CELL BIOLOGY

A LABORATORY HANDBOOK

Third Edition
Volume 1

CELL BIOLOGY

A LABORATORY HANDBOOK

Third Edition
Volume 1

Edited by

JULIO E. CELIS

Institute of Cancer Biology, Danish Cancer Society,
Copenhagen, Denmark

ELSEVIER
ACADEMIC
PRESS

AMSTERDAM • BOSTON • HEIDELBERG • LONDON
NEW YORK • OXFORD • PARIS • SAN DIEGO
SAN FRANCISCO • SINGAPORE • SYDNEY • TOKYO

Elsevier Academic Press
30 Corporate Drive, Suite 400, Burlington, MA 01803, USA
525 B Street, Suite 1900, San Diego, California 92101-4495, USA
84 Theobald's Road, London WC1X 8RR, UK

This book is printed on acid-free paper. ∞

Library of Congress Cataloging-in-Publication Data
Application Submitted

British Library Cataloguing in Publication Data
A catalogue record for this book is available from the British Library

ISBN 13: 978-0-12-164731-5
ISBN 10: 0-12-164731-5
Set ISBN 13: 978-0-12-164730-8
Set ISBN 10: 0-12-164730-7

For all information on all Elsevier Academic Press publications
visit our Web site at www.books.elsevier.com

Printed in China
05   06   07   08   09   10   9   8   7   6   5   4   3   2   1

Contents of Volume 1

Contents of other Volumes

PART E. APPENDIX

Section 16. Appendix

Contributors

Numbers in parenthesis indicate the volume (bold face) and page on which the authors' contribution begins.

Mads Aaboe (**4**: 83)   Clinical Biochemical Department, Molecular Diagnostic Laboratory, Aarhus University Hospital, Skejby, Brendstrupgaardvej, Aarhus N, DK-8200, DENMARK

Tanja Aarvak (**1**: 239)   Dynal Biotech ASA, PO Box 114, Smestad, N-0309, NORWAY

Harindra R. Abeysinghe (**3**: 345)   Department of Pathology and Laboratory Medicine, University of Rochester School of Medicine, 601 Elmwood Ave., Rm 1-6337, Rochester, NY 14642

Ruedi Aebersold (**4**: 437)   The Institute for Systems Biology, 1441 North 34th Street, Seattle, WA 98103-8904

Ueli Aebi (**3**: 233, 241)   ME Muller Institute for Microscopy, Biozentrum, University of Basel, Klingelbergstr. 50/70, Basel, CH-4056, SWITZERLAND

Cheol-Hee Ahn (**4**: 29)   School of Materials Science and Engineering, Seoul National University, Seoul, 151-744, SOUTH KOREA

Natalie G. Ahn (**4**: 443)   Department of Chemistry & Biochemistry, University of Colorado, 215 UCB, Boulder, CO 80309

Ramiro Alberio (**4**: 45)   School of Biosciences, University of Nottingham, Sutton Bonington, Loughborough, Leics, LE12 5RD, UNITED KINGDOM

Donna G. Albertson (**3**: 445)   Cancer Research Institute, Department of Laboratory Medicine, The University of California, San Francisco, Box 0808, San Francisco, CA 94143-0808

Heiner Albiez (**1**: 291)   Department of Biology II, Ludwig-Maximilians University of Munich, Munich, GERMANY

Terence Allen (**3**: 325)   CRC Structural Cell Biology Group, Paterson Institute for Cancer Research, Christie Hospital NHS Trust, Wilmslow Road, Withington, Manchester, M20 4BX, UNITED KINGDOM

Noona Ambartsumian (**1**: 363)   Department of Molecular Cancer Biology, Danish Cancer Society, Institute of Cancer Biology, Strandboulevarden 49, Copenhagen, DK-2100, DENMARK

Øystein Åmellem (**1**: 239)   Immunosystems, Dynal Biotech ASA, PO Box 114, Smestad, N-0309, NORWAY

Patrick Amstutz (**1**: 497)   Department of Biochemistry, University of Zürich, Winterthurerstr. 190, Zurich, CH-8057, SWITZERLAND

Jens S. Andersen (**4**: 427)   Protein Interaction Laboratory, University of Southern Denmark— Odense, Campusvej 55, Odense M, DK-5230, DENMARK

Mads Hald Andersen (**1**: 97)   Tumor Immunology Group, Institute of Cancer Biology, Danish Cancer Society, Strandboulevarden 49, Copenhagen, DK-2100, DENMARK

Helena Andersson (**4**: 63)   Bioscience at Novum, Karolinska Institutet, Halsovagen 7-9, Huddinge, SE-141 57, SWEDEN

Peter W. Andrews (**1**: 183)   Department of Biomedical Science, The University of Sheffield, Rm B2 238, Sheffield, S10 2TN, UNITED KINGDOM

Elsa Anes (**2**: 57)   Faculdade de Farmacia, Universidade de Lisboa, Av. Forcas Armadas, Lisboa, 1649-019, PORTUGAL

James M. Angelastro (**1**: 171)   Department of Pathology and Center for Neurobiology and Behavior, Columbia University College of Physicians and Surgeons, 630 West 168th Street, New York, NY 10032

Sergey V. Anisimov (4: 103)   Molecular Cardiology Unit, National Institute on Aging, NIH, 5600 Nathan Shock Drive, Baltimore, MD 21224

Celia Antonio (2: 379)   Department of Biochemistry & Molecular Biophysics, College of Physicians & Surgeons, Columbia University, 701 W 168ST HHSC 724, New York, NY 69117

Shigehisa Aoki (1: 411)   Department of Pathology & Biodefence, Faculty of Medicine, Saga University, Nebeshima 5-1-1, Saga, 849-8501, JAPAN

Ron D. Appel (4: 207)   Swiss Institute of Bioinformatics, CMU, Rue Michel Servet 1, Geneva 4, CH-1211, SWITZERLAND

Rolf Apweiler (4: 469)   EMBL Outstation, European Bioinformatics Institute, Wellcome Trust Genome Campus, Hinxton, Cambridge, CB10 1SD, UNITED KINGDOM

Nobukazu Araki (2: 147)   Department of Histology and Cell Biology, School of Medicine, Kagawa University, Mki, Kagawa, 761-0793, JAPAN

Christopher M. Armstrong (4: 295)   Dana Faber Cancer Institute, Harvard University, 44 Binney Street, Boston, MA 02115

Anthony J. Ashford (2: 155)   Antibody Facility, Max Planck Institute of Molecullar Cell Biology and Genetics, Pfotenhauerstrsse 108, Dresden, D-01307, GERMANY

Daniel Axelrod (3: 19)   Dept of Physics & Biophysics Research Division, University of Michigan, Ann Arbor, MI 48109-1055

Sheree Bailey (1: 475)   Dept of Immunology, Allergy and Arthritis, Flinders Medical Centre and Flinders University, Bedford Park, Adelaide, SA, 5051, SOUTH AUSTRALIA

Nathalie Q. Balaban (2: 419)   Department of Physics, The Hebrew University-Givat Ram, Racah Institute, Jerusalem, 91904, ISRAEL

William E. Balch (2: 209)   Department of Cell and Molecular Biology, The Scripps Research Institute, 10550 North Torrey Pines Road, La Jolla, CA 92037

Debabrata Banerjee (1: 315)   Department of Medicine, Cancer Institute of New Jersey, 195 Little Albany Street, New Brunswick, NJ 08903

Jiri Bartek (4: 253)   Department of Cell Cycle and Cancer, Danish Cancer Society, Strandboulevarden 49, Copenhagen, DK-2100, DENMARK

Werner Baschong (3: 5)   ME Muller Institute for Microscopy, Biozentrum, University of Basel, Klingelbergstrasse 50/70, Basel, CH-4056, SWITZERLAND

Philippe I. H. Bastiaens (3: 153)   Cell Biology and Cell Biophysics Program, European Molecular Biology Laboratory, Meyerhofstrasse 1, Heidelberg, 69117, GERMANY

Jürgen C. Becker (1: 103)   Department of Dermatology, University of Würzburg, Sanderring 2, Würzburg, 97070, GERMANY

Martin Béhé (4: 149)   Department of Nuclear Medicine, Philipp's-University of Marburg, Baldingerstraße, Marburg/Lahn, D-35043, GERMANY

Thomas M. Behr (4: 149)   Department of Nuclear Medicine, Philipp's-University of Marburg, Baldingerstraße, Marburg, D-35043, GERMANY

Stefanie Benesch (2: 399)   Department of Cell Biology, Gesellschaft fur Biotechnoogische Forschung, Mascheroder Weg 1, Braunschweig, D-38124, GERMANY

Aaron Bensimon (3: 429)   Laboratoire de Biophysique de l'ADN, Departement des Biotechnologies, Institut Pasteur, 25 rue du Dr. Roux, Paris Cedex 15, F-75724, FRANCE

John J. M. Bergeron (2: 41)   Department of Anatomy and Cell Biology, Faculty of Medicine, McGill University, STRATHCONA Anatomy & Dentistry Building, Montreal, QC, H3A 2B2, CANADA

Michael W. Berns (3: 351)   Beckman Laser Institute, University of California, Irvine, 1002 Health Sciences Road E, Irvine, CA 92697-1475

Joseph R. Bertino (1: 315)   The Cancer Institute of New Jersey, 195 Little Albany Street, New Brunswick, NJ 08901

Paulo Bianco (1: 79)   Dipartimento di Medicina Sperimentale e Patologia, Universita 'La Sapienza', Viale Regina Elena 324, Roma, I-00161, ITALY

Hans Kaspar Binz (1: 497)   Department of Biochemistry, University of Zürich, Winterthurerstr. 190, Zürich, CH-8057, SWITZERLAND

R. Curtis Bird (1: 247)   Department of Pathobiology, Auburn University, Auburn, AL 36849

Mina J. Bissell (1: 139)   Life Sciences Division, Lawrence Berkeley National Laboratory, 1 Cyclotron Road, Bldg 83-101, Berkeley, CA 94720

Stephanie Blackwood (**3**: 445)   Cancer Research Institute, University of California San Francisco, PO Box 0808, San Francisco, CA 94143-0808

Blagoy Blagoev (**4**: 427)   Protein Interaction Laboratory, University of Southern Denmark—Odense, Campusvej 55, Odense M, DK-5230, DENMARK

Kenneth R. Boheler (**4**: 103)   Laboratory of Cardiovascular Science, National Institute on Aging, NIH, 5600 Nathan Shock Drive, Baltimore, MD 21224-6825

Michelle A. Booden (**1**: 345)   Lineberger Comprehensive Cancer Center, University of North Carolina at Chapel Hill, Chapel Hill, NC 27599-7295

Gary G. Borisy (**3**: 277)   Department of Cell and Molecular Biology, Northwestern University Medical School, Chicago, IL 6011-3072

Elliot Botvinick (**3**: 351)   Beckman Laser Institute, University of California, Irvine, 1002 Health Sciences Road, East, Irvine, CA 92697-1475

Gérard Bouchet (**4**: 207)   Swiss Institute of Bioinformatics (SIB), CMU, rue Michel-Servet 1, Genève 4, CH-1211, SWITZERLAND

Rosemary Boyle (**4**: 437)   The Institute for Systems Biology, 1441 North 34th St., Seattle, WA 98109

Susanne Brandfass (**1**: 563)   Department of Biochemistry and Cell Biology, Max Planck Institute of Biophysical Chemistry, Am Faßberg 11, Gottingen, D-37077, GERMANY

Pascal Braun (**4**: 73)   Department of Chemistry and Chemical Biology, Harvard University, 12 Oxford Street, Cambridge, MA 02138

Steven A. Braut (**4**: 121)   Department of Anatomy and Structural Biology, Golding # 601, Albert Einstein College of Medicine of Yeshiva University, 1300 Morris Park Avenue, Bronx, NY 10461

Alvis Brazma (**4**: 95)   EMBL Outstation—Hinxton, European Bioinformatics Institute, Wellcome Trust Genome Campus, Hinxton, Cambridge, CB10 1SD, UNITED KINGDOM

J. David Briley (**3**: 471)   Department of Genomic Sciences, Glaxo Wellcome Research and Development, 5 Moore Drive, Research Triangle Park, NC 27709-3398

Simon Broad (**1**: 133)   Keratinocyte Laboratory, London Research Institute, 44 Lincoln's Inn Fields, London, WC2A 3PX, UNITED KINGDOM

Nicholas H. Brown (**3**: 77)   Wellcome Trust/Cancer Research UK Institute and Departmentt of Anatomy, University of Cambridge, Tennis Court Road, Cambridge, CB2 1QR, UNITED KINGDOM

Heather L. Brownell (**2**: 329, 341)   Office of Technology Licensing and Industry Sponsored Research, Harvard Medical School, 25 Shattuck Street, Gordon Hall of Medicine, Room 414, Boston, MA 02115

Damien Brunner (**3**: 69)   Cell Biology and Cell Biophysics Programme, European Molecular Biology Laboratory, Meyerhofstrasse 1, Heidelberg, D-69117, GERMANY

Suzannah Bumpstead (**3**: 463)   Genotyping / Chr 20, The Wellcome Trust Sanger Institute, The Wellcome Trust Genome Campus, Hinxton, Cambridge, CB10 1SA, UNITED KINGDOM

Deborah C. Burford (**3**: 403)   Wellcome Trust, Sanger Institute, The Wellcome Trust Genome Campus, Hinxton, Cambridge, CB10 1SA, UNITED KINGDOM

Gerald Burgstaller (**2**: 161)   Department of Cell Biology, Institute of Molecular Biology, Austrian Academy of Sciences, Billrothstrasse 11, Salzburg, A-5020, AUSTRIA

Ian M. Caldicott (**1**: 157)

Angelique S. Camp (**1**: 457)   Gene Therapy Centre, University of North Carolina at Chapel Hill, 7119 Thurston-Bowles (G44 Wilson Hall), Chapel Hill, NC 27599-7352

Keith H. S. Campbell (**4**: 45)   School of Biosciences, Sutton Bonington, Loughborough, Leics, LE12 5RD, UNITED KINGDOM

Yihai Cao (**1**: 373)   Microbiology & Tumor Biology Center, Karolinska Institute, Room: Skrivrum (G415), Box 280, Stockholm, SE-171 77, SWEDEN

Maria Carmo-Fonseca (**2**: 277, **3**: 419)   Institute of Molecular Medicine, Faculty of Medicine, University of Lisbon, Av. Prof. Egas Moniz, Lisbon, 1649-028, PORTUGAL

T. Carneiro (**3**: 419)   Faculty of Medicine, Institute of Molecular Medicine, University of Lisbon, Av. Prof. Egas Moniz, Lisboa, 1649-028, PORTUGAL

Nigel P. Carter (**2**: 133)   The Wellcome Trust, Sanger Institute, The Wellcome Trust, Genome Campus, Hinxton, Cambridge, CB10 1SA, UNITED KINGDOM

Célia Carvalho (**3**: 419)   Faculty of Medicine, Institute of Molecular Medicine, University of Lisbon, Av. Prof. Egas Moniz, Lisboa, 1649-028, PORTUGAL

Lucy A. Carver (**2**: 11)   Cellular and Molecular Biology Program, Sidney Kimmel Cancer Center, 10835 Altman Row, San Diego, CA 92121

Doris Cassio (**1**: 231, **3**: 387)   INSERM U-442: Signalisation cellulaire et calcium, Bat 443, Universite Paris-Sud, Street George Clemenceau Pack, 444, Orsay, Cedex, F-91405, FRANCE

Chris Catton (**3**: 207)   Department of Zoology, University of Oxford, South Parks Road, Oxford, OX1 3PS, UNITED KINGDOM

Julio E. Celis (**1**: 527, **4**: 69, 165, 219, 243, 289) Danish Cancer Society, Institute of Cancer Biology and Danish Centre for Translational Breast Cancer Research, Strandboulevarden 49, Copenhagen O, DK-2100, DENMARK

Pierre Chambon (**3**: 501)   Institut de Génétique et de Biologie Moléculaire et Cellulaire, 1 rue Laurent Fries, B.P.10142, Illkirch CEDEX, F-67404, FRANCE

Francis Ka-Ming Chan (**2**: 355)   Department of Pathology, University of Massachusetts Medical School, Room S2-125, 55 Lake Avenue North, Worcester, MA 01655

Ming-Shien Chang (**3**: 87)   Department of Physics, Duke University, 107 Physics Bldg, Durham, NC 27708-1000

Samit Chatterjee (**2**: 241)   Margaret M. Dyson Vision Research Institute, Department of Ophthalmology, Weill Medical College of Cornell University, 1300 York Avenue, New York, NY 10021

Sandeep Chaudhary (**1**: 121)   Veterans Affairs Medical Center, San Diego (V111G), 3350 La Jolla Village Drive, San Diego, CA 92161

Jingwen Chen (**3**: 471)   Department of Genomic Sciences, Glaxo Wellcome Research and Development, 5 Moore Drive, Research Triangle Park, NC 27709

Yonglong Chen (**1**: 191)   Institute for Biochemistry and Molecular Cell Biology, University of Goettingen, Justus-von-Liebig-Weg 11, Göttingen, D-37077, GERMANY

Yong Woo Cho (**4**: 29)   Akina, Inc., Business & Technology Center, 1291 Cumberland Ave., #E130, West Lafayette, IN 47906

Juno Choe (**1**: 269)   Institute for Systems Biology, 1441 N. 34th St, Seattle, WA 98103

Claus R. L. Christensen (**1**: 363)   Department of Molecular Cancer Biology, Danish Cancer Society, Institute of Cancer Biology, Strandboulevarden 49, Copenhagen, DK-2100, DENMARK

Theodore Ciaraldi (**1**: 121)   Veterans Affairs Medical Center, University of California, San Diego, 9500 Gilman Drive, La Jolla, CA 92093-9111

Aaron Ciechanover (**4**: 351)   Center for Tumor and Vascular Biology, The Rappaport Faculty of Medicine and Research Institute, Technion-Israel Institute of Technology, POB 9649, Efron Street, Bat Galim, Haifa, 31096, ISRAEL

Mark S. F. Clarke (**2**: 233, **4**: 5)   Department of Health and Human Performance, University of Houston, 3855 Holman Street, Garrison—Rm 104D, Houston, TX 77204-6015

Martin Clynes (**1**: 335)   National Institute for Cellular Biotechnology, Dublin City University, Glasnevin, Dublin, 9, IRELAND

Philippe Collas (**1**: 207)   Institute of Medical Biochemistry, University of Oslo, PO Box 1112 Blindern, Oslo, 0317, NORWAY

Kristen Correia (**4**: 35)   Krumlauf Lab, Stowers Institute for Medical Research, 1000 East 50th Street, Kansas City, MO 64110

Pascale Cossart (**2**: 407)   Unite des Interactions Bacteries-Cellules/Unité INSERM 604, Institut Pasteur, 28, rue du Docteur Roux, Paris Cedex 15, F-75724, FRANCE

Thomas Cremer (**1**: 291)   Department of Biology II, Ludwig-Maximilians University of Munich, Munich, 80333, GERMANY

Robert A. Cross (**2**: 371)   Molecular Motors Group, Marie Curie Research Institute, The Chart, Oxted, Surrey, RH8 0TE, UNITED KINGDOM

Matthew E. Cunningham (**1**: 171)   Hospital for Special Surgery, New York Hospital, 520 E. 70th Street, New York, NY 10021

Noélia Custódio (**3**: 419)   Faculty of Medicine, Institute of Molecular Medicine, University of Lisbon, Av. Prof. Egas Moniz, Lisboa, 1649-028, PORTUGAL

Zbigniew Darzynkiewwicz (**1**: 279)   The Cancer Research Institute, New York Medical College, 19 Bradhurst Avenue, Hawthorne, NY 10532

Ilan Davis (3: 187)   Wellcome Trust Centre for Cell Biology, Institute of Cell and Molecular Biology, The University of Edinburgh, Michael Swann Building, The King's Buildings, Mayfield Road, Edinburgh, EH9 3JR, SCOTLAND

Stephen C. De Rosa (1: 257)   Vaccine Research Center, National Institutes of Health, 40 Convent Dr., Room 5610, Bethesda, MD 20892-3015

Nicholas M. Dean (3: 523)   Functional Genomics, GeneTrove, GeneTrove (a division of Isis Isis Pharmaceuticals, Inc.), 2292 Faraday Avenue, Carlsbad, CA 92008

Anne Dell (4: 415)   Department of Biological Sciences, Biochemistry Building, Imperial College of Science, Technology & Medicine, Biochemistry Building, London, SW7 2AY, UNITED KINGDOM

Panos Deloukas (3: 463)   The Wellcome Trust, Sanger Institute, Hinxton, Cambridge, CB10 1SA, UNITED KINGDOM

Nicolas Demaurex (3: 163)   Department of Cell Physiology and Metabolism, University of Geneva Medical Center, 1 Michel-Servet, Geneva, CH-1211, SWITZERLAND

Chris Denning (4: 45)   Division of Animal Physiology, School of Biosciences, Institute of Genetics Room C15, University of Nottingham, Queens Medical Centre, Nottingham, NG7 2UH, UNITED KINGDOM

Ami Deora (2: 241)   Margaret M. Dyson Vision Research Institute, Department of Ophthalmology, Weill Medical College of Cornell University, 1300 York Avenue, New York, NY 10021

Julien Depollier (4: 13)   Centre de Recherche en Biochimie Macromoléculaire (UPR 1086), Centre National de la Recherche Scientifique (CNRS), 1919 Route de Mende, Montpellier Cedex 5, F-34293, FRANCE

Channing J. Der (1: 345)   Department of Pharmacology, University of North Carolina at Chapel Hill, Lineberger Comprehensive Cancer Center, Chapel Hill, NC 27599

Bart Devreese (4: 259)   Department of Biochemistry, Physiology and Microbiology, University of Ghent, K.L. Ledeganckstraat 35, Ghent, B-9000, BELGIUM

Alberto Diaspro (3: 201)   Department of Physics, University of Genoa, Via Dodecaneso 33, Genoa, I-16146, ITALY

James Fred Dice (4: 345)   Department Physiology, Tufts University School of Medicine, 136 Harrison Ave, Boston, MA 02111

Thomas J. Diefenbach (4: 307)   Department of Physiology, Tufts University School of Medicine, 136 Harrison Avenue, Boston, MA 02111

Chris Dinant (2: 121)   Biomolecular Sciences, UMIST, PO Box 88, Manchester, M60 1QD, UNITED KINGDOM

Da-Qiao Ding (3: 171)   Structural Biology Section and CREST Research Project, Kansai Advanced Research Center, Communications Research Laboratory, 588-2 Iwaoka, Iwaoka-cho, Nishi-ku, Kobe, 651-2492, JAPAN

Gilles Divita (4: 13)   Centre de Recherche en Biochimie Macromoléculaire (UPR 1086), Centre National de la Recherche Scientifique (CNRS), 1919 Route de Mende, Montpellier Cedex 5, F-34293, FRANCE

Eric P. Dixon (1: 483)   TriPath Oncology, 4025 Stirrup Creek Drive, Suite 400, Durham, NC 27703

Bernhard Dobberstein (2: 215)   Zentrum fur Molekulare Biologie, Universitat Heidelberg, Im Neuenheimer Feld 282, Heidelberg, D-69120, GERMANY

Lynda J. Donald (4: 457)   Department of Chemistry, University of Manitoba, Room 531 Parker Building, Winnipeg, MB, R3T 2N2, CANADA

Wolfgang R. G. Dostmann (2: 299)   Department of Pharmacology, University of Vermont, Health Science Research Facility 330, Burlington, VT 05405-0068

Adam Douglass (3: 129)   Department of Cellular and Molecular Pharmacology, The University of California, San Francisco, School of Medicine, Medical Sciences Building, Room S1210, 513 Parnassus Avenue, San Francisco, CA 94143-0450

Kate Downes (3: 463)   Genotyping / Chr 20, The Wellcome Trust, Sanger Institute, The Wellcome Trust Genome Campus, Hinxton, Cambridge, CB10 1SA, UNITED KINGDOM

Harry W. Duckworth (4: 457)   Department of Chemistry, University of Manitoba, Room 531 Parker Building, Winnipeg, MB, R3T 2N2, CANADA

Derek M. Dykxhoorn (3: 511)   CBR Institute for Biomedical Research, Harvard Medical School, 200 Longwood Ave, Boston, MA 02115

Lars Dyrskjøt (4: 83)   Clinical Biochemical Department, Molecular Diagnostic Laboratory, Aarhus University Hospital, Skejby, Brendstrupgaardvej, Aarhus N, DK-8200, DENMARK

Christoph Eckerskorn (4: 157)   Protein Analytics, Max Planck Institute for Biochemistry, Klopferspitz 18, Martinsried, D-82152, GERMANY

Glenn S. Edwards (3: 87)   Department of Physics, Duke University, 221 FEL Bldg, Box 90305, Durham, NC 27708-0305

Andreas A. O. Eggert (1: 103)   Department of Dermatology, Julius-Maximillians University, Josef-Schneider-Str. 2, Würzburg, 97080, GERMANY

Maria Ekström (4: 63)   Bioscience at Novum, Karolinska Institutet, Huddinge, SE-141 57, SWEDEN

Andreas Engel (3: 317)   Maurice E. Müller Institute for Microscopy at the Biozentrum, University of Basel, Klingelbergstrasse 70, Basel, CH-4056, SWITZERLAND

Anne-Marie Engel (1: 353)   Bartholin Institutte, Bartholinsgade 2, Copenhagen K, DK-1356, DENMARK

José A. Enríquez (2: 69)   Department of Biochemistry and Molecular and Cellular Biology, Universidad de Zaragoza, Miguel Servet, 177, Zaragoza, E-50013, SPAIN

Rachel Errington (1: 305)   Department of Medical Biochemistry and Immunology, University of Wales College of Medicine, Heath Park, Cardiff, CF14 4XN, UNITED KINGDOM

Virginia Espina (3: 339)   Microdissection Core Facility, Laboratory of Pathology, National Cancer Institute, 9000 Rockville Pike, Building 10, Room B1B53, Bethesda, MD 20892

H. Dariush Fahimi (2: 63)   Department of Anatomy and Cell Biology II, University of Heidelberg, Im Neuenheimer Feld 307, Heidelberg, D-69120, GERMANY

Federico Federici (3: 201)   Department of Physics, University of Genoa, Via Dodecaneso 33, Genoa, I-16146, ITALY

Daniel L. Feeback (2: 233, 4: 5)   Space and Life Sciences Directorate, NASA-Johnson Space Center, 3600 Bay Area Blvd, Houston, TX 77058

Patricio Fernández-Silva (2: 69)   Dept of Biochemistry and Molecular and Cellular Biology, Univeridad de Zaragoza, Miguel Servet 177, Zaragoza, E-50013, SPAIN

Erika Fernández-Vizarra (2: 69)   Dept of Biochemistry and Molecular and Cellulary Biology, Universidad de Zaragoza, Miguel Servet, 177, Zaragoza, E-50013, SPAIN

Patrick F. Finn (4: 345)   Department of Physiology, Tufts University School of Medicine, 136 Harrison Ave, Boston, MA 02111

Kevin L. Firth (2: 329, 2: 341)   ASK Science Products Inc., 487 Victoria St, Kingston, Ontario, K7L 3Z8, CANADA

Raluca Flükiger-Gagescu (2: 27)   Unitec— Office of Technology Transfer, University of Geneva and University of Geneva Hospitals, 24, Rue Général-Dufour, Geneva 4, CH-1211, SWITZERLAND

Leonard J. Foster (4: 363, 427)   Protein Interaction Laboratory, University of Southern Denmark, Odense, Campusvej 55, Odense M, DK-5230, DENMARK

Dimitrios Fotiadis (3: 317)   M. E. Müller Institute for Microscopy at the Biozentrum, University of Basel, Klingelbergstrasse 70, Basel, CH-4056, SWITZERLAND

Patrick L. T. M. Frederix (3: 317)   M. E. Müller Institute for Microscopy at the Biozentrum, University of Basel, Klingelbergstrasse 70, Basel, CH-4056, SWITZERLAND

Marcus Frohme (4: 113)   Functional Genome Analysis, German Cancer Research Center, Deutsches Krebsforschungszentrum, Im Neuenheimer Feld 580, Heidelberg, D-69120, GERMANY

Masanori Fujimoto (4: 197)   Department of Biochemistry and Biomolecular Recognition, Yamaguchi University School of Medicine, 1-1-1, Minami-kogushi, Ube, Yamaguchi, 755-8505, JAPAN

Margarida Gama-Carvalho (2: 277)   Faculty of Medicine, Institute of Molecular Medicine, University of Lisbon, AV. Prof. Egas Moniz, Lisbon, 1649-028, PORTUGAL

Henrik Garoff (1: 419, 4: 63)   Unit for Cell Biology, Center for Biotechnology. Karolinska Institute, Huddinge, SE-141 57, SWEDEN

Susan M. Gasser (2: 359)   Friedrich Miescher Institute fr Biomedical Research, Maulbeerstrasse 66, Basel, CH-1211, SWITZERLAND

Kristine G. Gaustad (1: 207)   Institute of Medical Biochemistry, University of Oslo, PO Box 1112 Blindern, Oslo, 0317, NORWAY

Benjamin Geiger (2: 419)   Dept. of Molecular Cell Biology, Weizman Institute of Science, Wolfson Building, Rm 617, Rehovot, 76100, ISRAEL

Kris Gevaert (4: 379, 4: 457)   Dept. Medical Protein Research, Flanders Interuniversity Institute for Biotechnology, Faculty of Medicine and Health Sciences, Ghent University, Instituut Rommelaere— Blok D, Albert Baertsoenkaai 3, Gent, B-9000, BELGIUM

Jilur Ghori (3: 463)   Genotyping / Chr 20, The Wellcome Trust, Sanger Institute, The Wellcome Trust, Genome Campus, Hinxton, Cambridge, CB10 1SA, UNITED KINGDOM

Alasdair J. Gibb (1: 395)   Department of Pharmacology, University College London, Gower Street, London, WC1E GBT, UNITED KINGDOM

Mario Gimona (1: 557, 2: 161, 4: 145)   Department of Cell Biology, Institute of Molecular Biology, Austrian Academy of Sciences, Billrothstrasse 11, Salzburg, A-5020, AUSTRIA

David A. Glesne (1: 165)   Biosciences Division, Argonne National Laboratory, 9700 South Cass Avenue, Argonne, IL 60439-4844

Martin Goldberg (3: 325)   Science Laboratories, University of Durham, South Road, Durham, DH1 3LE, UNITED KINGDOM

Kenneth N. Goldie (3: 267)   Structural and Computational Biology Programme, EMBL, Meyerhofstrasse 1, Heidelberg, D-69117, GERMANY

Jon W. Gordon (3: 487)   Geriatrics and Adult Development, Mount Sinai School of Medicine, One Gustave L. Levy Place, New York, NY 10029

Angelika Görg (4: 175)   Fachgebiet Proteomik, Technische Universität München, Am Forum 2, Freising Weihenstephan, D-85350, GERMANY

Martin Gotthardt (4: 149)   Department of Nuclear Medicine, Philipp's-University of Marburg, Baldingerstraße, Marburg/Lahn, D-35043, GERMANY

Frank L. Graham (1: 435)   Department of Biology, McMaster University, Life Sciences Building, Room 430, Hamilton, Ontario, L8S 4K1, CANADA

Claude Granier (1: 519)   UMR 5160, Faculté de Pharmacie, 15 Av. Charles Flahault, Montpellier Cedex 5, BP 14491, 34093, FRANCE

Lloyd A. Greene (1: 171)   Department of Pathology and Center for Neurobiology and Behavior, Columbia University, College of Physicians and Surgeons, 630 W. 168th Street, New York, NY 10032

Susan M. Gribble (3: 403)   Sanger Institute, The Wellcome Trust, The Wellcome Trust Genome Campus, Hinxton, Cambridge, CB10 1SA, UNITED KINGDOM

Gareth Griffiths (2: 57, 3: 299)   Department of Cell Biology, EMBL, Postfach 102209, Heidelberg, D-69117, GERMANY

Sergio Grinstein (3: 163)   Cell Biology Program, Hospital for Sick Children, 555 University Avenue, Toronto, Ontario, M5G 1X8, CANADA

Pavel Gromov (1: 527, 4: 69, 165, 243, 289)   Institute of Cancer Biology and Danish Centre for Translational Breast Cancer Research, Danish Cancer Society, Strandboulevarden 49, Copenhagen, DK-2100, DENMARK

Irina Gromova (4: 219)   Department of Medical Biochemistry and Danish Centre for Translational Breast Cancer Research, Danish Cancer Society, Strandboulevarden 49, Copenhagen, DK-2100, DENMARK

Dale F. Gruber (1: 33)   Cell Culture Research and Development, GIBCO/Invitrogen Corporation, 3175 Staley Road, Grand Island, NY 14072

Markus Grubinger (4: 145)   Institute of Physics and Biophysics, University of Salzburg, Hellbrunnerstr. 34, Salzburg, A-5020, AUSTRIA

Jean Gruenberg (2: 27, 201)   Department of Biochemistry, University of Geneva, 30, quai Ernest Ansermet, Geneva 4, CH-1211, SWITZERLAND

Stephanie L. Gupton (3: 137)   10550 North Torrey Pines Road, CB 163, La Jolla, CA 92037

Cemal Gurkan (2: 209)   Department of Cell and Molecular Biology, The Scripps Research Institute, 10550 North Torrey Pines Road, La Jolla, CA 92037

Martin Guttenberger (**4**: 131)   Zentrum für Molekulariologie der Pflanzen, Universitat Tübingen, Entwicklungsgenetik, Auf der Morgenstelle 3, Tübingen, D-72076, GERMANY

Thomas Haaf (**3**: 409)   Institute for Human Genetics, Johannes Gutenberg-Universität Mainz, 55101, Mainz, D-55131, GERMANY

Christine M. Hager-Braun (**1**: 511)   Health and Human Services, NIH National Institute of Environmental Health Sciences, MD F0-04, PO Box 12233, Research Triangle Park, NC 27709

Anne-Mari Håkelien (**1**: 207)   Institute of Medical Biochemistry, Institute of Medical Biochemistry, University of Oslo, PO Box 1112 Blindern, Oslo, 0317, NORWAY

Fiona C. Halliday (**1**: 395)   GlaxoSmithKline, Greenford, Middlesex, UB6 OHE, UNITED KINGDOM

Gerald Hammond (**2**: 223)   Molecular Neuropathobiology Laboratory, Cancer Reasearch UK London Research Institute, 44 Lincoln's Inn Fields, London, WC2A 3PX, UNITED KINGDOM

Klaus Hansen (**4**: 253)

Hironobu Harada (**1**: 367)   Department of Neurosurgery, Ehime University School of Medicine, Shitsukawa, Toon-shi, Ehime, 791-0295, JAPAN

Robert J. Hay (**1**: 43, 49, 573)   Viitro Enterprises Incorporated, 1113 Marsh Road, PO Box 328, Bealeton, VA 22712

Izumi Hayashi (**1**: 151)   National Medical Center and Beckman Research Institute, Division of Neurosciences, City of Hope, 1500 E. Duarte Rd, Duarte, CA 91010-3000

Timothy A. Haystead (**4**: 265)   Department of Pharmacology and Cancer Biology, Duke University Medical Center, Box 3813 Med Ctr, Durham, NC 27710

Rebecca Heald (**2**: 379)   Molecular and Cell Biology Department, University of California, Berkeley, Berkeley, CA 94720-3200

Florence Hediger (**2**: 359)   Department of Molecular Biology, University of Geneva, 30, Quai Ernest Ansermet, Geneva, CH-1211, SWITZERLAND

Rainer Heintzmann (**3**: 29)   Randall Division of Cell and Molecular Biophysics, King's College London, Guy's Campus, London, SE1 1UL, UNITED KINGDOM

Frederic Heitz (**4**: 13)   Centre de Recherche en Biochimie Macromoléculaire (UPR 1086), Centre National de la Recherche Scientifique (CNRS), 1919 Route de Mende, Montpellier Cedex 5, F-34293, FRANCE

Johannes W. Hell (**2**: 85)   Department of Pharmacology, University of Iowa, 2152 Bowen Science Building, Iowa City, IA 52242

Kai Hell (**4**: 269)   Adolf-Butenandt-Institut fur Physiologische Chemie, Lehrstuhl: Physiologische Chemie, Universitat Munchen, Butenandtstr. 5, Gebäude B, Munchen, D-81377, GERMANY

Robert R. Henry (**1**: 121)   Veterans Affairs Medical Center, San Diego (V111G), 3350 La Jolla Village Drive, San Diego, CA 92161

Johan Hiding (**2**: 45)   Göteborg University, Institute of Medical Biochemistry, PO Box 440, Göteborg, SE-403-50, SWEDEN

Yasushi Hiraoka (**3**: 171)   Structural Biology Section and CREST Research Project, Kansai Advanced Research Center, Communications Research Laboratory, 588-2 Iwaoka, Iwaoka-cho, Nishi-ku, Kobe, 651-2492, JAPAN

Mary M. Hitt (**1**: 435)   Department of Pathology & Molecular Medicine, McMaster University, 1200 Main Street West, Hamilton, Ontario, L8N 3Z5, CANADA

Julie Hodgkinson (**3**: 307)   School of Crystallography, Birkbeck College, Unversity of London, Malet Street, London, WC1E 7HX, UNITED KINGDOM

Klaus P. Hoeflich (**2**: 307)   Division of Molecular and Structural Biology, Ontario Cancer Institute, Department of Medical Biophysics, University of Toronto, 610 University Avenue, 7-707A, Toronto, Ontario, M5G 2M9, CANADA

Tracy L. Hoffman (**1**: 21)   ATCC, P.O. Box 1549, Manassas, VA 20108

Jörg D. Hoheisel (**4**: 113)   Functional Genome Analysis, German Cancer Research Center, Deutsches Krebsforschungszentrum, Im Neuenheimer Feld 580, Heidelberg, D-69120, GERMANY

Thomas Hollemann (**1**: 191)   Institute for Biochemistry and Molecular Cell Biology, University of Goettingen, Justus-von-Liebig-Weg 11, Göttingen, D-37077, GERMANY

Caterina Holz (**4**: 57)   PSF biotech AG, Huebnerweg 6, Berlin, D-14059, GERMANY

Akira Honda (2: 299)   Department of Pharmacology, University of Vermont, Health Science Research Facility 330, Burlington, VT 05405-0068

Masanori Honsho (2: 5)   Max Planck Institute of Molecular Cell Biology and Genetics, Pfotenhauerstrasse 108, Dresden, D-01307, GERMANY

Andrew N. Hoofnagle (4: 443)   School of Medicine, University of Colorado Health Sciences Center, Denver, CO 80262

Eliezer Huberman (1: 165)   Gene Expression and Function Group, Argonne National Laboratory, 9700 South Cass Avenue, Argonne, IL 60439-4844

M. Shane Hutson (3: 87)   Department of Physics, Duke University, 107 Physics Bldg, Durham, NC 27708-1000

Andreas Hüttmann (1: 115)   Abteilung für Hämatologie, Universitäkrankenhaus Essn, Hufelandstr. 55, Essen, 45122, GERMANY

Anthony A. Hyman (2: 155)   Max Planck Institute of Molecular Cell Biology and Gene Technology, Pfotenhauerstrasse 108, Dresden, D-01307, GERMANY

Sherrif F. Ibrahim (1: 269)   Institute for Systems Biology, 1441 N. 34th St, Seattle, WA 98103

Kazuo Ikeda (1: 151)   National Medical Center and Beckman Research Institute, Division of Neurosciences, City of Hope, 1500 East Duarte Road, Duarte, CA 91010-3000

Elina Ikonen (2: 181)   The LIPID Cell Biology Group, Department of Biochemistry, The Finnish National Public Health Institute, Mannerheimintie 166, Helsinki, FIN-00300, FINLAND

Pranvera Ikonomi (1: 49)   Director, Cell Biology, American Type Culture Collection (ATCC), 10801 University Blvd., Manassas, VA 20110-2209

Mitsuhiko Ikura (2: 307)   Division of Molecular and Structural Biology, Ontario Cancer Institute, Department of Medical Biophysics, University of Toronto, 610 University Avenue 7-707A, Toronto, Ontario, M5G 2M9, CANADA

Arup Kumar Indra (3: 501)   Institut de Génétique et de Biologie Moléculaire et Cellulaire (IGBMC), 1 rue Laurent Fries, B.P.10142, Illkirch CEDEX, F-67404, FRANCE

Takayoshi Inoue (4: 35)   National Institute for Neuroscience, 4-1-1 Ogawahigashi, Kodaira, Tokyo, 187-8502, JAPAN

Kumiko Ishii (2: 139)   Supra-Biomolecular System Research Group, RIKEN (Institute of Physical and Chemical Research), 2-1, Hirosawa, Wako-shi, Saitama, 351-0198, JAPAN

Dean A. Jackson (2: 121)   Department of Biomolecular Sciences, UMIST, PO Box 88, Manchester, M60 1QD, UNITED KINGDOM

Reinhard Jahn (2: 85)   Department of Neurobiology, Max-Planck-Institut for Biophysical Chemistry, Am Faßberg 11, Gottingen, D-37077, GERMANY

Kim D. Janda (1: 491)   Department of Chemistry, BCC-582, The Scripps Research Institute, 10550 N. Torrey Pines Road, La Jolla, CA 92037

Harry W. Jarrett (4: 335)   Department of Biochemistry, University of Tennessee Health Sciences Center, Memphis, TN 38163

Daniel G. Jay (4: 307)   Dept. Physiology, Tufts University School of Medicine, 136 Harrison Avenue, Boston, MA 02111

David W. Jayme (1: 33)   Cell Culture Research and Development, GIBCO/Invitrogen Corporation, 3175 Staley Road, Grand Island, NY 14072

Ole Nørregaard Jensen (4: 409)   Protein Research Group, Department of Biochemistry and Molecular Biology, University of Southern Denmark, Campusvej 55, Odense M, DK-5230, DENMARK

Jae Hyun Jeong (4: 29)   Department of Chemical & Biomolecular Engineering, Center for Ultramicrochemical Process Systems, Korea Advanced Institute of Science and Technology, Daejeon, 305-701, SOUTH KOREA

Jeff A. Jones (2: 233)   Space and Life Sciences Directorate, NASA-Johnson Space Center, TX 77058

Gloria Juan (1: 279)   Research Pathology Division, Room S-830, Memorial Sloan-Kettering Cancer Center, 1275 York Avenue, New York, NY 10021

Melissa S. Jurica (2: 109)   Molecular, Cell & Developmental Biology, Center for Molecular Biology of RNA, UC Santa Cruz, 1156 High Street, Santa Cruz, CA 95064

Eckhart Kämpgen (**1**: 103)   Department of Dermatology, Friedrich Alexander University, Hartmannstr. 14, Erlangen, D-91052, GERMANY

Roger Karlsson (**2**: 165)   Department of Cell Biology, The Wenner-Gren Institute, Stockholm University, Stockholm, S-10691, SWEDEN

Fredrik Kartberg (**2**: 45)   Göteborg University, Institute of Medical Biochemistry, PO Box 440, Gothenburg, SE, 403-50, SWEDEN

Irina N. Kaverina (**3**: 111)   Institute of Molecular Biotechnology, Austrian Academy of Sciences, Dr. Bohrgasse 3-5, Vienna, A-1030, AUSTRIA

Ralph H. Kehlenbach (**2**: 267)   Hygiene-Institut-Abteilung Virologie, Universitat Heidelberg, Im Neuenheimer Feld 324, Heidelberg, D-69120, GERAMNY

Daniel P. Kiehart (**3**: 87)   Department of Biology, Duke University, B330g Levine Sci Bldg, Box 91000, Durham, NC 27708-1000

Katherine E. Kilpatrick (**1**: 483)   Senior Research Investigator, TriPath Oncology, 4025 Stirrup Creek Drive, Suite 400, Durham, NC 27703

Jong-Duk Kim (**4**: 29)   Department of Chemical & Biomolecular Engineering, Center for Ultramicrochemical Process Systems, Korea Advanced Institute of Science and Technology, Daejeon, 305-701, SOUTH KOREA

Maurice Kléber (**1**: 69)   Institute of Cell Biology, Department of Biology, Swiss Federal Institute of Technology, ETH—Hönggerberg, Zurich, CH-8093, SWITZERLAND

Toshihide Kobayashi (**2**: 139)   Supra-Biomolecular System Research Group, RIKEN (Institute of Physical and Chemical Research) Frontier Research System, 2-1, Hirosawa, Wako-shi, Saitama, 351-0198, JAPAN

Stefan Kochanek (**1**: 445)   Division of Gene Therapy, University of Ulm, Helmholtz Str. 8/I, Ulm, D-89081, GERMANY

Anna Koffer (**2**: 223)   Physiology Department, University College London, 21 University Street, London, WC1E 6JJ, UNITED KINGDOM

Antonius Koller (**4**: 383)   Department of Cell Biology, Torrey Mesa Research Institute, 3115 Merryfield Row, San Diego, CA 92121

Erich Koller (**3**: 523)   Functional Genomics, GeneTrove, Isis Pharmaceuticals, Inc., 2292 Faraday Ave., Carlsbad, CA 92008

Robert L. Kortum (**1**: 215)   The Eppley Institute for Research in Cancer, The University of Nebraska Medical Center, 986805 Nebraska Medical Center, Omaha, NE 68198-6805

Irina Kratchmarova (**4**: 427)   Protein Interaction Laboratory, University of Southern Denmark—Odense, Campusvej 55, Odense M, DK-5230, DENMARK

Geri E. Kreitzer (**2**: 189)   Cell and Developmental Biology, Weill Medical College of Cornell University, LC-300, New York, NY 10021

Florian Kreppel (**1**: 445)   Division of Gene Therapy, University of Ulm, Helmholtz Str. 8/I, Ulm, D-89081, GERMANY

Mogens Kruhøffer (**4**: 83)   Molecular Diagnostic Laboratory, Clinical Biochemical Department, Aarhus University Hospital, Skejby, Brendstrupgaardvej, Aarhus N, DK-8200, DENMARK

Robb Krumlauf (**4**: 35)   Stowers Institute for Medical Research, 1000 East 50th Street, Kansas City, MO 64110

Michael Kühl (**1**: 191)   Development Biochemistry, University of Ulm, Albert-Einstein-Allee 11, Ulm, D-89081, GERMANY

Mark Kühnel (**2**: 57)   Department of Cell Biology, EMBL, Postfach 102209, Heidelberg, D-69117, GERMANY

Anuj Kumar (**3**: 179)   Dept. of Molecular, Cellular, and Developmental Biology and Life Sciences Institute, University of Michigan, 210 Washtenaw Avenue, Ann Arbor, MI 48109-2216

Thomas Küntziger (**1**: 207)   Institute of Medical Biochemistry, Institute of Medical Biochemistry, University of Oslo, PO Box 1112 Blindern, Oslo, 0317, NORWAY

Yasuhiro Kuramitsu (**4**: 197)   Department of Biochemistry and Biomolecular Recognition, Yamaguchi University School of Medicine, 1-1-1 Minami-kogushi, Ube, Yamaguchi, 755-8505, JAPAN

Sergei A. Kuznetsov (**1**: 79)   Craniofacial and Skeletal Disease Branch, NIDCR, NIH, Department of Health and Human Services, 30 Convent Drive MSC 4320, Bethesda, MD 20892

Joshua Labaer (4: 73)   Harvard Institute of Proteomics, 320 Charles Street, Boston, MA 02141-2023

Frank Lafont (2: 181)   Department of Biochemistry, University of Geneva, 30, quai Ernest-Ansermet 1211, Geneva 4, CH-1211, SWITZERLAND

Yun Wah Lam (2: 103, 115)   Wellcome Trust Biocentre, MSI/WTB Complex, University of Dundee, Dow Street, Dundee, DD1 5EH, UNITED KINGDOM

Angus I. Lamond (2: 103, 115)   Wellcome Trust Biocentre, MSI/WTB Complex, University of Dundee, Dow Street, Dundee, DD1 5EH, UNITED KINGDOM

Lukas Landmann (3: 5)   Institute for Anatomy (LL), Anatomisches Institut, University of Basel, Pestalozzistrasse 20, Basel, CH-4056, SWITZERLAND

Helga B. Landsverk (1: 207)   Institute of Medical Biochemistry, Institute of Medical Biochemistry, University of Oslo, PO Box 1112 Blindern, Oslo, 0317, NORWAY

Christine Lang (4: 57)   Department of Microbiology and Genetics, Berlin University of Technology, Gustav-Meyer-Allee 25, Berlin, D-13355, GERMANY

Paul LaPointe (2: 209)   Department of Cell and Molecular Biology, The Scripps Research Institute, 10550 North Torrey Pines Road, La Jolla, CA 92037

Martin R. Larsen (4: 371)   Department of Biochemistry and Molecular Biology, University of Southern Denmark, Campusvej 55, Odense M, DK-5230, DENMARK

Pamela L. Larsen (1: 157)   Department of Cellular and Structural Biology, University of Texas Health Science Center at San Antonio, San Antonio, TX 78229-3900

Eugene Ngo-Lung Lau (1: 115)   Leukaemia Foundation of Queensland Leukaemia Research Laboratories, Queensland Institute of Medical Research, Royal Brisbane Hospital Post Office, Brisbane, Queensland, Q4029, AUSTRALIA

Sabrina Laugesen (4: 371)   Department of Biochemistry and Molecular Biology, University of Southern Denmark, Campusvej 55, Odense M, DK-5230, DENMARK

Daniel Laune (1: 519)   Centre de Pharmacologie et Biotechnologie pour la Santé, CNRS UMR 5160, Faculté de Pharmacie, Avenue Charles Flahault, Montpellier Cedex 5, F-34093, FRANCE

Andre Le Bivic (2: 241)   Groupe Morphogenese et Compartimentation Membranaire, UMR 6156, IBDM, Faculte des Sciences de Luminy, case 907, Marseille cedex 09, F-13288, FRANCE

Ronald Lebofsky (3: 429)   Laboratoire de Biophysique de l'ADN, Departement des Biotechnologies, Institut Pasteur, 25 rue du Dr. Roux, Paris Cedex 15, F-75724, FRANCE

Chuan-PU Lee (2: 259)   The Department of Biochemistry and Molecular Biology, Wayne State University School of Medicine, 4374 Scott Hall, 540 E. Canfield, Detroit, MI 48201

Eva Lee (1: 139)   Life Sciences Division, Lawrence Berkeley National Laboratory, 1 Cyclotron Road, Bldg 83-101, Berkeley, CA 94720

Joon-Hee Lee (4: 45)   School of Biosciences, University of Nottingham, Sutton Bonington, Loughborough, Leics, LE12 5RD, UNITED KINGDOM

Kwangmoon Lee (1: 215)   The Eppley Institute for Research in Cancer, The University of Nebraska Medical Center, 986805 Nebraska Medical Center, Omaha, NE 68198-6805

Thomas Lee (4: 443)   Dept of Chemistry and Biochemistry, Univ of Colorado, 215 UCB, Boulder, CO 80309-0215

Margaret Leversha (3: 395)   Memorial Sloan Kettering Cancer Center, 1275 York Avenue, New York, NY 10021

Jeffrey M. Levsky (4: 121)   Department of Anatomy and Structural Biology, Golding # 601, Albert Einstein College of Medicine of Yeshiva University, 1300 Morris Park Avenue, Bronx, NY 10461

Alexandre Lewalle (3: 37)   Randall Centre, New Hunt's House, Guy's Campus, London, SE1 1UL, UNITED KINGDOM

Chung Leung Li (1: 115)   Experimental Haematology Laboratory, Stem Cell Program, Institute of Zoology/Genomics Research Center, Academia Sinica, Nankang 115, Nankang, Taipei, 11529, R.O.C.

LiQiong Li (3: 345)   Department of Pathology and Laboratory Medicine, University of Rochester School of Medicine, 601 Elmwood Ave., Rm 1-6337, Rochester, NY 14642

Mei Li (3: 501)   Institut de Génétique et de Biologie Moléculaire et Cellulaire (IGBMC), 1 rue Laurent Fries, B.P.10142, Illkirch CEDEX, F-67404, FRANCE

Siming Li (4: 295)   Dana Faber Cancer Institute, Harvard University, 44 Binney Street, Boston, MA 02115

Lih-huei Liaw (3: 351)   Beckman Laser Institute, University of California, Irvine, 1002 Health Sciences Road E, Irvine, CA 92697-1475

Antonietta M. Lillo (1: 491)   Department of Chemistry, BCC-582, The Scripps Research Institute, 10550 N. Torrey Pines Road, La Jolla, CA 92037

Uno Lindberg (2: 165)   Department of Cell Biology, Stockholm University, The Wenner-Gren Institute, Stockhølm, S-10691, SWEDEN

Christian Linden (1: 103)   Department of Virology, Julius-Maximillins University, Versbacher Str. 7, Würzburg, D-97080, GERMANY

Robert Lindner (2: 51)   Department of Cell Biology in the Center of Anatomy, Hannover Medical School, Hannover, D-30625, GERMANY

Lance A. Liotta (3: 339)   Chief, Laboratory of Pathology, National Cancer Institute Building 10, Room 2A33, 9000 Rockville Pike, Bethesda, MD 20892

Adam J. Liska (4: 399)   Max Planck Institute of Molecular Cell Biology and Genetics, Pfotenhauerst 108, Dresden, D-01307, GERMANY

Hong Liu (1: 139)   Life Sciences Division, Lawrence Berkeley National Laboratory, 1 Cyclotron Road, Bldg 83-101, Berkeley, CA 94720

Silvia Lommel (2: 399)   Department of Cell Biology, German Research Center for Biotechnology (GBF), Mascheroder Weg 1, Braunschweig, D-38124, GERMANY

Giuseppe S. A. Longo-Sorbello (1: 315)   Centro di Riferimento Oncologico, Ospedale "S. Vincenzo", Taormina, Contradra Sirinam, 08903, ITALY

Lovisa Lovmar (3: 455)   Department of Medical Sciences, Uppsala University, Akademiska sjukhuset, Uppsala, SE-75185, SWEDEN

Eugene Lukanidin (1: 363)   Department of Molecular Cancer Biology, Institute of Cancer

Biology, Danish Cancer Society, Strandboulevarden 49, Copenhagen, DK-2100, DENMARK

Jiri Lukas (4: 253)   Department of Cell Cycle and Cancer, Danish Cancer Society, Strandboulevarden 49, Copenhagen, DK-2100, DENMARK

Peter J. Macardle (1: 475)   Department of Immunology, Allergy and Artritis, Flinders Medical Centre and Flinders University, Bedford Park, Adelaie, SA, 5051, SOUTH AUSTRALIA

Peder S. Madsen (4: 69)   Institute of Medical Biochemistry, University of Aarhus, Ole Worms Alle, Building 170, Aarhus C, DK-8000, DENMARK

Nils E. Magnusson (4: 83)   Clinical Biochemical Department, Molecular Diagnostic Laboratory, Aarhus University Hospital, Skejby, Brendstrupgaardvej, Aarhus N, DK-8200, DENMARK

Asami Makino (2: 139)   Supra-Biomolecular System Research Group, RIKEN (Institute of Physical and Chemical Research) Frontier Research System, 2-1, Hirosawa, Wako-shi, Saitama, 351-0198, JAPAN

G. Mike Makrigiorgos (3: 477)   Department of Radiation Oncology, Dana Farber-Brigham and Women's Cancer Center, 75 Francis Street, Level L2, Boston, MA 02215

Matthias Mann (4: 363, 427)   Protein Interaction Laboratory, University of Southern Denmark, Odense, Campusvej 55, Odense M, DK-5230, DENMARK

Edward Manser (4: 285)   Glaxo-IMCB Group, Institute of Molecular and Cell Biology, Singapore, 117609, SINGAPORE

Ahmed Mansouri (3: 491)   Department of Molecular Cell Biology, Max-Planck-Institute of Biophysical Chemistry, Am Fassberg 11, Göttingen, D-37077, GERMANY

Alan D. Marmorstein (2: 241)   Cole Eye Institute, Weill Medical College of Cornell Cleveland Clinic, 9500 Euclid Avenue, i31, Cleveland, OH 44195

Bruno Martoglio (2: 215)   Institute of Biochemistry, ETH Zentrum, Building CHN, Room L32.3, Zurich, CH-8092, SWITZERLAND

Susanne E. Mason (1: 407)   Department of Physiology, University of Maryland School of Medicine, 655 W. Baltimore St., Baltimore, MD 21201

Stephen J. Mather (1: 539)   Dept of Nuclear Medicine, St Bartholomews Hospital, London, EC1A 7BE, UNITED KINGDOM

Arvid B. Maunsbach (3: 221, 289) Department of Cell Biology, Institute of Anatomy, Aarhus University, Aarhus, DK-8000, DENMARK

William Hayes McDonald (4: 391) Department of Cell Biology, The Scripps Research Institute, 10550 North Torrey Pines Rd, La Jolla, CA 92370

Kathleen M. McKenzie (1: 491) Department of Chemistry, BCC-582, The Scripps Research Institute, 10550 N. Torrey Pines Road, La Jolla, CA 92037

Alexander D. McLellan (1: 103) Department of Microbiology & Immunology, University of Otago, PO Box 56, 720 Cumberland St, Dunedin, NEW ZEALAND

Scott W. McPhee (1: 457) Department of Surgery, University of Medicine and Dentistry of New Jersey, Camden, NJ 08103

Jill Meisenhelder (4: 139) Molecular and Cell Biology Laboratory, The Salk Institute, 10010 North Torrey Pines Road, La Jolla, CA 92037

Paula Meleady (1: 13) National Institute for Cellular Biotechnology, Dublin City University, Glasnevin, Dublin, 9, IRELAND

Nicholas T. Mesires (4: 345) Department of Physiology, Tufts University School of Medicine, 136 Harrison Ave, Boston, MA 02111

Daniel Metzger (3: 501) Institut de Génétique et de Biologie Moléculaire et Cellulaire (IGBMC), Institut Clinique de la Souris (ICS), 1 rue Laurent Fries, B.P.10142, Illkirch CEDEX, F-67404, FRANCE

Martina Mirlacher (3: 369) Division of Molecular Pathology, Institue of Pathology, University of Basel, Schonbeinstrasse 40, Basel, CH-4031, SWITZERLAND

Suchareeta Mitra (4: 335) Department of Biochemistry, University of Tennessee Health Sciences Center, Memphis, TN 38163

Atsushi Miyawaki (2: 317) Laboratory for Cell Function and Dynamics, Advanced Technology Center, Brain Science Institute, Institute of Physical and Chemical Research (RIKEN), 2-1 Horosawa, Wako, Saitama, 351-0198, JAPAN

Dejana Mokranjac (4: 269) Adolf-Butenandt-Institut fur Physiologische Chemie, Lehrstuhl: Physiologische Chemie, Universitat Munchen, Butenandtstr. 5, Gebäude B, Munchen, D-81377, GERMANY

Peter L. Molloy (4: 325) CSIRO Molecular Science, PO Box 184, North Ryde, NSW, 1670, AUSTRALIA

Richard A. Moravec (1: 25) Promega Corporation, 2800 Woods Hollow Road, Madison, WI 53711-5399

José M. A. Moreira (1: 527) Institute of Cancer Biology and Danish Centre for Translational Breast Cancer Research, Danish Cancer Society, Strandboulevarden 49, Copenhagen O, DK-2100, DENMARK

May C. Morris (4: 13) Centre de Recherche en Biochimie Macromoléculaire (UPR 1086), Centre National de la Recherche Scientifique (CNRS), 1919 Route de Mende, Montpellier Cedex 5, F-34293, FRANCE

Robert A. Moxley (4: 335) Department of Biochemistry, University of Tennessee Health Sciences Center, Memphis, TN 38163

Anne Muesch (2: 189) Margaret M. Dyson Vision Research Institute, Department of Opthalmology, Weill Medical College of Cornell University, New York, NY 10021

Peggy Müller (1: 325) Zentrum für Angewandte Medizinische und Humanbiologische Forschung, Labor für Molekulare Hepatologie der Universitätsklinik und Polikinik für Innere Medizin I, Martin Luther University Halle-Wittenburg, Heinrich-Damerow-Street 1, Saale, Halle, D-06097, GERMANY

Steve Murray (3: 325) CRC Structural Cell Biology Group, Paterson Institute for Cancer Research, Christie Hospital NHS Trust, Wilmslow Road, Withington, Manchester, M20 4BX, UNITED KINGDOM

Connie Myers (1: 139) Life Sciences Division, Lawrence Berkeley National Laboratory, 1 Cyclotron Road, Bldg 83-101, Berkeley, CA 94720

Kazuyuki Nakamura (4: 197) Department of Biochemistry and Biomolecular Recognition, Yamaguchi University School of Medicine, 1-1-1 Minami-kogushi, Ube, Yamaguchi, 755-8505, JAPAN

Maithreyi Narasimha (3: 77) Wellcome Trust/Cancer Research UK Institute and Dept of Anatomyaraki, University of Cambridge, Tennis Court Road, Cambridge, CB2 1QR, UNITED KINGDOM

Dobrin Nedelkov (4: 279) Intrinsic Bioprobes, Inc., 625 S. Smith Road, Suite 22, Tempe, AZ 85281

Randall W. Nelson (4: 279)   Intrinsic Bioprobes Inc., 625 S. Smith Road, Suite 22, Tempe, AZ 85281

Frank R. Neumann (2: 359)   Department of Molecular Biology, University of Geneva, 30, Quai Ernest Ansermet, Geneva, CH-1211, SWITZERLAND

Walter Neupert (4: 269)   Adolf-Butenandt-Institut fur Physiologische Chemie, Lehrstuhl: Physiologische Chemie, Universitat Munchen, Butenandtstr. 5, Gebäude B, Munchen, D-81377, GERMANY

Axl Alois Neurauter (1: 239)   Immunosystem R & D, Dynal Biotech ASA, PO Box 114, Smestad, N-0309, NORWAY

Phillip Ng (1: 435)   Dept of Molecular and Human Genetics, Baylor College of Medicine, One Baylor Plaza, Houston, TX 77030

Garth L. Nicolson (1: 359)   The Institute for Molecular Medicine, 15162 Triton Lane, Huntington Beach, CA 92649-1041

Trine Nilsen (4: 275)   Department of Biochemistry, Institute for Cancer Research, The Norwegian Radium Hospital, Montebello, Oslo, N-0310, NORWAY

Tommy Nilsson (2: 45)   Göteborg University, Institute of Medical Biochemistry, PO Box, Göteborg, SE-403 50, SWEDEN

Lars Norderhaug (1: 239)   Dynal Biotech ASA, PO Box 114, Smestad, N-0309, NORWAY

Robert O'Connor (1: 5, 13, 335)   National Institute for Cellular Biotechnology, Dublin City University, Glasnevin, Dublin, 9, IRELAND

Lorraine O'Driscoll (1: 5, 335)   National Institute for Cellular Biotechnology, Dublin City University, Glasnevin, Dublin, 9, IRELAND

Martin Offterdinger (3: 153)   Cell Biology and Cell Biophysics Program, European Molecular Biology Laboratory, Meyerhofstrasse 1, Heidelberg, D-69117, GERMANY

Philip Oh (2: 11)   Cellular and Molecular Biology Program, Sidney Kimmel Cancer Center, 10835 Altman Row, San Diego, CA 92121

Takanori Ohnishi (1: 367)   Department of Neurosurgery, Ehime University School of Medicine, Shitsukawa, Toon-shi, Ehime, 791-0295, JAPAN

Sjur Olsnes (4: 19, 275)   Department of Biochemistry, The Norwegian Radium Hospital, Montebello, Oslo, 0310, NORWAY

Shao-En Ong (4: 427)   Protein Interaction Laboratory, University of Southern Denmark— Odense, Campusvej 55, Odense M, DK-5230, DENMARK

Akifumi Ootani (1: 411)   Department of Internal Medicine, Faculty of Medicine, Saga University, Nebeshima 5-1-1, Saga, 849-8501, JAPAN

Valerio Orlando (4: 317)   Dulbecco Telethon Institute, Institute of Genetics & Biophysics CNR, Via Pietro Castellino 111, Naples, I-80131, ITALY

Torben Faeck Ørntoft (4: 83)   Clinical Biochemical Department, Molecular Diagnostic Laboratory, Aarhus University Hospital, Skejby, Brendstrupgaardvej 100, Aarhus N, DK-8200, DENMARK

Mary Osborn (1: 549, 563)   Department of Biochemistry and Cell Biology, Max Planck Institute of Biophysical Chemistry, Am Faßberg 11, Gottingen, D-37077, GERMANY

Lawrence E. Ostrowski (2: 99)   Cystic Fibrosis/ Pulmonary Research and Treatment Centre, University of North Carolina at Chapel Hill, Thurston-Bowles Building, Chapel Hill, NC 27599-7248

Hendrik Otto (2: 253)   Institut für Biochemie und Molekularbiologie, Universität Freiburg, Hermann-Herder-Str. 7, Freiburg, D-79104, GERMANY

Kerstin Otto (1: 103)   Department of Dermatology, Julius-Maximillians University, Josef-Schneider-Str. 2, Würzburg, 97080, GERMANY

Michel M. Ouellette (1: 215)   Department of Biochemistry and Molecular Biology, Eppley Institute for Research in Cancer, The University of Nebraska Medical Center, 986805 Nebraska Medical Center, Omaha, NE 68198-6805

Jacques Paiement (2: 41)   Département de pathologie et biologie cellulaire, Université de Montréal, Case postale 6128, Succursale "Centre-Ville", Montreal, QC, H3C 3J7, CANADA

Patricia M. Palagi (4: 207)   Swiss Institute of Bioinformatics, CMU, 1 Michel Servet, Geneva 4, CH-1211, SWITZERLAND

Kinam Park (4: 29)   Department of Pharmaceutics and Biomedical Engineering, Purdue University School of Pharmacy, 575 Stadium Mall Drive, Room G22, West Lafayette, IN 47907-2091

Helen Parkinson (4: 95)   EMBL Outstation—Hinxton, European Bioinformatics Institute, Wellcome Trust Genome Campus, Hinxton, Cambridge, CB10 1SD, UNITED KINGDOM

Richard M. Parton (3: 187)   Wellcome Trust Centre for Cell Biology, Institute of Cell and Molecular Biology The University of Edinburgh, Michael Swann Building, The King's Buildings, Mayfield Road, Edinburgh, EH9 3JR, SCOTLAND

Bryce M. Paschal (2: 267)   Center for Cell Signaling, University of Virginia, 1400 Jefferson Park Avenue, West Complex Room 7021, Charlottesville, VA 22908-0577

Wayne F. Patton (4: 225)   Perkin-Elmer LAS, Building 100-1, 549 Albany Street, Boston, MA 02118

Staffan Paulie (1: 533)   Mabtech AB, Box 1233, Nacha Strand, SE-131 28, SWEDEN

Rainer Pepperkok (3: 121)   Cell Biology and Cell Biophysics Programme, European Molecular Biology Laboratory (EMBL), Meyerhofstrasse 1, Heidelberg, D-69117, GERMANY

Xomalin G. Peralta (3: 87)   Department of Physics, Duke University, 107 Physics Bldg, Durham, NC 27708-1000

Martha Perez-Magallanes (1: 151)   National Medical Center and Beckman Research Institute, Division of Neurosciences, City of Hope, 1500 E. Duarte Rd, Duarte, CA 91010

Stephen P. Perfetto (1: 257)   Vaccine Research Center, National Institutes of Health, 40 Convent Dr., Room 5509, Bethesda, MD 20892-3015

Hedvig Perlmann (1: 533)   Department of Immunology, Stockholm University, Biology Building F5, Top floor, Svante Arrhenius väg 16, Stockholm, SE-10691, SWEDEN

Peter Perlmann (1: 533)   Department of Immunology, Stockholm University, Biology Building F5, Top floor, Svante Arrhenius väg 16, Stockholm, SE-10691, SWEDEN

Timothy W. Petersen (1: 269)   Institute for Systems Biology, 1441 N. 34th St, Seattle, WA 98103

Patti Lynn Peterson (2: 259)   Department of Neurology, Wayne State University School of Medicine, 5L26 Detroit Receiving Hospital, Detroit Medical Center, Detroit, MI 48201

Nisha Philip (4: 265)   Department of Pharmacology and Cancer Biology, Duke University, Research Dr. LSRC Rm C115, Box 3813, Durham, NC 27710

Thomas Pieler (1: 191)   Institute for Biochemistry and Molecular Cell Biology, University of Goettingen, Humboldtallee 23, Göttingen, D-37073, GERMANY

Daniel Pinkel (3: 445)   Department of Laboratory Medicine, University of California San Francisco, Box 0808, San Francisco, CA 94143-0808

Javier Pizarro Cerdá (2: 407)   Unite des Interactions Bacteries-Cellules/Unité INSERM 604, Institut Pasteur, 28, rue du Docteur Roux, Paris Cedex 15, F-75724, FRANCE

Andreas Plückthun (1: 497)   Department of Biochemistry, University of Zürich, Winterthurerstrasse 190, Zürich, CH-8057, SWITZERLAND

Helen Plutner (2: 209)   Department of Cell and Molecular Biology, The Scripps Research Institute, 10550 North Torrey Pines Road, La Jolla, CA 92037

Piotr Pozarowski (1: 279)   Brander Cancer Research Institute, New York Medical College, Valhalla, NY 10595

Johanna Prast (1: 557)   Institute of Molecular Biology, Austrian Academy of Sciences, Billrothstrasse 11, Salzburg, A-5020, AUSTRIA

Brendan D. Price (3: 477)   Department of Radiation Oncology, Dana Farber-Brigham and Women's Cancer Center, 75 Francis Street, Level L2, Boston, MA 02215

Elena Prigmore (3: 403)   Sanger Institute, The Wellcome Trust, The Wellcome Trust Genome Campus, Hinxton, Cambridge, CB10 1SA, UNITED KINGDOM

Gottfried Proess (1: 467)   Eurogentec S.A., Liege Science Park, 4102 Seraing, B-, BELGIUM

David M. Prowse (1: 133)   Centre for Cutaneous Research, Barts and The London Queen Mary's School of Medicine and Dentistry, Institute of Cell and Molecular Science, 2 Newark Street, Whitechapel London, WC2A 3PX, UNITED KINGDOM

Manuela Pruess (4: 469)   EMBL outstation—Hinxton, European Bioinformatic Institute, Welcome Trust Genome Campus, Hinxton, Cambridge, CB10 1SD, UNITED KINGDOM

Eric Quéméneur (4: 235)   Life Sciences Division, CEA Valrhô, BP 17171 Bagnols-sur-Cèze, F-30207, FRANCE

Leda Helen Raptis (2: 329, 341)   Department of Microbiology and Immunology, Queen's University, Room 716 Botterell Hall, Kingston, Ontario, K7L3N6, CANADA

Anne-Marie Rasmussen (1: 239)   Dynal Biotech ASA, PO Box 114, Smestad, N-0309, NORWAY

Andreas S. Reichert (4: 269)   Department of Physiological Chemistry, University of Munich, Butenandtstr. 5, München, D-81377, GERMANY

Siegfried Reipert (3: 325)   Ordinariat II, Institute of Biochemistry and Molecular Biology, Vienna Biocenter, Dr. Bohr-Gasse 9, Vienna, A-1030, AUSTRIA

Guenter P. Resch (3: 267)   Institute of Molecular Biology, Dr. Bohrgasse 3-5, Vienna, A-1030, AUSTRIA

Katheryn A. Resing (4: 443)   Dept of Chemistry and Biochemistry, University of Colorado, 215 UCB, Boulder, CO 80309-0215

Donald L Riddle (1: 157)   Division of Biological Sciences, University of Missouri, 311 Tucker Hall, Columbia, MO 65211

Mara Riminucci (1: 79)   Department of Experimental Medicine, Universita' dell' Aquilla, Via Vetoio, Coppito II, L'Aquila, I-67100, ITALY

Terry L. Riss (1: 25)   Promega Corporation, 2800 Woods Hollow Road, Madison, WI 53711-5399

Pamela Gehron Robey (1: 79)   Craniofacial and Skeletal Disease Branch, NIDCR, NIH, Department of Health and Human Services 30 Convent Dr, MSC 4320, Bethesda, MD 20892-4320

Linda J. Robinson (2: 201)

Philippe Rocca-Serra (4: 95)   EMBL Outstation—Hinxton, European Bioinformatics Institute, Wellcome Trust Genome Campus, Hinxton, Cambridge, CB10 1SD, UNITED KINGDOM

Alice Rodriguez (3: 87)   Department of Biology, Duke University, Durham, NC 27708-1000

Enrique Rodriguez-Boulan (2: 189, 241)   Margaret M Dyson Vision Research Institute, Department of Opthalmology, Weill Medical College of Cornell University, New York, NY 10021

Mario Roederer (1: 257)   ImmunoTechnology Section and Flow Cytometry Core, Vaccine Research Center, National Institute for Allergy and Infectious Diseases, National Institutes of Health, 40 Convent Dr., Room 5509, Bethesda, MD 20892-3015

Peter Roepstorff (4: 371)   Department of Biochemistry and Molecular Biology, University of Southern Denmark, Campusvej 55, Odense M, DK-5230, DENMARK

Manfred Rohde (2: 399)   Department of Microbial Pathogenicity, Gesellschaft fur Biotechnoogische Forschung, Mascheroder Weg 1, Braunschweig, D-38124, GERMANY

Norbert Roos (3: 299)   Electron Microscopical Unit for Biological Sciences, University of Oslo, Blindern, Oslo, 0316, NORWAY

Sabine Rospert (2: 253)   Institut für Biochemie und Molekularbiologie, Universität Freiburg, Hermann-Herder-Str. 7, Freiburg, D-79104, GERMANY

Klemens Rottner (3: 111)   Cytoskeleton Dynamics Group, German Research Centre for Biotechnology (GBF), Mascheroder Weg 1, Braunschweig, D-38124, GERMANY

Line Roy (2: 41)   Department of Anatomy and Cell Biology, Faculty of Medicine, McGill University, STRATHCONA Anatomy & Dentistry Building, Montreal, QC, H3A 2B2, CANADA

Sandra Rutherford (3: 325)   CRC Structural Cell Biology Group, Paterson Institute for Cancer Research, Christie Hospital NHS Trust, Wilmslow Road, Withington, Manchester, M20 4BX, UNITED KINGDOM

Beth Rycroft (1: 395)   Department of Pharmacology, University College London, Gower Street, London, WC1E GBT, UNITED KINGDOM

Patrick Salmon (1: 425)   Department of Genetics and Microbiology, Faculty of Medicine, University of Geneva, CMU-1 Rue Michel-Servet, Geneva 4, CH-1211, SWITZERLAND

Paul M. Salvaterra (1: 151)   National Medical Center and Beckman Research Institute, Division of Neurosciences, City of Hope, 1500 E. Duarte Rd, Duarte, CA 91010-3000

R. Jude Samulski (1: 457)   Gene Therapy Centre, Department of Pharmacology, University of North Carolina at Chapel Hill, 7119 Thurston Bowles, Chapel Hill, NC 27599-7352

Susanna-Assunta Sansone (**4**: 95) EMBL Outstation—Hinxton, European Bioinformatics Institute, Wellcome Trust Genome Campus, Hinxton, Cambridge, CB10 1SD, UNITED KINGDOM

Ugis Sarkan (**4**: 95) EMBL Outstation—Hinxton, European Bioinformatics Institute, Wellcome Trust Genome Campus, Hinxton, Cambridge, CB10 1SD, UNITED KINGDOM

Moritoshi Sato (**2**: 325) Department of Chemistry, School of Science, University of Tokyo, 7-3-1 Hongo, Bunkyo-Ku, Tokyo, 113-0033, JAPAN

Guido Sauter (**3**: 369) Institute of Pathology, University of Basel, Schonbeinstrasse 40, Basel, CH-4003, SWITZERLAND

Carolyn L. Sawyer (**2**: 299) Department of Pharmacology, University of Vermont, Health Science Research Facility 330, Burlington, VT 05405-0068

Guray Saydam (**1**: 315) Department of Medicine, Section of Hematology, Ege University Hospital, Bornova Izmir, 35100, TURKEY

Silvia Scaglione (**3**: 201) BIOLab, Department of Informatic, Systemistic and Telematic, University of Genoa, Viale Causa 13, Genoa, I-16145, ITALY

Lothar Schermelleh (**1**: 291, 301) Department of Biology II, Biocenter of the Ludwig-Maximilians University of Munich (LMU), Großhadernerstr. 2, Planegg-Martinsried, 82152, GERMANY

Gudrun Schiedner (**1**: 445) CEVEC Pharmaceuticals GmbH, Gottfried-Hagen-Straße 62, Köln, D-51105, GERMANY

David Schieltz (**4**: 383) Department of Cell Biology, Torrey Mesa Research Institute, 3115 Merryfield Row, San Diego, CA 92121

Jan E. Schnitzer (**2**: 11) Sidney Kimmel Cancer Center, 10835 Altman Row, San Diego, CA 92121

Morten Schou (**1**: 353) Bartholin Institute, Bartholinsgade 2, Copenhagen K, DK-1356, DENMARK

Sebastian Schuck (**2**: 5) Max Planck Institute of Molecular Cell Biology and Genetics, Pfotenhauerstrasse 108, Dresden, D-01307, GERMANY

Herwig Schüler (**2**: 165) Department of Cell Biology, The Wnner-Gren Institute, Stockholm University, Stockholm, S-10691, SWEDEN

Michael Schuler (**3**: 501) Institut de Génétique et de Biologie Moléculaire et Cellulaire (IGBMC), 1 rue Laurent Fries, B.P.10142, Illkirch CEDEX, F-67404, FRANCE

Ulrich S. Schwarz (**2**: 419) Theory Division, Max Planck Institute of Colloids and Interfaces, Potsdam, 14476, GERMANY

Antonio S. Sechi (**2**: 393) Institute for Biomedical Technology-Cell Biology, Universitaetsklinikum Aachen, RWTH, Pauwelsstrasse 30, Aachen, D-52057, GERMANY

Richard L. Segraves (**3**: 445) Comprehensive Cancer Center, University of California San Francisco, Box 0808, 2400 Sutter N-426, San Francisco, CA 94143-0808

James R. Sellers (**2**: 387) Cellular and Motility Section, Laboratory of Molecular Cardiology, National Heart, Lung and Blood Institute (NHLBI), National Institutes of Health, 10 Center Drive, MSC 1762, Bethesda, MD 20892-1762

Nicholas J. Severs (**3**: 249) Cardiac Medicine, National Heart and Lung Institute, Imperial College, Faculty of Medicine, Royal Brompton Hospital, Dovehouse Street, London, SW3 6LY, UNITED KINGDOM

Jagesh Shah (**3**: 351) Laboratory of Cell Biology, Ludwig Institute for Cancer Research, University of California, 9500 Gilman Drive, MC 0660, La Jolla, CA 92093-0660

Norman E. Sharpless (**1**: 223) The Lineberger Comprehensive Cancer Center, The University of North Carolina School of Medicine, Lineberger Cancer Center, CB# 7295, Chapel Hill, NC 27599-7295

Andrej Shevchenko (**4**: 399) Max Planck Institute for Molecular Cell Biology and Genetics, Pfotenhauerstrasse 108, Dresden, D-01307, GERMANY

David M. Shotton (**3**: 207, 249, 257) Department of Zoology, University of Oxford, South Parks Road, Oxford, OX1 3PS, UNITED KINGDOM

David I. Shreiber (**1**: 379) Department of Biomedical Engineering, Rutgers, the State University of New Jersey, 617 Bowser Road, Piscataway, NJ 08854-8014

Snaevar Sigurdsson (3: 455)   Department of Medical Sciences, Uppsala University, Akademiska sjukhuset, Uppsala, SE-751 85, SWEDEN

Stephen Simkins (1: 483)   TriPath Oncology, 4025 Stirrup Creek Drive, Suite 400, Durham, NC 27703

Ronald Simon (3: 369)   Division of Molecular Pathology, Institute of Pathology, University of Basel, Schonbeinstrasse 40, Basel, CH-4031, SWITZERLAND

Kai Simons (1: 127, 2: 5, 181)   Max Planck Institute of Molecular Cell Biology and Genetics, Pfotenhauerstrasse 108, Dresden, D-01307, GERMANY

Jeremy C. Simpson (3: 121)   Cell Biology and Cell Biophysics Programme, European Molecular Biology Laboratory (EMBL), Meyerhofstrasse 1, Heidelberg, D-69117, GERMANY

Robert H. Singer (4: 121)   Department of Anatomy and Structural Biology, Golding # 601, Albert Einstein College of Medicine of Yeshiva University, 1300 Morris Park Avenue, Bronx, NY 10461

Mathilda Sjöberg (1: 419)   Department of Biosciences at Novum, Karolinska Instituet, Huddinge, SE-141-57, SWEDEN

Camilla Skiple Skjerpen (4: 275)   Department of Biochemistry, Institute for Cancer Research, The Norwegian Radium Hospital, Montebello, Oslo, N-0310, NORWAY

John Sleep (3: 37)   Randall Division, Guy's Campus, New Hunt's House, London, SE1 1UL, UNITED KINGDOM

J. Victor Small (1: 557)   Department of Cell Biology, Institute of Molecular Biology, Austrian Academy of Sciences, Billrothstrasse 11, Salzburg, A-5020, AUSTRIA

Joél Smet (4: 259)   Department of Pediatrics and Medical Genetics, University Hospital, De Pintelaan 185, Ghent, B-9000, BELGIUM

Kim Smith (3: 381)   Director of Cytogenetic Services, Oxford Radcliffe NHS Trust, Headington, Oxford, OX3 9DU, UNITED KINGDOM

Paul J. Smith (1: 305)   Dept of Pathology, University of Wales College of Medicine, Heath Park, Cardiff, CF14 4XN, UNITED KINGDOM

Antoine M. Snijders (3: 445)   Comprehensive Cancer Center, Cancer Research Institute, The University of California, San Francisco, Box 0808, 2340 Sutter Street N-, San Fransisco, CA 94143-0808

Michael Snyder (3: 179)   Department of Molecular, Cellular and Developmental Biology, Yale University, P. O. Box 208103, Kline Biology Tower, 219 Prospect St., New Haven, CT 06520-8103

Irina Solovei (1: 291)   Department of Biology II, Anthropology & Human Genetics, Ludwig-Maximilians University of Munich, Munich, GERMANY

Marion Sölter (1: 191)   Institute for Biochemistry and Molecular Cell Biology, University of Goettingen, Justus-von-Liebig-Weg 11, Göttingen, D-37077, GERMANY

Lukas Sommer (1: 69)   Institute of Cell Biology, Department of Biology, Swiss Federal Institute of Technology, ETH—Hönggerberg, Zurich, CH-8093, SWITZERLAND

Simon Sparks (3: 207)   Department of Zoology, University of Oxford, South Parks Road, Oxford, OX1 3PS, UNITED KINGDOM

Kenneth G. Standing (4: 457)   Department of Physics and Astronomy, University of Manitoba, 510 Allen Bldg, Winnipeg, MB, R3T 2N2, CANADA

Walter Steffen (3: 37, 307)   Randall Division, Guy's Campus, New Hunt's House, London, SE1 1UL, UNITED KINGDOM

Theresia E. B. Stradal (3: 111)   Department of Cell Biology, German Research Centre for Biotechnology (GBF), Mascheroder Weg 1, Braunschweig, D-38124, GERMANY

Per Thor Straten (1: 97)   Tumor Immunology Group, Institute of Cancer Biology, Danish Cancer Society, Strandboulevarden 49, Copenhagen, DK-2100, DENMARK

Hajime Sugihara (1: 411)   Department of Pathology & Biodefence, Faculty of Medicine, Saga University, Nebeshima 5-1-1, Saga, 849-8501, JAPAN

Chung-Ho Sun (3: 351)   Beckman Laser Institute, University of California, Irvine, 1002 Health Sciences Road E, Irvine, CA 92697-1475

Mark Sutton-Smith (4: 415)   Department of Biological Sciences, Imperial College of Science, Technology and Medicine, Biochemistry Building, London, SW7 2AY, UNITED KINGDOM

Tatyana M. Svitkina (3: 277)   Department of Cell and Molecular Biology, Northwestern University Medical School, Chicago, IL 60611

Ann-Christine Syvänen (3: 455)   Department of Medical Sciences, Uppsala University, Forskningsavd 2, ing 70, Uppsala, SE-751 85, SWEDEN

Masako Tada (1: 199)   ReproCELL Incorporation, 1-1-1 Uchisaiwai-cho, Chiyoda-ku, Tokyo, 100-0011, JAPAN

Takashi Tada (1: 199)   Stem Cell Engineering, Stem Cell Research Center, Institute for Frontier Medical Sciences, Kyoto University, 53 Kawahara-cho Shogoin, Sakyo-ku, Kyoto, 606-8507, JAPAN

Angela Taddei (2: 359)   Department of Molecular Biology, University of Geneva, 30, Quai Ernest Ansermet, Geneva 4, CH-1211, SWITZERLAND

Tomohiko Taguchi (2: 33)   Department of Cell Biology, Yale University School of Medicine, 333 Cedar Street, PO Box 208002, New Haven, CT 06520-8002

Kazusuke Takeo (4: 197)   Department of Biochemistry and Biomolecular Recognition, Yamaguchi University School of Medicine, 1-1-1, Minami-kogushi, Ube, Yamaguchi, 755-8505, JAPAN

Nobuyuki Tanahashi (2: 91)   Laboratory of Frontier Science, Core Technology and Research Center, The Tokyo Metropolitan Institute of Medical Sciences, 3-18-22 Honkomagome, Bunkyo-ku, Tokyo, 113-8613, JAPAN

Keiji Tanaka (2: 91)   Laboratory of Frontier Science, Core Technology and Research Center, The Tokyo Metropolitan Institute of Medical Sciences, 3-18-22 Honkomagome, Bunkyo-ku, Tokyo, 133-8613, JAPAN

Chi Tang (2: 121)   Dept of Biomolecular Sciences, UMIST, PO Box 88, Manchester, M60 1QD, UNITED KINGDOM

Kirill V. Tarasov (4: 103)   Molecular Cardiology Unit, National Institute on Aging, NIH, 5600 Nathan Shock Drive, Baltimore, MD 21224

J. David Taylor (3: 471)   Department of Genomic Sciences, Glaxo Wellcome Research and Development, 5 Moore Drive, Research Triangle Park, NC 27709-3398

Nancy Smyth Templeton (4: 25)   Department of Molecular and Cellular Biology & the Center for Cell and Gene Therapy, Baylor College of Medicine, One Baylor Plaza, Alkek Bldg., Room N1010, Houston, TX 77030

Kenneth K. Teng (1: 171)   Department of Medicine, Wiell Medical of Cornell University, 1300 York Ave., Rm-A663, New York, NY 10021

Patrick Terheyden (1: 103)   Department of Dermatology, Julius-Maximillians University, Josef-Schneider-Str. 2, Würzburg, D-97080, GERMANY

Scott M. Thompson (1: 407)   Department of Physiology, University of Maryland School of Medicine, 655 W. Baltimore St., Baltimore, MD 21201,

John F. Timms (4: 189)   Department of Biochemistry and Molecular Biology, Ludwig Institute of Cancer Research, Cruciform Building 1.1.09, Gower Street, London, WC1E 6BT, UNITED KINGDOM

Shuji Toda (1: 411)   Department of Pathology & Biodefence, Faculty of Medicine, Saga University, Nebeshima 5-1-1, Saga, 849-8501, JAPAN

Yoichiro Tokutake (3: 87)   Department of Physics, Duke University, 107 Physics Bldg, Durham, NC 27708-1000

Evi Tomai (2: 329)   Department of Microbiology and Immunology, Queen's University, Room 716 Botterell Hall, Kingston, Ontario, K7L3N6, CANADA

Kenneth B. Tomer (1: 511)   Mass Spectrometry, Laboratory of Structural Biology, National Institute of Environmental Health Sciences NIEH/NIH, 111 Alexander Drive, PO Box 12233, Research Triangle Park, NC 27709

Derek Toomre (3: 19)   Department of Cell Biology, Yale University School of Medicine, SHM-C227/229, PO Box 208002, 333 Cedar Street, New Haven, CT 06520-8002

Sharon A. Tooze (2: 79)   Secretory Pathways Laboratory, Cancer Research UK London Research Institute, 44 Lincoln's Inn Fields, London, WC2A 3PX, UNITED KINGDOM

David Tosh (1: 177)   Centre for Regenerative Medicine, Department of Biology and Biochemistry, University of Bath, Claverton Down, Bath, BA2 7AY, UNITED KINGDOM

Yusuke Toyama (3: 87)   Department of Physics, Duke University, 107 Physics Bldg, Box 90305, Durham, NC 27708-0305

Robert T. Tranquillo (1: 379)   Department of Biomedical Engineering and Department of Chemical Engineering and Materials Science, University of Minnesota, Biomedical Engineering, 7-112 BSBE. 312 Church St SE, Minneapolis, MN 55455

Signe Trentemølle (4: 165)   Institute of Cancer Biology and Danish Centre for Translational Breast Cancer Research, Danish Cancer Society,

Strandboulevarden 49, Copenhagen, DK-2100, DENMARK

Didier Trono (1: 425) Department of Genetics and Microbiology, Faculty of Medicine, University of Geneva, CMU-1 Rue Michel-Servet, Geneva 4, CH-1211, SWITZERLAND

Kevin Truong (2: 307) Division of Molecular and Structural Biology, Ontario Cancer Institute, Department of Medical Biophysics, University of Toronto, 610 University Avenue, 7-707A, Toronto, Ontario, M5G 2M9, CANADA

Jessica K. Tyler (2: 287) Department of Biochemistry and Molecular Genetics, University of Colorado Health Sciences Center at Fitzsimons, PO Box 6511, Aurora, CO 80045

Aylin S. Ulku (1: 345) Department of Pharmacology, University of North Carolina at Chapel Hill, Lineberger Comprehensive Cancer Center, Chapel Hill, NC 27599-7295

Yoshio Umezawa (2: 325) Department of Chemistry, The School of Science, University of Tokyo, 7-3-1 Hongo, Bunkyo-ku, Tokyo, 113-0033, JAPAN

Ronald Vale (3: 129) Department of Cellular and Molecular Pharmacology, The Howard Hughes Medical Institute, The University of California, San Francisco, N316, Genentech Hall, 1600 16th Street, San Francisco, CA 94107

Jozef Van Beeumen (4: 259) Department of Biochemistry, Physiology and Microbiology, University of Ghent, K.L. Ledeganckstraat 35, Ghent, B-9000, BELGIUM

Rudy N. A. van Coster (4: 259) Department of Pediatrics and Medical Genetics, University Hospital, University of Ghent, De Pintelaan 185, Ghent, B-9000, BELGIUM

Ger van den Engh (1: 269) Institute for Systems Biology, 1441 North 34th Street, Seattle, WA 98103-8904

Peter van der Geer (4: 139) Department of Chemistry and Biochemistry, University of California, San Diego, 9500 Gilman Dr., La Jolla, CA 92093-0601

Joël Vandekerckhovr (4: 379, 457) Department of Medical Protein Research, Flanders Interuniversity Institute for Biotechnology, KL Ledeganckstraat 35, Gent, B-9000, BELGIUM

Charles R. Vanderburg (4: 5) Department of Neurology, Massachusetts General Hospital, 114 Sixteenth Street, Charlestown, MA 02129

John Venable (4: 383) Department of Cell Biology, Scripps Research Institute, 10550 North Torrey Pines Road, La Jolla, CA 92037

Isabelle Vernos (2: 379) Cell Biology and Cell Biophysics Programme, European Molecular Biology Laboratory, Meyerhofstrasse 1, Heidelberg, D-69117, GERMANY

Peter J. Verveer (3: 153) Cell Biology and Cell Biophysics Program, European Molecular Biology Laboratory, Meyerhofstrasse 1, Heidelberg, D-69117, GERMANY

Marc Vidal (4: 295) Cancer Biology Department, Dana-Farber Cancer Institute, 44 Binney Street, Boston, MA 02115

Emmanuel Vignal (2: 427) Department Genie, Austrian Academy of Sciences, Billrothstrasse 11, Salzbourg–Autriche, A-520, 5020, AUSTRIA

Sylvie Villard (1: 519) Centre de Pharmacologie et Biotechnologie pour la Santé, CNRS—UMR 5094, Faculté de Pharmacie, Avenue Charles Flahault, Montpellier Cedex 5, F-34093, FRANCE

Hikka Virta (1: 127) Department of Cell Biology, European Molecular Biology Laboratory, Cell Biology Programme, Heidelberg, D-69012, GERMANY

Alfred Völkl (2: 63) Department of Anatomy and Cell Biology II, University of Heidelberg, Im Neuenheimer Feld 307, Heidelberg, D-69120, GERMANY

Sonja Voordijk (4: 207) Geneva Bioinformatics SA, Avenue de Champel 25, Geneva, CH-1211, SWITZERLAND

Adina Vultur (2: 329, 341) Department of Microbiology and Immunology, Queen's University, Room 716 Botterell Hall, Kingston, Ontario, K7L3N6, CANADA

Teruhiko Wakayama (1: 87) Center for Developmental Biology, RIKEN, 2-2-3 Minatojima-minamimachi, Kobe, 650-0047, JAPAN

Daniel Walther (4: 207) Swiss Institute of Bioinformatics (SIB), CMU, rue Michel-Servet 1, Genève 4, 1211, SWITZERLAND

Gang Wang (3: 477)   Department of Radiation Oncology, Dana Farber-Brigham and Women's Cancer Center, 75 Francis Street, Level L2, Boston, MA 02215

Nancy Wang (3: 345)   Department of Pathology and Laboratory Medicine, University of Rochester School of Medicine, 601 Elmwood Ave., Rm 1-6337, Rochester, NY 14642

Xiaodong Wang (2: 209)   Department of Cell and Molecular Biology, The Scripps Research Institute, 10550 North Torrey Pines Road, La Jolla, CA 92037

Yanzhuang Wang (2: 33)   Department of Cell Biology, Yale University School of Medicine, 333 Cedar Street, PO BOX 208002, New Haven, CT 06520-8002

Yu-Li Wang (3: 107)   Department of Physiology, University of Massachusetts Medical School, 377 Plantation St., Rm 327, Worcester, MA 01605

Graham Warren (2: 33)   Department of Cell Biology, Yale University School of Medicine, 333 Cedar Street, PO Box 208002, New Haven, CT 06520-8002

Clare M. Waterman-Storer (3: 137)   10550 North Torrey Pines Road, CB 163, La Jolla, CA 92037

Jennifer C. Waters (3: 49)   Department of Cell Biology, Department of Systems Biology, Harvard Medical School, 240 Longwood Ave, Boston, MA 02115

Fiona M. Watt (1: 133)   Keratinocyte Laboratory, London Research Institute, 44 Lincoln's Inn Fields, London, WC2A 3PX, UNITED KINGDOM

Gerhard Weber (4: 157)   Protein Analytics, Max Planck Institute for Biochemistry, Klopferspitz 18, Martinsried, D-82152, GERMANY

Peter J. A. Weber (4: 157)   Proteomics Division, Tecan Munich GmbH, Feldkirchnerstr. 12a, Kirchheim, D-, 85551, GERMANY

Paul Webster (3: 299)   Electron Microscopy Laboratory, House Ear Institute, 2100 West Third Street, Los Angeles, CA 90057

Jürgen Wehland (2: 399)   Department of Cell Biology, Gesellschaft fur Biotechnoogische Forschung, Mascheroder Weg 1, Braunschweig, D-38124, GERMANY

Dieter G. Weiss (3: 57)   Institute of Cell Biology and Biosystems Technology, Department of Biological Sciences, Universitat Rostock, Albert-Einstein-Str. 3, Rostock, D-18051, GERMANY

Walter Weiss (4: 175)   Fachgebiet Proteomik, Technische Universität Muenchen, Am Forum 2, Freising Weihenstephan, D-85350, GERMANY

Adrienne R. Wells (3: 87)   Department of Biology, Duke University, Durham, NC 27708-1000

Jørgen Wesche (4: 19)   Department of Biochemistry, The Norwegian Radium Hospital, Montebello, Oslo, 0310, NORWAY

Pamela Whittaker (3: 463)   Genotyping / Chr 20, The Wellcome Trust Sanger Institute, The Wellcome Trust Genome Campus, Hinxton, Cambridge, CB10 1SA, UNITED KINGDOM

John Wiemann (3: 87)   Department of Biology, Duke University, B330g Levine Sci Bldg, Box 91000, Durham, NC 27708-1000

Sebastian Wiesner (2: 173)   Dynamique du Cytosquelette, Laboratoire d'Enzymologie et Biochimie Structurales, UPR A 9063 CNRS, Building 34, Bat. 34, avenue de la Terrasse, Gif-sur-Yvette, F-91198, FRANCE

Ilona Wolff (1: 325)   Prodekanat Forschung, Medizinnishce Fakultät, Martin Luther University Halle-Wittenburg, Magdeburger Str 8, Saale, Halle, D-06097, GERMANY

Ye Xiong (2: 259)   The Department of Biochemistry and Molecular Biology, Wayne State University School of Medicine, 4374 Scott Hall, 540 E. Canfield, Detroit, MI 48201

David P. Yarnall (3: 471)   Department of Metabolic Diseases, Glaxo Wellcome Inc, 5 Moore Drive, Research Triangle Park, NC 27709-3398

Hideki Yashirodas (2: 91)   Laboratory of Frontier Science, Core Technology and Research Center, The Tokyo Metropolitan Institute of Medical Sciences, 3-18-22 Honkomagome, Bunkyo-ku, Tokyo, 133-8613, JAPAN

John R. Yates III (4: 383, 391)   Department of Cell Biology, Scripps Research Institute, 10550 North Torrey Pines Road, La Jolla, CA 92037

Charles Yeaman (2: 189)   Department of Cell and Developmental Biology, Weill Medical College of Cornell University, New York, NY 10021

Robin Young (2: 41)   Département de pathologie et biologie cellulaire, Université de Montréal, Case postale 6128, Succursale "Centre-Ville", Montreal, QC, H3C 3J7, CANADA

Christian Zahnd (1: 497) Department of Biochemistry, Univerisity of Zürich, Winterthurerstr. 190, Zürich, CH-8057, SWITZERLAND

Zhuo-shen Zhao (4: 285) Glaxo-IMCB Group, Institute of Molecular and Cell Biology, Singapore, 117609, SINGAPORE

Huilin Zhou (4: 437) Department of Cellular and Molecular Medicine, Ludwig Institute for Cancer Research, University of California, San Diego, 9500 Gilman Drive, CMM-East, Rm 3050, La Jolla, CA 92093-0660

Timo Zimmermann (3: 69) Cell Biology and Cell Biophysics Programme, EMBL, Meyerhofstrasse 1, Heidelberg, D-69117, GERMANY

Chiara Zurzolo (2: 241) Department of Cell Biology and Infection, Pasteur Institute, 25,28 rue du Docteur Roux, Paris, 75015, FRANCE

Preface

Scientific progress often takes place when new technologies are developed, or when old procedures are improved. Today, more than ever, we are in need of complementary technology platforms to tackle complex biological problems, as we are rapidly moving from the analysis of single molecules to the study of multifaceted biological problems. The third edition of *Cell Biology: A Laboratory Handbook* brings together 236 articles covering novel techniques and procedures in cell and molecular biology, proteomics, genomics, and functional genomics. It contains 165 new articles, many of which were commissioned in response to the extraordinary feedback we received from the scientific community at large.

As in the case of the second edition, the *Handbook* has been divided in four volumes. The first volume covers tissue culture and associated techniques, viruses, antibodies, and immunohistochemistry. Volume 2 covers organelles and cellular structures as well as assays in cell biology. Volume 3 includes imaging techniques, electron microscopy, scanning probe and scanning electron microscopy, microdissection, tissue arrays, cytogenetics and in situ hybridization, genomics, transgenic, knockouts, and knockdown methods. The last volume includes transfer of macromolecules, expression systems, and gene expression profiling in addition to various proteomic technologies. Appendices include representative cultured cell lines and their characteristics, Internet resources in cell biology, and bioinformatic resources for in silico proteome analysis. The Handbook provides in a single source most of the classical and emerging technologies that are essential for research in the life sciences. Short of having an expert at your side, the protocols enable researchers at all stages of their career to embark on the study of biological problems using a variety of technologies and model systems. Techniques are presented in a friendly, step-by-step fashion, and gives useful tips as to potential pitfalls of the methodology.

I would like to extend my gratitude to the Associate Editors for their hard work, support, and vision in selecting new techniques. I would also like to thank the staff at Elsevier for their constant support and dedication to the project. Many people participated in the realization of the *Handbook* and I would like to thank in particular Lisa Tickner, Karen Dempsey, Angela Dooley, and Tari Paschall for coordinating and organizing the preparation of the volumes. My gratitude is also extended to all the authors for the time and energy they dedicated to the project.

Julio E. Celis
Editor

CELL AND TISSUE CULTURE: ASSORTED TECHNIQUES

General Techniques

1

Setting up a Cell Culture Laboratory

Robert O'Connor and Lorraine O'Driscoll

I. INTRODUCTION

Over the past three decades, the continuous culture of eukaryotic cells has become a mainstay technique in many different forms of biological, biochemical, and biomedical experimentation. While at first, to the uninitiated, the techniques, methodology, and equipment can appear daunting, clear specification of the experimental requirements can help make choices straightforward. This article does not purport to be an exhaustive guide but rather aims to prompt the researcher to plan and make choices appropriate to their experimental, environmental, and financial resources.

II. WHERE TO START

Cell culture needs a commitment of energy and resources to be undertaken in a professional manner for any continuous period. Therefore, the biggest decision to be made before going down this experimental road is whether there will be an ongoing need for culture facilities or whether for short periods of work it might be more economical to collaborate with an established laboratory or sub contract work. Assuming that there is an agreed need to set up a cell culture facility, there are several *fundamental considerations*.

A. Environment

Purpose-built facilities are optimal for ergonomic and experimental reasons but often this choice is not available. Small cell culture facilities can be engineered

(with inherent limitations) without making significant modifications to laboratory rooms. Before going into the detailed choices available, it is perhaps timely to go through some of the basics.

There are two fundamental considerations that govern most choices available to the would-be cell culture researcher: contamination and safety.

The fundamentals of cell culture owe much to the basic methodologies developed by microbiologists over the last two centuries. However, microorganisms reproduce several orders of magnitude more rapidly than eukaryotic cells and, in direct competition, bacteria and fungi will rapidly reproduce more biomass than eukaryotic cells. Eukaryotic cells are also very sensitive to the primary and secondary metabolic products of microbes. Bacteria and fungi are therefore the biggest problem, for those growing eukaryotic cells. As mentioned in the article by *Meleady and O'Connor*, there is also potential for cross contamination of one cell line by another if proper procedures are not observed (a major problem with the first cultured human cell lines).

Eukaryotic cells can potentially harbour subcellular microbes that could cause disease to human beings or animals. More specifically, most human cell cultures are derived from human cancers. Being derived from human beings who could be harbouring several known (and potentially, as yet, unknown) pathogens, appropriate steps must be taken to ensure that cells do not pose a risk of passing disease on to human beings, including the researcher or others including visitors or cleaning staff. In practice, the majority of pathogenic organisms are quite fragile and do not survive well under general culture conditions. The risk of disease is therefore greatest when working with primary biological material, i.e., material recently removed from another (human) being. However, one should assume that any eukaryotic cells could potentially harbour microbes and/or viruses (including oncogenic viruses) and/or prion-contaminated material. As a general rule, taking the maximum amount of reasonable precautions (in procedures and equipment) gives the best margin of protection for staff, provides peace of mind for operators, and limits the culpability of supervisors/managers. The Centers for Disease Control (CDC) in the United States stipulate that general cell cultures should be undertaken in biosafety level 2 containment facilities (CDC, 1999).

B. Location

Having briefly outlined these fundamentals, it should be clear that correctly locating a cell culture facilities is of paramount importance. The *working environment* needs to be clean, free from dust, and easy to disinfect. The immediate area should have limited/restricted access, with no passing traffic. Consideration should be given to proper ergonomics in the area, e.g., correct heights of equipment, nothing requiring bending down, suitable chairs, and so on, to reduce the chance of chronic or acute injuries to staff (see the excellent laboratory website by the NIH for further details, specifications, and illustrations). Thought should also be given to how large and often heavy equipment, particularly laminar flow cabinets, can be brought in and out without major disruptions. If a building is being planned, this may include provisions for large lifts with fully opening doors or direct door access to higher floors with high loading equipment. Door openings need to be extra wide, typically at least 1 m full clearance, and corners designed that large equipment can get by. Provision must also be made for the movement and storage of consumables and bulk items. In practice, it is usually better to pool certain sets of equipment together in individual rooms, e.g., several biosafety cabinets being located with an incubator and other ancillary equipment. This reduces the overall equipment/cost necessary and can also make for a more "communal" working environment. Quarantine areas are an obvious exception to this suggestion.

C. Gases

Many cell cultures can be maintained in HEPES-buffered medium, which utilises carefully controlled incubators that do not need a separate carbon dioxide gas supply. However, some cells do not grow optimally in HEPES and need the buffering provided by the equilibrium of 5% CO_2 with bicarbonate in medium. This can be important for some specific cell lines, primary culture, and hybridoma work (Freshney, 2000). At a minimum, such incubators should have two separate CO_2 sources supplying them. Direct connection of a single cylinder means that the incubators are vulnerable every time a cylinder is changed and cylinders may fail to provide an adequate pressure of gas as they near an empty state. Cylinder changeover units permit, for example, one main line of gas and one backup cylinder, which means that incubators can be left for weeks or months (depending on use) without needing gas replacement and, when cylinders are replaced, there is no interruption in supply to the incubator (see NTC services website). Cylinders in a laboratory environment must always be fully fixed to an immovable object to lessen the risk of them toppling and doing serious injury and damage. Appropriate automatic changeover units can also be incorporated into external supplies of CO_2. It is always better not to have large cylinders of gas in a culture room for safety, practical, and aesthetic reasons. Building regulations in some areas may also legislate against the use of large cylinders in enclosed rooms.

D. Ventilation

Ventilation and airflow in the cell culture environment are critical to operation. At its simplest, there must be no disruption to the laminar airflow pattern in the biological safety cabinet and no undue circulation of dust and dirt that could occur with, for example, significant staff movement around the unit or location near drafts or vents. Ideally, there should be no openable windows; if so, they should be sealed to prevent drafts, insects, and dust entering (Freshney, 2000). However, in a purpose-built facility, if possible, it is ergonomically and aesthetically desirable that there be a source of natural light. When planning, one should also make provisions in case there is a need to fumigate the room or equipment (Doyle, 1998). Cabinets will usually need to be fumigated in advance of any filter changes, although this can now usually be performed on single cabinets (using a cabinet bag system) rather than whole rooms.

Clearly, whatever the room design, there must be some replenishment of air in a room and such ventilation must not interfere with the operation of the cabinet. If liquid nitrogen is being utilised in the same location as the cabinet, the ventilation must be adequate to remove the continuous evaporation of liquid and the ventilation/air space sufficient to ensure that if there is an acute spillage of liquid or a rupture of the vacuum vessel, there will still be sufficient oxygen concentration to support life and permit evacuation, i.e., that instant evaporation of all the liquid nitrogen will not reduce the oxygen concentration below 14%. If this cannot be guaranteed at all times, oxygen monitoring and alarm equipment will be required (Angerman, 1999).

High efficiency particulate air (HEPA) filtration within the safety cabinet will ensure a good quality working environment. However, if the air around the unit is dusty, the cabinet filters will age and clog quite rapidly. HEPA filtration is a statistical process (99.999%–99.97% efficient depending on the manufacturer) and unduly contaminated air may also reduce the air quality inside the cabinet. HEPA filtration of air into a cell culture room should improve operation and cabinet filter longevity and is also a requirement of biocontainment-classified rooms. However, if the room itself is not maintained properly, expensively ducted HEPA-filtered air may merely be clean air being utilised to circulate dust and microbes. Depending on the biocontainment level required for operation, HEPA filtration may also be required on exhaust vents for a cell culture room (class 2+ biocontainment and above).

In larger cell culture facilities, careful balancing of air pressures in culture and anterooms may be useful or may be required for operation to appropriate biocontainment levels, specifically the positive and negative air pressures required for class 2 and greater biocontainment (CDC, 1999). Balancing and filtration require careful installation and validation from the outset and continued regular maintenance and monitoring to ensure appropriate function. Ventilation systems will also need to be tied into building fire management systems with, for example, automatic smoke dampeners to prevent the ventilation system from fanning or spreading smoke and fire.

E. Basic Cell Culture Requirements

To perform a basic range of cell culture procedures for any prolonged period the following list of equipment is required (Freshney, 2000).

Laminar flow cabinet
Incubator
Centrifuge
Refrigerator
Freezer
Microscope
Haemocytometer
Pipette boy
Micropipette

Outside immediate culture environment
    Autoclave
    Selection of appropriate consumables and cultureware

General laboratory facilities for the storage of chemicals, provision of standard reagents such as clean water, and controlled temperature water baths

F. Ideal Layout

1. Purpose-Built Facility

There are significant advantages in having the resources to custom design and implement a *purpose-built facility*. A purpose-built facility should be strongly considered (and will likely be more economical in the long term) if cell culture is performed on a significant scale (involving many researchers) for an established organisation and/or if this work is likely to be undertaken on a commercial basis. Aside from the support departments, which are mentioned later, the standard accepted laboratory scheme for cell culture involves a general laboratory room, an anteroom (physically and methologically separating the general laboratory area from the cell culture room), and a specific, self contained, cell culture room.

2. Equipment

a. Biosafety Cabinets. While basic cell culture can be performed with sterile equipment and good techniques, there is no question that the development of modern biological cabinets has greatly facilitated routine culture procedures and the prolonged manipulation of cells required for many experimental procedures. Laminar flow cabinets come in many sizes, capacities, and with several variants appropriate for different types of cell culture. Most manufacturers have excellent technical schematics describing the appropriate operation, dimensions, and so on. For examples, see the websites by Heto-Holten or Baker. Any laminar flow cabinet should be designed to an internationally recognised specification and biological safety standard appropriate to the work being undertaken or likely to be performed in at least the next 5 years. [See CDC (1995, 2000) and the list of international specifications in the Appendix for further details.] Laminar flow cabinets represent a significant investment and will usually give decades of service if adequately maintained. A careful choice is therefore important. The precise operation of a laminar flow cabinet is beyond the scope of this review except to say that they fundamentally consist of a large air-pumping motor, pumping air through a HEPA filter onto a working surface (CDC, 2000). Cabinet HEPA filters are designed to remove particulates (from proportionately large dust down to submicroscopic viral particles). Forcing large volumes of air through these filters needs large motors, which inherently imply significant weight in their own right, and a heavy chassis to support this weight and to prevent vibration. Before locating/installing, one therefore needs to consider how the unit is to be placed in the required position. Typical weights can vary from 200 kg for new models to an excess of 500 kg for older units (which may still be very usable). More modern concrete-fabricated buildings will usually be designed to take the weight of one or more cabinets on a limited floor space, but this may need to be checked with an appropriate engineer and will certainly cause limitations in many lighter prefabricated or older buildings. A clear path from the point where a delivery vehicle may drop off the cabinet to its final resting place is required. Doors will need to be wide enough for the delivery pallet or, at least, the unpacked unit. Lift size and weight restrictions may be a particular problem, as can sharp corners on the route. Lift specifications should always be in excess of the largest envisaged unit that they might be expected to transport. In addition or if an appropriate lift cannot be provided, direct access to corridors on upper floors should be provided so that low loaders or cranes can directly insert heavy equipment to each floor of the building. Although one should plan for the unit to be in place for many years, one should also consider possibilities of bringing in new units and/or removing units as a particular group of scientists may expand or move their operation. Careful planning is therefore essential. Some manufacturers also have built-in modifications or supply units in subsections, which allow their units to take up a smaller space during movement than required for the operational unit.

Which Unit to Purchase/Use? Several international bodies have developed standard specifications for laminar flow cabinets (*see specification list in Appendix*). General cell culture requires a class II cabinet for an adequate protection of cells and operator (see Section A). These units recirculate HEPA-filtered air and exhaust a portion of that air back into the room through a HEPA filter. Where hazardous agents may be used, higher specification units such as external venting cabinets may be required (e.g. class II type B2/B3). These require very specific and costly exhaust ducting (CDC, 2000).

A careful choice of cabinet size should be made. In our centre, many researchers prefer a 4-ft (1.2 m)-wide laminar with plenty of space between the inner top of the cabinet and the work surface. Smaller cabinets may suit more restricted spaces but also restrict the amount of work that can be undertaken in comfort inside. Larger units may be useful for bulkier operations such as batch media production. The operator should be able to get their legs fully under the working space to permit appropriate posture and to reduce stretching and bending, which can cause repetitive strain injuries. The unit and the culture room should be well lit with low flicker lighting of an appropriate intensity (preferably supplemented with natural light). The motor size of safety cabinets causes the units to give off a significant amount of heat. A very significant heat load may be generated in areas where there are several (nonducted) units in a confined space with incubators. The air-handling system must be able to cope with such a loading and maintain the temperature at a comfortable level. Optional extras, such as ultraviolet lights, are often dispensable and may have limitations or be inappropriate. If laboratory benches are used to support the cabinet, they must be sufficiently sturdy to support the full weight of the unit for the duration of use. Legs must also be sufficiently broadly spaced to prevent any risk of toppling.

b. Incubators. Most mammalian cells are maintained at 37°C, whereas insect cell cultures typically grow at 28°C. As mentioned previously, there may also be a need for a regulated use of CO_2 and possibly other gas mixtures in the cabinet. Many different-sized *incubators* are available to suit the needs and requirements of the researcher. Temperature control and accuracy usually to 0.1°C is an obvious requirement, and units that have heated doors and glass inner doors, permitting limited observation without letting all the heat out, are preferable. In our laboratory, units of 200- to 300-liter internal capacity are capable of supporting the culture output of several researchers. Roller bottle and spinner flask culture vessels may also require special incubators or adapters. The incubator should ideally be located close to the laminar cabinet to reduce temperature changes, which could affect cultured cells. General cell culture incubators require the ambient temperature to be usually 5–10°C colder than the target temperature. If the ambient temperature is too close to the target temperature, standard units can overheat, making temperature control in the room critically important for reliable operation. The incubator should also be at standing height to prevent the need for bending down. Temperature-controlled warm rooms may be useful in specific circumstances; however, larger incubators and particularly warm rooms are very difficult to maintain at a homogeneous temperature all through the space, which can adversely affect growth.

All internal surfaces should be polished metal and accessible for cleaning purposes. Any surface or component that cannot be cleaned but is inside the air space of the incubator is likely to see significant microbial growth as residue builds up. Where vented culture flasks and CO_2 are being used, the unit must also have a bath of clean water inside and the incubator must monitor relative humidity. Vented flasks and other unsealed culture vessels, such as 96-well culture plates, will rapidly dry out if humidity drops.

Most units available now are very straightforward to calibrate the level of CO_2 and temperature. Sensors that use infrared measurement of CO_2 are the easiest to calibrate as opposed to older electrochemical sensors. It is advisable to periodically check the calibration of such equipment using external measuring devices.

c. Centrifuges. Many different models of centrifuge are available for cell culture. For ergonomic reasons, the unit should be located near the biosafety cabinet and at an appropriate height. All *centrifuges* will vibrate to some degree, and measures should be taken to ensure that such vibrations cannot damage other equipment, cause unnecessary noise, or allow the centrifuge to creep and potentially fall. The centrifuging of tubes containing media and cells can induce aerosols, and all centrifuges used for cell culture and biological procedures must have some form of seal either above the rotor head or over the buckets, which prevents aerosol leaking, particularly where tubes may rupture or leak during operation. A common operational centrifugation rate is appproximately $110 \times g$, and units must capable of fractions and low multiples of this rate to cope with different research requirements. Temperature control, particularly the ability to spin at refrigerated temperatures, can be a very useful add on to standard units. The buckets in the rotor should have adapters that snugly fit the standard consumables used for culture in that facility, e.g., 25- and 50-ml universal tubes. Control of the acceleration and deceleration rates can also be useful when working with certain very sensitive cell lines.

d. Refrigerator/Freezer Units. Storage of media, media components, buffers, and so on needs *refrigeration and freezing facilities*. Domestic-type units are often used for this purpose; however, domestic units can have limitations in their ability to maintain a steady temperature and must also never be used if flammable liquids are to be stored inside (an externally thermostat-controlled unit must be used for such purposes). It can also be useful if the door has a lockable latch, as this can help ensure that the unit is closed properly after every use. In practice, careful monitoring of the storage and use of refrigerators/freezers are

necessary to prevent space wastage. Fridges and freezers should be cleaned out regularly. When this is not done, it is common to see gradual increases in the number of such units, which can greatly add to heat loads in the laboratory. In larger scale facilities, it may be prudent to combine the use of fridges/freezers with longer term, volume storage in central cold/freezer rooms. In such laboratories it may also be useful to use cooling/freezing units that have condenser units located outside the laboratory. Although not usually required for cell culture operations, freezers that can operate below $-60°C$, usually $-80°C$, are often necessary for the storage of molecular biological enzymes and reagents.

e. Liquid Nitrogen. Maintenance of cultured cells for any prolonged period requires that cell stocks can be kept well below the glass temperature of water (approximately $-60°C$). In practice, *liquid nitrogen storage* is often a convenient general storage environment for stocks of cultured cells. However, electrical freezers operating at $-120°C$ can perform a similar function. Liquid nitrogen boils at $-196°C$, and because the resultant nitrogen gas is an asphyxiant in high concentrations, proper handling and ventilation procedures are necessary (*see article by Meleady and O'Connor*). Proper inherently safe methods for transport and storage must be used to get the liquid gas into the cell storage containers. As the storage vessels usually require topping up at least once a week (more regularly as the containers age), the route from the gas delivery/production area to the laboratory must be as short as possible and the surface and equipment appropriate for the transport involved. Storage of cells by multiple users for prolonged periods necessitates careful inventory management to ensure that cells are maintained optimally and economically. It is good practice to have two independently stored stocks of cells: one vessel containing master stocks of important cells and the other, in the laboratory environment, containing working stocks.

Modern incubators, refrigerators, freezers, and liquid nitrogen vessels typically have alarms that can be set if parameters, such as temperature, drop below a critical values. However, such alarms are only of use if there is someone nearby to hear them. Where critical or expensive procedures are being utilised, it is wise to have such equipment linked into a monitoring system that can alert a researcher day or night. Consideration should also be given to the provision of a backup electricity generator that can automatically restore supply for a finite period in the event of an electrical blackout.

f. Autoclave. Modern single-use plastic consumables have reduced the dependence on a laboratory *autoclave* for sterilising equipment and reagents; however, an appropriately sized, robust autoclave is still vital for continued cell culture work. Although appropriately sourced plastic consumables and presterilised reagents can eliminate the potential for contamination, there can still be huge cost savings by utilising specific pieces of glass equipment that can be repetitively autoclaved, used, and recycled, particularly reagent and media bottles, and by sterilising one's own general reagents, such as water and phosphate-buffered saline. Small autoclaves can generate significant amounts of foul odour, steam, and heat and should not be used in a general laboratory environment. Because of the biohazardous nature of cell culture, autoclaving of all materials and reagents that come into contact with cells is also required. For larger facilities, it is strongly recommended to have a centralised autoclave facility away from an individual laboratory. A distinction of clean and waste autoclaves is necessary to prevent cross contamination, particularly of odours and volatile substances. A backup autoclave is also good insurance against maintenance and unanticipated downtime. As such facility autoclaves are large, need regular maintenance and inspection, and can have a limited life span, provision for easy access and removal/replacement is advisable. The operational characteristics of the autoclave should also be regularly checked with spore strips, for example, to validate that the autoclave is operating effectively.

3. General Environmental Recommendations

Ideally, there should be a limited facility for the temporary storage of small amounts of culture consumables next to the biosafety cabinet. However, cardboard can be a significant source of fungal spores, and large-scale storage in the culture area causes clutter. Consumables are best stored in a central location, and if cell culture is being undertaken on a larger scale, it is far more economical to bulk purchase supplies and distribute them as necessary to each laboratory. Provision also needs to be made for the storage of flammable materials, especially disinfectant alcohols, which are used for local disinfection and "swabbing" (disinfectant wiping of consumables and reagents as they enter the working space inside safety cabinets).

Ideally, all flooring, walls, and other surfaces in the culture environment should be readily accessible, chemically resistant, nonadsorbent, and easy to clean. Shelves should not be used, and open flat surfaces should be minimised as these will need to be cleaned and can be a source for the buildup of dirt and dust.

Walls should be smooth skimmed and coated with epoxy paint. The floor should also be resistant and bonded to the wall so that there are no crevices or corners that cannot be cleaned and any spillages can be easily isolated and cleaned. Sinks, coat hooks, and so on should all be kept in an anteroom and not in the culture room.

Regular training, standard operating procedures, and centralised management of all aspects of cell culture, particularly technique and safety, ensure that there is an economical, continuously monitored, high standard of operation. The function of vital equipment should be continuously monitored and recorded. Particular attention should be paid to the monitoring of environmental microbial levels throughout the facility. For example, "settle" plates should be periodically left in laminar flow cabinets to check for sterility, as well as in all laboratory areas to validate the cleanliness of such areas.

The microbial quality of cell stocks should also be monitored as part of this process. All cells coming into a facility, regardless of the source, especially if supplied by noncommercial sources, should be quarantined and initially cultured in isolation from the general culture environment. *Mycoplasma* is the main microbial contaminant of concern because such contamination can go unnoticed for long periods and is very easy to pick up and cross contaminate stocks of cells. Ideally, a facility should have provision for routine *Mycoplasma* detection. Experience and expertise are required to do this job reliably, and where such expertise is not available in-house, commercial testing facilities exist.

Industrial disinfectants such as Virkon and/or Tego need to be employed (at recommended dilutions) as part of the routine cleaning of equipment and areas, in addition to diluted industrial methylated spirits or isopropyl alcohol, which is used for local disinfection and "swabbing."

Other Areas. A successful laboratory will have a two-way flow of people and materials. To manage and control these, it is useful to have a centrally located reception area to administer all incoming deliveries of consumables, cells, chemicals, and so on. Provision needs to made for effective communication within the organisation, including phone access in all but the most specialist of rooms, internet, and e-mail access and storage of records and scientific literature. If the unit is fully self-contained, the human environment must also be considered with an adequate provision of locker and toilet facilities and appropriate locations for taking breaks and eating completely away from the laboratory.

III. SUMMARY

The reader will note the preference in this text for larger centralised, purpose-built facilities. In individual situations, it may be appropriate to represent the areas mentioned in a smaller more general way; however, even at its simplest, maintenance of a small cell culture environment necessitates that time be spent validating and monitoring the quality of that environment and the various flow paths contained therein. This obviously reduces the time available for research. However, in larger facilities, such critical validatory tasks can become specific jobs enabling the researcher to concentrate with confidence on biological research. Where such facilities are properly established, it is common that they continue to expand. The final thought in setting up a cell culture environment is to look as much as possible to the future and to allow for changes in use and facility expansion from the very start.

APPENDIX

The following relevant international standards are used for biosafety cabinets.

American National Standards Institute (National Sanitation Foundation). NSF/ANSI—49 (1992). Class II (Laminar Flow) Biohazard Cabinetry. Now superseded by NSF/ANSI49 (2002) and NSF/ANSI 49-02e (2002).

Australian standards AS2252.1, Biological Safety Cabinets (Class I) for personal protection (1980). AS2252.2 Biological Safety Cabinets (Class II) for personal protection (1981). Standards Association of Australia.

Canadian Standards Association, CSA Z316.3-95 (1995). Biological Containment Cabinets: Installation and Field Testing.

European standard EN12469:2000 (2000). Performance criteria for microbiological safety cabinets. Supersedes EU member state standards such as BS 5726 (British), DIN 12950 (German), and NF X44-201 (French).

Japanese Industrial Standard JIS K 3800 (JACA) for Class II biological safety cabinets.

South African standards, SABS 0226:2001 (2001). The installation, postinstallation tests, and maintenance of microbiological safety cabinets (2001) and VC 8041:2001, microbiological safety cabinets (Classes I, II, and III) (2001).

References

Angerman, D. (1999). "Handbook of Compressed Gases," 4th Ed. Kluwer Academic, New York.

Baker Biosafety cabinet website. http://bakercompany.com/products/ 161 Gatehouse Road, Sanford, Maine 04073 USA.

CDC (1995). "Biosafety Cabinets; Primary Containment for Biohazards: Selection, Installation and Use of Biological Safety Cabinets," 1st Ed. U.S. Department of Health and Human Services Public Health Service Centers for Disease Control and Prevention and National Institutes of Health. U.S. Government Printing Office, Washington. Web edition http://www.niehs.nih.gov/odhsb/biosafe/bsc/bsc.htm

CDC (1999). U.S. Department of Health and Human Services Public Health Service Centers for Disease Control and Prevention and National Institutes of Health. "Biosafety in Microbiological and Biomedical Laboratories," 4th Ed. U.S. Government Printing Office, Washington. Web edition http://www.cdc.gov/od/ohs/biosfty/bmbl4/bmbl4toc.htm. See also the CDC Office of Health and Safety biosafety general website: http://www.cdc.gov/od/ohs/biosfty/biosfty.htm

CDC (2000). "Primary Containment for Biohazards: Selection, Installation and Use of Biological Safety Cabinets." U.S. Department of Health and Human Services Public Health Service Centers for Disease Control and Prevention and National Institutes of Health. U.S. Government Printing Office, Washington. Web edition http://www.cdc.gov/od/ohs/biosfty/bsc/bsc.htm

Doyle, A. (1998). "Cell and Tissue Culture: Laboratory Procedures" (A. Doyle, J. B. Griffiths, and D. G. Newell, eds.), Chap. 1. Wiley, Chichester.

Freshney, R. I. (2000). "Culture of Animal Cells: A Manual of Basic Technique," 4th Ed. Wiley-Liss, New York.

Heto Holten Biosafety cabinet website. http://www.heto-holten.com/prod-holten.htm. Heto-Holten now part of Jouan Nordic, Gydevang 17–19, DK-3450, Allerød, Denmark.

National Institutes of Health (NIH) laboratory safety website http://www.nih.gov/od/ors/ds/ergonomics/lab3.html

NTC services limited. Northern Technical & Chemical Services. Unit D44, Brunswick Business Centre, Liverpool, L3 4BD.UK. http://www.merseyworld.com/ntcs/

2

General Procedures for Cell Culture

Paula Meleady and Robert O'Connor

I. INTRODUCTION

A. Background

Mammalian cell culture emerged as a valuable research tool in the 1950s when the first cell line, HeLa, was successfully cultured from a human cervical cancer (Gey *et al.*, 1952). However, it is only since the mid-1980s that reproducible and reliable large-scale culture of mammalian cells has been achieved. The development of cell culture led to new experimental approaches to cellular physiology in which isolated, functionally differentiated cells could be maintained in culture under conditions that allowed direct manipulations of the environment and measurement of the resulting changes in the function of a single cell type. Today many aspects of research and development involve the use of animal cells as *in vitro* model systems, substrates for viruses, and in the production of diagnostic and therapeutic products in the pharmaceutical industry.

The process of initiating a culture from cells, tissues, or organs taken directly from an animal and cultured either as an explant culture or following dissociation into a single cell suspension by enzyme digestion is known as primary culture. Certain primary cultures may be passaged for a finite number of population doublings before senescence occurs, but usually the number of doublings is limited in adult-derived or differentiated cell types. However, these cells are still invaluable as they retain many of the differentiated characteristics of the cell *in vivo*. After a number of subcultures a cell line will either die out, referred to as a finite cell line (and is usually diploid), or a population of cells can transform to become a continuous cell line.

Lines of transformed cells can also be obtained from normal primary cell cultures by infecting them with oncogenic viruses or treating them with carcinogenic chemicals. It is often very difficult to obtain a normal human cell line from a culture of normal tissue. In contrast, neoplasms from humans have been generated into many cell lines. It appears that the possession of a cancerous phenotype allows the easier adaptation to cell culture, which may be due in part to the fact that cancer cells are aneuploid. Transformed cell lines have the advantage of almost limitless availability; however, they often retain very little of the original *in vivo* characteristics.

Cell cultures *in vitro* take one or two forms, either growing in suspension (as single cells or small clumps) or as an adherent monolayer attached to the surface of the tissue culture flask. It is necessary that cell culture medium is produced so that it mimics the physiological conditions within tissues. *In vitro* growth of cell lines requires a sterile environment in which all the nutrients for cellular metabolism can be provided in a readily accessible form at the optimal pH and temperature for growth. Media formulations vary in complexity and have been developed to support a wide variety of cell types, including Eagle's minimum essential medium (MEM), Dulbecco's modified Eagle's medium (DMEM), RPMI 1640, and Ham's F12. Cell culture media essentially consist of a number of factors required for the growth of the cells, including amino acids (essential and nonessential), lipids (essential fatty acids, glycerides, etc.), trace elements, vitamins, and cofactors. Carbohydrates such as glucose or fructose are usually added as an energy source. Other essential components include inorganic salts, which provide buffering capacity and osmotic balance (260–320 mOsm/kg) to counteract the effects of carbon

13

dioxide and lactic acid produced during cellular metabolism. The pH of medium should ideally be between 7.2 and 7.4 (however, fibroblasts prefer a pH range of 7.4–7.7 whereas transformed cells prefer a pH range of 7.0–7.4). When phenol red is included in the medium, the medium turns purple above pH 7.6 and yellow below pH 7.0. Buffering of culture media is usually provided by sodium bicarbonate and cells are usually maintained in vented tissue culture flasks in an atmosphere of 5% CO_2. Synthetic buffers such as HEPES can also help maintain correct pH levels in closed, nonvented tissue culture flasks, although it may be toxic to some cell types. Serum, which is a complex mixture of albumins, growth promoters, and growth inhibitors, may also be incorporated into the growth medium at concentrations from 5 to 20%, although certain production processes and experimental procedures require the use of serum-free conditions. The majority of cell lines require the addition of serum to defined culture medium to stimulate growth and cell division but can be subject to significant biological variation. The most common source is bovine and this may be of adult, newborn, or foetal origin.

A number of other books give more detailed and comprehensive treatment of procedures for mammalian cell culture and the reader is recommended to refer to these books for additional information (Doyle *et al.*, 1998; Freshney, 1992, 2000; Shaw, 1996).

B. General Safety when Working with Mammalian Cells in Culture

In general, because of the potential risks that may be associated with material of biological origin, standard and specific laboratory regulations should always be adhered to.

- Antibodies, sera, and cells (particularly but not exclusively those of human and nonhuman primate origin) may pose a potential threat of infection or other biological hazard (e.g., prion disease).
- Many animal cells contain C-type particles, which may be retrovirus related. All such materials may harbour pathogens and should be handled as potentially infectious material in accordance with local guidelines.
- Laboratory coats are essential. The Howie-type coat is the only recommended coat for biological work. Coats should be used only in the culture area and should be laundered frequently.
- Protective glasses should be worn at all times while contact lenses should never be worn in the laboratory area.

- No eating, drinking, or smoking should be permitted.
- No mouth pipetting of any solutions.
- Operators must make sure that any cuts, especially on the hands, are covered. Wearing gloves is strongly recommended, particularly for manipulations involving cells and biological material. Gloves should be of a standard appropriate to the risk of the agent being handled. Nitrile gloves may provide superior protection with lower allergy potential than traditional latex gloves.
- Thorough washing of hands before and after cell work with appropriate laboratory soap is essential.
- Immunisation against hepatitis B may be recommended if working with primary human material. In certain countries, where tuberculosis (TB) vaccination is not a standard (or staff are employed from such countries and where the material being handled may have a TB hazard) BCG vaccination may also be recommended (Richmond and McKinney, 1999). However, this should be at the discretion of the individual operator, and regular follow-up and paperwork, including titre estimation, are necessary to have an effective vaccination policy.
- To comply with current safety regulations, a cell culture laboratory should be fully ventilated, preferably with high efficiency particulate air (HEPA) filters on the inlets, and equipped with HEPA-filtered workstations where the airflow is directed away from the operator. A class II (type A) downflow recirculating laminar flow biological safety cabinet will provide a safe working environment for standard hazard material and should be checked yearly (or as recommended by manufacturer) for containment, airflow velocity, and efficiency. A horizontal flow cabinet should never be used, as it can, even in the absence of viruses, possibly increase exposure to allergens.
- In addition to standard alcohol-based disinfection for routine work and introduction of consumables, primary equipment and work areas should be regularly disinfected with laboratory disinfectants, e.g., Virkon, Tego, or equivalent. Manufacturer guidelines should be followed as some disinfectants may corrode materials.

II. MATERIALS AND INSTRUMENTATION

A. Materials

The following are from Sigma Aldrich: DMEM (Cat. No. D5648), Ham's F12 (Cat. No. N2650), RPMI (Cat.

No. R6504), NaHCO$_3$ (Cat. No. S5761), HEPES (Cat. No. H4034), dimethyl sulphoxide (DMSO, Cat. No. D5879), foetal calf serum (FCS, Cat. No. F7524), EDTA (Cat. No. E5134), soybean trypsin inhibitor (Cat. No. T6522), haemocytometers (Neubauer improved chamber) (Cat. No. Z35, 962–9), and replacement coverslips (Cat. No. Z37, 535–7).

The following are from Invitrogen (GIBCO brand): 200 mM (100×) L-glutamine (Cat. No. 25030-024), penicillin/streptomycin (5000 IU/500 μg/ml) (Cat. No. 15070-063), 2.5% trypsin (10×) (Cat. No. 15090-046), and Trypan blue (Cat. No. 15250-061).

The following are from Corning (Costar brand): 10-ml (Cat. No. 4101CS) and 25-ml (Cat. No. 4251) pipettes, 25-cm^2 (Cat. No. 3055) and 75-cm^2 (Cat. No. 3375) nonvented tissue culture flasks, and 25-cm^2 (Cat. No. 3056) and 75-cm^2 (Cat. No. 3376) vented tissue culture flasks.

The following are from Greiner: 30-ml (Cat. No. 2011-51) and 50-ml (Cat. No. 210161G) sterile containers/centrifuge tubes, cryovials (Cat. No. 122278G), and autoclave bags (Cat. No. Bag1).

The following are from Millipore: 0.22-μm low protein-binding filters for small volumes (Cat. No. SLGVR25KS) and 0.22-μm filters for large volumes (Cat. No. SPGPM10RJ).

Phosphate-buffered saline (PBS) (Cat. No. BR14A) is from Oxoid. Laboratory disinfectants such as Virkon (Cat. No. CL900.05) and Tego (Cat. No. 2000) are from Antec and Goldschmidt, respectively.

B. Instrumentation

Automatic pipette aids (Accu-jet, Model No. Z33,386–7) are from Sigma Aldrich, the inverted microscope with phase-contrast optics (Model No. DM 1L) is from Leica Microsystems, and the centrifuge is from Eppendorf (Model No. 5810). The 37°C incubator (Model No. 310) is from Thermo Forma. The laminar flow cabinet (Model No. NU-425-600) is from Nuaire.

Other standard laboratory apparatus includes refrigerators, –20°C and –80°C freezers, and a liquid nitrogen freezer. Clean autoclaves should be available for solutions, glassware, and other items that require sterilisation by moist heat. A separate waste autoclave should be available for general biological laboratory waste, including plastics and waste media. Dishwashers should also be available to ensure thorough cleaning of all glassware used in cell culture procedures.

III. PROTOCOLS

A. Good Practice and Safety Considerations

Equipment in the designated area for cell culture should be kept to the minimum required for the job. There should be proper entry facilities and internal surfaces must be easy to clean and dust free. When setting up a cell culture laboratory the following equipment is essential:

- Class II downflow recirculating laminar flow cabinet
- Low-speed biological centrifuge
- CO$_2$ incubator
- Inverted microscope with phase-contrast capabilities
- Refrigeration and freezing facilities
- Cell storage (liquid nitrogen) facilities

Prevention of contamination by bacteria, fungi (especially yeast), mycoplasma, or viruses is absolutely necessary in cell culture. Good laboratory practice requires that the following standard procedures are followed:

- Cell culture should be performed in a designated area that is easy to clean and free from clutter. Equipment used should broadly be designated for that purpose to prevent potential chemical or biological contamination by or due to other laboratory processes and operation. Ensure that all equipment are cleaned and serviced regularly.
- Cell culture by definition involves the handling of biological material. As such, all biological material can potentially harbour infective agents. Therefore, routine precautions to prevent infection should be exercised. Waste media and items coming into contact with biological material should be disinfected; autoclaving is probably the most broadly useful method. Where material of primary origin is in use and in the absence of specific legal guidelines, validation of the inactivation of biological material is vital.
- Although the majority of common culture lines are characterised as biosafety level 1, the Centers for Disease Control and Prevention (CDC) in the United States suggest that handling procedures be of biosafety level 2 standard, where material is of mixed origin, including some primary material. Biosafety level 2 or 2+ will be mandatory unless prior knowledge indicates the need for even higher standards of safe handling (Richmond and McKinney, 1999).
- It is important that cell lines are obtained from a reputable source, preferably the laboratory of origin or an established cell repository (e.g., ATCC or ECACC).

Cells from all sources should be handled in quarantine until all quality control checks are completed, particularly microbial (and especially *Mycoplasma*) contamination should be checked.

- Only sterile, wrapped items (i.e., pipettes, culture flasks) should enter the room, and discarded media and waste should be removed each day. All used materials should be disposed of safely, efficiently, and routinely in accordance with local regulatory requirements. Keep cardboard packaging to a minimum in all cell culture areas.

- Stock cultures of two cell lines should never be worked on at the same time in a laminar flow cabinet. When working with different cell lines, a thorough cleaning of surfaces with a suitable disinfectant is required and a minimum of 15 min between handling different cell lines is essential to prevent cross contamination. Use of pipettes, medium/waste bottles, and so on for more than one cell line is another possible source of cross contamination and must be avoided.

- When setting up a large frozen stock line, aliquots should be thawed to test for viability, growth, and absence of contamination (including *Mycoplasma*). They should also be characterised and authenticated by some appropriate criteria (e.g., DNA fingerprinting and cytogenetic analysis), both for comparison to the parent cell line and as a standard for comparison of future stocks.

- It is advisable not to use antibiotics continuously in culture medium as this will inevitably lead to the appearance of antibiotic-resistant strains.

- Quality control all of the reagents used in tissue culture to ensure all materials used provide reproducible growth characteristics and are free from contamination prior to use with cells.

- Ensure laboratory coats are changed regularly.

- A disinfectant, e.g., 70% isopropyl alcohol [or alternatives such as 70% industrial methylated spirits (IMS) or ethanol] should be used liberally to disinfectant work surfaces, bottles, plastics, and so on.

- Cell culture is often combined with other methodologies, including chemical and drug methodologies. These may necessitate additional consideration of the chemical/toxicological hazards involved (e.g., use of cytotoxic laminar flow cabinets, cytoguards).

B. Subculturing Cells

In order to maintain cell cultures in optimum conditions it is essential to keep cells in the log phase of growth as far as it is practicable. There are two general types of culture method appropriate for adherent and suspension cells.

1. Subculturing Adherent Cells

Adherent cells (usually of epithelial, endothelial, and fibroblastic origin) will continue to grow *in vitro* until either they have covered the surface available for growth or they have depleted the nutrients in the surrounding medium. Cells kept for prolonged periods in the stationary phase of growth will lose plating efficiency and may become senescent, lose viability and other characteristics, or even die. The frequency of subculture is dependent on a number of factors, including inoculation density, growth rate, plating efficiency, and saturation density. These factors will vary between cell lines.

Protocol 1. Subculturing of Adherent Cells

The following procedure is described for the adherent cell line A549 (human lung adenocarcinoma, purchased from the ATCC) of epithelial origin, which can be subcultured indefinitely. These cells can be grown in HEPES-buffered medium and can therefore be routinely grown in nonvented tissue culture flasks in a normal 37°C incubator.

Media and Solutions

100× L-glutamine stock (200 mM): Thaw stock solution and aliquot into sterile 5-ml amounts and store at −20°C.

100× penicillin/streptomycin solution: Aliquot stock solution into sterile 5-ml amounts and store at −20°C.

FCS: Thaw stock solution and aliquot into sterile 20- to 50-ml amounts and store at −20°C.

1× phosphate-buffered saline (PBS): Add 1 tablet to 500 ml of ultrapure water, according to manufacturer's instructions, and autoclave. Store sterile solution at 4°C.

1% EDTA solution: Add 1 g of EDTA to 100 ml of ultrapure water and autoclave. Store sterile solution at 4°C.

0.25% trypsin/EDTA solution: Thaw the 2.5% (10×) stock trypsin solution. To 440 ml of sterile PBS solution, add 10 ml of sterile 1% EDTA and 50 ml of 2.5% trypsin stock solution. Mix resultant solution by gentle inversion and aliquot into sterile 20 ml amounts. Store aliquots at −20°C.

1× DMEM: Prepare 1× DMEM from 10× powder according to the manufacturer's instructions, supplement the medium with $NaHCO_3$ and HEPES (adjust pH to 7.2–7.4), and filter sterilise using a 0.22-μm filter capable of handling large volumes or, alternatively, purchase premade 1× medium from a relevant supplier. Store basal media at 4°C.

1× Ham's F12: Prepare 1× Ham's F12 from 10× powder according to the manufacturer's instructions, supplement the medium with $NaHCO_3$ and HEPES (adjust pH to 7.2–7.4), and filter sterilise using a 0.22-μm filter capable of handling large volumes or, alternatively, purchase premade 1× medium from a relevant supplier. Store basal media at 4°C.

Complete DMEM/Ham's F12 1:1 medium: To prepare 100 ml of complete medium, mix 47 ml of 1× DMEM with 47 ml of 1× Ham's F12 and supplement with 5 ml of FCS and 1 ml of 100× L-glutamine. One milliliter of antibiotic solution (e.g., 100× penicillin/streptomycin solution) may also be supplemented to the medium if absolutely necessary.

Procedure

1. Examine the condition of the cells using an inverted microscope with phase-contrast capabilities. Ensure that the cells are healthy and subconfluent (i.e., in the exponential phase of growth) and free of contamination.

2. Sanitise the laminar flow cabinet by wiping the surface of the working area with 70% IMS (or equivalent alternative). Wipe the surfaces of any materials prior to starting work, including gloves, media bottles, and pipettes.

3. Remove the spent growth medium from the flask using a pipette and wash the monolayer with a sufficient volume of prewarmed trypsin/EDTA solution (0.25% trypsin/EDTA solution in PBS) to ensure the removal of all media from the flask.

4. Add an appropriate volume of the 0.25% trypsin/EDTA solution to the flask (2 ml for a 25-cm² flask, 4 ml for a 75-cm² flask, 7–8 ml for a 175-cm² flask) and incubate at 37°C to allow the cells to detach from the inside surface of the flask (the length of time depends on the cell line, but usually this will occur within 2–10 min). Examine the cells with an inverted microscope to ensure all the cells are detached and in suspension. Gently tap the flask with the palm of the hand a couple of times to release any remaining detached cells.

5. Inactivate the trypsin by adding an equal volume of serum-containing (complete) media to the flask.

6. Remove the cell suspension from the flask and place in to a sterile container. Centrifuge typically at 1000 rpm for 5 min.

7. Pour off the supernatant from the container and resuspend the pellet in complete medium. Perform a viable cell count (see article by Hoffman) and reseed a flask with an aliquot of cells at the required density. The size of culture flask used depends on the number of cells required. An appropriate volume of complete medium is added to the flask (5–7 ml for a 25-cm² flask, 10–15 ml for a 75-cm² flask, or 20–30 ml for a 175-cm² flask). A cell count may not always be necessary if the cell line has a known split ratio (refer to ATCC or ECACC data sheets for the required seeding density). Label each flask with cell line name, passage number, and date.

Notes

• It is recommended to prepare fresh complete medium 1 week in advance of use, which gives time for a 5-day sterility check.

• L-Glutamine is more labile in cell culture solution than any other amino acid and the rate of degradation is dependent on storage temperatures, age of product, and pH. It is usually added in excess to culture medium, as it can be a limiting factor during cell growth. However, the degradation of L-glutamine causes a buildup of ammonia, which can have a detrimental effect on some culture suspensions. It is important to be very cautious when exceeding the formulated level of L-glutamine originally in the medium. Its degradation occurs at a faster rate at 37°C compared to 4°C; therefore it is recommended to keep medium containing L-glutamine at 4°C.

• To maintain serum-free conditions for specific cultures, trypsin may be neutralised with soybean trypsin inhibitor where an equal volume of inhibitor at a concentration of 1 mg/ml is added to the trypsinised cells. Cells can then be processed as described in steps 6 and 7 described earlier. For more details, refer to the article by Gayme and Gruber.

• An alternative to proteases is cell-scraping methods.

• Although most cells will detach in the presence of trypsin alone, EDTA is added to enhance the activity of the enzyme.

• Cells should only be exposed to trypsin long enough to detach the cells. Prolonged exposure can damage surface receptors on cells.

2. Subculturing Suspension Cells

Once established in culture, many tumour cell lines do not produce attachment factors and remain in suspension either as single cells or as clumps of cells. Examples of this cell type are lymphoblasts and cells of haematopoietic origin. Lymphoblastoids and many other tumour cell lines do not require a surface on which to grow and therefore do not require proteases such as trypsin for subculturing. Nonadherent cells may be subcultured by a number of methods, including subculture by direct dilution and subculture by sedimentation followed by dilution. The disadvantage of subculturing suspension cells by dilution is that during growth, cells produce metabolic by-products, which become toxic if allowed to accumulate in the medium. Initially, sufficient waste metabolites will be removed by diluting cell cultures, allowing growth to continue but the viability of the cells will gradually decline after a few subcultures. Surviving cells may also undergo selection pressures, resulting in altered characteristics. Subculture by sedimentation overcomes this problem of toxic metabolites but one must be careful not to overdilute the cells, which will increase the lag phase of the cells and may prevent them from dividing and also removes any growth factors produced by the cells.

Protocol 2. Subculture of Suspension Cells by Sedimentation

The following procedure is described for the human haematopoietic cell line HL60 (purchased from the ATCC), which can be subcultured indefinitely in suspension culture. HL60 cells cannot be grown in medium containing HEPES as a buffering agent. These cells need to be grown in vented tissue culture flasks in an atmosphere of 5% CO_2. HEPES is known to be toxic to some cell lines, so this needs to be checked before culturing the cell line of interest.

Media and Solutions

1× RPMI: Prepare 1× RPMI from 10× powder according to the manufacturer's instructions and supplement the medium with $NaHCO_3$ (adjust pH to 7.2–7.4) and filter sterilise using a 0.22 µm filter capable of handling large volumes or, alternatively, purchase premade 1× medium from a relevant supplier. Store basal media at 4°C.

Complete RPMI medium: To prepare 100 ml of complete medium, aliquot 89 ml of 1× RPMI and supplement with 10 ml of FCS and 1 ml of 100× L-glutamine. One milliliter of antibiotic solution (e.g., 100× penicillin/streptomycin) may also be supplemented to the medium if absolutely necessary.

Procedure

1. Examine the cultures microscopically for signs of cell deterioration and high-density growth. Cells in the exponential phase of growth will appear bright, round, and refractile, whereas dying cultures show cell lysis and shrunken cells. Hybridomas may be sticky and adhere loosely to the surface of the flask. These may require a gentle tap to the flask to detach the cells into suspension. Examination of the colour of the media may also indicate the growth stage of the cells (pH indicator in media).

2. Remove a volume of the cells for a viable cell count (refer to the article by Hoffman for details). The viability of suspension cell cultures should not be allowed to fall below 90%.

3. Transfer the cells to a sterile container and centrifuge the cells at 1000 rpm for 5 min to form a pellet.

4. Discard the supernatant and resuspend the pellet with fresh media.

5. Set up the required number of vented tissue culture flasks and add an appropriate volume of prewarmed growth medium to the flasks. Add the appropriate number of cells to the flasks and incubate the flasks at 37°C in a CO_2 incubator. Refer to data sheets for recommended seeding densities of the cell line of interest (a typical growing density range for many suspension cells is 10^5–10^6 cells/ml).

C. Cryopreservation of Cells

Cell lines can change properties over time such as growth rate, antigen expression, and isoenzyme profile. It is therefore essential to set up a large frozen stock of each cell line, i.e., a working cell bank, and to work only within a defined number of cell doublings (or passages at a defined dilution/split ratio). A 10-passage period is satisfactory for many cell lines to maintain characteristic profiles.

1. Good Practice and Safety Considerations

Liquid nitrogen is the main refrigerant used for the long-term storage of biological material on a laboratory scale. The nitrogen boils above −196°C (−320°F) and thus the liquid is intensely cold and can pose several forms of safety hazard (Angerman, 1999).

• Cold of this magnitude will almost instantly burn skin and tissue. Such burns will also be produced by contact with material that has been exposed to liquid nitrogen. In some cases, items may actually adhere to skin and tissue that they contact, exacerbating burn damage.

• Liquid nitrogen can only be transported and stored in appropriately insulated containers. Most materials become extremely brittle in contact with such temperatures. For small quantities, special polystyrene containers can be used for transport. For larger quantities, purpose-built vacuum-insulated dewars are vital.

• The constant boiling of liquid nitrogen, particularly the instantaneous boiling that can occur if objects at room temperature are encountered, necessitates that all exposed areas of skin are covered. Hands, in particular, will need specific thermal-insulating gloves designed for low temperatures and these must never be used if damp or wet. The face should also be protected by a full face visor. This becomes particularly important when cryovials are being removed from liquid phase storage. The liquid gas can enter such vials during storage. The gas phase of nitrogen occupies approximately 700-fold the space of the liquid. Hence boiling of the liquid in a closed vial generates unbearable pressure that can result in explosive projections.

• The huge expansion of liquid nitrogen also causes boiling of the liquid to displace air if ventilation is insufficient. As nitrogen is an asphyxiant gas, careful consideration must be given to storage and transport. Ventilation must be adequate at all times to prevent buildup of the boiled gas produced under normal storage and the amount of gas that could be immediately produced in the event of a vessel rupture. Storage of large liquid nitrogen volumes needs specialist attention, including automatic oxygen level monitoring in the area. There have unfortunately been notable cases of death by asphyxiation where storage and monitoring have not been appropriate.

DMSO, which is often used in cryopreservation, must be handled with care. The chemical is flammable, although high temperatures are required to generate a risk of explosion. DMSO readily permeates some types of gloves and the skin. In itself DMSO is not hugely toxic to the body; however, DMSO may carry other chemicals and allergens through the skin barrier (Freshney, 2000).

2. Cryopreservation of Adherent and Suspension Cells

It is impractical to maintain cell lines in culture indefinitely as cell cultures will undergo genetic drift with continuous passage and risk losing their differentiated characteristics. It therefore becomes necessary to store cell stocks for future use and as a backup in case of contamination or alterations in the characteristics of the cells. There should be a limit placed on the number of passages any cell line can undergo before being discarded and replaced with new cells from the cryopreserved stock (e.g., a limit of 10 passages). Nearly all cell lines can be cryopreserved indefinitely in liquid nitrogen at −196°C. The cryopreservation process is based on slow freeze and fast thaw, together with a high protein concentration (foetal calf serum) and the presence of a cryopreservative agent that increases membrane permeability (e.g., DMSO or glycerol).

Cryopreservatives may be classed as being penetrative or nonpenetrative. Penetrative cryopreservatives, such as DMSO and glycerol, protect the cells against freezing damage caused by intracellular ice crystals and osmotic effects. Nonpenetrative cryopreserva-

tives, such as serum, protect the cells from damage by extracellular ice crystals. Cells should only be cryopreserved when in the exponential phase of growth to increase the chances of good recovery. It is important to check each batch of cryopreserved vials for viability, sterility, and maintenance of specific cell characteristics.

Protocol 3. Cryopreservation of Adherent and Suspension Cells

Media and Solutions

2× freezing medium (20% DMSO): To prepare 20 ml of freezing medium, add 4 ml of DMSO to 16 ml of FCS. Filter sterilise the solution using a 0.22-μm low-protein-binding filter. Aliquot the final solution into 5-ml amounts and store at −20°C.

Procedure

1. Check cultures using an inverted microscope to assess the degree of cellular growth and to ensure that the cells are free of contamination. Adherent or suspension cells are harvested for cryopreservation in the exponential phase of growth (refer to Protocols 1 and 2) and are counted as described in the article by Hoffman. Ideally, the cell viability should be greater than 90% in order to achieve a good recovery after freezing.

2. Thaw the 2× freezing medium and leave at 4°C until required.

3. Resuspend the cells in a suitable volume of serum. Slowly, in a dropwise manner, add an equal volume of the cold 20% DMSO/serum solution (or alternative freezing medium) to the cell suspension while at the same time gently swirling the cell suspension to allow the cells to adapt to the presence of DMSO. Note that DMSO is toxic to cells if added too quickly. The final concentration of DMSO should be 10%.

4. Place a total volume of 1 ml of this suspension, which ideally should contain between 5×10^6 and 1×10^7 cells, into each cryovial. Ensure that each cryovial is clearly marked with the cell line name, passage number, and date of freezing.

5. Place these vials in the vapour phase of a liquid nitrogen container, which is equivalent to a temperature of −80°C for a minimum of 3 h (or overnight).

6. Remove the vials from the vapour phase of the liquid nitrogen container and transfer them to the liquid phase for storage (−196°C).

Notes

• DMSO is a good universal cryoprotectant and most cell lines can be frozen down in a final concentration of 5–10% DMSO. However, it may not be suitable for every cell line

   i. Alternative freeze medium (e.g., glycerol): check the suitability on individual cell lines.

   ii. It may have differentiation effects on certain cell lines (Freshney, 2000).

• Programmable freezers: the European Collection of Animal Cell Cultures (ECACC) recommends a cooling rate of −3°C per minute for most cell types.

• With serum-free conditions, nonpenetrative cryoprotectants such as methylcellulose and polyvinylpyrrolidone (Merten *et al.*, 1995; Ohno *et al.*, 1988) can be used as alternatives to serum to maintain serum-free conditions (Keenan *et al.*, 1998).

Protocol 4. Thawing of Cryopreserved Cells

1. Aliquot a volume of 9 ml of medium to a sterile container. Remove the cryopreserved cells from the liquid nitrogen and thaw at 37°C. It is important to thaw the cryopreserved cells rapidly to ensure minimal damage to the cells by the thawing process and the cryopreservation agent DMSO, which is toxic above 4°C.

2. Allow the contents to thaw until a small amount of ice remains in the cryovial.

3. Wipe the outside of the cryovial with a tissue moistened with 70% IMS (or alternative) and transfer the cell suspension to the previously aliquoted media.

4. Centrifuge the resulting cell suspension at 1000 rpm for 5 min to remove the cryoprotectant. Remove the supernatant and resuspend the pellet in fresh culture medium.

5. Assess the viability of the cells as described in the article by Hoffman to ensure that the viability is above 90%.

6. Add cells to an appropriately sized tissue culture flask (usually a 25- or 75-cm² flask) with a suitable volume of growth medium. Allow adherent cells to attach overnight and refeed the next day to remove any dead or floating cells. It is advisable to start cultures at between 30 and 50% of their final maximum cell density as this allows the cells rapidly to condition the medium and enter the exponential phase of growth.

IV. COMMENTS

• The procedures described in this article are applicable to many different continuous cell lines. However, some specialised cell types require special conditions for growth and differentiation (e.g., defined media, growth factors, coated flasks/plates) and are described elsewhere within this volume.

• A number of other basic cell culture procedures are also applicable to this article but are described elsewhere. These include cell counting (Hoffman) proliferation assays (Riss and Moravec), cell line authentication (Hay), and testing of cell cultures for microbial and viral contamination, including *Mycoplasma* (Hay and Ikonomi).

References

Angerman, D. (ed.) (1999). "*Handbook of Compressed Gases: Compressed Gas Association*," 4th Ed. Kluwer Academic, New York.

Doyle, A., Griffiths, J. B., and Newell, D. G. (eds.) (1998). "*Cell and Tissue Culture: Laboratory Procedures*." Wiley, New York.

Freshney, R. I. (ed.) (1992). "*Animal Cell Culture, a Practical Approach*," 2nd Ed. IRL Press, Oxford.

Freshney, R. I. (ed.) (2000). "*Culture of Animal Cells: A Manual of Basic Techniques*," 4th Ed. Wiley-Liss, New York.

Gey, G., Coffman, W. D., and Kubiech, M. T. (1952). *Cancer Res.* **12**, 264.

Keenan, J., Meleady, P., and Clynes, M. (1998). Serum-free media. *In* "*Animal Cell Culture Techniques*" (M. Clynes, ed.), pp. 54–66. Springer-Verlag, New York.

Merten, O. W., Peters, S., and Couve, E. (1995). A simple serum free freezing medium for serum free cultured cells. *Biologicals* **23**, 186–189.

Ohno, T., Kurita, K., Abe, S., Eimori, N., and Ikawa, Y. (1988). A simple freezing medium for serum-free cultured cells. *Cytotechnology* **1**, 257–260.

Richmond, J. Y., and McKinney, R. W. (eds.) (1999). "*Biosafety in Microbiological and Biomedical Laboratories*." U.S. Department of Health and Human Services, Public Health Service Centers for Disease Control and Prevention and National Institutes of Health. 4th Ed. U.S. Government Printing Office, Washington. 1999 Web edition: http://www.cdc.gov/od/ohs/biosfty/bmbl4/bmbl4toc.htm.

Shaw, A. J. (ed.) (1996). "*Epithelial Cell Culture: A Practical Approach*." IRL Press, Oxford.

3

Counting Cells

Tracy L. Hoffman

I. INTRODUCTION

Cell culture is the technique of growing cell lines or dissociated cells extracted from normal tissues or tumors. The *cell count* measures the status of the culture at a given time and is essential when subculturing or assessing the effects of experimental treatments on cells. The cell count can be expressed as the number of cells per milliliter of medium (for suspension cultures) or per centimeter squared area of attachment surface (for monolayer cultures).

II. MATERIALS AND INSTRUMENTATION

Dulbecco's phosphate-buffered salire (PBS; DPBS) and *exclusion dyes* are available from ATCC (DPBS, Cat. No. 30-2200; 0.1% erythrosin B stain solution, Cat. No. 30-2404; 0.4% trypan blue stain solution, Cat. No. 30-2402). Citric acid and crystal violet are available from Sigma Chemical Co. (citric acid, Cat. No. C-0759; crystal violet, Cat. No. C-0775). The hemocytometer set and tally counter can be obtained from VWR (double Neubauer chamber and cover glass, Cat. No. 15170-173 or Bright-Line chamber with Neubauer ruling and cover glass, Cat. No. 15170-172; tally counter, Cat. No. 15173-252).

The Innovatis "Cedex" cell analysis system and associated reagents are available within the United States from Biomedical Resources International, Inc. (Cedex System, Cat. No. 702 00 00 00; 20-position automatic sampler, Cat. No. AS20, autosampler cups, Cat. No. 600 05 00 00; detergent concentrate, Cat. No.

702.00.125 or 702.00.250; density reference beads, Cat. No. 7 60 00 00 17). The Vi-CELL cell viability analyzer is available from Beckman Coulter (1-position carousel model, Part No. 6605617; 12-position carousel model, Part No. 6605587). The NucleoCounter mammalian cell counting system is available from New Brunswick Scientific (NucleoCounter, Cat. No. M1293-0000; NucleoCassettes, Cat. No. M1293-0100; starter kit, Cat. No. M1293-0020).

III. PROCEDURES

Cell culture measurements can be divided into three major categories: (1) visual methods using light microscopy and a hemocytometer; (2) indirect visual methods, which count cell nuclei using light microscopy and a hemocytometer; and (3) electronic systems, which automate the visual method.

A. Visual Methods

This procedure is taken from those of Absher (1973), Freshney (2000), Merchant *et al.* (1964), and Phillips (1973).

The most practical method for routine determination of the number of cells that are in suspension or that can be put into suspension is a direct count in a *hemocytometer*. The hemocytometer is a modified glass slide engraved with two counting chambers of known area. Each chamber grid is composed of nine squares (subgrids), with each square being 1 mm² (see Fig. 1). The hemocytometer is supplied with a glass coverslip of precise thickness that is supported 0.100 mm above the ruled area.

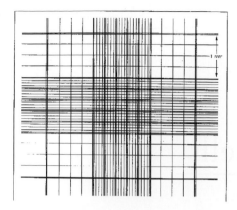

FIGURE 1 Schematic of one chamber grid of a hemocytometer with Neubauer ruling.

When the coverslip is positioned correctly, the volume of liquid over each square (subgrid) is a known constant, $0.1\,mm^3$ or $1.0 \times 10^{-4}\,mL$. If cells are counted in each of the corner squares and in the central squares of each of the two chamber grids, a total volume of $1 \times 10^{-3}\,mL$ ($10 \times 0.1\,mm^3 = 1.0\,mm^3$) will have been examined, and the cells per milliliter (concentration) of the sample added to the hemocytometer can be determined by multiplying the count by 1000. It should be noted that when cells are diluted in stain or buffer during preparation, then multiplication by the appropriate dilution factor is also required to obtain the concentration of the sample. *Determination of viability* with the hemocytometer method involves the dye exclusion test. This test relies on the ability of living cells to exclude certain stains from crossing the cell membrane. Dead cells are permeable and will take up the stain.

1. Dilution of Cell Suspension

a. Place 0.3 mL of sterile DPBS into a polypropylene tube capable of holding at least a 1.0-mL volume.

b. Aseptically collect a sample from the suspension culture or detached monolayer culture. Prior to collection, mix the cell culture to be counted by pipetting up and down two to five times.

c. Transfer 0.1 mL of the mixed culture sample to the tube containing the DPBS (Step 1a). Mix by pipetting up and down two to five times.

d. Add 0.1 mL exclusion dye (0.1% erythrosin B or 0.4% trypan blue) to the tube containing the cell suspension (step 1c). Mix by pipetting up and down two to five times.

2. Use of Hemocytometer and Counting of Cells

a. Obtain a hemocytometer and coverslip from storage (70% alcohol bath). Wipe both pieces dry and put coverslip in place. With a Pasteur pipette or pipetman and pipette tip, introduce enough homogeneous stained cell suspension (Step 1d) to the V-shaped filling troughs for each of the two chambers to fill the hemocytometer.

b. Place the hemocytometer on the microscope stage and focus with an objective that will permit one subgrid (one of the nine squares) to be visible (usually $100 \times$ magnification).

c. Using a tally counter with at least two registers, count viable (nonstained) cells and nonviable (stained) cells in each grid by counting cells in each of 10 subgrids (5 subgrids per chamber grid). The subgrids to be counted are the 4 outer, corner subgrids and the middle subgrid in each of the two chambers.

d. Remove cells from hemocytometer and return it to storage.

3. Calculation of Cell Count

a. Calculate the viable cell concentration using the formula $C = N \times D \times 10^3$, where C is the viable cells/mL, N is the number of viable cells counted in 10 subgrids ($1.0\,mm^3$), D is the dilution factor, and 10^3 is the hemocytometer correction factor.

b. Calculate percentages cell viability using the formula $V = N/T \times 100\%$, where V is the percentage viability, N is the number of viable cells counted per 10 subgrids ($1\,mm^3$), and T is the number of total cells counted per 10 subgrids ($1\,mm^3$).

B. Indirect Visual Methods

This procedure is modified from that of Patterson (1979).

Some adherent cells, such as human diploid cells, are difficult to count directly. In such cases, a procedure can be used to stain and count cell nuclei rather than the cells themselves.

1. Preparation of Solutions

a. Prepare 0.1 M citric acid by dissolving 1.9212 g in 100 mL distilled water.

b. Prepare 0.1 M citric acid containing 0.01% (w/v) crystal violet by dissolving 0.005 g crystal violet (also known as basic violet 3 or gentian violet; C.I.42555) in 50 mL of the 0.1 M citric acid (prepared in step B1b)

2. Procedure

a. Aseptically transfer a sample from the suspension culture or detached monolayer culture containing approximately 5×10^5 cells to a centrifuge tube. Prior to collection, mix cell culture to be counted by pipetting up and down two to five times.

b. Centrifuge at $500 \pm 50\,g$ for 5 to 10 min.

c. Decant the supernatant. Add 1.0 mL 0.1 M citric acid solution to the cell pellet. Mix well and incubate at 35° to 37°C for 1 to 2 h.

d. Separate nuclei by violent shaking followed by centrifugation at $1000 \pm 100\,g$ for 20 to 25 min.

e. Discard supernatant. Resuspend the cell pellet in 0.5 to 1.0 mL citric acid-crystal violet solution.

3. Use of Hemocytometer and Counting of Cells

a. Obtain a hemocytometer and coverslip from storage (70% alcohol bath). Wipe both pieces dry and put coverslip in place. With a Pasteur pipette or pipetman and pipette tip, introduce enough homogeneous stained cell suspension (step 1d) to the V-shaped filling troughs for each of the two chambers to fill the hemocytometer.

b. Place the hemocytometer on the microscope stage and focus with an objective that will permit one subgrid (one of the nine squares) to be visible (usually 100X magnification).

c. Using a tally counter, count stained nuclei in each grid by counting nuclei in each of 10 subgrids (5 subgrids per chamber grid). The subgrids to be counted are the 4 outer, corner subgrids and the middle subgrid in each of the two chambers.

d. Remove cells from hemocytometer and return it to storage.

4. Calculation of Cell Count

Calculate the cell concentration using the formula $C = N \times D \times 10^3$, where C is the cells/mL, N is the number of nuclei counted in 10 subgrids (1.0 mm^3), D is the dilution factor, and 10^3 is the hemocytometer correction factor.

C. Electronic Systems

The visual method can be user dependent, tedious, and time-consuming. Over the course of the last few years, systems that eliminate these problems through the automation of the process have become available commercially. There are two major types of system: (1) those that use trypan blue to arrive at a viable cell count and (2) those that use propidium iodide (PI) for counting cell nuclei.

1. Automated Trypan Blue Method

Both the Cedex system (Innovatis) and the Vi-CELL system (Beckman Coulter) incorporate image analysis technology and fluidics management to automate the mixing, staining, and hemocytometer counting process. As a result, the staining process is executed with high precision and reproducibility. Automation of the sample preparation process eliminates human errors in connection with pipette handling.

a. Set up the system with the appropriate reagents following the manufacturer's instructions.

b. Calibrate the system as per the manufacturer's schedule and instructions.

c. Aseptically collect a sample from the suspension culture or detached monolayer culture. Prior to collection, mix the cell culture to be counted by pipetting up and down two to five times. Transfer at least 1.0 mL to a sample cup.

d. Load sample cup onto the unit.

e. Log in pertinent information about the sample and initiate the counting sequence.

f. When the count cycle is complete (typically about 3 min), the viable cell density, total cell density, and percentage viability will appear on the screen. The results can then be printed and saved as appropriate.

g. After use, the sample cup(s) can be disposed of as biological waste. The system flow path is flushed and cleaned automatically, following each count, as part of the count cycle.

2. Automated Propidium Iodide Method

Similar to the Cedex and Vi-CELL systems, the NucleoCounter mammalian cell counting system (New Brunswick Scientific) also incorporates image analysis technology and fluidics management to automate the mixing, staining, and hemocytometer counting process. Unlike these systems, however, the NucleoCounter stains nuclei rather than intact cells. Because of this, two separate counts are needed to obtain culture viability.

a. Aseptically collect two samples from the suspension culture or detached monolayer culture. Prior to collection, mix the cell culture to be counted by pipetting up and down two to five times.

b. Pretreat the first sample with lysis and stabilization buffers following the manufacturer's instructions. Load the pretreated sample into the NucleoCassette. Place into the unit and press "run."

c. A total cell count will be displayed when the run is complete (typically about 30 s).

d. Load the second, untreated sample into another NucleoCassette. Place into the unit and press "run."

e. A total cell count will be displayed when the run is complete (typically about 30 s).

f. Divide the total count obtained from the pretreated sample (step C2b) by the total count obtained from the untreated sample (step C2d) and multiply by 100% to obtain the percentage viability of the culture.

g. After use, the NucleoCassette(s) can be disposed of as biological waster, with the PI dye safely enclosed. The system itself requires no cleaning.

IV. PITFALLS

A. The visual method is best applied when only a few samples are to be counted. If numerous cultures are to be counted on a routine basis, electronic systems should be considered.

B. The dye exclusion test gives information about the integrity of the cell membrane, but does not necessarily indicate how the cell is functioning.

C. Trypan blue has a greater affinity for serum proteins than for cellular protein. It is recommended that cells be suspended in DPBS or serum-free medium before counting when using this exclusion dye. When using trypan blue as the exclusion dye, mix the sample well and allow it to stand for 5 to 10 min prior to performing the cell count.

D. Among the major errors in all visual counts are:

1. *Nonuniform cell suspensions.* The total volume over a chamber grid is assumed to be a random sample. This assumption is invalid unless the cell suspension is uniform and free of cell clumps. Also, cells settle from suspensions rapidly. A uniform suspension can only be obtained if cells are mixed thoroughly before sampling.

2. *Improper filling of chamber grids.* The chamber must be filled by capillary action, without overflowing. Both pipettes and the hemocytometer set must be scrupulously clean.

3. *Failure to adopt a routine method for counting cells in contact with boundary lines.* All such conventions are arbitrary, but are essential to obtain reproducible results by counting comparable areas. The most commonly used method is to count only those cells that touch the left and/or upper boundary lines. Those cells that touch the right and/or lower boundary lines are not counted.

4. *Statistical error.* The cell sample should be diluted so that no fewer than 10 and no more than 50 cells are over each 1-mm^2 square. Because the cells follow a Poisson distribution in the hemocytometer chamber grids, the count error will be approximated by the square root of the count. Using the best techniques, an experienced individual can generally obtain counts with a total error of 10 to 15% (Absher, 1973).

E. The dynamic range for the Cedex and Vi-CELL systems is 1×10^4 to 1×10^7 cells/mL. For the Nucleo-Counter, it is 5×10^3 to 4×10^6, with the optimum being 10^5 to 2×10^6 cells/mL. Samples should be diluted or concentrated as appropriate and rerun if the values reported by the system fall outside these ranges.

F. Due to the small bore of the flow cell path in the Cedex and Vi-CELL systems, routine maintenance of the instruments as outlined by the manufacturer is critical. Failure to perform routing priming and/or flushing could result in clogging.

References

Absher, M. (1973). Hemocytometer counting. *In "Tissue Culture: Methods and Applications"* (P. F. Kruse, Jr., and M. K. Patterson, Jr., eds.), pp. 395–397. Academic Press, New York.

Freshney, R. I. (2000) Quantitation. *In "Culture of Animal Cells: A Manual of Basic Technique,"* 4th Ed., pp. 309–312. Wiley-Liss, New York.

Merchant, D. J., Kahn, R. H., and Murphy, W. H., Jr. (1964). *"Handbook of Cell and Organ Culture,"* 2nd Ed., pp. 155–157. Burgess, Minneapolis.

Patterson, M. K. (1979). Measurement of growth and viability of cells in culture. *In "Cell Culture"* (W. B. Jakoby and I. H. Pastan, eds.), pp. 141–149, Academic Press, New York.

Phillips, H. J. (1973). Dye exclusion test for cell viability. *In "Tissue Culture: Methods and Applications"* (P. F. Kruse, Jr., and M. K. Patterson, Jr., eds.), pp. 406–408. Academic Press, New York.

Cell Proliferation Assays: Improved Homogeneous Methods Used to Measure the Number of Cells in Culture

Terry L. Riss and Richard A. Moravec

I. INTRODUCTION

Over the last decade several improvements have been made in assay technology to enable miniaturization and more efficient measurement of the number of cells present in microwell plates. A variety of different methods have been optimized for convenient use in multiwell formats, making it easier to do large numbers of assays. The most significant improvement in efficiency has been the development of homogeneous "add, mix, and measure" assay formats compatible with robotic automation for high-throughput screening (HTS) of test compounds.

Making a choice among available assay formats often depends on the preference for which marker is measured or the level of sensitivity required. Homogeneous assays are now available to measure total cell number, viable cell number, the number of dead cells present in a population, or the number of cells undergoing apoptosis. For many experimental systems the most useful information is the number of viable cells at the end of a treatment period. The parameter used most conveniently to determine the number of viable cells in culture is measurement of an indicator of active metabolism.

This article describes four options for measuring cell number that are based on assaying different aspects of cellular metabolism. The example assays chosen include ATP quantitation, tetrazolium reduction using [3-(4,5-dimethylthiazol-2-yl)-5-(4-sulfophenyl)-2H-tetrazolium, inner salt (MTS), resazurin reduction, and total lactate dehydrogenase (LDH) activity measurement. The four examples are all homogeneous methods sensitive enough to detect cell numbers typically used in automated high-throughput 96- and 384-well plate formats. The methods are equally suitable for measuring just a few samples processed manually. In addition, all of these assays have been shown to be reproducible and exhibit good Z'-factor values (Zhang *et al.*, 1999) desirable for automated HTS applications. Each assay has its own set of advantages and disadvantages that contribute to the decision of which one to choose.

II. MATERIALS

RPMI 1640 culture medium containing 15 mM HEPES (Cat. No. R-8005), 2-mercaptoethanol (Cat. No. M-7154), trypan blue solution (0.4% Cat. No. T-8154), and phenazine ethosulfate (PES; Cat. No. P-4544) are from Sigma. Fetal bovine serum (Cat. No. SH30070) is from Hyclone. Ninety-six-well plates with opaque white walls and clear bottoms (Cat. No. 3610) are from Corning.

The CellTiter-Glo luminescent cell viability assay (Cat. No. G7571) for determining ATP content, the CytoTox-ONE homogeneous membrane integrity assay (Cat. No. G7891) for determining total LDH activity, the CellTiter 96 AQ$_{ueous}$ one solution cell proliferation assay (Cat. No. G3580) for measuring MTS tetrazolium reduction, the CellTiter-Blue cell viability assay (Cat. No. G8080) for measuring resazurin reduction, and recombinant human interleukin (IL)-6 (Cat. No. G5541) are from Promega.

III. INSTRUMENTATION

The Model MTS-4 plate shaker obtained from IKA Works, Inc. is used to mix the contents of the 96-well plates. Absorbance is recorded using a Molecular Devices V_{max} plate reader spectrophotometer. Luminescence is recorded using a Dynex MLX plate-reading luminometer. Fluorescence is recorded using a Labsystems Fluoroscan Ascent fluorescence plate reader fitted with a 560 (excitation) and 590 (emission) filter pair.

IV. PROCEDURES

The assay plates for all four procedures are prepared in an identical manner using stock cultures of the IL-6-dependent B9 hybridoma cell line. B9 cells are cultured in RPMI 1640 containing $15 \, mM$ HEPES and $50 \, \mu M$ 2-mercaptoethanol supplemented with 10% fetal bovine serum (assay medium) and $2 \, ng/ml$ of IL-6 (specific activity $168 \, U/ng$ protein, assigned by direct comparison to the interim reference standard #88/514 from the National Institute of Biologic Standards and Controls). Cell cultures are maintained at densities between 2×10^4 and $3 \times 10^5/ml$ in a humidified chamber at 37°C with 5% CO_2.

Stock cultures of cells used to prepare the proliferation assay are seeded at $2 \times 10^4/ml$ in standard T-75 flasks containing assay medium supplemented with $2 \, ng/ml$ IL-6 and are allowed to expand for 3 days. Cells are harvested using centrifugation (4 min at 200 g) and washed twice using assay medium without IL-6. Cells are suspended in assay medium without IL-6, treated with trypan blue solution, counted using a hemocytometer, and adjusted to 6.25×10^4 viable cells/ml. Eighty microliters of cell suspension (5000 cells) is dispensed into each well of multiple 96-well clear-bottom white opaque walled plates with replicates of four for each sample.

Serial two-fold dilutions of IL-6 are prepared in assay medium without IL-6 so 20-μl/well additions would contain a final concentration range of 0–2.0 ng/ml. An equivalent volume of assay medium without IL-6 is added to sets of negative control wells. Assay plates are cultured for 72 h at 37°C with 5% CO_2 before processing with each of the cell number assays.

A. MTS Tetrazolium Reduction Assay

Viable cells reduce tetrazolium compounds into intensely colored formazan products that can be detected as an absorbance change with a spectrophotometer. The amount of formazan color produced is directly proportion to the number of viable cells in standard culture conditions. Cells rapidly lose the ability to reduce tetrazolium compounds shortly after death, which enables tetrazolium reduction to be used as an indicator of viable cell number. The first MTT tetrazolium reduction cell proliferation assay was described two decades ago (Mosmann, 1983). Upon cellular reduction, the MTT tetrazolium reagent results in the formation of a formazan precipitate and requires the addition of a solubilizing agent to generate a solution suitable for recording absorbance.

The chemical properties of the MTS tetrazolium compound provide an improvement over the MTT assay. The formazan product resulting from bioreduction of MTS is directly soluble in cell culture medium, thus eliminating the solubilization step required for the MTT assay. The chemical properties of MTS that contribute to the formation of a soluble formazan product also restrict entry of MTS into viable cells. As a result, cell-permeable, electron-coupling agents (such as PES) are used in combination with MTS to shuttle in and out of cells to pick up reducing equivalents from molecules such as NADH. The reduced PES can move from the cytoplasm into the culture medium and reduce MTS into the soluble formazan product.

The combination of MTS + PES provides an assay that requires only a single reagent addition to the cell culture wells and results in a homogeneous "add, incubate, and measure" assay format. An additional advantage of using aqueous soluble tetrazolium assays is that data can be recorded from the same plate at various intervals after the addition of MTS + PES, which simplifies the optimization of the incubation period during characterization of the effects of particular compounds on cells (Fig. 1). The sensitivity of the MTS assay is dependent on cell type, but it is usually adequate for detecting the number of cells used commonly in microwell plates. Typically, the MTS assay can detect fewer than 1000 viable cells/well in the 96-well plate format or fewer than 250 cells/well in 384-well plates. For additional background information, refer to Promega Technical Bulletin #245.

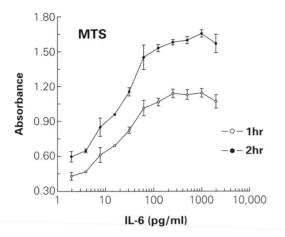

FIGURE 1 The effect of different concentrations of IL-6 on cell number was determined by measuring MTS reduction to the colored formazan product. Absorbance at 490 nm was recorded after 1 h of incubation with the MTS + PES reagent. Plates were returned to the cell culture incubator for an additional hour before recording the 490-nm absorbance after a total of 2 h of incubation. The 490-nm absorbance from control wells of cells without IL-6 (not shown on the log scale) was 0.39 for 1 h of incubation and 0.54 for 2 h of incubation. Values represent the mean ± standard deviation from four replicate wells.

Steps

1. Warm the CellTiter 96 AQ$_{ueous}$ one solution reagent to 37°C. (*Note*: The reagent contains 2 mg/ml MTS tetrazolium and 300 μM PES electron transfer reagent in a solution of phosphate-buffered saline. This solution is photolabile and should be stored protected from light.)

2. Remove the multiwell assay plate from the cell culture incubator and transfer to a laminar flow hood.

3. Add 20 μl of CellTiter 96 AQ$_{ueous}$ one solution reagent to the culture medium in each well using a multichannel pipette and aseptic conditions.

4. Mix the contents of the culture wells using a plate shaker to ensure uniform suspension of MTS reagent in the culture medium.

5. Return assay plate to a 37°C cell culture incubator for a period of 1 to 4 h, depending on the number and metabolic activity of cells being used.

6. Transfer assay plate to a multiwell plate-reading spectrophotometer, shake plate for 10 s (using the instrument's on-board plate shaking function) to ensure a uniform solution in the assay wells, and record absorbance at 490 nm.

7. Plot 490 nm absorbance vs cell number.

B. ATP Assay

The measurement of ATP has become widely accepted as a valid indicator of the number of viable cells present in culture (Ekwall *et al.*, 2000). Under cell culture experimental conditions that do not alter metabolism drastically, the amount of ATP is directly proportional to the number of viable cells (Crouch *et al.*, 1993). Historically, sample preparation for ATP assays has been a multistep process requiring inactivation of endogenous ATPases (known to interfere with measurement of ATP) and neutralization of the acidic extract prior to addition to a luciferase-containing reaction mixture (Lundin *et al.*, 1986; Stanley, 1986). Firefly luciferase purified from *Photinus pyralis* has been used most often as a reagent for ATP assays (Lundin *et al.*, 1986; Crouch *et al.*, 1993). Unfortunately, the native form of luciferase has only moderate stability *in vitro* and is sensitive to its chemical environment, e.g., pH and detergents, thus limiting its usefulness for developing a robust homogeneous ATP assay.

A stable form of luciferase has been developed from a different firefly, *Photuris pennsylvanica*, using an approach of directed evolution to select for characteristics that improve performance in ATP assays. The development strategy included selection for increased thermostability and resistance to degradation products of luciferin, which inhibit luciferase activity. The unique characteristics of this mutant (Luc*Ppe2*^{m}) enabled design of a homogeneous single-step reagent approach for performing ATP assays on cultured cells that overcomes the problems caused by factors such as ATPases that reduce the level of ATP in cell extracts. The CellTiter-Glo reagent is physically robust and provides a sensitive and stable luminescent output that is ideal for automated HTS cell proliferation and cytotoxicity assays. The homogeneous "add, mix, and measure" format results in cell lysis, inhibition of endogenous ATPases, and generation of a luminescent signal proportional to the amount of ATP present. In addition, the CellTiter-Glo assay conditions generate a "glow-type" luminescent signal, having a half-life of greater than 5 h, providing flexibility for recording data.

For most situations, the ATP assay is the method of choice because it has a simple homogeneous "add–mix–measure" procedure, it provides the quickest way to collect data (i.e., it avoids the 1- to 4-h incubation step required for tetrazolium or resazurin assays), and it has the best detection sensitivity among all the available methods. The ATP-based detection of cells has been shown to be more sensitive than other methods (Petty *et al.*, 1995). Assay sensitivity and range of responsiveness are typically between 50 and 50,000 cells/well in 96-well plates, but sensitivities of as few as 4 cells/well have been achieved using 384-well plates. For additional background information, refer to Promega Technical Bulletin #288.

Preparation of CellTiter-Glo Reagent

1. Allow a vial of the lyophilized substrate and a bottle of frozen assay buffer to equilibrate to ambient temperature (22°C).

2. Add 10 ml of assay buffer to the substrate to reconstitute the lyophilized cake and form the CellTiter-Glo reagent and mix gently to dissolve. The reagent is a buffered solution containing detergents to lyse cells, ATPase inhibitors to stabilize the ATP released from the lysed cells, luciferin as a substrate, and luciferase to generate a bioluminescent signal proportional to the amount of ATP present in the cell lysate.

Steps

1. Remove the multiwell assay plate from the cell culture incubator and equilibrate to ambient temperature for approximately 20–30 min. (*Note*: Transferring eukaryotic cells from 37°C to ambient temperature for the length of time required for temperature equilibration has little affect on cell viability or ATP content.)

2. Add a volume of CellTiter-Glo reagent equal to the volume of cell culture medium present in each well using a multichannel pipette. (*Note*: For 384-well plates containing 25 µl of culture medium, add 25 µl of reagent.)

3. Mix the contents of the assay wells to ensure uniform distribution of the reagent in the culture medium and to speed cell lysis.

4. Allow the assay plate to stand at ambient temperature for 10 min.

5. Record luminescence using a multiwell plate-reading luminometer.

6. Plot luminescence vs cell number (Fig. 2).

C. LDH Assay

Lactate dehydrogenase is a cytoplasmic enzyme that has been used as a marker of cell damage *in vitro* because the enzymatic activity is relatively stable in cell culture medium and can be measured easily after leakage out of cells with a compromised membrane (i.e., nonviable cells). The LDH assay is performed most commonly as a cytotoxicity assay by measuring the enzymatic activity from a sample of culture medium removed from the treated population of cells. Most assay methods require transfer of an aliquot of culture medium (without cells) into a separate assay vessel because the reagent formulation would damage living cells, resulting in the release of additional LDH (Korzeniewski and Callewaert, 1983). Reactions often proceed for 30 min and result in a colorimetric signal (Decker and Lohmann-Matthes, 1988).

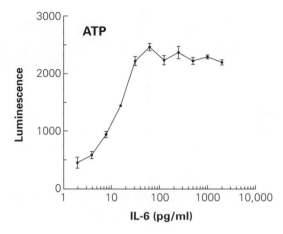

FIGURE 2 The effect of different concentrations of IL-6 on cell number was determined by measuring the total amount of ATP. Luminescence was recorded 10 min after addition of the CellTiter-Glo reagent. The luminescence of control wells of cells without IL-6 (not shown on the log scale) was 260 relative light units. Values represent the mean ± standard deviation from four replicate wells.

Recent improvements in LDH assay performance have been accomplished by using more sensitive fluorescent reporter molecules and by formulating the assay reagents in a physiologically balanced buffer that is not harmful to viable cells. These improvements enabled the development of a rapid homogeneous cytotoxicity assay format to detect the number of damaged cells directly in cell culture wells containing a mixed population of viable and nonviable cells. The reagent used to perform the coupled enzymatic assay is a buffered solution containing lactate as a substrate and NAD^+ as a cofactor to drive the LDH reaction. The reagent also contains the enzyme diaphorase to catalyze the NADH-driven reduction of resazurin into the fluorescent resorufin product.

As illustrated in the following example procedure, the total number of cells in culture (i.e., viable and nonviable) also can be estimated using an LDH assay by measuring total enzymatic activity from the entire population of cells. The cytotoxicity assay format is modified to detect total LDH in cultures by including a cell lysis step in the procedure utilizing a detergent that is compatible with the LDH assay chemistry. The detection sensitivity and linear range of the fluorescent LDH assay are typically 800–50,000 cells/well in the 96-well plate format, with sensitivity improving to 200 cell/well in 384-well plates. For additional background information, refer to Promega Technical Bulletin #306.

Preparation of CytoTox-ONE Reagent

1. Allow a vial of the lyophilized substrate and a bottle of frozen assay buffer to equilibrate to ambient temperature (22°C).

2. Add 11 ml of assay buffer to the substrate to reconstitute the lyophilized cake and form the CytoTox-ONE reagent and mix gently to dissolve. Protect the reagent from direct light.

Steps

1. Remove the multiwell assay plate from the cell culture incubator.

2. Add 2 µl of the lysis solution component of the kit (i.e., 9% Triton X-100) to each 100 µl of culture medium to lyse the cells and release LDH into the surrounding medium. (*Note*: If it is inconvenient to pipette very small volumes, a 1:5 dilution of lysis solution may be prepared using water so 10 µl can be dispensed into each well.)

3. Shake the plate for approximately 10 s to ensure uniform distribution of the contents of the wells and to ensure complete cell lysis.

4. Equilibrate the plate to ambient temperature (approximately 20–30 min).

5. Add a volume of CytoTox-ONE reagent equal to the volume of cell culture medium present in each well using a multichannel pipette. (*Note*: For 384-well plates containing 25 µl of culture medium, add 25 µl of reagent to each well.)

6. Mix the contents of the assay wells for 30 s to ensure uniform distribution of the reagent in the culture medium.

7. Allow the assay plate to incubate at ambient temperature for 10–15 min for the LDH reaction to proceed.

8. Add 50 µl to each well of the stop solution [3% (w/v) sodium dodecyl sulfate] provided as a component of the CytoTox-ONE assay kit. (*Note*: Add 12.5 µl stop solution for the 384-well plate format containing 25 µl of cells in culture medium and 25 µl of CytoTox-ONE reagent.)

9. Record the fluorescence of resorufin using a multiwell plate reading fluorometer fitted with a filter set for 560-nm excitation and 590-nm emission wavelengths. (*Note*: The spectra of resorufin will enable a variety of filter sets to be used. Data can be collected using excitation filters in the range of 530–570 nm and emission filters in the range of 580–620 nm.)

10. Plot fluorescence vs cell number (Fig. 3).

D. Resazurin Reduction Assay

Resazurin is a redox dye that can be reduced by cultured cells to form resorufin. Resazurin is dark blue

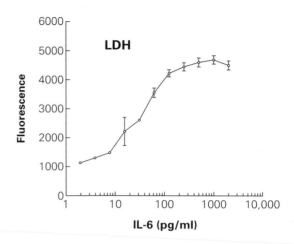

FIGURE 3 The effect of different concentrations of IL-6 on cell number was determined by measuring the total amount of LDH from the lysed population of cells. Cells were lysed by the addition of 2 µl/well of 10% (v/v) Triton X-100 in phosphate-buffered saline prior to addition of the CytoTox-ONE reagent. The reaction was stopped by addition of the stop solution, and fluorescence (560 excitation/590 emission) was recorded. The fluorescence value of control wells of cells without IL-6 (not shown on the log scale) was 1081 relative fluorescence units. Values represent the mean ± standard deviation from four replicate wells.

and has little intrinsic fluorescence until it is reduced to the pink resorufin product. The spectral properties of resorufin allow the molecule to be detected using either fluorescence or absorbance; however, fluorescence is the preferred method because it provides greater sensitivity.

The resazurin reduction assay is based on the ability of metabolically active living cells to convert a redox dye (resazurin) into a fluorescent end product (resorufin). Resazurin can enter living cells where it becomes reduced, and the resazurin product, which is also permeable, can be found in the cell culture medium (O'Brien *et al.*, 2000). The specific cellular mechanisms responsible for the reduction of resazurin are unknown (Gonzales and Tarloff, 2001), but probably involve reactions generating reducing equivalents such as NADH. Nonviable cells lose metabolic capacity rapidly and do not reduce resazurin to generate a fluorescent signal. The resazurin substrate is soluble in phosphate-buffered saline compatible with preparation of a reagent for direct addition to cell cultures. The homogeneous assay procedure involves addition of a single resazurin-containing reagent directly to cells cultured in serum-supplemented medium. After an incubation step, data are recorded using either a plate-reading fluorometer (preferred method) or a spectrophotometer. The reagent is generally nontoxic to cells, allowing extended incubation periods in some

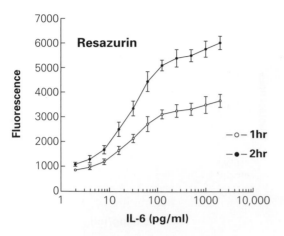

FIGURE 4 The effect of different concentrations of IL-6 on cell number was determined by measuring the ability of viable cells to reduce resazurin into the fluorescent resorufin. Fluorescence was recorded after 1 h of incubation with the CellTiter-Blue reagent. Plates were returned to the cell culture incubator for an additional hour before recording fluorescence a second time after a total of 2 h of incubation. The average fluorescence from control wells of cells without IL-6 (not shown on the log scale) was 814 for 1 h of incubation and 1017 for 2 h of incubation. Values represent the mean ± standard deviation from four replicate wells.

situations. Fluorescence data can be recorded at various intervals after the addition of resazurin, which simplifies optimization of the incubation period during characterization of the effects of particular compounds on cells (Fig. 4). Longer incubation periods may result in increased detection sensitivity; however, there may be a loss in the linear range of response. Assay sensitivity and range depend on cell type and metabolic capacity, but are typically between 200 and 50,000 cells/well in 96-well plates. For additional background information, refer to Promega Technical Bulletin #317.

Steps

1. Thaw CellTiter-Blue reagent containing resazurin and warm to 37°C.

2. Remove the multiwell assay plate from the cell culture incubator and transfer to a laminar flow hood.

3. Add 20 μl CellTiter-Blue reagent to each well containing 100 μl culture medium using a multichannel pipette and aseptic conditions. (*Note*: For 384-well plates containing 25 μl of culture medium, add 5 μl/well of CellTiter-Blue reagent.)

4. Mix the contents of the assay wells for 10 s to ensure uniform distribution of the reagent in the culture medium.

5. Return assay plate to a 37°C cell culture incubator for a period of 1 to 4 h to develop the fluorescent resorufin product.

6. Record the fluorescence of resorufin using a multiwell plate-reading fluorometer fitted with a filter set for 560-nm excitation and 590-nm emission wavelengths. (*Note*: The spectra of resorufin will enable a variety of filter sets to be used. Data can be collected using excitation filters in the range of 530–570 nm and emission filters in the range of 580–620 nm.)

7. Plot fluorescence vs cell number.

V. COMMENTS

Each assay format has its own set of advantages and disadvantages. The tetrazolium assay is currently the most widely used method of estimating the number of viable cells in multiwell plates and is the most often cited in the scientific literature. The resazurin reduction assay is functionally similar to the tetrazolium assay, except it has the optional advantage of using fluorescent detection methods. In contrast to the tetrazolium and resazurin reduction assays that require 1–4 h of incubation to obtain meaningful results, data from ATP and LDH assays can be obtained after a 10-min incubation. The ATP and LDH assays lyse cells and thus provide "a snapshot" of the condition of the cells at time of lysis. This advantage provides a quicker assay and avoids any toxic effects of the assay reagents that may occur during the incubation period. For many applications, the ATP assay may be the best choice. It is the most sensitive, provides results faster than any of the other assays, and is the easiest to use. However, one of the limitations of the ATP assay is that it requires a multiwell plate-reading luminometer that may not be available in all laboratories. The choice of which particular assay format to use may depend on the availability of instruments to record data, the detection sensitivity required, the number of samples to be measured, and whether total cell number, viable cell number, or nonviable cell number is chosen as an end point for measurement.

Multiplexing of two assays to gather more than one type of data from the same experimental well may help eliminate the possibility of artifacts. For example, an LDH cytotoxicity assay and an ATP viability assay can be done using the same sample of cells. A small aliquot of culture supernatant can be used for estimating the number of dead cells by measuring the amount of LDH released into the culture medium. Because the sample of cells remains intact, an ATP assay (or any of the other methods) can be used to measure viable or total cell number.

VI. PITFALLS

Temperature is a factor that affects the performance of the aforementioned assays because of its effect on enzymatic rates. It is critical to run the assays at a uniform temperature to ensure reproducibility across a single plate or among stacks of several plates. For assays developed at room temperature, it is important to ensure adequate equilibration of samples after the removal of assay plates from a 37°C incubator to avoid differential temperature gradients resulting in "edge effects." Stacking large numbers of assay plates together in close proximity should be avoided to ensure complete temperature equilibration.

Proper negative and positive controls are required to test whether compounds being measured have an effect on the assay chemistry or result in artifacts. For example, strong reducing compounds may interfere with procedures using redox dyes such as the tetrazolium or resazurin reduction assays. Culture medium supplemented with pyruvate will slow the rate of the LDH reaction and thus will require longer incubation periods to generate an adequate fluorescent signal in the CytoTox-ONE assay. In addition, different animal sera have different amounts of LDH activity that will influence background fluorescence. To correct for many of these factors, use of a "no treatment" negative control and a positive control to show maximum effect on each multiwell plate is recommended for all assays.

References

CellTiter 96 AQ_ueous One Solution Cell Proliferation Assay *Technical Bulletin #TB245*, Promega Corporation.

CellTiter-Blue Cell Viability Assay *Technical Bulletin #TB317*, Promega Corporation.

CellTiter-Glo Luminescent Cell Viability Assay *Technical Bulletin #TB288*, Promega Corporation.

Cory, A. H., *et al.* (1991). Use of an aqueous soluble tetrazolium/formazan assay for cell growth assays in culture. *Cancer Commun.* **3**(7), 207–212.

Crouch, S. P., *et al.* (1993). The use of ATP bioluminescence as a measure of cell proliferation and cytotoxicity. *J. Immunol. Methods* **160**, 81–88.

CytoTox-ONE Homogeneous Membrane Integrity Assay *Technical Bulletin #TB306*, Promega Corporation.

Decker, T., and Lohmann-Matthes, M. L. (1988). A quick and simple method for the quantitation of lactate dehydrogenase release in measurements of cellular cytotoxicity and tumor necrosis factor (TNF) activity. *J. Immunol. Methods* **115**, 61–69.

Ekwall, B., *et al.* (2000). MEIC Evaluation of acute systemic toxicity. VIII. Multivariat partial least squares evaluation, including the selection of a battery of cell line tests with a good prediction of human acute lethal peak blood concentrations for 50 chemicals. *Altern. Lab. Anim.* **28**(Suppl. 1), 201–234.

Gonzales, R. J., and Tarloff, J. B. (2001). Evaluation of hepatic sub-cellular fractions for Alamar Blue and MTT reductase activity. *Toxicol In Vitro* **15**, 257–259.

Korzeniewski, C., and Callewaert, D. M. (1983). An enzyme-release assay for natural cytotoxicity. *J. Immunol. Methods* **64**, 313–320.

Lundin, A., *et al.* (1986). Estimation of biomass in growing cell lines by ATP assay. *Methods Enzymol.* **133**, 27–42.

Mosmann, T. (1983). Rapid colorimetric assay for cellular growth and survival: Application to proliferation and cytotoxicity assay. *J. Immunol. Methods* **65**, 55–63.

O'Brien, J., *et al.* (2000). Investigation of the Alamar Blue (resazurin) fluorescent dye for the assessment of mammalian cell cytotoxicity. *Eur. J. Biochem.* **267**, 5421–5426.

Petty, R. D., *et al.* (1995). Comparison of MTT and ATP-based assays for measurement of viable cell number. *J. Biolumin. Chemilumin.* **10**, 29–34.

Stanley, P. E. (1986). Extraction of adenosine triphosphate from microbial and somatic cells. *Methods Enzymol.* **133**, 14–22.

Zhang, H.-U., *et al.* (1999). A simple statistical parameter for use in evaluation and validation of high throughput screening assays. *J. Biomol. Screen.* **4**, 67–73.

Development of Serum-Free Media: Optimization of Nutrient Composition and Delivery Format

David W. Jayme and Dale F. Gruber

I. INTRODUCTION

In a previous edition, we introduced many of the basic formulation concerns to be considered when developing a serum-free basal medium (Jayme and Gruber, 1998). Since that publication, societal and safety concerns have prompted additional scientific requirements that ultimately impact the construction of serum-free medium. Safety concerns, to be discussed at some length later in this article, now recommend all medium components to be animal-origin free (Jayme, 1999). Consistent with these guidelines, every medium constituent must document a traceable history free of primary or secondary contact with animal-origin products. These regulatory requirements apply to all scales of production from developmental to production formats, and are also extended to bioreactor supplementation and to downstream purification processes. This article reviews the impetus for these new requirements and improvements in serum-free medium development consistent with these directives.

Although many practitioners of tissue culture consider their field to be an art, others characterize it as a science. Both positions may be appropriate, considering the circumstances surrounding the ontogeny of medium development. The advantages of cultivating cells and tissues in a defined nutrient medium were first recognized over eighty years ago. However, since its inception, maturation of the tissue culture field may be characterized as a series of fits and starts predicated

mostly upon perceived societal, economic, or experimental needs of the time. Tissue culture has now inarguably gained legitimacy as a research tool and has enabled many scientific investigations to reach successful conclusions. Yet investigators have continued to press for more fully defined cell cultivation environments. Theoretically, a fully defined cell culture medium system would eliminate confounding variables and allow the most accurate assessment and description of dependent variables. Although relatively simple in both concept and statement, practical elimination of tissue culture variables has proved to be no trivial undertaking. As the field has matured and cell-based applications have become more complex, there has been a transition from simple salt solutions to simple basal media and to more complex, nutrient enhanced media, all of which required some level of supplementation by animal sera.

The ultimate goal—to eliminate serum and develop effective serum-free or even protein-free or chemically defined media—proved more challenging. Merely eliminating the serum additive proved ineffective due to the broad diversity of serum functions within a cell culture environment. The presumptive serum-free medium formulation must contain all requisite materials to support the synthesis of new cells (proliferation), as well as all the vitamins, minerals, energy substrates, lipids, and inorganic ions essential to maintain all normal and/or genetically engineered physiological functions.

As such, the field of tissue culture may be generally described as always being in the discovery mode. As specific cellular requirements are identified, appropriate changes are made in media formulae to support that particular cellular function. Certain cell types, important to a thorough understanding of normal and dysfunctional physiology, have been nutritionally fastidious and difficult to propagate *in vitro*. Often, these cell types were more effectively cultured and more persistently retained differentiated cellular function in the absence of serum. Finally, the target end point evolved from supporting cell proliferation or increasing total biomass to optimizing the bioreactor yield of a biomolecule (monoclonal antibody, recombinant protein, virus) of interest.

Innovations in proteomics, gene therapy, and bio- and chemoinformatics have also added momentum to serum-free medium development. Societal issues, such as the global concerns regarding bioterrorism (e.g., anthrax, smallpox), immunodeficiency virus (HIV), and transmissible spongiform encephalopathy (prion) contamination, have prompted an amalgamation of scientific and regulatory positions. Public attention has focused on practices relevant to pharmaceutical and vaccine development. As the entire body of tissue culture knowledge focused on these biotechnology applications, analytical techniques facilitated acceleration in the design and development of numerous next-generation serum-free media. As biotechnology era products matured from discovery phases through production and advanced clinical trials and, eventually, into commercialization phases, more regulatory fervor has been engendered, encompassing all portions of cell culture from cell growth and expansion to gene amplification and expression, and biological product harvest and purification.

A. History

In its infancy (1950s), tissue culture was conducted primarily in glass vessels, the matrix of choice at that time. Commensurate with increased popularity in the 1960s, tissue culture systems scaled up to larger formats to include microcarrier and stir tank reactor formats and practices. Although representing a significant step in scale, the tissue culture field nonetheless languished under the daunting sterility practices and requirements associated with reuseable glass vessels. In the 1970s, sterile disposable plastic culture flasks were developed and, after 30 years, remain the mainstay of today's practitioners. In the 1980s, the tissue culture landscape was transfigured by the introduction of tissue culture-derived bioproducts into the commercial marketplace. In the 1990s, tissue culture expanded directly into therapeutic products, including skin replacement for burn patients, cartilage for focal

articular surface repair, and experimental cell and gene therapy regimens.

During the last decade and in parallel to the maturation of tissue culture there has been explosive growth in the knowledge and practices pertaining to molecular and cellular biology. As pathological situations have been ascribed to the presence or absence of specific genes or proteins, the fields of genomics and proteomics have assumed both academic and therapeutic importance. The desire to produce specific proteins in culture evoked the ability to select, clone, and express the gene encoding that specific target protein of interest. Once gene expression for the specific protein was verified in a stable cell bank, it became the responsibility of the process development function to maximize product yield through optimized upstream and downstream activities. Optimization of the *in vitro* cultivation conditions typically involves a careful balance between maintaining the narrowly defined *in vivo* environment of normal cells and perturbing that environment (physicochemically and nutritionally) to stress the cells into differentially producing copious quantities of the desired bioproduct.

B. Serum as a Culture Additive

This audience is familiar with the evolution of tissue culture beginning with simple salt solutions to simple and more complex media formulations supplemented with serum, extracts, hydrolysates, or peptones. Technical considerations and disadvantages to the use of serum in cell culture applications have been reviewed previously (Jayme and Greenwold, 1991) and included physiological variability, quality control complexity, cellular specificity, downstream processing artifacts, adventitious agent contamination, cost and availability, growth inhibitors, and proteolytic enzyme activity (Freshney, 2000).

A curious paradox concerning the use of serum as a cell culture additive is that cells cannot proliferate in serum alone. The specialized microenvironment required by cultured cells is mimicked *in vitro* by replacing serum with a mixture of metabolic precursors, macromolecules, and biophysical elements (Ham, 1982; Bottenstein *et al.*, 1979). Serum functions in cell culture (Freshney, 2000; Jayme and Blackman, 1985) and factors to consider in selecting serum for specific applications (Hamilton and Jayme, 1998) have been reviewed previously.

C. Regulatory Impacts

In 2001, the council of Europe adopted regulation (EC) 999/2001 known as the transmissible spongiform encephalopathies (TSE) regulation. Its purpose was to establish principles and guidelines to minimize the risk of transmission of a TSE via human or veterinary medicinal products. This regulation applied to all materials used in the preparation of active and excipient substances and included all source materials and reagents (including culture medium) used in production of an end product. The U.S. Food and Drug Administration (FDA), although not in regulation form as yet, issued letters (1991, 1993, and 1996) and a guidance document (1997), issuing strong recommendations toward the reduction or elimination of all animal-origin components used in the manufacture of FDA-regulated products. These strong regulatory positions have resulted in concerted efforts toward the elimination of serum and other animal-derived factors (e.g., serum albumin, transferrin) and performance qualification of substitute serum-free formulations.

II. CURRENT ISSUES IN NUTRIENT MEDIUM DEVELOPMENT

Failure in transitioning to a presumptive serum-free medium may be attributable to a variety of factors, including ineffective cellular adaptation and cultivation protocols or suboptimal nutrient composition. Three general approaches may be implemented to achieve a serum-free culture environment: (1) replacing serum with recombinant cytokines and with nonanimal transport or adhesion factors necessary for proliferation or production; (2) adapting or genetically modifying parental cells to reduce or eliminate their requirement for serum specific factors; or (3) developing or supplementing a basal formulation with low molecular weight constituents to yield an enriched, protein-free, biochemically defined nutrient medium optimized for the target application.

A useful exercise at the outset is an analysis of the primary motivation for eliminating serum (Jayme and Greenwold, 1991). Is it because serum is ill-defined and variable from lot to lot so that you are uncertain of all factors that impact your culture environment? Are you concerned with cost and availability of qualified serum additives that "work" in your system? Is your cell type unable to grow with serum supplementation because of overgrowth by contaminating fibroblasts or inhibitory or differentiating serum factors? Does serum contain other elements that mask or inhibit a normal biological function you desire to study? Are you concerned with potential adventitious contaminants or degradative enzymes? Is your project

exclusively a laboratory research study or will its results ultimately be transferred to pilot or production-scale environments for diagnostic or biopharmaceutical applications? Results from this self-directed analysis may lead the investigator along different paths toward the development and optimization of serum-free culture medium (Waymouth, 1984; Jayme, 1991).

This section, focuses exclusively on three categories: (1) problematic constituents of basal nutrient media, (2) manufacturing process issues and, (3) concerns regarding cell maintenance under serum-free conditions. We will comment on several emerging trends in a subsequent section.

A. Nutrient Medium Constituents

1. Raw Material Definition and Standardization

Efforts to design effective serum-free formulations are often thwarted by the improper selection of basal components. In many instances, compendial specifications do not exist for certain medium ingredients. Unlike many constituents of classical medium formulations that may have compendial specifications for pharmaceutical use, novel growth and attachment additives frequently used as components of serum-free media remain unstandardized. Some additives are quantitated based upon protein content, whereas others are described in units of activity determined by bioassay performance results. Although percentage purity is frequently reported, there may be substantial variation among suppliers regarding the percentage and type of impurity. The presence or absence as raw material contaminants of certain micronutrients or cytotoxic elements may be amplified under serum-free cultivation conditions.

Some serum-free media formulations contain substantial levels of ill-defined hydrolysates, peptones, or extracts that exhibit lot-to-lot variability in biochemical composition and biological performance properties. For example, plant hydrolysates exhibit variable performance as a function of inherent seasonal growth variations, maturity at harvest, mechanism of harvest, or processing differences. These same additives may also create regulatory concerns regarding adventitious contaminants if primary or secondary processing involved contact with animal-origin materials. If possible without compromising biological activity, substitution of protein-free or chemically defined formulations devoid of macromolecular constituents may offer both technical and regulatory benefits. Protein additives to serum-free media should be synthetic in origin or be treated by processes validated to eliminate adventitious contaminants. Ensure that all medium components will be available from at least one reliable supplier at a cost and consistent quality commensurate with projected application needs.

Even where such compendial descriptions exist, specifications and analytical tests may vary significantly among global sources and suppliers. Despite efforts toward international harmonization of specifications, there remain United States (USP), European, and Japanese pharmacopoeias and an U.S. National Formulary (NP). The USP contains legally recognized standards of identity, strength, quality, purity, packaging, storage, and labeling. The NF includes standards for excipients, botanicals, and other "nondrug" ingredients that could feasibly be present in media formulations. Although there are similarities among the three international pharmacopoeias, there remain procedural or reporting differences in component identification, strength, stability, sterility, or endotoxin determination that may lend themselves to significant formulation or performance differences within serum-free culture environments.

2. Water

An often overlooked and undervalued component is water, the principal constituent of liquid cell culture medium. Given the wide variation in source materials, processing and storage methods, and quality parameters, water could readily qualify as a key variable component of the cell culture environment. Without the protective benefits of elevated serum proteins, variations in water quality could introduce variables that critically impact serum-free cultures, such as beneficial or cytotoxic trace metals, organic materials, bioburden, or bacterial endotoxin. The effects of these various contaminants will vary by cell type and by concentration, but each must be rigorously considered, controlled, and evaluated for effect. A useful guideline might be for investigators to use water that meets "water for injection" (WFI) standards to minimize these variables. We recommend that WFI should be produced fresh using validated procedures and should conform to published quality standards and specifications established through a routine testing program (Freshney, 2000).

3. Dissociating Enzymes

Trypsin is the most common enzyme used for tissue disaggregation or dissociation, as it is well tolerated by a wide range of cell culture applications. However, with the transition to serum-free culture environments without animal-origin constituents, the origin of trypsin materials is problematic. Trypsin is a naturally occurring protein in the pancreas of most mammals. The traditional sources of trypsin have been bovine or porcine pancreatic tissues, neither of which is acceptable under the current regulatory definitions as primarily and secondarily animal-origin free. Raw materials are routinely treated with gamma irradiation to destroy parvovirus and other likely adventitious contaminants.

Various protease alternatives to pancreatic trypsin have been commercialized from fungal, plant, and other nonanimal sources. These dissociating enzyme preparations appear to function equivalently to serine proteases derived from animal tissues and may exhibit other practical advantages for small-scale and production-scale cell culture applications. It is anticipated that these nonanimal trypsin substitutes will reduce the risk of introducing adventitious virus to adherent cultures and will serve as a superior alternative to porcine trypsin for regulated applications.

B. Manufacturing Process Issues

1. Storage and Stability

The functional shelf life of a particular nutrient medium may vary, depending on the professional perspective. A quality assurance professional with an analytical chemistry background might be challenged by the reality that medium formulations do not exhibit a "potency" analogous to a pharmaceutical agent. Nutrient media are formulated to deliver a reasonably homogeneous range of 30–70 different constituents. However, process variables, such as formulation water temperature, speed and duration of mixing, and filtration media, will differentially impact the postprocessing active concentration of the various ingredients.

The postfiltration stability of medium constituents is also highly variable. Some ingredients, such as ascorbic acid, break down quite rapidly in aqueous medium, whereas other components are quite stable with refrigerated storage. Most nutrient media formulated without glutamine and stored in the dark at 2–8°C should perform stably for 6–12 months. Once L-glutamine or other relatively unstable constituents have been added to the medium, performance should be monitored for individual applications. Spontaneous deamidation of glutamine to yield pyroglutamate and ammonia can be problematic due both to glutamine limitation and to sensitivity of serum-free cultures to ammonia.

Perhaps the most reliable determinations of medium shelf life are derived from cell-specific applications. Such analysis should not be limited to the observation of cellular growth kinetics, but should also include the investigation of biological function, as the production of biomolecules and other cellular functions may be impaired earlier than the proliferative rate.

2. Light Sensitivity

Deterioration of tissue culture medium components by exposure to high-intensity light has been documented since the late 1970s (Wang, 1976; Wang and Nixon, 1978; Spierenburg et al., 1984; Parshad and Sanford, 1977). Fluorescent lighting caused the deterioration of riboflavin, tryptophan (Wang, 1976; Lee and Rogers, 1988), and HEPES buffer (Zigler et al., 1985; Lepe-Zuniga et al., 1987) in medium. Riboflavin (vitamin B_2) is broken down by visible and ultraviolet light exposures below 540 nm. These photo effects on riboflavin and HEPES are mediated through the generation of superoxide radicals. These free radicals are relatively short lived, but are highly reactive and deleterious to most hydrocarbon moieties in their immediate vicinity, particularly ringed heterocycles. The resultant peroxides are also cytotoxic, particularly in serum-free environments. Of course, cell types exhibit varying sensitivities to light-induced medium cytotoxicity (Spierenburg et al., 1984), but it is generally recommended that medium, chemstocks, and cell cultures be protected from light to minimize the possibility of deleterious photo effects.

3. Lipid Delivery Mechanisms

Lipids play an integral cellular role in signal transduction, cellular communication, and intermediary metabolism. Lipids are an essential, albeit underinvestigated, ingredient of serum-free medium formulations. From a medium development perspective, lipids present unique challenges (Darfler, 1990; Gorfien et al., 2000).

Traditionally, lipid constituents were obtained from animal sources, similar to the lipid elements present in human intravenous feeding solutions. Sourcing biologically active sterols from nonanimal sources to meet customer and regulatory requirements was problematic due to biochemical differences that resulted in a significant loss of bioactivity. Traditionally, ovine cholesterol derived from lanolin has been the primary sterol additive source for cell culture. Synthetic cholesterol and plant-derived sterols have exhibited encouraging performance as substitutes for ovine cholesterol in serum-free culture.

The ability to solubilize lipids in aqueous medium and to maintain them stably following filter sterilization has raised additional challenges. Historically, lipid delivery was accomplished through attachment to albumin or other serum-derived proteins. With the demand to eliminate animal-derived proteins, ethanol

dissolution was initially investigated, but it proved less desirable due to limited lipid dissolution capacity and stability. Pluronic-based microemulsions overcame some of these issues (Stanton, 1957; Schmolka, 1977), but the production scale was limited by the capacity of the microfluidization apparatus. Pluronic-based emulsions also exhibited variable stability, and sterile filtration of single-strength nutrient medium following the addition of concentrated lipid emulsion effectively removed all supplemental lipid. Advances in cyclodextrin-based technology appear to have overcome many of these practical limitations to lipid delivery (Walowitz et al., 2003).

4. Vendor Audits

As noted earlier, under serum-free or protein-free environments, cultured cells are exquisitely sensitive to fluctuations in medium quality associated with any component of the manufacturing conversion process. Absent the protective and detoxifying contribution of serum proteins, raw material impurities may exert a greater impact on culture performance. Manufacturing protocols should be consistent with current good manufacturing practices (cGMP). Prompted by quality compliance and regulatory issues, most media manufacturers have instituted a raw material qualification program, including routine vendor audits to assess and control the quality and consistency of individual medium components. Quality documentation for all raw materials and process components should be carefully scrutinized during routine audits of nutrient medium suppliers to ensure compliance with appropriate standards and specifications.

C. Cell Maintenance under Serum-Free Cultivation Condition

Unique constraints exist for cell cultivation in serum-free environments that have been reviewed exhaustively elsewhere (Bottenstein et al., 1979; Ham, 1982; Jayme and Blackman, 1985). Three critical elements are noted briefly here.

1. Adaptation

Experimental procedures for adapting cultures to serum-free medium have been described previously (Jayme and Gruber, 1998). Our global interactions with investigators attempting to transition to serum-free culture suggest that many failures may be attributed to inadequate or inappropriate efforts to adapt cells to a novel exogenous environment. Key concerns include the growth state of the cellular inoculum, cell seeding density, subcultivation techniques, and biophysical attributes of the cell culture system.

Typically, cultures may be transitioned from serum-supplemented medium to serum-free medium over a period of 3–6 weeks, following a weaning protocol that sequentially adapts cells in a proportionate mixture of conditioned and fresh media over a period of multiple subcultures. Following adaptation, cells should be recloned in serum-free medium to establish both master and working cell banks. Adapted cell banks should be verified for consistent biological performance properties and absence of adventitious agents for cGMP applications.

2. Cryopreservation and Recovery

Two principal criteria for successful cell cryopreservation and recovery are to initiate the freeze with a healthy cell population and to ensure that both cryopreservation and recovery procedures minimize cellular insult. Log-phase cultures should be maintained with normal proliferative characteristics for a minimum of three passages in the selected serum-free medium prior to cryopreservation.

Because historical cryopreservation protocols included serum or albumin in the freezing medium, we are frequently asked if the addition of animal origin proteins is required. Our experience indicates that it is not necessary to include these additives to achieve high viability recovery of cryopreserved cells. Cells previously adapted to serum-free medium conditions have been cryopreserved successfully in a formulation consisting of equal portions of conditioned and fresh medium, supplemented with 5–10% (v/v) dimethyl sulfoxide (DMSO) as a cryoprotectant. The cryoprotectant agent is necessary to minimize the disruption of cellular and organellar membranes by ice crystals that form during freezing. To accommodate cell-specific variations, titration of DMSO may be required to determine the optimal cryoprotectant concentration for viable cell recovery.

Recovery protocols remain controversial and may vary by cell type as alternative optimization schedules are developed. A generally recommended procedure is a rapid thawing of cryovials in a 37°C water bath and an immediate dilution of cryoprotectant by inoculating cells into prewarmed nutrient medium. During recovery from a cryopreserved state, cells are particularly sensitive to mechanical disruption. Consequently, vigorous trituration, centrifugation, and other physical stresses should be avoided or minimized. When cellular recovery is evidenced by adequate observable increases in cell density, spin down cells gently and remove as much as possible of the medium containing the residual cryoprotectant and replace on a volume-for-volume basis with fresh prewarmed medium.

3. Adherent Cultures

Attachment-dependent cultures pose additional challenges to the development of serum-free media. Many adherent cell types have the capacity to deposit complex extracellular matrices, utilizing exogenous attachment factors and synthesized glycosaminoglycans for cell-to-matrix attachment purposes. In serum-supplemented medium, such attachment factors (e.g., fibronectin, vitronectin) were contributed by the serum additive. Within a serum-free environment, there are three general approaches to improving cell attachment: (1) selection or genetic modification of the cell line to augment native adherence, (2) modification of substratum properties to facilitate attachment and spreading (Griffith, 2000; Han and Hubbell 1996), and (3) nutrient medium supplementation with attachment-promoting factors or precursors. Although some progress has been made with medium modification, many potential factors are commercially unattractive from cost and animal origin perspectives and from their propensity to adsorb to container surfaces.

III. EMERGING TRENDS

We have chosen to focus on four trends that are exerting a significant impact on the development of serum-free culture environments: (1) format options for delivering nutrient medium, (2) outsourcing of biological production activities, (3) alimentation options to improve bioreactor productivity, and (4) expanded cell culture applications.

A. Format Evolution

To achieve production-scale economies, biotechnology manufacturers have transitioned to larger and more efficient bioreactors. Previous nutrient medium format options were limited to single-strength liquid medium in bulk containers or to ball-milled powder configurations. However, both of these historical options presented technical and logistical challenges for large-scale biological production applications.

Development of liquid media concentrates (50× subgroupings) facilitated stable solubilization of complex constituents of serum-free medium into a format that was readily reconstituted in either batch or continuous mode to yield production volumes of nutrient medium (Jayme et al., 1992). Liquid media concentrates have been utilized commercially for vaccine and recombinant protein. Combining this technology with an in-line mixing device facilitated commercial production of >30,000 liter batches of liquid medium dispensed directly into bulk containers from common ingredients (Jayme *et al.*, 1996).

To minimize the shipment of water, many production-scale applications prefer a dry format, but encounter performance variations, incomplete solubilization, and hygienic concerns with the hydration of serum-free formulations produced as a ball-milled powder. Milling within a stainless steel hammer mill (FitzMill) overcame technical concerns regarding thermal inactivation of heat-labile components and regulatory concerns regarding sanitization associated with ceramic ball-milling processes.

A novel approach to producing a dry-form nutrient medium was introduced by the application of fluid bed granulation to serum-free formulations, yielding a granular medium format (Fike *et al.*, 2001). This alternative approach, termed advanced granulation technology (AGT), resulted in the homogeneous distribution of trace elements and labile components onto granules that dispersed and dissolved rapidly within a medium formulation tank. Upon hydration, medium granules yield a single-strength nutrient medium with superior biological performance relative to the ball-milled powder format of the identical formulation and with specified pH and osmolality without requirement for manual titration (Radominski *et al.*, 2001).

B. Outsourcing

Given the extended developmental lead time and associated expense from identifying a lead candidate, through process development, clinical investigation and regulatory submission, and eventually culminating in an approved biological product, many companies are choosing to defer capital investment in production capacity or hiring of manufacturing personnel pending regulatory approval and commercial success. Such companies may develop mutually beneficial partnerships with contract manufacturing organizations with the ability to produce gram-to-kilogram quantities of a biological product.

Similarly, many companies have identified core competencies that they uniquely possess and have chosen to outsource noncritical capabilities, finding it financially advantageous to hire, buy, or acquire source technology. Various instances of consolidation within the biotechnology industry have resulted in acquisition by the pharmaceutical industry of entrepreneurial firms with intellectual property assets. Other biotechnology companies have opted to outsource significant requirements to contract research organizations. Others have chosen to defer capital investment in media or buffer kitchens or cell banking capabilities by contracting these ancillary services or purchasing ready-to-use materials.

C. Bioreactors

Inevitably, the transition from bench-scale shake flasks or spinner cultures to pilot or production-scale stirred tank or airlift bioreactors encounters various scale-up challenges. In addition to the classical engineering issues of agitation, gas transfer, and control of temperature and pH, there may be qualitative or quantitative adjustments in nutrient formulation or delivery schedule. Nutrient medium qualified for the batch culture of specific cells may benefit from additional optimization if cultures are to be expanded in a bioreactor that has a linear nutrient flow, such as a hollow fiber or horizontal plate-style bioreactor. Variations in nutrient consumption kinetics dependent on the bioreactor environment, such as mechanical stress, dissolved oxygen content, and gas sparging, may profoundly impact cellular energetics and predispose cells to alternative metabolic pathways that alter cell functionality or product quality.

To extend bioreactor longevity and provide more consistent product quality, process development groups often augment traditional batch bioreactor regimens by supplementing exhausted nutrients in fed-batch or perfusion mode (Mahadevan *et al.*, 1994). While a concentrated nutrient feed cocktail may be developed solely by eliminating inorganic salts and buffering components from the base medium, superior cell viability and biological productivity may be achievable through analysis of spent medium (Fike *et al.*, 1993). Determination of component exhaustion kinetics by quantitative analysis of spent medium produced by high-density cultures can yield valuable information regarding metabolite reduction or enrichment to optimize culture productivity. Nutrient modifications derived from iterative analysis, resulting either in adjustment of the initial formulation or inperiodic or continuous addition of concentrated nutrient supplements, have resulted in enhanced bioreactor longevity and specific productivity (Jayme, 1991). Although initially more laborious, this method often yields a simple, customized nutrient cocktail that permits optimized adjustment of individual nutrients and avoids inhibitory effects resulting from excessive additives or metabolic by-products.

D. Applications to Cell Therapy and Tissue Engineering

Tissue engineering applies engineering and life sciences techniques to develop biological substitutes that restore, maintain, or improve tissue or organ function. Integration of biomaterials and biological scaffolding (Griffith, 2000; Han and Hubbell, 1996) with a suitable cell type(s) and a bathing nutrient fluid can expand progenitor cells along a desirable lineage or repopulate a deficient tissue with healthy cells possessing the desired biological function. Cells may be obtained from autologous, allogeneic, or xenogeneic sources or be derived from immortalized cell lines or stem cell progenitors. Given the intended therapeutic application, nutrient media for each specialized application will ultimately need to be acceptable by both technical and regulatory criteria.

IV. CONCLUSIONS

The justifications for eliminating serum from cell culture are numerous, and serum-free media have now been developed for a broad array of biotechnological applications. The past few years have addressed regulatory concerns regarding the potential contamination of biopharmaceuticals by adventitious agents introduced via medium constituents of animal origin or defects in the manufacturing process. Eliminating all animal-origin components has proved daunting, but numerous protein-free nutrient formulations have been specifically designed and developed to eliminate questionable components without sacrificing biological performance. Coincident to the removal of animal-origin components has been the concomitant increase in the biochemical definition of medium composition. Biochemical definition will ultimately translate to enhanced production consistency for prospective biopharmaceuticals synthesized in animal cell-based bioreactors and for biomaterials designed for tissue repair or regeneration, drug delivery, or genetic therapy. Defined serum-free nutrient media and scaffolding matrices compatible with technical and regulatory requirements are under active investigation for the three-dimensional regeneration of bone, ligaments, skin, blood vessels, nerves, and organ functions. These applications extend into many of today's problematic areas, such as arthritis, diabetes, cancer, cardiovascular disease, congenital defects, or sports injuries.

Over the past decade, a striking series of trends has transformed the landscape of tissue culture. The electronic availability of information and data has reduced response times and established new international standards. Biotechnology and tissue culture have become globalized in both needs and concerns. The synergy of molecular biology techniques into functional genomics and proteomics has revolutionized and condensed the developmental pipeline for cell-based products. Enhanced social and safety consciousness has redefined our fundamental responsibility for generating products and research capabilities that will meet global criteria for scientific, economic, regulatory, and ethical acceptability.

References

Bottenstein, J., Hayashi, I., Hutchings, S., Masui, H., Mather, J., McClure, D. B., Ohasa, S., Rizzino, A., Sato, G., Serrero, G., Wolfe, R., and Wu, R. (1979). The growth of cells in serum-free hormone-supplemented media. *Methods Enzymol.* **LVIII**, 94–109.

Darfler, F. J. (1990). Preparation and use of lipid microemulsions as nutritional supplements for culturing mammalian cells. *In Vitro* **26**, 779–783.

Fike, R., Dadey, B., Hassett, R., Radominski, R., Jayme, D., and Cady, D. (2001). Advanced granulation technology (AGT): An alternate format for serum-free, chemically-defined and protein-free cell culture media. *Cytotechnology* **36**, 33–39.

Fike, R., Kubiak, J., Price, P., and Jayme, D. (1993). Feeding strategies for enhanced hybridoma productivity: Automated concentrate supplementation. *BioPharm.* **6**(8), 49–54.

Freshney, R. I. (2000). "Culture of Animal Cells: A Manual of Basic Technique," 4th Ed. Wiley-Liss, New York.

Gorfien, S. F., Paul, B., Walowitz, J., Keem, R., Biddle, W., and Jayme, D. (2000). Growth of NS0 cells in protein-free, chemically-defined medium. *Biotechnol. Prog.* **16**(3), 682–687.

Griffith, L. G. (2000). Polymeric biomaterials. *Acta Mater.* 48, 263–277.

Ham, R. G. (1982). Importance of the basal nutrient medium in the design of hormonally defined media. *In "Growth of Cells in Hormonally Defined Media"* (G. H. Sato, A. B. Pardee, and D. A. Sirbasku, eds.), pp. 39–60. Cold Spring Harbor Laboratory, New York.

Hamilton, A. O., and Jayme, D. W. (1998). Fetal bovine serum lot testing. *In "Cell Biology: A Laboratory Handbook"* (J. E. Celis, ed.), Vol. 1, pp. 27–34. Academic Press, New York.

Han, D. K., and Hubbell, J. A. (1996). Lactide-based poly(ethylene glycol) polymer networks for scaffolds in tissue engineering. *Macromolecules* **29**, 5233–5235.

Jayme, D. W. (1991). Nutrient optimization for high density biological production applications. *Cytotechnology* 5, 15–30.

Jayme, D. W. (1999). An animal origin perspective of common constituents of serum-free medium formulations. *Dev. Biol. Stand.* **99**, 181–187.

Jayme, D. W., and Blackman, K. E. (1985). Review of culture media for propagation of mammalian cells, viruses and other biologicals. *In "Advances in Biotechnological Processes"* (A. Mizrahi and A. L. van Wezel, eds.), Vol. 5, pp. 1–30. Al. R. Liss, New York.

Jayme, D. W., DiSorbo, D. M., Kubiak, J. M., and Fike, R. M. (1992). Use of nutrient medium concentrates to improve bioreactor productivity. *In "Animal Cell Technology: Basic and Applied Aspects"* (H. Murakami, S. Shirahata, and H. Tachibana, eds.), Vol. 4, pp. 143–148. Kluwer, New York.

Jayme, D. W., and Greenwold, D. J. (1991). Media selection and design: Wise choices and common mistakes. *Bio/Technology* **9**, 716–721.

Jayme, D. W., and Gruber, D. F. (1998). Development of serum-free media and methods for optimization of nutrient composition. *In "Cell Biology: A Laboratory Handbook"* (J. E. Celis, ed.), Vol. 1, pp. 19–26, Academic Press, New York.

Jayme, D. W., Kubiak, J. M., Battistoni, and Cady, D. J. (1996). Continuous, high capacity reconstitution of nutrient media from concentrated intermediates. *Cytotechnology* **22**, 255–261.

Lee, M. G., and Rogers, C. M. (1988). Degradation of tryptophan in aqueous solution. *J. Parenteral Sci. Tech.* **42**, 20–22.

Lepe-Zuniga, J. L., Zigler, J. S., Jr., and Gery, I. (1987). Toxicity of light exposed HEPES media. *J. Immunol. Methods* **103**, 145.

Mahadevan, M. D., Klimkowsky, J. A., and Deo, Y. M. (1994). Media replenishment: A tool for the analysis of high-cell density perfusion systems. *Cytotechnology* **14**, 89–96.

Parshad, R., and Sanford, K. K. (1977). Proliferative response of human diploid fibroblasts to intermittent light exposure. *J. Cell Physiol.* **92**, 481–485.

Radominski, R., Hassett, R., Dadey, B., Fike, R., Cady, D., and Jayme, D. (2001). Production-scale qualification of a novel cell culture medium format. *BioPharm* **14**(7), 34–39.

Schmolka, I. R. (1977). A review of block polymer surfactants. *J. Am. Oil Chem. Soc.* **54**, 110–116.

Spierenburg, G. T., Oerlemans, F. T., van Laarhoven, J. P., and de Bruyn, C. H. (1984). Phototoxicity of N-2-hydroxyethylpiperazine-N2-ethanesulfonic acid-buffered culture media for human leukemic cell lines. *Cancer Res.* **44**, 2253–2254.

Stanton, W. B. (1957). Polymeric nonionic surfactants. *Soap Chem. Spec.* **33**, 47–49.

Walowitz, J. L., Fike, R. M., and Jayme, D. W. (2003). Efficient lipid delivery to hybridoma culture by use of cyclodextrin in a novel granulated dry-form medium technology. *Biotechnol. Prog.*

Wang, R. J. (1976). Effect of room fluorescent light on the deterioration of tissue culture medium. *In Vitro* **12**, 19–22.

Wang, R. J., and Nixon B. T. (1978). Identification of hydrogen peroxide as a photoproduct toxic to human cells in tissue culture medium irradiated with daylight fluorescent light. *In Vitro* **14**, 715–722.

Waymouth, C. (1984). Preparation and use of serum-free culture media. *In* "Methods for Preparation of Media, Supplements, and Substrata for Serum-Free Animal Cell Culture," pp. 23–68. A. R. Liss, New York.

Zigler, J. S., Jr., Lepe-Zuniga, J. L., Vistica, B., and Gery, I. (1985). Analysis of the cytotoxic effects of light-exposed HEPES-containing culture medium. *In Vitro Cell Dev. Biol.* **21**, 282–287.

6

Cell Line Authentication

Robert J. Hay

I. INTRODUCTION

This article provides a strategy and summarizes steps for the authentication of cell line stocks. It serves as a preface for the following article and for others in this series offering detail on the characterization of cells and cell lines. For those working with serially propagated cells, it is absolutely critical that quality control tests be applied periodically. Rationale and pertinent key references are included here.

Literally hundreds of instances of cross-contamination in cell culture systems have been documented (Nelson-Rees, 1978: Nelson-Rees *et al.*, 1981.; Hukku *et al.*, 1984; MacLeod *et al.*, 1999). Many others have gone unreported. The novice technician or student using cell culture techniques soon is made painfully aware of the potential for bacterial and fungal infection. Generally, however, one must be alerted to the more insidious problems of animal cell cross-contaminations, the presence of mycoplasma, and especially the potential for latent or otherwise inconspicuous viral infection. The financial losses in research and production efforts resulting from the use of contaminated cell lines are certainly equivalent to many millions of dollars. Accordingly, frequent reiteration of the details of cell culture contaminations and of precautionary steps to avoid and detect such problems clearly is warranted.

This article includes a review of quality control steps applied to authenticate cell lines, i.e., to ensure absence of microbial, viral, and cellular contamination, as well as potential tests to verify the identity of human cells. The approach suggested has been developed during the establishment of a national cell repository. Specific rationales for applying the tests indicated are included in this volume and are discussed in more detail elsewhere (Hay *et al.*, 2000).

Most established cell lines have been characterized by the originator and collaborators well beyond the steps essential for quality control. Specific details are provided in subsequent chapters of this series and include, for example, phase-contrast and ultrastructural morphologies; detailed cytogenetic analysis; definition of protooncogene, oncogene, or oncogene product presence, nature, and location; detailed evaluation of intermediate-filament proteins; and demonstration of tissue-specific antigens or production of other specific products. These characterizations obviously increase the value of each line for research and for production work. However, cell resource organizations need not repeat all of these tests before distributing the stock cultures. Decisions must be made to establish the most acceptable authentication steps, consistent with maintaining the lowest possible cost, to provide a high-quality cell stock. Authentication can be considered the act of confirming or verifying the identity and critical feature of a specific line, whereas characterization is the definition of the many traits of the cell line, some of which may be unique and also may serve to identify or reauthenticate that line specifically. Essential steps for quality control will vary with the type of cell resource constructed. Minimal descriptive data frequently will be supplemented with a much broader characterization base for each particular cell line.

II. SEED STOCK CONCEPT

Definitions of public repository seed stocks may vary from those used for specific applications such as the production of vaccines or other biologicals. A scheme illustrating the steps involved in developing seed stocks is presented as Fig. 1.

Generally, starter cultures or ampules are obtained from the originator, and progeny are propagated according to the instructions to yield the first "token" freeze. Cultures derived from such token material are then tested for bacterial, fungal, and mycoplasmal contamination. The species of each cell line is verified. These quality control steps are the minimum ones that must be performed before eventual release of a line. If these steps confirm that further efforts are warranted, the material is expanded to produce the seed and distribution stocks. Note that, under ideal conditions, additional major quality control and characterization efforts are applied to cell populations from seed stock ampules. Test results refer to *specific numbered stocks*. The distribution stock consists of ampules that are distributed on request to investigators. The reference seed stock, however, is retained to generate further distribution stocks as the initial distribution stock becomes depleted. The degree of characterization applied to master cell banks or master working cell banks in production facilities is generally most rigorous. The seed stock here, like the master cell bank, is used as a reservoir to replenish depleted distribution lots over the years. By adherence to this principle, one can avoid problems associated with genetic instability, cell line selection, or transformation.

III. MICROBIAL CONTAMINATION

Microbial contamination in cell culture systems remains a serious problem. Cryptic contaminants, even of readily isolatable bacteria and fungi, are missed by many laboratories. The American Type Culture Collection (ATCC) still receives cultures, even for the patent depository, that contain yeast, filamentous fungi, and/or mycoplasma contaminants.

A. Bacteria and Fungi

Microscopic examination is not sufficient for the detection of gross contaminations; even some of these cannot be detected readily by simple observations. Therefore, an extensive series of culture tests is also required to provide reasonable assurance that a cell line stock or medium is free of fungi and bacteria. Details are given in the following article.

B. Mycoplasma Infection

Contamination of cell cultures by mycoplasma can be a much more insidious problem than that created by the growth of bacteria or fungi. Although the presence of some mycoplasma species may be apparent because of the degenerative effects induced, other mycoplasma metabolize and proliferate actively in the

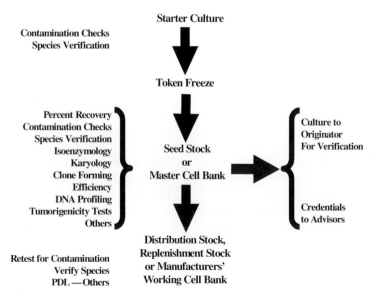

FIGURE 1 Accessioning Scheme.

culture without producing any overt morphological change in the contaminated cell line. Thus, cell culture studies relating to metabolism, surface receptors, virus–host interactions, and so forth are certainly suspect to interpretation, if not negated in interpretation entirely, when conducted with cell lines that harbor mycoplasma. The seriousness of these problems has been documented through published data from testing services and cell culture repositories.

The high incidence of mycoplasma contamination from human operators is supported by the fact that *Mycoplasma orale* and others of human origin (*Mycoplasma hominus*, *M. salivarium*, and *M. fermentas*) are among those most frequently isolated. In the study of Del Giudice and Gardella (1984) of the 34,697 lines tested, 3955 (11%) were positive; 36% of these isolates were mycoplasmas of human origin. A high incidence of isolation of *Mycoplasma hyorhinus* was noted that may have resulted from using contaminated sera or by culture-to-culture spread in laboratories working with infected biologicals. After a more recent study, Uphoff *et al.* (1992) reported that 84 (33%) of 253 cell lines submitted for their developing cell repository in Germany were infected with mycoplasma. Data showing that these are not unusual findings are presented in Table I. Results from seven laboratories published since 1980 indicate clearly that mycoplasma infection is still a very major problem in the cell culture field. Some 5 to 20% of cultures tested were positive. This is the range even today.

Protocols to test for mycoplasma infection are included in the following chapter.

Four general recommendations can be offered to avoid the problem. The implementation of an effective regimen to monitor cell lines for mycoplasma is one critical step. Quarantining all new, untested lines and the use of mechanical pipetting aids are others. Most experts also strongly suggest that the use of antibiotics be eliminated when possible. Antibiotic-free systems permit overgrowth by bacteria and fungi to provide ready indication whenever a lapse in aseptic technique occurs. When a primary tissue is used, e.g., a human tumor sample, antibiotics may be employed initially, but after the primary population has been grown out and cyropreserved, reconstituted cells may be propagated further in antibiotic-free medium.

C. Viruses

Verification of the absence of viruses in cell lines is recognized as a most significant problem. That these may coexist as noncytopathic entities (e.g., the c-type retroviruses) or in a latent form (e.g., papilloma viruses and some herpes viruses) compounds difficulties in detection. Judicious choices are necessary not only to select appropriate methods available for recognizing viruses associated with cell lines, but also to identify the offending species. The nature of the cell resource, its users, and budget available, plus the intended purposes for which the lines will be needed, all affect decisions on testing. More complete detail and protocols are provided in the following article.

IV. CELLULAR CROSS-CONTAMINATION

Wherever cells are grown in culture, serious risk exists for the inadvertent addition and subsequent overgrowth by cells of another individual or species. One most certainly cannot rely on morphologic criteria alone to recognize or identify cell lines. Data-documenting problems have been collected over the years by groups offering identification services for cell culture laboratories in the United States and elsewhere.

In one study, 466 lines from 62 laboratories were examined. Of these, 75 (16%) were found to be identified incorrectly. A total of 43 lines (9%) were not of the species expected, whereas 32 lines (7%) were either incorrect mixtures of two or more lines or were not the individual line as stated (Nelson-Rees, 1978). Hukku *et al.* (1984) examined 275 lines over a period of 18 months. Results of their analyses are summarized in Table II. A total of 96 lines (35%) were not as indicated by the donor laboratories. For purported human lines, 36% were not as expected, 25% were a different species, and 11% a different human individual. More recently, Drexler *et al.* (1994), while developing a resource of cell lines, reported that 10% of those provided by outside investigators contained cells different from those expected, probably due to misidentification or cross-contamination.

TABLE I Mycoplasma Infection of Cell Lines (1982–2002)

Reference test laboratory	Percentage positive	Total
McGarrity (1982)	4.7	16,197
Del Giudice and Gardella (1984)	11.4	34,697
Uphoff *et al.* (1992)	33	233
Takeuch *et al.* (1993)	21	2,332
Lundin and Lincoln (1994)	10.2	1,000
Hay *et al.* (1996b)	15.5	5,362
Uphoff and Drexler (2002)	17	549

TABLE II   Summary of Cell Line Cross-Contamination[a]

Reported cell species	Cultures received	Interspecies	Intraspecies	Percentage of total
Human	160	40 (25%)	18 (11%)	36
Others	115	38	—	33
Total	275	78	18	35

[a] Adapted from Hukku et al. (1984)

To minimize the risk of cellular cross-contamination, culture technicians require a laminar flow hood for ideal operation. These individuals must be instructed periodically to work with only one cell line at any given time, use one reservoir of medium for each line, and avoid introducing pipettes that have been used to dispense or mix cells into any medium reservoir. Technicians must be reminded repeatedly to legibly label each and every cell culture with designations, passages, and dates. Labels of differing colors can be used to more readily distinguish one cell line from another during expansion.

Technicians must also be instructed to allow at least 5 min of hood clearance time, with ultraviolet lights and blower on, between cell lines when working on more than one line during a particular period. The inner surfaces of each hood should be swabbed with 70% ethanol between such uses.

The studies outlined earlier illustrate the severity of the problem of cellular cross-contamination and provide a strong rationale for vigilance in careful handling, characterization, and authentication of cell lines.

A. Species Verification

Species of origin can be determined for cell lines by a variety of immunological tests, by isoenzymology, and/or by cytogenetics (Nelson-Rees, 1978; Hukku et al., 1984; Hay et al., 2000). The indirect fluorescent antibody-staining technique is used in many laboratories to verify the species of a cell line (for details, see Hay et al., 1992). Isozyme analyses performed on homogenates of cell lines from over 25 species have demonstrated the utility of these biochemical characteristics for species verification (O'Brien et al., 1977). By determining the mobilities of three isozyme systems—glucose-6-phosphate dehydrogenase, lactic acid dehydrogenase, and nucleoside phosphorylase—using vertical starch gel electrophoresis, the species of origin of cell lines can be identified with a high degree of certainty. Alternatively, a standardized kit employing agarose gels and stabilized reagents may be obtained for this purpose (AuthentiKit, Innovative Chemistry, Inc., Marshfield, MA).

Karyologic techniques have long been used informatively to monitor for interspecies contamination among cell lines. In many instances, the chromosomal constitutions are so dramatically different that even cursory microscopic observations are adequate. In others, for example, in comparisons among cell lines from closely related primates, careful evaluation of banded preparations is required (Nelson-Rees, 1981; Hukku et al., 1984). Cytogenetics has the advantage of detecting even very minor contaminants, on the order of 1% or less in some circumstances. Furthermore, it can provide precise identification of specific lines in cases where marker chromosomes are known or detected (e.g., Drexler et al., 2002). However, cytogenetic analyses are time-consuming and interpretation requires a high degree of skill. The karyotype is constructed by cutting chromosomes from a photomicrograph and arranging them according to arm length, position of centromere, presence of secondary constrictions, and so forth. Automated analytical systems are available but expensive. The "Atlas of Mammalian Chromosomes" (Hsu and Benirschke, 1967–1975) illustrates examples of conventionally stained preparations from over 550 species. Detailed protocols are available elsewhere (Hay et al., 2000).

B. Intraspecies Cross-Contamination

With the dramatic increase in numbers of cell lines being developed, especially from human tissues, the risk of intraspecies cross-contamination rises proportionately. The problem is especially acute in laboratories requiring work with the many different cell lines of human and murine origin available today.

Methods for verifying cell line species employing enzyme mobility studies have been mentioned. Using similar technology, but with different enzyme systems, one can also screen for intraspecies cellular cross-contamination. Cell lines from various individuals of the same species often show different codominant alleles for a given enzyme locus, the products of which are polymorphic and electrophoretically resolvable. In most cases, the phenotype for these allelic enzymes (allozymes) is extremely stable. Consequently, when allozyme phenotypes are determined over a suitable spectrum of loci, they can be used effectively to provide an allozyme genetic signature for each line under study (O'Brien *et al.*, 1977; Hukku *et al.*, 1984).

The application of recombinant DNA technology and cloned DNA probes to identify and quantitate allelic polymorphisms provides additional powerful means for cell line identification. These polymorphisms can be recognized as extremely useful markers, even if they are not expressed through transcription and translation to yield structural enzymatically active proteins.

Hybridization probes to regions of the human genome that are highly variable have been produced for DNA profiling applications, including cell line individualization. Profiles derived from human cell lines can be interpreted best using scanning devices as the patterns are complicated. For protocols and examples, see Jeffreys *et al.* (1985), Gilbert *et al.* (1990), and Hay *et al.* (1996, 2000).

Original procedures included fingerprints using larger (10–20 bp) minisatellite DNA and Southern blotting. With application of the polymerase chain reaction (PCR), loci can be typed in hours rather than days. Smaller microsatellite (2–6 bp) loci have been identified as well. Edwards *et al.* (1992) demonstrated the usefulness of these "short tandem repeat" (STR) loci in recognizing individuals at the DNA level. One significant advantage of STR loci over minisatellite repeats is their small size. The small size of STR loci allows multiplex PCR reactions to be developed in which many loci are examined simultaneously in a single reaction. Commercially available multiplexed STR systems are available for routine screening of new cell lines for authenticity as well as for validating any subsequent distribution of replenishment cell lines. More detail on STR profiling of human DNA cell lines is available elsewhere (Hay *et al.*, 2000). A comprehensive database can be accessed via the ATCC website at http://www.atcc.org/cultures/str.cfm. STR loci are among the most informative polymorphic markers in the genome. The profiling process used at ATCC involves simultaneous amplification of eight STR loci and the amelogenin gene in a multiplex PCR reaction (Promega PowerPlex 1.2 system). This allows for discrimination of fewer than 1 in 10^8 individuals. The amplicons are separated by electrophoresis and are analyzed using Genotyper 2.0 software from Applied Biosystems. Each peak in the resulting electropherogram represents an allele that is alphanumerically scored and entered into our database.

The classical method for intraspecies cell line individualization involves karyotype analysis after treatment with trypsin and Giemsa stain (Giemsa or G-banding). The banding patterns made apparent by this technique are characteristic for each chromosome pair and permit recognition, by an experienced cytogeneticist, of even comparatively minor inversions, deletions, or translocations. Many lines retain multiple marker chromosomes, readily recognizable by this method, that identify the cells specifically and positively (Chen *et al.* 1987; Drexler *et al.* 2002). If readily recognized marker chromosomes are present, contaminations at less than 1% can be recognized with careful scrutiny. This technique even permits discrimination among lines from the same individual that cannot be identified as different by DNA. These lines have similar marker chromosomes, indicating a common source, but each also has markers unique in type and copy number. In contrast, Gilbert *et al.* (1990) noted no distinct differences among the nine HeLa derivatives examined using the minisattelite probe 33.6 after *Hae*III designation. Metabolic differences among the HeLa derivatives (Nelson-Rees *et al.*, 1980) will ultimately be traced to genetic differences among the lines, as reflected by the unique cytogenetic profiles. On this basis, the importance of documenting the precise cytogenetics of cell lines used for production purposes is very clear. Protocols are provided elsewhere in this series.

V. ORIGIN AND FUNCTION

The markers used for verification of the source tissues for cell lines are probably as numerous as the types of metazoan cells. Major methods of demonstration include an analysis of fine structure, immunological tests for cytoskeletal and tissue-specific proteins, and, of course, an extremely broad range of biochemical tests for specific functional traits of tissue cells. Ultrastructural features such as desmosomes or Weibel–Palade bodies identify epithelia and endothelia, respectively. The nature of intermediate-filament proteins, demonstrated using monoclonal antibodies, permits differentiation among epithelial subtypes, mesenchymal, and neurological cells. Tissue- and tumor-specific antigens can be used when reagents are reliable and available. In addition, tissue-specific biochemical reactions or syntheses may be used for absolute identification if these features are retained by the cell line in question. One excellent example is the cell line NCI-H820 (ATCC—HTB 181) isolated from a metastic lesion of a human papillary lung adenocarcinoma. Cells of this line reportedly retain multilamellar bodies suggestive of type 2 pneumocytes and express the three surfactant-associated proteins SP-A (constitutively), SP-B, and SP-C (after dexamethasone stimulation; A. Gazdar, personal communication). For more examples, see other articles of this series and Hay (1992).

VI. CONCLUSIONS

The overall utility of any bank of cultured cell lines depends on the degree of characterization of the holdings that has been performed by the originators, the banking agency, and other individuals within the scientific community. Ready availability at a reasonable cost of the lines and such data, as well as the ability to track distribution of the biologicals, are additional critical considerations. Documenting the verification of species and identity of each cell line, when possible, is considered essential. Freedom from bacterial, fungal, and mycoplasmal infection must be assured. However, from the cell banking perspective, applying all possible characterizations to every seed or master cell stock developed is neither essential nor practical. At ATCC, for example, screens for particular viruses have been applied when a specific program support is available for such testing. Similarly, the definition of ultrastructural features, tumorigenicity, and functional traits, for example, is performed with appropriate external support and adequate rationale. The central responsibility is to produce reference stocks, authenticated and well characterized for multiple purposes, and to return to those preparations over the years for the development of working stocks for distribution or other specific applications. Each replacement distribution stock requires reauthentication prior to distribution to intended users.

References

Chen, T. R. (1988). Re-evaluation of HeLa, HeLaS3 and Hep-2 karyotypes. *Cytogenet. Cell Genet.* **48**, 19–24.

Chen, T. R., Drabkowski, D., Hay, R. J., Macy, M. L., and Peterson, W., Jr. (1987). WiDr is a derivative of another colon adenocarcinoma cell line, HT-29. *Cancer Genet. Cytogenet.* **27**, 125–134.

Del Giudice, R. A., and Gardella, R. S. (1984). Mycoplasma infection of cell culture: Effects, incidence and detection. *In "In Vitro* Monograph 5: Uses and Standardization of Vertebrate Cell Cultures," pp. 104–115. Tissue Culture Association, Gaithersburg, MD.

Drexler, H. G., Gignac, S. M., Minowada, J. (1994). Hematopoietic cell lines. *In "Atlas of Human Tumor Cell Lines* (R. J. Hay, R. Gazdar, and J. G. Park, eds.), pp. 213–250. Academic Press, Orlando.

Drexler H. G., Quentmeier, H., Dirks, W. G., Uphoff, C. C., and MacLeod, R. A. (2002). DNA profiling and cytogenetic analysis of cell line WSU-CLL reveal cross-contamination with cell line REH (pre B-ALL). *Leukemia* 1868–70.

Edwards, A., Hammond, H. A., Jin, L., Caskey, C. T., and Chakraborty, R. (1992). Genetic variation at five trimeric and tetrameric tandem repeat loci in four human population groups. *Genomics* 241–253.

Gilbert, D.A, Reid, Y. A., Gail, M. H., Pee, D., White, C., Hay, R. J., and O'Brien, S. J. (1990). Application of DNA fingerprints for cell line individualization. *Am. J. Hum. Genet.* **47**, 499–514.

Hay, R. J. (1992). Cell line preservation and characterization. *In "Animal Cell Culture: A Practical Approach"* (R. I. Freshney, ed.), 2nd Ed., pp.95–148. IRL Press, New York.

Hay, R. J., Caputo, J., and Macy, M. L. (1992). "ATCC Quality Control Methods for Cell Lines," 2nd Ed. ATCC, Rockville, MD.

Hay, R. J., Reid, Y. A., McClintock. P. R., Chen, T. R., and Macy, M. L. (1996). Cell line banks and their role in cancer research. *J. Cell Biochem. Suppl.* **27**, 1–22.

Hay, R. J., Reid, Y. A., and Miranda, M. (1996b). Advances in methodologies for cell line authentication. *In "Culture Collections to Improve the Quality of Life"* (R. A. Sampson, J. A. Staplers, D. vander Mei, and A. H. Stouthamer, eds.), pp. 131–137. CVS, Baarm, The Netherlands.

Hay, R. J., Cleland, M. M., Durkin, S., and Reid, Y. A. (2000). Cell line preservation and authentication *In "Animal Cell Culture: A Practical Approach"* (J. R. W. Masters, ed.), 3rd Ed., pp. 69–103. Oxford University Press, New York.

Hsu, T. C., and Benirschke, K. (1967–1975). "At Atlas of Mammalian Chromosomes." Springer-Verlag, New York.

Hukku, B., Halton, D. M., Mally, M., and Peterson, W. D., Jr. (1984). Cell characterization by use of multiple genetic markers. *In "Eukaryotic Cell Cultures"* (R. T. Acton and J. D., Lynn, eds.), pp. 13–31. Plenum, New York.

Jeffreys, A. J., Wilson, L., and Thein, S. L. (1985). Hypervariable minisatellite regions in human DNA. *Nature* **314**, 67–73.

Lundin, D. J., and Lincoln, C. K. (1994). Mycoplasmal testing of cell cultures by a combination of direct culture and DNA-fluorochrome staining. *In Vitro* **30A**, 111.

MacLeod R. A. F., Dirks, W. G., Kaufmann, M., Matsuo Y., Milch, H., and Drexler, H. G. (1999). Widespread intraspecies cross-contamination of human tumor cell line arising at source. *Int. J. Cancer* **83**:555–563.

McGarrity, G. J. (1982). Detection of mycoplasmal infection of cell cultures. *Adv. Cell Culture* **2**, 9–131.

Nelson-Rees, W. A. (1978). The identification and monitoring of cell line specificity. *Progr. Clin. Biol. Res.* **20**, 25–79.

Nelson-Rees, W. A., Daniels, D. W., and Flandermeyer, R. R. (1981). Cross-contamination of cell lines. *Science* **212**, 446–452.

Nelson-Rees, W. A., Hunter, L., Darlington, G. J., and O'Brien, S. J. (1980). Characteristics of HeLa strains: Permanent vs. variable features. *Cytogenet. Cell Genet.* **27**, 216–231.

O'Brien, S. J., Kleiner, G., Olson, R., and Shannon, J. E. (1977). Enzyme polymorphisms as genetic signatures in human cell culture. *Science* **195**, 1345–1348.

Reid, Y. A., Gilbert, D. A., and O'Brien, S. J. (1990). The use of DNA hypervariable probes for human cell line identification. *Am. Type Cult. Collect. Newslett.* **10**(4), 1–3.

Reid, Y. A., and Lou, X. (1993). The use of PCR-amplified hypervariable regions for the identification and characterization of human cell lines. *In Vitro* **29A**, 120A.

Takeuchi, M., Yoshida, T., Satoh, M., Kuno, H., and Ohno, T. (1993). Survey of mycoplasmal contamination in animal cell lines collected by three cell banks in Japan. *Bull. JFCC* **9**, 13–18.

Uphoff, C. C., Brauer, S., Grunicke, D., Gignae, S. M., MacLeod, R. A. F., Quentmeier, H., Steube, K., Tummler, M., Voges, M., Wagner, B., and Drexler, H. G. (1992). Sensitivity and specificity of the different mycoplasma detection assays. *Leukemia* **6**, 335–341.

Uphoff, C. C., and Drexler, H. G. (2002). Detection of mycoplasma in leukemia-lymphoma cell lines using polymerase chain reaction. *Leukemia* **16**, 289–293.

Detection of Microbial and Viral Contaminants in Cell Lines

Robert J. Hay and Pranvera Ikonomi

I. INTRODUCTION

The presence of microbial contaminants—bacteria, fungi, mycoplasma, or protozoa—in cell cultures seriously compromises virtually all research or production work involving culture technology. Although many contamination events are overt and readily apparent, others are insidious and more difficult to detect. Similarly, viral infection may be obvious if cytopathogenesis is affected, but many viruses do not induce a drastic alteration in host cells and some are present in latent forms.

This article provides representative test protocols suitable for detecting most microbes and many viruses that might be expected in cell culture systems. The perspective is that of staff operating a national cell culture resource.

II. MATERIALS

The following media and reagents are from Difco: Bacto Sabouraud dextrose-broth (Cat. No: 0382-1); Bacto fluid thioglycollate medium (Cat. No. 0256-01); beef extract (Cat. No. 0131); brain–heart infusion (BHI) broth (Cat. No. 0037-016); neopeptone (Cat. No. B119); proteose peptone (Cat. No. 3); YM broth (Cat. No. 0711-01); nutrient broth (Cat. No. 003-01); Bacto yeast extract (Cat. No. 012701); trypsin, 1:250 Difco certified (Cat. No. 0152-15); and blood agar base (Cat. No. 0045-01). Trypticase soy broth powder (Cat. No. 01-162), trypticase (Cat. No. B11770), mycoplasma broth base (Cat. No. 11458), and mycoplasma agar base (Cat. No. 11456) are from Becton-Dickinson Microbiology

Systems (formerly BBL). Fresh defibrinated rabbit blood (Cat. No. 82-8614) and sheep blood are from Editek, and North American Biologicals provided Diamond's TP-S-1 broth base powder (Cat. No. 73-9502) and Diamond's TP-S-1 vitamin solution (40X, Cat. No. 72-2315). American Hoechst is the source of bisbenzamide fluorochrome stain (Cat. No. 33217), and oil-free, dry annealed aluminum foil is from Reynolds Aluminum (Cat. No. 1235-0). Unless specified otherwise, cell culture media (various) and sera are from ATCC, Sigma, GIBCO-BRL, or Hyclone.

Template-primers $poly(rA) \cdot poly(dT)_{12-18}$ (Cat. No. 7878) and $poly(dA)–poly(dT)_{12-18}$ (Cat. No. 7868) are from P-L Biochemicals, and [*methyl*-^{3}H]thymidine triphosphate ([3H]TTP, carrier-free, specific activity 50–60 Ci/mmol, 1.0 mCi/ml, Cat. No. NET 221X) is from Dupont New England Nuclear. The scintillation counter fluid (Betafluor, Cat. No. LS-151) was purchased from National Diagnostics.

The following items are required in screening for c-type viruses such as HIV and HTLV:

Rneasy minikit, Qiagen (Cat. No. 74104)
SuperScript double-stranded cDNA synthesis kit, Invitrogen (Cat. No. 11917-010)
Platinum Taq DNA polymerase, Invitrogen (Cat. No. 0966034)
Sybr Green I, Molecular Probes, Inc. (Cat. No. S7563)
DNA standard and HIV and HTLV positive cell lines, ATCC (Cat. No. 53069 CRL-8993 and CRL 8543)

Standards, primers, and probes:

A pUC18 vector containing a 9.9-kb insert from HIV-standard DNA is used as standard DNA for HIV (ATCC, Manassas, VA, ATCC No. 53069). For GAPDH and HTLV, polymerase chain reaction

(PCR) products containing the primer and probe sequence for both targets have been cloned into the pUC18 vector.

Primers and probes include:

HIV	Forward primer	5' TCCACCTATCCCAG TAGGAGAAAT 3'
	Reverse primer	5' GGTCCTTGTCTTAT GTCCAGAATG 3'
	TaqMan probe	5' GATTAAATAAAATAG TAAGAATGTATAGC 3'
HTLV	Forward primer	5' CAATCACTCATACA ACCCCCAA 3'
	Reverse primer	5' CTGGAAAAGACAGG GTTGGG 3'
	TaqMan probe	5' TCCTCCAGGCCATG CGCAAATACTCG 3'

In most cases, Sigma or VWR supplied general laboratory chemicals and solvents plus such items as instruments and bacteriological or cell culture glass- and plasticware.

The following additional specialty items are required: Leighton tubes (Cat. No. 3393) from Costar; cellulose filters, 0.45 and 0.22 μm (Cat. Nos. HATF 14250 and GSTF 14250, respectively) from Millipore; GasPak anaerobic systems (Cat. No. 60465) from Becton-Dickinson Microbiology Systems; embryonated chicken eggs from SPAFAS; egg candlers (Cat. No. C6372N-50001) from Nasco; egg drills, cutters, and moto tool (Cat. No. 9826-00) from Cole-Parmer; stainless-steel sterilizing pans (Cat. No. 2065-5) from Orem; and adjustable microliter pipettes and Pipetman (Cat. No. P-20 D/P-200D) from Rainin. Reference microbes, cell lines, and viruses are from the American Type Culture Collection (ATCC): *Pseudomonas aeruginosa* (e.g., ATCC 14502), *Micrococcus salivarious* (e.g., ATCC 14344), *Escherichia coli* (e.g., ATCC 4157), *Bacteroides distasonis* (e.g., ATCC 8503), *Penicillium notatum* (e.g., ATCC 8537), *Aspergillus niger* (e.g., ATCC 34467), *Candida albicans* (e.g., ATCC 10231), influenza virus (e.g., ATCC VR-95 or VR-810), Newcastle disease virus (e.g., ATCC VR-108 or VR-109), and Rous sarcoma virus (e.g., ATCC VR-140 or VR-724). The Gen-Probe kit is available from Fisher (Cat. No. GP1591) and the mycoplasma PCR kit is from ATCC (Cat. No. 90-1001K).

III. PROCEDURES

A. Bacteria and Fungi

Tests for sterility are performed routinely at ATCC on all culture media used, on cultures submitted from the community, on cultures at various stages during the accessioning process, and on all seed and distribution freezes. *Pseudomonas* species, micrococci, and *E. coli* are common bacterial isolates, whereas *Penicillium*, *Aspergillus*, and *Candida* species are common fungal and yeast contaminants.

1. Preparation of Media Solutions

1. *Sabouraud dextrose broth:* Dissolve 30 g dehydrated powder in 1000 ml distilled water and dispense 10-ml aliquots into each of one hundred 16 × 150-mm test tubes. Cap each tube loosely and sterilize in an autoclave for 15 min at 15 lb pressure (121°C) on slow exhaust. After removing the tubes from the autoclave, press down caps securely and store at room temperature until used. Caps of differing colors may be used to permit ready identification.

2. *Nutrient broth with 2% yeast extract:* Dissolve 8 g of nutrient broth powder plus 20 g of Bacto yeast extract in 1000 ml distilled water and dispense 10-ml aliquots into each of one hundred 16 × 150-mm test tubes. Cap each tube loosely, sterilize, and store as described for solution 1.

3. *Thioglycollate medium:* Suspend 29.8 g dehydrated powder in 1000 ml distilled water in a 3-liter flask and heat to boiling to dissolve the powder completely. Dispense 10-ml aliquots of the thioglycollate medium into each of one hundred 16 × 150-mm test tubes; cap each tube loosely. Sterilize in the autoclave as described for solution 1. After removing the tubes from the autoclave, press down caps securely and store in the dark at room temperature. This medium changes color in processing. As it dissolves it turns red or gold depending on the amount of dissolved oxygen. After autoclaving it is clear and gold in color, like nutrient broth. After cooling, the top layer of medium oxidizes and the indicator in the upper portion of the tube turns pink or red. The fluid should not be used if the indicator has changed to a red color in the lower third of the tube.

4. *Trypticase soy broth:* Suspend 30 g powder in 1000 ml distilled water and mix thoroughly and warm gently until solution is complete. Dispense 10-ml aliquots of die trypticase soy booth into each of one hundred 16 × 150-mm test tubes; cap each tube loosely and sterilize and store as described for solution 1.

5. *BHI broth:* Suspend 37 g powder in 1000 ml of distilled water, dissolve completely, and dispense 10-ml aliquots of BHI into 16 × 150-mm test tubes. Cap each tube loosely. Sterilize in the autoclave for 15 min at 15 lb pressure on slow exhaust. After autoclaving, press down caps securely, cool, and store at 4°C.

6. *YM broth:* Dissolve 21 g powder in 1000 ml distilled water and dispense 10-ml aliquots of the broth into each of one hundred 16 × 150-mm test tubes; cap each tube loosely. Sterilize and store as described for solution 1.

7. *Blood agar plates:* Suspend 40 g blood agar base in 950 ml cold distilled water and heat to boiling to dissolve the powder completely. Sterilize in the autoclave for 15 min at 15 lb pressure on slow exhaust. When the sterile blood agar base is cooled to 50°C, add 5% (50 ml) of pretested, fresh, defibrinated rabbit blood and mix by swirling. Dispense aseptically to 9-cm plates and store at 4°C. The rabbit blood is pretested for sterility by inoculating 0.5-ml aliquots into BHI broth and YM broth and onto blood agar base plates with subsequent incubation at 25° and 37°C. Negative results in 48 to 72 h are usually sufficient to permit use of the tested fluid.

2. Examination

Steps

1. Using an inverted microscope, equipped with phase-contrast optics if possible, examine cell culture vessels individually. Scrutiny should be especially vigorous in cases where large-scale production is involved.

2. Check each culture first using low power. After moving the cultures to a suitable isolated area, remove aliquots of fluid from cultures that are suspect; retain these for further examination. Alternatively, autoclave and discard all such cultures.

3. Prepare wet mounts using drops of the test fluids and observe under high power.

4. Prepare smears, heat fix, and stain by any conventional method using filtered solutions.

5. Examine under oil immersion for microbial contaminants.

6. Consult Freshney (2000) for further details.

3. Inoculation and Incubation of Test Samples

Steps

1. After cryopreservng stocks of cells (Hay *et al.*, 2000), retrieve and thaw about 5% of the ampoules from liquid nitrogen or vapor storage. Pool and mix the contents of the ampoules from each cryopreserved lot using a sterile 1-ml disposable pipette. It is recommended that antibiotics not be included in media used to prepare stocks of cells for preservation. If antibiotics are used, the pooled suspension should be centrifuged at 2000 g for 20 min and the pellet should be resuspended in antibiotic-free medium. A series of three such washes with antibiotic-free medium prior to testing eliminates traces of antibiotics that could obscure contamination.

2. From each pool, inoculate each of the following with a minimum of 0.3 ml of the test cell suspension: (a) two blood agar plates, (b) two tubes of thioglycollate broth, (c) two tubes of trypticase soy broth, (d) two tubes of BHI broth, (e) two tubes of Sabouraud broth, (f) two tubes of YM broth, and (g) two tubes of nutrient broth with 2% yeast extract.

3. Incubate test plates and broths as follows. (a) Blood agar plates: one at 37°C under aerobic conditions and one at 37°C anaerobically (a BBL Gaspak anaerobic system is convenient for the latter). (b) Tubes of thioglycollate broth, trypticase soy broth, BHI broth, Sabouraud broth, YM broth, and nutrient broth with yeast extract: one each at 37°C and one each at 26°C under aerobic conditions. (c) Incubate and examine periodically for 14 days the tubes of thioglycollate, trypticase soy broth, BHI broth, and blood agar plates. (d) Observe the tubes of Sabouraud broth, YM broth, and nutrient broth with yeast extract for 21 days before concluding that the test is negative. Contamination is indicated if colonies appear on solid media or if any of the liquid media become turbid.

4. Repeat any components of the test series that are positive initially to confirm the presence of a contaminant.

5. Autoclave and discard any contaminated cultures or ampule lots.

Comments

Of the seven media employed, trypticase soy, BHI, blood agar, and thioglycollate are suitable for detecting a wide range of bacterial contaminants. Sabouraud broth, YM broth, and nutrient broth with yeast extract will support growth of fungal contaminants. Stock media and incubation conditions used can be tested with the following ATCC control strains: *P. aeruginosa*, *M. salivarius*, *E. coli*, *B. distasonis*, *P. notatum*, *A. niger*, and *C. albicans*. Table I summarizes this recommended test regimen.

Pitfalls

Although this test regimen permits detection of most common bacterial and fungal organisms that grow in cell cultures, we have experiences with at least one very fastidious bacterial strain that initially escaped observation. This was present in nine different cell cultures from a single clinical laboratory in the United States submitted for testing and expansion under a government contract. The organism grew extremely slowly but could be detected after 3 weeks of incubation with cell cultures without antibiotics and with no fluid changes. Samples so developed were inoculated into sheep blood agar plates and New York City broth (ATCC medium 1685). The organism could

TABLE I   Suggested Regimen for Detecting Bacterial or Fungal Contamination

Test medium	Temperature (°C)	Gas phase	Observation time (days)
Blood agar with fresh defibrinated rabbit blood (5%)	37	Aerobic	14
	37	Anaerobic	14
Thioglycollate broth	37	Aerobic	14
	26		
Trypticase soy broth	37	Aerobic	14
	26		
Brain–heart infusion broth	37	Aerobic	14
	26		
Sabouraud broth	37	Aerobic	21
	26		
YM broth	37	Aerobic	21
	26		
Nutrient broth with 2% yeast extract	37	Aerobic	21

be observed during a subsequent 6-week incubation period at 37°C.

The bacteriology department at ATCC determined suitable culture conditions for this microorganism and tentatively identified it as a *Corynebacterium*. Antibiotic sensitivity tests revealed bacteriostasis with some compounds, but no bactericidal antibiotics have yet been found.

This incident emphasizes the critical importance of diligent testing of cell cultures for contaminant microorganisms. By combining protocols such as those described here with procedures discussed later (e.g., fluorescent or nucleic acid probes for mycoplasma and viruses), one can be more certain that clean cell cultures are available for experimentation.

B. Mycoplasma

Mycoplasmal contamination of cell cultures has been established as a common occurrence that is capable of altering normal cell structure and function. Mycoplasmas have been shown to inhibit cell metabolism and growth, alter nucleic acid synthesis, affect cell antigenicity, induce chromosomal alterations, interfere with virus replication, and mimic viral actions. Basically, the growth of mycoplasma in cell cultures can be detected either by a direct microbiological agar culture procedure or by indirect procedures using staining, biochemical methods, or nucleic acid hybridization techniques (McGarrity, 1982; Hay et al., 1989, 1992, 2000).

In testing cell cultures for contamination, both direct and indirect procedures should be employed. The indirect method employed most frequently at the ATCC was originally described by Chen (1977). It

requires the bisbenzamide DNA fluorochrome staining procedure plus an indicator cell. This adaptation is described here with slight modifications.

Duplicate screening techniques are generally recommended for rigorous cell line testing. Alternative methods to those included here include nucleic acid hybridization and a new technique involving the polymerase chain reaction. Details are available elsewhere (Hu and Buck, 1993; ATCC website http://www.atcc.org).

In screening cell lines for mycoplasma contamination, it is important to include positive controls in order to be assured that the test systems being used are optimal. Special precautions, however, are necessary when working with such material. The handling of mycoplasma cultures should be done at the end of a particular test and, when possible, in an isolated area using a biohazard-type hood. All equipment used in manipulations involving control cultures should be collected and sterilized immediately by autoclaving. More detailed accounts of the measures necessary to detect and prevent the spread of contamination can be found elsewhere (Uphoff and Drexler, 2001; Freshney, 2000).

1. Direct Method

The procedures used in preparing and pretesting the following culture media should be standardized, and the final pH should be adjusted to 7.2 to 7.4. Both media are prepared in quantities to be utilized within 3 to 4 weeks. The quality of the major components of the media may vary from batch to batch in the degree of toxicity and in their ability to support the growth of mycoplasma. The growth-promoting properties of each new lot of freshly prepared broth and agar media are determined by making inoculations using ATCC 23206, *Acholeplasma laidlauii*, and ATCC 23838, *Mycoplasma arginini*. In addition, the horse serum, like all sera employed for cell culture work, is screened for mycoplasmal contamination. Briefly, a 100-ml aliquot of the serum being tested is used as the serum supplement for 400 ml of broth medium. The cultures are incubated aerobically at 37°C for 4 weeks and are observed for turbidity and change in pH. Subcultures to agar plates and inoculations onto Vero indicator cultures for the indirect test are performed weekly during the incubation period. In testing samples in which bacterial and/or fungal contaminations may be prevalent, penicillin and thallium acetate are added to the basic medium. Penicillin is added to the stock solution (step a below) to provide a final concentration of 500 U per milliliter. Thallium acetate is added to the basic media (step b below) to provide a final concentration of 1:2000.

Steps

a. Preparation of Stock Solution. To 900 ml freshly distilled water, add 50 g dextrose and 10 g L-arginine HC1. Mix the ingredients at 37°C until dissolved. Bring the final volume up to 1000-ml. Sterilize the solution by filtration using a 0.22-µm filter, dispense into 100-ml aliquots, and store at −70°C until needed.

b. Preparation of Mycoplasma Broth Medium. Add 14.7 g mycoplasma broth base and 0.02 g phenol red to 600 ml water, heat to dissolve. Sterilize the solution by autoclaving for 15 min at 121°C using a slow exhaust cycle and allow the broth mixture to cool to room temperature. Aseptically add 200 ml horse serum, 100 ml yeast extract (15%), and 100 ml thawed stock solution (step a). Mix the solution completely. Dispense 10-ml aliquots of the broth medium into sterile test tubes and cap. Store broth tubes at 5°C and use within 3 to 4 weeks.

c. Preparation of Mycoplasma Agar Medium. Add 23.8 g mycoplasma agar base to 600 ml water. To dissolve, bring solution to a boil and sterilize the solution by autoclaving for 15 min at 121°C using a slow exhaust cycle. Place the sterilized medium in a water bath at 50°C. Place 200 ml horse serum, 100 ml yeast extract (15%), and 100 ml stock solution (step a) in a water bath at 37°C. Allow the components to equilibrate at these respective temperatures. Aseptically add the horse serum, yeast extract, and stock solution to the medium; mix well. Proceed immediately to dispense 10-ml aliquots in 60 × 15-mm petri dishes. Add the fluid as quickly as possible in order to eliminate

the problem of the agar solidifying before the medium is dispensed completely. Stack the agar plates into holding racks, wrap in autoclave bags to minimize dehydration, and store at 5°C. Use within 3 to 4 weeks.

NOTE This procedure involves a total incubation time of about 35 days for both broth and agar cultures. This schedule is advisable for detecting lower levels of mycoplasma contamination that otherwise might be scored as false negatives.

d. Inoculation of Test Sample. Select a cell culture that is near confluency and has not received a fluid renewal within the last 3 days. Remove and discard all but 3 to 5 ml of the culture medium. Scrape a portion of the cell monolayer into the remaining culture medium using a sterile disposable scraper. For suspension culture systems, take the test sample directly from a heavily concentrated culture that has not received a fresh medium supplement or renewal within the last 3 days. Samples for testing can also be taken directly from thawed ampules that have been stored in the frozen state.

Inoculate 1.0 ml of the test cell culture suspension into a mycoplasma broth culture. Inoculate 0.1 ml of the test sample onto an agar culture plate. Incubate the broth culture aerobically at 37°C. Observe daily for the development of turbidity and/or shift in pH. Incubate the agar plate anaerobically at 37°C in a humidified atmosphere of 5% CO_2–95% nitrogen. After 5 to 7 days of incubation, and again after 10 to 14 days, remove a 0.1-ml sample from the broth culture and inoculate a new agar plate. Incubate these plates anaerobically at 37°C.

Microscopically examine the agar plates weekly for at least 3 weeks for mycoplasma colony formation and growth before considering them to be negative. Observe the plates at 100 to 300 magnification using an inverted microscope.

The positive differentiation of mycoplasma colonies on agar plates, as opposed to air bubbles, tissue culture cells, and pseudocolonies, can be accomplished by subculturing a small section (1 cm^2) of the suspicious area of the agar culture into a new broth culture (for examples, see Freshney, 2000).

2. Indirect Method (Staining for DNA)

The bisbenzamide stain concentrate (step a below) should be examined routinely for contamination. Sterilization by filtration diminishes the quality of fluorescence, and fresh stock needs to be prepared periodically. The pH of the mounting medium (step b below) is critical for optimal fluorescence and should also be monitored routinely.

Continuous cells lines, such as ATCC.CCL 81, Vero, African green monkey kidney or ATCC.CCL 96, 3T6 mouse fibroblast, have been used very effectively as indicator cells in the indirect DNA-staining procedure. The use of transformed cell lines is not recommended because they produce large amounts of nuclear background fluorescence, which interfere with the interpretation of the results.

Utilization of the indicator cell with the DNA-staining procedure provides two major advantages. First, the indicator cell line supports the growth of the more fastidious mycoplasma species. Second, both positive and negative control cultures are readily available for direct comparisons with the culture samples being tested.

Steps

a. Preparation of Stain Concentrate. To 100 ml of Hanks' balanced salt solution without sodium bicarbonate or phenol red, add 5.0 mg bisbenzamide fluorochrome stain and 10 mg thimerosol. Mix thoroughly using a magnetic stirrer for 30 to 45 min at room temperature. The stain is heat and light sensitive. Prepare the concentrate in a brown amber bottle wrapped completely in aluminum foil. Store aliquots at −20°C. These are stable for about 1 year.

b. Preparation of Mounting Medium. Combine 22.2 ml *0.1 M* citric acid, 27.8 ml 0.2 *M* disodium phosphate, and 50 ml glycerol and adjust pH of mixture to 5.5. Store in a cold room at 5°C.

c. Preparation of Indicator Cell Cultures and Inoculation of Test Samples. Aseptically place a glass coverslip (previously sterilized) into each 60 × 15-mm culture dish. Dispense 3 ml Eagle's minimum essential medium with Earle's salts, 100 U/ml penicillin, and 100 µg/ml streptomycin plus 10% bovine calf serum into each culture dish. Make certain that each glass coverslip is totally submerged and not floating on top of the medium.

Prepare a single cell suspension of ATCC.CCL 81, the African green monkey kidney cell line Vero, in this medium at a concentration of 1.0×10^5 cells/ml. The 3T6 murine line (ATCC.CCL 96) can be used instead. Inoculate 1 ml of the cell suspension into each culture dish and incubate the cultures overnight in a 5% CO_2 95% air incubator at 37°C. Examine the cultures microscopically to verify that the cells have attached to the glass coverslip. Number the top of each culture dish for identification purposes to record the test sample inoculated. Add 0.5 ml of culture medium to each of two cultures for negative controls and 0.2 to 0.5 ml of each test sample to each of two culture dishes. Add 0.5 ml ATCC 29052, *M. hyorhinis*, to each of two cultures for positive controls. Alternatively, a known infected cell line can be used. Return the cultures to the CO_2 incubator and allow to incubate undisturbed for 6 days.

d. Fixing, Staining, and Mounting Coverslips. To prepare the staining solution:

1. Add 1.0 ml of stock concentration (step a) to 100 ml Hanks' balanced salt solution without sodium bicarbonate and phenol red.
2. Prepare in a brown amber bottle wrapped in aluminum foil.
3. Mix thoroughly for 30 to 45 min at room temperature using a magnetic stirrer.

Remove cultures from the incubator and aspirate the medium from each dish. Add 5 ml of a 1:3 mixture of acetic acid:methanol to each culture dish for 5 min. Do not allow the culture to dry between removal of the culture medium and addition of the fixative. Aspirate each culture dish and repeat the fixation step for 10 min. Aspirate the fixative and let the cultures air dry. Add 5 ml of the staining solution [step d (1–3)] to each culture dish; cover and let stand at room temperature for 30 min. Aspirate the stain and rinse each culture three times with 5 ml distilled water.

After the third rinse, aspirate well so that the glass coverslip is completely dry. Let air dry if necessary. Place a drop of mounting fluid (step b) on a clean glass slide. Use forceps to remove the glass coverslip containing the specimen from the culture dish and place face up on the top of the mounting fluid. Add a second drop of mounting fluid onto the top of the specimen coverslip and cover with a larger clean coverslip. Lower both coverslips onto the mounting fluid in such a way as to eliminate trapped air bubbles. Label each slide to identify the specimen being tested and record results.

Observe each specimen under oil immersion, including both the positive and the negative controls, by fluorescence microscopy at 500X. A blue glass excitation filter (BG12 for Zeiss microscopes) is used in combination with a No. 50 barrier filter. Small fluorescing particles indicate mycoplasmal DNA and infection.

Alternative molecular methods readily available in kit form should also be considered. The PCR-based method for mycoplasma detection (Harasawa *et al.*, 1993; Hu and Buck, 1993) in use at the ATCC requires primers based on the DNA sequences in 16S and 23S mycoplasmal rRNA. These amplify DNA from all of the common mycoplasma found in cell cultures to levels easily detected after gel electrophoresis and ethidium bromide staining. Advantages of the method include speed and sensitivity, as well as the ability to detect and identify species of most of the common mycoplasma known to infect cell cultures. Furthermore, it does not suffer from interpretation difficulties associated with some of the Hoechst or DAPI-stained preparations. Levels of sensitivity compare favorably with the Hoechst stain. Sample sizes need consideration. Detailed methodologies are provided with the kits (see the ATCC website for more details).

C. Protozoa

The overall frequency of infection of cell cultures with protozoans is low but the incidence may be higher if one is working with tissues such as human clinical material and animal tissues such as kidney or colon. The small limax amoebae belonging to the *Acanthamoeba* (or *Hartmanella*) genus are ubiquitous in nature and have been isolated from cells and tissues in culture in a significant number of laboratories. Jahnes *et al.* (1957) first reported spontaneous contamination of monkey kidney cells in culture by such free-living amoebae. The organisms have also been detected as occasional contaminants in such diverse cell lines as dog lymphosarcoma (LS30), HeLa, chick embryo fibroblast-like, and Chang liver cells (Holmgren, 1973). In some cases, protozoans are demonstrably cytopathic in cell culture.

Observation, cytological examination, and attempts at isolation are required in the detection of protozoan contaminants. These techniques are suitable for detecting many of the most common flagellates and amoeboid protozoans, including species of the genera *Acanthamoeba, Giardia, Leishmania, Naegleria,* and *Trypanosoma* (for more details, see Hay *et al.,* 1992).

1. Preparation of Solutions and Protozoan Media Steps

1. *Trypsin-EDTA:* Combine 2.5 g trypsin (1:250 Difco certified), 0.3 g EDTA, 0.4 g KCl, 8.0 g NaCl, 1.0 g glucose, 0.58 g NaHCO3, and 0.01 g phenol red in 1 liter double-distilled water, sterilize by filtration (0.22-μm Millipore filter), and store at −40°C.

2. *Hanks' balanced salt solution without divalent cations:* Combine 8.0 g NaCl, 0.4 g KCl, 0.05 g Na_2HPO4, 0.06 g KH_2PO4, and 0.02 g phenol red in 50 ml double-distilled water to dissolve chemicals; then bring volume to 100 ml. Autoclave on slow exhaust for 15 min, adjust pH to 7.2 to 7.4 with sterile 0.4 N NaOH, and store at 4°C.

3. *Giemsa stock solution:* For stock solution of stain, combine 40 ml glycerol, 65 ml absolute methanol, and

1.0 g Giemsa powder. Filter two or three times and store at 4°C.

4. *Price's buffer (IOX):* Combine 6.0 g Na_2HPO_4, 5.0 g KH_2PO_4, and 1.0 liter distilled water. Before use dilute buffer with distilled water to 1X.

5. *Price's Giemsa stain:* Dilute Giemsa stain stock 3:97 with 1X buffer. After staining, discard unused portion.

6. *ATCC medium No. 400, Diamond's TP-S-1 medium for axenic cultivation of Entamoeba (ATCC medium No. 400):* Dissolve one packet of Diamond's TP-S-1 broth base powder in 875 ml distilled water, adjust pH to 7.0 with 0.4 N NaOH, and filter through Whatman No. 1 paper. Sterilize at 120°C for 15 min. Aseptically add 100 ml inactivated (56°C for 30 min) bovine serum and 25 ml Diamond's TP-S-1 vitamin solution (40X, North American Biologicals), and aseptically dispense 13 ml per sterile test tube. Some commercial lots of Diamond's TP-S-1 medium have been shown to be toxic to *Entamoeba. To* test for toxicity, subculture *Entamoeba* through three to five passages.

7. *Locke's solution:* Combine 8.0 g NaCl, 0.2 g NaCl, 0.2 CaCl₂, 0.3 g KH_2PO_4, 2.5 glucose, and 1.0 liter distilled water and autoclave the solution for 20 min at 121°C.

8. *Diphasic blood agar medium (ATCC medium No. 1011):* Infuse 25.0 g beef extract in 250 ml distilled water by bringing to a rapid boil for 2 to 3 min while stirring constantly. Filter through Whatman No. 2 filter paper and add 10.0 g Difco neopeptone, 2.5 g NaC₁, and 10.0 g agar. Heat to boiling and filter through Whatman No. 2 paper, make up volume to 500 ml with distilled water, and adjust pH to 7.2 to 7.4. Autoclave for 20 min at 121°C, cool mixture to 50°C, aseptically add 30% sterile, defibrinated rabbit blood (Editek) to whole mixture, and dispense in sterile tubes and slant. After the slants have set, cover with 3.0 ml sterile Locke's solution.

9. *PYb medium (ATCC medium No. 711):* Combine 1.0 g Difco proteose peptone, 1.0 g yeast extract, 20.0 g agar, and 900.0 ml distilled water. Prepare and sterilize separately each of the following stock solutions and add to the basal medium as indicated to avoid precipitation: $CaCl_2$ (0.05 M), 4.0 ml; $MgSO_4 \cdot 7H_2O$ (0.4 M), 2.5 ml; Na_2HPO_4 (0.25 M), 8.0 ml; and KH_2PO_2 (0.25 M), 32 ml. Make the volume to 1 liter, check that the pH is at 6.5, and sterilize by autoclaving for 25 min at 120°C. Pour into petri dishes and allow to solidify.

10. *Brain-heart infusion blood agar (ATCC medium No. 807):* For the agar component, dissolve 37.0 g Difco BHI broth and 18.0 g agar in 1 liter boiling water. Dispense 5.0 ml solution per tube (16 × 125 mm) and sterilize for 25 min at 121°C. Cool to 48°C. Add 0.5 ml per tube of sterile, defibrinated rabbit blood and slant. After slants

have set, cover with 0.5 ml BHI broth (1.0 liter distilled water and 37.0 g BHI broth) with sterilization by autoclaving at 121°C for 25 min.

11. *Leishmania medium (ATCC medium No. 811):* Combine 1.2 g sodium citrate, 1.0 g NaCl, and 90.0 ml distilled water. Dispense 1.0 ml per tube, autoclave for 25 min at 121°C, and cool. Add 1.0 ml defibrinated, lysed rabbit blood solution (prepare by mixing equal parts of whole rabbit blood and sterile distilled water and freezing and thawing twice).

12. *NTYG medium (ATCC medium No. 935):* Combine 5.0 g trypticase, 5.0 g yeast extract, 10.0 g glucose, and 1.0 liter distilled water. Dispense 10.0 ml per test tube and sterilize. Just before use, add 0.2 ml dialyzed, heat-activated bovine serum and 0.1 ml defibrinated sheep blood. Protozoan growth media retain stability for at least 3 months if maintained at 4°C, with the exception of ATCC medium 400, which maintains stability for 2 to 4 weeks.

2. Preparation of Cell Culture Samples for Inoculation into Protozoan Media

Steps

1. Rapidly thaw a frozen ampoule of the sample in a water bath at 37°C.

2. Aseptically open the ampoule. Continue to use sterile techniques.

3. Transfer 0.8 ml of the concentrated cell suspension from the ampoule into a T-25 flask. Save 0.2 ml of the suspension for Giemsa staining (step 3 below).

4. Add 7 ml of the appropriate cell culture medium to maintain the culture.

5. Incubate at 37°C until the monolayer becomes confluent (3 to 5 days depending on the cell line). Examine the culture microscopically during this incubation period for the presence of (a) movement (i.e., motile cells), (b) intracellular contaminants, and (c) cytopathology.

6. Transfer the supernate from the confluent test cell culture to a sterile 15-ml plastic centrifuge tube and retain at room temperature for use in step 12.

7. Rinse the cell monolayer (T-25 flask, step 5) with 5 ml Ca"- and Mg"- free Hanks' saline and discard saline solution.

8. Add 2 ml 0.25% trypsin-EDTA solution to the T-25 flask and incubate at 37°C for 10 min.

9. Add 7 ml of cell culture medium to the T-25 flask and aspirate gently to obtain a single-cell suspension.

10. Dispense aliquots (0.5 ml) of the trypsinized single-cell suspension to the following ATCC protozoan growth media: (a) ATCC medium No. 400 (for *Entamoeba, Giardia*); (b) ATCC medium No. 711 with *Enterobacter aerogenes* (for *Acanthamoeba*) [use a wire loop to streak medium No. 711 with *E. aerogenes* (*ATCC* 15038) 48 h before use]; (c) ATCC medium Nos. 807, 811, and 1011 (for trypanosomatids); and (d) ATCC medium No. 935 (for *Naegleria*).

11. Incubate samples for 7 to 10 days at 35°C and examine microscopically for the presence of flagellate, cyst, and trophozoite forms of protozoa.

12. Prepare five wet mounts for each test cell monolayer using the supernate collected in step 6. Examine microscopically with phase contrast for the presence of motile and nonmotile protozoans.

3. Preparation of Culture Cells for Giemsa Staining

Steps

1. Aseptically add 1.5 ml of the appropriate culture medium to a sterile Leighton tube containing a coverslip.

2. Dispense 0.2 ml of the original cell suspension (sample preparation just earlier) into the Leighton tube and incubate at 37°C until the culture is confluent.

3. Remove the coverslip from the Leighton tube, fix with absolute methanol for 1 min, and airdry.

4. Stain for 10 min with Price's Giemsa, rinse with tap water, and mount the coverslip to a glass slide using Aquamount.

5. Examine the slide; use low power (20X) for scanning and high power (100X) for close examination.

Controls

It is recommended that positive controls be included. For example, if cultured cells of the upper respiratory tract are being used, *Acanthameba castellanii* (ATCC 30010) or *Naegleria lovaniensis* (ATCC 30569) can be used as positive controls. *A. castellanii* was isolated from human clinical material. *N. lovaniensis* strain TS, another nonpathogenic strain of amoebae, was isolated from a Vero cell culture at passage 120. *Entamoeba histolytica* (ATCC 30042), the common pathogen causing amoebic dysentery, or the nonpathogenic *Entamoeba invadens* (ATCC 30020) can be used for positive controls if cells are being isolated from the intestinal tract. *E. histolytica* is a human isolate, and *E. invadens* strain PZ is a snake isolate.

Comments

The methods described are suitable for the detection of most common protozoan genera (i.e., limax amoebae) that could survive in association with cells in culture. Because cysts and trophozoites closely resemble damaged tissue cells, their presence as occasional contaminants can remain unnoticed. However, cells in cultures infected productively with amoebae of the genus *Acanthamoeba* frequently become granular and gradually progress to complete disintegration. The time elapsed depends on the inoculum size and whether cysts or the motile trophozoites predominate in the inoculum. The cytopathic effect of amoebic contaminants has been reported, and in some cases the responsible agent has been mistakenly identified as viral in origin. Therefore, frequent observation of the cell culture is particularly stressed when examining for parasitic protozoan contaminants.

The possible presence of other genera (i.e., *Entamoeba* or trypanosomatids) should be considered not only in experimental studies involving primary tissues, but also with work requiring development or utilization of cell lines. The only known case of an isolation other then an amoeboid protozoan occurred in the isolation of a trypanosomatid from liver tissue.

The particular animal and tissue employed provide valuable clues as to the type of protozoan contaminant, the specific media, and staining procedures required.

D. Viruses

Of the various tests applied for detection of adventitious agents associated with cultured cells, those for endogenous and contaminant viruses are the most problematical. Table II lists representative problem viruses. Development of an overt and characteristic cytopathogenic effect (CPE) will certainly provide an early indication of viral contamination; however, the absence of a CPE definitely does not indicate that the culture is virus free. In fact, persistent or latent infections may exist in cell lines and remain undetected until the appropriate immunological, cytological, ultrastructural, and/or biochemical tests are applied. Unfortunately, separate tests are necessary for each class of virus and for specific viruses. Additional host systems or manipulations, e.g., treatment with halogenated nucleosides, may be required for virus activation and isolation (Aaronson *et al.*, 1971). Common

screening methods or tests for specific virus classes are listed in Table III.

Without such screens, latent viruses and viruses that do not produce an overt CPE or hemadsorption will escape detection. Some of these could be potentially dangerous for the cell culture technician. For example, Hantaan virus, the causative agent of Korean hemorrhagic fever, replicates in tumor and other cell lines. Outbreaks of the disease in individuals exposed to infected colonies of laboratory rats have been reported separately in five countries. An incident of transmission during passage of a cell line was confirmed in Belgium. As a result of these findings, cell lines expanded in this laboratory were screened using an indirect immunofluorescent antibody assay (LeDuc *et al.*, 1985) and were found to be negative.

Substantial concern over laboratory transmission of the human immunodeficiency viruses is also evident. Cases of probable infection during processing in U.S. laboratories have been described. One, for example, was presumed due to parenteral exposure and another to work with highly concentrated preparations (Weiss *et al.*, 1988). In the latter circumstance, strict adherence to biosafety level 3 containment and practices is essential. A more detailed discussion of safety precautions for work with cell lines in general is provided elsewhere (Caputo, 1988).

ATCC cell lines from selected groups have been screened for HIV-1 using PCR amplification followed by a slot-blot test for envelope and GAG sequences (Ou *et al.*, 1988). The oligonucleotide primer pairs SK 38/39 and SK 68/69 *plus* SK 19 or SK 70 probes were used. Human cell lines of T-cell, monocyte–macrophage, brain and nervous system, B-cell, and gastrointestinal origin plus an array of other primate lines have been examined to date. Only those already known to be infected with HIV-1 have been positive. Additional viruses that could present a substantial health hazard to cell culture technicians include, for example, hepatitis and cytomegaloviruses. Rapid PCR-based tests for these have been described [e.g.,

TABLE II Representative Viruses of Special Concern in Cell Production Work

Human	Other
Human immunodeficiency viruses	Hantavirus
Human T-cell leukemia viruses	Lymphocytic choriomeningitis virus
Other endogenous retroviruses	Ectromelia virus
Hepatitis viruses	Murine hepatitis
Human herpesvirus 6	Simian viruses
Cytomegalovirus	Sendai virus
Human papillomavirus	Avian leukosis virus
Epstein–Barr virus	Bovine viral diarrhea virus

TABLE III Common Methods for Detection of Viruses in Cell Line Stocks[a]

Cytopathogenic effect observation	Reverse transcriptase assays
Chorioallantoic membrane inoculation	Nucleic acid hybridization
Hemagglutination	Fluorescent antibody staining
Hemadsorption	Electron microscopic fine structure
Cocultivation	Animal inoculation

[a] See IABS (1989) and Hsuing *et al.* (1994) for more details.

Ulrich *et al.* (1989) and Cassol *et al.* (1989), respectively].

Other viruses that may present problems generally in cell culture work include ectromelia virus, the causative agent of bovine viral diarrhea (BVDV), and Epstein–Barr virus (EBV). [See also Bolin *et al.* (1994), Harasawa *et al.* (1994), and Hay *et al.* (2000) for testing methodology and further discussion.]

It should be emphasized at the outset that the following protocols represent an expedient compromise established at ATCC to monitor for readily detectable viruses associated with cell lines. Egg inoculations plus select cocultivations and hemadsorption tests were included in addition to routine examinations for CPE using phase-contrast microscopy. Similar general tests are recommended by government agencies in cases where cell lines are to be used for biological production work (Code of Federal Regulations on Animals and Animal Products, 9 CFR *113.34-113.52*, revised Jan. 1, 1978; Code of Federal Regulations on Food and Drugs, Subchapter F on Biologics, *21* CFR *630.13* b-c, revised April 1, 1979; IABS, 1989; Lubiniecki and May, 1985). Procedures for reverse transcriptase assays to detect oncogenic viruses are also being applied at the ATCC for selected cell lines.

Because endogenous and most exogenous retroviruses produce no morphological transformation or cytopathology in infected cells, the production of such viruses by cell cultures is generally undetectable except by serological or biochemical means. At ATCC the concentration of particulate material from culture supernates and assay for viral RNA-directed DNA polymerase (RDDP) provide sensitive and reliable means for detecting retrovirus production by cultured cells.

One or more of the following procedures is currently being applied to all cell lines accessioned for the ATCC repository. Tests for specific viruses may be applied through collaborations as described earlier.

1. Examination of Established Cultures for Overt Cytopathogenic Effect or Foci

Steps

1. Hold each flask or bottle so that light is transmitted through the monolayer and look for plaques, foci, or areas that lack uniformity. If frozen stocks of cells are to be examined, pool and mix the contents of about 5% of the ampoules from each lot using a syringe with a cannula. Establish cultures for morphological examinations and for tests in the following sections using progeny from such pooled populations.

2. Using an inverted microscope equipped with phase-contrast optics wherever possible, examine cell culture vessels individually, paying special attention to any uneven areas in gross morphology observed in step 1. Check first using low power. If the cell line is suspect, subculture taking the appropriate safety precautions. Prepare coverslip cultures for further examination. Alternatively, autoclave and discard all suspect cultures. (Stainless-steel collection and sterilizing pans for this purpose can be obtained from the Orem Medical Company.)

3. Remove fluid from coverslip cultures that require additional study. Treat with neutral buffered formalin or other suitable fixative. Prepare a wet mount and examine under high power. [Consult Rovozzo and Burke (1973), Hsuing *et al.* (1994), and Yolken *et al.* (1999) for examples of cytopathogenic effects and further details.]

2. Application of the Hemadsorption Test

Steps

1. Establish test cultures in T-25 flasks using an inoculation density such that the monolayers become confluent in 48 to 72 h.

2. Prepare washed red blood cell suspensions on the day the test is to be performed. Pack the erythrocytes from 5 ml of the purchased suspensions by centrifugation at $100 g$ and resuspend in 35 ml Hanks' saline without divalent canons. Repeat three times and resuspend the final pellet to yield a 0.5% suspension (v/v) of red blood cells in saline.

3. Remove the culture fluid and rinse the test monolayers with 5 ml Hanks' saline minus divalent cations.

4. Add 0.5 ml each of the suspensions of chick, guinea pig, and human type O erythrocytes from step 2. Then place the flask with monolayer down at 4°C for 20 min.

5. Observe macroscopically and microscopically under low power for clumping and adsorption of red blood cells to the monolayer.

6. Repeat steps 2–4 on all test cultures not exhibiting hemadsorption before recording a negative result. [A suitable positive control can be established by infecting a flask of rhesus monkey kidney cells with 0.2 ml of undiluted ATCC VR-95 (influenza virus strain A/PR/8/34) 48 to 72 h before testing.]

3. Egg Preparation

Steps

1. Drill a small hole in the egg air sac (blunt end) using the electric drill (Cole-Parmer) and a 1/16-in. burr-type bit or an 18-gauge needle in this and subsequent operations; work with sterile instruments. Swab areas of the shell to be drilled with 70% ethanol before

and after each manipulation. The drill bits may be placed in 70% ethanol before use.

2. Using the candling lamp (Nasco), locate the area of obvious blood vessel development and, at a central point, carefully drill through the shell, leaving the shell membrane intact.

3. Place 2 or 3 drops of Hanks' saline on the side hole and carefully pick through the shell membrane with a 26-gauge syringe needle. The saline will seep in and over the chorioallantoic membrane (CAM) to facilitate its separation from the shell membrane.

4. Apply gentle suction to the hole in the air sac using a short piece of rubber tubing with one end to the mouth and the other pressed to the blunt end of the egg. Use the candling lamp to monitor formation of the artificial air sac over the CAM.

5. Seal both holes with squares of adhesive or laboratory tape and incubate the eggs horizontally at 37°C. Standard cell culture incubators and walk-in rooms are entirely adequate for egg incubations. High-humidity or air/CO_2 boxes are not satisfactory.

4. Egg Inoculations

Steps

1. Obtain suspensions of test cells in the appropriate growth medium and adjust the concentration such that 0.2 ml contains 0.5 to 1×10^7 cells.

2. Remove the seal from side holes in the embryonated eggs and inject 0.2 ml of the cell suspension onto the CAM of each of 5 to 10 eggs.

3. Using the candling lamp, examine the embryos 1 day after adding the cell suspension; discard any embryos that have died. Repeat the examination periodically for 8 to 9 days.

4. If embryos appear to be viable at the end of the incubation period, open the eggs over the artificial air sac and examine the CAM carefully for edema, foci, or pox. Check the embryo itself for any gross abnormalities such as body contortions or stunting.

5. In cases in which viral contamination is indicated, repeat steps 1–4 both with a second aliquot of the suspect cells and with fresh fluid samples from eggs in which the embryos have died or appear abnormal. Positive controls may be established by inoculating eggs with influenza virus, Newcastle disease virus, and/or Rous sarcoma virus.

5. Cocultivation Trials

Steps

1. Select two appropriate cell lines for cocultivation with each cell line to be tested. The lines chosen will depend on the species from which the test cell line originated. For example, for a human cell line, one could cocultivate with ATCC CCL 75 (WI-38), ATCC CCL 171 (MRC-5), or primary human embryonic kidney (HEK) cells. A cell line from a second species of choice in this example could be ATCC CCL 81 (Vero) originating from the African green monkey.

2. Inoculate a T-75 flask with 10^6 cells from each line in a total of 8 ml of an appropriate growth medium. In some cases, the inocula may have to be adjusted in an attempt to maintain both cell populations during the cocultivation period. For example, if a very rapidly proliferating line is cocultivated with a test line that multiplies slowly, the initial ratio of the former to the latter could be adjusted to 1:10. Similarly, the population that multiplies slowly might have to be reintroduced to the cocultivation flasks if it were being overgrown by the more rapidly dividing cells.

3. Change the culture fluid twice weekly and subcultivate the population as usual soon after it reaches confluence.

4. Examine periodically for CPE and hemadsorption over a 2- to 3-week period at minimum, using procedures described earlier.

Viral isolates may be identified through standard neutralization (hemadsorption inhibition, plaque inhibition, hemagglutination inhibition) or complement fixation tests. The ATCC virology department retains and distributes antisera to many viral serotypes, and identification can be accomplished readily.

6. Reverse Transcriptase Assays

Positive serological assays for retrovirus antigens in cells and cell packs indicate that a retrovirus genome is present, but these assays do not indicate whether release of progeny virus particles is occurring. It has been found that the concentration of particulate material from culture supernates and the assay for viral RDDP (Baltimore, 1970; Temin and Mizutani, 1970) provide a sensitive and reliable means for detecting retrovirus production by cultured cells.

a. Preparation of Cell Cultures. Cell cultures to be examined for the production of retrovirus should be cultured by the methods and in the media that are optimal for the particular cells. It is important that the cells be in good condition and not undergoing degeneration and autolysis.

Steps

1. When adherent cell cultures are about 50 to 60% confluent, or when suspension cultures are at a cell density about 50% of the maximum, completely replace the medium and reincubate the cultures.

2. Harvest fluid approximately 24 h after feeding.

b. Processing of Culture Fluid

Steps

1. Collect culture medium aseptically.

2. Clarify medium by centrifugation at 1000 to $3000\,g$ for 10 min at 4°C. Decant and save the clarified supernates and discard sedimented materials.

3. The clarified medium contains $0.15\,M$ NaCl; add $5.0\,M$ NaCl to a final concentration of $0.5\,M$ NaCl. Calculate the volume of $5.0\,M$ NaCl according to the formula $0.15(V_1) + 5.0(V_2) = 0.5\,(V_1 + V_2)$, where V_1 is the volume of clarified culture fluid and V_2 is the volume of $5.0\,M$ NaCl to be added. Mix well. If the medium becomes cloudy after adding NaCl, centrifuge at $10,000\,g$ for 10 min and save the supernate.

4. To 2 volumes of clarified supernate containing $0.5\,M$ NaCl, add 1 volume 30% PEG 6000 in $0.5\,M$ NaCl. Mix well.

5. Allow precipitation to occur for at least 1 h while holding in wet ice. At this point samples may be held overnight at 4°C if necessary.

6. Centrifuge at $7000\,g$ for 10 min.

7. Decant and discard the supernates.

8. Drain the pellets thoroughly while holding at 4°C.

9. Resuspend the pellets in 50% (v/v) buffer A [0.05 M Tris–HCl, pH 7.5, $0.1\,M$ KCl, 0.5 mM EDTA, 10 mM dithiothreitol, 0.05% (v/v) Triton X-100, and 50% glycerol]. Care must be taken to ensure that pellets are completely resuspended.

10. Store resuspended pellets at –20°C. Aliquots are used for RDDP assays.

c. Assay of RNA-Directed DNA Polymerase Activity.
This procedure is based on that of Gallagher and Gallo (1975).

Solutions

1. *Stock mix:* 0.5% (v/v) Triton X-100, $1.13\,M$ *KCl.*

2. *Template-primer solutions (P-L Biochemicals).* Mix A: Combine 1 mg/ml poly(rA)–poly(dT)$_{12-18s}$ with 0.01 M Tris–HCl, pH 7.5, and $0.1\,M$ NaCl. Mix B: Combine 1 mg/ml poly(dA)–poly(dT)$_{12-18}$ with $0.1\,M$ NaCl and $0.01\,M$ Tris–HCl, pH 7.5.

3. *Working mixtures of template-primer solutions:* Mix stock mix and template-primer solutions in 3:2 (v/v) ratio.

4. *Reaction cocktail:* Evaporate 250 μl [3H]TTP (carrier-free, New England Nuclear 221-X) to dryness under vacuum. Redissolve in 720/μl H_2O before adding the following components (volumes given are for 10 tubes): $1.0\,M$ Tris–HCl, pH 7.8 (40 μl), $0.2\,M$ dithiothreitol (40 μl), $0.01\,M$ MnCl$_2$ (50 μl). Add MnCl$_2$ last (just before initiating reactions).

Steps

1. Distribute culture medium concentrates and positive and negative control samples into siliconized 10×75-mm assay tubes. (a) For positive controls, use concentrates prepared from culture media of cell cultures known to be producing retroviruses. (b) For negative controls, use buffer A-glycerol (Table IV).

2. Add appropriate template-primer mix to each tube and mix with a vortex mixer. Hold tubes in wet ice for 15 min.

3. Initiate reactions by adding reaction cocktail to each tube and mixing. Allow reactions to proceed for 30 min in a 37°C water bath.

4. At the end of incubation period, remove tubes to an ice bath; terminate reactions by adding 25 μl $0.1\,M$ EDTA per tube (Sethi and Sethi, 1975).

TABLE IV   Contents of RNA-Directed Polymerase Assays for Three Samples

Tube No.	Medium concentrate	Sample volume	Mix A	Mix B	Cocktail
1	1	5 μl	10 μl	—	85 μl
2	1	5 μl	—	10 μl	85 μl
3	2	5 μl	10 μl	—	85 μl
4	2	5 μl	—	10 μl	85 μl
5	3	5 μl	10 μl	—	85 μl
6	3	5 μl	—	10 μl	85 μl
7	Negative control	5 μl[a]	10 μl	—	85 μl
8	Negative control	5 μl[a]	—	10 μl	85 μl
9	Positive control	5 μl[b]	10 μl	—	85 μl
10	Positive control	5 μl[b]	—	10 μl	85 μl

[a] Buffer A-glycerol (1:1) as used for resuspending PEG precipitates.

[b] PEG-precipitated particles from medium collected from known retrovrus (e.g., murine leukemia virus)-producing cells.

5. Spot 100 μl from each tube onto appropriately numbered DE-81 filters. Allow liquid to soak into filters.

6. Wash batches of filters with gentle manual swirling in at least 10 ml (per filter) of 5% (w/v) $Na_2HPO_4 \cdot 7H_2O$. Repeat for a total of six washes (Sethi and Sethi, 1975).

7. Wash twice with distilled H_2O and twice with 95% ethanol and arrange filters on cardboard covered with absorbent paper. Dry thoroughly under a heat lamp.

8. Place each filter in a separate numbered scintillation vial, add 10 ml PPO-POPOP scintillation cocktail (Betafluor) to the vial, and count in a liquid scintillation counter (Beckman LS-3133) using a tritium window.

Comments

A number of precautions must be observed in the interpretation of the results obtained in this assay. If the cultures to be tested are very heavy and undergoing autolysis, a large amount of cellular DNA-directed DNA polymerase may be associated with microsomal particles in the culture medium. These particles are concentrated by the polyethylene glycol procedure just as virus particles are. Because cellular DNA polymerases do not exhibit an absolute specificity for a DNA template, a certain level of [3H]TMP ([³H]thymidylate) incorporation directed by an RNA template will result from cellular polymerase activity. If a high level of DNA-directed cellular polymerase activity is present in medium concentrates, the (sometimes high) degree of incorporation by these enzymes can mask true RDDP activity, which may be present. Consequently, it is important that media be collected from healthy, actively growing cultures.

It must be remembered that enzymes that catalyze the polymerization or terminal addition of [3H]TMP with a poly(rA) · oligo(dT) template-primer are not exclusively viral (Harrison *et al.*, 1976). poly(rA) DD–CC oligo(dT) is generally employed because the activity of retroviral RDDP is usually greater with that template-primer than it is when measured by the incorporation of dGMP (deoxyguanylate) directed by poly(rC) · oligo(dG1) or by methylated derivatives of the poly(rC) template; however, DNA synthesis directed by poly(rC) · oligo(dG) is more specific for viral enzyme. Consequently, medium concentrates that show incorporation of [3H]TMP with the poly(rA) template should be tested for the incorporation of [³H]dGMP directed by a poly(rC) template.

Incorporation of isotopic precursors into macromolecular form is generally detected by the precipitation of macromolecules with trichloroacetic acid after the enzymatic reaction is terminated. Although background levels of radioactivity may be somewhat higher by the use of adsorption to and elution from ion-exchange filter paper, the ion-exchange procedure obviates the need for a filtration manifold, which is generally employed for acid precipitation. Also, the batch method employed allows many more samples to be processed efficiently.

7. A Rapid PCR-Based Procedure for Detecting the Presence of HIV and HTLV RNA in ATCC Cell Lines

An additional high-throughput method has been adopted to screen for HIV and HTLV RNA in ATCC cell lines using quantitative PCR. Priorities include lines, which might be expected to support growth of these viruses, namely T cells, macrophages and monocytes, lines from the brain and nervous system, lines of gastrointestinal origin, selected human hybridomas, and others.

Steps

1. Total RNA is extracted from 10^6 cells of selected cell lines (Qiagen, Valencia, CA).

2. After quantitative and qualitative analyses using the Bioanalyser 2100, 500 ng of RNA is reverse transcribed in the presence of a mixture of random hexamers and oligo(dT). For real-time PCR (ABI PRISM 700, sequence detector, Foster City, CA), 12.5 ng of initial RNA (0.5 μl of cDNA reaction) is used.

3. Specific primers and TaqMan probes are used individually to measure levels of HIV-1, HTLV-1, and GAPDH transcripts. For HIV-1 and HTLV-1, the Sybr Green I dye assay is also used.

Notes and Results

Both the SybrGreenI dye (Molecular Probes, Eugene, OR) and TaqMan assays for HIV-1 and HTLV-1 show an amplification signal. GAPDH *transcripts*, used as an internal control, are detectable in all the cell lines tested. Every sample is run on triplicates, and an average is calculated. Absolute quantities of HIV, HTLV, and GADPH are calculated using the standard DNA method as described in User Bulletin 2 (PE Applied Biosystems, Foster City, CA).

We have used two different detection methods: Sybr Green I dye or the TaqMan probe method. Sybr Green I is a double-stranded DNA-binding dye, which, when added into the PCR mix, binds to the amplicon, the double-stranded DNA fragment produced during PCR. As the PCR progresses, more amplicons are created. Due to the binding of Sybr Green I dye to all double-stranded DNA, the increase in fluorescence is proportionate to the amount of PCR product. In addition to forward and reverse primers, TaqMan technology uses a sequence-specific oligonucleotide labeled with fluorescent dye at the 5′ end and a quencher dye at the 3′ end. While the probe is intact, the proximity of the quencher dye reduces the fluorescence emitted by the reporter dye by fluorescence resonance energy transfer (FRET) through space. If the target sequence is present, the probe anneals downstream from one of the primer sites and is cleaved by the 5′ nuclease activity of *Taq* DNA polymerase during primer extension (Figs. 1A and 1B). The cleavage of the probe separates the reporter dye from the quencher dye, increasing the reporter dye signal (Fig. 1C). Additional cleavage of the probe occurs at every cycle, resulting in an increase of the fluorescent signal proportional to the amount of the amplicon. Thus the presence of the TaqMan probe enables detection of the specific amplicon as it accumulates during PCR cycles.

HIV- and HTLV-specific primers and probes were designed to avoid false-negative samples due to nucleotide variation in the target sequence (Desire 2001; Schutten, 2000; Bisset, 2001). Blast search of GenBank indicated that the probe and primer set used in these analyses will detect the majority of HIV-1 subtypes and HTLV-I. Total RNA was extracted from frozen cell pellets for all the cell lines used in the test. For CRL-8993, total RNA was used from both frozen

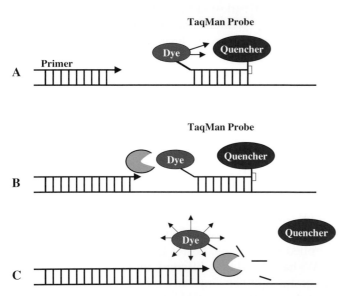

FIGURE 1 Description of 5′ nuclease assay. (A) Sequence-specific primers and dual-fluorescence TaqMan probe anneal to complementary sequences in the DNA template. Due to frequency resonance energy of transfer emission of the fluorescence dye (reporter) is reduced significantly by the presence of the proximal quencher. (B) Due to *Taq* polymerase activity, primer extension and synthesis of a complementary strand occur. (C) While extending, due to its 5′ nuclease activity, the *Taq* polymerase cleaves the TaqMan probe and enables the release of a fluorescent signal by the reporter.

and fresh cells as a comparison. During RNA extraction, DNase I was added to avoid contamination of RNA extracts with genomic DNA. Extracted RNA was evaluated using the 2100 Bioanalyzer (Agilent Technologies, Wilmington, DE). Although the amount of total RNA extracted from frozen cells is significantly lower than the total RNA extracted from fresh cells, the RNA extracted from frozen cells is sufficient for several RT-PCR assays and no RNA degradation was observed. Total RNA (300 ng) was reverse transcribed in a 20-µl reaction using Superscript (Invitrogen, Carlsbad, CA) and 1 µl of the cDNA product was used for real-time PCR. We have used both Sybr Green and TaqMan methods to quantitate and detect HIV-1 and HTLV-I viruses in human cell lines. For absolute quantitation, known amounts of plasmid DNA containing full-length or partial sequences of HIV and HTLV were used in parallel with the unknown templates. PCR was performed using the same set of primers and the TaqMan probe for both standard and sample cDNA. Amplification plots and standard curves were generated for HIV (Fig. 2A) and HTLV (data not shown). The absolute amounts of HIV and HTLV for each sample were calculated based on the standard curves (Figs. 2B and 2C). We have also analyzed levels of GAPDH as an endogenous control, and standard curves for GAPDH were generated using plasmid DNA containing a GAPDH cDNA (Fig. 3). Absolute amounts of GAPDH were also calculated based on the standard curve method.

Quantitation of copies of viral HIV and HTLV, as well as quantitation of endogenous GAPDH, is shown in Table V and Fig. 4. Sybr Green assays were run for the quantitation of HIV, HTLV, and GAPDH, analyses were performed using the standard curve method, and similar results were obtained (data not shown). Therefore, if primers are well designed and PCR conditions are such that do not generate nonspecific amplicons or primer dimers, testing for the presence of viral RNA using Sybr Green I dye could be less expensive and as effective as the TaqMan assay. As shown in Figs. 3 and 4, HIV and HTLV viral RNA was detected only in CRL-8993 and CRL-8543 cell lines. CRL-8993 is a lymphoid cell line reportedly positive for HIV, and CRL-8543 is also reported to contain HIV and HTLV viruses. CRL-8993 was used as a control, and total RNA was extracted from both fresh culture and frozen cells. As shown in Fig. 4, the difference in absolute copies of either GAPDH or HIV transcripts between the two RNA extracts is insignificant, suggesting that total RNA extracted from frozen cells could be used effectively for RT-PCR analyses.

In summary, we have established a quick and accurate method for the detection of HIV-1 and HTLV-I

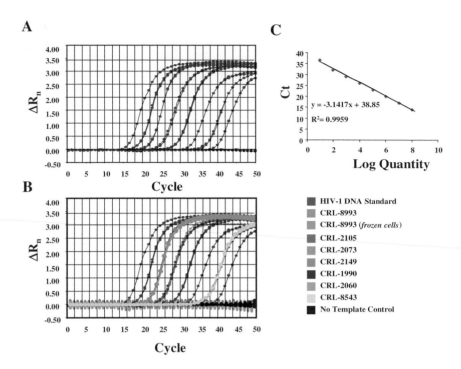

FIGURE 2 Real-time PCR assays for the detection and quantitation of HIV. Amplification plots (A) and standard curves (B) generated for the quantitation of HIV. (C) Amplification plots of unknown templates (total RNA from human cell lines).

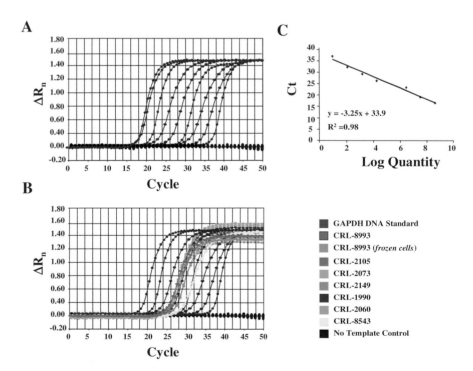

FIGURE 3 Real-time PCR assays for the quantitation of endogenous GAPDH transcripts as an internal control. Amplification plots (A) and standard curves (B) generated for the quantitation of GAPDH. (C) Amplification plots of endogenous GAPDH from each specific cell line.

HIV, HTLV (copies/ng RNA)

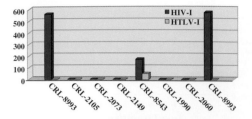

GAPDH (copies/ng RNA)

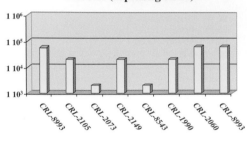

FIGURE 4 Detection and quantitation of HIV, HTLV, and GAPDH transcripts based on real-time PCR analyses.

TABLE V　HIV and HTLV Contaminants in Selected Cell Lines[a]

Cell line	Designation	GAPDH	HIV	HTLV
CRL-8993	8E5(CEM)	82,000	560	0
CRL-2105	HH	20,000	0	0
CRL-2073	NCCIT	2,000	0	0
CRL-2149	SK-N-DZ	20,000	173	51
CRL-8543	H9/HTLV-111B	2,000	0	0
CRL-1990	J45–01	20,000	0	0
CRL-2060	PFSK-1	60,000	0	0
CRL-8993	8E5(CEM)	60,000	572	0

[a] Results are presented as copies of HIV and HTLV transcripts per nanogram of total RNA. Quantitation of GAPDH transcripts is also provided.

viruses. Given the high sensitivity of real-time PCR, very low amounts of total RNA are needed for PCR analyses. Therefore, RNA can be extracted directly from frozen cell pellets and particular cell lines can be tested for the presence of viral RNA prior to cultivation. Real-time PCR offers the possibility of detecting and quantitating as low as 10 copies of viral DNA.

IV. GENERAL COMMENTS

The microbial contamination and viral infection of cell lines are still extremely serious problems.

Mycoplasmal infection has been especially well studied, and the incidence of problems has been documented through government-funded programs. Screening results reported within the past two decades showed that as many as 4 to 33% of cultures tested were infected with one or more species of mycoplasma (Hay *et al.*, 1989). It *is absolutely imperative* that cell lines used in research or production work be tested routinely for such adventitious infection. The comparative cost in time and materials is extremely small. Rewards in terms of research or production reliability are substantial.

Testing for viral infection is more problematical in that it is expensive and multiple tests are required to provide even a limited degree of assurance on freedom from infection. We recommend consideration of screening on a case-by-case basis depending on anticipated use for the line, funding available, and a risk-versus-benefit analysis. Of course, a potential health hazard for cell culture technicians is a major concern.

Finally, it is also critically important to verify the identity of cell lines employed. Hukku *et al.* (1984) documented the incidence of cross-contamination of cell lines reporting misidentifications in excess of 35%.

Thus, a reasonably rigorous authentication pro-gram must include not only reliable tests to ensure an absence of microbial infections (including mycoplasma), but also cell species verification. Methods are detailed elsewhere (Hay *et al.*, 2000, 1992) as are precautions required to avoid operator-induced contamination during routine processing (Hay, 1991).

References

Aaronson, S. A., Todaro, G. J., and Scolnick, E. M. (1971). Induction of murine C-type viruses from clonal lines of virus-free BALB/3T3 cells. *Science* **174**, 157–159.

Baltimore, D. (1970). RNA-dependent DNA polymerase in virions of RNA tumor viruses. *Nature* (*Lond.*) **226**, 1209–1211.

Bisset L. R. (2001). Quantification of *in vitro* retroviral replication using a one-tube real time RT-PCR system incorporating direct RNA preparation. *J. Virol. Methods.* **91**(2), 149–155.

Bolin, S. R., Ridpath, I. F., Black, J., Macy, M., and Roblin, R. (1994). Survey of cell lines in the American Type Culture Collection for bovine viral diarrhea virus. *J. Virol. Methods* **48**, 211–221.

Caputo, J. (1988). Biosafety procedures in cell cultures. *J. Tissue Cult Methods* **11**, 223–228.

Cassol, S. A., Poon, M.-C., Pal, R., Naylor, M. J., Culver-lames, J., Bowen, T. J., Russel, J. A., Krawetz, S. A., Pon, R. T., and Hoar, I. (1989). Primer mediated enzymatic amplification of cytomegalovirus (CMV) DNA. *J. Clin. Invest.* **83**, 1109–1115.

Chen, T. R. (1977). *In situ* demonstration of mycoplasma contamination in cell cultures by fluorescent Hoechst 33258 stain. *Exp. Cell Res.* **104**, 255–262.

Desire, N. (2001). Quantification of HIV type 1 proviral load by TaqMan real time PCR. *J. Clin. Microbiol.* **39**(4), 1303–1310.

Fogh, J. (1973). Contaminants demonstrated by microscopy of living tissue cultures or of fixed and stained tissue culture preparations.

In *"Contamination in Tissue Culture"* (J. Fogh, ed.), pp. 65–106. Academic Press, New York.

Freshney, R. I. (2000). "Culture of Animal Cells: A Manual of Basic Technique," 4th Ed., Wiley-Liss, New York.

Gallagher, R. E., and Gallo, R. C. (1975). Type C RNA tumor virus isolated from cultured human acute myelogenous leukemia cells. *Science* **187**, 350–353.

Harasawa, R., Kazumasa, H., Tanabe, H., Takada, Y., and Mizusawa, H. (1994). Detection of adventitious pestivirus in cell cultures by polymerase chain reaction using nested-pair primers. *Jpn. Tissue Cult. Assoc. J.* **12**, 215–220.

Harrison, T. A., Barr, R. D., McCaffrey, R. P., Sarna, G., Silverstone, A. F., Perry, S., and Baltimore, D. (1976). Terminal deoxynucleotidyl transferase in AKR leukemia cells and lack of relation of enzyme activity to cell cycle phase. *Biochem. Biophys. Res. Commun.* **69**, 63–67.

Hay, R. J. (1991). Operator-induced contamination in cell cultures systems. *Aeres-Sorono Symp. Dev. Biol. Stand.* **75**, 193–204.

Hay, R. J., Caputo, J., and Macy, M. (1992). "ATCC Quality Control Methods for Cell Lines," 2nd Ed. ATCC, Rockville, MD.

Hay, R. J., Cleland, M. M., Durkin, S., and Reid, Y. A. (2000). Cell Line Preservation and Authentication in "Animal Cell Culture" (J. R. W. Masters, ed.) Oxford Univ. Press, New York.

Hay, R. J., Macy, M. L., and Chen, T. R. (1989). Mycoplasma infection of cultured cells. *Nature* **339**, 487–488.

Holmgren, N. B. (1973). Contamination in tissue culture by parasites. *In "Contamination in Tissue Culture"* (J. Fogh, ed.), pp. 195–203. Academic Press, New York.

Hsiung, G. D., Fong, C. K. Y., and Landry, M. L. (1994). "Hsuing's Diagnostic Virology," 4th Ed. Yale Univ. Press, London.

Hu, M., and Buck, C. (1993). Application of polymerase chain reaction technique for detection of mycoplasma contamination. *J. Tissue Cult. Methods* **15**, 155–160.

Hukku, B., Halton, D. M., Mally, M., and Peterson, W. D., Jr. (1984). Cell characterization by use of multiple genetic markers in eukaryotic cell cultures. *In "Eukaryotic Cell Cultures, Basics and Applications"* (R. T. Acton and J. D. Lyn, eds.), pp. 13–31. Plenum Press, New York.

IABS (1989). "Continuous Cell Lines as Substrates for Biologicals." IABS Symposium on Developments in Biological Standardization, Vol. 70. Karger, Basel.

Jahnes, W. G., Fullmer, H. M., and Li, C. P. (1957). Free-living amoebae as contaminants in monkey kidney tissue culture. *Proc. Soc. Exp. Biol. Med.* **96**, 484–488.

LeDuc, J. W., Smith, G. A., Macy, M. L., and Hay, R. J. (1985). Certified cell lines of rat origin appear free of infection with hantavirus. *J. Infect. Dis.* **152**, 1081–1082.

Lubiniecki, A. S., and May, L. H. (1985). Cell bank characterization for recombinant DNA mammalian cell lines. *Dev. Biol. Stand.* **60**, 141–146.

McGarrity, G. J. (1982). Detection of mycoplasmal infection of cell cultures. *Adv. Cell Cult.* **2**, 99–131.

Ou, C.-Y., Kwok, S., Mitchell, S. W., Mack, D. H., Sninsky, J. J., Krebs, J. W., Feorino, P., Warfield, D., and Schochetman, G. (1988). DNA amplification for direct detection of HIV-1 in DNA of peripheral blood mononuclear cells. *Science* **239**, 295–297.

Rovozzo, G. C., and Burke, C. N. (1973). "A Manual of Basic Virological Techniques." Prentice-Hall, Englewood Cliff's, NJ.

Schutten, T. (2000). Development of real-time Q RT-PCR for detection of HIV-2 RNA in plasma. *J. Virol. Methods* **88**, 81–87.

Sethi, V. S., and Sethi, M. L. (1975). Inhibition of reverse transcriptase activity of RNA tumor viruses by fagaronine. *Biochem. Biophys. Res. Commun.* **63**, 1070–1076.

Temin, H. M., and Mizutani, S. (1970). RNA-dependent DNA polymerase in vMons of ROUS sarcoma virus. *Nature (Lond.)* **226**, 1211–1213.

Ulrich, P. P., Bhat, R. A., Seto, B., Mack, D., Sninsky, J., and Yvas, G. N. (1989). Enzymatic amplification of hepatitis B virus DNA in serum compared with infectivity testing in chimpanzees. *J. Infect. Dis.* **160**, 37–43.

Uphoff, C. C., and Drexler, H. G. (2001). Prevention of mycoplasma contamination in leukemia-lymphoma cell lines. *Hum. Cell* **14**, 244–247.

Weiss, S. H., Goedert, J. J., Gartner, S., Popovic, M., Waters, D., Markham, P., Veronese, F. M., Gail, M. H., Barkley, W. E, Gibbons, J., Gill, F. A., Leuther, M., Shaw, G. M., Gallo, R. C., and Blattner, W. A. (1988). Risk of human immunodeficiency virus (HIV-1) infection among laboratory workers. *Science* **239**, 68–71.

Yolken, R. H., Lennette, D. A., Smith, T. F., and Warner, J. L. (1999). "Manual of Clinical Microbiology," 7th Ed. ASM Press, Washington, DC.

Culture of Specific Cell Types: Stem Cells

Neural Crest Stem Cells

Maurice Kléber and Lukas Sommer

I. INTRODUCTION

Cellular diversity in the vertebrate peripheral nervous system is achieved by the differentiation of neural crest stem cells (NCSCs) in a spatially and temporally regulated fashion. During embryonic development, neural crest cells detach from the neuroepithelium of the dorsal neural tube and migrate to their sites of terminal differentiation (Le Douarin and Kalcheim, 1999). At least a subpopulation of these cells are multipotent and able to give rise to neuronal, glial, and nonneural derivatives, as has been shown by grafting experiments and by clonal analysis in culture and *in vivo* (Anderson *et al.*, 1997; Ziller *et al.*, 1983). Moreover, some crest cells display features of stem cells that not only generate multiple cell types, but also have the capacity to self-renew (Morrison *et al.*, 1999; Stemple and Anderson, 1992). In migrating neural crest and in target tissues of the neural crest, multipotent crest cells coexist with cells that have a more restricted developmental potential (Sommer, 2001). Cell intrinsic differences between crest cells from different regions of the peripheral nervous system are involved in the generation of neural diversity. In addition, the decision of a NCSC to survive, self-renew, or differentiate depends on the combinatorial activity of multiple environmental signals (Sommer, 2001). To identify these signals, NCSCs have to be challenged by altering both their extracellular environment and their intrinsic genetic programs. This is facilitated greatly by the availability of neural crest culture systems. Because signals required to maintain undifferentiated, multipotent NCSCs for an extended period of time in culture have not yet been identified, various cell culture conditions have been established by different laboratories. This article describes methods for culturing rat and mouse neural crest stem cells, largely based on articles by Stemple and Anderson (1992) (rat NCSCs), Sommer *et al.* (1995) (mouse NCSCs), Greenwood *et al.* (1999) (culture conditions permissive for sensory neurogenesis), and Morrison *et al.* (1999) (postmigratory NCSCs).

II. MATERIALS AND INSTRUMENTATIONS

A. Instruments and Plasticware

Modular incubator chamber (Billups-Rothenberg Inc., www.brincubator.com); Dumont #3 and #5 forceps (Fine Science Tools, Cat. Nos. 11231-30 and 11251-10); Vannas style iris spring scissors (Fine Science Tools, Cat. No. 15000-02); cell culture dishes, 35 × 10-mm style (Corning Cat. No. 430165); 6-well cell culture dishes (Corning Cat. No. 430166); tissue culture dishes, 60 × 15-mm style (Nunclon DSI Cat. No. 064194); Omnifix, 50 ml (Braun Cat. No. 459785OF); Millex syringe-driven filter unit, 0.22 μm (Millipore Cat. No. SLGPO33RB); and Steritop 500 GP express plus membrane, 0.22 μm (Milipore Cat. No. SCGPT05RE).

B. General Buffers and Reagents

Hanks' balanced salts (HBSS) without Ca^{2+} and Mg^{2+}, 95.18 g (Amimed Cat. No. 3-02P30-M); Hanks' balanced salts without phenol red, 97.5 g (Amimed Cat. No. 3-02P32-M); phosphate-buffered saline (PBS) Dulbecco (D-PBS) 50 l (Biochrom KG Cat. No. L-182-

50); formaldehyde solution, 250 ml (Fluka Cat. No. 47608); and potassium hydroxide (Fluka Cat. No. 60375).

C. Enzymes

Dispase 1 (neutral protease) 10 × 5 mg (Roche Cat. No. 1 284 908); collagenase types 1, 3, and 4 (Worthington Biochemical Coorporation); 0.25% trypsin–EDTA, 100 ml (Invitrogen-GIBCO Cat. No. 25200-056); 2.5% trypsin (10×), 100 ml (Invitrogen-GIBCO Cat. No. 25090-028); hyaluronidase type IV-S, 50 mg (Sigma H-4272); and deoxyribonuclease type 1 (Sigma Cat. No. D-4263).

D. Substrates

Fibronectin (0.1% solution), 5 mg (Sigma Cat. No. F1141), and poly-D-lysine (pDL), 5 mg, (Sigma Cat. No. P-7280).

E. Media Components

Dulbecco's modified Eagle medium (DMEM)-low glucose, 500 ml (Invitrogen-GIBCO Cat. No. 11880-028); DMEM, 500 ml (Invitrogen-GIBCO Cat. No. 41966-029); minimum essential medium (MEM), 500 ml (Invitrogen-GIBCO Cat. No. 31095-029); Leibovitz's L15 medium powder (Invitrogen-GIBCO Cat. No. 41300-021); dimethyl sulfoxide (DMSO) 25 l (Aldrich Cat. No. 27,043-1); N2-supplement (100×), 5 ml (Invitrogen-GIBCO Cat. No. 17502-048); 2-mercaptoethanol (Sigma Cat. No. M-7522); B-27 supplement (50×), 10 ml (Invitrogen-GIBCO Cat. No.17504-048); forskolin, 10 mg (Sigma Cat. No. F-6886); fetal bovine serum (FBS), 500 ml (different suppliers); water, cell culture tested, 500 ml (Sigma Cat. No. W-3500); phenol red solution, 100 ml (Sigma Cat. No. P-0290); imidazole, 1 g (Sigma Cat. No. I-0250); hydrochloride acid solution, 100 ml (Sigma Cat. No. H-9892); sodium bicarbonate, 500 g (Sigma Cat. No. S-5761); dexamethasone, 25 mg (Sigma Cat. No. D-4902); bovine albumin crystalline, 5 g (Sigma Cat. No. A-4919); 99.5% glycerol, 500 ml (Invitrogen-GIBCO Cat. No. 15514-011); transferrin, holo, bovin plasma, 100 mg (Calbiochem Cat. No. 616420); putrescine (Sigma Cat. No. P-7505); (+/−)-α-tocopherol (vitamin E), 5 g (Sigma Cat. No. T-3251); insulin, 100 mg (Sigma Cat. No. I-6634); human epidermal growth factor (hEGF), 200 µg (R&D Systems Cat. No. 236-EG-200); human nerve growth factor (β-NGF) (R&D Systems Cat. No. 256-GF-100); selenious acid (Aldrich Cat. No. 22,985-7); basic fibroblast growth factor (bFGF), 25 µg (R&D Systems Cat. No. 233-FB-025); progesterone, 1 g (Sigma Cat. No. P-8783);

human neurotrophin 3 (NT3), 10 µg (BioConcept Cat. No. 1 10 01862); human brain-derived neurotrophic factor (BDNF), 10 µg (BioConcept Cat. No. 1 10 11961); insulin-like growth factor (IGF1), 250 µg (R&D Systems Cat. No. 291-G1-250); and retinoic acid (Sigma Cat. No. R-2625).

F. Stable Vitamin Mix

Aspartic acid, 100 g (Sigma Cat. No. A-4534); L-glutamic acid, 100 g (Sigma Cat. No. G-8415); L-proline, 25 g (Sigma Cat. No. P-4655); L-cystine (Sigma Cat. No. C-7602); p-aminobenzoic acid (Aldrich Cat. No. 42,976-7); 3-aminoproprionic acid, 100 g (Sigma Cat. No. A-9920); vitamin B_{-12}, 1 g (Sigma Cat. No. V-6629); myo-inositol, 50 g (Sigma Cat. No. I-7508); choline chloride, 100 g (Sigma Cat. No. C-7527); fumaric acid, 100 g (Sigma Cat. No. F-8509); coenzyme A, 100 mg (Sigma Cat. No. C-4282); D-biotin (Sigma Cat. No. B-4639); and DL-α-lipoic acid, 5 g (Sigma Cat. No. T-1395).

G. 1:1:2

Dextrose [D-(+)-glucose] (Sigma Cat. No. G-7021); L-glutamine (Sigma Cat. No. G-6392); and penicillin–streptomycin, 100 ml (Invitrogen-GIBCO Cat. No. 15140-015).

H. Mix7

DL-β-Hydroxybutyric acid sodium salt (Sigma Cat. No. H-6501); cobalt chloride, 25 g (Sigma Cat. No. C-8661); oleic acid, 250 mg (Sigma Cat. No. O-7501); α-melanocyte-stimulating hormone (α-MSH), 5 mg (Sigma Cat. No. M-4135); prostaglandin$_{E1}$, 1 mg (Sigma Cat. No. P-5515); and 3,3',5- triiodo L-thyronine (T3), 100 mg (Sigma Cat. No. T-6397).

I. FVM

DMPH B grade (Calbiochem Cat. No. 31636); L-glutathione (Sigma Cat. No. G-6013); and L-ascorbic acid, 25 g (Sigma Cat. No. A-4544).

III. PROCEDURES

A. Solutions and Stocks

1. Chicken Embryo Extract (CEE)

Incubate white chicken eggs for 11 days at 38°C in a humidified atmosphere. Wash eggs with 70% ethanol, open the top of each shell, and remove the embryos and place them into a petri dish containing MEM at 4°C. Macerate the embryos by pressing through a 50-ml syringe into a 50-ml centrifuge tube (Falcon) (approximately 25 ml of homogenate per tube). Add 25 ml of MEM per 25 ml of chicken homogenate. Shake the tubes at 4°C for 1 h. Add 100 μl (800 U) sterile hyaluronidase to 50 ml of chicken homogenate and centrifuge the mixture for 6 h at 30,000 g at 4°C. Collect the supernatant, filtrate through a 0.22-μm Steritop filter, and distribute in 5-ml aliquots. Store at −80°C until use.

2. Stable Vitamin Mix (SVM)

Collect 198 ml water into a detergent-free beaker. To the 198 ml water add 0.6 g aspartic acid, 0.6 g L-glutamic acid, 0.6 g L-proline, 0.6 g L-cystine, 0.2 g p-aminobenzoic acid, 0.2 g 3-aminoproprionic acid, 80 mg vitamin B_{-12}, 0.4 g myo-inositol, 0.4 g choline chloride, 1.0 g fumaric acid, and 16 mg coenzyme A. Suspend 0.4 mg D-biotin and 100 mg DL-α-lipoic acid in 10 ml water and add 2 ml of this solution to the solution in the beaker. Mix the solutions, prepare 1.5-ml aliquots and store at −20°C until use.

3. L-15CO2

Add 3.675 g L-15 powder, 0.019 g imidazole, and 1.6 ml SVM to 288 ml water in a 500-ml beaker. Mix the solution until dissolved and add 240 μl of 1 M HCl to adjust the pH between 7.35 and 7.40. In a 100-ml beaker, mix 0.8 g sodium bicarbonate, 120 μl phenol red, and 59 ml water. Apply CO_2 directly to this solution using a Pasteur pipette until it turns yellow and no further color change can be observed. Then mix the sodium bicarbonate solution with the L-15 solution and apply CO_2 again for a short time. Determine the pH, which should range between 7.15 and 7.25. Filter the solution through a 0.22-μm filter into a 500-ml tissue culture bottle and store at 4°C until use.

4. 1:1:2

Slowly dissolve 60 g dextrose in 160 ml water by stirring. Adjust the volume to 200 ml after dextrose is dissolved. Add 100 ml glutamine (200 mM) and 100 ml penicillin–streptomycin, filter the solution through a 0.22-μm filter, and distribute in 2-ml aliquots. Store at −20°C until use.

5. Fresh Vitamin Mix (FVM)

Dissolve 5 mg DMPH, 25 mg glutathione, and 500 mg L-ascorbic acid in 80 ml water by stirring. After all chemicals are dissolved, raise the pH to 5–6 with 1 M potassium hydroxide. Then adjust the volume to 100 ml, filter the solution through a 0.22 μm filter, and store in 550-μl aliquots at −20°C until use.

6. Mix7

Dissolve 630 mg DL-β-hydroxybutyrate in 10 ml water (1000× stock). Dissolve cobalt chloride to 10 mg/ml in water and then add 25 μl from that solution to 10 ml of L15CO2 to obtain a stock of 25 μg/ml (1000×). Dissolve biotin to 10 mg/ml in DMSO and then dilute to 1 mg/ml in L15CO2 (1000×). Dissolve oleic acid to 2.8 mg/ml in water and then add 37.5 μl of this to 10 ml of L15CO2 to 10 μg/ml (1000×). Dissolve αMSH to 1 mg/ml in water and dilute to 0.1 mg/ml in L15CO2 (1000×). Dissolve prostaglandin to 1 mg/ml in 95% ethanol and dilute 1:100 in L15CO2 to 10 μg/ml (1000×). Dissolve T_3 to 10 mg/ml in DMSO and add 67.5 μl of this to 10 ml L15CO2 to 67.5 μg/ml (1000×). Add 5 ml of each of the aforementioned solutions to 15 ml of L15CO2, filter the solution through a 0.22-μm filter, and store in 550-μl aliquots at −20°C until use.

7. Additives

Dissolve most additives in H_2O, except for the following: dissolve retinoic acid to 17.5 mg/ml in DMSO and then dilute this solution 1:500 in equal volumes of 95% ethanol and L15CO2 to 35 μg/ml (1000×). Dissolve vitamin E to 50 mg/ml in DMSO and dilute this 1:10 to 5 mg/ml (1000×). Dissolve 3.93 mg dexamethasone in 10 ml 95% ethanol to 1 mM stock; for use, dilute stock 1:100 in L15CO2. Dissolve 50 mg insulin in 10 ml 5 mM HCl solution. Dissolve 31.5 mg progesterone in 10 ml 95% ethanol to 10 mM and dilute 1:100 in 95% ethanol to a 0.1 mM stock. Dissolve 100 mg transferrin in 2 ml 1× D-PBS. Dissolve 100 μg β-NGF in 2 ml L15CO2(+1 mg/ml BSA) to 50 μg/ml. Dissolve 200 μg hEGF in 2 ml L15CO2(+1 mg/ml BSA) to 100 μg/ml. Dissolve 25 μg bFGF in 1 ml L15CO2(+1 mg/ml BSA) to 25 μg/ml. Dissolve 10 μg NT-3 in 400 μl L15CO2 (+1 mg/ml BSA) to 25 μg/ml. Dissolve 10 μg BDNF in 400 μl L15CO2(+1 mg/ml BSA) to 25 μg/ml. Dissolve 250 μg IGF-1 in 2 ml L15CO2(+1 mg/ml BSA) to 125 μg/ml. Store these additives at −80°C until use. Dissolve 80 mg putrescine in 10 ml water to 8 mg/ml. Dissolve 1.29 g selenious acid in 10 ml water and dilute to a stock of 0.1 mM. Store these additives at 4°C until use.

B. Media

1. Standard Medium (SM)

To prepare 50 ml of defined medium (DM) (Stemple and Anderson, 1992) take 46.3 ml L15CO2, 50 mg BSA (1 mg/ml), 2 ml 1:1:2, 500 μl FVM, 315 μl glycerol, 100 μl putrescine (16 μg/ml), 100 μl transferrin (100 μg/ml), 50 μl vitamin E (5 μg/ml), 50 μl EGF (100 ng/ml), 50 μl insulin (5 μg/ml), 20 μl NGF (20 ng/ml), 15 μl selenious acid (30 nM), 8 μl bFGF (4 ng/ml), 10 μl progesterone (20 nM), 0.5 µl dexamethasone (100 nM), 500 µl Mix7. To prepare 50 ml standard medium, add to 45 ml DM 5 ml CEE and 50 µl retinoic acid (35 ng/ml). Filter the medium through a 0.22-µm filter and store at 4°C until use. (Final concentrations are in parentheses.)

A simplified SM has been used by Morrison and colleagues (Bixby *et al.*, 2002; Morrison *et al.*, 1999). To prepare 50 ml of standard medium take 38.9 ml DMEM low glucose, 2 ml 1:1:2, 25 µl retinoic acid (17.5 ng/ml), 500 µl N2 salt supplement (1%), 15 µl selenious acid, 1 ml B27 supplement (1:50), 2.5 µl 2-mercaptoethanol (50 µM), 8 µl IGF1 (20 ng/ml), and 7.5 ml CEE (15%). For SM1, add 40 µl bFGF (20 ng/ml). After 6 days of culture incubation, use SM2 to allow differentiation: reduce bFGF levels to 20 µl (10 ng/ml) and CEE to 500 µl (1%) and fill up to 50 ml with DMEM low glucose. Filter the medium through a 0.22-µm filter and store at 4°C until use.

Comment

In our hands, the simplified medium according to Morrison *et al.* (1999) was less efficient in supporting neural crest cultures than the more complex standard medium (Stemple and Anderson, 1992). To increase cell survival, IGF1 has been added in a more recent study by Bixby *et al.* (2002).

2. Medium Supporting Early Neural Crest Stem Cells and Sensory Neurogenesis (SN1 + SN2)

To prepare 50 ml of SN medium (Greenwood *et al.*, 1999) take 47.1 ml L15CO$_2$, 50 mg BSA (1 mg/ml), 2 ml 1:1:2, 50 µl insulin (5 µg/ml), 100 µl putrescine (16 µg/ml), 10 µl progesterone (20 nM), 15 µl selenious acid (30 nM), 0.5 µl dexamethasone, 143 µl glycerol, 50 µl vitamin E (5 µg/ml), and 500 µl Mix7.

To culture NCSCs for 2–3 days, make SN1 by adding 20 µl bFGF (10 ng/ml) to SN. For further differentiation, use SN2, which is prepared by adding the following reagents to SN: 8 µl bFGF (4 ng/ml), 25 µl EGF (50 ng/ml), 25 µl retinoic acid (17.5 ng/ml), 25 µl NGF (25 ng/ml), 25 µl BDNF (12.5 ng/ml), 25 µl NT3 (12.5 ng/ml), and 250 µl CEE (0.05%). Filter the medium through a 0.22-µm filter and store at 4°C until use. In brackets, (Final concentrations are in parentheses.)

C. Isolation

1. Migratory Neural Crest Stem Cells from Neural Tube Explant Cultures

Mouse NCSCs are isolated at embryonic day 9 (E9) (Sommer *et al.*, 1995), whereas rat NCSCs are isolated

at E10.5 (Stemple and Anderson, 1992) (Fig. 1). The isolation of trunk neural crest is described here.

1. Sacrifice time-mated females by CO$_2$ asphyxiation in accordance with National Institutes of Health guidelines.

2. Remove the uterus into a 10-cm petri dish containing sterile HBSS without phenol red.

3. With a pair of fine spring scissors, cut an opening along the length of each uterus, being careful not to cut into the embryo and yolk sac.

4. Under a dissecting microscope, remove the embryos by squeezing the uterus gently with a Dumont #3 forceps while cutting the surface of the decidua and amnionic sac with another #3 forceps. After every embryo has been removed from the uterus, transfer the embryos to a new 10-cm petri dish containing sterile HBSS without phenol red.

5. Use an L-shaped electrolytically sharpened tungsten needle and a Dumont #5 forceps to dissect a block of tissue from a region corresponding to the region caudal to the heart to the most caudal somite. Pool and place the trunks into a new 3-cm petri dish. They can be stored for an hour at 4°C.

6. Prepare digestion mix using 12 ml HBSS without Ca^{2+} and Mg^{2+} and one vial (5 mg) of dispase 1. Distribute digestion mix to three 3-cm petri dishes. Transfer the trunks with a Pasteur pipette to the dispase mix and transfer them from the first to the second and then from the second to the third petri dish. Triturate slowly for 2 min at room temperature. Place the dish at 4°C for 6 min.

7. Triturate the trunks gently and patiently until the neural tubes are free of other tissues. Transfer every tube to DMEM+10% FBS to stop the digestion reaction. Then transfer the tubes to the appropriate media.

8. Coat 35-mm Corning tissue culture dishes with fibronectin (FN) as described later and preincubate with appropriate media. Withdraw media and plate three to four neural tubes directly onto the dish. Monitor each step carefully under a dissecting microscope.

9. Allow the tubes to attach for 30 min at 37°C in a 5% CO$_2$ atmosphere and then flood the dish gently with 1 ml medium. If the tubes do not attach, withdraw the media gently and repeat the step until the tube is properly attached to the dish. Incubate the dishes in medium appropriate to the experiment.

2. Isolation of Postmigratory Neural Crest Stem Cells

Postmigratory NCSCs from dorsal root ganglia (DRG) (Fig. 1), from sciatic nerve, and from gut have been isolated at various stages of development, both

Isolation of NCSCs from neural tube explant cultures

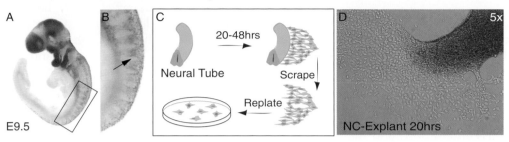

Isolation of Postmigratory NCSCs from DRG

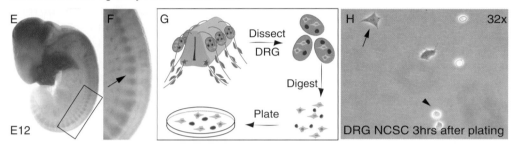

FIGURE 1 To illustrate the localization of migratory and postmigratory neural crest cells (arrows), X-gal staining was performed on mouse embryos at E9.5 (A) and at E12 (E), in which neural crest cells had been marked by Wnt1Cre-mediated recombination (Danielian *et al.*, 1998) of the ROSA26 reporter gene (Soriano, 1999). Boxes in A and E represent areas enlarged in B and F, respectively. (C) Scheme of the explant culture system of a trunk neural crest and (D) phase-contrast picture of a neural crest explant at 5× magnification. (G) Isolation scheme of postmigratory NCSCs from DRG and (H) phase-contrast picture of DRG-derived postmigratory NCSCs (arrow) and neuronal cells (arrowhead).

from mouse and from rat embryos (Bixby *et al.*, 2002; Hagedorn *et al.*, 1999; Lo and Anderson, 1995; Morrison *et al.*, 1999; Paratore *et al.*, 2002; Pomeranz *et al.*, 1993). Studies have also shown that NCSCs can be isolated from postnatal and adult gut (Kruger *et al.*, 2002). Note that the plating efficiency is very low for adult gut NCSCs.

a. NCSCs from Embryonic DRG.

1. Sacrifice time-mated females as described. After removal from the uterus, transfer the embryos to a new 10-cm petri dish containing sterile HBSS without phenol red.

2. Use an L-shaped electrolytically sharpened tungsten needle and a Dumont #5 forceps to dissect a block of tissue from a region rostral to the heart to the most caudal somite. Pool and place the trunks into a new 3-cm petri dish. They can be stored for an hour at 4°C.

3. Gently drive the tungsten needle between the cartilage primordium of the vertebral bodies and the neural tube while stabilizing the trunk with the forceps. Take care not to damage the neural tube. Dissect the cartilage primordium by pulling the needle

ventrally. Tear apart the tissue lateral to the neural tube to display the ventral part of the neural tube.

4. Hold the neural tube with a Dumont #3 forceps and separate from dorsal muscle and epithelial tissue using another forceps.

5. Collect the DRGs, which remain attached to the neural tube, using the tungsten needle. Pool the DRGs in ice-cold HBSS without phenol red.

6. Centrifuge the DRGs for 2 min at 2000 rpm and withdraw the HBSS. Digest the DRGs by incubation in 0.25% trypsin and 3.5 mg collagenase type 1 in HBSS without Ca^{2+} and Mg^{2+} for 20 min at 37°C.

7. Stop the reaction by adding FBS to 10%, centrifuge the cells for 2 min at 2000 rpm, resuspend the cells in the appropriate medium, and plate the cells onto culture dishes that have been precoated with either fibronectin or pDL/fibronectin (see Section III,D,1).

b. Sciatic Nerve NCSCs.

1. Isolate embryos as described earlier.

2. Fix embryos on a wax support. Cut an opening dorsolateral to the hind limb, proximal to the spinal

cord. Nerve and nerve plexus are revealed underneath muscle tissue.

3. Fix the hind limb with a Dumont #5 forceps. With another #5 forceps, pull out sciatic nerve running into the hind limb. Dissect sciatic nerves into ice-cold HBSS without Ca^{2+} and Mg^{2+}. Centrifuge cells at 2000 rpm for 2 min, withdraw HBSS, and resuspend the pellet in a solution containing 0.025% trypsin and 1 mg/ml type 3 collagenase.

4. Incubate for 4 min at 37°C and then quench the digestion with 2 volumes of $L15CO_2$ containing 1 mg/ml BSA, penicillin/streptomycin, and 25 µg/ml deoxyribonuclease type 1.

5. Centrifuge the cells at 2000 rpm for 2 min and slowly triturate them in medium.

c. NCSCs from the Enteric Nervous System

1a. To prepare enteric NCSCs from embryos, dissect the entire gut distal to the stomach and digest in 1 mg/ml collagenase type I in HBSS without Ca^{2+} and Mg^{2+} for 20 min at 37°C. Preparations from older embryos are digested for 45 min in a solution that, in addition to the collagenase, contains 0.01% trypsin.

1b. Stop the digestion by adding FBS to 10%. Subsequently, centrifuge the cells for 2 min at 1800 rpm, triturate, and resuspend.

2a. To isolate and culture early postnatal gut NCSCs, separate the small intestine from the attached mesentry and place into ice-cold HBSS without Ca^{2+} and Mg^{2+}. Peel free the outer muscle/plexus layers of the underlying epithelium, mince, and dissociate in 0.025% trypsin/EDTA plus 1 mg/ml type 4 collagenase in HBSS without Ca^{2+} and Mg^{2+} for 8 min at 37°C. Quench the digestion with two volumes medium, centrifuge the cells, and triturate.

2b. Filter the cells through a nylon screen to remove clumps of cells and undigested tissue. Before plating, resuspend the cells in medium.

3. Flow Cytometry

Isolation of prospectively identified NCSCs by FACS avoids contamination by nonneural cells and allows enrichment of the NCSC population. Suspend dissociated cells in antibody-binding buffer, add the primary antibody (or mixture of antibodies) at the appropriate concentration, and incubate for 20–25 min on ice. Wash three times in antibody-binding buffer and incubate with fluorophore-conjugated secondary antibody. Wash cells and resuspend in buffer containing 2 µg/ml of the viability dye 7-aminoactinomycin D (7-AAD; Molecular Probes, Eugene, OR). This step allows exclusion of 7-AAD-positive dead cells during the FACS procedure. To isolate NCSCs from sciatic nerve, cells have been sorted that express the neu-

rotrophin receptor p75 but not P0, a PNS myelin component (Morrison et al., 1999). For the isolation of gut NCSCs, a selection for p75/α4 integrin double-positive cells has been performed (Bixby et al., 2002; Kruger et al., 2002). Prior to and after sorting, it is recommended to keep tissue culture plates in sealed plastic bags gassed with 5% CO_2 to maintain the pH in the medium.

D. Culture of NCSCs

1. Substrate Preparation

Dishes coated with fibronectin: dilute 5 ml (one vial; 5 mg) fibronectin in 20 ml sterile 1× D-PBS. Apply 1 ml fibronectin to 35-mm Corning tissue culture dishes and withdraw it immediately and add the appropriate media. It is possible to reuse the fibronectin solution several times. Dishes coated with poly-D-lysine/fibronectin: resuspend 5 mg poly-D-lysine in 10 ml cell culture water. Rinse each 35 mm Corning tissue culture dish with 1 ml poly-D-lysine solution. Allow plates to air dry. Subsequently, wash twice with tissue culture water and air dry plates again. Apply fibronection to poly-D-lysine-coated plates as described earlier.

2. Culturing NCSCs from Neural Tube Explants in SM

In the absence of instructive growth factors (see Section III,D,5), the following conditions are permissive for the generation of autonomic neurons, peripheral glia, and nonneural smooth muscle-like cells. After neural tube isolation, the culture dishes are incubated at 37°C, 5% CO_2 and 20% O_2 (from air) for 24 h (rat neural crest) to 48 h (mouse neural crest) in SM. At this stage, most of the emigrated neural crest cells coexpress the transcription factor Sox10 and the low-affinity neurotrophin receptor p75 as markers for undifferentiated NCSCs (Paratore et al., 2001; Stemple and Anderson, 1992). For further incubation, it is possible to scrape away the neural tube from the neural crest cells that migrated onto the substrate using an L-shaped tungsten needle and an inverted phase-contrast microscope equipped with a 5 or 10× objective lens (Fig. 1). Differentiated cell types become apparent after a few days of culture in SM (Stemple and Anderson, 1992). Differentiation is promoted by the addition of 10% FBS and 5 µM forskolin (Sommer et al., 1995).

Comments

In the presence of the neural tube and upon addition of NT3, BDNF, and LIF, the generation of sensory neurons is observed proximal to the neural tube, in addition to the autonomic neurons that are found scattered throughout the outgrowth after 8 days in culture (Greenwood *et al.*, 1999).

3. Culturing NCSCs from Neural Tube Explants in SN Medium

In the absence of instructive growth factors (see Section III,D,5), neural tube explants in SN medium consist of early NCSCs that can generate sensory neurons. After plating, neural tube explants are incubated in SN1 medium for 20 h. Outgrowth of NCSCs occurs during this period. To allow sensory neuronal differentiation, withdraw the SN1 medium from the plates after 48 h of explant incubation and add SN2 medium for another 2 days.

Comments

Twenty hours after having plated the isolated neural tubes, the early neural crest explants cultured in SN1 express neither the sensory marker Brn-3A nor NF160, whereas virtually all neural crest cells express p75 and Sox10 (Figs. 2A and 2B) (Hari *et al.*, 2002). Sensory neurons obtained after prolonged incubation are characterized by coexpression of the POU domain transcription factor Brn-3A and NF160 (Fedtsova and Turner, 1995) (Figs. 2C and 2D).

4. Replating NCSCs from Neural Tube Explants and Cloning Procedure

Allow rat and mouse NCSCs to emigrate for 24 and 48 h, respectively. After scraping away the neural tubes, carefully wash plates once with DMEM. Detach the NCSCs by treatment with a 0.05% trypsin solution for 2 min at 37°C. Quickly resuspend the cells in DMEM+10% FBS to abolish the reaction. Centrifuge the cells at 2000 rpm for 2 min and resuspend the pellet in 1 ml fresh medium. Count the cells in a Neubauer counting chamber. Plate the cells at low density (100–300 cells/35-mm plate). Let the cells settle down for approximately 3 h. Single NCSCs are mapped by labeling the surface antigen p75 on living cells (Stemple and Anderson, 1992).

Staining is performed in SM for 30 min using a rabbit antimouse p75 antibody (1:300 dilution, Chemikon International). Wash the cells three times in DMEM and visualize the staining using a Cy3-coupled goat antirabbit IgG (Jackson Laboratories) in SM for 30 min at room temperature. Wash the cells three times in DMEM and add 1 ml fresh SM medium to each plate. Detect p75-expressing NCSCs with an inverted fluorescence microscope at 10× magnification. Mark single founder cells by inscribing them with a 3- to 4-mm circle using a grease pencil on the bottom of the dish.

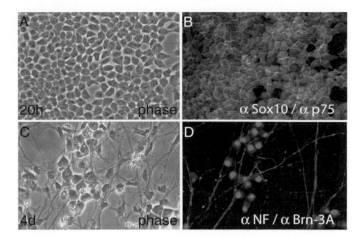

FIGURE 2 NCSCs are identified by coexpression of the transcription factor Sox10 and the low-affinity neurotrophin receptor p75 (B). Neural tube explant cultures cultured in SN conditions for 20 h were fixed with 3.7% formaldehyde in D-PBS for 10 min. Cells were treated for 10 min at room temperature with blocking buffer containing 10% goat serum, 0.3% Triton X-100, and 0.1% BSA in D-PBS and were stained with rabbit antimouse p75 (1:300 dilution, Chemikon International) for 1 h at room temperature and with the monoclonal anti-Sox10 antibody (1:10 dilution; Paratore *et al.*, 2001) for 2 h at room temperature. Within 4 days in culture, neural crest cells differentiate into sensory neurons identified by coexpression of the POU transcription factor Brn-3A and NF160 (D). Immunocytochemistry with the polyclonal rabbit anti-Brn-3A antibody [1:300 dilution (Fedtsova and Turner, 1995)] and monoclonal anti-NF160 antibody NN18 (1:300 dilution, IgG, Sigma-Aldrich) were carried out at room temperature for 1 h. Immunostainings were visualized by incubation for 1 h at room temperature using the following secondary antibodies at 1:200 dilution: Cy3-conjugated goat antimouse IgG, Cy3-conjugated goat antirabbit IgG, FITC-coupled donkey antirabbit IgG (Jackson Immuno Research Laboratories), and FITC-coupled horse antimouse IgG (Vector Laboratories). (A and C) Corresponding phase-contrast pictures.

Comments

For reasons not entirely clear, mouse NCSCs display a low survival capacity at clonal density. Clonal experiments with mouse NCSCs are therefore only possible under certain conditions (such as in the presence of fetal bovine serum) (Paratore *et al.*, 2001). In the rat, clonogenic culture systems allowed assessment of the state of commitment of neural crest cells by exposing individual cells to changing environmental cues. Using such experiments, instructive growth factors have been identified that are able to promote the differentiation of NCSCs to specific lineages *in vitro* (Morrison *et al.*, 1999, 2000b; Shah *et al.*, 1994, 1996). Bone morphogenic protein 2 (BMP2) promotes a neuronal and, to a lesser extent, a smooth muscle-like fate, whereas single neural crest cells are instructed by transforming growth factor-β (TGFβ) to adopt a nonneural fate. Furthermore, individual neural crest cells choose a glial fate upon either Notch signal activation or treatment with GGF, an isoform of neuregulin 1 (NRG1). Finally, canonical Wnt signaling instructively promotes sensory neurogenesis in NCSCs (Lee et al., 2004). However, the response of NCSCs to instructive growth factors is modulated by short-range cell–cell interactions termed community effects and other signals (Hagedorn et al., 1999).

Moreover, serial subcloning experiments demonstrated the self-renewal capacity of NCSCs (Morrison et al., 1999; Stemple and Anderson, 1992). The signals promoting self-renewal and maintenance of NCSCs have not yet been discovered.

5. Culturing Postmigratory NCSCs

Cells are cultured either in 35-mm or in 6-well plates that have been precoated with poly-D-lysine and fibronectin (see earlier discussion). Both the traditional SM according to Stemple and Anderson (1992) and a simplified SM (Bixby et al., 2002; Morrison et al., 1999) have been used successfully to culture postmigratory NCSCs (Bixby et al., 2002; Hagedorn et al., 1999; Morrison et al., 1999). When using the simplified SM, incubate cells in SM1 for 6 days and then add SM2 for another 8 days to favor differentiation. For clonal analysis, directly plate cells at low density after dissociation of postmigratory neural crest target tissues (100–300 cells/35-mm plate; fewer than 30 cells per well of a 6-well plate).

Comments

Although neural crest cells isolated both from neural tube explant cultures and from various neural crest-derived tissues have been shown to be multipotent and responsive to instructive growth factors, cell-intrinsic differences between NCSCs from different origins affect fate decisions by changing the sensitivity of the cells to specific extracellular signals (Bixby et al., 2002; Kruger et al., 2002; Paratore et al., 2001; White et al., 2001).

6. Culturing NCSCs at Reduced Oxygen Levels

Reduced levels of oxygen have been shown to influence the survival, proliferation, and cell fate decision of neural stem cells (Morrison et al., 2000a). To culture neural tube explants at reduced oxygen levels, put all the dishes after neural tube isolation into a gas-tight modular incubator chamber and flush the chamber for 3–5 min with a custom gas mixture of 1% O_2, 6% CO_2, and balance N_2 to generate an actual O_2 level of 3–6%. The gas-tight chamber is housed inside a normal incubator. Once cultures are established in the reduced oxygen chamber, minimize the opening to avoid reperfusion.

Comment

We observed that in SN1 NCSCs appeared healthier after 20 h when cultured at reduced oxygen levels (M. Kléber et al., unpublished results). Moreover, culturing NCSCs at reduced oxygen levels in the presence of BMP-2 and forskolin revealed that low oxygen levels can influence cell fate (Morrison et al., 2000a).

IV. PITFALLS

1. In order to establish NCSC cultures, follow the instructions carefully. Note that small differences in concentrations of media ingredients may influence outgrowth, proliferation, survival, and differentiation of NCSCs.

2. Different batches of CEE and FBS might have different effects on the cultures. When using a new batch of CEE or FBS, always compare it to an older batch.

3. During isolation, triturate the neural tubes slowly and patiently. Rapid trituration can damage the neural tubes, which can impair efficient neural crest outgrowth.

4. Do not exceed the time of the digestion during isolation.

5. Always use fresh media. Media should not be stored for more than 1 week.

Acknowledgments

The protocols described here have been developed over a number of years. They represent the work of many people, mainly from the laboratories of Dr. David J. Anderson and Dr. Sean J. Morrison and from our own laboratory. We thank Dr. Ned Mantei and Hye-Youn Lee for help with the manuscript and thank Drs. Andrew McMahon, Philippe Soriano, Eric Turner, and Michael Wegner for tools used to prepare the figures.

References

Anderson, D. J., Groves, A., Lo, L., Ma, Q., Rao, M., Shah, N. M., and Sommer, L. (1997). Cell lineage determination and the control of neuronal identity in the neural crest. Cold Spring Harb. Symp. Quant. Biol. 62, 493–504.

Bixby, S., Kruger, G. M., Mosher, J. T., Joseph, N. M., and Morrison, S. J. (2002). Cell-intrinsic differences between stem cells from different regions of the peripheral nervous system regulate the generation of neural diversity. Neuron 35, 643–656.

Danielian, P. S., Muccino, D., Rowitch, D. H., Michael, S. K., and McMahon, A. P. (1998). Modification of gene activity in mouse embryos in utero by a tamoxifen-inducible form of Cre recombinase. *Curr. Biol.* **8**, 1323–1326.

Fedtsova, N. G., and Turner, E. E. (1995). Brn-3.0 expression identifies early postmitotic CNS neurons and sensory neural precursors. *Mech. Dev.* **53**, 291–304.

Greenwood, A. L., Turner, E. E., and Anderson, D. J. (1999). Identification of dividing, determined sensory neuron precursors in the mammalian neural crest. *Development* **126**, 3545–3559.

Hagedorn, L., Suter, U., and Sommer, L. (1999). P0 and PMP22 mark a multipotent neural crest-derived cell type that displays community effects in response to TGF-ß family factors. *Development* **126**, 3781–3794.

Hari, L., Brault, V., Kléber, M., Lee, H. Y., Ille, F., Leimeroth, R., Paratore, C., Suter, U., Kemler, R., and Sommer, L. (2002). Lineage-specific requirements of beta-catenin in neural crest development. *J. Cell Biol.* **159**, 867–880.

Kruger, G. M., Mosher, J. T., Bixby, S., Joseph, N., Iwashita, T., and Morrison, S. J. (2002). Neural crest stem cells persist in the adult gut but undergo changes in self-renewal, neuronal subtype potential, and factor responsiveness. *Neuron* **35**, 657–669.

Le Douarin, N. M., and Kalcheim, C. (1999). "The Neural Crest." *Cambridge Univ. Press, UK.*

Lee, H. Y., Kléber, M., Hari, L., Brault, V., Suter, U., Taketo, M. M., Kemler, R., and Sommer, L. (2004). Instructive Role of Wnt/β-Catenin in Sensory Fate Specification in Neural Crest Stem Cells. *Science*, **303**, 1020–1023.

Lo, L., and Anderson, D. J. (1995). Postmigratory neural crest cells expressing c-RET display restricted developmental and proliferative capacities. *Neuron* **15**, 527–539.

Morrison, S. J., Csete, M., Groves, A. K., Melega, W., Wold, B., and Anderson, D. J. (2000a). Culture in reduced levels of oxygen promotes clonogenic sympathoadrenal differentiation by isolated neural crest stem cells. *J. Neurosci.* **20**, 7370–7376.

Morrison, S. J., Perez, S. E., Qiao, Z., Verdi, J. M., Hicks, C., Weinmaster, G., and Anderson, D. J. (2000b). Transient Notch activation initiates an irreversible switch from neurogenesis to gliogenesis by neural crest stem cells. *Cell* **101**, 499–510.

Morrison, S. J., White, P. M., Zock, C., and Anderson, D. J. (1999). Prospective identification, isolation by flow cytometry, and *in vivo* self-renewal of multipotent mammalian neural crest stem cells. *Cell* **96**, 737–749.

Paratore, C., Goerich, D. E., Suter, U., Wegner, M., and Sommer, L. (2001). Survival and glial fate acquisition of neural crest cells are regulated by an interplay between the transcription factor Sox10 and extrinsic combinatorial signaling. *Development* **128**, 3949–3961.

Paratore, C., Hagedorn, L., Floris, J., Hari, L., Kléber, M., Suter, U., and Sommer, L. (2002). Cell-intrinsic and cell-extrinsic cues regulating lineage decisions in multipotent neural crest-derived progenitor cells. *Int. J. Dev. Biol.* **46**, 193–200.

Pomeranz, H. D., Rothman, T. P., Chalazonitis, A., Tennyson, V. M., and Gershon, M. D. (1993). Neural crest-derived cells isolated from gut by immunoselection develop neuronal and glial phenotypes when cultured on laminin. *Dev. Biol.* **156**, 341–361.

Shah, N., Groves, A., and Anderson, D. J. (1996). Alternative neural crest cell fates are instructively promoted by TGFß superfamily members. *Cell* **85**, 331–343.

Shah, N. M., Marchionni, M. A., Isaacs, I., Stroobant, P., and Anderson, D. J. (1994). Glial growth factor restricts mammalian neural crest stem cells to a glial fate. *Cell* **77**, 349–360.

Sommer, L. (2001). Context-dependent regulation of fate decisions in multipotent progenitor cells of the peripheral nervous system. *Cell Tissue Res.* **305**, 211–216.

Sommer, L., Shah, N., Rao, M., and Anderson, D. J. (1995). The cellular function of MASH1 in autonomic neurogenesis. *Neuron* **15**, 1245–1258.

Soriano, P. (1999). Generalized lacZ expression with the ROSA26 Cre reporter strain. *Nature Genet.* **21**, 70–71.

Stemple, D. L., and Anderson, D. J. (1992). Isolation of a stem cell for neurons and glia from the mammalian neural crest. *Cell* **71**, 973–985.

White, P. M., Morrison, S. J., Orimoto, K., Kubu, C. J., Verdi, J. M., and Anderson, D. J. (2001). Neural crest stem cells undergo cell-intrinsic developmental changes in sensitivity to instructive differentiation signals. *Neuron* **29**, 57–71.

Ziller, C., Dupin, E., Brazeau, P., Paulin, D., and Le Douarin, N. M. (1983). Early segregation of a neural precursor cell line in the neural crest as revealed by culture in a chemically defined medium. *Cell* **32**, 627–638.

9

Postnatal Skeletal Stem Cells: Methods for Isolation and Analysis of Bone Marrow Stromal Cells from Postnatal Murine and Human Marrow

Sergei A. Kuznetsov, Mara Riminucci, Pamela Gehron Robey, and Paolo Bianco

I. INTRODUCTION

Skeletal stem cells are found among the adherent and clonogenic subset of bone marrow stromal cells. It is of utmost importance to realize that the very existence of a skeletal stem cell is established through a complex sequence of *ex vivo* isolation and expansion, and *in vivo* transplantation. Through the *ex vivo* expansion of a single cell-derived strain and the subsequent *in vivo* transplantation, a complete heterotopic bone/bone marrow organ, containing a hematopoiesis-supporting stroma, must be established in order to prove that the single, originally cloned cell was indeed a stem cell. In addition, stromal cells isolated from the heterotopic organ must be able to transfer the hematopoietic microenvironment and have the potential to establish bone tissue *in vivo* upon serial transplantation. *In vivo* transplantation of stroma cells from a variety of species usually does not lead to the formation of cartilage. This is due to the conditions of a relatively high oxygen tension established in open transplantation systems. Hence, the chondrogenic potential of the cell strain under examination must be probed separately using *in vitro* micromass cultures (Bianco and Robey, 2004).

It is also important to realize that monolayers of stromal cells, established through clonal or nonclonal cultures, are not per se homogeneous populations of "stem" cells, but an uncontrolled mixture of cells and progenitors of highly diverse differentiation potential. This is the natural consequence of the natural asymmetric kinetics of stromal stem cell growth in culture, which inherently leads to a progressive dilution of the original stem cells present in the explanted cells, unless some degree of expansion of the stem cells is allowed, during culture, through the stochastic reversal of asymmetric kinetics to symmetric expansion.

In vitro and *in vivo* assays are necessarily complementary to one another in the study of the biology and pathology of skeletal stem cells and cannot be sensibly used in isolation. Most *in vitro* assays of differentiation potential do not necessarily predict the behavior of the same test strain upon *in vivo* transplantation.

II. MATERIALS AND INSTRUMENTATION

Mice of any strain, including transgenic lines, can be used as a source of bone marrow stromal cells (BMSCs). Guinea pigs (Hartley Davis), used to create irradiated feeder cells, are obtained from Charles River Laboratories. Human bone fragments are collected as surgical waste, and human bone marrow aspirates are obtained from normal volunteers, both under internal review board-approved protocols for the use of human subjects in research. For *in vivo* transplantation experiments, female Bg Nu/Nu-Xid mice, between 6 weeks and 6 months of age, are from Harlan. Standard tissue culture supplies (tubes, pipettes, dishes, flasks) and solutions [α-MEM, Coon's modified Ham's F-12, Hanks' balanced salt solution (HBSS), trypsin/EDTA, glutamine, penicillin–streptomycin] are not vendor specific. Cell strainers are from Becton–Dickinson (Cat. No. 2350). Lot-selected fetal bovine serum is obtained from a number of vendors (see Section V). Chondroitinase ABC is obtained from Seikagaku America (Cat. No. 100330-1A). Cloning cylinders were obtained from Sigma (Cat. No. C3983). Recombinant human TGFβ1 is obtained from Austral Biologics (Cat. No. GF-230-2). Hydroxyapatite/tricalcium ceramic, particle size 0.5–1 mm, is obtained from Zimmer by a material transfer agreement, and Gelfoam is from Upjohn (dental packs, size 4, 2 × 2 cm, Cat. No. NDC 0009-0396-04). Mouse fibrinogen and thrombin are from Sigma (Cat. Nos. F-4385 and T-8397, respectively). Ketamine hydrochloride, xylazine hydrochloride, and acepromazine (Cat. Nos. K2753, X1251, and A6908, respectively) are from Sigma. All other standard chemicals and reagents are from Sigma. Polymerase chain reaction is performed using a commercially available kit from Roche Diagnostics (Cat. No. 1 636 103). Standard equipment for use in dissection (scissors, forceps, blade knives) are sterilized by autoclaving prior to use and can be obtained from any vendor. Cell number enumeration is determined by use of a standard hemocytometer. Cell cultures are viewed by standard inverted and dissecting microscopes, and tissue sections are viewed by standard bright-field microscopes.

III. PROCEDURES

Bone marrow stromal cells can be prepared from bone specimens or bone marrow aspirates from any animal species using a variety of procedures. While there have been a number of modifications whereby single cell suspensions of marrow are subfractionated by density gradient centrifugation, the original assay described by Friedenstein (and described later) relies on the rapid adherence of BMSCs to plastic and avoids loss of these cells during fractionation (Friedenstein, 1980; Friedenstein *et al.*, 1992; Kuznetsov *et al.*, 1997a,b). When working with human cells, BMSCs can be isolated by FACS using the mouse monoclonal antibody Stro-1. However, there is significant contamination by hematopoietic cells such that adherence to tissue culture plastic is necessary to purify them further. It must also be noted that culture conditions must be

optimized for each animal species that is used, particularly, in selecting appropriate lots of serum. The procedures presented here focus on the preparation of mouse and human cells in particular due to the fact that they are used most frequently, and to highlight some of the differences between establishing cultures from these two different species.

A. Collection and Preparation Single Cell Suspensions of Bone Marrow

Solutions

1. *Nutrient medium*: α-MEM
2. *Heparinized nutrient medium*: α-MEM containing 100 U/ml sodium heparin.

Steps

1. For preparation of mouse and guinea pig marrow, animals are euthanized by CO_2 inhalation in compliance with institutionally approved protocols for the use of animals in research. Femora, tibiae, and humeri are removed aseptically, and the entire bone marrow content of medullary cavities is flushed with nutrient medium and combined. From human surgical specimens, trabecular bone fragments are scraped with a steel blade into the nutrient medium and washed until the bone became marrow free. In other cases, a 0.5-ml aspirate is collected and mixed with 5 ml of ice-cold nutrient medium containing 100 U/ml sodium heparin. The cells are centrifuged at $135g$ for 10 min, and the pellet is resuspended in fresh nutrient medium.

2. To prepare single cell suspensions (from all animal species), marrow preparations are pipetted up and down several times, passed through needles of decreasing diameter (gauges 16 and 20) to break up aggregates, and subsequently filtered through a cell strainer. Excessive pressure, both positive and negative, should be avoided while passing cell suspensions through the needles. Mononuclear cell concentrations are determined with a hemocytometer.

3. Guinea pig marrow suspensions, used as feeder cells in mouse cultures, are γ-irradiated with 6000 cGy to prevent the proliferation of adherent guinea pig cells.

B. Determination of Colony-Forming Efficiency (enumeration of CFU-F)

The concentration of CFU-F in bone marrow is usually expressed as the colony-forming efficiency (CFE), or number of BMSC colonies per 1×10^5 marrow nucleated cells in the original marrow cell suspension. In animals under physiological conditions, CFE remains relatively stable; it is, however, somewhat age dependent and can be altered significantly by experimental procedures, such as acute bleeding, irradiation, or curettage (Friedenstein, 1976, 1990). In humans, CFE is also relatively constant; in normal bone marrow not diluted with peripheral blood, as occurs when aspirated, it is between 20 and 70 per 1×10^5 marrow cells (Kuznetsov and Gehron Robey, 1996).

Solutions

1. *Serum-containing medium (SM)*: SM consists of α-MEM, glutamine (2 m*M*), penicillin (100 U/ml), streptomycin sulfate (100 µg/ml), and 20% lot-selected fetal bovine serum.
2. *Hanks' balanced salt solution*
3. *100% methanol*
4. *Saturated methyl violet*

Steps

1. Mouse cells (6–15×10^5 nucleated cells) or human cells (1–6×10^5 nucleated cells) are plated into 25-cm² plastic culture flasks in 5 ml of SM. If significantly abnormal CFE can be expected, as in some human pathologies, numbers of nucleated cells per flask should be adjusted accordingly. In problematic cases, it is recommended that several groups of flasks, containing, for example, 1×10^4, 1×10^5, and 1×10^6 nucleated cells, are prepared.

2. After 2–3 hr of adhesion, unattached cells are removed by aspiration, and cultures are washed vigorously three times with SM. No more than several hundred nonadherent cells remain after the washing step.

3. Each flask receives 5 ml of SM. For mouse cultures, irradiated guinea pig feeder cells (1.0–1.5×10^7 nucleated cells per flask) are added. If no more than 1×10^5 human cells per 25-cm² flask are plated, steps 2 and 3 can be omitted.

4. Cultivation is performed at 37°C in a humidified atmosphere of 5% CO_2 with air. On days 10–14, cultures are washed with HBSS, fixed with methanol, and stained with an aqueous solution of saturated methyl violet.

5. Colonies containing 30 or more cells are counted using a dissecting microscope, and colony-forming efficiency (number of colonies per 1×10^5 marrow cells plated) is determined (See Fig. 1A).

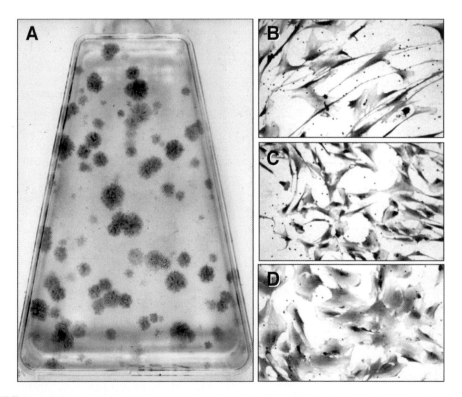

FIGURE 1 (A) When single cell bone marrow suspensions are plated at low density, the colony-forming unit-fibroblast present in the cell population adheres and forms colonies of bone marrow stromal cells. Their enumeration at low density is the basis of the colony-forming efficiency assay. Colonies are heterogeneous and grow at different rates, generating a broad range of colony sizes that are composed of cells with different morphologies, ranging from elongated, spindle-shaped cells (B), cells with a more compact morphology (C), and very flat and extended cells (D).

C. Preparation of Multicolony-Derived Strains of BMSCs

Solutions

1. *Hanks' balanced salt solution*
2. *Chondroitinase ABC*: Dissolve powdered chondroitinase ABC in HBSS to achieve a concentration of 20mU/ml, aliquot, and store at −20°C.
3. *Serum-containing medium*: SM is prepared as described earlier.
4. *Trypsin/EDTA*: 0.05% trypsin with 0.53 mM EDTA
5. *Fetal bovine serum (FBS)*

Steps

1. Suspensions of mouse cells are prepared as described earlier. Contents of six mouse bones (two each of femora, tibiae, and humeri, approximately $6–8 \times 10^7$ nucleated cells) are plated per 75-cm^2 flask. Suspensions from human surgical specimens are plated at 5×10^6 to 5×10^7 nucleated cells, and suspensions from aspirates are plated at 5×10^6 to 20×10^7 nucleated cells in 75-cm^2 flasks or 150-mm^2 dishes containing 30–50 ml of SM.

2. Cells are cultured at 37°C in a humidified atmosphere of 5% CO_2 with air, and medium is replaced on day 1 for human aspirates and at day 7 for all cultures. Passage generally is performed on days 12 to 14.

3. The resulting mouse cultures are passaged by (a) washing twice with HBSS, (b) incubating with chondroitinase ABC for 25–35 min at 37°C, (c) washing with HBSS, (d) treatment with trypsin/EDTA for 25–30 min at room temperature, (e) a second treatment with trypsin/EDTA for 25–30 min at 37°C, and (f) a final wash with SM. Steps b and c are omitted after passages greater than two. Human cultures are washed twice with HBSS and treated with two consecutive applications of trypsin/EDTA for 10 to 15 min each at room temperature, followed by a wash with SM.

4. Cold FBS is added to each fraction as collected (final concentration 1%) to inhibit enzymatic activity. Fractions are combined, pipetted vigorously to break up cell aggregates, and centrifuged at 135 g for 10 min, and the cell pellet is resuspended in fresh SM. Mouse cells are plated at $2–10 \times 10^6$ cells per 75-cm^2 flask depending on hematopoietic cell numbers. The next passage is performed when cultures approach confluency. Human cells are plated at 2×10^6 cells per 75-cm^2 flask or 150-mm dish. Upon reaching approximately 70% confluency, cells are passaged using the same procedure.

D. Establishment of Single Colony-Derived Strains of BMSCs

While multicolony derived strains of BMSCs take on a homogeneous appearance after passaging, and their differentiation potential can be characterized en masse, examination of colonies that form when bone marrow suspensions are plated at low density (as in the CFE assay, Fig. 1A) shows a great deal of heterogeneity in the starting population. There is a marked difference in the growth rate of the colonies, as demonstrated by colonies of different size. Furthermore, colonies are formed by cells with varying morphologies, ranging from a long, spindle shape (Fig. 1B) to a more compact shape (Fig. 1C) and to a very flat and spread morphology (Fig. 1D). Based on this heterogeneity, a number of studies have focused on the characterization of single colony-derived strains, prepared as described later (Kuznetsov *et al.*, 1997b), in order to better understand the hierarchy of BMSCs and their differentiation potential.

Solutions

1. *Serum-containing medium*: SM is prepared as described earlier.
2. *Hanks' balanced salt solution*
3. *Trypsin/EDTA*

Steps

1. Mouse cells, $6–15 \times 10^5$ nucleated cells, are plated in 150-mm petri dishes in order to prepare single colony-derived strains. From human surgical specimens, $0.007–3.5 \times 10^3$ nucleated cells/cm^2, and from aspirates, $0.14–14.0 \times 10^3$ nucleated cells/cm^2, are plated in 150-mm dishes containing 30–50 ml of SM. Cells may also be plated by limiting dilution in 96-well microtiter plates.

2. After adhesion for 2–3 hr, cultures are washed vigorously, and irradiated guinea pig cells are added to mouse cultures as described earlier.

3. After 14 to 16 days, cultures are inspected visually, and well-separated colonies of perfectly round shape are identified for cloning. The medium is removed, cultures are washed with HBSS, and individual colonies are surrounded by a cloning cylinder attached to the dish with sterilized high vacuum grease.

4. Cells inside the cylinder are treated with two consecutive aliquots of trypsin/EDTA for 5–10 min each at room temperature. In both cases, the released cells are transferred to individual wells of 6-well plates containing SM.

5. Subsequent passage is performed before cells reach confluence, usually 5 to 10 days later. Each strain

is passaged consecutively to a 25-cm^2 flask (second passage) and to a 75-cm^2 flask (third passage).

E. Cartilage Formation by BMSCs in Micromass (Pellet) Cultures

Cartilage formation by BMSCs is generally performed *in vitro* using high-density "pellet" cultures, which generate a relatively anaerobic environment that is conducive for chondrogenesis, along with a chondrogenic medium. The following procedure is essentially as described by Johnstone *et al.* (1998), although it has been suggested that cells grown from day 0 with basic fibroblast growth factor (FGF-2) display more chondrogenic potential (Muraglia *et al.*, 2003).

Solution

1. *Chondrogenic medium*: Coon's modified Ham's F-12 medium is supplemented with $10^{-6}M$ bovine insulin, $8 \times 10^{-8}M$ human apo-transferrin, $8 \times 10^{-8}M$ bovine serum albumin, $4 \times 10^{-6}M$ linoleic acid, $10^{-3}M$ sodium pyruvate, $10\,ng/ml$ rhTGFβ1, $10^{-7}M$ dexamethasone, $2.5 \times 10^{-4}M$ ascorbic acid, or media with similar formulation.

Steps

1. Either multicolony-derived or single colony-derived BMSCs (2.5×10^5) are centrifuged at $500\,g$ in 15-ml polypropylene conical tubes in 5 ml of chondrogenic medium.

2. Cultures are incubated with caps partially unscrewed for 3 weeks at 37°C in 5% CO_2, with a medium change at 2- to 3-day intervals. Pellets should not be attached to the tubes.

3. At harvest, pellets are washed with phosphate-buffered saline (PBS), fixed in 4% neutral-buffered formaldehyde for 2 h, and embedded in paraffin for histological analysis (Figs. 3A and 3B).

F. Preparation of BMSC and Hydroxyapatite/Tricalcium Phosphate Constructs

In vivo transplantation of BMSCs has become the gold standard by which to measure their multipotentiality (see Fig. 2). In conjunction with hydroxyapatite/tricalcium phosphate (HA/TCP) or collagen sponges as described by Krebsbach *et al.* (1997), BMSCs have the ability to form bone, myelosupportive stroma, and adipocytes, thereby recreating an ectopic bone/marrow organ (ossicle) when transplanted subcutaneously into immunocompromised mice. It

bone marrow single cell suspension

multi- or single colony derived strains

micro-mass culture

attach to ceramic particles or collagen sponges

cartilage

bone, myelosupportive stroma, adipocytes

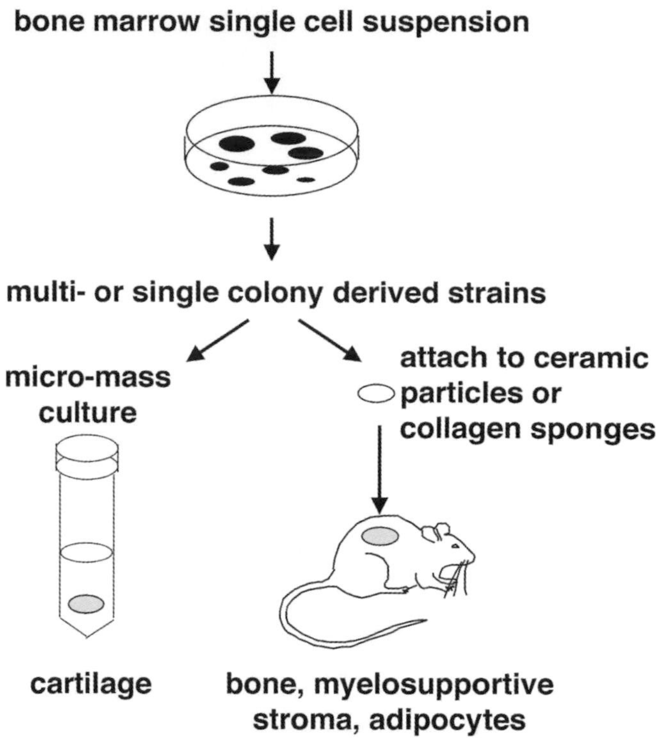

FIGURE 2 BMSC colonies can be passaged together to form multicolony-derived strains or isolated individually to form single colony-derived strains. Both types of cultures can be used to form cartilage in micromass (pellet) cultures in the presence of a chondrogenic medium and to demonstrate the ability to form bone, myelosupportive stroma, and adipocytes by *in vivo* transplantation in association with hydroxyapatite/tricalcium phosphate particles or collagen sponges.

should be noted that while human BMSCs are not as sensitive to culture conditions as murine BMSCs, they are more sensitive to the substrates used for *in vivo* transplantation. To date, HA/TCP particles appear to provide the best substrate for ossicle formation by human BMSCs (Figs. 3C and 3D).

Solutions

1. *Serum-containing medium*: SM is prepared as described earlier.

2. *Mouse fibrinogen*: Mouse fibrinogen is reconstituted in sterile PBS at 3.2 mg/ml, aliquoted, and kept at −80°C.

3. *Mouse thrombin*: Mouse thrombin is reconstituted in sterile 2% $CaCl_2$ at 25 U/ml, aliquoted, and kept at −80°C. (Because the salt is $CaCl_2 \cdot 2H_2O$, in order to prepare 2% solution, 2.65 g of the salt should be diluted in 100 ml of water.)

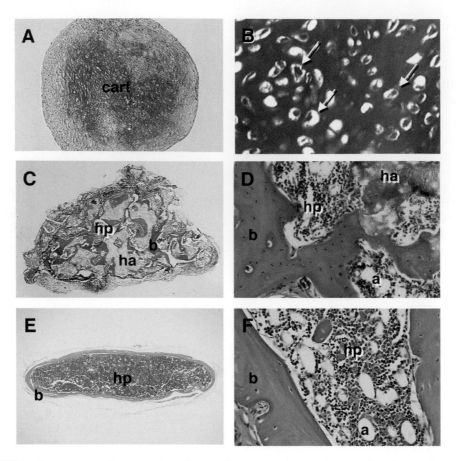

FIGURE 3 In micromass (pellet) cultures in chondrogenic medium, human BMSCs form a dense extracellular matrix that is metachromatic when stained with toluidine blue (A) and features chondrocyte-like cells embedded in this matrix (arrows in B). When transplanted in conjunction with hydroxyapatite/tricalcium phosphate particles (ha), human BMSCs form bone (b) that is actively deposited on the surface of the particles (C) and, with time, establish a fully functional hematopoietic marrow (hp), including adipocytes (a) as shown (D). Murine BMSCs are also able to form a cortex of bone (E) that surrounds a complete hematopoietic marrow with adipocytes when transplanted in collagen sponges (F).

Steps

1. Sterilization of HA/TCP particles is achieved by placing the particles into a glass bottle sealed with foil and heating at 220°C overnight (minimum 8 h). The particles are then aliquoted (40-mg/sterile round-bottomed centrifuge tube) in a hood using a sterile balance, sterile weighing spatula, and sterile filter paper.

2. Passaged BMSCs (see earlier discussion) are counted, pelleted at 135 g for 10 min, and resuspended in SM so that the volume in milliliters is equal to the number of transplants to be prepared.

3. HA/TCP powder is washed twice with SM, and the medium is discarded.

4. BMSCs designated for an individual transplant ($1-2 \times 10^6$ cells in 1 ml of SM) are transferred into a tube with HA/TCP powder. The powder is mixed with the cell suspension and is incubated at 37°C for 70–100 min with slow rotation (25 rpm).

5. Particles with adherent BMSCs are collected by a brief centrifugation (135 g for 1 min), and the supernatant is removed carefully.

6. Mouse fibrinogen (15 µl) is added to the cell/particle mixture, and the components are mixed gently. Mouse thrombin (15 µl) is added to the cell/particle mixture, and the components are mixed gently.

7. The tubes are left for a few minutes at room temperature. After a clot is formed, the cap is closed tightly to prevent the transplant from drying. The resulting fibrin clot with HA/TCP powder and BMSCs has the consistency of a gel; it can be easily taken out of the tube with a spatula and placed into a recipient animal as described later.

G. Preparation of BMSC and Collagen Sponge Constructs

Murine BMSCs generate complete ossicles in conjunction with HA/TCP and also when transplanted in collagen sponges (Gelfoam) (Figs. 3E and 3F), whereas human BMSCs only generate limited amounts of bone in the latter substrate.

Solution

1. *Serum-containing medium*: SM is prepared as described earlier.

Steps

1. Sterile Gelfoam sponges are cut into cubes of the desired size or into any other shape, placed into SM, and squeezed with forceps so that they regain their full size but with no air bubbles left inside.

2. BMSCs designated for individual transplants are transferred ($1–2 \times 10^6$ cells/1 ml of SM) into an individual Eppendorf tube, pelleted at $135\,g$ for $10\,min$, and most of the supernatant is discarded. The volume left should be smaller than the volume of the sponge (usually about $50\,\mu l$). The pellet is resuspended in this small volume.

3. Individual sponges are blotted between two sheets of sterile filter paper and are placed immediately into freshly resuspended cells in an Eppendorf tube where it expands, capturing the cell suspension. The tube is sealed tightly to prevent drying.

4. The sponges are incubated at $37°C$ for $90\,min$.

5. Sponges with cells are transplanted as described later.

H. In Vivo Transplantation of BMSC Constructs

Solutions

1. *Anesthesia*: Combine $225\,\mu l$ ketamine, $69\,\mu l$ of xylazine, $75\,\mu l$ of acepromazine, and $231\,\mu l$ of H_2O (total volume = $600\,\mu l$). Use $100\,\mu l$/mouse ($25\,g$). If mice are smaller, less anesthesia should be used.

2. *Betadine*

3. *70% ethanol*

Steps

1. Anesthetize the mouse and clean the skin with betadine and 70% ethanol. A single 3-cm-longitudinal incision is made in the skin along the dorsal surface.

2. The tip of the dissecting scissors is used to make a pocket for the transplant by inserting the scissor subcutaneously and then opening the scissors approximately 1 cm. A sterile spatula is used to insert HA/TCP

transplants, and forceps are used to insert collagen sponge transplants. The procedure is repeated twice on each side of the incision for a total of four pockets. The incision is closed with several autoclips. Autoclips are not removed due to the fact that it causes extensive bleeding in immunocompromised mice.

3. Transplants are harvested at various time points, fixed with 4% neutral-buffered formaldehyde overnight, decalcified, and embedded in paraffin for standard histological analyses.

4. Because both HA/TCP and collagen sponge transplants are open systems, there is no barrier to prevent host cells from invading transplants. Determination of cells of donor origin in transplants generated by murine cells requires that the donor bear a marker (e.g., lacZ, green fluorescent protein, or a transgene) (Krebsbach *et al.*, 1997). When human cells are used, antibodies that recognize human proteins but not mouse analogs are commonly used (Krebsbach *et al.*, 1997). *In situ* hybridization using human-specific Alu sequences as the probe has been particularly useful in characterizing transplants generated by human cells (Kuznetsov *et al.*, 1997).

IV. COMMENTS

The methods of *ex vivo* expansion presented here are expressly for maintaining BMSCs in an uncommitted state. Subsequently, it is possible to demonstrate that these cells, either multicolony-derived or some single colony-derived strains, have the ability to generate multiple phenotypes. Their chondrogenic potential can be demonstrated *in vitro* by the establishment of high-density pellet cultures with a chondrogenic medium, and their ability to form bone, myelosupportive stroma, and adipocytes is best assessed by transplantation *in vivo*. These results demonstrate that a true postnatal stem cell exists within the population of BMSCs, with well-known implications in the emerging field of regenerative medicine. Furthermore, *in vivo* transplantation of BMSCs bearing gene defects, either occurring naturally or created by molecular engineering, provides the opportunity to study the specific role of a gene in the process of establishing a bone/marrow organ (Bianco and Robey, 1999).

While the differentiation capacity of BMSCs is best evaluated by *in vivo* transplantation, there are many methods for inducing an osteogenic phenotype *in vitro*. However, it must be noted that these methods result in the formation of a tissue that does not have the structural organization of bone that is formed *in vivo*; in many cases, mineralization is due to dystrophic calcification as opposed to true bone formation. Adipogenesis can also be induced *in vitro* by a variety of different culture modifications (reviewed in Bianco *et al.*, 1999), but again, the adipocytes that are formed tend to be multivacuolar (immature), whereas mature adipocytes in marrow are univacuolar. Nonetheless, cultures of BMSCs provide the opportunity to study the effects of extrinsic signals that cause these cells to shift from one phenotype to another and to analyze the resultant changes in the pattern of gene expression.

V. PITFALLS

1. The specific lot of fetal bovine serum used is critical not only for the determination of CFE, but also for the *ex vivo* expansion of BMSCs. It is not well recognized that fetal bovine serum must be tested extensively to select lots that are suitable for one animal species or another.

2. In murine BMSC cultures (and from other rodents), macrophages represent a major contaminant in the adherent population. In low-density cultures, clusters of macrophages are often mistaken by the untrained eye for a colony established by a CFU-F. In cultures established by high-density plating, ~11% of the adherent population are macrophages (as identified by α-naphthyl acetate esterase activity) at the second passage, and further passage or cell sorting is needed to eliminate them (Krebsbach *et al.*, 1997). The contamination by macrophages in murine cultures not only has an impact on determination of CFE, but also on the interpretation of a variety of *in vitro* studies intending to determine the direct effect of factors on BMSCs and *in vivo* studies aimed at identifying where BMSCs engraft and what cell types they form after systemic injection.

3. In establishing BMSCs from bone marrow aspirates, the volume should not exceed 0.5 cc from any one site, or without repositioning the needle. Drawing larger volumes results in contamination of the marrow by peripheral blood, which influences the determination of CFE. Furthermore, peripheral blood has a negative influence on the growth of BMSCs.

References

Bianco, P., Riminucci, M., Kuznetsov, S., and Robey, P. G. (1999). *Crit. Rev. Eukaryot. Gene Expr.* **9**, 159–173.

Bianco, P., and Robey, P. (1999). *J. Bone. Miner. Res.* **14**, 336–341.

Bianco, P., and Robey, P. G. (2004). Skeletal stem cells. *In* "Handbook of Adult and Fetal Stem Cells" (R. P. Lanza, ed.). Academic Press, Sam Diegs.

Friedenstein, A. J. (1976). *Int. Rev. Cytol.* **47**, 327–359.

Friedenstein, A. J. (1980). *Hamatol. Bluttransfus.* **25**, 19–29.

Friedenstein, A. J. (1990). Osteogenic stem cells in bone marrow. *In* "Bone and Mineral Research" (J. N. M. Heersche, J. A. Kanis, eds.), pp. 243–72. Elsevier, New York.

Friedenstein, A. J., Latzinik, N. V., Gorskaya Yu, F., Luria, E. A., and Moskvina, I. L. (1992). *Bone Miner.* **18**, 199–213.

Johnstone, B., Hering, T. M., Caplan, A. I., Goldberg, V. M., and Yoo, J. U. (1998). *Exp. Cell Res.* **238**, 265–272.

Krebsbach, P. H., Kuznetsov, S. A., Satomura, K., Emmons, R. V., Rowe, D. W., and Robey, P. G. (1997). *Transplantation* **63**, 1059–1069.

Kuznetsov, S., and Gehron Robey, P. (1996). *Calcif. Tissue Int.* **59**, 265–270.

Kuznetsov, S. A., Friedenstein, A. J., and Robey, P. G. (1997a). *Br. J. Haematol.* **97**, 561–570.

Kuznetsov, S. A., Krebsbach, P. H., Satomura, K., Kerr, J., Riminucci, M., *et al.* (1997b). *J. Bone Miner. Res.* **12**, 1335–1347.

Muraglia, A., Corsi, A., Riminucci, M., Mastrogiacomo, M., Cancedda, R., *et al.* (2003). *J. Cell Sci.* **116**, 2949–2955.

10

Establishment of Embryonic Stem Cell Lines from Adult Somatic Cells by Nuclear Transfer

Teruhiko Wakayama

I. INTRODUCTION

Usually, embryonic stem (ES) cells are established from fertilized embryos. Recent studies have shown that it is possible to produce cloned embryos from adult somatic cell nuclei; even with a low success rate, some of those embryos can develop to full term (Wakayama *et al.*, 1998, 2000a). Now, using this technique, it has become possible to create a new embryonic stem (ES) cell line via nuclear transfer (ntES cell line) from adult somatic cells (Munsie *et al.*, 2000; Wakayama *et al.*, 2001). These ntES cells differentiate not only all three embryonic germ layers *in vitro*, but also germ cells *in vivo*, and some ntES cell nuclei can develop to full term after a second nuclear transfer (Wakayama *et al.*, 2001). The ntES cell techniques could be useful for research tools used in reprogramming, imprinting, and gene modification (Wakayama *et al.*, 2002). "Whereas the ntES technique is expected to have applications in regenerative medicine without immune rejection, we demonstrated that it is also applicable to the preservation of genetic resources of mouse strains instead of an embryo, oocyte, or spermatozoon. At present, this technique is the only one available for the preservation of valuable genetic resources from mutant mouse without germ cells (Wakayama et al., 2005a,b)."

II. MATERIALS AND INSTRUMENTS

A. Mouse Cloning Medium

KSOM (long-term culture medium) from Specialty media (#MR-106-D). All others are obtained from Sigma unless otherwise mentioned. NaCl (S7653), KCl (P9327), $CaCl_2 \cdot 2H_2O$ (C5080), $MgSO_4 \cdot 7H_2O$ (M5921), KH_2PO_4 (P5379), D-glucose (G8769), $NaHCO_3$ (S5761), sodium lactate 60% (ml/liter) (L1375), HEPES·Na (H3784), EDTA·2Na (ED2SS), phenol red (P0290), cytochalasin B (C6762), $SrCl_2$ (S0390), polyvinylpyrrolidone (PVP; 360kD) (PVP-360), polyvinyl alcohol (PVA) (P8136), bovine serum albumin (BSA, A3311), dimethyl sulfoxide (DMSO, D8779), hyaluronidase (H4272), and mineral oil (M5310).

B. ES Cell Establishment and Medium Maintenance

Acidic tyrode solution (MR-004-D), DMEM (SLM-120-B), phosphate-buffered saline (PBS) Ca/Mg free (BSS-1006-B), 0.1% gelatin (ES-006-B), nucleosides (ES-008-CD), nonessential amino acid (TMS-001-C), 2-

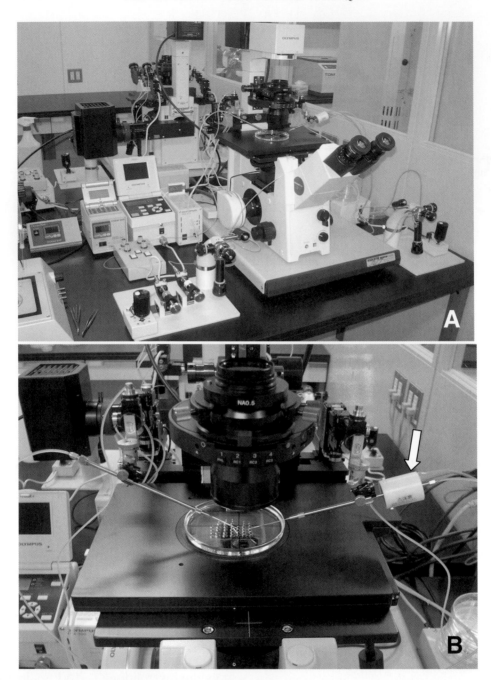

FIGURE 1 Micromanipulator and piezo system. (A) Equipment includes an inverted microscope with Hoffman optics, two injectors, air cushion, and warming plate on the stage of the microscope. (B) Piezo impact unit attached to the injection pipette holder (arrow).

mercaptoethanol (ES-007-E), penicillin–streptomycin solution (TMS-AB2-C), trypsin/EDTA solution (SM-2000-C), mitomycin C (Sigma, 0503): fetal calf serum (FCS, ES-009-B), KSR (invitrogen, 10828-028) and Lif (GIBCO #13275-029); Specialty media (http://www.specialitymedia.com) can provide most of these solutions.

C. Instruments

Inverted microscope with Hoffman optics is from Olympus (IX71) or Nikon (TE2000). Micromanipulator set is from Narishige (MMO-202ND) (Fig. 1A). Piezo impact drive system is from Prime Tech (PMM-150FU) (Fig. 1B). Pipette puller is from Sutter Instrument

TABLE I Formulations of Mouse Cloning Medium (mg/100 ml)[a]

Type of medium	CZB–HEPES Nuclear injection	CZB–PVP Donor cell diffusion	CZB–CB Oocyte enucleation	CZB–Sr Oocyte activation
NaCl	479	479	479	479
KCl	36	36	36	36
CaCl$_2$·2H$_2$O	25	25	25	—
KH$_2$PO$_4$	16	16	16	16
MgSO$_4$·7H$_2$O	29	29	29	29
D-Glucose	100	100	100	100
Sodium lactate 60% syrup	0.53 ml	0.53 ml	0.53 ml	0.53 ml
EDTA2Na	4	4	4	4
NaHCO$_3$	42	42	42	210
HEPESNa	520	520	520	—
Polyvinyl alcohol	10	10	10	10
Phenol red	1	1	1	1
Polyvinyl pyrrolidone		12000		
Cytochalasin B			0.5	
SrCl$_2$				2.7
BSA				500

[a] CZB medium is used with slight modifications for nuclear transfer.

Company (P-97). CO$_2$ incubator, centrifuge, pipettes, and tissue culture dishes/flasks are also needed.

III. PROCEDURES

A. Preparation of Medium

1. KSOM

Thaw the freezing KSOM medium inside a refrigerator overnight. After thawing, aliquot in 5-ml disposable tube and keep at 4°C. Make about 10 tubes and use a new tube every day. Tubes can be used 2 weeks after thawing.

2. CZB–HEPES (Embryo Handling and Nuclear Injection Medium)

Basically, the CZB medium (Chatot et al., 1990) is used with slight modifications during in vitro manipulations. As shown in Table I, all drugs must mix into 99 ml distilled water (finally becoming 100 ml). After everything is mixed adjust pH to about 7.4–7.6 using 1 N HCl and 1 N NaOH. Sterilize with 0.45-μm filters and aliquot into 5-ml tubes the same as KSOM.

3. CZB–PVP (Donor Cell Diffusion and Pipette-Washing Medium)

Add PVP in CZB–HEPES medium and keep in the refrigerator overnight and then filtrate and aliquot in a 1-ml tube.

4. CZB–CB (Oocyte Enucleation Medium)

Dissolve 1 mg cytochalasin B in 2 ml DMSO as stock solution (100 times concentration) and store at –20°C. Take 10 μl of this stock solution and mix with 990 μl of CZB–HEPES.

5. CZB–Sr (Oocyte Activation Medium)

As shown in Table I, this medium is similar to CZB but includes SrCl$_2$ and BSA instead of calcium and HEPES. After mixture, filtrate and aliquot in a 1-ml tube.

6. For ES Cell Establishment

All media can be obtained from Specialty media.
"After mixing all solutions, add FCS (for culture, 20%) or knockout serum (KSR; for ntES cell establishment, 20%) and filtrate."

B. Preparation of Micromanipulation Pipette and Manipulation Chamber

1. *Holding pipette*: The diameter of the outer pipette should be smaller than the oocyte. If the holding pipette is larger than the oocyte, you will lose the oocyte metaphase II spindle when the oocyte is caught by the holding pipette.

2. *Enucleation/injection pipette*: Break the tip of the pipette at flat, vertical and blunt ends using a microforge (Narishige Cat. No. MF-900) (Fig. 2A). A notched tip often kills the oocyte during enucleation and injection. The inner diameter of the enucleation pipette is about 8 μm and the injection pipette depends on the cell type (e.g., for cumulus cell: 5–6 μm, for fibroblast or ES cell: 6–7 μm, for G2/M phase cell: 7–8 μm).

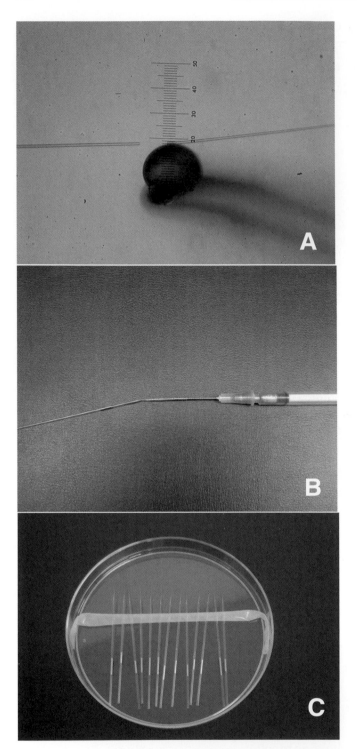

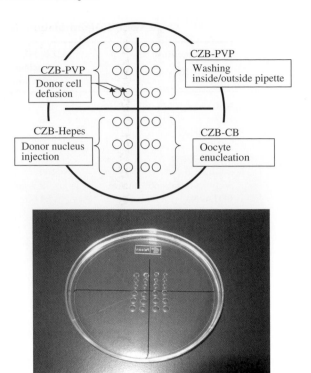

FIGURE 3 Manipulation chamber. Right above the line is CZB–PVP medium for washing pipettes. Right under the line is CZB–CB medium for oocyte enucleation. Left above the line is CZB–PVP medium. Top two drops and pick up nucleus are washing and last drop is donor cell diffusion. Left under the line is CZB–HEPES medium for donor nuclei injection into enucleated oocytes.

enhances the piezo impact power and control (Fig. 2B).

5. *Manipulation chamber*: As shown in Fig. 3, different media were put on the dish and then covered with mineral oil. This chamber can be use for both enucleation and injection.

6. Attach these pipettes to the pipette holder of the piezo impact drive unit of the micromanipulator (Fig. 1B).

7. Expel the air, oil, and mercury from the enucleation/ injection pipette under CZB–PVP medium. Wash inside/outside of the pipette with the PVP medium, until inside wall of pipette becomes smooth. Without this washing, the enucleation/injection pipette becomes dirty quickly and needs to be changed.

C. Collection of Oocytes and Enucleation

1. Collect the oocyte/cumulus cell complex from ampullae of the oviduct 14–15 h after hCG injection and move that complex into hyaluronitase (0.1 %) containing CZB–HEPES medium. About 10 min later, pick up oocytes and keep in KSOM in the incubator.

2. Place about 20 oocytes into CZB–CB (enucleation medium) and keep in this medium at least 5 min before starting enucleation.

FIGURE 2 Injection pipette and storage. (A) Breaking the tip of a pipette using a microforge. The inner diameter of the tip depends on donor cell size. (B) Inject a small mount of mercury into the pipette using a 1-ml syringe with a 26-gauge needle. (C) Storage of the pipettes in a 10-cm dish. All pipettes are attached softly on sticky tape.

3. Bend the pipette very close to the tip at 15–20° by the microforge.

4. Load a small mount of mercury into the enucleation/injection pipette, which is not essential but

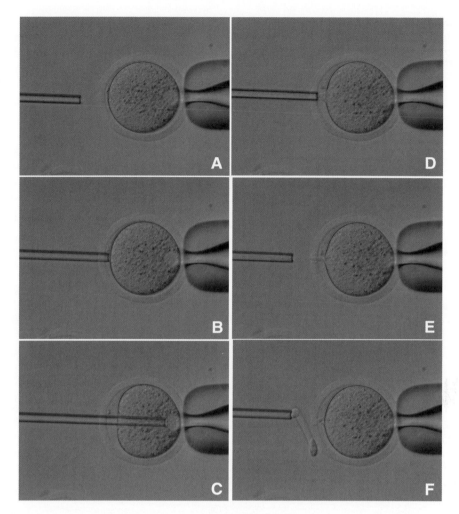

FIGURE 4 Oocyte enucleation. (A) Rotate and find the metaphase II spindle and place it in the 8 to 10 o'clock position. Then hold this oocyte on the holding pipette. (B) Cut the zona pellucida using an enucleation pipette with piezo pulses. (C) Insert enucleation pipette until it reaches the spindle. (D) Remove the spindle by suction without breaking the plasma membrane and gently pull away the pipette from oocytes. (E) The stretched cytoplasmic bridge from oocytes to pipette was pinched off. (F) Push out the spindle in order to check enucleation, which is harder than cytoplasm.

3. Find the metaphase II spindle inside the oocyte. Without any staining, the spindle can be recognized under Nomarski or Hoffman optics. Locate the spindle between the 2 and 4 o'clock positions and then hold on the holding pipette (Fig. 4A).

4. Using a few piezo pulses, cut the zona pellucida (Fig. 4B). To avoid damaging the oocyte by piezo pulses, make a large space between the zona pellucida and the oolemma.

5. Insert the enucleation pipette into the oocyte without breaking the oolemma (Fig. 4C) and remove the metaphase II spindle by aspirating with a minimal volume of oocyte cytoplasm. Oocyte membrane and spindle must be pinched off slowly, do not apply piezo pulses to cut the membrane (Figs. 4D and 4E).

6. Wash the enucleated oocytes several times to remove cytochalasin B completely and keep in KSOM medium in the incubator until used for donor cell injection.

D. Donor Cell Preparation

1. Choose the cell type: Cumulus cell (Wakayama *et al.*, 1998; Wakayama and Yanagimachi, 2001a), tail tip cell (probably fibroblast; Wakayama and Yanagimachi, 1999b; Ogura *et al.*, 2000a), Sertoli cell (Ogura *et al.*, 2000b), fetus cell (Wakayama and Yanagimachi, 2001b; Ono *et al.*, 2001), and ES cell (Wakayama *et al.*, 1999) have been successful in producing cloned mice. The methods are slightly different among those cell types, as shown in Table II. Cumulus cells are easy to prepare as donors. However, tail tip cells must be prepared at least 1 week before nuclear transfer, and then single cell treatment is required.

TABLE II   Type of Donor Cell and Preparation Method
of Single Cells

Donor cell type	Culture period	Single cell treatment
Cumulus cell	Fresh	No need
Sertoli cell	Fresh	Collagenase and washing
Tail tip cell	1–2 weeks	Tripsin and washing
Fetus cell	Fresh or 1–2 weeks	Tripsin and washing
ES cell	Any time	Tripsin and washing

2. To make single cells, remove cells from culture flask or dish by trypsin treatment. Because trypsin is very toxic at the time of nuclear injection, donor cells must be centrifuged at least three times to remove trypsin completely. However, cumulus cells can be used immediately without any treatment.

3. For diffusing donor cells, pick up 1–3 µl donor cell suspension and introduce into a CZB–PVP drop on the micromanipulation chamber (Fig. 3). Mix the donor cells and PVP medium gently using sharp tweezers. If there is not enough mixture, donor cells stick to each other and it becomes difficult to pick up a single cell (Fig. 5A). Do not scratch the bottom of the chamber.

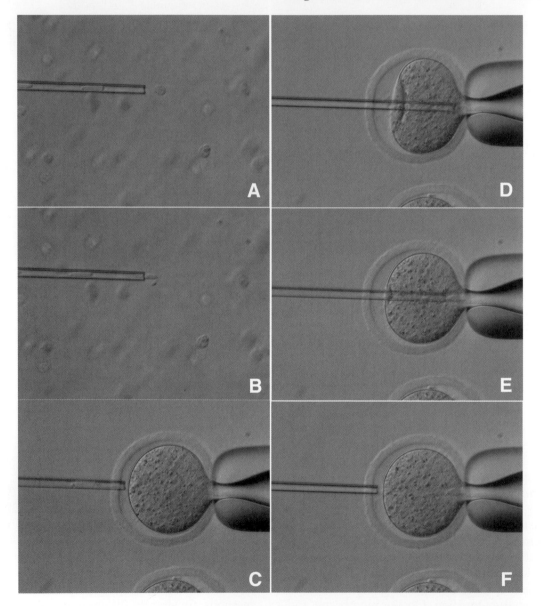

FIGURE 5   Donor nucleus injection. (A and B) Donor nuclei are aspirated in and out of the injection pipette gently until their nuclei are largely devoid of visible cytoplasmic material. (C) Hold the enucleated oocyte and cut the zona pellucida using piezo pulses. (D) Insert the injection pipette into enucleated oocyte. (E) Apply a single piezo pulse to break the membrane and immediately inject the donor nucleus. (F) Pull the pipette away gently.

E. Donor Cell Nucleus Injection

1. Place about 20 enucleated oocytes into CZB–HEPES (injection medium, Fig. 3).

2. Remove donor nuclei from cells and gently aspirate in and out of the injection pipette until the nucleus is devoid of visible cytoplasmic material (Figs. 5A and 5B). Put a few nuclei within the injection pipette.

3. Hold an enucleated oocyte using the holding pipette.

4. Cut the zona pellucida using a few piezo pulses as described in the enucleation method (Fig. 5C).

5. Push one nucleus forward until it is near the tip of the pipette and then advance the pipette through the enucleated oocyte until it almost reaches the opposite side of the oocyte cortex (Fig. 5D).

6. Apply one piezo pulse to puncture the oolemma at the pipette tip, which is indicated by a rapid relaxation of the oocyte membrane (Fig. 5E). Expel a donor nucleus into the enucleated oocyte cytoplasm immediately with a minimal amount of PVP medium. Gently withdraw the injection pipette from the oocyte (Fig. 5F).

7. After injection, leave the injected oocyte for at least 10 min at room temperature. Then transfer the oocytes into KSOM medium and store in the incubator until they are activated with strontium.

F. Oocyte Activation and Culture

1. Nuclear transferred oocytes should be cultured in KSOM medium for at least 30 min before activation. During this time, donor nuclei may be reprogrammed under oocyte cytoplasm (Wakayama et al., 1998, 2000b).

2. Activate the oocytes in CZB-Sr medium for 6 h with medium containing 2–10 mM strontium chloride and 5 μg/ml cytochalasin B (Fig. 6). Strontium activates oocytes, and cytochalasin B prevents extrusion of the donor chromosomes as a polar body (Wakayama et al., 1998).

3. Following activation, transfer all embryos and wash several times in KSOM medium to remove cytochalasin B completely (Wakayama and Yanagimachi, 2001b). Then culture until cloned embryos develop to the blastocyst stage. "Rarely, cytochalasin B seeps from activation drops and prevents cloned embryo development. In this case, it is better to culture all cloned embryo in another dish." These embryos may develop to full term when transferred into a pseudo-pregnant mother with about a 1–2% success rate (Wakayama and Yanagimachi, 2001a).

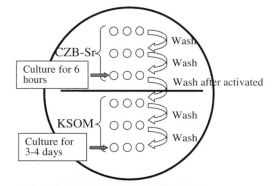

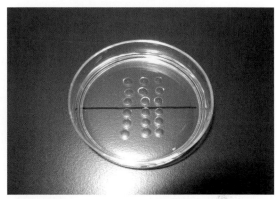

FIGURE 6 Reconstructed oocyte activation medium. Above the line is CZB–Sr medium, which activate the oocytes. The top two drops are used for washing oocytes to remove calcium from medium. Under the line is KSOM medium. The top two drops are used for washing activated oocytes and to remove A strontium and cytochalasin B.

G. Establishment of ntES Cell Line from Cloned Blastocysts

1. To make an embryonic feeder cell collect day 12.5 to 13.5 dpc fetuses from pregnant mother and then remove the head and internal organs on a 10-cm petri dish containing PBS. Place the embryos into a new 10-cm dish and mince the embryos into very small pieces with sterile scissors. Add 25 ml DMEM medium and plate into a larger size (175-cm²) tissue culture flask. One or 2 days later, split the cells 1:5 by trypsinization and allow them to grow to confluence.

2. Mitomycin C treatment. When the cells become confluent, treat with 10 μg/ml mitomycin C for 2 h in an incubator. Wash the flask several times with PBS to remove mitomycin C and then collect the cells by trypsinization. Pellet the cells by centrifugation (1000 rpm for 10 min). Aspirate the supernatant and gently resuspend the cell pellet in freezing medium (final concentration about 1×10^6 cells/ml) and divide into cryovials. Place the vials in a −80°C freezer overnight; the next day, vials can be transferred to liquid nitrogen for long-term storage.

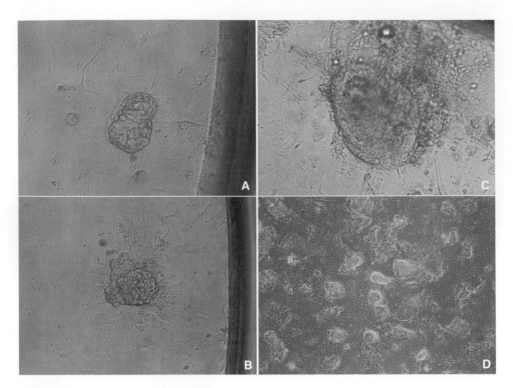

FIGURE 7 Establishment of ntES cell line. (A) A cloned blastocyst was attached to the feeder cell in the 96-well dish. (B) Spreading the trophoblast cells and an inner cell mass (ICM) appearing. (C) ICM grew up almost 5 to 10 times as large. (D) After trypsinization, some wells show a nearly established ntES cell line.

3. Thaw the vial quickly and transfer cells into a 15-ml tube filled with ES medium. Pellet the cells by centrifuging at 1000 rpm for 10 min and then aspirate the supernatant and resuspend with fresh ES medium. Plate on a 96-well multidish and culture in an incubator until use. This feeder cell preparation should be done at least 1 day before plating the cloned blastocysts.

4. Remove the zona pellucida from the cloned blastocysts using acidic tyrode solution. The zona pellucida dissolves quickly within 30 s.

5. Wash the cloned blastocysts several times in KSOM medium and then plate them into a 96-well multidish containing a performed feeder layer. Culture the multidish about 10 days. During this period, cloned blastocysts attach to the surface of the feeder layer and the inner cell mass (ICM) can be seen to grow (Figs. 7A and 7B).

6. Some wells of the multidish may have clumps of large ICM (Fig. 7C). If these clumps appear, treat with trypsin and disaggregate the clump using a 200 μl pipettor (Nagy *et al.*, 2003). Then replate the suspension into another well of the same multidish. Several days later, if ICM-like clumps appear on the dish, replate again the same as before.

7. When ES-like cell colonies dominate the well (Fig. 7D), the cells should be expanded gradually into 48-, 24-, and 12-wells and then into 12.5-, 25-, and 75-cm^2 flasks. After the cell number has increased greatly, freeze and store the cells the same as feeder cells.

IV. COMMENTS

1. So far many types of somatic cell have been used as donors, but only a few cell types have demonstrated the capacity of full-term developments.

2. For nuclear transfer, only 100 to 150 cells were used as donors. However, to pick up better cells, donor cells must be prepared with high concentrations (more than 10^5 cells) on the manipulation chamber.

3. When using cumulus cells, these cells can be picked up immediately after collection without washing, as hyaluronidase is not toxic for oocyte injection.

4. Nuclear injection by piezo is a very new method and very difficult. However, this is almost the same method as sperm injection into oocytes. You should begin with the sperm injection experiment before starting the nuclear transfer experiment. Sperm injection will help you understand the mechanisms of the piezo system.

V. PITFALLS

1. Washing the pipette by PVP is very important. This process affects not only the survival rate, but also development after nuclear transfer.

2. During strontium treatment, some eggs die and the medium becames dirty. This is a normal process and surviving oocytes suffer no damage by strontium treatment.

3. It will probably take a few months to generate data due to incomplete technical skills (Perry *et al.*, 2002). Without hard practice, the establishment of a ntES cell line is impossible.

References

Chatot, C. L., Lewis, J. L., Torres, I., and Ziomek, C. A. (1990). Development of 1-cell embryos from different strains of mice in CZB medium. *Biol. Reprod.* **42**, 432–440.

Munsie, M. J., Michalska, A. E., O'Brien, C. M., Trounson, A. O., Pera, M. F., and Mountford, P. S. (2000). Isolation of pluripotent embryonic stem cells from reprogrammed adult mouse somatic cell nuclei. *Curr. Biol.* **10**, 989–992.

Nagy, A., Gertsenstein, M., Vintersten, K., and Behringer, R. (2003). Manipulating the Mouse Embryo, 3rd Ed. Cold Spring Harbor Laboratory Press, Cold Spring Harbor, NY.

Ogura, A., Inoue, K., Ogonuki, N., Noguchi, A., Takano, K., Nagano, R., Suzuki, O., Lee, J., Ishino, F., and Matsuda, J. (2000a). Production of male cloned mice from fresh, cultured, and cryopreserved immature sertoli cells. *Biol. Reprod.* **62**, 1579–1584.

Ogura, A., Inoue, K., Takano, K., Wakayama, T., and Yanagimachi, R. (2000b). Birth of mice after nuclear transfer by electrofusion using tail tip cells. *Mol. Reprod. Dev.* **57**, 55–59.

Ono, Y., Shimozawa, N., Ito, M., and Kono, T. (2001). Cloned mice from fetal fibroblast cells arrested at metaphase by a serial nuclear transfer. *Biol. Reprod.* **64**, 44–50.

Perry, A. C., and Wakayama, T. (2002). Untimely ends and new beginnings in mouse cloning. *Nature Genet.* **30**, 243–244.

Rideout, W. M., III, Hochedlinger, K., Kyba, M., Daley, G. Q., and Jaenisch, R. (2002). Correction of a genetic defect by nuclear transplantation and combined cell and gene therapy. *Cell* **109**, 17–27.

Wakayama, S., Kishigami, S., Van, Thuan, N., Ohta, H., Hikichi, T., Mizutani, E., Yanagimachi, R. and Wakayama, T. (2005) Propagation of an Infertile Hermaphrodite Mouse Lacking Germ Cells, Using Nuclear Transfer and Embryonic Stem Cell Technology. *Proc. Natl. Acad. Sci. USA* **102**, 29–33.

Wakayama, S., Ohta, H., Kishigami, S., Van, Thuan, N., Hikichi, T., Mizutani, E., Miyake, M. and Wakayama, T. (2005) Establishment of Male and Female Nuclear Transfer Embryonic Stem Cell Lines from Different Mouse Strains and Tissues. *Biol. Reprod.* In press.

Wakayama, T., and Perry, A. C. F. (2002). Cloning mice: Perspective and prospective. *In* "Principles of Cloning" (J. Cibelli, R. P. Lanza, K. H. S. Campbell, and M. D. West, eds.), pp. 301–341. Academic Press, San Diego.

Wakayama, T., Perry, A. C. F., Zuccotti, M., and Yanagimachi, R. (1998). Full term development of mice from enucleated oocytes injected with cumulus cell nuclei. *Nature* **394**, 369–374.

Wakayama, T., Rodriguez, I., Perry, A. C. F., Yanagimachi, R., and Mombaerts, P. (1999). Mice cloned from embryonic stem cells. *Proc. Natl. Acad. Sci. USA* **26**, 14984–14989.

Wakayama, T., Shinkai, Y., Tamashiro, K. L. K., Niida, H., Blanchard, D. C., Blanchard, R. J., Ogura, A., Tanemura, K., Tachibana, M., Perry, A. C. F., Colgan, D. F., Mombaerts, P., and Yanagimachi, R. (2000a). Cloning of mice to six generations. *Nature* **407**, 318–319.

Wakayama, T., Tabar, V., Rodriguez, I., Perry, A. C. F., Studer, L., and Mombaerts, P. (2001). Differentiation of embryonic stem cell lines generated from adult somatic cells by nuclear transfer. *Science* **292**, 740–743.

Wakayama, T., Tateno, H., Mombaerts, P., and Yanagimachi, R. (2000b). Nuclear transfer into mouse zygotes. *Nature Genet.* **24**, 108–109.

Wakayama, T., and Yanagimachi, R. (1999). Cloning of male mice from adult tail-tip cells. *Nature Genet.* **22**, 127–128.

Wakayama, T., and Yanagimachi, R. (2001a). Mouse cloning with nucleus donor cells of different age and type. *Mol. Reprod. Dev.* **58**, 376–383.

Wakayama, T., and Yanagimachi, R. (2001b). Effect of cytokinesis inhibitor, DMSO and the timing of oocyte activation on mouse cloning using cumulus cell nuclei. *Reproduction* **122**, 49–60.

T-Cell Isolation and Propagation *in vitro*

Mads Hald Andersen and Per thor Straten

I. INTRODUCTION

Cellular immunity is largely based on T-lymphocytes. Like B cells, T cells also arise from the bone marrow. However, unlike B cells, they migrate to the thymus for maturation. A T-cell expresses a unique antigen-binding molecule called the T-cell receptor (TCR) on the cell surface. In contrast to membrane-bound antibodies on B cells, which can recognize the antigen alone, the majority of TCR recognizes a complex ligand, comprising an antigenic peptide bound to a protein called the major histocompatibility complex (MHC) [known in humans as human leukocyte antigen (HLA)] molecule (Moss *et al.*, 1992). When a T-cell encounters an antigen in the context of an HLA molecule, it undergoes clonal expansion and differentiates into memory and various effector T-cells: CD4+ T-helper cells and CD8+ cytotoxic T-lymphocytes (CTL). Activation of both humoral and cell-mediated parts of the immune response requires cytokines produced by T-helper cells. The activation of T-helper cells is carefully regulated, and naïve cells only become activated when they recognize an antigen presented by class II HLA molecules in context with the appropriate costimulatory molecules on the surface of professional antigen-presenting cells (macrophages, B cells, and dendritic cells) (Stockwin *et al.*, 2000).

Although antigen-presenting cells encounter and incorporate an antigen in many different compartments, the interaction with T-helper cells is largely confined to secondary lymphoid organs (Fu *et al.*, 1999). While the T-helper cells provide help to activate B-cells, antigen-presenting cells, and CTL, only the latter has a vital function in monitoring the cells of the body and in eliminating cells that display antigen, primarily virus-infected cells. However, in addition to providing protection against infectious agents, CTL are thought to provide some degree of protection against spontaneous tumors by virtue of their ability to detect quantitative and qualitative antigenic differences in transformed cells (Castelli *et al.*, 2000). Tumorigenic alterations result in an altered protein repertoire inside the cell. Class I MHC molecules sample peptides from protein degradation inside the cell and present these at the cell surface to CTL. Hence, this enables CTL to scan for cellular alterations.

Until recently, measurements of the levels of cellular immune responses, i.e., those mediated by CD4+ and CD8+ T-lymphocytes, have depended largely on culture *in vitro* and the subsequent measurement of specific functions (such as cytolysis). More recently, new technologies based around tetrameric class I peptide complexes (tetramers) have allowed immunologists to isolate and measure CD8+ T-lymphocyte levels directly *ex vivo*. This article describes measures used to generate and clone specific T-cells in culture as well as measures to isolate specific T-cells by means of recombinant HLA/peptide complexes. Finally, it describes the conventional chrome release assay and the ELISPOT assay for the measurement of specific T-cell immunity. It should be noted, however, that T-cells behave very differently and that a T-cell protocol should not be considered as the definitive receipt but rather as a guideline that can be altered depending on the target or the donor. Finally, the making of dendritic cells has been described in detail elsewhere in this volume. However, this is an important first step of making primary T-cells and is consequently mentioned shortly here.

II. MATERIALS AND INSTRUMENTATION

A. Tissue Culture Medium

X-vivo medium (Cambrex Bio Science, Cat. No. 04-418Q). Store at 4°C.

Human serum (Sigma, Cat. No. H1513). Store at −20°C.

Standard medium: X-vivo, 5% human serum. Store at 4°C.

RPMI 1640 medium (GIBCO, Cat. No. 61870-010). Store at 4°C.

Fetal calf serum (FCS) (Sigma, Cat. No. F7524). Store at 4°C.

R10 medium: RPMI 1640 + 10% FCS. Store at 4°C.

B. Cytokines

Interleukin (IL)-2 [Apodan, Cat. No. 004184]. Store at −80°C. Aliquot 20 units/µl.

IL-4 [Peprotech (trichem), Cat. No. 200-04]. Store at −20°C. Aliquot 10 units/µl.

IL-7 [Peprotech (trichem), Cat. No. 200-07]. Store at −20°C. Aliquot 5 ng/µl.

IL-12 [Peprotech (trichem), Cat. No. 200-12]. Store at −20°C. Aliquot 20 units/µl.

GM-CSF [Peprotech (trichem), Cat. No. 300-03]. Store at −20°C. Aliquot 800 units/µl.

TNF-α [Peprotech (trichem), Cat. No. 300-01A]. Store at −20°C. Aliquot 10 ng/µl.

PHA (Sigma, Cat. No.). Store at −20°C. Aliquot 0.9 mg/µl.

C. Antibodies

Anti-CD28 (eBioscience, Clone 28.8, Cat. No. 16-0288-81). Store at 4°C.

Anti-CD3 (eBioscience, Clone OKT3, Cat. No. 14-0037-82). Store at 4°C.

Tricolor-anti-CD8 (Caltag, Burlingame, CA, Cat. No.). Store at 4°C.

D. Sterile Plastic Plates

96-well plates [Boule (Corning Costar), Cat. No. 3799]
48-well plates [Boule (Corning Costar), Cat. No. 3548]
24-well plates [Boule (Corning Costar), Cat. No. 3526]
6-well plates [Boule (Corning Costar), Cat. No. 3516]

E. Additional Materials

Lymphoprep/Ficoll (Medinor, Cat. No. 30066.03). Store in the dark at 4°C.

[51]Crom (Amersham, Cat. No. CJS1) 5 mCi/1 ml. Store at −20°C. Dilute 1:5 in phosphate-buffered saline (PBS) before use.

Biotinylated monomers or PE-labeled tetramer (Proimmune, Oxford, UK). Store at 4°C.

Streptavin-coated magnetic beads (Dynabeads M-280, Dynal A/S, Cat. No. M-280). Store at 4°C.

F. Materials for ELISPOT

Nitrocellulose plates (Millipore MAIPN 4550)

Coating antibody: Mab anti-hIFN γ clone 1-D1K, 1 mg/ml, MABTECH 3420-3. Store at 4°C. Dilute to 7.5 µg/ml in PBS before use.

Secondary antibody: Biotinylated Mab anti-hIFN-γ, 7-b6-1, 1 mg/ml, MABTECH 3420-6. Store at 4°C. Dilute to 0.75 µg/ml in diluting buffer before use.

Streptavidin AP: Calbiochem CAL 189732, 2 ml. Add 2 ml H$_2$O and 2 ml glycerol (85%). Store at 4°C. Dilute 1:1000 in diluting buffer before use.

NBT/BCIP substrate system. DAKO, Cat. No. K 0598. Store at 4°C. Dilute 1:5 in substrate buffer before use.

In addition to standard cell culture instruments:

Gamma counter (Cobra 5005, Packard Instruments, Meriden, CT)

ELISPOT counter [ImmunoSpot Series 2.0 analyzer (CTL Analyzers, LLC, Cleveland, OH]

Magnetic isolator (Dynal A/S, Oslo, Norway)

FACSVantage (Becton-Dickinson, Mountain View, CA)

III. PROCEDURES

A. Induction of Specific T Cells as Primary Responses

The following is a description on how to grow antigen-specific T-cells using peptide-loaded dendritic cells as stimulator cells (Pawelec, 2000). This protocol can also be used to grow tumor-specific T cells if tumor lysate-loaded dendritic cells are used as stimulator cells as described.

1. Stimulator Cell (Dendritic Cell) Culture

Day −7

Dilute 50 ml blood 1:1 with RPMI medium and separate on lymphoprep by centrifugation for 30 min at 1200 rpm. Harvest mononuclear cells, mix with an equal volume of RPMI, and centrifuge at 1500 rpm for 10 min, followed by two washes in R10 (1200 rpm, 5 min). Resuspend cells to 20×10^6 cells/ml and plate out at 3 ml/well in 6-well plates. Incubate the cells for 2 h at 37°C. Remove nonadherent cells (lymphocytes) by gentle suction. If necessary, the lymphocytes may be frozen. Add 2.5 ml standard medium containing 800 units/ml GM-CSF and 500 units/ml IL-4 to each well.

Day –5

Add 2.5 ml standard medium containing 1600 units/ml GM-CSF and 1000 units/ml IL-4.

Day –3

Remove 2.5 ml of medium from each well and replace with 2.5 ml of fresh medium containing 1600 units/ml GM-CSF and 1000 units/ml IL-4.

Day –1

Remove 1 ml of medium from each well and replace with 1 ml of fresh medium containing 10 ng/ml of TNF-α.

Day 0

Harvest the cultured DC and wash twice in RPMI medium. Resuspend in 1 ml RPMI medium containing 50 μg/ml peptide and 3 μg/ml β_2m. Incubate the cells for 4 h at 37°C; gently resuspend every hour. Irradiate at 25 Gy, wash twice with RPMI medum, and resupend cells (stimulator cells) at 3×10^5/ml in standard medum.

2. Initiation of Primary T-Cell Culture

Mix 3×10^6 freshly isolated lymphocytes and 3×10^5 peptide-pulsed stimulator cells in a 24-well plate at 2 ml/well.

Day 1

Add IL-7 to a final concentration of 5 ng/ml and 100 pg/ml IL-12.

Day 7

Remove 1 ml of medium and replace with 1 ml of standard medium containing 20 ng/ml of IL-7.

Day 12

Harvest responder cells, separate over Ficoll, wash once, and count viable cells. Resuspend at 1.5×10^6/ml in standard medium and keep tube at 37°C. Thaw 2×10^6 autologous peripheral blood mononuclear cells (PBMC) per 1.5×10^6 responder cells. Wash the PBMC once in RPMI medium and irradiate at 60 Gy. Wash again in RPMI and incubate 2 h at 37°C with 20 μg/ml peptide and 2 μg/ml β_2m. Remove medium and gently wash once in RPMI. Mix 1.5×10^6/ml and 2×10^6 peptide-pulsed autologous PBMC in 2 ml of standard medium. Alternatively, instead of peptide-pulsed PBMC, use peptide-pulsed DC prepared as on Day 0.

Day 14

Remove 1 ml of medium and replace with 1 ml of fresh medium containing 40 units/ml IL-2 to each well.

Day 19

Restimulate as on day 12. Add IL-2 to the culture as on day 14. Restimulation is needed four or five times before a primary response is measurable.

B. Cloning of T Cells by Limiting Dilution

Day 0

Wash autologous PBMC once in RPMI medium and irradiate at 30 Gy. Wash again in RPMI and incubate 2 h at 37°C with 20 μg/ml peptide and 2 μg/ml β_2m. Plate T-cells at limiting dilution (featuring 10; 3; 1; 0.3 cells/well) in 96-well round-bottom microtiter plates containing 10^5 irradiated, autologous peptide-pulsed PBL, 10 units/ml IL-4, and 40 units/ml IL-2 in 100 μl standard medium.

Day 4

Add 50 μl standard medium containing IL-2 and IL-4 to a final concentration of 10 and 40 units/ml, respectively.

Day 8

Add 50 μl standard medium containing IL-2 and IL-4 to a final concentration of 10 and 40 units/ml, respectively.

Day 12

Inspect cells for growing cells microscopically. Transfer growing cells to 48-well plates containing peptide-pulsed autologous feeder cells and antigen. At the same time, test clones for antigen specificity by the chrome release assay or ELISPOT. Incubate plates 7 days at 37°C in a 5% CO_2 incubator adding IL-2 and IL-4 every third day and transfer antigen-specific T-cells to 24-well plates.

An Alternative Cloning Protocol Using Anti-CD3 and Anti-CD28 Antibodies

This procedure is modified from Oelke *et al.* (2000).

Day –1

Coat a 96-well plate with 100 ng/ml of anti-CD3 and anti-CD28 antibodies in PBS for 24 h at room temperature.

Day 0

Plate T cells at limiting dilution in precoated 96-well plates containing 10^5 irradiated, autologous PBL, 10 units/ml IL-4, and 40 units/ml IL-2 in 100 µl standard medium. Continue as described in previously.

C. Expansion of T-Cell Clones

This is a protocol for the expansion of already established clones modified from Dunbar *et al.* (1998). The cloning mix described in the following can, however, also be used as stimulators to clone T-cells instead of autologous PBL as described previously.

Day 1

For *preparing the cloning mix* isolate fresh lymphocytes from at least three individuals, resuspend them in standard medium, and irradiate (20 Gy). Count the lymphocytes and mix them together to give a final total concentration of 1×10^6/ml. Add PHA to a final concentration of 1 µg/ml and leave in the incubator while the clone is thawed and counted. Thaw the clone, count, and resuspend in prewarmed cloning mix at between 10^5 and 5×10^5 cellsml. Plate out into a U-bottomed, 96-well plate at 100 µl/well.

Day 3

Add 100 ml medium and 40 units/ml IL-2 to each well.

Day 7

Prepare fresh cloning mix, plate 1 ml per well into 24-well plates and prewarm in an incubator.

Transfer 1 well of the 96-well plate into 1 well of the prewarmed 24-well plate.

Day 10

Add 1 ml of medium and 40 units/ml IL-2 to each well.

Day 14

Pool the wells for each individual clone, count, and remove the required amount of cells for a chrome release assay to check if the clones have maintained antigen specificity. Freeze the remaining cells in aliquots of 2 to 7 million per vial.

D. Isolation of Peptide-Specific T Cells

Momomeric or tetrameric MHC/peptide complexes can be bought commercially, e.g., proimmune (Oxford, UK). They can, however, also be made as described by Pedersen *et al.* (2001). However, these rather complex procedures are not the scope of this review.

1. Isolation of Specific T-Cells Using MHC Class I/Peptide Complexes Coupled to Magnetic Beads

This procedure is performed according to Andersen *et al.* (2001; Schrama *et al.*, 2001). Incubate 2.5 µg biotinylated monomer of a given HLA/peptide complex with 5×10^6 streptavin-coated magnetic beads in 40 µl PBS for 20 min at room temperature. Wash the magnetic complexes three times in PBS using a magnetic field.

Add 10^7 PBL or lymphocytes in 100 µl PBS with 5% bovine serum albumin (BSA) and rotate very gently for 1 h. Wash the antigen-specific T-cells associating with the magnetic complexes gently three times in PBL. Incubate for 2 h at 37°C and resuspend several times in standard medium to release the cells from the magnetic beads.

Assay the antigen specificity of isolated antigen-specific CD8⁺ T-cells in an ELISPOT or chrome release assay after at least 5 days in culture.

2. Isolation of Specific T-Cells by FACS

This protocol is modified from Dunbar *et al.* (1999). Culture PBL overnight in standard medium before sorting. Alternatively, pulse PBL with 10 µM peptide in standard medium, plus 10 U/ml IL-7 and culture for 10 days before sorting. Stain cells with PE-labeled tetrameric HLA/peptide complex for 15 min at 37°C before the addition of tricolor-anti-CD8 for 15 min on ice. Wash the cells six times in PBS before analysis on FACS.

Gate specific T-cells according to tetramer/CD8 double staining by forward and side scatter profile. Sort single cells directly into U-bottom, 96-well plates and stimulate these as described in Section III,B.

E. Examination of Antigen Specificity

1. Crome Release Assay

Conventional ^{51}Cr release assays for CTL-mediated cytotoxicity can be used to test the specificity of CTL lines against relevant target cells, e.g., autologous EBV-transformed B-cell lines or cancer cell lines. This procedure is performed according to Brunner *et al.* (1968).

Label 10^6 target cells in 50 µl R10 medium with [^{51}Cr] (100 µCi) in a round-bottomed well of a 96-well plate at 37°C for 60 min. If necessary, add 4 µg of peptide. Wash the target cells four times and plate out in 96-well plates in 100 µl R10 medium. Add T-cells at various effector:target ratios in another 100 µl R10 and incubate at 37°C for 4 h. Aspirate 100 µl of medium and count [^{51}Cr] release in a gamma counter. Determine the maximum [^{51}Cr] release in separate wells by the addition of 100 μl 10% Triton X-100 and the spontaneous release by the addition of 100 μl R10 only to target cells.

Calculate the specific lysis using the following formula:

$$[(\text{experimental release} - \text{spontaneous release})/$$
$$(\text{maximium release} - \text{spoontaneous release})] \times 100.$$

2. ELISPOT

For the measurement of specific T-cell immunity, ELISPOT analysis, involving the incubation of primary PBMC with one or more peptide epitopes, is probably the most sensitive and reliable assay (Pittet *et al.*, 1999). ELISPOT is based on the detection of the peptide-induced release of cytokines such as interferon (IFN)-γ by single T-cells upon stimulation with a peptide antigen (Scheibenbogen *et al.*, 1997).

Single T-cells can be detected and quantified as cytokine spots on special nitrocellulose filter plates. Cytokine-specific antibodies are coated to the filters to capture the secreted cytokine; peptide-pulsed target cells are added together with cells containing the pre-

cursor T-cells. If a T-cell recognizes the peptide epitope examined, the cell releases cytokine. This can be detected as a spot by a colorimetric reaction with secondary antibodies, which represents the cytokine after secretion by a single activated cell. The principle of ELISPOT is illustrated in Fig. 1. If case responses are weak, a single round of *in vitro* stimulation can be used. For cytotoxic T-cells, IFN-γ, or granzyme B, ELISPOT can be performed; for T-helper cells, IFN-γ, or IL-4 (for T-helper 1-type and T-helper 2-type immunity, respectively) can be performed.

Solutions

PBS

Washing buffer: PBS and 0.05% Tween 20; store at room temperature

Diluting buffer: PBS, 1% BSA, and 0.02% NaN₃; store at room temperature

Substrate buffer: 0.1 M NaCl, 50 mM MgCl₂, and 0.1 M Tris–HCl, pH 9.50; store at room temperature

Coating of Plates

Add 75 μl 7.5 ug/ml coating antibody. The antibody concentration may change depending on the cytokine target. Leave overnight at room temperature (if coating

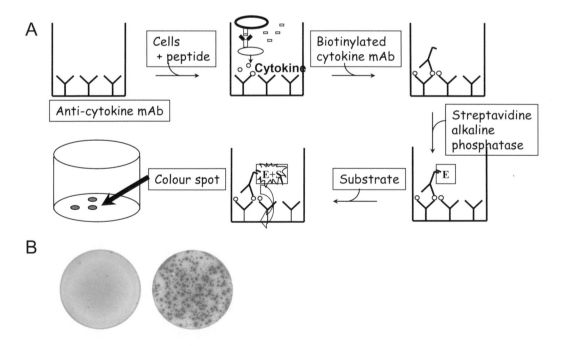

FIGURE 1   (A) Schematic illustration of the ELISPOT assay. Cytokine-specific antibodies are coated onto nitrocellulose filter plates to capture the secreted cytokine; a peptide-pulsed target cell is added together with cells containing the precursor T-cells. If a T-cell recognizes the peptide epitope examined, the cell releases cytokine. This can be detected as a spot by a colorimetric reaction with secondary antibodies. Thus, the spot represents the cytokine after secretion by a single activated cell. (B) ELISPOT wells after incubation with T-cells that were either nonreactive (*left*) or reactive (*right*) against the antigen examined.

3–5 days in advance, leave at 4°C). Wash the plate with 6 × 200 μl PBS. Block the plate with 200 μl media. Leave in incubator for 2 h.

Setting up the ELISPOT

Prepare the serial dilutions of the cells (sterile) at relevant concentrations in order to add the cells to each well in 200 μl media. Poor off the blocking media, and add 200 μl cells and 0.5 μl peptide (2 mM) to each well.

Leave overnight in incubator. (The plate must not shake during incubation.) Wash the plate with 6 × 200 μl washing buffer. Add 75 μl of the secondary antibody (1 μg/ml) to each well. The antibody concentration may change depending on the cytokine target. Incubate for 2 h at room temperature. Wash the plate with 6 × 200 μl washing buffer.

Add 75 μl streptavidin (diluted 1 : 1000) to each well. Incubate for 1 h at room temperature.

Wash the plate with 6 × 200 μl washing buffer and 1 × 200 μl substrate buffer. Mix the substrate: 10 ml substrate buffer + 44 μl NBT + 33 μl BCIP.

Add 75 μl fresh substrate to each well.

Leave at room temperature for 2–20 min.

Stop the reaction with tap water when spot development is satisfactory.

Count the spots using the ImmunoSpot Series 2.0 analyzer and calculate the peptide-specific T-cell frequency from the number of spot-forming cells.

IV. COMMENTS AND PITFALLS

It is always optimal to prepare T-cell cultures from fresh material. T-cells behave very differently, and it is always important to carefully inspect T-cell cultures daily in a microscope. Thus, it is always possible to alter the T-cell protocol depending on the target or the donor, e.g., the concentration of cytokines, the addition of new cytokines, the amount of antigen-presenting cells, the day of restimulation, and the addition of antibodies such as anti-CD28 anti-CD3, or other costimulatory factors. Additionally, T-cells should never be kept at too low cell densities, as cell-to-cell contact is very important. This is also why round-bottom wells are used instead of flat-bottom wells during the cloning of T-cells. Furthermore, when T-cells are transferred from small to larger wells, it is important to carefully inspect the cells microscopically to ensure that the cells are in good condition. In this regard, feeder cells (e.g., irradiated, autologous PBL) may be added if needed.

References

Andersen, M. H., Pedersen, L. O., Capeller, B., Brocker, E. B., Becker, J. C., and thor, S. P. (2001). Spontaneous cytotoxic T-cell responses against survivin-derived MHC class I-restricted T-cell epitopes in situ as well as ex vivo in cancer patients. *Cancer Res.* **61**, 5964–5968.

Brunner, K. T., Mauel, J., Cerottini, J. C., and Chapuis, B. (1968). Quantitative assay of the lytic action of immune lymphoid cells on 51-Cr-labelled allogeneic target cells *in vitro*; inhibition by isoantibody and by drugs. *Immunology* **14**, 181–196.

Castelli, C., Rivoltini, L., Andreola, G., Carrabba, M., Renkvist, N., and Parmiani, G. (2000). T-cell recognition of melanoma-associated antigens. *J. Cell Physiol.* **182**, 323–331.

Dunbar, P. R., Chen, J. L., Chao, D., Rust, N., Teisserenc, H., Ogg, G. S., Romero, P., Weynants, P., and Cerundolo, V. (1999). Cutting edge: Rapid cloning of tumor-specific CTL suitable for adoptive immunotherapy of melanoma. *J. Immunol.* **162**, 6959–6962.

Dunbar, P. R., Ogg, G. S., Chen, J., Rust, N., van der Bruggen, P., and Cerundolo, V. (1998). Direct isolation, phenotyping and cloning of low-frequency antigen-specific cytotoxic T-lymphocytes from peripheral blood. *Curr. Biol.* **8**, 413–416.

Fu, Y. X., and Chaplin, D. D. (1999). Development and maturation of secondary lymphoid tissues. *Annu. Rev. Immunol.* **17**, 399–433.

Moss, P. A., Rosenberg, W. M., and Bell, J. I. (1992). The human T-cell receptor in health and disease. *Annu. Rev. Immunol.* **10**, 71–96.

Oelke, M., Moehrle, U., Chen, J. L., Behringer, D., Cerundolo, V., Lindemann, A., and Mackensen, A. (2000). Generation and purification of CD8$^+$ melan-A-specific cytotoxic T-lymphocytes for adoptive transfer in tumor immunotherapy. *Clin. Cancer Res.* **6**, 1997–2005.

Pawelec, G. (2000). New methods to approach immunotherapy of cancer–and strategies of tumours to avoid elimination. Conference report, on behalf of EUCAPS. European Cancer Research Consortium. *Cancer Immunol. Immunother.* **49**, 276–280.

Pedersen, L. O., Nissen, M. H., Nielsen, L. L., Lauemøller, S. L., Hansen, N. J. V., Blicher, T., Hansen, A., Hviid, C. S., Thomsen, A. R., and Buus, S. (2001). Efficient assembly of recombinant major histocompatibility complex class I molecules with preformed disulfide bonds. *Eur. J. Immunol.* **31**, 2986–2996.

Pittet, M. J., Valmori, D., Dunbar, P. R., Speiser, D. E., Lienard, D., Lejeune, F., Fleischhauer, K., Cerundolo, V., Cerottini, J. C., and Romero, P. (1999). High frequencies of naive Melan-A/MART-1-specific CD8($^+$) T-cells in a large proportion of human histocompatibility leukocyte antigen (HLA)-A2 individuals. *J. Exp. Med.* **190**, 705–715.

Scheibenbogen, C., Lee, K. H., Mayer, S., Stevanovic, S., Moebius, U., Herr, W., Rammensee, H. G., and Keilholz, U. (1997). A sensitive ELISPOT assay for detection of CD8$^+$ T-lymphocytes specific for HLA class I-binding peptide epitopes derived from influenza proteins in the blood of healthy donors and melanoma patients. *Clin. Cancer Res.* **3**, 221–226.

Schrama, D., Andersen, M. H., Terheyden, P., Schroder, L., Pedersen, L. O., thor Straten, P., and Becker, J. C. (2001). Oligoclonal T-cell receptor usage of melanocyte differentiation antigen-reactive T-Cells in stage IV melanoma patients. *Cancer Res.* **61**, 493–496.

Stockwin, L. H., McGonagle, D., Martin, I. G., and Blair, G. E. (2000). Dendritic cells: Immunological sentinels with a central role in health and disease. *Immunol. Cell Biol.* **78**, 91–102.

Generation of Human and Murine Dendritic Cells

Andreas A.O. Eggert, Kerstin Otto, Alexander D. McLellan, Patrick Terheyden, Christian Linden, Eckhart Kämpgen, and Jürgen C. Becker

I. INTRODUCTION

Dendritic cells (DC) are professional antigen-presenting cells (APC) that serve as sentinels for the induction and regulation of immune responses. Because they are potent stimulators for B and T lymphocytes and activate natural killer cells, they link the innate and the acquired immune system (Fernandez *et al.*, 1999). Dendritic cells are migratory leucocytes originating from the bone marrow, specialized for the uptake, processing, and presentation of antigens. In peripheral tissues, they are found in an immature stage. After contact with antigenic material in the context of danger signals, they mature and upregulate major histocompatibility complexes and costimulatory molecules required for effective interaction with T cells. They are considered a powerful tool to manipulate the immune system (Banchereau and Steinman, 1998).

In humans, by virtue of several *in vitro* techniques, monocyte-derived DC, stem cell-derived DC, and DC isolated directly from peripheral blood can be distinguished depending on the origin of progenitor cells. Directly isolated peripheral blood DC can be subdivided into myeloid and plasmacytoid DC (Cella *et al.*, 2000). The physiological counterpart or *in vivo* relevance of each DC type is not entirely clear. This article focuses on protocols that allow the reliable and reproducible generation of the respective cell types.

In mice, DC subsets in different tissues can be distinguished (Shortman and Liu, 2002), e.g., Langerhans cells in skin, three subsets of spleen DC characterized by their CD8 and CD4 expression (Vremec *et al.*, 2000; McLellan and Kämpgen, 2000), and at least three subsets of lymph node DC (Ruedl *et al.*, 2000; Vremec

et al., 1997). In addition, dendritic cells can be generated from precursors in bone marrow or from peripheral blood monocytes (Inaba *et al.*, 1992; Schreurs *et al.*, 1999).

II. MATERIALS AND INSTRUMENTATION

A. Human Dendritic Cells

Acid citrate dextrose (ACD-A, Prod. No. 70010035) can be obtained from Fresenius, Bad Homburg, Germany, and Lymphoprep. (Cat. No. 1053980) from Axis-Shield PoC AS, Oslo, Norway. Phosphate-buffered saline (PBS) (Cat. No. BE17-512F); EDTA, 0.2 mg/ml (Cat. No. 17-711E), RPMI 1640 (Prod. No. BE12-167F); Hanks' balanced salt solution (HBSS) (Cat. No. BE10-547F); and L-glutamine, 200 mM (Cat. No. BE17-605E), are available from Bio Whittaker, Cambrex Company, Verviers, Belgium, 20% human serum albumin (HSA) from Aventis Behring, Marburg, Germany, glucose 40% (Prod. No. 2357542) from Braun, Melsungen, Germany, and dimethyl sulfoxide (DMSO) (Prod. No. D-2650), as well as bovine serum albumin (5% stock solution, Prod. No. A-4128), are from Sigma-Aldrich Chemie, Taufkirchen, Germany. For cytokine use, GM-CSF (Leucomax 400) from Essex Pharma, München, Germany, interleukin (IL)-4 (Cat. No. 9511231), TNF-α (Cat. No. 9511512), IL-1β (Cat. No. 9511180), and IL-6 (Cat. No. 9511260) from Strathmann, Hamburg, Germany, can be used. PGE$_2$ (Minprostin E2 10 mg/ml) is produced by Pharmacia, Erlangen, Germany. Refobacin 10 mg is provided by Merck, Darmstadt, Germany, penicillin/streptomycin

(PenStrep, Cat. No. 15140-122) by Invitrogen Corporation, Karlsruhe, Germany, and Liquemin N 10000 by Hoffmann-La Roche AG, Grenzach-Wyhlen, Germany. Fetal calf serum (FCS, Cat. No. 14-801E) can be obtained from Bio Whittaker. Syringes, 10 ml (Prod. No. 4606108) and 20 ml (Prod. No. 4606205), can be obtained from Braun. Culture flasks, 80 cm³ (Prod. No. 201045) and 175 cm³ (Prod. No. 200573), as well as cryotubes, 1.8 ml (Prod. No. 363401) and 3.6 ml (Prod. No. 366524), are available from Nunc GmbH & Co. KG, Wiesbaden, Germany. Tubes, 15 ml (Prod. No. 25315-15) and 50 ml (Cat. No. 430829), are from Corning GmbH, Bodenheim, Germany. Disposable filter units (red rim, Ref. No. 10462200) are from Schleicher & Schuell, Dassel, Germany. Pipettes, 2 ml (Cat. No. 4486), 5 ml (Cat. No. 4487), 10 ml (Cat. No. 4488), and 25 ml (Cat. No. 4489), are from Corning GmbH. Pipette tips, 10 µl (Cat. No. 790011), are available from Biozym, Hessisch-Oldendorf, Germany; 100 µl (Cat. No. 0030 010.035) and 1000 µl (Cat. No. 0030 010.051) are from Eppendorf. Tissue culture dishes (Prod. No. 35-3003) are from Falcon, Becton-Dickinson, New Jersey, syringe needles, 0.9 × 40 mm (Prod. No. 4657519), are from Braun, trypan blue (Cat. No. S11-004) is from PAA Laboratories, Linz, Austria, and antibodies for FCM analysis are from Becton-Dickinson. Antibodies, antibody-conjugated diamagnetic beads, separation columns, and magnets for the isolation of peripheral blood dendritic cells can be obtained from Miltenyi Biotech GmbH, Bergisch-Gladbach, Germany (blood dendritic cell isolation kit, Cat. No. 130-046-801). For further reference, we recommend the frequently updated online protocols of Miltenyi (http://www.miltenyibiotech.com).

B. Murine Dendritic Cells

Bovine serum albumin (BSA) (Cat. No. B 4287), β-mercapthoethanol (Cat. No. M 6250), and EDTA (Cat. No. 6511) are from Sigma-Aldrich, Deisenhofen, Germany. Cytokines are from Strathmann Biotech, Hamburg, Germany: rmGM-CSF (Cat. No. mGMCSF-10) and rmIL-4 (Cat. No. mIL4-10). Collagenase type II (Cat. No. LS04176) or type III (Cat. No. LS04182) is from Cell Systems Biotech GmbH, St. Katharinen, Germany. DNase I is from Roche Diagnostics GmbH, Mannheim, Germany (20 mg/ml, Cat. No. 92566700). Fetal calf serum (FCS) (Cat. No. 14-801E) was tested before usage and obtained as IMDM (Cat. No. 12-726F) from Cambrex/Bio Witthaker Europe, Verviers, Belgium. Hank's (Cat. No. 14175-053), HBSS (Cat. No. 14175-053), L-glutamine (Cat. No. 25030-024), penicillin/streptomycin (Cat. No. 15140-122), PBS (Cat. No. 14190094), and RPMI 1640 (Cat. No. 31870-025) are

from Invitrogen, Karlsruhe, Germany. Heparin (Cat. No. L6510) and HEPES are from Biochrom AG Seromed, Berlin, Germany (Cat. No. L 1603). KCl (Cat. No. 104933) and NaCl (Cat. No. 106404) are from Merck, Darmstadt, Germany. Trypsin (Cat. No. 59227) is from JRH Biosciences, Shawnee Mission, Kansas. Nycodenz powder (Nycoprep) is from Nycomed Pharma AS, Asker, Norway. Plastic materials are obtained from Becton-Dickinson, Heidelberg, Germany: cell culture dish (Cat. No. 353003), petri dish (Cat. No. 351001), 70-µm cell stainer (Cat. No. 352350), 6-well plate (Cat. No. 353224), 24-well plate (Cat. No. 353226), syringe needle (Cat. No. 300400), and 50-ml tube (Cat. No. 352070). Instruments are from NeoLab, Heidelberg, Germany: sieve (diameter 70 mm, Cat. No. 2-7530) and tweezers (Cat. No. 2-1034). Cy5-PE-conjugated anti-CD4 monoclonal antibody (mab) (clone H129.19; Cat. No. 553654) is from Becton-Dickinson. Multiple species Ig-absorbed, fluorescein isothiocyanate (FITC)-conjugated antihamster Ig (Cat. No. 127-096-160) and PE antirat Ig (Cat. No. 112-116-143) are both from Dianova, Hamburg, Germany.

III. PROCEDURES

Solutions for the Generation of Human Dendritic Cells

1. *PBS/0.1 M EDTA buffer:* 2.5 ml 2 M EDTA and 497.5 ml PBS. Store at 4°C and use within 1 week.

2. *Freezing medium:* 20% HSA, 20% DMSO, and 6% glucose. To prepare 200-ml, pipette 130 ml HSA 20% into a sterile 200-ml flask and add 40 ml DMSO and 30 ml 40% glucose. Store at 4°C and use within 1 week.

3. *R0 culture medium (R0):* To one bottle of 500 ml RPMI 1640, add 5 ml glutamine and 2 ml 10 mg Refobacin. Store at 4°C and use within 1 week.

4. *Complete medium (CM):* To one bottle of 500 ml RPMI 1640, add 5 ml glutamine, 2 ml 10 mg Refobacin, and 5 ml sterile-filtered autologous heat-inactivated plasma. Store at 4°C and use within 1 week.

5. *PBS/ACD buffer:* Add 50 ml ACD-A to one bottle (500 ml) PBS and use within 1 day.

6. *PBS/heparin solution:* Add 500 µl Liquemin N 10000 (5000 IE sodium heparin) to 500 ml PBS buffer. Store at 4°C.

7. *R10 culture medium (R10):* 1% glutamine, 1% PenStrep, and 10% FCS in RPMI 1640. To one bottle of 500 ml RPMI 1640, add 5 ml glutamine, 5 ml PenStrep, and 50 ml heat-inactivated FCS. Store at 4°C.

8. *Adherence medium (AM):* To one bottle of 500 ml RPMI 1640, add 5 ml PenStrep, 5 ml glutamine, 15 ml sterile-filtered plasma surrogate, and 20,000 U GM-CSF. Store at 4°C and use within 1 day.

9. *BSA/EDTA buffer (BEB):* 0.5% bovine serum albumin and 2 mM EDTA. For preparation of 100 ml buffer, add 10 ml of 5% bovine serum albumin stock solution and 3.2 ml 2% EDTA to 86.8 ml PBS. *Note:* it is necessary to degas buffer prior to use. Store and use at 4°C.

A. Generation of Mature Dendritic Cells from Precursor Cells Obtained from Leukapheresis Products and Whole Blood

The development of techniques to generate large numbers of homogeneous DC populations *in vitro* from human progenitors allowed their use for research purposes as well as for clinical applications. Protocols established by several different groups describe the generation of DC either from rare, proliferating CD34[+] cells or from more frequent, but nonproliferating CD 14[+] monocytic cells, which can be both isolated from peripheral blood. Currently, monocyte-derived DC are used more frequently, as no special pretreatment (e.g., systemic application of G-CSF) of the donor is required; moreover, DC yielded from this progenitor population still seem to be more homogeneous and easier to generate with reproducible characteristics; hence we will restrict the provided protocol to this population.

Monocyte-derived DC from CD14[+] precursors are generated by use of GM-CSF and IL-4 as key cytokines, whereas the generation of DC from CD34[+] progenitors requires GM-CSF and TNF-α. The following protocols are adapted from procedures reported by Feuerstein *et al.* (2000) and Thurner *et al.* (1999) for the generation of mature DC from CD14[+] precursors isolated from leukapheresis products and buffy coats/whole blood. It should be noted that not only the used progenitor cell population but also the method of their isolation influence the subsequent generation process, e.g., upon culture in the presence of GM-CSF and IL-4, CD14[+] monocytes differentiate within 5–6 (leukapheresis cells) or 7 (buffy coat/whole blood cells) days to immature DC, which are characterized phenotypically as large adherent cells with irregular outlines possessing only rarely longer processes or veils. The Addition of cytokine mix at the respective day, consisting of TNF-α, PGE2, IL-1β, and IL-6, however, results in fully mature DC, which are nonadherent and veiled. Flow cytometry using antibodies recognizing typical DC surface molecules

additionally allows quality control of generated cells (Fig. 1).

1. Protocol to Start from Leukapheresis Products

Day 0

1. Fill leukapheresis product (200–250 ml) into a 1000-ml culture flask and add warm (room temperature) PBS-ACD buffer to a final volume of 480 ml.

2. Fill 15 ml Lymphoprep. into each of sixteen 50-ml tubes.

3. Overlay Lymphoprep. carefully with 30 ml of the cell suspension.

4. Spin in a warm centrifuge for 30 min (22°C, 300 g). Make sure that the centrifuge runs off without brake.

5. While the cells are spinning, fill autologous plasma in 50-ml tubes and incubate for 30 min in a 56°C hot water bath. Thereafter, spin tubes for 10 min (22°C, 600 g), aliquot supernatants into 15-ml tubes, and freeze at −20°C.

6. Take leukapheresis product-containing tubes carefully out of the centrifuge and harvest the interphase into 50-ml tubes containing 15 ml of cold PBS/EDTA. Add cold PBS/EDTA buffer to a final volume of 45 ml.

7. Spin tubes for 10 min (4°C, 200 g).

8. Remove supernatant, resuspend pellet with cold PBS/EDTA buffer by observing a 2:1 transfer, and spin tubes for 5 min (4°C, 300 g).

9. Repeat step 8.

10. Remove supernatant and resuspend with 40 ml cold R0.

11. Take an aliquot of all four tubes and dilute each 1:50 for cell counting using a Neubauer chamber. While counting cells, put 50-ml tubes containing cells on wet ice.

12. After calculating the amount of cells, devide cells in fractions for immediate replating and for freezing. Spin cells in cold centrifuge for 5 min (4°C, 300 g).

13. Cells that are not used for immediate generation of DC should be frozen in 20% HSA at 120×10^6 cells per 1.8 ml and freezing medium (1:1) in cold 3.6-ml vials using a freezing device on wet ice before they are transferred to a nitrogen-freezing machine to cool down to −150°C. Store frozen cells in liquid nitrogen.

14. For immediate generation of DC, spin cells in a cold centrifuge for 5 min (4°C, 300 g), remove supernatant, and resuspend at $15–25 \times 10^6$ cells/ml of CM.

15. Preload dishes with 8 ml CM. Add 2 ml of cell suspension per dish, swing slightly, and transfer dishes into the incubator for at least 30 min.

16. Subsequently, control adherence. If sufficient (close layer of adherent cells), wash away the nonad-

herent fraction with warm PBS (room temperature) twice. Add 10 ml of warm CM per dish and retransfer into an incubator overnight.

Day 1

1. Take dishes out of the incubator and control cell adherence under a microscope.
2. Remove CM from dishes by pipetting carefully.
3. Add 9 ml of warm CM (room temperature).
4. Fill 1 ml CM per dish into a 50-ml tube and add 8000 U GM-CSF and 10,000 U IL-4 per dish.
5. Add diluted cytokine mix to dishes, swing slightly, and transfer dishes into an incubator.

Day 5

1. Take dishes out of the incubator and control cell adherence under a microscope.
2. Carefully collect 5 ml of culture supernatant from each dish into a 50-ml tube.
3. Add 4 ml of room-temperatured CM to each dish.
4. Spin 50 ml tubes for 5 min (room temperature 300 g).
5. Remove supernatant, add 1 ml CM, 8000 U GM-CSF, and 10,000 U IL-4 per dish and resuspend.
6. Redistribute 1 ml of this suspension to each dish, swing slightly, and transfer dishes into an incubator.

Day 6

1. Take dishes out of incubator and control cell morphology under a microscope.
2. Remove 1 ml from all dishes by pipetting into a 50-ml tube.
3. Add TNF-α (10 ng/ml cell culture volume), PGE$_2$ (1 μg/ml), IL-1β (2 ng/ml), and IL-6 (5 ng/ml), resuspend, and add equal amounts into each dish. Swing slightly and retransfer dishes to an incubator. Incubate for at least 24 h.

Day 7

1. Take dishes out of incubator and control cell morphology under a microscope. Mature DC appear veiled and nonadherent.
2. Harvest cells into 50-ml tubes, spin for 5 min (room temperature 300 g), remove supernatant, and resuspend cells in CM. Remove an aliquot to count cells.
3. Determine the phenotype of the cells by FCM analysis (Fig. 1).

2. Protocol to Start from Frozen PBMC Isolated from Leukapheresis Products

Day 0

1. Fill 5 ml of cold HBSS in 15-ml tubes.

2. Thaw vials with frozen cells in warm water to an extent that a frozen core of cells remains in each tube.
3. Dump frozen cells into the 15-ml tubes preloaded with 5 ml HBSS. Use one vial of frozen cells per 15-ml tube.
4. Rinse the storage tube once with cold HBSS, pipette suspension into the 15–ml tubes, and fill up with HBSS to a final volume of 13–14 ml.
5. Spin in a cold centrifuge for 10 min (4°C, 240 g).
6. Remove supernatants, resuspend pellets with a small amount of cold R0, and pool pellets into one 50-ml tube.
7. Spin in a cold centrifuge for 5 min (4°C, 300 g).
8. Remove supernatant and resuspend pellet with 40 ml of cold R0.
9. Take an aliquot and count cells.
10. Follow steps 14–16 of the protocol described in the previous protocol.

For day 1 to day 7 proceed as described in the previous protocol.

3. Protocol to Start from Buffy Coats or Whole Blood

Day 0

1. Transfer buffy coat or whole blood into a culture flask and dilute 1:3 with PBS/heparin solution.
2. Fill 15 ml Lymphoprep. into 50-ml tubes.
3. Overlay Lymphoprep. carefully with 30 ml of the cell suspension.
4. Spin in a warm centrifuge for 30 min (22°C, 300 g). Make sure that the centrifuge runs off without brake.
5. Collect supernatant, further called plasma surrogate, into a 50-ml tube and incubate for 30 min in a 56°C hot water bath. Thereafter, centrifuge for 10 min (room temperature 600 g). Save supernatant in a 50-ml tube for preparing AM.
6. Harvest interphase carefully to a tube containing 15 ml cold PBS/EDTA buffer. Fill up to a final volume of 40 ml using PBS/EDTA buffer.
7. Centrifuge for 10 min (4°C, 200 g).
8. Remove supernatant and resuspend pellet in a small volume of PBS/EDTA buffer. Pool pellets (2:1 transfer) and add cold PBS/EDTA buffer to a final volume of 40 ml per tube.
9. Centrifuge for 5 min (4°C, 300 g).
10. Repeat steps 8 and 9.
11. Remove supernatant, resuspend pellets in a small volume of AM, and pool cells into one 50-ml tube.
12. Remove an aliquot and count cells.
13. Load dishes with AM and cell suspension by observing 30–50 × 10^6 cells per dish. Swing slightly

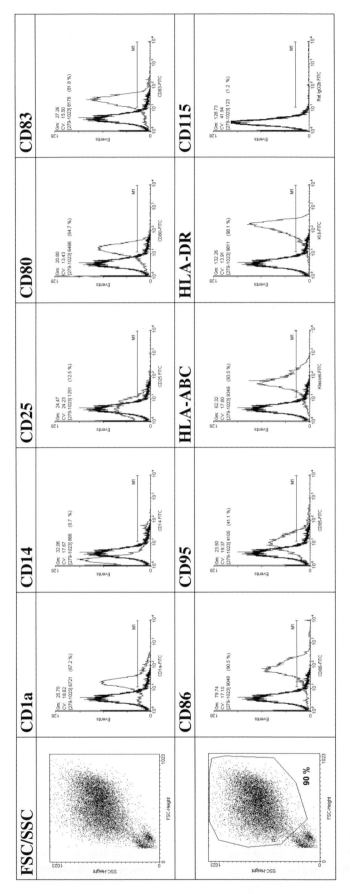

FIGURE 1 Phenotypic characterization of day 7 cells generated by FCM analysis.

and transfer dishes into the incubator for 60 min.

14. If cells adhere sufficiently, gently wash away nonadherent cells with warm PBS (room temperature). Repeat once. Add 10 ml warm R10 supplemented with 10,000 U IL-4 and 8000 U GM-CSF per dish and put into an incubator overnight.

Day 3

1. Take dishes out of the incubator and control cell adherence under a microscope.
2. Remove 5 ml R10 from dishes by pipetting carefully.
3. Add 4 ml of warm R10 (room temperature) to each dish.
4. Add 4000 U GM-CSF and 5000 IL-4 diluted in 1 ml of R10 to each dish.
5. Swing slightly and transfer dishes into the incubator.

Day 5

1. Follow steps 1–5 as described for day 3.

Day 7

1. Take dishes out of incubator and control cell morphology under a microscope.
2. Harvest cells into 50-ml tubes, spin for 5 min (4°C, 300 g), and collect supernatant. Resuspend cells in R10. Remove an aliquot to count cells.
3. Replate 5–7 × 10⁶ cells into each culture dish containing 10 ml final volume (1:1 saved supernatant and R10). Add 4000 U GM-CSF and 5000 IL-4 per dish, as well as TNF-α (10 ng/ml cell culture volume), PGE₂ (1 μg/ml), IL-1β (2 ng/ml), and IL-6 (5 ng/ml), and pipette equal portions into the dishes.
4. Swing dishes slightly and transfer them into incubator.

Day 10

1. Take dishes out of the incubator and control cell morphology under a microscope. Mature dendritic cells look veiled and nonadherent.
2. Harvest cells into 50-ml tubes, spin for 5 min (4°C, 300 g), remove supernatant, and resuspend cells in R10.
3. Determine the phenotype of cells by FCM analysis.

B. Direct Isolation of Myeloid and Plasmacytoid DC from Peripheral Blood

In principle, peripheral blood DC are obtained by isolation of lineage marker negative, CD4⁺ cells (O'Doherty et al., 1994). This basic procedure is described in the following text. However, advanced procedures were developed for the isolation of DC subtypes based on blood DC antigen (BDCA) expression; therefore, myeloid DC can be isolated by depletion of B cells from PBMC and positive selection of BDCA-1+ cells; plasmacytoid cells are isolated directly by the selection of BDCA-4 expressing cells. Techniques for the isolation of peripheral blood DC are based on positive and/or negative selection of primary and secondary antibody-conjugated magnetic bead-binding subpopulations in PBMC.

Steps

1. Isolate PBMC from heparinized whole blood by standard gradient centrifugation with Lymphoprep; about 2 × 10⁸ cells are usually needed. Note: Due to large cell need, i.e., about 1% cell yield at the end of the procedure, we recommend buffy coats.

2. Resuspend cells in 300 μl of buffer per 1 × 10⁸ cells. Add 100 μl human IgG (FcR blocking reagent) and label cells with 100 μl haptenized murine anti-CD3, anti-CD11b, and anti-CD16 antibodies (Hapten-Antibody cocktail) per 10⁸ cells. Incubate at 4°C for 10 min.

3. Wash cells twice with 20-fold labeling volume with buffer and centrifuge at 300 g for 10 min. Finally, resuspend the pellet in 900 μl buffer/10⁸ cells.

4. Add 100 μl cells antihapten antibodies attached to diamagnetic beads (antihapten microbeads), per 10⁸ mix well, and incubate for 15 min at 4°C.

5. Separate magnetic (T cells, monocytes, and natural killer cells) cells from the nonmagnetic fraction using a magnetic separation column. To do this, a depletion column must be placed into a magnet (we usually use VarioMACS) and must be assembled with a flow resistor (an injection needle of 20–22 gauge is appropriate; follow the manufacturer's instructions) together with a side syringe and a three-way stopcock. Fill the column with degassed cool buffer and rinse. Cool the magnet until use; the procedure itself can be performed at room temperature.

6. Apply the cell suspension on top of the depletion column and allowed it to enter for 5 min. Wash the column six times and collect the flow through, which contains the desired cells.

7. Wash cells at 300 g for 10 min. Remove the supernatant and resuspend the cell pellet in 100 μl buffer.

8. To separate B cells from the DC, which are enriched in the resulting cell fraction, label cells with 100 μl anti-CD4-conjugated magnetic beads (MACS CD4 MicroBeads) and incubate for 30 min at 4°C.

9. Wash cells once by adding 4 ml of buffer and centrifuge at 300 g for 10 min. Resuspend in 500 μl of buffer.

10. For the next depletion step, a column must be prepared, again following the manufacturer's instruc-

tion (we use a MS+ column and a miniMACS magnet). Prepare the column by washing with 500 μl buffer.

11. Apply and allow cells to penetrate the column.

12. Rinse the column with 3 × 500 μl buffer (flow through contains nondendritic cells).

13. Remove the column from the magnet, place it on a 15-ml plastic tube, and elute cells by rinsing with 1 ml buffer. Use an appropriate plunger to press fluid through the column. Fill the eluate into a fresh column and repeat the procedure.

14. Assess DC purity by flow cytometry, with DC defined as HLA-DR$^+$ cells lacking expression of CD3, CD4, CD14 , CD16, and CD19.

C. Preparation of Murine Langerhans Cells

This method was published by Kämpgen et al. (1994) and is described here with some important modifications to improve LC yield and purity.

Solutions

1. *Nycodenz gradient (Vremec and Shortman, 1997)*

Solution A (Shortman buffer) (500 ml of 308 mOsm (EDTA-SS)): 0.154 M NaCl (4.5 g), 4 mM KCl (0.1491 g), 14.8 mM HEPES (2.96 ml of 2.5 M stock, pH 7.2), and 5 mM EDTA (5 ml of 0.5 M) and make up to 500 ml volumetrically with dH$_2$O (Osm = 308 mOsM)

Solution B (230.78 ml of 30.55% Nycodenz (d = 1.16)): 70.5 g Nycodenz powder; make up to 230.78 ml volumetrically with dH$_2$O (Osm = 308 mOsM)

For 14.1% Nycoprep (d~1.077, 308 mOsm): Add 269.22 ml of solution A to 230.78 ml of solution B to a final volume of 500 ml. Both solutions and the remaining EDTA-SS should be sterile filtered.

2. *Langerhans cell culture medium (I10)*: IMDM, 10% FCS, 50 μM β-mercapthoethanol, 200 μM L-glutamine, 100 μg/ml penicillin, and 50 μg/ml streptomycin

Steps

1. Kill 15–30 mice by CO$_2$ inhalation, cut off both ears right above the ring cartilage, and place them in a petri dish (Falcon 1001) filled with 10 ml 70% alcohol for 3 min. Hold ears under alcohol and strike out blood and air with a round side of a bent tweezers. Place ears on a sterile 10 × 10-cm swab and dry them for 20 min at room temperature under a hood.

2. Prepare one petri dish containing 6 ml HDSS for the inner ear (ventral side with cartilage) and a dish with 9 ml HBSS for the soft outer dorsal half of the ear. Split ears by using two Edson forceps and place the corresponding side on top of the HBSS and add 9 ml 2.5% trypsin into the dish with the ventral side of the ears and 3.5 ml 2.5% trypsin for the outer halves of the ears. Incubate ventral ear halves for

90 min and the dorsal ear halves for 45 min at 37°C in 5% CO$_2$.

3. Place a metal sieve in a petri dish containing 15 ml of cold I10. Peel off the epidermis using two bent curved watchmaker's forceps and lay the epidermal sheets onto the medium. The sheets will tend to spread out upon reaching the surface of the medium. Knock against the sieve for 3 min. Wash cells once prior to an incubation at a density of 1–2 × 10^7 per 10-ml cell culture dish or 1–2 × 10^6/ml in a 6-well plate in I10 and 10 ng/ml rmGM-CSF at 37°C, 5% CO$_2$ in humidified air for 2–3 days.

4. After 2–3 days, cells are harvested and resuspended in 10 ml 14.1% Nycodenz gradient (p = 1.077) before transfer to two 15-ml tubes. This suspension is overlayed with 2 ml of Shortman buffer and spun at 4°C at 600 g for 20 min without the brake. Harvest low-density cells and wash once with medium prior to use.

D. Preperation of Murine Spleen Dendritic Cells

This procedure is performed according to McLellan et al., (2002) with minor modifications.

Solutions

1. *Nycodenz gradient*: See Section A
2. *Würzburger wash buffer*: PBS, 1% BSA, 5 mM EDTA, and 20 μg/ml DNase
3. *FACS buffer*: PBS and 0.2% BSA
4. *Dendritic cell culture medium (R10)*: RPMI 1640, 10% FCS, 50 μM β-mercapthoethanol, 200 μM L-glutamine, 100 μg/ml penicillin, and 50 μg/ml streptomycin

Steps

1. Kill five to six mice by CO$_2$ inhalation, disinfect skin with 70% alcohol for 1 min, cut abdominal skin using sterile siccors below left ribs, and tear away the skin. Cut the peritoneum directly above the spleen, which is isolated by removal of all other tissues. Under laminar air flow, place spleens into 5 ml Hanks buffered salt solution in a petri dish. Add 120 μg/ml DNase I, pierce splenic capsule at both ends, and with curved watchmaker's forceps squeeze out all cellular contents. Tear the splenic capsule into small pieces, add 400 μl FCS to the dish, and transfer the cell suspension and fragments to a 50-ml polypropylene tube and add collagenase to a concentration of 1 mg/ml. Wash plate once with 5 ml Hanks'. Store petri dish after adding another 5–10 ml of Hanks' on ice. In a 50-ml tube, pipette cells up and down ten times with a 10-ml pipette to break up loose aggregates. Subsequently,

incubate the tube with constant but gentle swirling for 25 min at 37°C.

2. After this incubation, add 200 µl 0.5 M EDTA (pH 7.2) and incubate for an additional 5 min. Gently press remaining fragments through a coarse metal sieve with a rubber syringe plunger and collect into the original petri dish.

3. Rinse sieve and petri dish with Hanks' and add this to the cell suspension. Filter the cell suspension through a 70-µm cell stainer into a 50-ml tube. Top up to 50 ml with cold washing buffer. Wash once.

4. Keep tube in ice and gently add to the cell pellet 10 ml 14.1% Nycodenz (308 mOsm; ρ = 1.077) for six spleens and slowly resuspend until all clumps are resolved. Transfer the suspension to two 15-ml polypropylene tubes, overlay with 2 ml 308 mOsm Shortman buffer, and spin at 4°C at 600 g for 20 min without any brake.

5. Harvest low-density cells from all layers without touching the cell pellet and wash once in ice-cold washing buffer. A normal yield should be around 5–10 $\times 10^6$ CD11c$^+$ DC/spleen at 30–50% purity.

6A. **Isolation of spleen DC subsets (CD11c$^+$/CD4$^-$/CD8α$^-$, CD11c$^+$/CD4$^-$/CD8α$^-$, CD11c$^+$/CD4$^-$/CD8α$^+$) by FACS.** Block nonspecific Ig-binding sites on low-density spleen cells by incubation for 10 min in 25 µl 10% goat and mouse serum before labeling with 500 µl N418 culture supernatant (hamster antimouse CD11c mab) and 500 µl 53-6.7 culture supernatant (rat antimouse CD8α mab) for 30 min and ice. After one wash with Würzburger buffer, incubate spleen cells with multiple species Ig-absorbed, FITC-conjugated antihamster Ig and PE antirat Ig for 30 min on ice in the dark. After an additional washing step, add 400 µl 10% rat serum and Cy5-PE-conjugated anti-CD4 mab (clone H129.19); incubate for 30 min, subsequently wash cells once, and resuspend at 5×10^7 cells/ml in PBS/1 mM EDTA. Stained cells can now be sorted into CD11c$^+$/CD4$^-$/CD8α$^-$, CD11c$^+$/CD4$^{+/}$CD8α$^-$, and CD11c$^+$/CD4$^-$/CD8α$^+$ subsets by FACS (e.g., FACS Vantage, Becton-Dickinson, Heidelberg, Germany).

6B. **Enrichment of spleen DC by adherence.** Incubate up to 50×10^6 Nycodenz-enriched, low-density cells per each Falcon 3003 dish in R10 supplemented with 10 ng/ml mGM-CSF for 2 h. For 3003 plates, wash gently against the wall of the dish with warm (37°C) R10 about five times. For 24-well plates, wash gently (5 ×) by removal of media with a R10 Pasteur pipette under vacuum and immediately refill each well with R10. Washes must be performed with extreme care to avoid dislodging too many DC. For the final wash, refill wells with 0.5 ml R10 with 10 ng/ml mGM-CSF (for 24-well plate). The next day, harvest nonadherent DC prior to use in experiments.

For serum-free culture conditions, substitute 0.5–1% mouse serum for BSA or FCS, including wash buffers.

E. Preperation of Murine Bone Marrow-Derived Dendritic Cells (BMDC)

This procedure is performed according to Inaba *et al.* (1992) with minor modifications.

Solution

BMDC culture medium: IMDM, 10% FCS, 50 µM β-mercaptoethanol, 200 µM L-glutamine, 100 µg/ml penicillin, and 50 µg/ml streptomycin

Steps

1. Kill needed number of mice by cervical dislocation or CO_2 inhalation, disinfect skin of the legs and lower abdomen with 70% alcohol for 1 min, cut the skin with a scissors at the inside of the leg upward, tear the skin away, and remove the muscle tissue. Disconnect the complete leg by cutting the ligaments holding the femur in the joint. Scrape the residual tissue with a scalpel from the bone, break it at the distal diaphysis, and transfer the femur and tibia to a tube with PBS.

2. Place bones in 70% alcohol for 1 min, wash them with PBS, and place them in a 2-cm petri dish. Cut off both ends of the bones with the scalpel with gently sawing movements; while placing the bone over a 50-ml tube with tweezers, flush the bone marrow with 1 ml of medium using a 10-ml syringe armed with a 0.5 × 24-mm cannula, turn the bone upside down, and repeat flushing.

3. Resuspend flushed bone marrow vigorously with a 10-ml pipette for 1 min to dislodge clusters. Transfer the cell suspension into a new 50-ml tube through a 70-µm filter to remove pieces of bone, pellet at 1500 rpm for 4 min, and resuspend cells. Incubate cells in one 6-well plate/mouse in 3 ml medium/well for overnight adherence.

4. The following day, harvest and count nonadherent cells. Up to 50% are lost upon adherence. Seed 0.8–1 $\times 10^6$ cells/well/6-well plate in 4 ml BMDC culture medium supplemented with 15 ng/ml rmGM-CSF and 15 ng/ml rmIL-4.

5. On day 3 of culture, add 1 ml medium containing 50 ng/ml rmGM-CSF and 50 ng/ml rmIL-4 to each well.

6. On day 6, after resuspending cells three times with a 5-ml pipette to dislodge DC clusters (adherent and nonadherent), harvest, pellet at 1500 rpm for 5 min, and replate cells in 6-well plates with new BMDC culture medium-supplemented cytokines as described earlier.

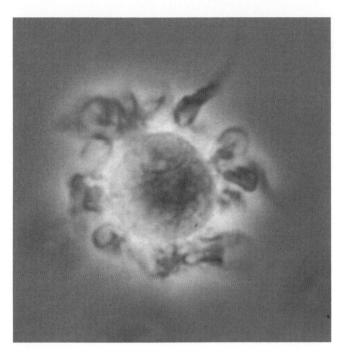

FIGURE 2 Murine bone marrow-derived dendritic cell in culture (400×) showing veils and dendrites spread homogeneously over the cell surface.

7. On day 8, swirl the plates gently, harvest nonadherent cells (Fig. 2) by gently washing the bottom of the well once, and count cells. The expected yield of BMDC is 10×10^6 cells per mouse, depending on the age of the animal.

F. Preperation of Murine Monocyte-Derived Dendritic Cells (MoDC)

This procedure is performed according to Schreurs *et al.* (1999) with little modification.

Solutions

1. *Transport medium*: ice-cold HBSS (without Ca, Mg) supplemented with 100 U/ml heparin
2. *Stock solution for mouse monocyte gradient*: Nine parts Percoll and one part 10× PBS. Do not use longer than 2 days.
3. *Working solution for mouse monocyte gradient*: Take 18.3 ml of the gradient stock solution and add 11.7 ml HBSS; mix well and store at 12°C.
4. *MoDC culture medium*: IMDM, 10% FCS, 50 μM β-mercapthoethanol, 200 μM L-glutamine, 100 μg/ml penicillin, and 50 μg/ml streptomycin
5. *MoDC adherence medium*: IMDM, 3% FCS, 50 μM β-mercapthoethanol, 200 μM L-glutamine, 100 μg/ml penicillin, and 50 μg/ml streptomycin

Steps

1. Anesthetize mice one by one, disinfect skin of the upper abdomen with 70% alcohol for 1-min, puncture heart with a 1-ml syringe harnessed with a 0.5 × 24-mm cannula containing 50 U heparin, and aspirate blood slowly. Expect 0.7–1 ml blood per mouse. Transfer blood to the tube containing transport medium. Immediately afterward kill the mouse by cervical dislocation. In general, 30 mice are needed to obtain sufficient numbers of MoDC precursors. In the following steps, all measures are described for this number of mice.

2. Dilute blood to 100 ml in HBSS and overlay carefully in four fractions on a 7-ml mouse monocyte gradient in four 50-ml tubes. Spin for 30 min at 2850 rpm, 12°C, without any brake. Pool two interphases, wash twice with ice-cold HBSS, pool the remaining pellets, and wash five times with ice-cold HBSS to remove all thrombocytes. Finally wash once with MoDC adherence medium and count cells.

3. Resuspend cells in 12 ml MoDC adherence medium and incubate for 90 min at 37°C in a 6-well plate at 2 ml per well for adherence. Discharge nonadherent cells, wash each well four times, and subsequently check the purity of the adherent cells using a microscope at 100× magnification. If satisfactory, add 3 ml DC medium/well supplemented with 20 ng/ml rmGM-CSF and 20 ng/ml rmIL-4.

4. After 3 days of culture, add 1 ml medium containing 50 ng/ml rmGM-CSF and 50 ng/ml rmIL-4 to each well.

5. On days 7–9 of culture, swirl the plates gently and harvest nonadherent cells. Gentle washing of the well, which should avoid the detachment of macrophages, improves the yield of cells. Expect 1.5 × 10^6 MoDC per 30 mice.

References

Banchereau, J., and Steinman, R. M. (1998). Dendritic cells and the control of immunity. *Nature* **392**, 245–252.

Cella, M., Facchetti, F., Lanzavecchia, A., and Colonna, M. (2000). Plasmacytoid dendritic cells activated by influenza virus and CD40L drive a potent TH1 polarization. *Nature Immunol.* **1**, 305–310.

Fernandez, N. C., Lozier, A., Flament, C., Ricciardi-Castagnoli, P., Bellet, D., Suter, M., Perricaudet, M., Tursz, T., Maraskovsky, E., and Zitvogel, L. (1999). Dendritic cells directly trigger NK cell functions: Cross-talk relevant in innate anti-tumor immune responses *in vivo*. *Nature Med.* **5**, 405–411.

Feuerstein, B., Berger, T. G., Maczek, C., Roder, C., Schreiner, D., Hirsch, U., Haendle, I., Leisgang, W., Glaser, A., Kuss, O., Diepgen, T. L., Schuler, G., and Schuler-Thurner, B. (2000). A method for the production of cryopreserved aliquots of antigen-preloaded, mature dendritic cells ready for clinical use. *J. Immunol. Methods* **245**, 15–29.

Inaba, K., Inaba, M., Romani, N., Aya, H., Deguchi, M., Ikehara, S., Muramatsu, S., and Steinman, R. M. (1992). Generation of large numbers of dendritic cells from mouse bone marrow cultures supplemented with granulocyte/macrophage colony-stimulating factor. *J. Exp. Med.* **176**, 693–1702.

Kämpgen, E., Koch, F., Heufler, C., Eggert, A., Gill, L. L., Gillis, S., Dower, S. K., Romani, N., and Schuler, G. (1994). Understanding the dendritic cell lineage through a study of cytokine receptors. *J. Exp. Med.* **179**, 1767–1776.

McLellan, A. D., and Kämpgen, E. (2000). Functions of myeloid and lymphoid dendritic cells. *Immunol. Lett.* **72**, 101–105.

McLellan, A. D., Kapp, M., Eggert, A., Linden, C., Bommhardt, U., Brocker, E. B., Kammerer, U., and Kampgen, E. (2002). Anatomic location and T-cell stimulatory functions of mouse dendritic cell subsets defined by CD4 and CD8 expression. *Blood* **99**, 2084–2093.

O'Doherty, U., Peng, M., Gezelter, S., Swiggard, W. J., Betjes, M., Bhardwaj, N., and Steinman, R. M. (1994). Human blood contains two subsets of dendritic cells, one immunologically mature and the other immature. *Immunology* **82**, 487–493.

Ruedl, C., Koebel, P., Bachmann, M., Hess, M., and Karjalainen, K. (2000). Anatomical origin of dendritic cells determines their life span in peripheral lymph nodes. *J. Immunol.* **165**, 4910–4916.

Schreurs, M. W., Eggert, A. A., de Boer, A. J., Figdor, C. G., and Adema, G. J. (1999). Generation and functional characterization of mouse monocyte-derived dendritic cells. *Eur. J. Immunol.* **29**, 2835–2841.

Shortman, K., and Liu, Y. J. (2002). Mouse and human dendritic cell subtypes. *Nature Rev. Immunol.* **2**, 151–161.

Steinman, R. M., and Pope, M. (2002). Exploiting dendritic cells to improve vaccine efficacy. *J. Clin. Invest.* **109**, 1519–1526.

Thurner, B., Roder, C., Dieckmann, D., Heuer, M., Kruse, M., Glaser, A., Keikavoussi, P., Kaempgen, E., Bender, A., and Schuler, G. (1999). Generation of large numbers of fully mature and stable dendritic cells from leukapheresis products for clinical application. *J. Immunol. Methods* **1**, 1–15.

Vremec, D., Pooley, J., Hochrein, H., Wu, L., and Shortman, K. (2000). CD4 and CD8 expression by dendritic cell subtypes in mouse thymus and spleen. *J. Immunol.* **164**, 2978–2986.

Vremec, D., and Shortman, K. (1997). Dendritic cell subtypes in mouse lymphoid organs: Cross-correlation of surface markers, changes with incubation, and differences among thymus, spleen, and lymph nodes. *J. Immunol.* **159**, 565–573.

Culture of Specific Cell Types: Haemopoietic, Mesenchymal, and Epithelial

Clonal Cultures *in vitro* for Haemopoietic Cells Using Semisolid Agar Medium

Chung Leung Li, Andreas Hüttmann, and Eugene Ngo-Lung Lau

I. INTRODUCTION

An exponential increase in knowledge of the regulation of haemopoiesis has been witnessed since the mid-1980s. Over 20 cytokines (colony-stimulating factors, erythropoietin, thrombopoietin, and interleukins) have now been identified, molecularly cloned, and expressed. Many of these recombinant haemopoietic growth factors are now being used in clinical situations where they have been found to be useful in correcting anaemia and white cell deficiencies either in chronic-inherited disease or following acute treatment (e.g., chemotherapy, bone marrow transplantation) (Atkinson, 1993; Clark and Kamen, 1987; Mertelsmann *et al.*, 1990; Sheridan *et al.*, 1989).

The discovery of haemopoietic growth factors was facilitated greatly by the ability to grow haemopoietic cells *in vitro*. These culture systems enable the undifferentiated haemopoietic precursors to proliferate and differentiate into various haemopoietic cell lineages. Especially valuable has been the development of clonal cultures using semisolid agar or methylcellulose culture medium for haemopoietic precursor cells. In the presence of appropriate growth factors, these precursor cells proliferate and produce a clonal colony of differentiated cells. This will allow biological, viral, biochemical, or molecular studies to be performed on individual cell clones. In addition, by counting colonies, it is also possible to infer the number of precursor cells in the starting cell population. This is possible because a linear relationship exists between the number of colonies formed and the number of cells cultured. By comparison, liquid suspension cultures of primary haemopoietic cells do not allow enumeration of precursor cell numbers as the progeny are intermingled in the culture dish. The second feature of clonal cultures is the dose–response relationship that exists between the amount of growth factor and the number of colonies stimulated. This dose–response relationship is sigmoid, having a linear phase and a plateau phase. The linear portion of the curve can be used to determine the amount of growth factor activity; in cultures described here, 50 units of growth factor activity correspond to the amount of activity stimulating 50% of maximal colony numbers. More detailed information on haemopoietic colony formation and cytokines can be found in Metcalf (1984, 1985, 1986, 1991) and Nicola (1991).

II. MATERIALS AND INSTRUMENTATION

A. Semisolid Agar Medium Cultures

Iscove's modified Dulbecco's medium powder, with 4 mM L-glutamine and 25 mM HEPES buffer (IMDM, Cat. No. 12200-085 for a 5-litre batch), is from GIBCO-Invitrogen. DEAE-dextran hydrochloride (Cat. No. D9885), L-asparagine monohydrate (Cat. No. A4284), 2-mercaptoethanol (Cat. No. M7522), penicillin G (Cat. No. P7794), and streptomycin (Cat. No. S9137) are from Sigma-Aldrich. Bacto agar is from Bacto Laboratories (Cat. No. 0140-17). Recombinant haemopoietic growth factors and cytokines can be purchased from a number of commercial suppliers (e.g., PeproTech EC Ltd, R&D Systems). For routine cultures, conditioned medium can be prepared from a number of tissues (see later). Foetal calf serum (FCS) can be obtained from a number of suppliers, but requires pretesting (see later). Bacteriological graded petri dishes (35–36 and 100 mm diameter) can be obtained from a number of suppliers as a tissue culture grade dish is not required. For viable cell count, eosin Y from Sigma-Aldrich (Cat. No. E4009) is required. Distilled Milli-Q endotoxin-free water (H_2O) is routinely used.

B. Conditioned Medium

IMDM (Cat. No. 12200-085) and pokeweed mitogen (PWM, Cat. No. 15360-019) are from GIBCO-Invitrogen. 2-Mercaptoethanol (2ME, Cat. No. M7522), sodium bicarbonate ($NaHCO_3$, Cat. No. S6014), penicillin G (Cat. No. P7794), and streptomycin (Cat. No. S9137) are from Sigma-Aldrich. FCS can be obtained from various suppliers, and batches can only be selected by prior testing. If a batch is available that is known to support good colony formation, then this can be used to prepare conditioned media. A variety of flasks can be used to prepare PWM-stimulated spleen cell-conditioned media (PWM-SCM) for murine cultures, although we routinely use 1- to 2-litre glass flasks fitted with a cotton plug to allow gas diffusion. For preparation of human placental conditioned media, a variety of flasks can also be used, but we routinely use disposable 75-cm^2 tissue culture flasks (Cat. No. 25110-75, Corning).

C. *In Situ* Colony Staining

Glutaraldehyde solution (grade II, 25% in water, Cat. No. G6257), ethanol (Cat. No. 02862), urea (Cat. No. U6504), acetylthiocholine iodide (Cat. No. A5751), sodium phosphate dibasic (Na_2HPO_4, Cat. No. 71629), sodium phosphate monobasic (NaH_2PO_4, Cat. No. 71492), sodium citrate (Cat. No. S1804), copper(II) sulphate ($CuSO_4$, Cat. No. C1297), potassium ferricyanide [$K_3Fe(CN)_6$, Cat. No. 60310], and Mayer's haematoxylin staining solution (Cat. No. MHS-16) are from Sigma-Aldrich. Luxol fast blue MBS (Code Pack No. 10732) is from NBS Biologicals. Hardened ashless 541 filter paper of 5.5-cm diameter (Cat. No. 1541 055) is from Whatman, and DePex (D.P.X) neutral mounting medium (Cat. No. 3197) is from Bacto Laboratories. Plain 2 × 2-in. microscope slides (Cat. No. 5075) and 45 × 50-mm cover glasses (Cat. No. 4550-1) can be obtained from Brain Research Laboratories.

D. Instrumentation

For colony examination and counting, a zoom stereomicroscope (Model SZ4045) fitted with a clear stage plate and a base illuminator from Olympus is used routinely. Many brands of CO_2 incubators are available and it is recommended that one with stainless-steel water jackets be obtained. In addition, to minimize the desiccation of cultures, an incubator without an inbuilt fan is preferred. A standard light microscope equipped with 10 and 40× objectives will be sufficient for routine cell counting and *in situ* colony typing. Graduated, glass blow-out pipettes in volumes of 1, 5, 10, and 25 ml are routinely used, although they can be replaced by disposable pipettes. For the concentration of conditioned media, we use an Amicon hollow-fibre concentrator (Model DC2A) fitted with a HIP10 membrane.

III. PROCEDURES

A. Method for Establishing and Scoring Agar-Medium Cultures

The fundamental steps in establishing agar-medium cultures are fourfold: (1) mix equal volumes of double-strength medium and double-strength agar solution, (2) add the cells to be cultured and mix, (3) pipette the cell suspension onto the culture dishes, and (4) after gelling the culture dishes are put in an incubator for colony formation.

Solutions

1. *IMDM for agar cultures (AIMDM)*: 88.3 g IMDM, 0.6 g penicillin G, 0.375 g streptomycin, 7.5 ml DEAE-dextran (50 mg/ml solution), 1.0 g L-asparagine, 24.5 g $NaHCO_3$, and 29.5 µl 2-mercaptoethanol. Dissolve the contents of a 5-litre package IMDM (88.3 g) in 1 litre of H_2O using a magnetic stirrer for mixing. Rinse the inside of the package to remove all traces of powder. Add afore-listed reagents while stirring continuously. Add H_2O to a final volume of 1.95 litres and gas medium with 100% CO_2 until it is a yellow–orange colour. The prepared endotoxin-free medium should then be filter sterilized and distributed in 100- or 250-ml aliquots, which are then tightly capped, stored at 4°C, and protected from light. Each preparation of AIMDM should be batch tested against one that is currently in use if applicable.

2. *Agar*: 0.6 g Bacto-agar and 100 ml H_2O. Weigh agar into a 100-ml flask, add 100 ml H_2O, and plug flask loosely. Bring to boiling for 2 min over a gas flame. Prepare immediately before setting up cultures

and maintain at 45°C in a water bath. Each new lot of Bacto-agar should be batch tested against one that is currently in use if applicable.

3. *Foetal calf serum*: FCS is used as a source of nutrients in cell cultures. Batches of FCS should be tested extensively prior to purchase, for optimal colony formation in colony number and colony size, in semisolid cultures. They should also be titrated to determine the optimal concentration (final concentration is usually 5–20%). The storage/shelf life of FCS at –20°C is at least 2 years and at –70°C it as long as 10 years. Centrifugation may be necessary to remove any sediment that forms after thawing. It is otherwise ready for use. Heat inactivation is optional but not necessary.

4. *Eosin for viable cell counts*: Prepare stock solution [10% eosin-yellow powder (w/v)] in normal saline and keep at 4°C. Mix 0.2 ml of stock eosin solution with 8.6 ml normal saline and 1.5 ml FCS to prepare working solution. Aliquots of the working solution should be store frozen.

5. *Hemopoietic growth factors*: If purchased commercially, they should be pretested to determine the amount required for optimal colony formation. If conditioned media are prepared (see later), they also require titration to determine the optimal concentration for maximal colony formation without evidence of high-dose inhibition. Stimuli should be divided into aliquots at a concentration at least 10-fold higher than that to be used finally in the culture dish. They can be stored frozen, but once thawed, they should not be frozen again as this can result in a loss of activity.

6. *Single-strength Iscove's modified Dulbecco's medium*: Dissolve the entire contents of a 5-litre package IMDM (88.3 g) with 4.8 litres of H_2O and mix with gentle stirring. Rinse the inside of the package to remove all traces of powder. Add 15.12 g $NaHCO_3$, 29.5 µl 2-mercaptoethanol, 150 mg penicillin G, and 100 mg streptomycin while stirring continuously. Add H_2O to a final volume of 5 litres and gas medium with 100% CO_2 until it is a yellow–orange colour. The prepared endotoxin-free medium should then be filter sterilized and distributed in 100- or 500-ml aliquots, which are then capped tightly, stored at 4°C, and protected from light.

7. *Pokeweed mitogen*: This should be prepared immediately prior to use. Any material not used should be discarded. Make up powder with 5 ml of double-distilled, deionized water. Remove from the vial and dilute 1:15 (v/v) with H_2O.

Steps

1. Warm AIMDM to room temperature.
2. Prepare agar solution.
3. Count viable cells using a haemocytometer and eosin as dye.

4. Draw culture layout, in a book, showing culture number, stimuli for each culture dish, and the number of cells for each culture dish.

5. Place required number of culture dishes on incubator trays and number lids individually according to the culture book.

6. Add required stimuli to appropriate culture dishes as described in the culture book. For each culture, the required amount of stimulus is usually added in 0.1 ml per culture dish. This amount can be less but should not exceed 0.2 ml as the agar may not gel properly.

7. For agar cultures with 20% FCS final concentration, mix AIMDM (three parts) and foetal calf serum (two parts) first and then add an equal volume of agar (five parts). Cells are added last to this single-strength agar medium, which should be now roughly at 36–37°C before plating. For each group of cultures, allow 1 ml of agar medium each per culture plus at least 1 ml extra for wastage in pipetting. If by pretesting, a lower concentration of FCS is sufficient for optimal colony formation, replace the leftover volume with H_2O so that the amount of AIMDM and FCS equals 0.5 ml for each 1 ml of culture.

8. Aliquot 1-ml volumes into petri dishes and swirl to mix stimuli and agar medium-containing cells.

9. Allow mixture to gel and place in a fully humidified containing 5–10% CO_2 in air.

10. After the required incubation period (normally 7 days for murine progenitor cells and 14 days for human progenitors and murine high-proliferative stem/progenitor cells), remove cultures from incubator and count colonies using an Olympus SZ stereomicroscope at 30–35× magnification. Murine haematopoietic colonies are defined routinely as clones greater than 50 cells and human colonies greater than 40 cells (some investigators count human colonies as having greater than 20 cells). Place the microscope on top of a black platform and adjust the concave side of the mirror until cells appear white against a black background (Fig. 1). To aid in the enumeration of colonies, we routinely put the 35-mm agar culture dishes on top of an inverted 60-mm culture dish marked with 6-mm² grids.

B. Preparation of Conditioned Media

Although specific stimuli may be required for many situations, and always for use as a positive control, a variety of conditioned media containing mixtures of haemopoietic growth factors can be prepared. The following examples provide descriptions for the preparation of two conditioned media: one suitable for human cultures and the other for murine cultures.

FIGURE 1 Photomicrograph of a dispersed granulocyte–macrophage colony (×100).

1. Preparation of Human Placenta-Conditioned Medium

Steps

1. Placenta should be obtained within 9 h of birth. Place it on a large sterile tray in a biological safety cabinet.

2. Using sterile instruments, remove outer layer of placenta. Assume that this portion is not sterile. Instruments can be kept sterile by periodically returning them to boiling water.

3. Cut a portion (1 cm³) of exposed placenta and place in a 100-mm petri dish containing 10 ml IMDM. Limit each petri dish to 8–10 pieces.

4. Having removed sufficient pieces of placenta, rinse each piece through three changes of IMDM in 100-mm petri dishes to remove most of the blood.

5. Place 18–20 pieces of placenta into tissue culture flasks (75 cm², Cat. No. 25110-75, Corning) in 60 ml IMDM with 5% (v/v) FCS.

6. Place the placenta cultures, with the caps sealed loosely, in a 37°C fully humidified incubator containing 5–10% CO_2 in air.

7. After 5–7 days of incubation, harvest the medium free of placenta by pouring the contents of each flask through cotton gauze into a collection flask. Centrifuge the medium at 3000 g and store the supernatant at −20°C until 4–5 litres has accumulated.

8. Concentrate the placenta-conditioned medium approximately 10-fold using a hollow fibre concentrator (this type of concentrator is preferred because of the relatively large volumes involved).

9. Filter sterilize the concentrate and test by titration using human bone marrow semisolid agar cultures. Select the batches of conditioned media that display a sigmoid dose–response relationship with the number of colonies formed and without evidence of a high-dose inhibition.

2. Preparation of Murine Pokeweed Mitogen-Stimulated Spleen Cell-Conditioned Medium

Steps

1. Prepare a single-cell suspension of murine spleen cells, either by teasing the spleen tissue with needle or by forcing it through a fine stainless steel mesh.

2. Place the spleen cells in a tube and allow them to stand for 5 min to allow larger tissue fragments to sediment. Remove the supernatant and determine viable cell numbers.

3. Make up the cells to 2×10^6/ml in IMDM containing 10% FCS. (The concentration of FCS should be as low as possible and can be determined only by preliminary testing.) Add pokeweed mitogen (0.05 ml of a 1:15 dilution of freshly prepared stock is added for each millilitre of culture medium).

4. Incubate the cells in medium for 7 days at 37°C in a fully humidified incubator containing 5–10% CO_2 in air. The cells can be incubated in a variety of containers. We routinely use 2-litre flasks, with cotton plugs, containing 250 ml of medium.

5. After incubation, harvest the conditioned medium and centrifuge at 3000 g to remove cellular debris. Concentrate the medium 10-fold as described earlier for human placenta-conditioned medium.

6. Titrate the concentrated PWM-SCM using cultures of murine bone marrow cells to determine the concentration required to give plateau numbers of colonies.

7. The conditioned medium can then be diluted to a concentration 10 times that is required for maximal colony formation, divided into 20-ml aliquots, and stored at −20°C until required. Once thawed the PWM-SCM should be stored at 4°C.

III. COLONY TYPING

Colonies grown in agar cultures can be typed using an *in situ* whole plate staining sequentially with Luxol Fast Blue to detect eosinophil granules (Johnson & Metcalf, 1980), acetylcholinesterase to detect megakaryocytes (von Melchner and Lieschke, 1981) and Mayer's haematoxylin for nuclear morphology to detect neutrophils and monocyte/macrophages.

Preparation of Fixative and Staining Solutions

A 2.5% (v/v) glutaraldehyde in PBS is prepared by mixing 1 part of 25% glutaraldehyde with 9 parts of PBS. The luxol fast blue staining solution is prepared by dissolving 0.1 g powdered luxol fast blue MBS dye (NBS Biologicals Ltd., Hungtingdon, Cambs, England) in 100 ml of 70% ethanol saturated with urea. The substrate solution for acetylcholinesterase staining should be prepared fresh each time by dissolving 10 mg acetylthiocholine iodide (Sigma-Aldrich) in 15 ml of 100 mM sodium phosphate buffer, pH 6.0. One millilitre of 100 mM sodium citrate, 2 ml of 30 mM copper sulphate, and 2 ml of potassium ferricyanide solutions are then added sequentially with constant stirring.

Steps

1. Add 2 ml 2.5% glutaraldehyde into individual culture dishes and leave at room temperature overnight.

2. Transfer the gel onto a 3×2-in. glass slide with the aid of a water bath and then cover the gel with a wet 5.5-cm Whatman 541 filter paper and allow drying at room temperature in a fume hood.

3. Remove the filter paper, leaving a thin film of gel containing compressed colonies on the glass slides.

4. Slides are first stained for acetylcholinesterase by incubating the slides in the dark with the substrate solution at room temperature for 3 h.

5. After washing under running tap water for 10–15 min, transfer slides into the luxol fast blue staining solution and stain for 30 min at room temperature.

6. After another washing under running tap water for 30 min, counterstain slides with Mayer's haematoxylin for 1 min, wash, and "blue" in running tap water.

7. Dry slides at room temperature, mount with DePex, and examine under a light microscope after drying.

IV. PITFALLS

1. Numerous pitfalls are associated with these procedures. A major problem involves selection of a suitable batch of FCS. If possible, a known positive sample should be obtained from a colleague or a commercial source (e.g., StemCell Technologies) for use as a control when testing new batches.

2. The agar needs to be boiled to ensure that it is dissolved properly. For this reason, it is recommended to use a gas flame. The agar will initially bubble up when boiling, and care must be taken to prevent it from overflowing. Once the bubbling has subsided, the agar should be boiled for another minute. An autoclave should not be used to prepare the agar solution.

3. It is imperative that the incubator being used is fully humidified, as desiccation of the cultures will prevent colony growth. With satisfactory cultures, a small volume of liquid will be evident at the edge of the agar medium when the cultures are tilted. If desiccation has occurred, the surface of the agar medium will display irregularities instead of being smooth and shiny. To prevent desiccation, the incubator should contain one or more large open trays containing H_2O. Humidity can be improved by pumping the air–gas mixture into the incubator via a tube immersed in one of the trays of water. Many incubators are fitted with a fan to produce a uniform atmosphere with the closed incubator. This can also cause the desiccation of cultures and the fan may have to be disconnected.

4. The possibility of cultures drying out is increased by extending the incubation time. For cultures in excess of 7 days it is a good policy to place culture dishes in a 100-cm petri dish (two cultures per dish) containing a third open-lid, 35-mm culture dish with H_2O. It is also a good policy to minimize the number of openings of the incubator.

5. When establishing the cultures, problems can arise due to the temperature of the agar–medium mixture. If too cold, it will gel prematurely and the cells will not be immobilized properly. If too hot, it will kill the cells. It is recommended that the agar be maintained at 45°C and that the AIMDM and FCS be allowed to warm to room temperature (18–20°C). When agar, medium, and FCS are mixed, the temperature of the solution will be about 37°C, which will not kill the cells but will still be above the gelling temperature of the agar. Once cells have been added to the agar medium mixture and mixed, it should be dispensed to culture dishes as soon as possible. As 0.1 ml of stimulus is usually present in the culture dish, the mixture should be swirled gently to allow homogeneous mixing. All these actions should be executed before gelling starts to occur.

References

Atkinson, K. (1993). Cytokines in bone marrow transplantation. *Today's Life Sci.* **5**, 28–38.

Clark, S. C., and Kamen, R. (1987). The human hematopoietic colony-stimulating factors. *Science* **236**, 1229–1237.

Johnson, G. R., and Metcalf, D. (1980). Detection of a new type of mouse eosinophil colony by Luxol-Fast-Blue staining. *Exp. Hematol.* **8**, 549–561.

Mertelsmann, R., Herrman, F., Hecht, T., and Schulz, G. (1990). Hematopoietic growth factors in bone marrow transplantation. *Bone Marrow Transplant.* **6**, 73–77.

Metcalf, D. (1984). "The Hemopoietic Colony Stimulating Factors." Elsevier, Amsterdam.

Metcalf, D. (1985). The granulocyte-macrophage colony-stimulating factors. *Science* **229**, 16–22.

Metcalf, D. (1986) How reliable are *in vitro* clonal cultures? Some comments based on hemopoietic cultures. *Int. J. Cell Cloning* **4**, 287–294.

Metcalf, D. (1991). Control of granulocytes and macrophages: Molecular, cellular, and clinical aspects. *Science* **254**, 529–533.

Nicola, N. A. (1991). Receptors for colony stimulating factors. *Br. J. Haematol.* **77**, 133–138.

Sheridan, W. P., Morstyn, G., Wolf, M., Lusk, J., *et al.* (1989). Granulocyte colony-stimulating factor and neutrophil recovery after high dose chemotherapy and autologous bone marrow transplantation. *Lancet* **2**, 891–895.

von Melchner, H., and Lieschke, G. J. (1981). Regeneration of hemopoietic precursor cells in spleen organ cultures from irradiated mice: Influence of genotype of cells injected and of the spleen microenvironment. *Blood* **57**, 906–912.

14

Human Skeletal Myocytes

Robert R. Henry, Theodore Ciaraldi, and Sandeep Chaudhary

I. INTRODUCTION

Human muscle tissue consists of many cell types, including adipocytes, fibroblasts, nerve cells, and stromal-vascular components. In the past, biochemical studies of human muscle used organ culture or dissociated monolayers of primary cells. By the nature of theses techniques, both methods would result in contamination of the muscle cells with other cell types. In 1981, Blau and Webster developed a method to maximize proliferation and differentiation of muscle satellite cells to produce *cultures of pure human muscle cells.* Subsequently, serum-free media were developed to optimize the growth of human myocytes without differentiation (Ham *et al.,* 1988). It then became feasible to use a sequential two-media approach to grow and differentiate human myocytes. With these techniques, and more recent modifications, it is now possible to study human muscle myocytes without the confounding complication of contamination by other cellular components of muscle tissue.

There are many reasons why it is valuable to be able to study myocytes in isolation. Some investigators have considered the potential of human myoblasts in gene therapy. This exploits the important characteristic of muscle cells; the progeny of a single cell can be taken full circle from the animal to the culture dish and then back to the animal where they fuse into mature myofibers of the host (Blau *et al.,* 1993). In fact, this quality of myoblasts affords itself as a way to develop gene- and cell-based therapies for many genetic disorders, including Duchenne muscular dystrophy (Gussoni *et al.,* 1997). A number of investigators have been studying the effects of insulin on multiple aspects of metabolism in human muscle cells as it relates to type 2 diabetes mellitus and other insulin-resistant states (Borthwick *et al.,* 1995; Park *et al.,* 1998; Gaster *et al.,* 2002). Maintenance of muscle cells under controlled conditions permits evaluation of the contribution of the components of the type 2 diabetic environment, such as hyperglycemia, hyperinsulinemia, and hyperlipidemia, to the metabolic behaviors of muscle as compared to the intrinsic or genetic properties of muscle. The culture conditions can also be manipulated to reproduce acquired behaviors. As a final example of how human myoblast cultures can be employed, investigators have also looked at the development of factors during the maturation of human myocytes to myotubes (Gunning *et al.,* 1987). Therefore, the ability to investigate human muscle cells *in vitro* has a great importance in discovering clinically significant *in vivo* pathophysiology.

Investigators have discovered that the differentiation of myoblasts to myotubes while in culture involves changes in gene expression (Gunning *et al.,* 1987). Using the technique illustrated in this article, *characterization of the muscle cells* in terms of the biochemical markers they express during the fusion process has been described previously. Some of the changes in gene expression in major markers that are expressed include increases in sarcomeric specific α-actin protein (3.5-fold), the muscle-specific isoform of creatinine phosphokinase (CPK-M) mRNA (2-fold), and CPK-M enzyme activity (6-fold). The process of myoblast fusion into myotubes can also be observed structurally through fluorescent micrographs of cells stained for nuclei, as seen in Fig. 1 (Henry *et al.,* 1995). A more complete list of changes in biochemical and histologic markers following differentiation of myocytes into myotubules is presented in Table I. It is important to monitor a number of these markers when manipulating conditions or comparing cells from different individuals.

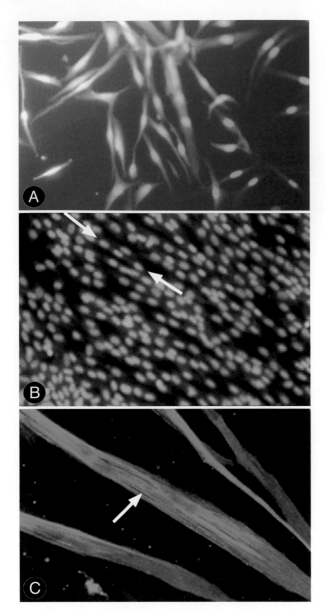

FIGURE 1 **Myocytes and myotubules in various stages of differentiation.** (A) Individual myocytes prior to differentiation. (B) Myotubules showing the classic "peas in a pod" appearance of multiple nuclei within single cells. Note that there are several mononucleated as well as multinucleated cells in the culture. (C) Striations in differentiated myotubules.

II. MATERIALS AND INSTRUMENTATION

Hams F10 media (Cat. No. 9056), custom ATV (Cat. No. 9920), α-MEM (Cat. No. 9142), Fungibact (Cat. No. 9350), penicillin/streptomycin (pen/strep) (Cat. No. 9366), and glutamine (Cat. No. 9317) are from Irvine Scientific (Santa Ana, CA). The SKBM and bullet kit

(Cat. No. 3160) are from BioWhittaker (Walkersville, MD). To make SKGM, the components of the bullet kit are added to one bottle of SKBM, along with 10 ml Fungibact and 5 ml glutamine. The insulin bullet can be added or omitted depending on the final purpose for which the cells will be utilized (see later).

Fetal bovine serum (FBS) is from Gemini (Calabasas, CA) (Cat. No. 100–106). The Hams F10 media, SKGM, and α-MEM should all be stored at 4°C. Stocks of custom ATV, fetal bovine serum, glutamine, Fungibact, and pen/strep should be stored at –20°C. To make fusion media, add fetal bovine serum to a final concentration of 2%, 10 ml of pen/strep, and 5 ml of glutamine to 500 ml of α-MEM.

The following were all obtained from Fisher Scientific: 6-well tissue culture plates (Cat. No. 08772-1G), 12-well tissue culture plates (Cat. No. 08772-3A), 24-well tissue culture plates (Cat. No. 07200-84), 100-mm culture dishes (Cat. No. 08772-E), sterile scalpel (Cat. No. 08927-5A), sterile pasteur pipettes (Cat. No. 13678-20B), sterile flasks (Cat. No. 100429-A), 15-ml conical polypropylene tubes (Cat. No. 145970C), 50-ml conical tubes (Cat. No. 1495949-A), 5-cc pipettes (Cat. No. 13675-22), 10-cc pipettes (Cat. No. 13675-20), 25-cc pipettes (Cat. No. 13655-30), p200 pipette tips (Cat. No. 02681422), and p100 pipette tips (Cat. No. 02681422). No special treatment or substrate is needed for the cells to attach to the plastic. Cells are grown and maintained in a humidified, 37°C incubator under 5% CO_2.

III. PROCEDURE

The following techniques for the growth of human skeletal muscle cultures were established through modifications of previously described methods (Blau and Webster, 1981; Sarabia *et al.*, 1990).

A. Cell Isolation

All steps are to be performed under sterile conditions.

1. Collect muscle tissue in a 50-ml conical tube in 20 ml cold (4°C) Hams F10 media: approximately 100–200 mg of tissue is needed.

2. Aspirate media. Wash tissue three times with chilled (on ice) Hams F10 to remove blood.

3. Transfer tissue in a small volume (~5 ml) of Hams F10 to a 100-mm culture dish.

4. Mince tissue with sterile scalpels. Make pieces as small as possible.

5. Transfer tissue and media back to centrifuge tube and aspirate media with a sterile Pasteur pipette.

TABLE I Markers of Differentiation of Human Myocytes to Myotubues

	Change during differentiation	Reference
Biochemical marker		
Sarcomeric-specific α-actin protein	Increase from 20 to 40% of actin produced	Blau and Webster (1981)
CPK gene expression as measured by mRNA	Increase 2-fold	Henry *et al.* (1995)
Acetylcholine receptor as measured by α-bungarotoxin staining	Increase 33-fold	Blau and Webster (1981)
CPK enzyme activity	Increase 6- to 18-fold	Blau and Webster (1981); Henry *et al.* (1995)
Sarcomeric myosin heavy Chain mRNA	Increase from 7 to 42% of maximum concentration during first day of differentiation and then increase slowly over 14 days	Gunning *et al.* (1987)
α- and β-tubulin mRNA	Peak at day 5 to 10 with a change in β:α ratio from 3 to greater than 8	Gunning *et al.* (1987)
Desimin protein	Increases	van der Ven *et al.* (1992)
Human myoglobin protein	Increases	Caviedes *et al.* (1992)
Total myosin	Increases	Caviedes *et al.* (1992)
Morphologic markers		
Multinucleation contained within myotubules	Rapid increase to 40% of nuclei in myotubes in the first 24 h, with a progressive doubling of multinucleation over the next 14 days	Gunning *et al.* (1987)
Striation	Appears	Blau and Webster (1981)
Myosin light chain staining	Increases	Blau and Webster (1981)

Because the tissue will settle out on its own, the sample does not need to be centrifuged.

6. Add 20 ml of trypsin/EDTA (custom ATV) and transfer to a small flask (e.g., 50-ml Erlenmeyer flask) containing a stir bar.

7. Stir 20–30 min at room temperature in a sterile hood. Collect supernatant and place on ice.

8. Add 20 ml ATV to flask and repeat step 7 two more times. Pool the supernatants together on ice.

9. To the pooled supernatants add FBS to a final concentration of 10%.

10. Centrifuge cells for 5 min at 1600 rpm (550 g) at room temperature.

11. Aspirate supernatant and add 20 ml of SKGM [SKBM + bullet kit (with or without insulin) + 10 ml Fungibact + 5 ml glutamine] to cell pellet. Pipette up and down gently to resuspend cells.

12. Transfer media with cells into two 100-mm dishes. Mark plates with subject identifier and place in incubator at 37°C, 5% CO_2.

13. Change SKGM media approximately every 3 days. Continue for 2–3 weeks.

14. During the next 2–3 weeks, the muscle cells will grow attached to the surface of the culture dish. Even-tually, the cells will form a confluent layer on the bottom of the dish.

B. Cell Culture

1. When the cells are 60–70% confluent, aspirate media and rinse the cells once with 5 ml of custom ATV and aspirate.

2. Place another 5 ml custom ATV on plates, aspirate after 30 s, and then incubate for 5 min at 37°C.

3. Check cells under a microscope to see if they are detached (cells will look rounded).

4. If cells are detached, add 5 ml SKGM to rinse plate, collect cells, and transfer to a 15-ml centrifuge tube. If they are only partially detached, remove the attached cells by pipetting up and down gently.

5. Add 5 ml SKGM media to trypsinized cells and centrifuge at room temperature for 5 min at 1600 rpm (550 g).

6. Aspirate supernatant and resuspend cells in up to 5 ml SKGM media (not more than 6 ml, depending on volume needed) by pipetting up and down gently.

7. Count cells in a hemocytometer chamber.

8. Plate cells:

100-mm dishes: 60,000 cells per plate (approximately 60 µl cell suspension)

6-well dishes: 20,000 cells per well (approximately 30 µl)

12-well dishes: 6000 cells/well (approximately 20 µl)

24-well dishes: 3000 cells/well (approximately 10 µl per well)

9. Cover wells with SKGM: 10 ml per 100-mm dish, 2 ml for each well per 6-well plate, and 1 ml for each well per 12-well and 24-well plates.

10. Change the SKGM every 2–3 days.

C. Cell Fusion/Differentiation

1. When cell cultures are 80–90% confluent, aspirate media and rinse two times with α-MEM/pen–strep/ 5 mM glutamine/2% FCS (fusion media).

2. Add fusion media to wells: 10 ml per 100-mm dish, 2 ml /well per 6-well plate, and 1 ml/well per 12-well and 24-well plates.

3. Culture at 37°C, 5% CO$_2$.

4. Change media every 48 h.

5. Fusion/differentiation is complete by 96 h.

IV. COMMENTS

The technique just described has been modified primarily to investigate the metabolic characteristics of human muscle cells. The system has been employed to study insulin action on glucose uptake and glycogen synthesis, fatty acid uptake and oxidation, insulin signaling, and regulation of gene expression. Other investigators have employed different techniques for the culture of human muscle cells; however, the focuses of those investigations were not on hormone action or metabolic activities (Rando and Blau, 1994; Webster and Blau, 1981), which is why the types of media employed differ from those described in this article. The major differences from the procedure described here are that growth media include Hams F-10 with 0.5% chick embryo extract and 20% fetal calf or horse serum. Furthermore, fusion media contain Dulbecco's modified Eagle's media with 2% horse serum. It is uncertain how the constituents of these other media would differ in their impact on human skeletal muscle metabolism when compared to the defined SKGM and fusion media described in this article.

In the system described here, cells are passed only a single time before terminal differentiation and are not maintained over multiple passages. Our experience is that both the extent of differentiation (percent-age multinucleated cells) and the insulin responsiveness for metabolic events are diminished in a passage-dependent manner. Other investigators have maintained cultures from a single subject for a greater number of passages (as high as 15) (Halse et al., 1998).

V. PITFALLS/CAUTIONS

1. It is critical to use sterile technique to avoid cell contamination. Bacteria or fungal contamination can lead to altered metabolic activity of the muscle cells.

2. Omit insulin from the SKBM if planning to investigate the effect of insulin on muscle cultures. If the insulin bullet is added, the final concentration (~30 µM) is high enough to produce a state of insulin resistance.

3. To avoid excessive cell damage from the trypsin, do not overincubate with ATV.

4. Change media every 48 h to avoid exhaustion of growth factors and glucose. This can alter the metabolic activity of the muscle culture.

5. Treatment of cells can be performed either during the fusion/differentiation period or at completion. If treatment is done during differentiation, then the extent of differentiation must be monitored for each new manipulation. This can be done with careful checking of differentiation by following muscle markers mentioned in Table I.

6. It is important to pass the cells at 60–70% confluency. Beyond that point spontaneous fusion may begin in adjacent cells, reducing subsequent proliferation.

7. If the fusion media and SKBM are used more than 10 days after assembly, it will be necessary to supplement media with additional glutamine.

References

Blau, H., Jyotsna, D., and Grace, P. (1993). Myoblasts in pattern formation and gene therapy. TIG **9**(8), 269–274.

Blau, H., and Webster, C. (1981). Isolation and characterization of human muscle cells. Proc. Natl. Acad. Sci. USA **78**(9), 5623–5627.

Borthwick, A., Wells, A., Rochford, J., Hurel, S., Turnbull, D., and Yeaman, S. (1995). Inhibition of glycogen synthase kinase-3 by insulin in cultured human skeletal muscle myoblasts. Biochem. Biophys. Commun. **210**(3), 738–745.

Caviedes, R., Liberona, J., Hidalgo, J., Tascon, S., Salas, K., and Jaimovich, E. (1992). A human skeletal muscle cell line obtained from an adult donor. Biochem. Biophys. Acta **1134**(3), 247–255.

Gaster, M., Petersen, I., Hojlund, K., Poulsen, P., and Beck-Nielsen, H. (2002). The diabetic phenotype is conserved in myotubules established from diabetic subjects: Evidence for primary defects in glucose transport and glycogen synthase activity. Diabetes **51**(4), 921–927.

Gunning, P., Hardeman, E., Wade, R., Ponte, P., Bains, W., Blau, H., and Kedes, L. (1987). Differential patterns of transcript accumulation during human myogenesis. *Mol. Cell. Biol.* **7**(11), 4100–4114.

Gussoni, E., Blau, H., and Kunkel, L. (1997). The fate of individual myoblasts after transplantation into muscles of DMD patients. *Nature Med.* **3**(9), 970.

Halse, R., Rochford, J., McCormack, J., Vandenheede, J., Hemmings, B., and Yeaman, S. (1998). Control of glycogen synthesis in cultured human muscle cells. *J. Biochem. Chem.* **274**(2), 776–780.

Ham, R., St. Clair, J., Webster, C., and Blau, H. (1988). Improved media for normal human muscle satellite cells: Serum-free clonal growth and enhanced growth with low serum. *In Vitro Cell. Dev. Biol.* **24**(8), 833–844.

Henry, R., Abrams, L., Nikoulina, S., and Ciaraldi, T. (1995). Insulin action and glucose metabolism in nondiabetic control and NIDDM subjects. *Diabetes* **44**, 936–946.

Park, K., Ciaraldi, T., Lindgren, K., Abrams-Carter, L., Mudaliar, S., Nikoulina, S., Tufari, S., Veerkamp, J., Vidal-Puig, A., and Henry, R. (1998). Troglitazone effects on gene expression in human skeletal myocytes of type II diabetes involve upregulation of peroxisome proliferator-activated receptor-γ. *J. Clin. Endrocrinol. Metab.* **83**(8), 2830–2835.

Rando, T., and Blau, H. (1994). Primary mouse myoblast purification, characterization and transplantation for cell-mediated gene therapy. *J. Cell Biol.* **125**(6), 1275–1287.

Sarabia, V., Lam, L., Burdett, E., Leiter, L.A., and Klip, A. (1990). Glucose uptake in human and animal muscle cells in culture. *Biochem. Cell Biol.* **68**, 536–542.

Van der Ven, P.F., Schaart, G., Jap, P.H., Sengers, R.C., Stadhounders, A.M., and Ramaekers, F.C. (1992). Differentiation of human skeletal muscle cells in culture: Maturation as indicated by titin and desmin striation. *Cell Tissue Res.* **270**(1), 189–198.

15

Growing Madin–Darby Canine Kidney Cells for Studying Epithelial Cell Biology

Kai Simons and Hilkka Virta

I. INTRODUCTION

Epithelial cells display a structural and functional polar organization (Simons and Fuller, 1985). In these cells, the plasma membrane can be divided into two distinct domains, the apical membrane and the basolateral membrane, each containing different sets of proteins. The apical membrane facing a secretory or an absorptive lumen is delimited by a junctional complex from the basolateral membrane. The tight junction (zonula occludens) is the most apical member of the complex. It is found at the intersection between the apical and the lateral plasma membranes and joins each cell to its neighbors, thus limiting the diffusion of molecules between the luminal and the serosal compartments (Gumbiner, 1987). This junction also prevents the lateral diffusion of membrane proteins from one domain to another, thus maintaining their unique composition. Immediately basal to the tight junctions is the intermediate junction (zonula adherens or belt desmosomes). The other more basal junctional elements are desmosomes (maculae adherentes) and gap junctions, which attach the lateral membranes of adjacent cells to each other. The junctional complex is involved in sealing the epithelium; it prevents molecules from diffusing between adjacent cells. The basolateral membrane faces the bloodstream and is involved in cell–cell contact and cell adhesion to the basement membrane.

For most studies on epithelial cell polarity, cultured cells have been used. These cells are superior to cells obtained from tissues because they can be grown under carefully controlled conditions and are easily manipulated. The cell population is homogeneous. Biosynthetic experiments using pulse–chase techniques with radioactive precursors can be accomplished at an analytical level with a short time resolution. Endocytosis and transcytosis can also be studied.

The most well-studied epithelial cell is the Madin–Darby canine kidney (MDCK) cell. This cell line is derived from normal dog kidney (McRoberts *et al.*, 1981). An unusual feature of these cells is that while in culture they retain many differentiated properties characteristic of kidney epithelial cells. Among these are an asymmetric distribution of enzymes and vectorial transport of sodium and water from the apical to the basolateral faces. The latter gives rise to "domes" or "blisters" in confluent cultures, which are transient areas where collected fluid has forced the monolayer to separate from the substratum. Morphologically, the cells resemble a typical cuboidal epithelium with microvilli on the apical side of the cells. Two different strains of the MDCK cell are known (Richardson *et al.*, 1981; Balcarova-Ständer *et al.*, 1984). Strain I cells are derived from a low-passage MDCK cell stock and these cells form a tight epithelium with transepithelial resistance above $2000\,\Omega\cdot cm^2$. Strain II cells form a monolayer of lower resistance of 100–$200\,\Omega\cdot cm^2$. MDCK strain II cells have been used primarily for studies of the cell biology of epithelial cells. Transcytosis is, however, studied more conveniently in MDCK strain I cells because of their high electrical resistance.

Several factors are important for optimal expression of the epithelial phenotype *in vitro* (Simons and Fuller, 1985). A primary consideration is the polarity of nutrient uptake. *In vivo*, many nutrients reach the epithelial sheet from the basolateral side, which faces the blood supply; however, when epithelial cells are cultured on glass or plastic, they are forced to feed from the apical surface, which faces the culture medium. Hence, the basolateral surface becomes isolated from the growth medium as the monolayer is sealed by the formation of tight junctions. To grow properly, the epithelial sheet must remain somewhat leaky or expose basolateral proteins responsible for the uptake of nutrients and binding of growth factors on the apical side.

These problems can be overcome simply by growing the epithelial cells on permeable supports, such as polycarbonate and nitrocellulose filters. Epithelial cells form monolayers with a higher degree of differentiation when the basolateral surface is directly accessible to the growth medium. This is evident from the morphology of the cells, their increased responsiveness to hormones, and the exclusion of basolateral proteins from their apical surfaces.

II. MATERIALS AND INSTRUMENTATION

Minimal essential medium with Earle's salt (MEM) is purchased as a powder (Cat. No. 11700-077) from Biochrom, mixed with Milli-Q-filtered H_2O, and sterile filtered. Glutamine (200 mM, Cat. No. 25030-024), penicillin (10,000 IU/ml)–streptomycin (10,000 mg/ml) (Cat. No. 15140-122), trypsin (0.05%)–EDTA (0.02%) (Cat. No. 25300-054), and phosphate-buffered saline (PBS, Cat. No. 041-04040H) are from GIBCO-BRL. The Transwell polycarbonate filters (2.45 cm, Cat. No. 3412, and 10 cm, Cat. No. 3419) are from Costar. Tissue culture flasks (75 cm², Cat. No. 156499) are from Nunc. The glass petri dishes (140 mm diameter and 30 mm high) for holding six 2.4-cm Transwell filters are from Schott Glasware. The laminar flow hood (Steril-Gard Hood Model VMB-600) is from Baker. The CO_2 incubator (Model 3330) is from Forma Scientific. The inverted Diavert microscope is from Leitz. The electrical resistance measuring device (EVOM) is from World Precision Instruments. The centrifuge (Type 440) is from Hereaus-Christ.

III. PROCEDURES

A. Growing Madin–Darby Canine Kidney Cells on Plastic

MDCK I and II cells are passaged every 3–4 days up to 25 passages. One flask is usually split into five new flasks. MDCK II cells usually form domes within 2 days of splitting, whereas MDCK I cells do not blister.

Solutions

1. *MEM growth medium*:

	Stock	Volume/liter
5% fetal calf serum (MDCK II)	100%	50 ml
10% fetal calf serum (MDCK I)	100%	100 ml
2 mM glutamine	200 mM	10 ml
100 IU/ml penicillin 100/μg/ml streptomycin	100X	10 ml

2. *Phosphate-buffered saline*
3. *0.05% trypsin/0.02% EDTA*

Steps

1. Wash hands and wipe laminar flow hood with 70% ethanol. Warm all solutions to 37°C. All manipulations are done in the laminar flow hood. When splitting the cells, remove the growth medium from the 75-cm² flasks containing the confluent layer of MDCK cells and add 10 ml PBS. Rinse and discard wash solution.

2. Add 5 ml trypsin–EDTA solution, seal flask, and incubate for 10–15 min at room temperature (until small patches of cells are rounded up but not yet detached from the flask).

3. Remove the trypsin–EDTA solution and add 1.5 ml of fresh trypsin–EDTA. Reseal the flask and incubate at 37°C for 10–15 min (MDCK II) or 25–30 min (MDCK I). At this point the cells should flow down to the bottom of the flask when the flask is turned up. Hit the flask hard against the palm of your hand.

4. Add 10 ml prewarmed MEM growth medium and resuspend the cells with a sterile 10-ml pipette (at least five times up and down). Using the inverted microscope, check that the cells are not sticking to each other.

5. Plate 2 ml of the cell suspension in a new 75-cm² flask containing 20 ml of MEM growth medium.

B. Seeding MDCK Cells on Polycarbonate Filters

Solutions

1. *MEM growth medium containing 10% fetal calf serum, penicillin-streptomycin, and 2 mM glutamine* (see solution A1).
2. *Phosphate-buffered saline*
3. *0.05% trypsin/0.02% EDTA*

Steps

1. Seed the cells on the filters at high density, higher than that achieved by confluent cells on plastic. The cells form tight junctions within 24 h and reach maximum tightness on the filters in 4 days. During this time cell density increases to more than five times that achieved on plastic. We place the filters in the petri dishes containing growth medium. For seeding we use one 75-ml flask, containing a confluent layer of MDCK I and II cells, for 2.4-cm-diameter filters (Transwell 3412). If you use large 10-cm-diameter filters (Transwell 3419), seed one 75-cm^2 culture flask of MDCK cells into each large filter.

2. Pour off medium from the culture flask and rinse cells with 10 ml of warm PBS. Pour off PBS.

3. Add 5 ml of warm trypsin–EDTA to cells. Leave in laminar flow hood.

4. After 15 min remove the trypsin–EDTA with a pipette, add 1.5 ml of trypsin–EDTA, and put the flask into a CO_2 incubator (37°C) for 10–15 min (MDCK II) or 25–30 min (MDCK I).

5. Remove flask (cells should be loose). Hit the flask hard against your palm. Add 10 ml of warm growth medium and suspend cells by pipetting up and down with a 10-ml pipette. Put suspension into a 50-ml Falcon tube and centrifuge for 5 min at 1000 rpm in a Hereaus-Christ centrifuge.

6. Remove supernatant and suspend cells in 9.5 ml of growth medium.

7. Pour 90 ml medium into glass petri dish containing six Transwell 3412 filters. Use 140 ml for one Transwell 3419 filter. The petri dishes contain filter holders specially made to fit either 3412 or 3419 filters (Fig. 1). Autoclave these units before use. Place filters into filter holders and allow filters to get wet from the bottom with medium. This should be done while the cells are in the centrifuge.

8. Add 1.5 ml of cell suspension to each filter in its holder. Use six Transwell 3412 filters or one Transwell 3419 filter per petri dish. Be careful not to spill cells over the edge of the filter holder.

9. Swirl petri dish gently to remove any trapped air from beneath filters.

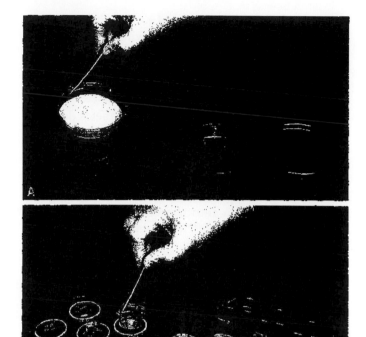

FIGURE 1 (A) The glass petri dish contains one filter holder for one Transwell 2419 filter. (B) The glass petri dish contains six filter holders for Transwell 3412 filters.

10. Place the petri dish with the filters in the CO_2 incubator.

11. Leave for 3–4 days in the incubator. No medium change is required during this time.

C. Transepithelial Resistance Measurement

Steps

1. Transepithelial resistance of filter-grown MDCK cells is measured with EVOM "chop-stick" electrodes. Each leaf has an outer and an inner electrode. The outside electrodes are small silver pads for passing current through the membrane sample. Inside the electrodes are small Ag/AgCl voltage sensors.

2. To test the instrument, switch the mode switch to R and turn the power on. Push the test R button. With the range switch in the 2000-V position, the meter will read 1000 (±1 digit). In the 20-k range, the meter will read 1.00. The meter is now ready for use.

3. To test the electrodes, insert the small telephone-type plug at the end of the chopsticks electrode cable into the jack on the front panel of the EVOM. Place the

tips of the electrodes into $0.1\,M$ KCl. Switch the mode switch to Volts. Turn the power switch on. The digital panel meter may read 1 or 2 mV due to the asymmetry of the voltage sensor pair. After 15 min, adjust this voltage to 0 mV with the screwdriver adjustment labeled "Zero V."

4. Measure resistance. The electrode set is designed to facilitate measurements of membrane voltage and resistance of cultured epithelia in culture cups by dipping one stick electrode inside the cup on top of the cell layer and the second stick electrode in the external bathing solution. To measure resistance, immerse the electrode pair again into the electrolyte and set the mode switch to "Ohm." The display should read zero; if not, adjust the display to zero with the Ohms Zero screwdriver adjustment. Push the measure R button. A steady ohm reading of the resistance should result.[1]

Example

a. Measure resistance R from solution + sample membrane support	109
b. Measure resistance R from solution + membrane support + tissue	189
c. Subtract (a) from (b)	$189 - 109 = 80$, R (tissue) $= 80\,V$
d. Calculate resistance × area product	resistance × area $= 1.2\,cm \times p \times r^2$ $= 80\,V \times 3.14 \times (1.2\,cm)^2$ $= 361.9\,V\,cm^2$

5. When moving the electrodes from one dish to another it is best not to rinse the electrodes with distilled water. If it is necessary to wash the electrodes between measurements, they should be rinsed with the membrane perfusate (e.g., PBS). Do not touch the cell layer with the internal electrode when making a measurement. Small differences in the apparent fluid resistance may occur if the depth to which the electrodes' tips are immersed varies. If the tips are unusually dirty, a light and very brief sanding with a fine nonmetallic abrasive paper will clean the sensor tip. For sterilization the electrodes may be soaked in alcohol or bactericides. After sterilization, the electrodes should be rinsed extensively with sterile perfusing solution or $1\,M$ KCl.

---

[1] If the electrode asymmetry potential difference exceeds the zero adjustment range, the central electrodes may be dirty or contaminated.

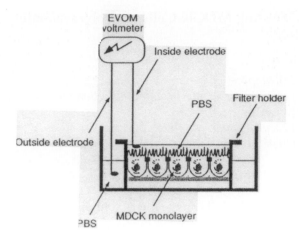

FIGURE 2 Scheme illustrating the setup for measuring electrical resistance. The stick designated for external use is slightly longer than its companion.

IV. COMMENTS

The cell layer on the filter cannot be observed in the inverted microscope because the filters are not transparent. Transparent filters are also available commercially' but they are more expensive than the polycarbonate ones; however, either the cells in one filter can be stained or the transepithelial resistance can be measured to ensure that the layer is intact. Our experience is that when one filter in the petri dish checks out, the other filters will also be fine.

It is recommended that MDCK cells not be used for more than 20–25 passages. New stock cells should then be thawed from liquid nitrogen storage.

We use filter holders for growing MDCK cells on either 2.4- or 10-cm polycarbonate filters. It is possible to grow cells in either the six-well plate for the Transwell 3412 filter or in the petri dish supplied with the Transwell 3419 filters. Under these latter culture conditions, growth media have to be changed every day, as the cells do not get enough nutrients and do not grow to optimal density. The problems with changing the medium every day are (1) the extra work involved and (2) the considerably increased risk of contamination. Therefore, we prefer to place filter holders in petri dishes into which one can add enough growth medium to last 4 days.

References

Balcarova-Stander, J., Pfeiffer, S. E., Fuller, S. D., and Simons, K. (1984). Development of cell surface polarity in the epithelial Madin-Darby canine kidney (MDCK) cell line. *EMBO J.* **3**, 2687–2694.

Gumbiner, B. (1987). Structure, biochemistry and assembly of epithelial tight junctions. *Am. J. Physiol.* **253**, C749–C758.

McRoberts, J. A., Taub, M., and Saier, M. H., Jr. (1981). The Madin-Darby canine kidney (MDCK) cell line. *In "Functionally Differentiated Cell Lines"* (G. Sato, ed.), pp. 117–139. A. R. Liss, New York.

Richardson, J. C. W., Scalera, V., and Simmons, N. L. (1981). Identification of two strains of MDCK cells which resemble separate nephron tubule segments. *Biochim. Biophys. Acta* **673**, 26–36.

Simons, K., and Fuller, S. D. (1985). Cell surface polarity in epithelia. *Annu. Rev. Cell. Biol.* **1**, 243–288.

16

Cultivation and Retroviral Infection of Human Epidermal Keratinocytes

Fiona M Watt, Simon Broad, and David M Prowse

I. INTRODUCTION

There are many techniques for culturing human epidermal keratinocytes, but the method described here is the one devised by Rheinwald and Green (1975). With this method, keratinocytes grow as multilayered sheets in which proliferation is confined to the basal layer and terminal differentiation takes place in the suprabasal layers, thus mimicking the spatial organization of normal interfollicular epidermis. These cultures have a range of applications in basic research and in the clinic. They are used to study the factors that regulate stem cell proliferation, terminal differentiation, and tissue assembly, as well as the events that take place during neoplastic transformation (Jones and Watt, 1993; Zhu and Watt, 1996; Levy *et al.*, 2000; Dajee *et al.*, 2003). Practical applications include the treatment of burns victims with cultured autografts (Compton *et al.*, 1989; Ronfard *et al.*, 2000) and the use of transduced keratinocytes as vehicles for gene therapy (Gerrard *et al.*, 1993; Dellambra *et al.*, 2001; Ortiz-Urda *et al.*, 2002).

What follows is a description of the procedures our laboratory uses to initiate and maintain cultures of keratinocytes from neonatal foreskins; it is based on the original Rheinwald and Green method, improved over the years as described by Rheinwald (1989). The key component of the culture system is the presence of a feeder layer of 3T3 cells that supports the growth of keratinocytes from clonal seeding densities (see Fig. 1). In the laboratory, our preferred method for manipulating gene expression in keratinocytes is by retroviral infection (Zhu and Watt, 1996; Levy *et al.*, 1998; Zhu and Watt, 1999) and we have therefore included our current protocol.

II. MATERIALS AND INSTRUMENTATION

FAD powder [three parts Dulbecco's modified Eagle's medium (DMEM) and one part Ham's F12 medium (F12) supplemented with $1.8 \times 10^{-4}M$ adenine] is custom made by Bio Whittaker. Alternatively, F12 (liquid, Cat. No. 41966-029) and DMEM (liquid, Cat. No. 21765-029) can be bought separately from Invitrogen, and adenine can be purchased from Sigma Aldrich Co. Ltd (Cat. No. A3159). Penicillin and streptomycin can be purchased from Invitrogen (Cat. No.15070-063). A suitable supplier of foetal calf serum (FCS) and bovine serum (BS) is Invitrogen. Hydrocortisone (Cat. No. 386698) and cholera enterotoxin (Cat. No. 227036) are purchased from Merck Biosciences Ltd. Cholera enterotoxin is from ICN Biomedicals Ltd. (Cat. No. 190329). Epidermal growth factor (EGF) is purchased from Peprotech EC Ltd. (Cat. No. 100 15). Insulin is from Sigma Aldrich Co. Ltd. (Cat. No. I5500). Mitomycin C is also obtained from Sigma Aldrich Co. Ltd (Cat. No. M0503). Trypsin (0.25%) (Cat. No. M0503) and 0.02% EDTA (Cat. No. 15040-033) can be obtained from Invitrogen. Mikrozid is purchased from Schulke & Mayer UK Ltd. Dimethyl sulfoxide (DMSO, Cat. No. 10323) is obtained from VWR International Ltd. CelStirs are made by Wheaton, USA (Cat. No. 356873; Wheaton products are available through Jencons Scientific Ltd.). Cryotubes are purchased from Alpha Laboratories Ltd. UK (Cat. No. LW 3435). Puromycin (Cat. No. P8833), chloroquine (Cat. No. C6628), polybrene (Cat. No. H9268), HEPES (Cat. No. H7006), and calcium chloride (Cat. No. C5080) are from Sigma-Aldrich Co. Ltd. Collagen-coated dishes are purchased from BD Biosciences. Hygromycin B

133

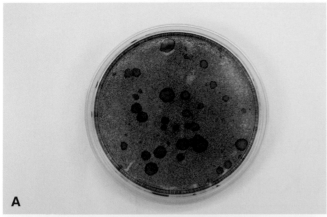

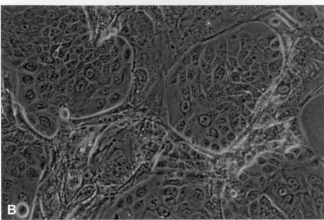

FIGURE 1   (A) Dish of cultured human keratinocytes plated on a feeder layer of 3T3 J2 cells and stained with rhodanile B. Individual clones of keratinocytes are seen as large round plaques. (B) Phase-contrast view of clones of cultured human keratinocytes on a feeder layer of 3T3 J2 cells.

(Cat. No. 10687 010) is from Invitrogen and diphtheria toxin (Cat. No. 322326) can be purchased from Merck Biosciences Ltd.

III. PROCEDURES

A. Feeder Layer

1. Solutions

1. *Culture medium for 3T3 J2 cells*: This consists of DMEM supplemented with 100 IU/ml penicillin, 100 µg/ml streptomycin, and 10% bovine serum. It is essential to batch test the serum for optimal growth of 3T3 J2.

2. *Trypsin–EDTA*: Mix one part 0.25% trypsin and four parts 0.02% EDTA. The same solution is used to harvest keratinocytes.

3. *Phosphate-buffered saline*: To make 1 litre, dissolve 0.2 g KCl, 0.2 g KH_2PO_4, 8.0 g NaCl, and 2.16 g $Na_2HPO_4 \cdot 7H_2O$ in 900 ml distilled water. Adjust pH

to 7.4, add distilled water to 1-litre final volume, autoclave, and store at room temperature.

4. *Mitomycin C in PBS*: Prepare a stock solution of 0.4 mg/ml in PBS. Filter sterilize and store in aliquots at –20°C.

1. Cells

The J2 clone of random-bred Swiss mouse 3T3 cells was selected to provide optimal feeder support of keratinocytes (Rheinwald, 1989). The cells are maintained by weekly passaging at 1:10 to 1:20 dilution. Fresh cells are thawed every 3 months, as with prolonged passaging the cells start to senesce or undergo spontaneous transformation.

2. Preparing the Feeder Layer

Steps

1. To irreversibly inhibit proliferation, add mitomycin C (final concentration, 4 µg/ml) to confluent flasks of 3T3 J2 and incubate for 2 h at 37°C.

2. Remove the medium, rinse the cells once with 0.02% EDTA, and then harvest in trypsin–EDTA. The optimal density of the feeder layer is one-third confluent (Rheinwald and Green, 1975; Rheinwald, 1989); hence each mitomycin C-treated flask is split 1:3.

3. The feeders can be used immediately (i.e., plated at the same time as the keratinocytes) or prepared 1–2 days before they are required; if prepared in advance, maintain in DMEM + 10% BS. The feeder layer should not be prepared more than 2 days before use because the feeder cells will start to degenerate.

B. Keratinocyte Culture Medium

Solutions

1. *Hydrocortisone*: Prepare a 5 mg/ml stock in absolute ethanol. Store at –20°C.

2. *Cholera enterotoxin*: Prepare a $10^{-5} M$ stock in distilled water. Store at 4°C.

3. *EGF*: Prepare at 100 µg/ml stock in FAD + FCS. Store at –20°C.

4. *Insulin*: Prepare a 5 mg/ml solution in 5 mM hydrochloric acid. Store at –20°C.

Steps

1. The basic medium consists of three parts DMEM and one part F12 supplemented with $1.8 \times 10^{-4} M$ adenine (FAD), 100 IU/ml penicillin, and 100 µg/ml streptomycin. Store at 4°C.

2. Supplement FAD with 10% FCS. It is essential to batch test the serum for its ability to support high colony-forming efficiency, rapid growth, and serial passage of keratinocytes. Serum batches optimal for keratinocytes tend to be completely unsuitable for

fibroblastic cells or hybridomas. Store FCS at −20°C before use.

3. Supplement FAD + FCS further with a HICE cocktail consisting of hydrocortisone (0.5 μg/ml), insulin (5 μg/ml), cholera enterotoxin ($10^{-10} M$), and EGF (10 ng/ml) (all final concentrations).

4. Store complete medium (FAD + FCS + HICE) at 4°C and use within 1 week.

C. Source of Keratinocytes

The usual source of keratinocytes is neonatal foreskin obtained, with ethical approval, from routine circumcisions. When handling any human tissue it is essential to take appropriate precautions against the transmission of infectious agents. Obtain the foreskin as soon as possible after circumcision and transfer to the laboratory dry in a sterile Bijou. If it cannot be used immediately, store overnight at 4°C.

D. Isolation of Keratinocytes

Solutions

1. *PBS containing 100 IU/ml penicillin and 100 μg/ml streptomycin.* Other solutions are as described in Sections III,A and III,B.

Steps

1. Rinse the foreskin thoroughly in PBS containing 100 IU/ml penicillin and 100 μg/ml streptomycin.
2. Transfer the tissue to a 100-mm-diameter petri dish, epidermis down. Remove as much connective tissue (muscle and dermis) as possible using sterile curved scissors.
3. Transfer the epidermis and remaining connective tissue to a fresh dish and chop into fine pieces (1–3 mm²) using scalpels.
4. Flood the dish with 10 ml trypsin–EDTA and transfer the solution containing pieces of skin with a wide-bore pipette to a sterile CelStir. A CelStir is an autoclavable glass vessel containing a magnet suspended by a rod from the lid. Solutions are introduced and removed via a side arm in the vessel.
5. Incubate the CelStir at 37°C for 30 min on a magnetic stirrer; allow the lumps of tissue to settle out and remove the supernatant.
6. Add fresh trypsin–EDTA and repeat the procedure.
7. Determine the number of cells in the supernatant with a haemocytometer. Collect the cells by centrifugation and resuspend in FAD + FCS + HICE. There are usually no cells from the first incubation and then 1–5 × 10⁶ cells per subsequent incubation.

8. After four or five incubations the cell yield starts to decline; discard any remaining lumps of tissue.
9. Pool cells from each incubation and seed onto the feeder layer at a density of 5 × 10⁵ per 75-cm² flask. The average yield per foreskin is approximately 5 × 10⁶ – 10⁷ cells.
10 After 2–3 days, small colonies of keratinocytes are visible, surrounded by 3T3 J2 feeder cells. Over the following weeks, individual colonies expand, displacing the feeder layer and merging with one another. At confluence, virtually no feeder cells remain. Feed the cells with fresh medium three times per week.

E. Passaging Keratinocytes

Solutions are as described in Sections III,A and III,B.

Steps

1. Passage keratinocytes prior to confluence (approximately 7–10 days after plating). Remove the medium and rinse the cultures once with 0.02% EDTA.
2. Add fresh EDTA and incubate the cultures at 37°C for about 5 min. Then selectively detach the feeders by gentle aspiration with a pipette (Rheinwald, 1989).
3. To detach the keratinocytes, add tryspin–EDTA (see Section III,A, solution 2) to the flasks and incubate at 37°C for about 10 min.
4. Transfer the cells to a centrifuge tube; use a small volume of culture medium to rinse the flask and then add it to the tube, inactivating the trypsin.
5. Recover the cells by centrifugation, count in a haemocytometer, and resuspend in complete medium.
6. Cells can be passaged at a density of 1–2 × 10⁵ per 75-cm² flask containing 3T3 J2 feeders.

F. Frozen Stocks of Keratinocytes and 3T3 J2 Cells

Solution

1. *10% DMSO/90% fetal calf serum*

Steps

1. Most of the primary keratinocyte cultures are used to prepare frozen stocks; most experiments are carried out on second and subsequent passages.
2. For freezing, harvest the keratinocytes as described earlier, but resuspend at 10⁶/ml in 10% DMSO/90% foetal calf serum.
3. Place 1 ml of cell suspension in each 1.8-ml CryoTube.

4. Place the tubes in a rubber rack wrapped in cotton wool overnight at –70°C and then transfer to liquid nitrogen

5. Freeze the 3T3 J2 cells in the same way.

G. Preparation of Stable Packaging Lines for Retroviral Vectors

Solutions

1. *Culture medium for packaging lines*: This consists of DMEM supplemented with 100 IU/ml penicillin, 100 µg/ml streptomycin, and 10% heat-inactivated foetal calf serum. The serum is heat inactivated in a water bath at 56°C for 45 min and can then be stored frozen in aliquots. Heat inactivation of FCS improves retroviral titres.

2. *Transfection solutions*: 50 mM chloroquine (2000×) in dH$_2$O, 2 M CaCl$_2$ in dH$_2$O, and 2xHBS, pH 7.0. Filter sterilise each solution, aliquot, and store at –20°C.

3. *2xHBS*: This is made up with 8.0 g NaCl, 6.5 g HEPES, and 10 ml Na$_2$HPO$_4$ stock solution (5.25 g in 500 ml of water). Adjust the pH to 7.0 using NaOH or HCl, bring the volume up to 500 ml, and check the pH again.

4. *Polybrene*: Prepare a 5 mg/ml stock solution in PBS. Filter sterilise, aliquot, and store at –20°C.

5. *Puromycin*: Prepare a 2 mg/ml stock solution in PBS. Filter sterilise, aliquot, and store at –20°C.

Cells

Phoenix E (ecotropic) packaging cells (Swift *et al.*, 1999) are obtained from ATCC with prior approval of G. Nolan (Stanford University, USA). The AM12 amphotropic packaging line was generated by Markowitz *et al.* (1988). Packaging cells are maintained by passage at 1:5 (Phoenix E) or 1:10 (AM12) dilution. For optimal virus production, Phoenix cells should initially be reselected sequentially for 4 weeks, 2 weeks in 1 µg/ml diphtheria toxin, and then two weeks in 500 µg/ml hygromycin to increase Envelope and Gag-Pol expression (http://www.stanford.edu/group/nolan/index.html). Packaging cells are frozen at 10^6/ml in 10% DMSO/90% foetal calf serum as described (in Section III, F,) and fresh cells are thawed every 3 months. J2 puro are 3T3 J2 cells that have been transfected with pBabe puro to render them resistant to puromycin (Levy *et al.*, 1998).

Generation of Stably Transduced AM12 Cells

Packaging lines that are generated by retroviral infection tend to have higher viral titres than those generated by transfection (Levy *et al.*, 1998). Therefore, we use a two-step procedure in which Phoenix E cells are transiently transfected with retroviral vector, and packaged virus released by Phoenix E cells is used to infect AM12 cells.

Steps

1. Twenty-four hours prior to transfection, plate 5 × 10^6 Phoenix E cells into one 10-cm-diameter petri dish in 10 ml medium.

2. Add chloroquine (final concentration 25 µM) to each plate 5 min before transfection. Add 20 µg retroviral vector DNA (we routinely use pBabe puro; Morgenstern and Land, 1990; Levy *et al.*, 1998) to 61 µl 2 M CaCl$_2$ and make up to a 500 µl total volume with ddH$_2$O. Add 500 µl 2xHBS quickly and then bubble vigorously with an automatic pipettor for 15 s. Add the HBS/DNA solution dropwise into the Phoenix E culture dish and then rock the plate gently a few times to distribute DNA/CaPO$_4$ particles evenly. Transfer the plate to a 37°C incubator.

3. Twenty-four hours posttransfection, change the culture medium. Gently add 6 ml of fresh medium per 10 cm plate. Leave cells overnight in a 32°C incubator; retroviruses are more stable at 32°C than at 37°C. In addition, seed 2 × 10^5 AM12 cells per 10 cm plate in 10 ml medium and incubate overnight at 37°C.

4. The next day (i.e., 48 h posttransfection) harvest virus by collecting the medium from Phoenix E cultures. Gently add fresh medium to cells and harvest virus again 8 h later. Add fresh medium for a third time, incubate the cells overnight at 32°C, and collect the medium. All three supernatants contain active virus. Each viral supernatant should be filtered using a 0.45-µm filter on collection and either used immediately (i.e., added to AM12 cells) or stored at –80°C (titre will halve on each freeze/thaw cycle).

5. Virus harvested from Phoenix E cells is used to infect AM12 cells. Remove 8 ml medium from the AM12 culture. Add 5 µl polybrene (5 mg/ml) and 3 ml viral supernatant to each plate (5 ml total volume), shake gently, and place in a 32°C incubator for 24 h.

6. Twenty-four to 48 h postinfection the AM12 cells are 60–80% confluent and are ready to be selected by adding 2 µg/ml puromycin to the medium. Change medium every 2–3 days. Once the cells are confluent, passage them, prepare frozen cell stocks, and use the rest to infect keratinocytes.

7. For some applications, AM12 are subjected to further selection to achieve maximal viral titres. This can be achieved by either cloning at limiting dilution or, if there is a suitable surface marker, FACS sorting of bulk populations. Both methods have been described previously (Levy *et al.*, 1998).

H. Retroviral Infection of Cultured Human Keratinocytes Using Viral Supernatant

J2 puro are 3T3 J2 cells that have been stably transfected with pBabe puro to render them resistant to puromycin (Levy *et al.*, 1998). They are handled in the same way as 3T3 J2 cells, except that the medium is supplemented with 2 μg/ml puromycin.

Steps

1. Grow infected AM12 packaging cells to 95% confluence and transfer to keratinocyte culture medium. Collect virus over a 24- to 48-h period at 32°C, harvesting virus every 8–12 h and replacing with fresh medium. The virus should be filtered through a 0.45-μm filter on collection and either used immediately or stored at −80°C (titre will halve on each freeze/thaw cycle).

2. Keratinocytes with the greatest proliferative potential (putative stem cells) can be enriched by rapid adhesion to collagen (Jones and Watt, 1993). Selection of keratinocytes on collagen prior to infection increases the retroviral infection efficiency to 60–80%. Harvest subconfluent keratinocytes, first removing the 3T3 J2 feeders. Plate 2×10^6 keratinocytes per 10-cm collagen-coated plate (or equivalent plating density if smaller dishes are used). Allow cells to adhere for 15–20 min. Wash gently with PBS and add fresh keratinocyte medium and mitomycin C-treated J2 puro cells. The final volume of medium should be 10 ml per 10-cm dish. Incubate cells overnight at 37°C.

3. The next day, remove 8 ml of medium and add 3 ml of retrovirus supernatant and 5 μl polybrene (5 mg/ml).

4. Incubate cells overnight at 32°C. The next day remove virus containing medium and add fresh keratinocyte medium.

5. Infected keratinocytes are ready to use or can be transferred to puromycin containing medium and passaged 24 h later.

Retroviral Infection of Cultured Human Keratinocytes by Coculture with Packaging Line

Steps

1. Prepare AM12 by treatment with mitomycin C in the same way as for 3T3 J2 cells (Section III,A).

2. Plate keratinocytes in the same way as when plating on 3T3 J2 cells (Section III,E). Use complete keratinocyte medium.

3. After 2–3 days remove AM12, with EDTA treament, as for 3T3 J2 cells (Section III,E). Add mitomycin C-treated J2 puro and supplement the culture medium with puromycin.

IV. COMMENTS

It is possible to obtain up to 100 population doublings of neonatal foreskin keratinocytes prior to senescence (Rheinwald, 1989). The number of passages prior to senescence varies between cell strains: 5 is the minimum, 10 is the average, and greater than 20 is the maximum.

The same basic culture procedure can be used to grow keratinocytes from adult epidermis (although the number of population doublings obtained will be somewhat reduced) and from other stratified squamous epithelia, such as the lining of the mouth and the exocervix (Rheinwald, 1989).

Methods for transfecting Phoenix cells and retrovirally infecting AM12 cells are based on those of Swift *et al.* (1999; http://www.stanford.edu/group/nolan/index.html). We routinely use the pBabe puro retroviral vector (Morgenstern and Land, 1990). Puromycin is the optimal drug for selection, as it is not toxic to transduced keratinocytes and kills nontransduced cells within 2–4 days (Levy *et al.*, 1998). Following culture and selection, the proportion of stably transduced cells obtained with the supernatant and coculture methods are equal (>90%). The advantage of the supernatant method is that large numbers of cells are infected simultaneously; however, with the coculture method, fewer keratinocytes are required initially.

V. PITFALLS

1. Under the conditions described, fibroblast contamination of keratinocytes is very rarely a problem because the feeder layer and culture medium suppress the growth of any human fibroblasts isolated from the skin at the same time as the keratinocytes (Rheinwald, 1975). We have noted that some strains of keratinocytes contain ndk-like (for nondifferentiating keratinocyte; see Adams and Watt, 1988) epithelial cells, but these cells are not abundant and can usually be removed with EDTA.

2. When keratinocytes are plated in culture there is selective attachment of the basal cells, but within 1 day the cultures consist of a mixture of basal and terminally differentiating keratinocytes (e.g., Jones and Watt, 1993). Seeding keratinocytes at high density appears, in our experience, to promote terminal differentiation and is not, therefore, the answer if you are in a hurry to obtain more cells.

3. Because keratinocytes are maintained in culture for long periods it is essential to be scrupulous in sterile technique and laboratory cleanliness to avoid

fungal or bacterial contamination. We spray our incubators three times a week with Mikrozid.

4. It is essential to be well organised and to plan your experiments in advance. It takes about 10 days from plating for keratinocytes to be ready for use and sufficient feeders must be available on the days when keratinocytes are ready for passaging.

5. Use of an amphotropic retrovirus is potentially hazardous because the virus can infect human cells. Special care should therefore be taken and appropriate local safety guidelines followed.

References

Adams, J. C., and Watt, F. M. (1988). An unusual strain of human keratinocytes which do not stratify or undergo terminal differentiation in culture. *J. Cell Biol.* **107**, 1927–1938.

Compton, C. C., Gill, J. M., Bradford, D. A., Regauer, S., Gallico, G. G., and O'Connor, N. E. (1989). Skin regenerated from cultured epithelial autografts on full-thickness burn wounds from 6 days to 5 years after grafting: A light, electron microscopic and immunohistochemical study. *Lab. Invest.* **60**, 600–612.

Dajee, M., Lazarov, M., Zhang, J. Y., Cai, T., Green, C. L., Russell, A. J., Marinkovich, M. P., Tao, S., Lin, Q., Kubo, Y., and Khavari, P. A. (2003). NF-KB blockade and oncogenic Ras trigger invasive human epidermal neoplasia. *Nature* **421**, 639–643.

Dellambra E., Prislei, S., Salvati, A. L., Madeddu, M. L., Golisano, O., Siviero, E., Bondanza, S., Cicuzza, S., Orecchia, A., Giancotti, F. G., Zambruno, G., and De Luca, M. (2001). Gene correction of integrin β4-dependent pyloric atresia-junctional epidermolysis bullosa keratinocytes establishes a role for β4 tyrosines 1422 and 1440 in hemidesmosome assembly. *J Biol Chem.* **276**, 41336–41342.

Gerrard, A. J., Hudson, D. L., Brownlee, G. G., and Watt, F. M. (1993). Towards gene therapy for haemophilia B using primary human keratinocytes. *Nature Genet.* **3**, 180–183.

Jones, P. H., and Watt, F. M. (1993). Separation of human epidermal stem cells from transit amplifying cells on the basis of differences in integrin function and expression. *Cell* **73**, 713–724.

Levy, L., Broad, S., Diekmann, D., Evans, R. D., and Watt, F. M. (2000). β1 integrins regulate keratinocyte adhesion and differentiation by distinct mechanisms. *Mol. Biol. Cell.* **11**, 453–466.

Levy, L., Broad, S., Zhu, A. J., Carroll, J. M., Khazaal, I., Péault, B, and Watt, F. M. (1998). Optimised retroviral infection of human epidermal keratinocytes: Long-term expression of transduced integrin gene following grafting on to SCID mice. *Gene Ther.* **5**, 913–922.

Markowitz, D., Goff, S., and Bank, A. (1988). Construction and use of a safe and efficient amphotropic packaging cell line. *Virology.* **167**, 400–406.

Morgenstern, J. P., and Land, H. (1990). Advanced mammalian gene transfer: High titre retroviral vectors with multiple drug selection markers and a complementary helper-free packaging cell line. *Nucleic Acids Res.* **18**, 3587–3596.

Ortiz-Urda, S., Thyagarajan, B., Keene, D. R., Lin, Q., Fang, M., Calos, M. P., and Khavari, P. A. (2002). Stable nonviral genetic correction of inherited human skin disease. *Nature Med.* **8**, 1166–1170.

Rheinwald, J. G. (1989). Methods for clonal growth and serial cultivation of normal human epidermal keratinocytes and mesothelial cells. *In* "Cell Growth and Division: A Practical Approach" (R. Baserga, ed.) pp. 81–94. IRL Press, Oxford.

Rheinwald, J. G., and Green, H. (1975). Serial cultivation of strains of human epidermal keratinocytes: The formation of keratinizing colonies from single cells. *Cell* **6**, 331–343.

Ronfard, V., Rives, J. M., Neveux, Y., Carsin, H., and Barrandon, Y. (2000). Long-term regeneration of human epidermis on third degree burns transplanted with autologous cultured epithelium grown on a fibrin matrix. *Transplantation* **15**, 1588–1598.

Swift, S. E., Lorens, J. B., Achacoso, P., and Nolan, G. P. (1999). Rapid production of retrovirus for efficient gene delivery to mammalian cells using 293T cell-based systems. *In* "Current Protocols in Immunology" (J. E. Coligan, A. M. Kruisbeek, D. H. Margulies, E. M. Shevach, and W. Strober, eds.), Vol 1. 17C, pp: 1–17. *Wiley, New York.*

Zhu, A. J., and Watt, F. M. (1996). Expression of a dominant negative cadherin mutant inhibits proliferation and stimulates terminal differentiation of human epidermal keratinocytes. *J. Cell Sci.* **109**, 3013–3023.

Zhu, A. J., and Watt, F. M. (1999). β-Catenin signalling modulates proliferative potential of human epidermal keratinocytes independently of intercellular adhesion. *Development* **126**, 2285–2298.

Three-Dimensional Cultures of Normal and Malignant Human Breast Epithelial Cells to Achieve *in vivo*-like Architecture and Function

Connie Myers, Hong Liu, Eva Lee, and Mina J. Bissell

I. INTRODUCTION

Apicobasal polarity, properly positioned cell–cell contacts, and attachment to basement membrane are fundamental characteristics of simple glandular epithelia, such as the mammary gland. The development and maintenance of this polarized structure are essential for the formation of tissue architecture, control of proliferation, and the differentiated function of epithelial cells (Roskelley *et al.*, 1995). Loss of architectural intactness and polarity is one of the pathological hallmarks of epithelial carcinoma (Bissell and Radisky, 2001). Although traditional studies of epithelial cells grown as a monolayer on tissue culture plastic remain a powerful tool to dissect and understand the molecular events of signaling machineries, tissue culture plastic does not recapitulate the microenvironment or morphology of glandular epithelium *in vivo*. Animal models have provided us with invaluable insights, which have led to a greater understanding of the events involved in mammary morphogenesis and tumorigensis *in vivo*, and aid in the translation from basic cellular research into clinical application. However, the complexity of animal model systems prevents us from precisely pinning down the specific biochemical and cell biological pathways involved in mammary morphogenesis and tumor formation. Therefore an *in vitro* cell-based model system that provides epithelial cells with an *in vivo*-like microenvironment, recapitulates both the three-dimensional

(3D) organization and multicellular complexity, and is conducive to systematic experimental pertubation is optimal to bridge the gaps between epithelial monolayer cultures and animal models.

We have developed an assay in which primary or phenotypically normal human breast epithelial cells cultured in a laminin-rich basement membrane (lrBM) undergo a three-dimensional reorganization to form structures that mimic *in vivo* acinar structures in culture (Petersen *et al.*, 1992). LrBM, available commercially as Matrigel, is a mixture of basement membrane proteins that include ~80% laminin, ~10% type IV collagen, entactin, and ~10% heparin sulfate proteoglycan, derived from Engelbreth–Holm–Swarm mouse tumor (Orkin *et al.*, 1977; Kleinman *et al.*, 1986, 1987). The malleability of this specialized gel and its ability to signal through integrins, as well as other ECM receptors, induce changes in the cells, which allow them to withdraw from the cell cycle, organize, and form a central lumen. The entire acinar structures achieve apical basal polarity, and cells within the acini express tissue-specific genes, processes that are not easily accomplished by the same cells cultured on conventional tissue-culture plastic. We refer to the assay as 3D BM assay. The following are advantages of 3D BM assays in studying normal and malignant cells. (1) Because the cells are able to form tissue-like structures, mechanisms of tissue specificity can be studied *ex vivo*. These cultures are amenable to a variety of perturbations and manipulations that can be used to understand how cells may signal *in vivo*. (2) The assay allows breast tumor cells to be distinguished easily from their nonmalignant counterparts because the tumor cells fail to stop proliferating and do not become organized into tissue-like structures (Petersen *et al.*, 1992; Weaver *et al.*, 1996). (3) Addition of specific signaling inhibitors or blocking antibodies or delivery of genes to the tumorgenic cells can be used to analyze signaling pathways involved in the acquisition of the nontumorgenic phenotype (Weaver *et al.*, 1997; Wang *et al.*, 1998, 2002). (4) The 3D BM cultures can be used to understand the novel roles of oncogenes and tumor supressors genes (Howlett *et al.*, 1994; Spancake *et al.*, 1999; Muthuswamy *et al.*, 2001; Debnath *et al.*, 2002). (5) These assays can assess the response of tumorigenic vs nontumorigenic cells to potential therapeutics and unravel novel mechanisms (Weaver *et al.*, 2002). (6) The assays (3–10 days) can be performed rapidly and in a high throughput manner relative to costly animal studies. (7) Last but not least, these assays allow us to study and manipulate human cellular responses in physiological context. For a brief review of the results obtained with these models and studies in rodents cells in 3D cultures and *in vivo*, see Bissell *et al.* (1999). For a history of development of 3D cultures in general see Schmeichel and Bissell (2003). For a review of more complex organotypic cultures, see Gudjonsson *et al.* (2003).

II. MATERIALS AND INSTRUMENTATION

Primary breast luminal cells from human reduction mammoplasty (Petersen *et al.*, 1992) as well as human mammary epithelial cell lines have been successfully cultured utilizing the 3D BM assay. These include the following.

The HMT3522 progression series (Briand *et al.*, 1987) consists of immortal human mammary epithelial cells originally isolated from fibrocystic breast tissue and includes the phenotypically normal S1 cells, as well as their tumorigenic derivative T4-2 cells, which were selected for their ability to grow in the absence of EGF (Briand *et al.*, 1996; Weaver *et al.*, 1996).

This article focuses mainly on the HMT3522 series but provides examples of the morphology of a few other nonmalignant cell lines, such as 184B5 (Walen and Stampfer, 1989); for a description of these series, see http://www.lbl.gov/~mrgs/review.html and MCF10A (Soule *et al.*, 1990). For further information and culture conditions, see http://www.atcc.org/ATCC Number: CRL-10317.

TABLE I   Growth Medium for HMT3522 Cells

Components	Growth medium
DMEM/F12	10 ml
Insulin (100 µg/ml stock)	25 µl (250 ng/ml final)
Transferrin (20 mg/ml stock)	5 µl (10 µg/ml final)
Sodium selenite (2.6 µg/ml stock)	10 µl (2.6 ng/ml)
Estradiol (10^{-7} M stock)	10 µl (10^{10} M final)
Hydrocortisone (1.4×10^{-3} M stock)	10 µl (1.4×10^{-6} M final)
Prolactin (1 mg/ml stock)	50 µl (5 µg/ml final)
EGF[a] (20 µg/ml stock)	5 µl (10 ng/ml final)

[a] S1 cells only.

A. Culture Media Composition

Media for the HMT3522 progression series of human mammary epithelial cells are composed of DMEM/F12 media (GibcoBRL Cat. No. 12400-024) supplemented with insulin (Sigma Cat. No. I-6634), human transferrin (Sigma Cat. No. T-2252), sodium selenite (Collaborative Research Cat. No. 40201), estradiol (Sigma Cat. No. E-2758), hydrocortisone (Sigma Cat. No. H-0888), prolactin (Sigma Cat. No. L-6520), and epidermal growth factor (EGF) (Roche Cat. No. 855731) as described in Briand *et al.* (1987) and Blaschke *et al.* (1994). Concentrations and stability are outlined in Table I.

B. Additional Reagents Required for Culture and Manipulation in Three-Dimensional Culture

Vitrogen (collagen I, Cohesion Technologies) can be made from rat tail collagen, phosphate-buffered saline (PBS) (any vendor), 0.25% trypsin–EDTA (any vendor), soybean trypsin inhibitor (Sigma Cat. No. T6522), trypan blue (Sigma Cat. No. T8154), growth factor reduced Matrigel (Becton-Dickenson-Collaborative Research Cat. No. 354230), AIIB2, blocking antibody against β1 integrin (Sierra BioSource Inc. Cat. No. SB959), LY294002 (Cell Signaling Technologies Cat. No. 9901), Mab225 (Oncogene Research Cat. No. GR13L), PP2 (Calbiochem Cat. No. 529573), PD98059 (Calbiochem Cat. No. 513000) and tyrphostin AG1478 (Calbiochem Cat. No. 658552). All other equipment and supplies can be purchased from a variety of vendors.

C. Minimal Equipment Required for Cell Culture

Hemacytometer, low-speed centrifuge, inverted light microscope equipped with 4, 10, 20, and 40×

objectives, pipetman p20, p200, p1000, pipette aid, biological safety cabinet, 37°C, 5% CO_2 humidified culture incubator, air-tight plastic container, 37°C water bath.

D. General Tissue Culture Supplies

Tissue culture plasticware including T75 flasks, 4-, 6-, 24-, 48-, and 96-well dishes, 5-, 10-, and 25-ml individually wrapped sterile pipettes, 15- and 50-ml sterile conical centrifuge tubes, sterile 1- to 200- and 500- to 1000-µl pipette tips, and sterile 9-in. Pasteur pipettes and cell lifters.

III. PROCEDURES

Successful growth of these cells requires special attention to the media, confluency, trypsinization, and feeding regime in order to maintain healthy cells capable of undergoing differentiation. Insufficient attention to any of the aforementioned parameters will quickly result in cells acquiring a fibroblast-like morphology and concomitant failure to undergo 3D organization when cultured in IrBM.

A. Cell Maintenance

1. Preparation of Collagen-Coated Tissue Culture Flasks

1. Make a working solution of collagen I [~65 µg/ml] by the addition of 1 ml Vitrogen (collagen I) to 44 ml of cold 1× PBS.

2. Add 3 ml of collagen solution for every 25-cm² surface area of flask; tilt flask to make sure entire surface is coated.
3. Store in an air-tight container, on even surface at 4°C, a minimum of overnight and a maximum of 3 weeks.
4. Before use, aspirate liquid in the flask, wash once with 2 ml DMEM/F12 per 25 cm² and add prewarmed growth media.

2. Passage of HMT3522 Cells

Change media every 2–3 days; it is important to ensure that fresh media be added 1 day before splitting. HMT3522 S1 cells should be passaged once the 2D colonies form rounded islands with the flask being approximately 75% confluent, as shown in Fig. 1A. This is usually occurs 8 to 10 days after the initial plating.

Passage HMT3522 T4-2 cells when the flask reaches 75% confluency, as shown in Fig. 1B. This usually occurs 3 to 5 days after initial plating. Maintain T4-2 cells on Vitrogen (collagen I) or rat tail collagen-coated flasks.

Perform the following steps in a tissue culture hood.

1. Aspirate medium.
2. Rinse cells in 0.5 to 1 ml of prewarmed 0.25% trypsin–EDTA for every 25 cm² of surface area.
3. Add 0.35 ml of 0.25% trypsin for every 25 cm² of surface area and cover cells by rotating the flask gently.
4. Place flask in 37°C incubator for 3 min.
5. Remove flask from incubator and check under the microscope to see if the cells have detached.

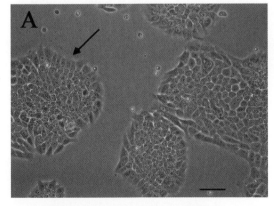
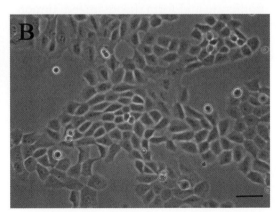

S1 (passage 57)   T4-2 (passage 37)

FIGURE 1 The morphology of S1 and T4-2 cultured on tissue culture plastic. (A) S1 cells, passage 57, at day 10 after plating. The colonies are essentially round with smooth edges and with the inner cells somewhat compacted. Cells at the edge should take on a polar-like appearance indicated by the arrows. (B) T4-2, passage 37, at day 5 after plating. Cells are larger than S1, the colonies are irregular, and cells at the edges rarely show any organization. Bar: 25 µm.

a. If cells are still adherent, redistribute the trypsin and return the flask to the 37°C incubator for 1 min. Repeat until cells are detached.

b. Once the majority of the cells are detached, knock the side of the flask gently; add ~3 ml of prewarmed DMEM/F12 and 30 µl of soybean trypsin inhibitor per 25 cm².

6. Gently pipette cells up and down three to five times to dissociate cell aggregates and transfer to a 15-ml conical centrifuge tube.

7. Pellet the cells by spinning at ~115 g for 5 min.

8. Aspirate media and resuspend the cell pellets in their appropriate growth media (~3.25 ml per 25 cm² of surface area of their original flask).

9. Determine the cell concentration by removing a 95-µl aliquot and placing into a 1.5-ml Eppendorf tube, add 5 µl of 0.4% trypan blue, mix gently with a pipetman, load onto a hemacytometer, count on an inverted microscope, and determine the number of viable (nonblue) cells per milliliter.

10. Plate

a. HMT 3522 S1 cells at a density of 2×10^4 cells/cm² on tissue culture plastic.

b. HMT 3522 T4-2 cells at a density of 1×10^4 cells/cm² on collagen I-coated flasks.

3. Three-Dimensional BM Assay Embedded in lrBM/EHS/(Matrigel)

Single phenotypically normal mammary epithelial cells embedded into lrBM will undergo several rounds of cell division, withdraw from cell cycle between 6 and 8 days (depending on the cell type), organize into polarized structures, and form acini-like structures, including a central lumen. Malignant cells, however, continue to proliferate and form disorganized, tumor-like structures (Fig. 2). Starting on day 4 but clearly by day 10, one can distinguish nonmalignant from malignant cells. Immunohistochemistry analysis for the Ki67 antigen on day 10 shows that ~60% of the malignant cells are still proliferating, whereas only <10% of the nonmalignant remain in the cell cycle.

Procedure

The following volumes and concentrations given are appropriate for a 35-mm tissue culture dish (9.6-cm² surface area). The volumes and concentrations must be adjusted to correct for the area of the plate used in the analysis. lrBM should be stored at −80°C. Thaw *immersed in ice* at 4°C overnight before use. Once thawed it should *not be refrozen* but can be kept *in ice* at 4°C for more than 1 month. Matrigel lots are tested before use and batches are purchased for experimental use. Alternatively, the basement membrane mixture (EHS) can be made from the Engelbreth-Holm–Swarm mouse tumor described in Kleinman *et al.* (1982); see Section III,B details.

Note: EHS or Matrigel should always be kept on ice as it will polymerize quickly at room temperature.

1. Chill plates on ice inside a tissue culture hood.

2. Add 120 µl of EHS per 35-mm dish spread in the center of the dish without introducing bubbles, spread EHS evenly across the surface of the dish using a sterile cell lifter or the back of a sterile pipette tip (make sure the entire surface is covered, including the edges), and place in a 37°C incubator for a minimum of 15 min to allow the EHS to polymerize. (Do not let sit more than 1 h in incubator as it will begin to dry out.)

3. While the EHS is polymerizing, trypinsize and count the cells as described earlier.

4. Place an aliquot containing the appropriate number of cells into a sterile centrifuge tube: 1.0×10^6 cells S1 (and other *non*tumorgenic cell lines) and 0.7×10^6 cells T4-2 (and other tumorgenic cell lines).

5. Centrifuge for 5 min at ~115 g to pellet the cells, aspirate media, and resuspend the pellet by flicking the tube.

6. Place the cell pellet on ice, add 1.2 ml EHS, and carefully pipette mixture up and down to distribute cells evenly but not introduce bubbles into the EHS.

7. Transfer to a precoated 35-mm dish.

8. Place in 37°C incubator for ~30 min until the EHS is polymerized.

9. Once a gel is formed, add 1.5 to 2.0 ml of the appropriate media to the dish.

10. Culture the cells for 10 days, changing media every 2 to 3 days.

Figures 3A–3D show cells cultured in Matrigel for 10 days.

4. Three-Dimensional BM Assay on Top of EHS/Matrigel

The culture of mammary epithelial cells, developed previously in our laboratory for functional studies of mouse cells (Barcellos-Hoff *et al.*, 1989; Roskelley *et al.*, 1994), has now been adopted to human cells as well. Making 3D BM cultures on top of EHS as opposed to inside the gel has some advantages and a few drawbacks. The proliferation and morphological differences between nonmalignant and malignant cells can be distinguished by 4 days in culture as opposed to the 8–10 days required for embedded cultures. The cell colonies can readily be imaged utilizing a live cell imager, allowing one to follow a single cell to an acini-like structure. Furthermore, cell colonies can be harvested easily by scraping the culture gently with a pipette tip and depositing the isolate on a glass slide for analysis

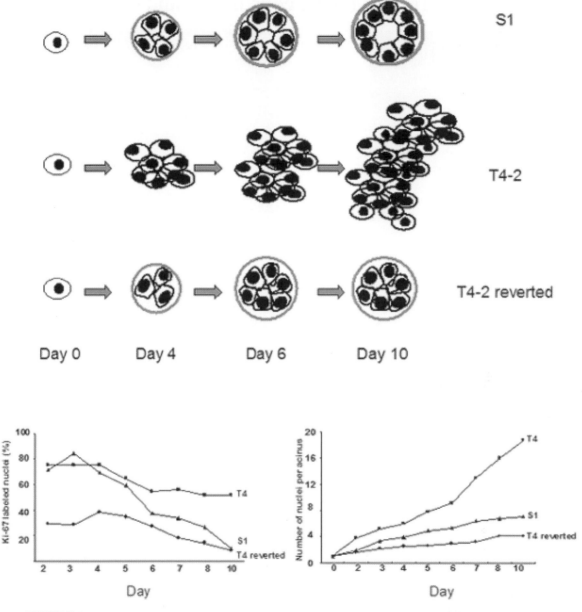

FIGURE 2 Morphology, size, and proliferation rate of S1 and T4-2 and T4-2 reverted cells embedded in 3D BM. (A) Schematic representations of the morphology of S1, T4-2, and T4-2 cells treated with signaling inhibitors as a function of time. (B) The percentage of Ki67 positive cells for S1, T4-2, and T4-2 cells treated with tyrophostin AG1478 from days 2 to 10. (C) The number of cells/acini for S1, T4-2, and T4-2 cells treated with tyrophostin AG1478 from days 2 to 10.

by immunohistochemistry. The advantage of this method of harvest is that the remaining culture can be harvested for DNA, RNA, or protein. However, neither the morphology nor the growth rate of the acini is as tightly controlled as the embedded cultures. An additional disadvantage of the *on-top* culture is that the amount of material collected per plate is less than the embedded cultures, as plating of the cells is restricted to one plane. Therefore, there are signifi-

cantly less cells per millimeter dish to be harvested compared to the embedded assay.

The following volumes and concentrations are appropriate for a 35-mm tissue culture dish (9.6-cm^2 surface area). The volumes and concentrations must be adjusted to correct for the area of the plate used in the analysis.

Before splitting the cells, prepare the EHS-coated plates.

Morphology of different breast cells inside and on top of Matrigel

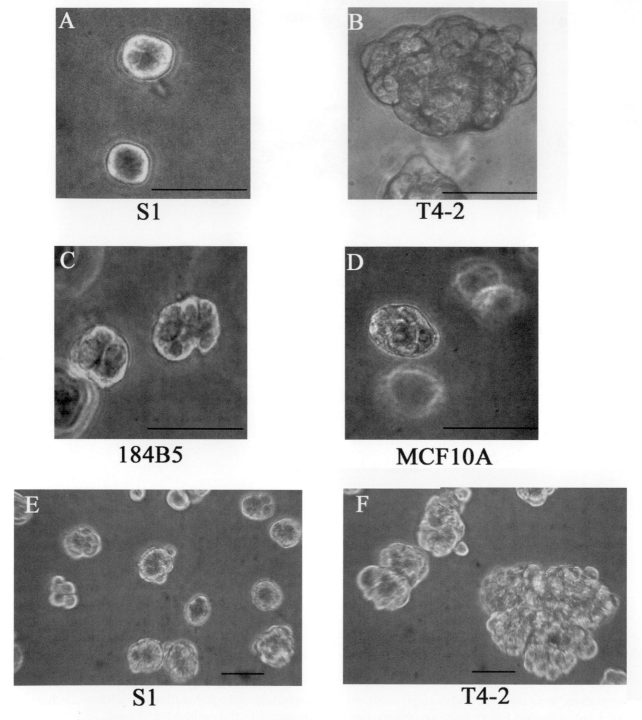

FIGURE 3   Morphologies of S1 (A), T4-2 (B), 184B5 (C), and MCF10 (D) cells embedded in 3D BM (all cells were cultured in Matrigel for 10 days) and S1 (E) and T4-2 (F) cells cultured on top (cells were cultured on top of Matrigel for 4 days). Photographs were taken using a phase-contrast microscope. Bar: 25μm.

1. Chill plates on ice in a tissue culture hood.

2. Add 0.5ml to the center of the dish and spread EHS evenly without creating bubbles as described earlier.

3. Place in 37°C incubator for ~30min until the EHS is polymerized.

4. Trypsinize cells as described previously.

5. Place an aliquot containing the appropriate

number of cells into a sterile centrifuge tube: 0.3×10^6 cells S1 (and other *non*tumorgenic cell lines) and 0.2×10^6 cells T4-2 (and other tumorgenic cell lines).

6. Centrifuge for 5 min at ~115 g to pellet the cells, aspirate media, and resuspend the pellet by flicking the tube.

7. Resuspend the cells in half of the total volume of the appropriate media for the cells (1 ml for 35 mm) according to the size of the plate (see later).

8. Plate the cells on top of the polymerized EHS-coated dishes, let sit in hood for 5 min, and then check under a microscope to see if cells are distributed evenly. If not, move plate on a flat surface, back to front and then side to side, to distribute and let sit for 5 min.

9. Return plates to the incubator for ~30 min until the cells have adhered.

10. Prepare media plus 10% EHS on ice.

11. Add the appropriate amount of media with 10% EHS to the cells to achieve a final concentration of 5% EHS.

12. Culture the cells for 4–6 days, changing media every 2–3 days.

Figures 3E and 3F show cells cultured on top of EHS for 4 days,

5. Reversion Assay

The reversion assay can determine the ability of tumor cells to become phenotypically normal or to undergo apoptosis as a function of treatments. In addition, the differential response of tumorigenic and non-malignant cells to a wide variety of perturbations can be analyzed. Analyses can be performed both on cells cultured embedded in or on top of EHS. This assay can be used as a screen for potential therapeutics by manipulating various genes, signaling pathways, and/or protein modifications and for determining their role in the establishment or maintenance of the tumorigenic phenotype. Cancer cells, cultured as described earlier, can be treated with a wide variety of agents, including inhibitory or stimulatory antibodies to integrins or growth factor receptors, small molecule inhibitors to different signaling pathways, gene delivery via transfection with expression vectors or transduction with virus overexpressing genes of interest, and RNAi for specific gene silencing.

Quantitative end points include morphology, proliferation index, degree of polarity, and level of expression of genes of interest using RT-PCR, Northern, or Western analysis, as well as immunofluorescence. Table II details a variety of molecules that have been shown to revert T4-2 cells. A typical reversion morphology of T4-2 cells is illustrated in Fig. 4.

TABLE II Reversion Agents in Three-Dimensional Assay

Reversion agent	Final concentration	Activity
AIIBII (6–10 mg/ml stock in PBS)	In Matrigel (0.16 mg/ml final)[a]	$\beta 1$ integrin inhibitory antibody
Mab225 (200 μg/ml stock in PBS)	In Matrigel (6–9 μg/ml final)[b]	EGF receptor inhibitory antibody
LY294002 (10 mM stock in DMSO)	In growth medium (10 μM final)	PI3 kinase inhibitor
Tyrophostin AG1478 (100 μM stock in DMSO)	In growth medium (100 nM final)	EGF receptor inhibitor
PD98059 (10 mM stock in DMSO)	In growth medium (10 μM final)	MEK1 inhibitor

[a] Antibodies may need slight adjustment in concentration from lot to lot.
[b] Add to IrBM/cell mix as well as to media.

6. Release of Cellular Structures from 3D BM

For molecular analysis of cell extracts, it is important to remove them from the gel. Apart from scraping the colonies as described earlier, which will still have minor EHS contamination, intact cellular structures can be released from EHS utilizing chelating agents, a procedure described in Weaver *et al.* (1997). A modified version is described here. Note that the solutions must be cold and the cells kept on ice for the entire harvest period. The maximum harvest should be 1 h as the cell–cell junctions will begin to come apart if left in these solutions for too long.

Perform the following steps on the day of harvest.

1. Wash cultures twice with cold PBS without Ca^{2+} or Mg^{2+} plus 0.005 M EDTA.

2. Scrape cultures into 3 ml cold PBS without Ca^{2+} or Mg^{2+} plus 0.005 M EDTA/25-cm^2 surface area, transfer into an appropriate size centrifuge tube, wash the plates twice with 3 ml each cold PBS, and bring final volume to 10 ml.

3. Incubate cells on ice for ~45 min, gently inverting the tube two or three times every 10 min.

4. Cell release from 3D BM is complete when the cellular structures begin to settle at the bottom of the centrifuge tube and the supernatant does not contain floating gel or flocculent substances.

5. Centrifuge gently for 2–4 min at 115 g until a loose pellet is obtained (the cells are very fragile at this point; therefore, centrifuging too hard or long will result in cell lysis).

8. Dialyze in cold room against the media of choice for 3 h.

9. Store on ice in thermos at 4°C if used within 2 months. Otherwise, freeze aliquots at −80°C and thaw on ice before use.

IV. COMMENTS

Using this assay system, one can screen a wide variety of breast tumor cell lines and assess their ability to revert or die when treated with various signaling inhibitors and/or potential therapeutics, as outlined in Wang *et al.* (2002).

V. PITFALLS

1. Use healthy cells that are no more than 75% confluent.

2. Make sure the cells have been dispersed into single cells and that you do not introduce bubbles into the lrBM for 3D BM-embedded cultures. Cell clumps will make it hard to interpret morphological characteristics at the end of the assay.

3. Feed cells every 2–3 days with fresh media when appropriate, include test compound.

Always include a cell line with a known positive response when performing the reversion assay.

Acknowledgments

Work from the authors' laboratory was supported by the United States Department of Energy, Office of Biological and Environmental Research (DE-AC03 SF0098 to M.J.B.). Additional funding was contributed by the Department of Defense Breast Cancer Research Program (#DAMD17-02-1-0438 to M.J.B.) and the National Cancer Institute (CA64786-02 to M.J.B., and CA57621 to Zena Werb and M.J.B). The authors thank Paraic Kenny for editorial comments.

References

Barcellos-Hoff, M. H., Aggeler, J., Ram, T. G., and Bissell, M. J. (1989). Functional differentiation and alveolar morphogenesis of primary mammary cultures on reconstituted basement membrane. *Development* 105, 223–235.

Bissell, M. J., and Radisky, D. (2001). Putting tumours in context. *Nature Rev. Cancer* 1, 46–54.

Bissell, M. J., Weaver, V. M., Lelievre, S. A., Wang, F., Petersen, O. W., and Schmeichel, K. L. (1999). Tissue structure, nuclear organization, and gene expression in normal and malignant breast. *Cancer Res.* 59, 1757–1763s; discussion 1763s–1764s.

Blaschke, R. J., Howlett, A. R., Desprez, P. Y., Petersen, O. W., and Bissell, M. J. (1994). Cell differentiation by extracellular matrix components. *Methods Enzymol.* 245, 535–556.

Briand, P., Nielsen, K. V., Madsen, M. W., and Petersen, O. W. (1996). Trisomy 7p and malignant transformation of human breast epithelial cells following epidermal growth factor withdrawal. *Cancer Res.* 56, 2039–2044.

Briand, P., Petersen, O. W., and Van Deurs, B. (1987). A new diploid nontumorigenic human breast epithelial cell line isolated and propagated in chemically defined medium. *In Vitro Cell Dev. Biol.* 23, 181–188.

Debnath, J., Mills, K. R., Collins, N. L., Reginato, M. J., Muthuswamy, S. K., and Brugge, J. S. (2002). The role of apoptosis in creating and maintaining luminal space within normal and oncogene-expressing mammary acini. *Cell* 111, 29–40.

Gudjonsson, T., Ronnov-Jessen, L., Villadsen, R., Bissell, M. J., and Petersen, O. W. (2003). To create the correct microenvironment: Three-dimensional heterotypic collagen assays for human breast epithelial morphogenesis and neoplasia. *Methods* 30, 247–255.

Howlett, A. R., Petersen, O. W., Steeg, P. S., and Bissell, M. J. (1994). A novel function for the nm23-H1 gene: Overexpression in human breast carcinoma cells leads to the formation of basement membrane and growth arrest. *J. Natl. Cancer Inst.* 86, 1838–1844.

Kleinman, H. K., Graf, J., Iwamoto, Y., Kitten, G. T., Ogle, R. C., Sasaki, M., Yamada, Y., Martin, G. R., and Luckenbill-Edds, L. (1987). Role of basement membranes in cell differentiation. *Ann. N. Y. Acad. Sci.* 513, 134–145.

Kleinman, H. K., McGarvey, M. L., Hassell, J. R., Star, V. L., Cannon, F. B., Laurie, G. W., and Martin, G. R. (1986). Basement membrane complexes with biological activity. *Biochemistry* 25, 312–318.

Kleinman, H. K., McGarvey, M. L., Liotta, L. A., Robey, P. G., Tryggvason, K., and Martin, G. R. (1982). Isolation and characterization of type IV procollagen, laminin, and heparan sulfate proteoglycan from the EHS sarcoma. *Biochemistry* 21, 6188–6193.

Muthuswamy, S. K., Li, D., Lelievre, S., Bissell, M. J., and Brugge, J. S. (2001). ErbB2, but not ErbB1, reinitiates proliferation and induces luminal repopulation in epithelial acini. *Nature Cell Biol.* 3, 785–792.

Orkin, R. W., Gehron, P., McGoodwin, E. B., Martin, G. R., Valentine, T., and Swarm, R. (1977). A murine tumor producing a matrix of basement membrane. *J. Exp. Med.* 145, 204–220.

Petersen, O. W., Ronnov-Jessen, L., Howlett, A. R., and Bissell, M. J. (1992). Interaction with basement membrane serves to rapidly distinguish growth and differentiation pattern of normal and malignant human breast epithelial cells. *Proc. Natl. Acad. Sci. USA* 89, 9064–9068.

Roskelley, C. D., Desprez, P. Y., and Bissell, M. J. (1994). Extracellular matrix-dependent tissue-specific gene expression in mammary epithelial cells requires both physical and biochemical signal transduction. *Proc. Natl. Acad. Sci. USA* 91, 12378–12382.

Roskelley, C. D., Srebrow, A., and Bissell, M. J. (1995). A hierarchy of ECM-mediated signalling regulates tissue-specific gene expression. *Curr. Opin. Cell Biol.* 7, 736–747.

Schmeichel, K. L., and Bissell, M. J. (2003). Modeling tissue-specific signaling and organ function in three dimensions. *J. Cell Sci.* 116, 2377–2388.

Soule, H. D., Maloney, T. M., Wolman, S. R., Peterson, W. D., Jr.,

Brenz, R., McGrath, C. M., Russo, J., Pauley, R. J., Jones, R. F., and Brooks, S. C. (1990). Isolation and characterization of a spontaneously immortalized human breast epithelial cell line, MCF-10. *Cancer Res.* **50**, 6075–6086.

Spancake, K. M., Anderson, C. B., Weaver, V. M., Matsunami, N., Bissell, M. J., and White, R. L. (1999). E7-transduced human breast epithelial cells show partial differentiation in three-dimensional culture. *Cancer Res.* **59**, 6042–6045.

Walen, K. H., and Stampfer, M. R. (1989). Chromosome analyses of human mammary epithelial cells at stages of chemical-induced transformation progression to immortality. *Cancer Genet. Cytogenet.* **37**, 249–261.

Wang, F., Hansen, R. K., Radisky, D., Yoneda, T., Barcellos-Hoff, M. H., Petersen, O. W., Turley, E. A., and Bissell, M. J. (2002). Phenotypic reversion or death of cancer cells by altering signaling pathways in three-dimensional contexts. *J. Natl. Cancer Inst.* **94**, 1494–1503.

Wang, F., Weaver, V. M., Petersen, O. W., Larabell, C. A., Dedhar, S., Briand, P., Lupu, R., and Bissell, M. J. (1998). Reciprocal interactions between beta1-integrin and epidermal growth factor receptor in three-dimensional basement membrane breast cultures: A different perspective in epithelial biology. *Proc. Natl. Acad. Sci. USA* **95**, 14821–14826.

Weaver, V. M., Fischer, A. H., Peterson, O. W., and Bissell, M. J. (1996). The importance of the microenvironment in breast cancer progression: Recapitulation of mammary tumorigenesis using a unique human mammary epithelial cell model and a three-dimensional culture assay. *Biochem. Cell Biol.* **74**, 833–851.

Weaver, V. M., Lelievre, S., Lakins, J. N., Chrenek, M. A., Jones, J. C., Giancotti, F., Werb, Z., and Bissell, M. J. (2002). beta4 integrin-dependent formation of polarized three-dimensional architecture confers resistance to apoptosis in normal and malignant mammary epithelium. *Cancer Cell* **2**, 205–216.

Weaver, V. M., Petersen, O. W., Wang, F., Larabell, C. A., Briand, P., Damsky, C., and Bissell, M. J. (1997). Reversion of the malignant phenotype of human breast cells in three-dimensional culture and in vivo by integrin blocking antibodies. *J. Cell Biol.* **137**, 231–245.

18

Primary Culture of *Drosophila* Embryo Cells

Paul M. Salvaterra, Izumi Hayashi, Martha Perez-Magallanes, and Kazuo Ikeda

I. INTRODUCTION

Primary cultures of *Drosophila* embryonic cells offer a unique system to study the transitions of undifferentiated cells into a variety of cell types. Coupled with the power of *Drosophila* classical and molecular genetics, the analyses that are possible *in vitro* using differentiating embryo cells will continue to contribute to a deeper understanding of development. Preparation of primary embryo cell cultures was developed independently in the laboratories of Seecof and colleagues (Seecof and Unanue, 1968) and Shields and Sang (1970). The procedures for preparing cultures are technically undemanding and can be adapted to solve a wide range of biological problems. Important variations include the culture of single embryos (Cross and Sang, 1978), as well as the development of techniques for isolating the precursors of different cell types such as neurons and muscle cells (Furst and Mahowald, 1985; Hayashi and Perez-Magallanes, 1994). Figure 1 shows a convenient way to collect embryos from large populations of egg-laying adults, and Fig. 2 illustrates a typical setup used for preparing primary embryo cultures.

Cultures are initiated from embryos in the early gastrula stage, making it possible to observe cell development from committed precursor to final differentiated cell type. The types of cells that differentiate in culture are primarily neurons (Seecof *et al.*, 1973a) and multinucleate myotubes (Seecof *et al.*, 1973b), but other epidermal and mesodermal derivatives are also present (Shields *et al.*, 1975). Cultures form a variety of neurons with highly differentiated phenotypes, including neurotransmitter systems (Salvaterra *et al.*, 1987), ion channels (Byerly and Leung, 1988; Germeraad *et al.*, 1992), axonal specializations, and even functional neuromuscular junctions (Seecof *et al.*, 1972) and synaptic activity (Lee and O'Dowd, 1999). Medium conditioned by embryo cells has also been shown to contain a variety of activities that can modulate the growth and differentiation properties of neurons (Salvaterra *et al.*, 1987; Hayashi *et al.*, 1992). Figure 3 shows some of the cell types observed in a differentiated *Drosophila* primary embryo culture and neurons marked with a fluorescent transgenes specific for different neurotransmitter phenotypes. Remarkably, cultured *Drosophila* embryo cells have also been shown to express a temporally specific transcriptional factor cascade thought to be important for establishing the developmental potential of subsequent cell lineages from primary neuroblasts (Brody and Odenwald, 2000, 2002).

II. MATERIALS AND INSTRUMENTATION

Penicillin and streptomycin are from GIBCO/BRL (Cat. No. 15140-122). Insulin is from Sigma (Cat. No. I-5500). 1X Schneider's *Drosophila* medium with glutamine is from GIBCO (Cat. No. 11720-034). Fetal bovine serum is from GIBCO (Cat. No. 10082-139).

Cells are usually cultured in 35-mm plastic dishes from Corning (Cat. No. 430165). For finer morphological observations, cells are plated directly onto round

FIGURE 1 Large plastic jars used for embryo collections. The jar on the left is empty and the one on the right has a grape juice agar embryo collection plate taped to the bottom and is stocked with about 3000 adult flies as described. Holes have been cut into the sides of the jars and are covered with nylon mesh to allow efficient air circulation.

FIGURE 2 Typical items used to prepare *Drosophila* gastrula stage embryo cultures are arranged in a small laminar flow hood.

glass coverslips placed in the bottom of a 35-mm plate (five per culture dish, Fisher Scientific, Cat. No. 12-545-801-1THK). Coverslips are presterilized using standard gas or autoclave procedures.

Dechorionated embryos are dissociated in 15-ml Dounce glass–glass homogenizers from Baxter Diagnostics (Cat. No. T4018-15). Nylon mesh (25–100 µm pour size) can be obtained from most fabric stores. All other equipment and reagents are commonly used for general laboratory work or tissue culture and can be obtained from any scientific supply company.

III. PROCEDURES

A. Egg Collection and Aging to Early Gastrula Stage

Solution and Egg Collection Plates

Yeast–proprionic acid paste: Dissolve dry yeast in 2% proprionic acid

Grape–juice–agar egg collection plates: 160 ml water, 160 ml grape juice (we use Welch's), 6.4 g Bacto-agar

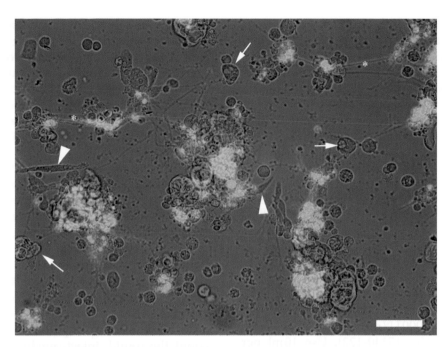

FIGURE 3 A *Drosophila* gastrula stage embryo cell culture after about 2 days of *in vitro* differentiation. The majority of cells differentiate into neurons organized in clusters. This image is a composite of a bright-field image overlaid with a false-colored fluorescent image of the same field. Green cells are neurons expressing a cholinergic-specific fluorescent transgene and are indicative of cells with a cholinergic neurotransmitter phenotype. White arrowheads show some multinuicleate myotubes, white and the asterisk indicates axonal-like processes emanating from neurons or neural clusters and are often connected to neurons in adjacent clusters. Scale bar: 50 μm.

(Becton, Dickinson, Cat. No. 214010), 0.8 g Nipagin [*p*-hydroxybenzoic acid methyl ester (methyl-praben)] (Sigma, Cat. No. H-5501). Combine water, grape juice, and Bacto-agar in a 1-liter flask. Microwave to dissolve agar, being careful not to let the solution boil over. Remove from the microwave, add Nipagin, mix, and pour into 12 large (14 cm) culture plates.

Steps

1. Prepare large egg-laying population by transferring 1.5-week-old adult flies from 8 to 12 ordinary culture bottles to fresh culture bottles supplemented with dry yeast sprinkled over the food. Allow the population to adapt for 1–2 days in a 25°C humidified room and transfer the adults to an embryo collection bottle on the day cultures will be prepared. Figure 1 shows an example of a homemade collection bottle or a similar one can be obtained commercially (Doc Frugal Scientific, Cat. No. 55-100). We generally stock each egg-laying population with about 3000 adult flies. For convenience, maintain the room on a reverse light–dark cycle so collections can be done in the investigators' morning (i.e., the flies' dusk).

2. Apply a small amount of yeast–proprionic acid paste on a small piece of filter paper and place in the center of a large (14-cm) grape juice–agar egg collection plate to attract females to lay eggs.

3. Invert the embryo collection bottles onto the food plate and collect eggs for 1 or 2 h.

4. If better developmental timing (synchronization) is desired, do a 30-min to 1-h precollection and discard plates. This may be necessary to induce females to lay eggs they have been storing for longer times.

5. Remove egg collection plates from the embryo collection bottles, cover, and allow embryos to develop to early gastrula stage at 25°C for 3.5 h.

B. Harvesting of Embryos

Solution

Deionized water: Prepare 2 to 3 liters of deionized water.

Steps

1. Embryos should be harvested 5–10 min before the 3.5-h aging period so the cultures can be initiated immediately at 3.5 h.

19

Laboratory Cultivation of *Caenorhabditis elegans* and Other Free-Living Nematodes

Ian M. Caldicott, Pamela L. Larsen, and Donald L. Riddle

I. INTRODUCTION

Nematodes have been cultured continuously in the laboratory since 1944 when Margaret Briggs Gochnauer isolated and cultured the free-living hermaphroditic species *Caenorhabditis briggsae*. Work with *C. briggsae* and other rhabditid nematodes, *C. elegans*, *Rhabditis anomala*, and *R. pellio*, demonstrated the relative ease with which they could be cultured (Dougherty, 1960; Vanfleteren, 1980). The culturing techniques described here were developed for *C. elegans*, but are generally suitable (to varying degrees) for other free-living nematodes. Whereas much of the early work involved axenic culturing, most of these techniques are no longer in common use and are not included here.

In the 1970s, *C. elegans* became the predominant research model due to work by Brenner and co-workers on the genetics and development of this species (Brenner, 1974). An adult *C. elegans* is about 1.5 mm long and, under optimal laboratory conditions, has a life cycle of approximately 3 days. There are two sexes, males and self-fertile hermaphrodites (Fig. 1), that are readily distinguishable as adults. The animals are transparent throughout the life cycle, permitting the observation of cell divisions in living animals using differential interference microscopy. The complete cell lineage and neural circuitry have been determined, and a large collection of behavioral and anatomical mutants has been isolated (Wood, 1988). *C. elegans* has six developmental stages: egg, four larval stages (L1–L4), and adult. Under starvation conditions or specific manipulations of the culture conditions, a

developmentally arrested dispersal stage, the dauer larva, can be formed as an alternative third larval stage (Golden and Riddle, 1984).

Many of the protocols included here and other experimental protocols have been summarized in Wood (1988). We also include a previously unpublished method for long-term chemostat cultures of *C. elegans*. General laboratory culture conditions for nematode parasites of animals have been described (Hansen and Hansen, 1978), but none of these nematodes can be cultured in the laboratory through more than one life cycle. Marine nematodes and some plant parasites have been cultured xenically or with fungi (Nicholas, 1975). Laboratory cultivation of several plant parasites on *Arabidopsis thaliana* seedlings in agar petri plates has also been reported (Sijmons *et al.*, 1991).

II. MATERIALS AND INSTRUMENTATION

Caenorhabditis elegans strains, as well as strains of other free-living nematodes, and bacterial food sources for them are available from the Caenorhabditis Genetics Center (250 Biological Sciences Center, University of Minnesota, 1445 Gourtner Ave., St. Paul, MN 55108). Most chemicals are obtained from general laboratory supply companies such as Fisher Scientific and Sigma; catalog numbers are for Fisher except where noted. Three sizes of polystyrene petri dishes are used in culturing nematodes: 35 × 10 mm (Cat. No. 8-757-100-A), 60 × 15 mm (Cat. No. 8-757-13-A), and 100 × 15 mm

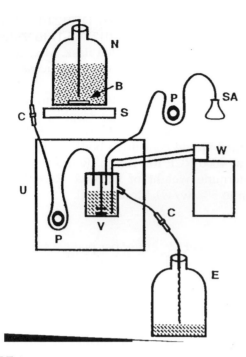

FIGURE 3 Chemostat assembly. N, nutrient reservoir; V, culture vessel; E, effluent tank; S, large stir plate; B, large stir bar (in reservoir to keep bacteria evenly suspended); W, refrigerated circulating water bath (set at 12°C) connected to cold finger of chemostat culture vessel; U, Bioflo I control unit; P, peristaltic pump; SA, selective agent; C, compression fitting (Swagelok Co.) used in the line for easy reservoir replacement (see Section III,C, step 12). Sterility is critical for the maintenance of long-term cultures.

3. When cool, add the other stock solutions, as to complete $1/5 \times S$ medium (see Section B), to the nutrient reservoir.

4. Add the concentrate from 12 liters of overnight OP50 cultures to the nutrient reservoir (final $OD_{600\,nm}$ = ~3.7).

5. Install culture vessel in chemostat control unit and make connections.

6. Start air flow and stir the culture vessel impeller at 220 rpm. Set the heat regulator to 20°C.

7. Run the feeding pump until culture vessel is full; then turn it off.

8. Inoculate culture vessel with sterile LI larvae prepared as described in Section III,E, steps 10 and 11.

9. Add OP50 concentrate from a 1-liter overnight culture to culture vessel. Add approximately 1 ml sterile antifoam A (Sigma, Cat. No. A-5758) as necessary to minimize foam.

10. Monitor the culture every 2 days by removing a sample and counting replicate aliquots spotted on plates (see Section III,A). Continue to add sterile antifoam A to culture vessel as needed. When the culture reaches ~50 animals per $100/A$, turn on the feeding pump (setting 9).

11. Monitor population density and adjust the speed of the feeding pump as necessary to maintain a reproducing culture. A dense population (~125 animals/10//I) that is mostly LI and L2 larvae and roughly one-fourth dauer larvae can be maintained.

12. Prepare a replacement nutrient reservoir for use when the first one is depleted. The effluent tank should be changed at the same time.

D. Freezing Strains for Long-Term Storage

The procedure is modified from that of Brenner (1974).

Solutions

1. *1M NaCl*: Dissolve 29.22 g of NaCl in a final volume of 500 ml distilled water and autoclave.

2. *S + glycerol*: Add 20 ml of $1M$ NaCl, 10 ml of KH_2PO_4 stock, and 60 ml of glycerol (Cat. No. G33) to 110 ml distilled water and autoclave.

Steps

1. Take three contaminant-free 60×15-mm plates 1 day after food is depleted and wash worms off plates with 2 ml M9 buffer.

2. Add equal volume of S + glycerol and mix by brief vortexing. Transfer in 0.5-ml aliquots to 2-ml cryovials (Vangard International, Cat. No. MS4502). Place vials in a styrofoam freezing box (a styrofoam block with holes the size of cryovials and a styrofoam lid) and immediately put at −70°C (cool at approximately l°C/min).

3. After 6 h the vials can be transferred to liquid nitrogen or to standard −70°C freezer boxes.

4. One vial should be thawed to check for viability, strain accuracy, and microbial contamination. Thaw vial by warming between hands until liquid; then pour contents onto seeded plate. Transfer young healthy worms to fresh plates the next day.

E. Isolation of Staged Animals

Solutions

1. *Dauer-inducing pheromone stock (modified from Golden and Riddle, 1984)*: Take 1 liter of starved liquid culture. Reduce volume 75% by evaporation under a stream of air at 100°C. Centrifuge at 10,000 RCF for 10 min. Dry the supernatant completely at 60°C. Extract four to six times with 50 ml of 95% ethanol until the extract is only slightly colored. Combine the extracts and dry under a stream of air at 60°C. Back-extract the resulting oily residue with 10 ml distilled water. Filter through Whatman 3 MM paper and store at 4°C.

2. *OP50 strep*: Transfer OP50 stock (see Section III, A) to preweighed sterile centrifuge bottles/tubes and centrifuge at 4000 RCF for 10 min. Remove supernatant and determine weight of bacteria. Resuspend in S medium to 5% (w/w). Add streptomycin (Sigma, Cat. No. S-6501) to 50/µg/ml final concentration. Store at 4°C for a maximum of 2 days.

Steps

1. Wash the worms off approximately five 60 × 15-mm plates containing a large number of gravid adults with approximately 2 ml M9 buffer per plate.

2. Combine in a 15-ml Corning polystyrene conical centrifuge tube and pellet nematodes. Remove all but 8 ml of the liquid.

3. Mix together 0.5 ml 5 *N* KOH and 1.2 ml 20% NaOCl (Cat. No. SS290) in a separate tube; combine with M9 and worms and vortex briefly.

4. Remove a small aliquot and monitor under a dissecting scope while agitating the remaining sample gently. When 50 to 75% of the adults in the sample have broken open, pellet nematodes.

5. Remove the supernatant, add 8 ml of fresh M9 buffer, and pellet. Repeat two more times leaving 0.5-ml volume after the last wash.

6. Resuspend the eggs and pipette onto plates. This method will generally leave some carcass parts.

Dauer Larvae

This procedure should give greater than 80% dauers.

1. Make NG agar without peptone (see Section III,A). Add approximately 25 ∧1/ml dauer-inducing pheromone stock just before pouring (make only as much as will be used immediately). Pour 2 ml per 35 × 10-mm plate.

2. After plates solidify, spot with 10 µl OP50 strep solution and allow to dry.

3. Add approximately 100 eggs or allow adults to lay 100 eggs; then remove the adults and incubate at 25°C for 48–60 h.

Other Stages

1. Follow the egg isolation procedure through step 5.

2. Bring volume to 10 ml with fresh M9 buffer; incubate on a rocker for 12 h or overnight.

3. Feed synchronized L1 larvae from the previous step. At 20°C mid-L1 larvae can be harvested after approximately 8 h, mid-L2 larvae at 18 h, mid-L3 larvae at 25 h, and L4 larvae at 37 h (Byerly *et al.*, 1976).

IV. COMMENTS

For genetic analysis and maintenance of strains in active use, the nematodes should be grown on 60 × 15-mm petri plates. When large numbers of worms are needed, 100 × 15-mm plates are used. When special additives are being used that are expensive or in limited supply, 35 × 10-mm plates are used. Long-term storage of all strains and those not in active use is best accomplished by freezing in liquid nitrogen. For biochemical purposes the nematodes should be grown in liquid culture. Chemostat culturing enables selection on a continuously reproducing population whose density is held constant (Dykhuizen and Haiti, 1983). Most commonly, selection is for an altered growth rate, whether due to induced or spontaneous mutations. Liquid or gaseous selective agents are compatible with the system described.

More precise synchronization for L4 larvae and adults can be accomplished by synchronizing through the dauer stage. If large numbers of dauers are needed, they can be obtained by slightly modifying the liquid culturing procedure. The culture should be allowed to continue for 2–3 days after clearing. After sucrose flotation, treat with a sterile solution of 1% sodium dodecyl sulfate (SDS) for 1 hr [resuspend in 22.5 ml of M9 and add 2.5 ml of 10% SDS (dissolve 10 g of SDS, Cat. No. S529, in a final volume of 100 ml distilled water)]. Wash twice with M9 buffer and then repeat liquid culture protocol starting with step 3.

References

Brenner, S. (1974). The genetics *of Caenorhabditis elegans. Genetics* **77**, 71–94.

Byerly, L., Cassada, R. C., and Russell, R. L. (1976). The life cycle of the nematode *Caenorhabditis elegans*. I. Wild type growth and reproduction. *Dev. Biol.* **51**, 23–33.

Dougherty, E. C. (1960). Cultivation of Aschelminths, especially Rhabditid nematodes. In "Nematology" (J. N. Sasser and W. R. Jenkins, eds.), pp. 297–318. Univ. of North Carolina Press, Chapel Hill.

Dykhuizen, D. E., and Haiti, D. L. (1983). Selection in chemostats. *Microbiol. Rev.* **47**(2), 150–168.

Golden, J. W., and Riddle, D. L. (1984). The *Caenorhabditis elegans* dauer larva: Developmental effects of pheromone, food, and temperature. *Dev. Biol.* **102**, 368–378.

Hansen, E. L., and Hansen, J. W. (1978). *In vitro* cultivation of nematodes parasitic on animals and plants. In "Methods of Cultivating Parasites in Vitro" (A. E. R. Taylor and J. R. Baker, eds.), pp. 227–278. Academic Press, London.

Nicholas, W. L. (1975). "The Biology of Free-Living Nematodes," pp. 74–91. Clarendon Press, Oxford.

Sijmons, P. C., Grundler, F. M. W., von Mende, N., Burrows, P. R., and Wyss, U. (1991). *Arabidopsis thaliana* as a new model host for plant-parasitic nematodes. *Plant J.* **1**, 245–254.

Sulston, J. E., and Brenner, S. (1974). The DNA of *Caenorhabditis elegans. Genetics* **77**, 95–104.

Vanfleteren, J. R. (1980). Nematodes as nutritional models. *In* "Nematodes as Biological Models" (B. M. Zuckerman, ed.). Vol. 2, pp. 47–79. Academic Press, New York.

Wood, W. B. (1988). "The Nematode *Caenorhabditis elegans*" (W. B. Wood, ed.), pp. 1–16. Cold Spring Harbor Laboratory Press, Cold Spring Harbor, NY.

Differentiation and Reprogramming of Somatic Cells

20

Induction of Differentiation and Cellular Manipulation of Human Myeloid HL-60 Leukemia Cells

David A. Glesne and Eliezer Huberman

I. INTRODUCTION

The human myeloid HL-60 leukemia cell line has enjoyed widespread use in studies of the molecular mechanisms involved in the control of cell growth, differentiation, and apoptosis. Following stimulation with specific agents, these cells acquire a granulocytic (Collins *et al.*, 1980), monocytic, or macrophage-like phenotype (Murao *et al.*, 1983). The maturation of these precursor cells into their mature progeny occurs at a high percentage and with some degree of predictable temporal uniformity. The acquisition of mature phenotypes can be demonstrated by a variety of differentiation markers. These include enzymatic markers such as an increase in the activities of nonspecific esterase or lysozyme; morphological changes such as the appearance of segmented nuclei, which are characteristic of a granulocyte-like phenotype (Collins *et al.*, 1980); or other markers such as cell attachment and spreading on culture dishes or increased phagocytosis, which are indicators of a macrophage phenotype (Tonetti *et al.*, 1994). Additionally, there are a variety of well-characterized immunological markers, some of which are lineage and/or stage specific. The use of these markers and forceful expression of specific cDNAs or inhibition of a particular gene's expression by pertinent antisense oligonucleotides has allowed for the identification of the temporal order of many components involved in the signal transduction processes that lead to HL-60 cell differentiation (Semizarov *et al.*, 1998; Laouar *et al.*, 1999; Xie *et al.*, 1998ab).

II. MATERIALS AND INSTRUMENTATION

HL-60 cells (Cat. No. CCL-240) are available from the American Type Culture Collection (ATCC). Culture medium RPMI 1640 (Cat. No. 11875-093), penicillin–streptomycin–glutamine (Cat. No. 10378-016), geneticin (Cat. No. 10131-035), and trypsin–EDTA (Cat. No. 15400-054) are from Invitrogen Life Technologies. Phorbol 12-myristate 13-acetate (PMA; Cat. No. P-1680) is from L.C. Laboratories, Inc. All-*trans* retinoic acid (ATRA; Cat. No. R-2625), L-α-lysophosphatidylcholine (Cat. No. L-4129), paraformaldehyde (Cat. No. P-1648), bovine serum albumin (BSA) fraction V (Cat. No. A-7906), anti-cd11b (Cat. No. C-0051), transferrin (Cat. No. T-2036), and insulin (Cat. No. I-6634) are available from Sigma Chemical Co. 1α,25-Dihydroxy-vitamin D_3 [1,25-$(OH)_2$vitD3; Cat. No. 679101] and hygromycin B (Cat. No. 400050) are from CalBiochem. PolyMount solution (Cat. No. 16866), 1.72-μm fluoresbrite beads (Cat. No. 17687), and hydroethidine (Cat. No. 17084) are from Polysciences Inc. 4',6-Diamidine-2'-phenylindole dihydrochloride (DAPI; Cat. No. 236276) is available from Boehringer Manneheim Biochemicals. Lab-Tek chamber slides (eight-well, Cat. No. 178599) are from Nunc, Inc. Fetal bovine serum (FBS; Cat. No. SH30071) is from HyClone. Dimethyl sulfoxide (DMSO) molecular biology grade (Cat. No. BP231-100) is from Fisher. Antibodies against CD14 (Cat. No. 30541A) and HLA-DR (Cat. No. 555562) are available from BD Pharmingen. Antibody against human glyceraldehde-

3-phosphate dehydrogenase (GAPDH, Cat. No. CR1093SP) is from Cortex Biochem. Antibody against actin (Cat. No. sc-1615) is from Santa Cruz Biotech. Vector pRL-null (Cat. No. E227A) is from Promega, and pIRES2-EGFP (Cat. No. 6029-1) is from BD Clontech.

III. PROCEDURES

A. Cell Growth and Differentiation Induction

Materials

1. *Culture medium*: Supplement RPMI 1640 culture medium with penicillin–streptomycin ($100\,\mu g/ml$), L-glutamine ($200\,mM$), and 10% FBS. If cells are to be used for viable immunostaining, inactivate serum complement by heat inactivation (65°C for 15 min followed by slow cooling to room temperature). Culture medium can be stored at 4°C for several months.

2. *Differentiation inducer stock solutions*: Dissolve PMA, 1,25-(OH)₂vitD3, and ATRA in DMSO at a concentration of $1\,mg/ml$. Store small aliquots in sterile microcentrifuge tubes at -70°C and avoid repeated freezing/thawing.

3. *Differentiation inducer working solutions*: The most effective concentration for the induction of differentiation depends on the potency of individual lots of the inducer and the endogenous sensitivity of the HL-60 cells being used. Examples of appropriate concentrations for some common inducers of HL-60 differentiation are shown in Table I. For new chemical inducers, a series of concentrations of the inducer should be tested, and conditions that impart an inhibition of cell multiplication and the appearance of one or more myelomonocytic differentiation markers should be

tested further. For many differentiation inducers, cells will exhibit maturation markers at inducer concentrations that inhibit about 50% or greater of cell multiplication rates.

4. *Cell culture*: The HL-60 cell line has a population doubling time of 20–24 h. Cell density should be maintained between 2×10^4 and 1×10^6 cells/ml with replacement of medium at high densities by pelleting of the cells by centrifugation if medium becomes acidified. Cultures should be maintained in either petri dishes or tissue culture plates at 37°C in an 8% CO_2, 95% water-humidified atmosphere.

Steps

1. Collect the cells and obtain cell density by hemocytometer chamber counting. Pellet the cells by gentle ($250\,g$) centrifugation. Resuspend in fresh growth medium and plate the cells at 2–10×10^4 cells/ml in culture dishes. Allow the cells to recover for several hours prior to addition of the inducing agent.

2. Dilute the differentiation-inducing agent in a minimal volume of culture medium and add to the experimental plates. Be sure to include a control culture to which the solution vehicle alone is added.

3. Culture cells at 37°C in an 8% CO_2 humidified atmosphere.

4. The appearance of differentiation markers occurs as early as 12 h and up to 6 days after treatment, depending on the concentration of the inducing agent and the marker to be assessed. To evaluate the phenotypic changes associated with cell differentiation, a series of morphological, biochemical, and immunological assays are performed (see Table II). We recom-

TABLE I Recommended Concentrations for Known Inducers of Differentiation in HL-60 Cells

Inducer	Concentration	End point	Reference
PMA	1–$10\,nM$	Macrophage	Murao *et al.* (1983)
1,25-(OH)₂vitamin D₃	10–$100\,nM$	Monocytic	Murao *et al.* (1983)
ATRA	100–$1000\,nM$	Granulocytic	Breitman *et al.* (1980)
Mycophenolic acid	3–$10\,\mu M$	Granulocytic	Collart and Huberman (1990)
DMSO	1–2%	Granulocytic	Collins *et al.* (1980)

TABLE II Examples of Markers Used to Characterize Differentiation in HL-60 Cells

Marker	Gran[a]	Mono[b]	Mac[c]	Reference
OKM1 MAb (CD11b)	+++	+++	++	Foon *et al.* (1982)
Banded and segmented nuclei	+++	−	−	Breitman *et al.* (1980)
Nonspecific esterase staining	Weak	+++	++	Yam *et al.* (1971)
Lysozyme activity	Weak	+++	+++	Murao *et al.* (1983)
Attachment and spreading	−	+	+++	Tonetti *et al.* (1994)
Phagocytosis	Weak	+	+++	Tonetti *et al.* (1994)
NM-6 MoAb	+++	+++	−	Murao *et al.* (1989)

[a] Granulocytic cells; treated with 1.2% DMSO.
[b] Monocytic cells; treated with $100\,nM$ 1,25-(OH)₂vitamin D₃.
[c] Macrophage-like cells; treated with $3\,nM$ PMA.

mend that two to three different assays be used to determine whether the cells acquire a granulocytic, monocytic, or macrophage-like phenotype. Many of the biochemical or histochemical markers can be assessed using commercially available kits.

B. Cell Adhesion and Spreading Assay

Material

Phosphate-buffered saline (PBS): To make 1 liter of stock solution (10×), dissolve 80.0 g of NaCl, 2.0 g of KCl, 2.0 g of KH_2PO_4, and 11.4 g of Na_2HPO_4 in high-quality (Milli-Q) water and sterilize by autoclaving. Dilute to 1× in water.

Steps

1. Treat HL-60 cells cultured in bacteriological petri dishes (see Section IV) with PMA and incubate at 37°C in an 8% CO_2, humidified atmosphere for 2 days.

2. Collect unattached cells by gentle pipetting and wash petri dishes twice with PBS, combining the washes with the initially collected cells.

3. Recover the remaining attached cells by treatment for 10 min with PBS supplemented with 0.05% trypsin–EDTA followed by forceful pipetting and/or use of a rubber policeman.

4. Count the number of attached and unattached cells using a hemocytometer chamber and calculate the percentage of attached cells as a function of the total recovered cell number.

5. In parallel, determine the number of spread cells by counting flattened cells (Fig. 1) in several microscopic fields of view. Count at least 200 cells/data point; the percentage cell spreading = (number of spread cells/total number of adherent cells) × 100.

C. Phagocytosis Assay

Materials

1. *Fluorescent beads*: Sterilize and opsonize fluores-brite beads by first adding three drops of beads to 10 ml of 70% ethanol. Mix and agitate at room temperature for 20 min. Recover the beads by centrifugation (400 g) and wash twice with PBS. After the second wash, add 5 ml of RPMI 1640 supplemented with 1 ml FBS (not heat inactivated). Mix and incubate overnight at 37°C with slow agitation. Pellet the beads at 1000 g and resuspend in 1 ml of culture medium at a concentration of 4×10^8 beads/ml.

2. *Working solutions*: Prepare the following solutions using PBS as solvent: 4% paraformaldehyde, 10 μg/ml L-α-lysophosphatidylcholine, 0.1 μg/ml DAPI, and 10 μg/ml hydroethidine.

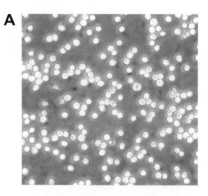

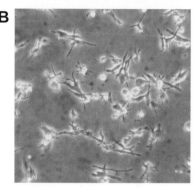

FIGURE 1 Bright-field microscopy images representing an example of HL-60 cells either untreated in suspension (A) or attached and spread following a 3-day treatment with 10 n*M* PMA (B).

Steps

1. Plate 1.2×10^4 cells in 400 μl of growth medium into each well of an eight-well Lab-Tek chamber slide. Treat the cells with PMA or other inducer as described previously.

2. After 30 h, add 100 μl of the bead suspension to each well.

3. Incubate for an additional 18 h, gently remove the growth medium, and air dry the cells for 30 s using a heated slide dryer.

4. Fix the cells by adding 100 μl of 4% paraformaldehyde to each well for 20 min and then washing each well several times with PBS.

5. Add 100 μl of L-α-lysophosphatidylcholine solution to each well for 20 min to permeabilize the cells, followed by several washes with PBS.

6. Cell nuclei are stained by incubation with DAPI for 5 min. Remove excess stain by washing with PBS and then counterstain the cytoplasm by incubating with hydroethidine for 10 min.

7. After three washes in PBS, remove the well partitions and mount the slide with Poly-Mount or other mounting medium and a coverslip.

8. Examine the cells using a fluorescent microscope equipped with a DAPI/UV-range filter set. DAPI-stained nuclei should appear blue, the stained cytoplasm red, and the fluoresbrite beads as small blue dots. A cell is considered positive for phagocytosis activity if it has engulfed more than 20 beads. Count >200 cells per point.

D. Immunofluorescence

Materials

1. *PBSA*: dissolve BSA (Cohn fraction V) to 1% in PBS.

2. *Primary antibodies*: Numerous monoclonal and polyclonal antibodies are available to characterize the phenotype of mature myeloid cells. These include but are not limited to the Mac-l/antiCDllb, CD14, or class II HLA-DR antigens. A more categorical listing is available online (http://www.bdbiosciences.com/pdfs/other/01-81024-3.pdf). These are available from many commercial sources (e.g., Immunotech, Coulter Immunology, Santa Cruz, Sigma, Pharmingen) and should be prepared and used according to the manufacturer's recommendations. For a new antibody, it is often prudent to perform a dilution series with both positive and negative controls to select an antibody concentration that provides a positive signal with a low background. For semiquantitative ratio imaging, useful standard antibodies are against GAPDH (sheep polyclonal, Cortex Biochem.) or actin (goat polyclonal, Santa Cruz). All antibody stock solutions are created by dilution in PBSA.

3. *Secondary antibodies*: There are numerous sources of labeled secondary antibodies. If one is to perform multiantigen imaging, it is critical that the secondary antibodies have been cross-absorbed against the other host organisms antibody repertoire. Such cross-absorbed antibodies are available tagged to a variety of fluorophores from both Jackson Immunoresearch and Molecular Probes. Stock solutions of these secondary antibodies are prepared at 1mg/ml in 50% glycerol/50% PBS. Store aliquots of the stock solution at −70°C. Prepare a working solution of the secondary antibody by diluting from 1:50 to 1:500 in PBSA using the furthest dilution that still provides significant signal.

4. *Microscopy*: A microscope equipped with multiple filter sets and/or cubes is required to visualize the fluorescence signal from multiple fluorophores. If one is to perform semiquantitative ratio imaging using multiple fluorescent signals, it is useful to have either an automated cube turret or automated excitation and emission filter wheels and a multipass dichroic cube.

Both this hardware and SlideBook software for processing such images are available from Intelligent Imaging Innovations (Denver, CO).

5. *Flow cytometery*: For the assessment of marker induction in granulocytic differentiation, flow cytometry has many advantages, particularly regarding the stastical significance attainable by scanning larger cell populations. However, in the case of monocytic and macrophage differentiation, significant clumping of cells is unavoidable and may lead to artifactual results. We therefore recommend fluorescence microscopy for such experiments and the examination of at least 200 cells per data point.

Steps

Reactions are performed on ice using either microcentrifuge tubes or, to facilitate the handling of numerous samples simultaneously, V-bottom 96-well plates.

1. Recover the cells by centrifugation and resuspend in PBS at 2×10^6/ml; add 200µl of the cells for each collection of antibodies to a well of the microtiter plate.

2. Centrifuge the plate ($400g$), remove the supernatant gently by inversion, and fix the cells by incubation with 100µl of 4% paraformaldehyde for 15min. Wash twice with PBS.

3. If the antigen to be visualized is not on the outside of the cell membrane, permeabilize the cells by the addition of either L-α-lysophosphatidylcholine solution or 0.1% Triton X-100 in PBS for 5min. Wash the cells three times with PBS.

4. Resuspend the cells in 100µl of PBSA and allow for blocking of nonspecific sites by incubation for 15min. Recover cells by centrifugation and mix the cells with 100µl of diluted primary antibody or mixture of antibodies (or a dilution series may be used) and incubate for 45min at room temperature.

5. After washing the cells twice with PBS, add 100µl of diluted secondary antibodies and continue incubating for 30min.

6. Wash the cells three times with PBS and then mount onto microscope slides using a small volume of a mounting solution. Overlay with a cover glass and seal with nail polish. Examine by fluorescence microscopy.

E. Transfection and Establishment of Stable Cell Lines

Materials

1. Most chemical approaches (e.g., charged lipids) for the transfection of HL-60 cells produce very poor results. Better transfection is achieved using electro-

poration, particularly with a square-wave device, and to a lesser degree by an exponential-decay device. We use a BTX T820 square-wave electroporator and achieve maximally 10–20% transfection rates.

2. For analysis of promoter activities, either CAT assays or luciferase activities can be monitored in HL-60 cells. However, standard firefly luciferase is unstable or degrades rapidly and the *Renilla* luciferase should be employed instead (pRL vectors, Promega).

3. Due to the low transfection efficiency of HL-60 cells, transient transfection assays should include a readily visualizable marker of transfection such as cotransfection with an EGFP plasmid, or preferably using a vector that expresses both your gene of interest and EGFP from the same plasmid (pIRES2-EGFP, Clontech).

Steps

1. If transient assays are to be performed, use 15 µg of supercoiled plasmid per transfection. If stable transfectants are to be isolated, linearize each plasmid with a restriction endonuclease that cleaves the vector but leaves your gene of interest, the selectable marker, and any regulatory elements intact. Solve either preparation in a small volume of PBS.

2. Collect HL-60 cells by centrifugation and resuspend at 1.0×10^6 in growth medium. Combine 400 µl of cells with DNA in a 0.4-cm-gap electroporation cuvette. Allow to stand for 5 min at room temperature.

3. Mix the cuvette by gentle tapping and insert into a square-wave electroporation device. Electroporate with three pulses of 1500 V each and 90 µs duration.

4. Allow the cells to recover for 5 min and then plate into 10 ml of growth medium and incubate at 37°C in 8% CO_2 in a humidified atmosphere.

5. For transient assays, recover the cells the next day and perform CAT or luciferase assays by standard protocols. For monitoring of fluorescence from EGFP expression, prepare the cells as in steps 1 and 2 of the immunofluorescence protocol, mount the cells, and examine for EGFP expression using a FITC filter set.

6. For stable transfectant recovery, wait 24 h and then add appropriate selective agent. Geneticin and hygromycin B are effective at 500 and 150 µg/ml, respectively, in HL-60 cells.

7. Allow for selection of stable transfectants by growth for 7–10 days. Cells can either be maintained at this stage as a pooled group of transfectants or individual clones can be isolated by single-cell dilution and expansion in 96-well plates.

8. Successful transfection in pools or individual clones should be monitored either by immunofluores-

cence or by Western blotting using either an antibody against the protein or an epitope tag if present.

F. Inhibition of Gene Expression Using Synthetic Oligonucleotides

Materials

1. Synthetic oligonucleotides should be synthesized as 15-mers using standard base chemistry. The exact concentration at which an oligonucleotide effectively disrupts translation or targets a given mRNA for RNase-mediated decay is a function of intracel-lular concentration, expression level of the target, and ability of the oligonucleotide to hybridize, i.e., secondary structure limitations. Therefore, several oligonucleotides and a range of oligonucleotide concentrations with a suitable scrambled control oligonucleotide need to be empirically tested. Initial experiments with siRNA indicate that effective gene knockdown by this approach can be performed transiently, but long-term expression of dsRNA may induce an interferon response (Bridge *et al.*, 2003) in HL-60 cells.

2. Prepare solutions of 5 mg/ml transferrin and 5 mg/ml insulin in unsupplemented RPMI 1640 medium. Prepare serum-free growth medium by supplementing RPMI 1640 with standard concentrations of penicillin–streptomycin (100 µg/ml) and L-glutamine (200 mM).

Steps

1. Collect HL-60 cells by centrifugation and wash three times in serum-free growth medium. Count the cells by hemocytometer chamber counting and plate the cells at 1×10^5 cells/ml in serum-free growth medium.

2. Supplement the cells with 5 µg/ml transferrin and 5 µg/ml insulin. Add a series of dilutions of your test and control oligonucleotides spanning a concentration range from 100 nM to 100 µM.

3. Allow uptake of the oligonucleotide to occur over a period of 6 h incubation at 37°C in an 8% CO_2 in a humidified atmosphere.

4. Restore FBS levels to 10% and allow overnight growth.

5. Monitor toxicity to the cells by trypan blue dye exclusion assays. Determine the effect of your test and control oligonucleotides on the expression of your gene of interest by either immunofluorescence or reverse transcriptase polymerase chain reaction (rtPCR) approaches.

6. Subsequent effects on differentiation induction and specific marker appearance can then be determined by following steps 1–4 and then adding the specific chemical inducer.

IV. COMMENTS AND PITFALLS

1. When thawing materials (antibodies, antigen, or samples), mix thoroughly before dilution or processing. Avoid multiple freeze–thaw cycles by aliquoting small volumes of reagents into multiple microcentrifuge tubes for storage.

2. When using all-*trans* retinoic acid, avoid exposure to light.

3. Fluorescent-activated cell sorting quantitation is not recommended when the differentiation inducer causes cell clumping, as is the case with PMA and related chemicals.

4. For some inducers that cause attachment, trypsin–EDTA treatment may not remove all cells from the surface of tissue culture dishes. Therefore, bacteriological grade petri dishes are utilized for procedures that require removal of the attached cells.

5. Use multiple differentiation markers when testing the effect of a new inducer.

6. In preparing cells for immunological analysis, use heat-inactivated serum to eliminate cell killing due to active complement.

Acknowledgments

This work was supported by the U. S. Department of Energy, Office of Health and Environmental Research, under Contract W-31-109-ENG-38, and the National Institutes of Health under Grant CA80826.

References

Breitman, T. R., Selonick, S. E., and Collins, S. I. (1980). Induction of differentiation of the human promyelocytic leukemia cell line (HL-60) by retinoic acid. *Proc. Natl. Acad. Sci. USA* **77**, 2936–2940.

Bridge, A. J., Pebernard, S., Ducraux, A., Nicoulaz, A. L., and Iggo, R. (2003). Induction of an interferon response by RNAi vectors in mammalian cells. *Nature Genet.* **34**, 263–264.

Collart, F., and Huberman, E. (1990). Expression of IMP dehydrogenase in differentiating HL-60 cells. *Blood* **75**, 570–576.

Collins, S. I., Bonder, A., Ting, R., and Gallo, R. C. (1980). Induction of morphological and functional differentiation of human promyelocytic leukemia cells (HL-60) by compounds which induce differentiation of murine leukemia cells. *Int. J. Cancer* **25**, 213–218.

Foon, K. A., Schroff, R. W., and Gale, R. P. (1982). Surface markers on leukemia and,lymphoma: Recent advances. *Blood* **60**, 1–19.

Laouar, A., Collart, F. R., Chubb, C. B., Xie, B., and Huberman, E. (1999). Interaction between alpha 5 beta 1 integrin and secreted fibronectin is involved in macrophage differentiation of human HL-60 myeloid leukemia cells. *J. Immunol.* **162**, 407–414.

Murao, S., Collart, F. R., and Huberman, E. (1989). A protein containing the cystic fibrosis antigens is an inhibitor of protein kinases. *J. Biol. Chem.* **264**, 8356–8360.

Murao, S., Gemmell, M. A., Callal1am, M. F., Anderson, N. L., and Huberman, E. (1983). Control of macrophage cell differentiation in human promyelocytic HL-60 leukemia cells by 1,25-dihydroxyvitamin D3 and phorbol-12-myristate-13-acetate. *Cancer Res.* **43**, 4989–4996.

Semizarov, D., Glesne, D., Laouar, A., Schiebel, K., and Huberman, E. (1998). A lineage-specific protein kinase crucial for myeloid maturation. *Proc. Natl. Acad. Sci. USA.* **95**, 15412–15417.

Tonetti, D. A., Henning-Chubb, C., Yamanishi, D. T., and Huberman, E. (1994). Protein kinase C-β is required for macrophage differentiation of human HL-60 leukemia cells. *J. Biol. Chem.* **269**, 23230–23235.

Xie, B., Laouar, A., and Huberman, E. (1998a). Autocrine regulation of macrophage differentiation and 92-kDa gelatinase production by tumor necrosis factor-alpha via alpha5 beta1 integrin in HL-60 cells. *J. Biol. Chem.* **273**, 11583–11588.

Xie, B., Laouar, A., and Huberman, E. (1998b). Fibronectin-mediated cell adhesion is required for induction of 92-kDa type IV collagenase/gelatinase (MMP-9) gene expression during macrophage differentiation: The signaling role of protein kinase C-beta. *J. Biol. Chem.* **273**, 11576–11582.

Yam, L., Li, C. Y., and Crosby, W. H. (1971). Cytochemical identification of monocytes and granulocytes. *Am. J. Clin. Pathol.* **55**, 283–290.

21

Cultured PC12 Cells: A Model for Neuronal Function, Differentiation, and Survival

Kenneth K. Teng, James M. Angelastro, Matthew E. Cunningham, and Lloyd A. Greene

I. INTRODUCTION

Since its initial description and characterization in 1976 (Greene and Tischler, 1976), the rat pheochromocytoma PC12 cell line has become a commonly employed model system for studies of neuronal development and function. In particular, PC12 cells have been a convenient alternative to cultured neurons for studying the trophic and differentiative actions of nerve growth factor (NGF; reviewed by Levi-Montalcini and Angeletti, 1968; Levi and Alemá, 1991). When cultured in serum-containing medium, PC12 cells adopt a round and phase-bright morphology and proliferate to high density. Under these conditions, PC12 cells display many of the properties associated with immature adrenal chromaffin cells and sympathicoblasts. When challenged with physiological levels of NGF, these cells cease division, become electrically excitable, extend long branching neurites, and gradually acquire many characteristics of mature sympathetic neurons. Under serum-free conditions, NGF promotes not only neuronal differentiation of PC12 cells, but also their survival (Greene, 1978; Rukenstein et al., 1991).

Several attributes of PC12 cells have led to their widespread popularity in neurobiological research. These include their relatively high degree of differentiation before and after NGF treatment, homogeneous response to stimuli, availability in large numbers for biochemical studies, and suitability for genetic manipulations. This article details experience gained with this cell line in terms of tissue culture requirements and treatment with NGF, as well as quantitative assessment of NGF actions. In addition, we describe some of the potential difficulties that one may encounter when culturing PC12 cells and suggest possible means to avoid or ameliorate these problems. The reader is referred to several prior articles (Greene and Tischler, 1982; Greene et al., 1987, 1991) for a more in-depth discussion of the properties and experimental exploitation of the PC12 cell line.

II. MATERIALS AND INSTRUMENTATION

Rosewell Park Memorial Institute 1640 (RPMI 1640) medium is purchased from Invitrogen (Carlsbad, CA; Cat. No. 23400062) in powder form. Donor horse serum (Cat. No. 12-44977P), fetal bovine serum (Cat. No. 12-10377P), and penicillin/streptomycin (Cat. No. 59-60277P) are obtained from JRH Biosciences (Lenexa, KA). It is recommended that sera be prescreened for their capacities to promote PC12 cell growth and maintenance The horse serum should be heat inactivated in a 56°C water bath for 30 min before use.

Tissue culture plasticwares are obtained from Falcon, Becton Dickson and Company (Lincoln Park, NJ). Freezing vials are purchased from Nunc, Denmark (Cat. No. 377267). Millipak-60 filters (0.22 μm, Cat. No. MPGL06SH2) are from Millipore (Bedford, MA). Filter units (0.45 μm, Cat. No. 245-0045) are obtained from Nalgene Company (Rochester, NY). Ethylhexadecyldimethylammonium bromide (Cat.

No. 117 9712) is purchased from Eastman Kodak Company (Rochester, NY).

NGF is prepared from adult male mouse submaxillary glands as described by Mobley *et al.* (1976). The glands can be purchased from Harlan Bioproducts for Science (Indianapolis, IN; Cat. No. 516) and stored at −80°C until use. NGF stocks (>100 μg/ml; in pH 5.0 acetate buffer, 0.4 M NaCl) are stored at −80°C and, once thawed, can be kept at 4°C for at least 1 month without significant loss of activity. Recombinant or purified NGF may also be purchased from a variety of commercial suppliers, including Harlan Bioproducts for Science, Roche Molecular, and Upstate Biotechnology, Inc.

Rat tail collagen is prepared in 0.1% acetic acid as described previously (Greene *et al.*, 1991) from the tendons of rat tails [see Fig. 14.2 of Kleitman *et al.* (1991) for a photographic illustration of the procedure for exposing and removing rat tail tendons]. Each large tail furnishes approximately 200 ml of stock collagen. Aliquots of collagen stock are stored at −20°C. Once thawed, the stock can be stored for up to several months at 4°C. Sterile technique should be employed throughout the preparation.

III. PROCEDURES

A. Routine Tissue Culture Techniques

Solutions

1. *Complete growth medium*: Prepare RPMI 1640 medium according to the supplier's protocol in reverse osmosis/Milli-Q or double-distilled water. After the addition of sodium bicarbonate (2 g/liter), penicillin (final concentration 25 U/ml), and streptomycin (final concentration 25 μg/ml), sterilize the medium by pressure filtration (driven by 90% air, 10% CO_2 mixture) through a Millipak-60 filter unit, dispense into 500-ml autoclaved bottles, and store in the dark at 4°C. The bottles should be dedicated to tissue culture only and should be cleaned by thorough rinsing without soap or detergent. To make up complete growth medium, add 50 ml of heat-inactivated horse serum and 25 ml of fetal bovine serum to 500 ml of RPMI 1640 medium. Store complete growth medium at 4°C.

2. *Medium for freezing of cells*: Mix 1 volume of dimethyl sulfoxide (DMSO) with 9 volumes of complete growth medium. This medium should be freshly prepared for immediate use only.

Steps

1. PC12 cells show optimal adherence to collagen-coated culture dishes. Before applying to dishes, freshly dilute the stock collagen solution with autoclaved reverse osmosis/Milli-Q water as noted later. The optimal final dilution of the collagen should be determined empirically by testing various concentrations for their capacities to foster cell attachment and NGF-promoted neurite outgrowth (Greene *et al.*, 1991). At too low a dilution, adhesion to substrate is poor, whereas at too high a concentration, neurite outgrowth is impeded and cells are difficult to dislodge for subculture. For application to plates without the necessity of spreading a thin layer, add the diluted collagen solution (typically a 1:50 dilution of the stock is optimal) to plastic tissue cultureware (10 ml/150-mm dish; 5 ml/100-mm dish; 1 ml/35-mm dish; 0.5 ml/well of 24-well culture plates). Leave dishes uncovered to air dry overnight in a tissue culture hood. Alternatively, for quicker drying, dilute the collagen by a factor of five less and add to cultureware in one-fifth the aforementioned volumes. Spread the collagen evenly over the surface of the culture dish with the use of an L-shaped glass rod. Dry collagen by leaving the plates uncovered for 1–2 h in a tissue culture hood. Store collagen-coated dishes at room temperature and use within 1 week after preparation.

2. Feed PC 12 cells three times a week with complete growth medium. Remove approximately two-thirds of the culture1 medium from each plate and replace with fresh complete growth medium (10 ml for 150-mm dishes; 5 ml for 100-mm dishes; 1.5 ml for 35-mm dishes). The medium should be added gently and from the side of the tissue culture dish. The feeding schedule should be kept rigid for maximum cell viability. Maintain PC12 cells in a 37°C incubator with a water-saturated, 7.5% COs atmosphere.

3. Passage PC12 cells (subcultured) when the cultures are 80–90% confluent. Dislodge the cells from the surface of the dish by repeated and forceful discharge of the culture medium directly onto the cells with a disposable glass Pasteur pipette. Forceful trituration of the cell suspension within the Pasteur pipette also decreases cell clumping. Mix the culture medium containing detached PC12 cells with fresh complete medium in a 1:3 or 1:4 ratio. Plate the PC12 cells onto collagen-coated dishes, and increase the passage number of the newly plated PC12 cells by one. As the cell doubling time is 3–4 days, subculture every 7–10 days. To avoid the potential accumulation of variants within the cultures, experiments should be carried out with cells that have undergone no greater than 50 passages.

4. Stock cultures of PC12 cells are frozen at high density (>5 × 10⁶ cells/ml; see later for cell counting procedure) as follows. Dislodge cells from tissue culture dish as described earlier, pellet by centrifuga-

tion at room temperature for 10 min at 500 g, and remove the medium. Add the appropriate volume of freezing medium (described earlier) and resuspend the cell pellet. Aliquot into a Nunc freezing vial (1 ml/vial) and transfer to a –80°C freezer for at least 1 day. For high-viability, long-term storage, the vials should be maintained in liquid nitrogen. The vials should not be permitted to warm up during the transfer to liquid nitrogen (e.g., transfer on dry ice).

5. Thaw frozen PC12 cell stocks (in freezing vials) rapidly in a 37°C water bath (2–3 min). Immediately transfer the cells into 9 volumes of complete growth medium. Pellet the cells by centrifugation at room temperature for 10 min at 500 g. Discard the supernatant. Resuspend the cell pellet in fresh, complete growth medium and plate cells on collagen-coated dishes.

B. Promotion and Assessment of NGF-Dependent Neurite Outgrowth

Solution

Low serum medium: Mix 1 ml of heat-inactivated horse serum per 100 ml of RPMI 1640 medium. Store at 4°C.

Steps

1. Dislodge PC12 cells from stock culture dishes and triturate well using a glass Pasteur pipette to break up cell clumps. Then plate cells at low densities (5 × 10^6 cells per 150-mm dish; 1–2 × 10^6 cells per 100-mm dish; 2–5 × 10^5 cells per 35-mm dish; see cell-counting procedure described later) on collagen-coated dishes in medium supplemented with NGF (50–100 ng NGF final concentration/ml of medium). Dilute NGF from the stock into the medium just before use. Because NGF binds to glass surfaces, use plasticware. Diluted solutions of NGF are not stable. Although neurite outgrowth is satisfactory in complete growth medium, low serum medium is recommended instead in order to economize on serum as well as to reduce cell clumping. Once plated, the cultures should be maintained in a 37°C incubator with a water-saturated, 7.5% CO_2 atmosphere and exchange the medium three times per week as described earlier. Neurite-bearing cells should be noticeable within 1–3 days of NGF treatment, and the number of PC12 cells with neurites should increase progressively with time of NGF exposure. By 7–10 days of treatment, at least 90% of the cells should generate neurites.

2. To determine the proportions of neurite-bearing PC12 cells after NGF treatment, observe cultures with a phase-contrast microscope under high magnification (e.g., 200×). Within random fields, score the propor-

tions of cells that possess at least one neurite greater than 20 µm (about two cell body diameters) in length. Continue counting until the total number of cells assessed exceeds 100. For consistent results, count only discernible and/or single cells, but not cell clumps. To determine mean neurite lengths and rates of neurite elongation, observe cultures using an eyepiece equipped with a calibrated micrometer. The latter is used to measure the entire length of randomly chosen neurites. At least 20–25 neurites are measured per culture.

3. Neurite regeneration experiments are carried out with NGF-pretreated PC12 cell cultures. This permits study of rapid neurite growth as well as a quantitative bioassay for NGF. Treat the cells first with NGF for 7–10 days as described in step 1. Then rinse the cultures five times with medium (without NGF) while the cells are still attached to the substrate. Mechanically dislodge the cells from the dish by trituration through a Pasteur pipette and wash them an additional five times in medium (without NGF) by repeated centrifugation at 500 g for 10 min at room temperature. Plate the washed cells at low density (about 10^5 cells/35-mm dish) in medium (complete or low serum) in the presence or absence of NGF (see step 1). Examine the cultures 24 h later and score for percentage of neurite-bearing cells or cell clumps. Because NGF-treated PC12 cells tend to aggregate, it is often necessary to score clumps rather than single cells. The ability of NGF to induce neurite regeneration from PC12 cells is determined by subtracting the number of neurite-bearing cells in culture medium without NGF from the number of neurite-bearing cells in NGF-containing culture medium. For well-washed cultures, 80–100% of the cells or cell clumps should regenerate neurites with NGF, whereas no more than 10–20% should regenerate neurites in the absence of NGF. The regeneration protocol can be used as a quantitative bioassay for NGF (Greene *et al.*, 1987).

C. Assessment of Survival-Promoting Actions of NGF and Other Substances

Solutions

1. *Nuclei counting solution stock (Soto and Sonnenschein, 1985):* Dissolve 5 g ethylhexadecyldimethylammonium bromide and 0.165 g NaCl in 80 ml of reverse osmosis/Milli-Q water. Add 2.8 ml glacial acetic acid and 1 drop bromphenol blue. Bring final volume to 100 ml and filter through a 0.45-µm filter unit. Store the solution at room temperature.

2. *Working nuclei counting solution (Soto and Sonnenschein, 1985):* Mix phosphate-buffered saline (10 ml),

10% Triton X-100 (5 ml), $1 M$ MgCl$_2$ (200 μl), and nuclei-counting solution stock (10 ml) with enough reverse osmosis/Milli-Q water so that the final volume is 100 ml. Pass through a 0.45 μm filter unit. Store the working nuclei-counting solution at room temperature.

Steps

1. To determine the numbers of PC12 cells suspended in culture or other medium, pellet the cells by centrifugation, aspirate to remove the medium, and resuspend the cells in a known volume of the working nuclei-counting solution. This solution provides a homogeneous suspension of intact nuclei, which are quantified using a hemocytometer. To count cells attached to a substrate, remove the medium and replace with a known volume of working nucleus-counting solution. Resuspend nuclei by trituration and quantify with a hemocytometer.

2. Wash PC12 cells (either naïve cells growing with serum or neuronally differentiated cells growing with serum and NGF) with serum-free RPMI medium five times while still attached to culture dishes and then, after detachment by trituration, wash another five times in serum-free RPMI medium by centrifugation/resuspension. Resuspend the cells in RPMI 1640 medium with or without NGF or other potential trophic agents. Plate the cells in collagen-coated, 24-well culture plates in 0.5 ml of medium (0.5–2×10^5 cells/well). Exchange the serum-free culture medium three times per week. Carry out cell counts by removing the medium, adding working nuclei-counting solution, and counting intact nuclei. Typically, without trophic substances such as NGF, 50% of the cell die by 24 h of serum deprivation and 90% by 3–4 days.

D. Cationic Lipid-Based Transfection Protocol for PC12 Cells

Solutions

OptiMEM-I and Lipofectamine2000: Both reagents are purchased from Invitrogen. OptiMEM-I is used for transfection without antibiotics.

Steps

1. As noted in Section I, an advantage of PC12 cells is that they can be subjected to genetic manipulation. A relatively high transient transfection efficiency (at least 20–30%) of PC12 cells is attainable by the use of lipid-based transfection reagents (Fig. 1). Following transfection, stable lines of PC12 transfectants can also be obtained by applying the appropriate selection pressure (e.g., G418 for mammalian expression

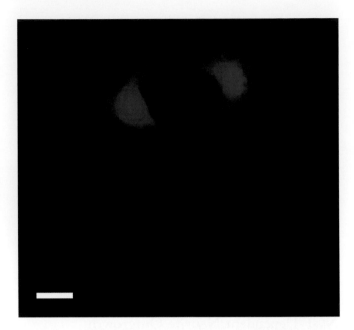

FIGURE 1   PC12 cells expressing green fluorescent protein (GFP). Photomicrograph of non-NGF-treated (naive) PC12 cells transfected with plasmid expressing GFP (green), followed by DAPI staining (blue). Bar: 50 μm.

plasmid carrying the neomycin resistance gene). The day before transfection, seed PC12 cells at high density (at least 90–95% confluency) on collagen-coated tissue culture plates and maintain in complete growth medium.

2. For PC12 cells plated on a 100-mm dish, add 8 μg DNA to 500 μl OptiMEM-I in one tube (tube 1) and add 20 μl Lipofectamine2000 to 500 μl OptiMEM-I in another (tube 2). The contents of tube 2 should be incubated at room temperature for at least 5 min before dropwise addition to tube 1. After which, incubat the mixture further at room temperature for 20 min.

3. During this time, remove complete growth medium from PC12 cells and replace with 4 ml of OptiMEM-I.

4. Add the Lipofectamine2000/DNA mixture to the PC12 cells and return the culture to the incubator. After 4–6 h, aspirate the Lipofectamine2000/DNA mixture (in OptiMEM-I) off completely and replace with complete growth medium. Expression of the transgene can be monitored 24–48 h afterward.

5. Note that the amount of DNA and Lipofectamine2000, as well as the volume of OptiMEM-I, can be adjusted proportionally to accommodate PC12 cells that are plated on smaller/larger size tissue culture plates. The aforementioned procedure can also been used to transfect PC12 cells that were already differentiated by NGF treatment. However, the transfection efficiency is significantly lower (1–2%) due to the need

to culture neuronally differentiated PC12 cells at lower density.

IV. COMMENTS

By adhering to the aforementioned protocols, our laboratory has been able to maintain (since 1977) PC12 cell stocks that are consistently responsive to NGF and that present a stable phenotype. However, a survey of the literature concerning the use of PC12 cells occasionally reveals conflicting or inconsistent results between laboratories. One possible cause for this may be the generation of variant "PC12 cell" lines. Like other continuous cell lines, PC12 cells are subjected to spontaneous mutations, and clonal PC12 cell variants have been identified from past studies. The introduction of nonstandard culture methods (e.g., changing sera, medium, substrate) can favor the selection of such variants over the wild-type population. Although the use of "variant" PC12 cell lines does not necessarily undermine the validity of data generated with them, it can give rise to uncertainty or confusion when one attempts to integrate/reproduce the finding from various reports. It is therefore our suggestion that a uniform standard of culturing PC12 cells be adopted for studies with this cell line.

Another cautionary note on the use of the PC12 cell line is that although it is a convenient model system for studying neuronal development and function, it is not a full substitute for "bona fide" nerve cells. Therefore, whenever feasible, experimental results obtained with PC12 cells should be verified or compared with representative neurons.

V. PITFALLS

1. Poor survival or growth of stock cultures has three probable causes: (i) the initial plating density is too low, (ii) the horse serum is not properly heat inactivated or is of insufficient quality (the latter is the usual cause for failure to thrive), and (iii) the culture medium is outdated (the glutamine has degraded).

2. The most probable cause of poor cell adherence is an insufficient level of collagen as the substrate.

3. A poor NGF response is indicated by the continuous proliferation of PC12 cells in NGF-containing medium and by the lack of neurite-bearing cells even after long-term NGF treatment. A possible cause is that the initial plating density is too high. Another is that the NGF may be inactive. Alternatively, the collagen

concentration on the dishes is too high or too low or the collagen has deteriorated. Finally, the cultures may contain a high proportion of nonresponsive variants (in this case, start with a new cell stock of lower passage number).

4. Spontaneously arising PC12 cell variants are indicated by the presence of flat (phase-dark), rapidly dividing, non-NGF responsive cells. Alternatively, contaminating variants may appear spiky in morphology even in the absence of NGF. The best solution to this is to replace the entire stock with PC12 cells from an earlier passage and to adopt culture conditions that do not favor selection of variants.

5. More than 50% of PC12 cells in serum-free medium without NGF should die within 24 h after plating. However, PC12 cells at high density are capable of conditioning the culture medium, retarding death. Therefore, if cultures exhibit a delay in serum-free cell death, the experiment should be repeated with a lower density of cells. Alternatively, a delay of cell death could be due to an insufficient washout of serum or, for neuronally differentiated cultures, of NGF. In this case, a more stringent washing procedure should be instituted.

6. Generally, it is prudent to discard contaminated cultures and replace with fresh PC12 cell stock. However, if it is necessary to rescue a nonreplaceable culture (such as cell line established from transfection experiments), the following treatments may be effective in removing common sources of contamination. (i) Yeast: treat the culture with 1% fungizone (final concentration) in complete medium. (ii) Mold: remove the contaminant by aspiration. Alternatively, use a Pasteur pipette to remove some of the PC12 cells from a small unaffected area of the dish and replate the cells onto a new dish. Treat the cells with 1% fungizone in complete medium. (iii) Bacteria: a combination of antibiotics and bacterial static agents [see Sambrook *et al.* (1989) for appropriate doses] may be added to the culture. PC12 cells can tolerate ampicillin, kanamycin, spectinomycin, tetracycline, and chloramphenicol.

References

Greene, L. A., Aletta, J. M., Rukenstein, A., and Green, S. H. (1987). PC12 pheochromocytoma cells: Culture, nerve growth factor treatment, and experimental exploitation. *Methods Enzymol.* **147B**, 207–216.

Greene, L. A., Sobeih, M. M., and Teng, K. K. (1991). Methodologies for the culture and experimental use of the PC12 rat pheochromocytoma cell line. In *"Culturing Nerve Cells"* (G. Banker and K. Goslin, eds.), pp. 207–226. *MIT Press*, Cambridge, MA.

Greene, L. A., and Tischler, A. S. (1976). Establishment of a noradrenergic clonal line of rat adrenal pheochromocytoma cells which respond to nerve growth factor. *Proc. Natl. Acad. Sci. USA* **73**, 2424–2428.

Greene, L. A., and Tischler, A. S. (1982). PC12 pheochromocytoma cultures in neurobiological research. *Adv. Cell. Neurobiol.* **3**, 373–414.

Kleitman, N., Wood, P. M., and Bunge, R. P. (1991). Tissue culture methods for the study of myelination. *In "Culturing Nerve Cells"* (G. Banker and K. Goslin, eds.), pp 337–377. *MIT Press*, Cambridge, MA.

Levi, A., and Alemá, S. (1991). The mechanism of action of nerve growth factor. *Annu. Rev. Pharmacol. Toxicol.* **31**, 205–228.

Levi-Montalcini, R., and Angeletti, P. U. (1968). Nerve growth factor. *Physiol. Rev.* **48**, 534–569.

Mobley, W. C., Schenker, A., and Shooter, E. M. (1976). Characterization and isolation of proteolytically modified nerve growth factor. *Biochemistry* **15**, 5543–5552.

Sambrook, J., Fritsch, E. F., and Maniatis, T. (1989). "Molecular Cloning: A Laboratory Manual," 2nd Ed. *Cold Spring Harbor Laboratory*, Cold Spring Harbor, NY.

Soto, A. M., and Sonnenschein, C. (1985). The role of estrogen on the proliferation of human breast tumor cells (MCF-7). *J. Steroid Biochem.* **23**, 87–94.

22

Differentiation of Pancreatic Cells into Hepatocytes

David Tosh

I. INTRODUCTION

The phenomenon of transdifferentiation is defined as the conversion of one differentiated cell type to another (Tosh and Slack, 2002). Generally, cells that have the potential to transdifferentiate arise from adjacent regions in the developing embryo. Therefore, transdifferentiation between adult cells probably reflects their close developmental relationship. Numerous examples of transdifferentiation have been described in literature (Eguchi and Kodama, 1993). Two examples are the transdifferentiation of pancreas to liver (reviewed in Tosh and Slack, 2002; Shen *et al.*, 2003) and the reverse, liver to pancreas conversion (Horb *et al.*, 2003). The liver and pancreas exhibit a close developmental relationship, as they arise from the same region of the endoderm (Wells and Melton, 1999). We have developed two different *in vitro* approaches for inducing the transdifferentiation of pancreatic cells to hepatocytes. The conversion of pancreatic cells to hepatocytes can be induced by culture of either the pancreatic cell line AR42J (Longnecker *et al.*, 1979; Christophe, 1994) or the mouse embryonic pancreas (Percival and Slack, 1999). The first system, AR42J cells, is a cell line originally isolated from a carcinoma of an azaserine-treated rat (Longnecker *et al.*, 1979); although a single cell type, they are considered to be amphicrine in nature, i.e., they possess both exocrine and neuroendocrine properties (Christophe, 1994). The expression of the exocrine enzyme amylase can be enhanced by short-term culture with $10\,nM$ dexamethasone (Logsdon *et al.*, 1985). The advantage of the second system for studying transdifferentiation, the cultured embryonic pancreas system, is that it is more physiological than the AR42J cell line, which has

been around for more than 20 years (Shen *et al.*, 2000). In addition, as the dorsal pancreatic organ grows as a flattened, branched structure, it is suitable for whole mount immunostaining. As well as being of interest to individuals who plan to investigate the transdifferentiation of pancreas to liver, the system is also relevant to those working on normal pancreas development. This article describes the use of AR42J cells and mouse embryonic pancreas as models for the transdifferentiation of pancreas to liver.

A. Models for Differentiation of Pancreatic Cells to Hepatocytes

Several protocols have been produced for inducing the *in vivo* appearance of hepatocytes in the pancreas. For example, administration of a methionine-deficient diet and exposure to a carcinogen (Scarpelli and Rao, 1981) can induce hepatocytes in the pancreas of hamsters. However, one of the most efficient means of converting pancreas to liver is to feed rats a copper-deficient diet in combination with a copper chelator, Trien (Rao *et al.*, 1988). After 7–9 weeks of copper deficiency, the animals are returned to their normal diet and hepatocytes begin to appear soon afterwards. Hepatocytes in the pancreas express a range of liver-specific proteins, e.g., albumin, and are able to respond to xenobiotics (Rao *et al.*, 1982, 1988).

One drawback to *in vivo* studies is that it is difficult to study individual changes at the cellular or molecular levels. An alternative approach is to use *in vitro* models, e.g., AR42J cells. The hepatocytes that are induced to differentiate from pancreatic AR42J cells express many of the proteins that are found in normal liver, e.g., albumin, glucose-6-phosphatase, transferrin, transthyretin, and proteins involved in

Preparation of Fibronectin-Coated Coverslips

Fibronectin comes as a lyophilised powder (Invitrogen, Cat. No. 33010-018).

1. Add 1 ml of sterilised water to a 1-mg vial of fibronectin and dissolve (1 mg/ml final concentration).
2. Aliquot 50 µl into 1-ml Eppendorfs and freeze at −20°C.
3. Prepare glass coverslips by baking for at least 3 h at 180°C.
4. Prior to adding fibronectin, coat the coverslips with 2% 3-aminoporpyltriethoxysilane (APTS) to enhance attachment of the bud to the coverslip.
5. To coat the coverslip with fibronectin, take one tube of the frozen 50 µl fibronectin and add 950 µl of sterilised water (make sure it is well dissolved).
6. In the tissue culture hood, pipette 50 µl of solution onto a sterilised coverslip that has been placed in a 35-mm culture dish. The final concentration of fibronectin is 50 µg/ml.
7. Allow the fibronectin to dry on the coverslips. This takes 2–3 h.

Preparation of Embryos

Mouse embryos are generated by timed matings. The day of the vaginal plug is taken as 0.5. For the purpose of the present study, we use 11.5-day embryos.

Mouse Embryonic Dissection and Culture

1. Kill mouse by cervical dislocation.
2. Test reflexes generally by gripping the paw. If reflexes are absent, then proceed.
3. Soak the fur in 70% ethanol. This is generally a source of infection.
4. Using blunt forceps and scissors, cut open the abdomen (first layer) by a small incision and then tear this back with the fingers and expose the peritoneal sack.
5. Cut the layer with the sharp scissors and forceps and expose one of the uterine horns by displacing the abdominal contents and fatty tissue to the side.
6. Find the furthest embryo from the base of the uterus in either horn and remove the whole string of embryos into a 90-mm petri dish containing ice-cold PBS.
7. Separate the embryos from each other using blunt forceps and scissors; alternatively, simply hold the embryo and uterus with blunt forceps in one hand and squeeze gently until the embryo pops out of the uterus. Place embryos in a clean petri dish of PBS.
8. Dissect the embryos from the sacs and remove their heads. Place in a new petri dish containing Hank's medium.

9. Dissect open the embryo and locate the stomach. At the lower end, just as the stomach empties into the intestine, there is a tissue lying over the surface of the organ. This is the dorsal embryonic bud. Separate the stomach (along with the dorsal pancreas) from the intestine and liver using the needle as a knife. Always remember to cut away from the tissue that you require to avoid damaging the delicate tissue. Finally, with the needle, prise the pancreas away from the stomach and place in a fresh dish of Hank's medium.

10. To culture the pancreatic bud, first place a cloning ring on top of the fibronectin-coated region of the coverslip. Fibronectin is visible as a dried "ring" on the coverslip so it is easy to locate. Add 2.5 ml of Earle's culture medium. Initially add some medium dropwise to the cloning ring but avoid bubbles, as this makes it difficult for the pancreas to settle and attach to the substratum.

11. For each pancreas culture, suck up tissue into a pipette and lower onto medium in the cloning ring. Although two to three pancreatic buds can be cultured within a single cloning ring, they sometimes grow and overlap so it is best to culture only one per dish.

12. Centre the pancreas with the tungsten needle and make sure that the cut surface of the pancreas bud is face down. It is important that the culture is placed in the centre, as occasionally the cloning ring can move slightly and could crush the bud, especially if it is too close to the edge of the cloning ring. Also, ensuring that the bud is placed cut surface down will increase the chances of the bud sticking to the fibronectin matrix.

13. Place the dish in the incubator at 37°C and culture overnight.

14. The following day, check the cultures under an inverted microscope. Generally, allow 24 h before changing the medium. Remove the cloning rings from the coverslips with forceps and then change the medium to fresh medium. Repeat medium changes every 1–2 days (Fig. 2).

The cultured pancreatic buds generally flatten out onto the substratum over the first 1–2 days, and mesenchymal cells spread rapidly out of the explant to form a monolayer of cells surrounding the epithelia in the centre. On the second or third day, branches begin to appear in the epithelium. Scattered cells expressing insulin or glucagon become evident during the first 3 days in culture. Over the next 3 days, the epithelium becomes an extended branched structure radiating from the original centre, and the development of exocrine cells can be recognized. Insulin cells become more numerous and strongly stained, and some islet-like architecture can be seen from day 6. These time

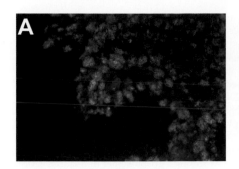

FIGURE 2 Immunostaining for amylase, insulin, and glucagon in normal pancreatic buds. Pancreatic buds were cultured for 7 days and then immunostained for amylase (A), insulin, (B), and glucagon (C).

courses show that the cell differentiation in the cultures resembles quite closely what occurs *in vivo*, although it is delayed about 1 day over a 4-day culture period.

Induction of Transdifferentiation

To induce transdifferentiation of pancreatic cells to hepatocytes, add $1 \mu M$ dexamethasone. This should be added at 1- to 2-day intervals from the $1 mM$ stock that is kept at $-20°C$. Dexamethasone can be added to the medium prior to pipetting onto the dishes or directly to the dish. Add an equivalent volume of ethanol to control dishes.

The embryonic buds can be analysed by immunofluorescent analysis, but the fixation and permeabilisation conditions are different from those for AR42J cells.

D. Immunofluorescence Analysis of Embryonic Pancreas

1. For immunofluorescent staining, fix the pancreata in MEMFA (10% formaldehyde, $0.1 M$ Mops, pH 7.4, $2 mM$, EGTA, $1 mM$ MgSO$_4$) for 30–45 min at room temperature.

2. Wash in PBS; they can be stored in PBS at $4°C$ for up to a few days.

3. Prior to immunostaining, treat the cultures with 1% Triton X-100 in PBS to permeabilise and then block in 2% blocking buffer (Roche), which contains 0.1% Triton X-100.

4. The buds can then be incubated sequentially with primary and secondary antibodies. We can perform triple labelling for the detection of amylase (FITC), insulin (TRITC), and glucagon (AMCA). Dilute the primary antibodies in blocking buffer and apply to the coverslip overnight at $4°C$. Dilute and obtain the antibodies as follows: rabbit polyclonal antiamylase (1/300, Sigma A8273), guinea pig polyclonal antiinsulin (1/300, Sigma, I6163), mouse monoclonal antiglucagon (1:100, Sigma G2654), goat polyclonal antirabbit IgG FITC conjugate (1/100; Vector Laboratories FI-1000), rabbit polyclonal antiguinea pig IgG TRITC conjugate (1/300; Sigma T7153), and horse antimouse IgG AMCA conjugate (1/100; Vector Laboratories CI-2000). The following day, wash the coverslips three times (15 min each) and apply the secondary antibody and leave on for 3 h. On the last day (day 4), wash the coverslip again (three times 15 min each) and mount the coverslip in an appropriate medium, e.g., gelvatol or mobiol. View specimens under a fluorescent microscope. After induction of transdifferentiation with dexamethasone, the hepatocytes can be detected in buds by using the liver antibodies described in Section III,B.

E. Imaging

For confocal imaging, we use a Zeiss LSM 510 confocal system on an inverted Zeiss fluorescent microscope fitted with a ×40/NA 1.30 or ×63/NA 1.40 oil immersion objective (Zeiss, Welwyn Garden City, UK). Alternatively, we use a Leica DMRB microscope fitted with a digital camera. Generally, when two or more antibodies are visualized through different fluorescent channels in the same specimen, the initial black-and-white CCD images are coloured and then recombined to form a single multicoloured image.

IV. COMMENTS

The differentiation of pancreatic AR42J cells to hepatocytes results in a marked morphological change. The cells become flattened and enlarged. These changes occur with 5 days of dexamethasone treatment. It is therefore possible to check the differentiation by noting the number and degree of flattening of cells.

V. PITFALLS

1. For the best results with the embryonic pancreatic cultures, ensure that the bud is placed cut surface down and in the centre of the cloning ring. If not, there is a chance that the cloning ring will move when the incubator door is closed, squashing the bud.

2. When immunostaining for liver proteins, do not be tempted to use serum as a blocking agent. Although good for conventional immunostaining, it is not suitable for some of the liver markers, which are serum proteins normally secreted from the organ, e.g., albumin.

3. Prior to inducing transdifferentiation, inoculate the AR42J cells at low density (e.g., we routinely inoculate at 10–15%). These cells have a high rate of cell division and although dexamethasone inhibits cell division, you may end up with too confluent a dish for optimal immunofluorescence.

Acknowledgment

This work was supported by the Medical Research Council.

References

Christophe, J. (1994). Pancreatic tumoral cell-line AR42J. An amphicrine model. *Am. J. Physiol.* **266**, G963–G971.

Eguchi, G., and Kodama, R. (1993). Transdifferentiation. *Curr. Opin. Cell Biol.* **5**, 1023–1028.

Herrera, P. L. (2000). Adult insulin and glucagon cells differentiate from two independent cell lineages. *Development* **127**, 2317–2322.

Horb, L. D., and Slack, J. M. W. (2000). Role of cell division in branching morphogenesis and differentiation of the embryonic pancreas. *Int. J. Dev. Biol.* **44**, 791–796.

Horb, M. E., Shen, C.-N., Tosh, D., and Slack, J. M. W. (2003). Experimental conversion of liver to pancreas. *Curr. Biol.* **13**, 105–115.

Logsdon, C. D., Moessner, J., William, J. A., and Goldfine, I. D. (1985). Glucocorticoids increase amylase mRNA levels, secretory organelles, and secretion in pancreatic acinar AR42J cells. *J. Cell Biol.* **100**, 1200–1208.

Longnecker, D. S., Lilja, H. S., French, J. I., Kuhlmann, E., and Noll, W. (1979). Transplantation of azeserine-induced carcinomas of pancreas in rats. *Cancer Lett.* **7**, 197–202.

Kamiya, A., *et al.* (1999). Fetal liver development requires a paracrine action of oncostatin M through gp130 signal transducer. *EMBO J.* **18**, 2127–2136.

Mashima, H., Shibata, H., Mine, T., and Kojima, I. (1996). Formation of insulin-producing cells from pancreatic acinar AR42J cells by hepatocyte growth factor. *Endocrinology* **137**, 3969–3976.

Percival, A. C., and Slack, J. M. W. (1999). Analysis of pancreatic development using a cell lineage label. *Exp. Cell Res.* **247**, 123–132.

Rao, M. S., Reddy, M. K., Reddy, J. K., and Scarpelli, D. G. (1982). Response of chemically induced hepatocyte-like cells in hamster pancreas to methyl clofenapate, a peroxisomal proliferator. *J. Cell Biol.* **95**, 50–56.

Rao, M. S., *et al.* (1988). Almost total conversion of pancreas to liver in the adult rat: A reliable model to study transdifferentiation. *Biochem. Biophys. Res. Commun.* **156**, 131–136.

Scarpelli, D. G., and Rao, M. S. (1981). Differentiation of regenerating pancreatic cells into hepatocyte-like cells. *Proc. Natl. Acad. Sci. USA* **78**, 2577–2581.

Shen, C.-N., Horb, M. E., Slack, J. M. W., and Tosh, D. (2003). Transdifferentiation of pancreas to liver. *Mech. Dev.* **120**, 107–116.

Shen, C.-N., Slack, J. M. W., and Tosh, D. (2000). Molecular basis of transdifferentiation of pancreas to liver. *Nature Cell Biol.* **2**, 879–887.

Tosh, D., Shen, C.-N., and Slack, J. M. W. (2002). Differentiated properties of hepatocytes induced from pancreatic cells. *Hepatology* **36**, 534–543.

Tosh, D., and Slack, J. M. W. (2002). How cells change their phenotype. *Nature Rev. Mol. Cell Biol.* **3**, 187–194.

Wells, J. M., and Melton, D. A. (1999). Vertebrate endoderm development. *Annu. Rev. Cell Dev. Biol.* **15**, 393–410.

TERA2 and Its NTERA2 Subline: Pluripotent Human Embryonal Carcinoma Cells

Peter W. Andrews

I. ORIGINS OF NTERA2

TERA2 is one of the oldest extant cell lines established from a human teratocarcinoma. It was derived from a lung metastasis of a testicular germ cell tumour and reported by Fogh and Tremp in 1975. However, its pluripotent embryonal carcinoma (EC) properties were not immediately obvious, in part because maintenance of its undifferentiated EC phenotype requires that cultures of these cells are passaged by scraping in order to retain small clumps of cells instead of using trypsinisation, which results in single cell suspensions. In an early study of human teratocarcinoma-derived cell lines we ourselves dismissed TERA2 as a potential EC cell line (Andrews *et al.*, 1980).

Initially, we failed to derive xenograft tumours from TERA2 (Andrews *et al.*, 1980). Nevertheless, after a number of attempts we did succeed in obtaining a single xenograft tumour after injecting a nude, athymic (*nu/nu*) mouse with TERA2 cells. Fortunately, we explanted some of this tumour back into culture, as well as fixing part for histology. To our surprise, histological examination then revealed that the tumour contained a variety of differentiated elements, as well as embryonal carcinoma components; indeed it was a teratocarcinoma. Meanwhile, it was evident that the cells in culture, now named NTERA2 to designate their passage through a nude mouse, did not resemble morphologically the cells that predominated in the culture of TERA2 that had been used to inoculate the host mouse (Andrews *et al.*, 1980). Indeed, the NTERA2 cells closely resembled other cells that we believed to represent human EC cells (Andrews *et al.*, 1982, 1984b).

From NTERA2 a number of cloned lines were obtained by isolating single cells and culturing clones from them. Detailed cytogenetic study and genetic analyses using isozyme markers demonstrated clearly that NTERA2 and its clones did indeed originate from TERA2 (Andrews *et al.*, 1984b). Subsequently, clones were also derived directly from earlier passages of TERA2 itself and these also exhibited EC cell properties (Thompson *et al.*, 1984; Andrews *et al.*, 1985). It was clear that the TERA2 line does contain EC cells but that these can be lost by differentiation and overgrowth if the cultures are maintained under suboptimal conditions. Nevertheless, remaining subpopulations of EC cells can be rescued from such cultures by cloning or by growing of xenograft tumours in immunosuppressed animals; evidently only undifferentiated EC cells are tumourigenic.

Several single cell clones of NTERA2 were initially studied, in particular NTERA2 clone B9, clone D3, and clone D1. However, there was no obvious difference among these and most subsequent studies have utilised the clone NTERA2 cl. D1, which has also been abbreviated NT2/D1 or sometimes simply NT2. The different clones do exhibit slightly different karyotypes, and it is certain that a small amount of genetic drift occurs upon prolonged culture. Nevertheless, the modal chromosome numbers of the NTERA2 clones, and of TERA2 itself, are very similar, about 61 chromosomes, including a variety of rearrangements. Many of these rearrangements are common to all clones, but new rearrangements continue to appear and characterise the individual clones (Andrews et al., 1984b, 1985).

II. UNDIFFERENTIATED NTERA2 EC CELLS

The features of undifferentiated human EC cells are best exemplified by another cell line, derived from a testicular germ cell tumour, 2102Ep (Andrews et al., 1982). 2102Ep cells, and indeed a number of other "nullipotent" human EC cell lines (Andrews et al., 1980; Andrews and Damjanov, 1994), tend to grow in tightly packed colonies of small cells with little cytoplasm and few prominent nucleoli within a nucleus that comprises most of the cell. They exhibit doubling times of about 20–24 h, and eventually form monolayers from which domes may form and floating vesicles bud off. The cells of these vesicles do not seem to be significantly different from cells that remain attached to the substrate. However, high-density cultures are required to retain these features—typically we seed cultures at densities of at least 5×10^6 cells per 75-cm^2 flask. At lower densities, the cells begin to flatten out and evidently some differentiate into trophectoderms (Andrews et al., 1982; Andrews, 1982; Damjanov and Andrews, 1983).

NTERA2 EC cells behave similarly, although there are differences. Maintenance at high cell densities is essential but, in this case, for passaging, cultures should be dispersed by scraping instead of using trypsin and EDTA—cell:cell contact seems to be required for the long-term retention of an undifferentiated phenotype. Like 2102Ep, NTERA2 cells at low density also flatten out and appear to differentiate, e.g., inducing expression of fibronectin, but they do not seem to produce trophectoderm (Andrews, 1982; Andrews et al., 1984b). Extraembryonic endoderm, evidenced by the production of laminin, may be formed (Andrews et al., 1983).

NTERA2 EC cells also share with other human EC, and indeed embryonic stem (ES) cell lines, the expression of characteristic marker genes, e.g., Oct4. However, of particular utility is their expression of surface antigens characteristic of undifferentiated human EC cells (Andrews et al., 1983, 1984a,b,c, 1996) (Table I). These include the glycolipid antigens stage-specific embryonic antigen (SSEA) -3 and SSEA4, but not SSEA1, the proteoglycan antigens TRA-1-60, TRA-1-81, and GCTM-2, the liver/bone/kidney alkaline phosphatase (L-ALP) associated antigens, TRA-2-49 and TRA-2-54, and also human Thy-1. This same surface antigen phenotype is also expressed by human

ES cells (Thomson et al., 1998; Reubinoff et al., 2000; Draper et al., 2002), and several of the same antigens (SSEA3, SSEA4, TRA-1-60, and TRA-1-81) are expressed by the inner cell mass of human blastocysts (Henderson et al., 2002).

In these respects, human EC cells differ from their murine counterparts, which do not express SSEA3 or SSEA4, or murine Thy-1 (Shevinsky et al., 1982; Kannagi et al., 1983; Martin and Evans, 1974), but do express SSEA1 (Solter and Knowles, 1978). Murine EC cells do express high levels of alkaline phosphatase (Bernstine et al., 1973), but that is not recognised by the antibodies TRA-2-49 and TRA-2-54. The antibodies TRA-1-60, TRA-1-81, and GCTM2 also do not appear to recognise epitopes expressed by mouse cells. A further difference between mouse and human EC cells is the expression of class 1 major histocompatibility complex (MHC) antigens, of which HLA is commonly expressed by human EC cells (Andrews et al., 1981), whereas H-2 is not expressed by mouse EC cells. In fact, undifferentiated NTERA2 cells only express low and variable levels of HLA-A,B,C, in contrast to other human EC cells (Andrews et al., 1984b), but these levels are increased markedly by exposure to interferon-γ (Andrews et al., 1987).

III. DIFFERENTIATION OF NTERA2 EC CELLS

NTERA2 EC cells differ from many other human EC cells by their susceptibility to differentiation induced by retinoic acid (Andrews, 1984) and to other inducing agents such as hexamethylene bisacetamide (HMBA) (Andrews et al., 1986, 1990) and the bone morphogenetic proteins (Andrews et al., 1994). These agents induce differentiation in distinct directions, although the best studied is retinoic acid-induced differentiation, which results in the formation of neurons as well as other cell types (Andrews, 1984). By contrast, many other human EC cells do not respond to retinoic acid (Matthaei et al., 1983).

Within 24–48 h after exposure to $10^{-5} M$ all-trans retinoic acid, NTERA2 cells commit to differentiate. During the following 2 weeks, expression of the key surface markers SSEA3, SSEA4, TRA-1-60, and TRA-1-81 is eliminated in most cells, while they lose the characteristic EC morphology, and begin expressing a range of new markers. Prominent among these induced markers are surface antigens SSEA1, A2B5, and ME311 (Table I), which appear to segregate to discrete subsets of cells (Fenderson et al., 1987). At the same time, expression of members of HOX gene clusters is induced (Mavilio *et al.*, 1988). This induction of *HOX* gene expression is dose dependent in a way that relates to the position of the genes in the *HOX* cluster (Simeone *et al.*, 1990; Bottero *et al.*, 1991). Thus, genes located at the 5' end of the clusters tend to require higher concentrations of retinoic acid for maximal induction than genes located at the 3' end of the clusters. This pattern appeared to relate to the expression pattern of the *HOX* genes along the anterior–posterior axis of the developing embryo and the postulated role for retinoic acid in establishing that axis. However, differentiation induced by HMBA, which causes a similar elimination of the EC surface marker antigens, was not accompanied by a comparable induction of the surface antigens, or the *HOX* genes, induced by retinoic acid (Andrews *et al.*, 1990).

Differentiation of NTERA2 cells induced by retinoic acid, and also HMBA, results in the appearance of susceptibility of the cells to replication of both human cytomegalovirus (HCMV) (Gönczöl *et al.*, 1984) and human immunodeficiency virus (HIV) (Hirka *et al.*, 1991). Neither virus is able to replicate in undifferentiated cells, although they can gain entry. In the case of HCMV, the block appears to lie directly in the inability of the cells to support transcription from the major immediate early promoter of the virus (LaFemina *et al.*, 1986; Nelson *et al.*, 1987). The nature of the block in the case of HIV is less certain. However, both viruses replicate readily in the differentiated derivatives of NTERA2 cells, yielding fully infectious virions. In the case of HIV, the entry of the virus into the cells does not involve CD4 as the receptor (Hirka *et al.*, 1991).

Among cells arising from the differentiation of NTERA2, the neurons that appear after retinoic acid induction are the most well defined and studied. These appear only rarely after induction with HMBA or BMPs (Andrews *et al.*, 1986, 1990, 1994). A variety of other cell types do appear in cultures induced with each of these agents. These are poorly defined and are not identified readily with specific cell types, although they may include mesenchymal cell types, including smooth muscle, fibroblasts, and chondrocytes (Duran *et al.*, 1992).

Neurons were initially identified in retinoic acid-induced cultures by their expression of neurofilaments and toxin receptors (Andrews, 1984). Subsequently, a detailed analysis of neurofilament expression confirmed their neural identity (Lee and Andrews, 1986), and electrophysiological studies indicated that they expressed ion channels appropriate to embryonic neurons (Rendt *et al.*, 1989). The neurons formed by the differentiation of NTERA2 cells can be purified readily and have been widely used in a variety of studies of human neural behaviour (Pleasure *et al.*, 1992; Pleasure

and Lee, 1993). Neural differentiation of NTERA2 cells appears to follow a sequence of events that parallel neural differentiation in the developing neural tube. Thus an early marker induced following initial exposure to retinoic acid is nestin, which is followed sequentially by the expression of neuroD1, a HLH transcription factor characteristic of postmitotic neuroblasts, and finally by markers of mature neurons such as synaptophysin (Pryzborski *et al.*, 2000). The appearance of the cell surface antigen A2B5 in differentiating cells appears to be related to the neural lineage and it seems that A2B5 expression is activated shortly before cells leave the cell cycle. A2B5 is expressed by the terminally differentiated NTERA2 neurons (Fenderson *et al.*, 1987).

IV. REAGENTS, SOLUTIONS, AND MATERIALS

Medium

All cells are cultured in Dulbecco's modified Eagle's medium (DMEM) (high glucose formulation), supplemented with 10% fetal calf serum (FCS). DMEM may be purchased from any reputable supplier of tissue culture reagents. Samples of FCS should be obtained from a number of suppliers and tested to identify a batch that supports optimal growth of NTERA2 cells and maintenance of an undifferentiated state, assayed by the expression of appropriate markers (see later). A batch of FCS that provides for good maintenance of undifferentiated NTERS2 cells is not necessarily the best for supporting differentiation. FCS should be batch tested separately for both purposes.

Dulbecco's Phosphate-Buffered Saline (PBS)

For the following procedures, PBS without Ca^{2+} and Mg^{2+} is used and may be purchased from any reputable supplier of tissue culture reagents.

Retinoic Acid

all-*trans* Retinoic acid can be purchased from Sigma-Aldrich or from Eastman-Kodak. It should be stored in the dark at $-70°C$, preferably under nitrogen. A stock solution of $10^{-2} M$ retinoic acid (3 mg/ml) in dimethyl sulphoxide (DMSO) should be prepared and also stored at $-70°C$. If prepared carefully under aseptic conditions, this stock solution may be assumed to be sterile.

Hexamethylene Bisacetamide (HMBA)

A $0.3 M$ stock solution of HMBA should be prepared in PBS and may be stored at $4°C$. It may be sterilised by passage through a $0.2-\mu m$ filter.

TABLE I Common Antibodies[a] Used to Detect Antigens Expressed by Human EC and ES Cells

Antibody	Antigen	Antibody species and Isotype	Reference
Surface Antigens of undifferentiated NTERA2 cells			
Globoseries glycolipids			
MC631	SSEA3	Rat IgM	Shevinsky *et al.* (1982); Kannagi *et al.* (1983)
MC813-70	SSEA4	Mouse IgG	Kannagi *et al.* (1983)
Keratan sulphate proteoglycans			
TRA-1-60	TRA-1-60	Mouse IgM	Andrews *et al.* (1984a); Badcock *et al.* (1999)
TRA-1-81	TRA-1-81	Mouse IgM	Andrews *et al.* (1984a); Badcock *et al.* (1999)
GCTM2	GCTM2	Mouse IgM	Pera *et al.* (1988); Cooper *et al.* (2002)
Liver/bone/kidney alkaline phosphatase			
TRA-2-49	L-ALP	Mouse IgG	Andrews *et al.* (1984c)
TRA-2-54	L-ALP	Mouse IgG	Andrews *et al.* (1984c)
Surface Antigens marking differentiated NTERA2 cells			
Lactoseries glycolipids			
MC480	SSEA1 [Le^x]	Mouse IgM	Solter and Knowles (1978); (Gooi *et al.* (1981); Kannagi *et al.* (1982)
Ganglioseries glycolipids			
A2B5	GT3	Mouse IgM	Eisenbarth *et al.* (1979)
ME311	9-*O*-AcetylGD3	Mouse IgG	Thurin *et al.* (1985)
VINIS56	GD3	Mouse IgM	Andrews *et al.* (1990)
VIN2PB22	GD2	Mouse IgM	Andrews *et al.* (1990)

[a] MC631, MC813-70, MC480, TRA-2-54, TRA-2-49, VNIS56, and VIN2PB22 are available from Chemicon Inc., Santa Cruz Inc., and the Developmental Studies Hybridoma Bank (University of Iowa). TRA-1-60 and TRA-1-81 are available from Chemicon Inc and Santa Cruz Inc. GCTM2 is available from ES Cell International, A2B5 from the American Type Culture Collection, and ME311 from the Wistar Institute of Anatomy (Philadelphia, PA).

a culture, supernatants can be removed by aspiration rather than dumping.

7. Wash the cells by adding 100 μl wash buffer to each well, seal, and agitate to resuspend the cells. Spin down as described earlier. After removing the supernatant, repeat with two further washes.

8. After the third wash, remove the supernatant and add 50 μl fluorescent-tagged antibody, previously titred and diluted in wash buffer to each well. FITC-tagged goat antimouse IgM or antimouse IgG, as appropriate to the first antibody, may be used. Antimouse IgM, but not antimouse IgG, usually works satisfactorily with MC631 (a rat IgM). Affinity-purified and/or F(ab)$_2$ second antibodies may be used if required to eliminate background.

9. Seal the plate as described earlier and repeat the incubation and washings as before.

10. Resuspend the cells at about 5×10^5/ml in wash buffer and analyze in the flow cytometer. The precise final cell concentration will depend upon local operating conditions and protocols.

Fluorescent-Activated Cell Sorting (FACS)

1. Harvest the hES cells using trypsin:EDTA as for analysis.

2. Pellet the cells and resuspend in primary antibody, diluted in medium without added azide, as determined by prior titration (100 μl per 10^7 cells). The primary and secondary antibodies should be sterilized using a 0.2-μm cellulose acetate filter.

3. Incubate the cells with occasional shaking at 4°C for 20–30 min.

4. Wash the cells by adding 10 ml medium and pellet by centrifugation at 200 g for 5 min; repeat this wash step once more.

5. Remove supernatant and flick gently to disperse the pellet. Add 100 μl of diluted secondary antibody per 10^7 cells and incubate, with occasional shaking, at 4°C for 20 min.

6. Wash the cells two times as just described. After the final wash, resuspend the cells in medium at 10^7 cells/ml. Sort cells using the flow cytometer according to local protocols.

Acknowledgments

This work was supported in part by grants from the Wellcome Trust, Yorkshire Cancer Research, and the BBSRC.

References

Ackerman, S. L., Knowles, B. B., and Andrews, P. W. (1994). Gene regulation during neuronal and non-neuronal differentiation of

NTERA2 human teratocarcinoma-derived stem cells. *Mol. Brain Res.* **25**, 157–162.

Andrews, P. W. (1982). Human embryonal carcinoma cells in culture do not synthesize fibronectin until they differentiate. *Int. J. Cancer* **30**, 567–571.

Andrews, P. W. (1984). Retinoic acid induces neuronal differentiation of a cloned human embryonal carcinoma cell line in vitro. *Dev. Biol.* **103**, 285–293.

Andrews, P. W., Banting, G., Damjanov, I., Arnaud, D., and Avner, P. (1984a). Three monoclonal antibodies defining distinct differentiation antigens associated with different high molecular weight polypeptides on the surface of human embryonal carcinoma cells. *Hybridoma* **3**: 347–361.

Andrews, P. W., Bronson, D. L., Benham, F., Strickland, S., and Knowles, B. B. (1980). A comparative study of eight cell lines derived from human testicular teratocarcinoma. *Int. J. Cancer* **26**, 269–280.

Andrews, P. W., Bronson, D. L., Wiles, M. V., and Goodfellow, P. N. (1981). The expression of major histocompatibility antigens by human teratocarcinoma derived cells lines. *Tissue Antigens* **17**, 493–500.

Andrews, P. W., Casper, J., Damjanov, I., Duggan-Keen, M., Giwercman, A., Hata, J. I., von Keitz, A., Looijenga, L. H. J., Millán, J. L., Oosterhuis, J. W., Pera, M., Sawada, M., Schmoll, H. J., Skakkebaek, N. E., van Putten, W., and Stern, P. (1996). Comparative analysis of cell surface antigens expressed by cell lines derived from human germ cell tumors. *Int. J. Cancer* **66**, 806–816.

Andrews, P. W., and Damjanov, I. (1994). Cell lines from human germ cell tumours. In *"Atlas of Human Tumor Cell Lines"* (R. J. Hay, J.-G. Park, and A. Gazdar, eds.), pp. 443–476. Academic Press, San Diego.

Andrews, P. W., Damjanov, I., Berends, J., Kumpf, S., Zappavingna, V., Mavilio, F., and Sampath, K. (1994). Inhibition of proliferation and induction of differentiation of pluripotent human embryonal carcinoma cells by osteogenic protein-1 (or bone morphogenetic protein-7). *Lab. Invest.* **71**, 243–251.

Andrews, P. W., Damjanov, I., Simon, D., Banting, G., Carlin, C., Dracopoli, N. C., and Fogh, J. (1984b). Pluripotent embryonal carcinoma clones derived from the human teratocarcinoma cell line Tera-2: Differentiation *in vivo* and *in vitro*. *Lab. Invest.* **50**, 147–162.

Andrews, P. W., Damjanov, I., Simon, D., and Dignazio, M. (1985). A pluripotent human stem cell clone isolated from the TERA-2 teratocarcinoma line lacks antigens SSEA-3 and SSEA-4 *in vitro* but expresses these antigens when grown as a xenograft tumor. *Differentiation* **29**, 127–135.

Andrews, P. W., Gönczöl, E., Plotkin, S. A., Dignazio, M., and Oosterhuis, J. W. (1986). Differentiation of TERA-2 human embryonal carcinoma cells into neurons and HCMV permissive cells: Induction by agents other than retinoic acid. *Differentiation* **31**, 119–126.

Andrews, P. W., Goodfellow, P. N., and Bronson, D. L. (1983). Cell-surface characteristics and other markers of differentiation of human teratocarcinoma cells in culture. In *"Teratocarcinoma Stem Cells"* (Silver, Martin, and Strickland, eds.), pp. 579–590. Cold Spring Harbor Press, Cold Spring Harbor, NY.

Andrews, P. W., Goodfellow, P. N., Shevinsky, L. H., Bronson, D. L., and Knowles, B. B. (1982). Cell-surface antigens of a clonal human embryonal carcinoma cell line: Morphological and antigenic differentiation in culture. *Int. J. Cancer* **29**, 523–531.

Andrews, P. W., Meyer, L. J., Bednarz, K. L., and Harris, H. (1984c). Two monoclonal antibodies recognizing determinants on human embryonal carcinoma cells react specifically with the liver isozyme of human alkaline phosphatase. *Hybridoma* **3**, 33–39.

Andrews, P. W., Nudelman, E., Hakomori, S., and Fenderson, B. A. (1990). Different patterns of glycolipid antigens are expressed following differentiation of TERA-2 human embryonal carcinoma

cells induced by retinoic acid, hexamethylene bisacetamide HMBA or bromodeoxyuridine BUdR. *Differentiation* **43**, 131–138.

Andrews, P. W., Trinchieri, G., Perussia, B., and Baglioni, C. (1987). Induction of class 1 major histocompatibility complex antigens in human teratocarcinoma cells by interferon without induction of differentiation, growth inhibition or resistance to viral infection. *Cancer Res.* **47**, 740–746.

Badcock, G., Pigott, C., Goepel, J., and Andrews, P. W. (1999). The human embryonal carcinoma marker antigen TRA-1-60 is a sialylated keratan sulfate proteoglycan. *Cancer Res.* **59**, 4715–4719.

Bernstine, E. G., Hooper, M. L., Grandchamp, S., and Ephrussi, B. (1973). Alkaline phosphatase activity in mouse teratoma. *Proc. Natl. Acad. Sci. USA.* **70**, 3899–3903.

Bottero, L., Simeone, A., Arcioni, L., Acampora, D., Andrews, P. W., Boncinelli, E., and Mavilio, F. (1991). Differential activation of homeobox genes by retinoic acid in human embryonal carcinoma cells. In *"Recent Results in Cancer Research"* (J. W. Oosterhuis, H. Walt, and I. Damjanov, eds.), Vol. 123, pp. 133–143. Springer-Verlag, New York.

Cooper, S., Bennett, W., Andrade, J., Reubinoff, B. E., Thomson, J., and Pera, M. F. (2002). Biochemical properties of a keratan sulphate/chondroitin sulphate proteoglycan expressed in primate pluripotent stem cells. *J. Anat.* **200**, 259–265.

Damjanov, I., and Andrews, P. W. (1983). Ultrastructural differentiation of a clonal human embryonal carcinoma cell line *in vitro*. *Cancer Res.* **43**, 2190–2198.

Draper, J. S., Pigott, C., Thomson, J. A., and Andrews, P. W. (2002). Surface antigens of human embryonic stem cells: Changes upon differentiation in culture. *J. Anat.* **200**, 249–258.

Duran, C., Talley, P. J., Walsh, J., Pigott, C., Morton, I., and Andrews, P. W. (2001). Hybrids of pluripotent and nullipotent human embryonal carcinoma cells: Partial retention of a pluripotent phenotype. *Int. J. Cancer* **93**, 324–332.

Eisenbarth, G. S., Walsh, F. S., and Nirenberg, M. (1979). Monoclonal antibody to a plasma membrane antigen of neurons. *Proc. Natl. Acad. Sci. USA* **76**, 4913–4917.

Fenderson, B. A., Andrews, P. W., Nudelman, E., Clausen, H., and Hakomori, S. (1987). Glycolipid core structure switching from globo- to lacto- and ganglio-series during retinoic acid-induced differentiation of TERA-2-derived human embryonal carcinoma cells. *Dev. Biol.* **122**, 21–34.

Fogh, J., and Trempe, G. (1975). New human tumor cell lines. In *"Human Tumor Cells in Vitro"* (J. Fogh, ed.), pp. 115–159. Plenum Press, New York.

Gönczöl, E., Andrews, P. W., and Plotkin, S. A. (1984). Cytomegalovirus replicates in differentiated but not undifferentiated human embryonal carcinoma cells. *Science* **224**, 159–161.

Gooi, H. C., Feizi, T., Kapadia, A., Knowles, B. B., Solter, D., and Evans, M. J. (1981). Stage specific embryonic antigen involves α1 → 3 fucosylated type 2 blood group chains. *Nature* **292**, 156–158.

Henderson, J. K., Draper, J. S., Baillie, H. S., Fishel, S., Thomson, J. A., Moore, H., and Andrews, P. W. (2002). Preimplantation human embryos and embryonic stem cells show comparable expression of stage-specific embryonic antigens. *Stem Cells* **20**, 329–337.

Hirka, G., Prakesh, K., Kawashima, H., Plotkin, S. A., Andrews, P. W., and Gönczöl, E. (1991). Differentiation of human embryonal carcinoma cells induces human immunodeficiency virus permissiveness which is stimulated by human cytomegalovirus coinfection. *J. Virol.* **65**, 2732–2735.

Kannagi, R., Cochran, N. A., Ishigami, F., Hakomori, S.-I., Andrews, P. W., Knowles, B. B., and Solter, D. (1983). Stage-specific embryonic antigens SSEA3 and -4 are epitopes of a unique globo-series ganglioside isolated from human teratocarcinoma cells. *EMBO J.* **2**, 2355–2361.

Kannagi, R., Nudelman, E., Levery, S. B., and Hakomori, S. (1982). A series of human erythrocyte glycosphingolipids reacting to the monoclonal antibody directed to a developmentally regulated antigen, SSEA1. *J. Biol. Chem.* **257,** 14865–14874.

Kohler, G., and Milstein, C. (1975). Continuous cultures of fused cells secreting antibody of predefined specificity. *Nature* **256,** 495–497.

LaFemina, R., and Hayward, G. S., (1986). Constitutive and retinoic acid-inducible expression of cytomegalovirus immediate-early genes in human teratocarcinoma cells. *J. Virol.* **58,** 434–440.

Lee, V. M-Y., and Andrews, P. W. (1986). Differentiation of NTERA-2 clonal human embryonal carcinoma cells into neurons involves the induction of all three neurofilament proteins. *J. Neurosci.* **6,** 514–521.

Martin, G. R., and Evans, M. J. (1974). The morphology and growth of a pluripotent teratocarcinomas cell line and its derivatives in tissue culture. *Cell* **2,** 163–172.

Matthaei, K., Andrews, P. W., and Bronson, D. L. (1983). Retinoic acid fails to induce differentiation in human teratocarcinoma cell lines that express high levels of cellular receptor protein. *Exp. Cell Res.* **143,** 471–474.

Mavilio, F., Simeone, A., Boncinelli, E., and Andrews, P. W. (1988). Activation of four homeobox gene clusters in human embryonal carcinoma cells induced to differentiate by retinoic acid. *Differentiation* **37,** 73–79.

Nelson, J. A., Reynolds-Kohler, C., and Smith, B. A., (1987). Negative and positive regulation by a short segment in the 5'-flanking region of the human cytomegalovirus major immediate-early gene. *Mol. Cell. Biol.* **7,** 4125–4129.

Pera, M. F., Blasco-Lafita, M. J., Cooper, S., Mason, M., Mills, J., and Monaghan, P. (1988). Analysis of cell-differentiation lineage in human teratomas using new monoclonal antibodies to cytostructural antigens of embryonal carcinoma cells. *Differentiation* **39,** 139–149.

Pleasure, S. J., and Lee, V. M. Y. (1993). NTERA-2 cells a human cell line which displays characteristics expected of a human committed neuronal progenitor cell. *J. Neurosci. Res.* **35,** 585–602.

Pleasure, S. J., Page, C., and Lee, V. M.-Y. (1992). Pure, post-mitotic, polarized human neurons derived from Ntera2 cells provide a system for expressing exogenous proteins in terminally differentiated neurons. *J. Neurosci.* **12,** 1802–1815.

Przyborski, S. A., Morton, I. E., Wood, A., and Andrews, P. W. (2000). Developmental regulation of neurogenesis in the pluripotent human embryonal carcinoma cell line NTERA-2. *Eur. J. Neurosci.* **12,** 3521–3528.

Rendt, J., Erulkar, S., and Andrews, P. W. (1989). Presumptive neurons derived by differentiation of a human embryonal carcinoma cell line exhibit tetrodotoxin-sensitive sodium currents and the capacity for regenerative responses. *Exp. Cell Res.* **180,** 580–584.

Reubinoff, B. E., Pera, M. F., Fong, C. Y., Trounson, A., and Bongso, A. (2000). Embryonic stem cell lines from human blastocysts: somatic differentiation *in vitro. Nature Biotechnol* **18,** 399–404.

Shevinsky, L. H., Knowles, B. B., Damjanov, I., and Solter, D. (1982). Monoclonal antibody to murine embryos defines a stage-specific embryonic antigen expressed on mouse embryos and human teratocarcinoma cells. *Cell* **30,** 697–705.

Simeone, A., Acampora, D., Arcioni, L., Andrews, P. W., Boncinelli, E., and Mavilio, F. (1990). Sequential activation of human HOX2 homeobox genes by retinoic acid in human embryonal carcinoma cells. *Nature* **346,** 763–766.

Solter, D., and Knowles, B. B. (1978). Monoclonal antibody defining a stage-specific mouse embryonic antigen SSEA1. *Proc. Natl. Acad. Sci. USA* **75,** 5565–5569.

Thompson, S., Stern, P. L., Webb., M., Walsh, F. S., Engström, W., Evans, E. P., Shi, W. K., Hopkins, B., and Graham, C. F. (1984). Cloned human teratoma cells differentiate into neuron-like cells and other cell types in retinoic acid. *J. Cell Sci.* **72,** 37–64.

Thomson, J. A., Itskovitz-Eldor, J., Shapiro, S. S., Waknitz, M. A., Swiergiel, J. J., Marshall, V. S., and Jones, J. M. (1998). Embryonic stem cell lines derived from human blastocysts. *Science* **282,** 1145–1147.

Thurin, J., Herlyn, M., Hindsgaul, O., Stromberg, N., Karlsson, K. A., Elder, D., Steplewski, Z., and Koprowski, H. (1985). Proton NMR and fast-atom bombardment mass spectrometry analysis of the melanoma-associated ganglioside 9-O-acetyl-GD3. *J. Biol. Chem.* **260,** 14556–14563.

24

Embryonic Explants from *Xenopus laevis* as an Assay System to Study Differentiation of Multipotent Precursor Cells

Thomas Hollemann, Yonglong Chen, Marion Sölter, Michael Kühl, and Thomas Pieler

I. INTRODUCTION

A mayor question in developmental biology is how cells or groups of cells are committed to their distinct fate during the elaboration of the body plan. An experimental system, which has been used to investigate the nature of inductive signals and therefore has led to the identification of components and interactions of various signal transduction pathways, is the so-called animal cap explant system of amphibian embryos. The animal cap consists of the ectodermal roof of a blastula stage embryo that can be kept in culture under the simplest conditions. An early use of this explant system was described by Johannes Holtfreter in an attempt to investigate the nature of inductive signals. In 1933, he established experimental tissue combination systems, referred to as "Umhüllungsversuch" (coating test) and "Auflagerungsversuch" (bedding test), to study the influence of factors emanating from mortified tissues on explanted ectoderm (Fig. 1) (Holtfreter, 1933).

Holtfreter showed that the mortified dorsal lip of a gastrula stage amphibian embryo led to the induction of neural tissue (neurale Blasen) in conjugated ectodermal explants (Fig. 1b). In 1971, Sudarwati and Nieuwkoop demonstrated that vegetal tissue is capable of inducing mesoderm in combined animal cap explants using the South African clawed frog *Xenopus laevis* as a source for animal cap explants (mesoderm induction assay). More recently, the animal cap system has been utilized to tackle a broad variety of developmental problems, including the specification of all germ layers, organ formation, and expression screenings for secreted proteins (Sudarwati and Nieuwkoop, 1971; Gurdon *et al.*, 1985; Slack, 1990; Smith *et al.*, 1990; Green *et al.*, 1992; Moriya *et al.*, 2000).

II. MATERIALS AND INSTRUMENTATION

Adult *X. laevis* frogs are from NASCO (Wisconsin). The reagents used are agarose (Cat. No. 15510-027, GibcoBRL); albumin bovine (BSA, Cat. No. A-8806, Sigma); chorionic gonadotropin (HCG, Cat. No. CG-10, Sigma); L-15 (Leibovitz, GibcoBRL); L-cysteine hydrochloride monohydrate (Cat. No. 30129, Fluka); HEPES (Cat. No. H-3375, Sigma); and penicillin/streptomycin solution (with 10,000 units penicillin and 10 mg streptomycin/ml, Cat. No. P-0781, Sigma). All other chemicals used are from Merck.

The gastromaster and replacement microsurgery tips are from XENOTEK Engineering (Belleville, IL). The microinjector for RNA injection is from Eppendorf (Microinjector 5242, Eppendorf).

III. BASIC PROCEDURES

The basic principles of the procedure have been described previously (Grunz and Tacke, 1989;

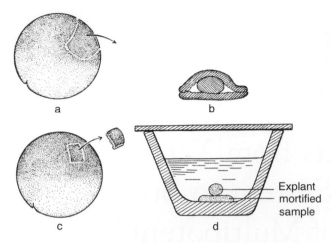

FIGURE 1 Schematic representation of the "Umhüllungsversuch" (coating test) and "Auflagerungsversuch" (bedding test) by Johannes Holtfreter, 1933. (a) Two pieces of ectoderm were isolated using a self-made cutting tool (eyebrow). (b) The mortified implant was placed between ectodermal tissues. (c) A small piece of animal ectoderm of a gastrula stage embryo was excised (c) and placed on the mortified sample (d). Adapted from Spemann, 1936.

Hollemann *et al.*, 1998). In addition to an update of the basic protocol, this article describes several readout systems that can be applied.

Solutions

Modified Barth's solution (MBS): 88 mM NaCI, 2.4 mM NaHCO$_3$, 1.0 mM KCI, 10 mM HEPES, 0.82 mM MgSO$_4$, 0.41 mM CaCl$_2$, and 0.33 mM Ca(NO$_3$)$_2$, pH 7.4

Ca^{2+}/Mg$^{2.+}$-free MBS: 88 mM NaCl, 2.4 mM NaHCO$_3$, 1.0 mM KCl, and 10 mM HEPES, pH 7.4

L-15/BSA solution: 65% L-15 and 0.1% BSA, pH 7.4

PCR mix *(one reaction; 25 µl)*: 2.5 µl 10× polymerase chain reaction (PCR) buffer containing 15 mM MgCl$_2$ (Roche), 0.2 µl 25 mM dNTPs (Biomol), 0.375 µl 10 µM forward primer, 0.375 µl 10 µM reverse primer, 21.425 µl PCR grade H$_2$O and 0.125 µl Ampli *Taq* DNA polymerase (5 U/µl) (Roche)

Histone H4 primer: H4-F: 5'-CGGGATAACATTC AGGGTATCACT-3' and H4-R: 5'-ATCCATGGCG GTAACTGTCTTCCT-3'

RT mix *(one reaction; 5 µl)*: 1 µl 25 mM MgCl$_2$ solution, 0.5 µl 10× PCR buffer II, 2 µl 2.5 mM dNTPs, 0.25 µl RNase inhibitor (20 U/µl), 0.25 µl MuLV reverse transcriptase (RT) (50 U/µl), 0.25 µl 50 µM random hexamers, and 0.75 µl (35 ng) RNA sample

PCR(RT) mix *(one reaction; 20 ml)*: 0.5 ml 25 mM MgCl$_2$ solution, 2 ml 10× PCR buffer II, 16.625 ml RNase-free H$_2$O, 0.375 ml 10 mM forward primer, 0.375 ml 10 mM reverse primer, and 0.125 ml Ampli *Taq* DNA polymerase (5 U/ml)

A. Preparation of Animal Cap Explants

At the late blastula stage (stage 8.5–9), *Xenopus* embryos show a relatively defined segregation of presumptive ectoderm (animal pole), mesoderm (equatorial area), and endoderm (vegetal pole); the blastocoel has attained its full size, and the inner surface of the blastocoel becomes smooth. The roof of the blastocoel, the animal cap (very top of the animal pole), consists of a single outer layer and two inner layers of cells. The pigmentation pattern and the cell size differences make it possible to roughly distinguish the presumptive ectoderm, mesoderm, and endoderm from outside of the embryo. However, the easiest way to identify the animal pole (animal cap) is to look at the embryos directly. What you see from the top is the animal pole, as the embryos are naturally floating inside the vitelline membrane with the animal pole up and the vegetal pole down due to gravity.

Preparation of Xenopus Embryos

Eggs from adult *Xenopus* are obtained from females 6–8 h after injection with human chorionic gonadotropin (500–1000 U/frog) and fertilized with minced testis in 0.1× MBS. Forty minutes after fertilization, dejelly the embryos in 2% cysteine (pH 8.0) and wash extensively in 0.1× MBS.

Isolation (Cutting) of Animal Cap Explants

As there is no marker to clearly delineate the boundary of presumptive ectoderm and mesoderm in a living embryo, it is difficult to isolate animal caps without presumptive mesodermal cell contamination. Many tools, including forceps, hair loops, fine glass needles, and tungsten needles, have been developed for cutting animal caps. The most advanced tool is the gastromaster. No matter which tool is used, the first step is to manually remove the vitelline membrane of stage 8.5–9 embryos (the best stage for capping) with a pair of watchmaker's forceps in a petri dish coated with 1% agarose. Although the manufacturer of the gastromaster demonstrates that it is possible to directly isolate animal caps without removing the vitelline membrane (see video clips at http://www.gastromaster.com/), we strongly recommend removing the membrane manually. The forceps do not have to be very sharp, but the two tips should be well matching. Digging a small pit into the surface of the coated agarose is helpful for fixing the embryo during manipulation. It is recommended to hold and remove the vitelline membrane with forceps from the equatorial area or from the vegetal pole, thus avoiding damage to the animal pole cells. Once the embryos are released from the vitelline membrane, they can no longer rotate freely according to gravity. Therefore, in order to isolate a clean animal cap from the right animal pole region, it is extremely important to properly orientate the nude embryo animal pole up and vegetal pole down (Fig. 2a). Dissect the animal cap (very top region of the animal pole, illustrated by the dashed circle in Fig. 2a) in 0.5× MBS using a gastromaster equipped with a yellow microsurgery tip of 400–500 mm in width (Figs. 2b–2e). This step is quite easy compared to the removal of vitelline membrane. We encourage visiting the gastromaster manufacturer's website for video clips that nicely demonstrate the dissecting process. Culture animal caps in 0.5× MBS containing penicillin (100 U/ml) and streptomycin (100 μg/ml) and harvest until control siblings have reached desired developmental stages for further analyses.

B. Manipulation of Animal Cap Cells

RNA Injection Techniques

One approach to investigate the activity of individual proteins is to introduce the corresponding mRNAs into the animal pole of early cleavage stage embryos followed by the isolation of animal caps for further analysis. Depending on the gene of interest, 5–2000 pg, e.g., FGF or XCYP26 (Isaacs et al., 1992; Hollemann et al., 1998), of in vitro-synthesized capped mRNA (Mega Message Machine, Ambion, USA) in 10 nl RNase-free water can be microinjected into the animal pole of both blastomeres of two-cell stage embryos, which are transferred into 1× MBS and kept at 13°C to slow down development during injection. Two hours after injection, culture embryos again in 0.1× MBS.

Application of Growth Factors

An alternative approach to monitoring the inductive activities of growth factors is to directly apply the active form of a given growth factor protein to the isolated animal caps with or without dissociation. The advantage of dissociation is that each cell is exposed to a uniform factor concentration.

Dissect animal caps from uninjected embryos as described earlier and keep in 1× Ca^{2+}/Mg^{2+}-free MBS in petri dishes coated with 0.7% agarose (about 50 caps per 60-mm petri dish). All the solutions used during disaggregation and reaggregation include penicillin (100 U/ml) and streptomycin (100 μg/ml). After approximately 15 min, cells start to dissociate. Mechanical sucking and blowing of the caps with a yellow pipette tip facilitates the disaggregation process. Caution should be taken to avoid damaging the cells. When the caps are completely disaggregated, replace the Ca^{2+}/Mg^{2+}-free MBS first with 1× MBS and

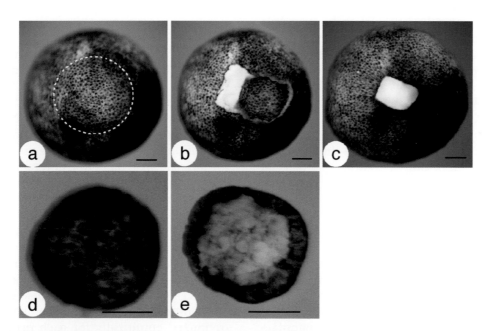

FIGURE 2 Cutting an animal cap with the gastromaster. (a) Animal pole view of a demembranated stage 9 Xenopus embryo. The dashed circle demarcates the very top region of the animal pole (the animal cap or presumptive ectoderm). (b) An animal cap freshly dissected with the gastromaster, still sitting on the mother embryo. (c) Wound after the isolation of the animal cap. (d) Outer layer view of an isolated animal cap. (e) Inner layer view of an isolated animal cap. Bars: 100 μm.

then with the L-15/BSA solution. After the treatment, wash and reaggregate cells in 1× MBS by gentle shaking of the petri dish. To reaggregate the cells, remove most of the solution from the petri dish with a vacuum pump and then swirl the cells together to the center of the dish by applying 1× MBS from one edge of the dish with the special apparatus invented by Horst Grunz that can supply continuous flow over a regulated tube connected to a glass bottle (for details of the apparatus, please visit Horst Grunz's website at http://www.uni-essen.de/zoophysiologie). Alternatively, aggregation can be done by shaking the petri dish gently. After one hour, cut the reaggregated cell mass (cake-like) into several pieces with a glass needle and transfer to sterilized larger glass culture bottles (100 ml) coated with 0.7% agarose. Freeze the tissues with liquid nitrogen when control siblings have developed to the desired stage of development.

C. Analysis (Readout Systems)

Effects on gene transcription can be analyzed by RT-PCR and whole mount *in situ* hybridization.

RT-PCR Analysis

The animal cap system can be used to elucidate which factors and signal transduction pathways regulate genes of interest or, vice-versa, which genes are downstream targets. Effects on gene transcription in explants can be analysed either by RNase protection or RT-PCR. We use a nonradioactive RT-PCR assay. It is normally sufficient to use this simple, semiquantitative approach to analyse strong effects on gene expression. A variety of other RT-PCR methods have also been described for *Xenopus*: quantitative, radioactive RT-PCR (Rupp and Weintraub, 1991) and quantitative, competitive RT-PCR (Onate *et al.*, 1992). The most sensitive and most reproducible technique based on fluorescence kinetics is quantitative real-time RT-PCR (Xanthos *et al.*, 2002), but it requires a special thermocycler.

RNA Isolation from Animal Cap Explants

We use the RNeasy minikit (Qiagen) to isolate RNA from animal caps and whole embryos for RT-PCR analysis following the RNeasy miniprotocol for the isolation of total RNA from animal tissues (see manufacturer's instructions with modifications detailed later). Digest genomic DNA using DNase I (Qiagen) for on column treatment according to the manufacturer's protocol.

Collect 10–12 animal cap explants (or 5 embryos) per sample at the appropriate time in a microcen-

trifuge tube, remove excess buffer, and freeze animal cap explants (embryos) with liquid nitrogen for storage at –80°C. For RNA isolation, add 350 μl (600 μl) buffer RLT/βME to each tube directly on the frozen sample. Carry out all further steps at room temperature. Homogenize animal cap explants (embryos) immediately by passing the lysate five times through a 24-gauge needle fitted to an RNase-free syringe. Elute RNA by 30 μl (70 μl) RNase-free water (60°C) and place on ice. After measuring the RNA concentration, adjust all samples to the same concentration with RNase-free water and store at –80°C. Control RNA integrity by use of a 1.5% TBE agarose gel; 75–100 ng of total RNA is normally obtained from a single animal cap explant.

Analysis of Genomic DNA Contamination

The exon–intron structure of most *Xenopus* genes is unknown. Therefore, primers used for the analysis that can target the same exon and genomic DNA contamination of isolated RNA would thus result in false-positive signals. In order to control for genomic DNA contamination, a regular PCR reaction for histone H4 genomic DNA is performed. Add 1 μl RNA (50 ng/μl) to 25 μl PCR mix containing the histone H4-specific primer. Use genomic DNA (5 ng) as a positive control and H_2O as a negative control, with the following cycling parameters: 94°C 2 min (94°C 45 s, 58°C 45 s, 72°C 30 s) × 35, 72°C 5 min. Analyze one-half of the PCR on a 2% TBE agarose gel.

RT-PCR

Carry out the reverse transcriptase reaction in 5-μl reactions (or a corresponding multiple of 5 μl) using the Gene Amp RNA PCR core kit (Roche) according to the manufacturer's instructions in a TRIO Thermoblock (Biometra). Use total RNA isolated from whole embryos collected at corresponding developmental stages and diluted to the same concentration as the animal cap explant RNA preparations as a positive control. As a negative control, perform the same reaction without adding reverse transcriptase (H_2O). Conditions for the RT reaction are 22°C 10 min, 42°C 30 min, 99°C 5 min, 5°C 5 min. Mix a 5-μl aliquot of the RT reaction with 20 μl PCR(RT) mix. We use the following cycling parameters: 94°C 2 min (94°C 45 s, X°C 45 s, 72°C 45 s) × Y cycles, 72°C 1 min. Analyze 10 μl of each RT-PCR on a 2% TBE agarose gel. The annealing temperature (X) and cycle number (Y) have to be optimized empirically for each primer pair. For that purpose, remove aliquots from a RT-PCR reaction using cDNA made from whole embryo RNA every second cycle over a 12-cycle range, usually starting at

20 cycles. The optimal cycle number is within the linear range of product accumulation. To control RNA quality, we perform histone H4-specific PCR; $X = 58°C$, $Y = 24$ cycles . Control contamination of animal cap explants by mesodermal tissue with primers for brachyury (early ACs) and muscle actin (later ACs). Sequences for these and other primers are available under http://www.xenbase.org.

Whole Mount in Situ Hybridization

Gene expression on the RNA level in animal cap explants can also be analyzed by whole mount *in situ* hybridization, originally adapted for *Xenopus* embryos by Harland (1991). The advantage of this method in comparison to RT-PCR is that individual explants can be controlled for homogeneity of gene expression on the level of individual cells. The disadvantage is that substantial numbers of explants are required if several genes are to be analyzed in parallel. We use the basic protocol for whole mount *in situ* hybridization described previously (Hollemann *et al.*, 1999) with the following minor modifications. In case of several samples, a transfer of fixed animal caps from glass vials to 24-well tissue culture plates before rehydration facilitates handling. Proteinase K treatment should be shortened to 5–8 min at room temperature.

D. Examples

RT-PCR Analysis of Animal Cap Explants

Many different genes are involved in the activation of the neuronal differentiation program, which drives naive ectodermal cells to become postmitotic neurons. In animal cap explants, X-ngnr-1 can also promote neurogenesis, as monitored by the activation of the neural determination marker N-tubulin. The density of cells expressing endogenous X-ngnr-1 is controlled by lateral inhibition mediated through Notch signalling (Ma *et al.*, 1996). In response to misexpression of the intracellular domain of Notch (ICD) that mimics active Notch signalling, the expression of target genes, such as Esr7, is induced (Fig. 3).

Whole Mount in Situ Hybridization in Animal Cap Explants

In addition to neural and mesodermal tissue, animal cap explants can also be converted into endoderm. For instance, VegT can synergize with β-catenin to induce a number of liver- and intestine-specific genes in animal cap explants. Figure 4 shows an example of whole mount *in situ* hybridization analysis for the induction of an endoderm marker gene in an animal cap upon VegT/β-catenin injection.

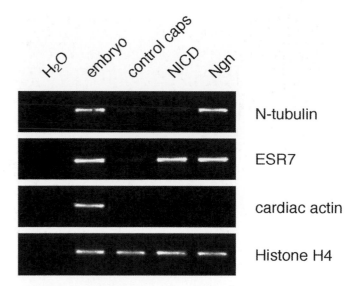

FIGURE 3 Induction of neuronal gene expression in animal caps as analyzed by RT-PCR. One hundred picograms X-Ngnr-1 mRNA (Ngn) or 300 pg Notch ICD mRNA (NICD) was injected in both cells of two-cell stage embryos, animal caps were dissected at early stage 9, cultured to stage 15, and then analyzed by RT-PCR. RNA isolated from uninjected total embryos of stage 15 was used as a positive control. X-Ngnr-1 induces the neuronal differentiation marker N-tubulin, as well as the Notch target gene ESR7 in animal caps. Histone H4-specific primers were used to control RNA input and quality. As shown by the absence of transcripts for cardiac actin, the isolated explants were not contaminated by mesodermal tissue (Sölter, *et al.*, unpublished results).

The Animal Cap to Study Cell Migration

Upon treatment with activin or FGF, the animal cap is not only induced to become mesoderm (see above), but as a consequence of mesoderm induction, the animal cap also changes its form (Fig. 5). Whereas untreated animal caps round up and look like spherical balls, induced animal caps elongate as a consequence of cell movements within the animal cap (Kuhl *et al.*, 2001). It is widely accepted that this kind of cell movement within the cap mimics mesodermal cell movements normally occurring during gastrulation of the *Xenopus* embryo. During gastrulation, cells from a more ventral position migrate towards the dorsal midline. The forces generated by this process *in vitro* not only push cells during gastrulation over the dorsal lip, but also lead to an elongation of the embryo along its anterior–posterior axis (for a review on cell movements during gastrulation, see Keller, 1986). With respect to gastrulation movements, the animal cap thus serves as an excellent system to study the molecular basis of cell migration. The elongation behaviour of injected or growth factor-treated caps can be studied and conclusions can be drawn on the function of a

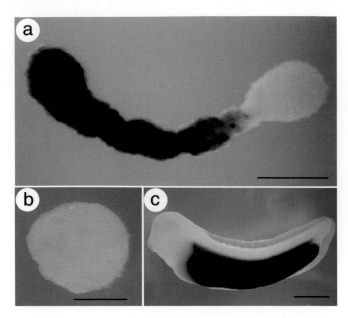

FIGURE 4 Induction of endodermal gene expression in animal cap explants as analysed by whole mount *in situ* hybridisation. Expression analysis of the activation of an endoderm-specific gene in VegT/β-catenin-injected animal cap explants. Cyl18 encodes a novel putative peptidase that demarcates the developing embryonic intestine. VegT (500 pg/embryo) and β-catenin (200 pg/embryo) mRNAs were injected into the animal pole of both blastomeres of two-cell stage embryos. Animal caps were isolated from stage 9 control uninjected and VegT/β-catenin-injected embryos and collected when control siblings had reached stage 34. (a) Cyl18 expression is activated in a subset of cells of an elongated animal cap injected with VegT/β-catenin. (b) Cyl18 is not expressed in the control cap. (b) Cyl18 is expressed exclusively in the presumptive intestine of stage 34 embryos (Chen *et al.*, 2003).

gene of interest with respect to cell migration. As an example, we show the effect of activin on animal caps (Fig. 5).

Recapitulation of Organ Formation in Animal Caps

a. Cement Gland Formation in Whole Animal Cap Explants. The homeobox-containing transcription factor PITX1 is expressed during all stages of cement gland formation (Hollemann and Pieler, 1999). The cement gland of young amphibian tadpoles is a specialised transient organ, which arises from the outer ectodermal layer (Fig. 6a). The function of the gland is to secrete sticky mucus that helps larvae adhere to solid surfaces, preventing larvae from being carried along by the water flow. To investigate the question if *Xpitx1* alone is able to induce cement gland formation, 100 pg of the corresponding synthetic messenger RNA was injected into one cell of a two-cell stage embryo. At the equivalent of embryonic stage 8, the animal cap was excised with the help of a gastromaster (yellow tip with 400 μm) and cultured in 0.5× MBS+P/S until nontreated siblings had reached stage 40. Whereas noninjected animal caps formed so-called artificial epidermis, a high number of animal caps injected with Xpitx1 developed groups of cells, which could be identified easily as typical cement gland cells based on their morphology (Fig. 6b).

b. Notochord from Dispersed Animal Cap Explant Cells. In response to different concentrations of activin, dispersed animal cap cells have been shown to exhibit sharp thresholds in respect to target gene activation. Activin A is added to the cells at a final concentration of 4 ng/ml for notochord differentiation and

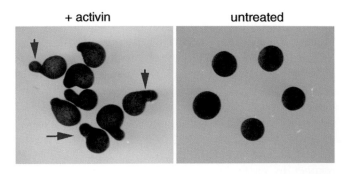

FIGURE 5 Elongation of animal caps mimics gastrulation movements. Animal cap explants of *Xenopus laevis* were isolated at stage 8 and treated with activin for 2 h. Treated caps elongate until stage 18 (left). Arrows indicate examples for elongation of the caps. In this example, all animal caps shown have indications of elongation. Untreated control caps remain spherical (right).

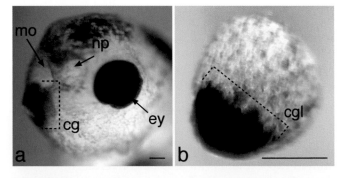

FIGURE 6 Cement gland formation in animal cap explants. (a) Anterior to the left, sloped lateral view of a head of a noninjected *Xenopus* larvae at NF stage 40. The cement gland is positioned ventrally immediately adjacent to the mouth opening. (b) Example of an animal cap explant that had been injected with Xpitx1 into one cell at the two-cell stage and cultured equivalent to stage 40. Bar: 50 μm. cg, cement gland; cgl, cement gland like; ey; eye; mo, mouth opening; np, nasal pit.

8 ng/ml for endoderm differentiation and incubated at room temperature for 1 h. In most cases, more than 90% of the reaggregated tissue mass from each batch of preparation differentiated into notochords. Some neural cells were formed concomitantly.

c. Inductive Processes: A Proteomics Approach. The animal cap assay has widely been used to study

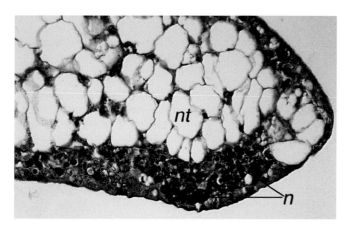

FIGURE 7 Histological staining of notochord tissues generated from animal cap cells treated with activin. In order to judge the inductive activities, one piece of the tissues from each batch of experiments was cultured further and finally fixed in Bouin's solution when control embryos reached stage 42. Differentiation of the tissues is assayed by histological analysis. nt, notochord cells; n, neural cells.

the inductive properties of growth factors such as activin or FGF. Molecular analysis of inductive events in general was on the level of mRNA expression. Changes in gene expression were analyzed by Northern blot or RT-PCR studies. This procedure resulted in the identification of a plethora of early and late response genes in the course of mesoderm or neural induction. However, posttranscriptional and posttranslational aspects of cellular regulation in this context (like protein translation, protein phosphorylation, and differential splicing events) were neglected. Two-dimensional (2D) protein analysis in conjunction with MALDI-TOF-based identification of proteins will be a useful tool to explore these processes (Shevchenko *et al.*, 1996). Figure 8 provides a 2D protein gel example of activin-treated animal caps in comparison to untreated animal cap explants, indicating that this rationale can be applied on the animal cap system in *Xenopus*. Thus, it can be expected that embryonic inductive processes will also be analyzed on the level of the proteome in the near future.

References

Chen, Y., Jurgens, K., Hollemann, T., Claussen, M., Ramadori, G., and Pieler, T. (2003). Cell-autonomous and signal-dependent expression of liver and intestine marker genes in pluripotent precursor cells from Xenopus embryos. *Mech. Dev.* **120**, 277–288.

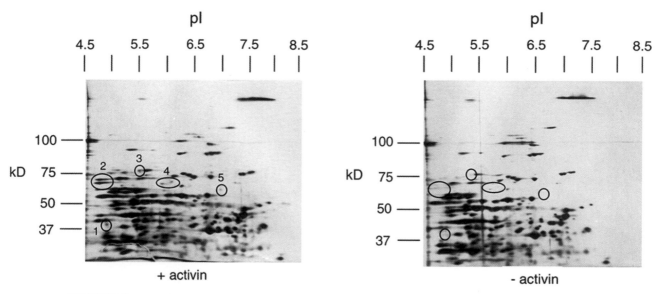

FIGURE 8 Two-dimensional gel analysis of proteins expressed in untreated and activin-treated animal caps. Embryos were injected at the two-cell stage with activin mRNA. Animal caps were isolated at stage 8 and cultured until stage 19. Equivalents of four animal caps were used for 2D protein gels. Circles indicate spots of different intensity upon treatment. Whereas spots 1 to 3 most likely represent proteins that are upregulated through activin treatment, spots 4 and 5 indicate two spots that are weakly affected and might also represent artefacts. Two-dimensional gels with a broad IP range were used. For a more detailed analysis of inductive processes, small range IP gels are required to increase the resolution. See detailed literature on 2D gel analysis for methods and detailed interpretation of the 2D gel. (Shevchenko *et al.*, 1996).

Green, J. B., New, H. V., and Smith, J. C. (1992). Responses of embryonic Xenopus cells to activin and FGF are separated by multiple dose thresholds and correspond to distinct axes of the mesoderm. *Cell* **71**, 731–739.

Grunz, H., and Tacke, L. (1989). Neural differentiation of Xenopus laevis ectoderm takes place after disaggregation and delayed reaggregation without inducer. *Cell Differ. Dev.* **28**, 211–217.

Gurdon, J. B., Fairman, S., Mohun, T. J., and Brennan, S. (1985). Activation of muscle-specific actin genes in Xenopus development by an induction between animal and vegetal cells of a blastula. *Cell* **41**, 913–922.

Harland, R. M. (1991). In situ hybridization: An improved wholemount method for Xenopus embryos. *Methods Cell Biol.* **36**, 685–695.

Hollemann, T., Chen, Y., Grunz, H., and Pieler, T. (1998). Regionalized metabolic activity establishes boundaries of retinoic acid signalling. *EMBO J.* **17**, 7361–7372.

Hollemann, T., Panitz, F., and Pieler, T. (1999). In situ hybridization techniques with Xenopus embryos. *In "A Comparative Methods Approach to the Study of Oocytes and Embryos"* (J. D. Richter, ed.), pp. 279–290. Oxford Univ. Press, New York.

Hollemann, T., and Pieler, T. (1999). Xpitx-1: A homeobox gene expressed during pituitary and cement gland formation of Xenopus embryos. *Mech. Dev.* **88**, 249–252.

Holtfreter, J. (1933). Nachweis der Induktionsfähigkeit abgetöteter Keimteile. Isolations- und Transplantationsversuche. *Roux' Arch.* **128**, 585–633.

Isaacs, H. V., Tannahill, D., and Slack, J. M. (1992). Expression of a novel FGF in the Xenopus embryo: A new candidate inducing factor for mesoderm formation and anteroposterior specification. *Development* **114**, 711–720.

Keller, R. E. (1986). The cellular basis of amphibian gastrulation. *Dev. Biol.* **2**, 241–327.

Kuhl, M., Geis, K., Sheldahl, L. C., Pukrop, T., Moon, R. T., and Wedlich, D. (2001). Antagonistic regulation of convergent extension movements in Xenopus by Wnt/beta-catenin and Wnt/Ca2+ signaling. *Mech. Dev.* **106**, 61–76.

Ma, Q., Kintner, C., and Anderson, D. J. (1996). Identification of neurogenin, a vertebrate neuronal determination gene. *Cell* **87**, 43–52.

Moriya, N., Komazaki, S., Takahashi, S., Yokota, C., and Asashima, M. (2000). *In vitro* pancreas formation from Xenopus ectoderm treated with activin and retinoic acid. *Dev. Growth Differ.* **42**, 593–602.

Onate, A., Herrera, L., Antonelli, M., Birnbaumer, L., and Olate, J. (1992). *Xenopus laevis* oocyte G alpha subunits mRNAs: Detection and quantitation during oogenesis and early embryogenesis by competitive reverse PCR. *FEBS Lett.* **313**, 213–219.

Rupp, R. A., and Weintraub, H. (1991). Ubiquitous MyoD transcription at the midblastula transition precedes induction-dependent MyoD expression in presumptive mesoderm of X. laevis. *Cell* **65**, 927–937.

Shevchenko, A., Jensen, O. N., Podtelejnikov, A. V., Sagliocco, F., Wilm, M., Vorm, O., Mortensen, P., Boucherie, H., and Mann, M. (1996). Linking genome and proteome by mass spectrometry: Large-scale identification of yeast proteins from two dimensional gels. *Proc. Natl. Acad. Sci. USA* **93**, 14440–14445.

Slack, J. M. (1990). Growth factors as inducing agents in early Xenopus development. *J. Cell Sci. Suppl.* **13**, 119–130.

Smith, J. C., Price, B. M., Van Nimmen, K., and Huylebroeck, D. (1990). Identification of a potent Xenopus mesoderm-inducing factor as a homologue of activin A. *Nature* **345**, 729–731.

Spemann, H. (1936). "Experimentelle Beiträge zu einer Theorie der Entwicklung." *Verlag von Julius Springer, Berlin.*

Sudarwati, S., and Nieuwkoop, P. D. (1971). Mesoderm formation in the anuran Xenopus laevis (Daudin). *Roux' Arch.* **166**, 189–204.

Xanthos, J. B., Kofron, M., Tao, Q., Schaible, K., Wylie, C., and Heasman, J. (2002). The roles of three signaling pathways in the formation and function of the Spemann organizer. *Development* **129**, 4027–4043.

25

Electrofusion: Nuclear Reprogramming of Somatic Cells by Cell Hybridization with Pluripotential Stem Cells

Masako Tada and Takashi Tada

I. INTRODUCTION

The technique of cell fusion, which was pioneered by Henry Harris (1965), has proved to be a powerful procedure with applications in cell biology, genetics, and developmental biology and in fields of practical concern such as medicine and agriculture. The spontaneous or induced cell fusion of two different types of cells (heterokaryons) generates intraspecific or interspecific hybrid cells. Genetically programmed spontaneous cell fusion is known to occur in the formation of polykaryons such as myotubes, osteoclasts, and syntrophoblasts *in vivo*. Under *in vitro* culture conditions, spontaneous cell fusion has been found to occur occasionally in some cell lines and malignant cells. Cell fusion due to membrane integrity between two different cells is induced by treatment with chemical agents such as calcium ions, lysolecithin, and polyethylene glycol; by mediation by viruses such as paramyxoviruses, including Sendai virus (HVJ), oncornavirus, coronavirus, herpesvirus and poxvirus; and by electrofusion.

In 1997, the successful production of the cloned sheep named Dolly demonstrated that committed animal somatic cell nuclei are able to reacquire totipotency as a result of nuclear transplantation into enucleated unfertilized oocytes and the subsequent embryonic development (Wilmut *et al.*, 1997). This nuclear reprogramming results from the resetting of the somatic cell-specific epigenetic program by *trans*-acting factors present in unfertilized oocytes. Nearly 20 years ago, genomic plasticity had already been examined by cell fusion experiments between differentiated cell types (Blau *et al.*, 1985; Baron and Maniatis, 1986; Blau and Baltimore, 1991). More recently, this approach has been used to study genomic reprogramming that occurs in X chromosome reactivation (Takagi *et al.*, 1983; Tada *et al.*, 2001; Kimura *et al.*, 2003) and switching of parental origin-specific marks of imprinted genes (Tada *et al.*, 1997).

An important finding is that pluripotential embryonic stem (ES) cells have an intrinsic capacity for epigenetic reprogramming of somatic genomes following cell fusion (Tada *et al.*, 2001, 2003; Kimura *et al.*, 2003). In hybrid cells between ES cells and adult thymocytes, nuclear reprogramming of the somatic genome has been demonstrated by (1) the contribution of ES hybrid cells to normal embryogenesis of chimeras, (2) reactivation of the silenced X chromosome derived from female somatic cells, (3) reactivation of pluripotential cell-specific genes, *Oct4*, *Xist*, and *Tsix*, which were derived from the somatic cell, (4) redifferentiation into a variety of cell types in teratomas, (5) tissue-specific gene expression from the reprogrammed somatic genome in addition to the ES genome in *in vivo*-differentiated teratomas and *in vitro*-differentiated neuronal cells, and (6) acquisition by reprogrammed somatic genomes of pluripotential cell-specific histone tail modifications. More interestingly, cell fusion experiments between somatic cells and embryonic germ (EG) cells derived from the gonadal primordial germ cells of mouse 11.5–12.5 dpc embryos have demonstrated that EG cells possess an additional potential for inducing the reprogramming of somatic cell-derived parental imprints accompanied by the disruption of parental origin-specific DNA methylation of imprinted genes (Tada *et al.*, 1997, 1998). Therefore, cell fusion with pluripotential stem cells is now recognized as a powerful approach for elucidating mechanisms of epigenetic reprogramming involving DNA and chromatin modifications.

More recent evidence has shown that neurosphere and bone marrow cells undergo nuclear reprogramming following spontaneous cell fusion with cocultured ES cells *in vitro* (Terada *et al.*, 2002; Ying *et al.*, 2002). Furthermore, experiments involving the *in vivo* transplantation of bone marrow cells have demonstrated that regenerated hepatocytes are derived from donor hematopoietic cells that undergo fusion with host hepatocytes, not from the transdifferentiation of hematopoietic stem cells or hepatic stem cells present in bone marrow (Vassilopoulos *et al.*, 2003; Wang *et al.*, 2003). Thus, the nuclear reprogramming of somatic cells by *in vivo* cell fusion is thought to play an important role in maintaining the homeostasis of some tissues by regeneration during defined self-renewal and following tissue damage.

This article describes a practical procedure for electrofusion to produce hybrid cells between pluripotential stem cells and committed somatic cells (mouse ES cells and lymphocytes isolated from the adult thymus) without the use of virus or chemicals to mediate the fusion. ES cells are adherent cells that undergo self-renewal by rapid cell division, whereas thymocytes are nondividing and nonadherent cells. In order to select the hybrid cells effectively, either thymocytes carrying the *neo* transgene or male ES cells deficient for the X-linked *Hprt* (*hyoxanthine phosphoribosyl transferase*) gene are used as the partner cells in the cell fusion. Consequently, only hybrid cell colonies are capable of surviving and growing in culture in the presence of antibiotic G418 or HAT (hypoxanthine, aminopterin, and thymidine).

II. MATERIALS AND INSTRUMENTATION

Cells: Adult mice, ES cells, and *neo*[r] feeder cells (see Section II,A)

Instruments: ECM 2001 AC/DC pulse generator (BTX), 1-mm gap microslide chambers (BTX P/N450-10WG), inverted microscope with 10 and 20× objectives, humidified incubator at 37°C, 5% CO_2, 95% air, 60-mm plastic tissue culture dishes, 60- and 100-mm bacterial dishes, 10- and 30-mm well plastic tissue culture plates, 15- and 50-ml conical tubes, 0.2-μm microfilters, 200- and 1000-μl capacity

adjustable pipetters with autoclaved tips, forceps, scissors, 2.5-ml syringes, 18-gauge needles

Compounds: Dulbecco's modified Eagle's medium/nutrient mixture F12 Ham (DMEM/F12) (Sigma D6421), Dulbecco's modified Eagle's medium (DMEM) (Sigma D5796), fetal bovine serum (FBS) (JRH Biosciences 12003-78P), recombinant leukemia inhibitory factor (LIF) (Chemicon ESG1107), 200 mM glutamine (GIBCO 320-5030AG), 2-mercaptoethanol (Sigma M7520), 10,000 IU/ml penicillin and 10 mg/ml streptomycin (penicillin–streptomycin 100×) (Sigma P-0781), 100 mM sodium pyruvate (Sigma S8636), 7.5% sodium bicarbonate (Sigma S8761), Ca^{2+}/Mg^{2+}-free phosphate-buffered saline (PBS) (GIBCO 10010-023), 0.25% trypsin/1 mM EDTA·4Na (GIBCO 25200-056), mytomycin C (Sigma M0503), gelatin from porcine skin, type A (Sigma G-1890), D-mannitol (Sigma M-9546), G418 (geneticin) (Sigma G-9516), HAT media supplement 50× (HAT) (Sigma H0262)

III. PROCEDURES

A. Mouse ES Cell and Feeder Cell Culture

One of the most important variables for cell fusion experiments is how stably ES cells ($2n = 40$) and hybrid cells ($2n = 80$) can be cultured without loss of the pluripotential competence and the full set of chromosomes derived from mouse ES cells and somatic cells through numerous cell divisions. The culture conditions are basically those described previously (Abbondanzo *et al.*, 1993). A crucial point is quality control of the FBS, which is added to make the ES cell culture medium cocktail. FBS certified for ES cell culturing has become available commercially. We strongly recommend the use of a suitable production lot of FBS that can support effective cell growth without inducing differentiation equivalently to the ES cell-certified FBS.

Solutions

1. *ES medium*: Mix 500 ml of DMEM/F12, 75 ml of FBS, 5 ml of 200 mM glutamine, 5 ml of penicillin–streptomycin (100×), 5 ml of 100 mM sodium pyruvate, 8 ml of 7.5% sodium bicarbonate, 4 μl of $10^{-4} M$ 2-mercaptoethanol, and 0.05 ml of 10^7 U/ml LIF (final 1000 U/ml). Store at 4°C.

2. *PEF medium*: Mix 500 ml of DMEM, 50 ml of FBS, 5 ml of 200 mM glutamine, 5 ml of 10,000 IU/ml penicillin, and 10 mg/ml streptomycin. Store at 4°C.

3. *0.25% trypsin/1 mM EDTA·4Na*: Dispense into aliquots and store at −20°C.

4. *Ca²⁺/Mg²⁺-free phosphate-buffered saline (PBS)*
5. *10 μg/ml mytomycin C*: 0.2 mg/ml in PBS, dispense into aliquots, and store at −20°C.
6. *0.1% gelatin*: 0.1% gelatin in distilled water. Sterilize by autoclaving and store at 4°C.

Steps

1. Coat 60-mm culture dishes with 0.1% gelatin for at least 30 min at room temperature.
2. Prepare mouse primary embryonic fibroblasts (PEFs) produced from day 13 embryos of ROSA26 transgenic mice carrying the ubiquitously expressed *neo/lacZ* gene (Friedrich and Soriano 1991). Treat *neoʳ* PEFs with 10 μg/ml mitomycin C (MMC) and incubate at 37°C in a CO_2 incubator for 2 h to produce mitotically inactivated feeder cells. Prepare frozen stocks of the MMC-treated PEFs at a concentration of 5×10^6 cells/ml and store in a cryotube in liquid nitrogen. The inactivated *neoʳ* PEFs are routinely used as feeder cells (1×10^6 cells/60-mm culture dish and 2.5×10^6 cells/100-mm culture dish) for culture of ES and hybrid cells, and also for selection of hybrid cell colonies with G418. For establishment of PEFs, see Abbondanzo *et al.* (1993).
3. Culture exponentially growing ES cells on the inactivated PEFs with changes of culture medium once or twice a day. Carry out subculturing of the ES cells every 2 days by a 1:4 split. ES cells at early passages are used for experiments. Before cell fusion, it should be verified that the karyotype of the ES cells is normal.
4. Prepare gelatin-coated 30-mm culture dishes (6-well-culture plates) containing the inactivated PEFs (4×10^5 cells/well) in 3 ml of ES medium 1 day before cell fusion experiments.

B. Pretreatment of ES and Somatic Cells for Cell Fusion

Solution

Fresh nonelectrolyte solution; 0.3 M mannitol buffer: To make 50 ml, dissolve 2.74 g of mannitol in distilled water. Filter through a 0.2-μm filter. Store at 4°C.

Steps

1. Trypsinize ES cells and remove excess trypsin quickly. Add 3 ml of ES medium to inactivate the trypsin and dissociate the cells into a single-cell suspension by gentle pipetting. Plate them on a new gelatin-coated 60-mm culture dish. Incubate the ES cells in a CO_2 incubator for 30 min to separate feeder cells from ES cells.

2. Collect unattached ES cells and harvest them by centrifugation at 1500 rpm for 5 min. Resuspend the cell pellet in 10 ml of DMEM, transfer them into a 15-ml conical tube, and place at room temperature.
3. Sacrifice a 6- to 8-week-old adult mouse humanely and dissect out the thymus in a clean room if a clean bench is not available. All of the dissection instruments should be sterilized by immersion in 70% ethanol, followed by flaming.
4. Wash the tissues with sterilized PBS twice in 60-mm petri dishes and place one lobe of the thymus in the barrel of a sterile 2.5-ml syringe with a sterile 18-gauge needle.
5. To dissociate the thymus into a single-cell suspension, gently expel the thymus through the tip of the needle into 2 ml of DMEM in a 50-ml conical tube. Draw up and expel the suspension several times. Allow to stand for several minutes at room temperature.
6. Transfer the supernatant excluding cell clumps to a 15-ml conical tube and add 10 ml of DMEM.
7. Spin down the ES cells and thymocytes separately in 15-ml conical tubes at 1500 rpm for 5 min.
8. Wash them with 10 ml of DMEM and spin down at 1500 rpm for 5 min and repeat again to remove FBS completely.
9. Add 5–10 ml of DMEM and adjust the density of ES cells and thymocytes each to 1×10^6 cells/ml.
10. Pellet a 1:5 mixture of ES cells and thymocytes (1 ml of the ES cell suspension and 5 ml of the thymocyte suspension made in step 9). Keep the remaining cells for control experiments.
11. Spin down and resuspend the cell pellet in 0.3 M mannitol buffer at 6×10^6 cells/ml. Usually, 1 ml of the mixture of ES cells and thymocytes is sufficient for the following fusion experiment. Use the cells immediately for electrofusion (Fig. 1).

C. Operation of ECM 2001 Pulse Generator and Electrofusion Protocol

Solution

70% ethanol

Steps

1. Sterilize the microslides by immersion in 70% ethanol, followed by flaming.
2. Set a microslide in a 100-mm plastic dish chamber.
3. For each electrofusion, apply 40 μl of cell mixture between the electrodes with a 1-mm gap on the microslide.

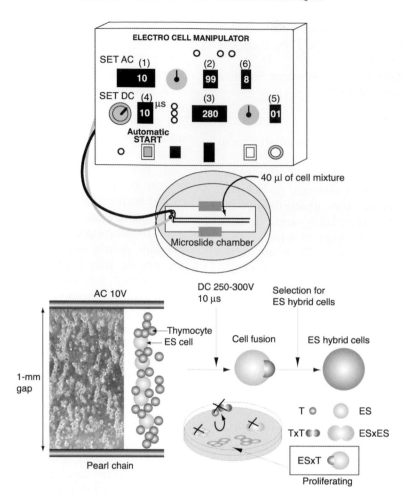

FIGURE 1 Schematic representation of the electrofusion system. A mixture of ES cells and somatic cells suspended in nonelectrolytic mannitol solution is placed in a 1-mm gap between electrodes on a microslide. Set parameters (1–6) of the AC/DC pulse generator and press the automatic start button. AC electric pulses induce pearl chain formation, and the subsequent DC electric pulse induces cell fusion. ES cells (ES) and ESxES hybrid cells are nonviable in the selection medium, whereas thymocytes (T) and TxT hybrid cells are nonadherent. ES hybrid cells (ESxT) rescued by the thymocyte genome are capable of surviving and proliferating in the selection medium.

4. Connect the microslide in the 100-mm plastic dish chamber with the ECM 2001 AC/DC pulse generator by electric cables. Set the chamber on an inverted microscope to allow for observation of cell alignment and the fusion processes. It is important to monitor the fusion process microscopically each time. The electrofusion may proceed somewhat differently depending on the density of cell preparations and on the cell type (cell size).

5. Set the optimized electrical parameters to fuse ES cells and thymocytes (Fig. 1): 10 V alternating current (AC), 99 s AC duration, and 250–300 V direct current (DC). Adjust the DC voltage according to the size of the gap between the electrodes. The appropriate electric field strength is 2.5–3.0 kV/cm. When microslides

with a 2-mm gap are used, the DC voltage should be almost 600 V.

DC pulse length: 10 μs
Number of DC pulses: 1
Postfusion AC duration: 8 s

6. Use the automatic operation switch to initiate AC followed by DC. AC is utilized to induce an inhomogeneous, or divergent electric field, resulting in cell alignment and pearl chain formation. DC is utilized to produce reversible temporary pores in the cytoplasmic membranes. When juxtaposed pores in the physically associated cells reseal, cells have a chance to be hybridized via cytoplasmic membrane fusion. AC application after the DC pulse induces compression of

the cells, which helps the process of fusion between the cell membranes.

7. Add 40 μl of DMEM to the fusion mixture between the electrodes to induce the recovery of membrane formation.

8. Place the cell mixture at room temperature for 10 min and transfer the cells to a 30-mm culture dish containing inactivated PEFs with 3 ml of the ES medium.

9. Repeat the cell fusion procedure sequentially using several microslides. Usually, cells recovered from three microslides (40 μl × 3) are plated into one 30-mm culture dish.

10. As a control, plate the untreated cell mixture and culture under the same conditions.

11. Change the medium to ES medium with appropriate supplements for selecting ES hybrid cells 24 h after cell fusion. The selection medium should be changed once a day (see Section III,D). During the 7-day selection treatment, unfused ES cells and hybrid cells between ES cells are killed and hybrid cells between thymocytes are nonadherent. Thus, only the hybrid cells between ES cells and somatic cells survive, proliferate, and form colonies. Several colonies of hybrid cells per 10^4 host ES cells are obtained under appropriate cell fusion and culture conditions.

12. Pick up the colonies with a micropipette and transfer each colony into a 10-mm well of a 24-well culture plate containing 1×10^5 inactivated PEFs per well and 0.8 ml of the ES medium with supplements for selection.

13. Subculture the cells every 2 or 3 days and gradually expand the number of cells in 30- and then 60-mm culture dishes with the inactivated PEFs and ES medium with supplements for selection. When the cells become nearly confluent in a 60-mm culture dish, it is considered that a hybrid cell line of passage 1 has been established.

14. Change the ES medium without selection supplements once or twice a day and subculture the hybrid cell line every 2 days under optimal culture conditions by splitting 1:4.

15. Subject hybrid cells to chromosome analysis soon after they are established.

D. Fusion Examples: Selection System for Hybrid Cells

This section describes one independent chemical selection system that can be used to select for hybrid cells between ES cells and somatic cells.

1. Normal ES cells are hybridized with thymocytes containing the bacterial neomycin resistance (*neo*^r) gene (Tada *et al.*, 1997, 2001). Thymocytes are derived from ROSA26 transgenic mice, which carry the ubiq-uitously expressed *neo/lacZ* transgene (Friedrich and Soriano, 1991). Only ES hybrid cells with the thymocytes can survive and grow in the ES medium supplemented with the antibiotic G418, a protein synthesis inhibitor. In this case, the ES hybrid cells and their derivatives are identified visually by their positive reaction with X-gal due to β-galactosidase activity, allowing one to analyze their contribution to the development of chimeric embryos and tissues (Tada *et al.*, 1997, 2001, 2003). Male ES cells deficient for the *Hprt* gene on the X chromosome are a powerful tool for producing hybrid cells with wild-type somatic cells. Electrofusion-treated cells are cultured in ES medium with the HAT supplement. In DNA synthesis, purine nucleotides can be synthesized by the *de novo* pathway and recycled by the salvage pathway. *Hprt* is a purine salvage enzyme, responsible for converting the purine degradation product hypoxanthine to inosine monophosphate, a precursor of ATP and GTP. In the presence of aminopterin, the *de novo* pathway is inhibited and only the salvage pathway functions. Consequently, dysfunction of *Hprt* induces cell death in cultures grown in HAT medium. Thus, the HAT medium proves fatal to *Hprt*-deficient ES cells, whereas ES hybrid cells, which are rescued by the thymocytes-derived wild-type *Hprt* gene, are able to survive and proliferate (Tada *et al.*, 2003; Kimura *et al.*, 2003).

Solutions

1. *ES medium with G418*: Reconstitute G418 with water (50 mg/ml). Sterilize through a 0.2-μm filter and store at 4°C. Add 50 μl of the G418 solution to 10 ml of ES medium, yielding a final concentration of 250 μg/ml.

2. *ES medium with HAT*: Reconstitute the HAT media supplement obtained from the supplier in a vial with 10 ml of DMEM (50× solution) and store at −20°C. Each vial contains $5 \times 10^{-3} M$ hypoxanthine, $2 \times 10^{-5} M$ aminopterin, and $8 \times 10^{-4} M$ thymidine. Add 200 μl of 50× solution to 10 ml of ES medium.

Steps

Selection with G418

1. Perform electrofusion between normal ES cells and thymocytes collected from the 6- to 8-week-old ROSA26 transgenic mice carrying the *neo/lacZ* transgene according to the procedure described earlier.

2. Culture the electrofusion-treated cells in ES medium for 24 h.

3. Change to ES medium supplemented with G418. ES hybrid cell colonies can be detected by 7–10 days.

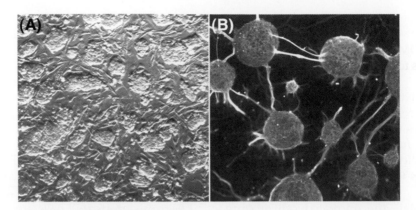

FIGURE 2   Pluripotential competence of ES hybrid cells with somatic cells. (A) Undifferentiated ES somatic hybrid cell colonies in culture on mitotically inactivated PEFs. (B) Neuronal cells differentiated *in vitro* from ES somatic hybrid cells on PA6 stromal feeder cells.

4. ES thymocyte hybrid cells are positive for X-gal staining and immunoreactive with the anti-β-galactosidase antibody.

Selection with HAT

1. Carry out electrofusion between ES cells (XY) deficient for *Hprt* and thymocytes collected from 6- to 8-week-old female mice (XX) according to the procedure described earlier.
2. Culture electrofusion-treated cells in ES medium for 24 h.
3. Change to ES medium containing the HAT supplement. ES hybrid cell colonies can be detected by 7–10 days.
4. ES thymocyte hybrid cells possess a karyotype of 4n = 80 with an XXXY sex chromosome constitution.

Figure 2A shows representative ES hybrid cells in culture on feeder cells in the ES medium. Figure 2B shows representative neuronal cells differentiated from ES hybrid cells. The ES hybrid cells are pluripotential and can differentiate into a variety of tissues *in vivo* and *in vitro*. Tissue-specific transcripts derived from the reprogrammed somatic genomes can be identified based on genetic polymorphisms found in inter-subspecific ES hybrid cells (*Mus musculus domesticus* × *M. m. molossinus*). The reprogrammed somatic cell genomes function similarly to the ES cell genomes in undifferentiated ES hybrid cells and also in ES hybrid cell derivatives differentiated *in vivo* and *in vitro* (Kimura *et al.*, 2003; Tada *et al.*, 2003).

IV. COMMENTS

ES hybrid cells can also be produced by 50% polyethylene glycol (PEG) treatment. Hybrid cells between embryonic carcinoma (EC) cells deficient for the *Hprt* gene and lymphocytes from the thymus or spleen are produced by cell fusion induced chemically by PEG (Takagi *et al.*, 1983). To produce ES hybrid cells using PEG, wash a mixture of ES cells and thymocytes with DMEM and pellet the cells by centrifugation. Prewarm 1 ml of a 50% PEG mixture (PEG4000/DMEM = 1:1) at 37°C and then add the PEG mixture to the cell pellet gradually using the tip of a pipette. Add 9 ml of DMEM gradually. Collect the cells by centrifugation, resuspend the cells in ES medium, and transfer them to a culture dish. Selection of hybrid cells is begun 1 day after the PEG-induced fusion treatment. Electrofusion has the following advantages over the PEG-induced cell fusion: (1) electrofusion is appropriate for *in vivo* applications of the hybrid cells, whereas PEG-induced fusion is not because PEG is toxic to cells; (2) electrofusion is more efficient and reproducible than PEG-induced cell fusion for producing ES hybrid cells; and (3) it is easier to produce hybrid cells by electrofusion than by PEG-induced fusion.

V. PITFALLS

If there are problems with the AC procedure, you may be able to solve the problems as follows. Pellet the mixed cells by centrifugation and resuspend the cells in a suitable amount of fresh mannitol buffer.

1. Adjust the cell density according to the size of the cells used as the fusion partner. Remove cell debris from the mixture of ES cells and somatic cells. Cell debris, which is irregular in size, sometimes makes the formation of pearl chains difficult.
2. Increase the cell density if the pearl chains form poorly.

3. Decrease the cell density if the cell movement is disturbed.

References

Abbondanzo, S. J., Gadi, I., and Stewart, C. L. (1993). Derivation of embryonic stem cell lines. In "Methods in Enzymology; Guide to Techniques in Mouse Development" (P. M. Wassarman and M. L. DePamphilis, eds.), pp. 803–823. Academic Press, San Diego.

Baron, M. H., and Maniatis, T. (1986). Rapid reprogramming of globin gene expression in transient heterokaryons. Cell 46, 591–602.

Blau, H. M., and Baltimore, D. (1991). Differentiation requires continuous regulation. J. Cell Biol. 112, 781–783.

Blau, H. M., Pavlath, G. K., Hardeman, E. C., Chiu, C. P., Silberstein, L., Webster, S. G., Miller, S. C., and Webster, C. (1985). Plasticity of the differentiated state. Science 230, 758–766.

Friedrich, G., and Soriano, P. (1991). Promoter traps in embryonic stem cells: A genetic screen to identify and mutate developmental genes in mice. Genes Dev. 5, 1513–1523.

Harris, H. (1965). Behaviour of differentiated nuclei in heterokaryons of animal cells from different species. Nature 206, 583–588.

Kimura, H., Tada, M., Hatano, S., Yamazaki, M., Nakatsuji, N., and Tada, T. (2003). Chromatin reprogramming of male somatic cell-derived Xist and Tsix in ES hybrid cells. Cytogenet. Genome Res. 99.

Tada, M., Morizane, A., Kimura, H., Kawasaki, H., Ainscough, J. F.-X., Sasai, Y., Nakatsuji, N., and Tada, T. (2003). Pluripotency of reprogrammed somatic genomes in ES hybrid cells. Dev. Dyn.

Tada, M., Tada, T., Lefebvre, L., Barton, S. C., and Surani, M. A. (1997). Embryonic germ cells induce epigenetic reprogramming of somatic nucleus in hybrid cells. EMBO J. 16, 6510–6520.

Tada, M., Takahama, Y., Abe, K., Nakatsuji, N., and Tada, T. (2001). Nuclear reprogramming of somatic cells by in vitro hybridization with ES cells. Curr. Biol. 11, 1553–1558.

Tada, T., Tada, M., Hilton, K., Barton, S. C., Sado, T., Takagi, N., and Surani, M. A. (1998). Epigenotype switching of imprintable loci in embryonic germ cells. Dev. Gene Evol. 207, 551–561.

Takagi, N., Yoshida, M. A., Sugawara, O., and Sasaki, M. (1983). Reversal of X-inactivation in female mouse somatic cells hybridized with murine teratocarcinoma stem cells in vitro. Cell 34, 1053–1062.

Terada, N., Hamazaki, T., Oka, M., Hoki, M., Mastalerz, D. M., Nakano, Y., Meyer, E. M., Morel, L., Petersen, B. E., and Scott, E. W. (2002). Bone marrow cells adopt the phenotype of other cells by spontaneous cell fusion. Nature 416, 542–545.

Vassilopoulos, G., Wang, P. R., and Russell, D. W. (2003). Transplanted bone marrow regenerates liver by cell fusion. Nature 422, 901–904.

Wang, X., Willenbring, H., Akkari, Y., Torimaru, Y., Foster, M., Al-Dhalimy, M., Lagasse, E., Finegold, M., Olson, S., and Grompe, M. (2003). Cell fusion is the principal source of bone-marrow-derived hepatocytes. Nature 422, 897–901.

Wilmut, I., Schnieke, A. E., McWhir, J., Kind, A. J., and Campbell, K. H. S. (1997). Viable offspring derived from fetal and adult mammalian cells. Nature 385, 810–813.

Ying, Q. L., Nichols, J., Evans, E. P., and Smith, A. G. (2002). Changing potency by spontaneous fusion. Nature 416, 545–548.

26

Reprogramming Somatic Nuclei and Cells with Cell Extracts

Anne-Mari Håkelien, Helga B. Landsverk, Thomas Küntziger, Kristine G. Gaustad, and Philippe Collas

I. INTRODUCTION

Genomic plasticity has generated considerable interest in the past two decades; nevertheless the mechanisms underlying nuclear programming remain poorly understood. We report a method that allows the processes of nuclear reprogramming to be investigated *in vitro*.

Nuclear reprogramming occurs in a variety of natural and experimental contexts. After fertilization, epigenetic alterations of the embryonic genome take place during successive stages of development. Epigenetic changes and alterations in gene expression also occur after the fusion of somatic cells with less differentiated cell types (Blau and Blakely, 1999; Tada *et al.*, 2001). The birth of clones and the production of embryonic stem cells by transplantation of nuclei into oocytes also provide evidence of complete reprogramming of somatic nuclei (Cibelli *et al.*, 1998; Gurdon *et al.*, 1979; Munsie *et al.*, 2000; Wilmut *et al.*, 1997). Based on these examples, reprogramming can be defined as an alteration of a differentiated nucleus into a totipotent or mutipotent state. Additional studies have shown that a somatic cell type could be, at least partially, turned into another somatic cell type. This was achieved in coculture conditions (Morrison, 2001) and, more recently, by exposing somatic nuclei or cells to an extract derived from another somatic cell type (Håkelien *et al.*, 2002; Landsverk *et al.*, 2002). These observations have led to the proposal of a simple definition of nuclear reprogramming. Reprogramming may not necessarily involve dedifferentiation or return to a more pluripotent state, but may refer to the dominance of the program of one cell type over another,

resulting in "the transformation of the pliant nucleus [in]to the dominant type" (Western and Surani, 2002).

We describe a procedure to redirect the nuclear program of a transformed human fibroblast cell line toward a T-cell program. The approach is outlined in **Fig. 1** and may, in principle, be applied to multiple cell types. The procedure involves the use of a nuclear and cytoplasmic extract derived from Jurkat T-cells in which reversibly permeabilized fibroblasts are incubated. At the end of incubation, the fibroblasts are resealed and cultured to assess the expression of T-cell-specific markers and the establishment of T-cell-specific functions. As large numbers of cells or nuclei can be processed simultaneously, and considering the ability of cell extracts to be manipulated biochemically, *in vitro* manipulation of nuclei and cells provides a potentially powerful system for analyzing the mechanisms of nuclear reprogramming.

II. MATERIALS AND INSTRUMENTATION

A. Materials

1. 293T human fibroblasts cultured on glass coverslips
2. Round, 12-mm glass coverslips, autoclaved
3. Poly-L-lysine (Cat. No. P8920, Sigma-Aldrich Co., St. Louis, MO)
4. Propidium iodide (Cat. No. P4170, Sigma). Make a 1-mg/ml stock solution in H_2O and store at $-20°C$ in the dark.
5. RPMI 1640 medium (Cat. No. R0883, Sigma) supplemented with 10% fetal calf serum

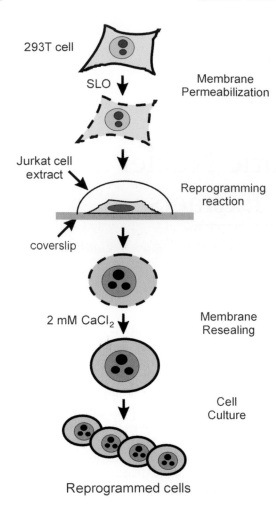

FIGURE 1 Strategy of *in vitro* cell reprogramming. 293T cells grown on coverslips are reversibly permeabilised with SLO. The permeabilised cells are incubated in a nuclear and cytoplasmic extract derived from Jurkat TAg cells for 1 h. The cells are resealed for 2 h in culture medium containing 2 mM $CaCl_2$. The $CaCl_2$-containing medium is replaced by regular complete culture medium, and the resealed, reprogrammed cells are cultured for assessment of reprogramming.

6. Hanks balanced salt solution (HBSS; Cat. No. 14170-088, Gibco-BRL; Paisley, UK)
7. Protease inhibitor cocktail (Cat. No. P2714, Sigma). This is a 100× stock solution.
8. Cell lysis buffer (50 mM NaCl, 5 mM $MgCl_2$, 20 mM HEPES, pH 8.2, 1 mM dithiothreitol, 0.1 mM phenylmethylsulphonyl fluoride and the protease inhibitor cocktail) at 4°C
9. Streptolysin O (SLO) (Cat. No. S5265, Sigma) at 100 μg/ml in H_2O, aliquoted, and stored at −20°C
10. 1 M $CaCl_2$ (Cat. No. C4901, Sigma) in sterile H_2O
11. ATP (Cat. No. A3377, Sigma) at 200 mM in H_2O, aliquoted, and stored at −20°C

12. Creatine kinase (Cat. No. C3755, Sigma) at 5 mg/ml in H_2O, aliquoted, and stored at −20°C
13. Phosphocreatine (Cat. No. P7936, Sigma) at 2 M in H_2O, aliquoted, and stored at −20°C
14. GTP (Cat. No. G8752, Sigma) at 10 mM in H_2O, aliquoted, and stored at −20°C
15. Nucleotide triphosphate (NTP) set (Cat. No. 1277057, Roche; Basel, Switzerland). Prepare a stock solution by mixing 20 μl of each NTP provided in the set at a 1:1:1:1 ratio on ice. Aliquot in 10 μl and store at −20°C. This makes an NTP mix at 25 mM of each NTP. Prepare more stock solution as needed.

B. Instrumentation

1. Sonicator fitted with a 2-mm-diameter probe (Model Labsonic M, B. Braun Biotech International; Melsungen, Germany)
2. Regular atmosphere incubator set at 37°C (for human cells)
3. 5% CO_2 incubator set at 37°C
4. 50- and 15-ml conical tubes (Corning; Corning, NY)
5. 1.5-ml centrifuge tubes
6. 24-well cell culture plates (Costar Cat. No. 3524, Corning)
7. Refrigerated centrifuge with swinging buckets suited for 15- and 50-ml conical tubes

III. PROCEDURES

The methods describe (1) the preparation of "donor" cells to be reprogrammed, (2) the preparation of the reprogramming extract, (3) the permeabilisation of the donor cells, (4) the reprogramming reaction, (5) the resealing of the reprogrammed cells, and (6) examples of assessments of nuclear and cell reprogramming. The procedures described are based on the reprogramming of a human fibroblast cell line (293T) in an extract derived from the human Jurkat TAg cell line (Håkelien *et al.*, 2002).

A. Seeding 293T Cells

On the day prior to reprogramming reaction, plate 293T cells onto 12-mm round, sterile, poly-L-lysine-coated glass coverslips at a density of 50,000 cells per coverslip. Each coverslip is placed in individual wells of a 24-well culture plate. Overlay coverslips with 500 μl of complete RPMI 1640 medium and place in a 5% CO_2 incubator at 37°C.

B. Preparation of the Reprogramming Extract

1. Cell Harvest

1. Transfer the Jurkat TAg cell suspension culture into 50-ml conical tubes and sediment the cells at $800\,g$ for 10 min at 4°C.
2. Wash the cells twice in ice-cold phosphate-buffered saline (PBS) by suspension and sedimentation at $800\,g$ for 10 min at 4°C. Cells can be pooled into a single tube after the first wash.

2. Cell Swelling

1. Resuspend the cells in 10 ml ice-cold cell lysis buffer (CLB). It is preferable to use a graduated 15-ml conical tube to estimate the cell volume after sedimentation.
2. Centrifuge at $800\,g$ for 10 min at 4°C.
3. Estimate the volume of the cell pellet. Resuspend the pellet into 2 volumes of ice-cold CLB.
4. Hold the cells on ice for 45 min to allow swelling. This makes it easier to lyse the cells during sonication. Keep the cells well suspended by occasional tapping of the tube during swelling. Note that the cells can be allowed to swell for longer than 45 min. This swelling step can be omitted for Jurkat TAg or primary T-cells as these cell types lyse promptly during sonication.

3. Extract Preparation

1. Aliquot the cell suspension into 200 μl in 1.5-ml centrifuge tubes previously chilled on ice. Sonicate each tube one by one (on ice) until all cells and nuclei are lysed. Lysis is assessed by complete disruption of the cells and nuclei, as judged by the sole appearance of cell "debris" by phase-contrast microscopy examination of 3-μl samples. Once lysis is achieved in a tube, keep the tube on ice and proceed with all other tubes. Power and duration of sonication varies with each cell type. For Jurkat TAg cells, sonication of each tube at 25% power and 0.5-s pulse cycle over 1 min 40 sec is recommended.
2. Pool all the lysates into one (or multiple, if needed) chilled 1.5-ml centrifuge tube. Sediment the lysate at $15,000\,g$ for 15 min at 4°C in a fixed-angled rotor. Note that a swing-out rotor can also be used.
3. Carefully collect the supernatant with a 200-μl pipette and transfer it into a new 1.5-ml tube chilled on ice. This is the reprogramming extract.
4. It is possible to aliquot the extract into 200-μl tubes such as those used for polymerase chain reaction, with 100 μl extract per tube. Snap-freeze each tube in liquid N_2 and store at −80°C. However, we recommend carrying out reprogramming with freshly made extract as the stability of the extract at −80°C may vary with cell types and batches.

5. Following sedimentation in step 3, remove 20 μl of extract to determine protein concentration and pH. The protein concentration should be between 20 and 25 mg/ml. The pH should be between 6.7 and 7.0 (see Comment 2).

4. Extract Toxicity Assay

Each new batch of Jurkat TAg cell extract requires a cell toxicity test.

1. Add 50,000 293T cells (or, in principle, HeLa cells or any other epithelial of fibroblast cell line growing in the laboratory) to 35 μl of extract on ice in a 1.5-ml centrifuge tube. The extract does not need to contain any additives (unlike for a reprogramming reaction; see Section III,E,1).
2. Incubate for 1 h at 37°C in a water bath.
3. Remove a 3-μl aliquot and assess cell morphology by phase-contrast microscopy. Fig. 2 illustrates primary rat fetal fibroblasts after a 30-min exposure to reprogramming extracts. In our hands, the morphology of the cells after a 30-min incubation in the extract reflects their survival in culture as judged 24 h after the toxicity assay. Cells shown in Fig. 2A survive the extract exposure, whereas cells in Fig. 2B have been damaged by the extract and do not survive in culture. Extract batches producing such cells should be discarded.
4. If so wished, replate the cells directly from the extract in complete RPMI 1640 for an overnight culture to assess survival further. There is no need to remove the extract prior to replating.

C. Permeabilisation of 293T Cells

In order for components from the reprogramming extract to enter 293T cells, the cells must be reversibly permeabilised. Permeabilisation is accomplished with the *Streptococcus pyogenes* toxin, streptolysin O. SLO is a cholesterol-binding toxin that forms large pores in the plasma membrane of mammalian cells (Walev et al., 2002).

1. Preparation of SLO Stock Solution

1. Dissolve SLO powder in sterile-filtered MilliQ H_2O to 100 μg/ml. Keep on ice while dissolving the SLO.
2. Aliquot 10 μl in 200-μl tubes and store at −20°C.
3. Discard all tubes after 1 month of storage at −20°C and prepare a new stock. Stock aliquots should be thawed only once. Note that because commercial batches of SLO vary in specific activity, a range of SLO concentrations (200, 500, 1000, 2000, and

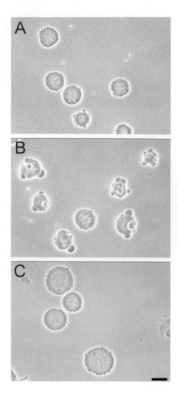

FIGURE 2 Results from toxicity assay. Rat embryo fibroblasts were exposed to a reprogramming extract for 1 h as described in the text. Cells were examined by phase-contrast microscopy. (A) Viable cells. (B) Nonviable cells. Reprogramming extracts producing such cells are discarded. (C) Control cells exposed for 30 min to the cell lysis buffer used to prepare the extract. Bar: 20 μm.

4000 ng/ml SLO) should be tested on the cell type to be reprogrammed after a new stock is prepared.

2. Cell Permeabilisation

1. Dilute the SLO stock in ice-cold HBSS to 230 ng/ml. This is the working solution. Note that this concentration is valid for 293T cells and should be adjusted for other cell types using a cell permeabilisation assay described in Section III,D.
2. Keep the SLO on ice until addition to 293T cells.
3. Remove the RPMI 1640 medium from wells containing 293T cells grown on coverslips and wash the cells four times with PBS at room temperature to remove all Ca^{2+} from the culture medium. This step is essential as Ca^{2+} inhibits SLO activity.
4. Add 250 μl SLO working solution to each well.
5. Incubate at 37°C in regular atmosphere for 50 min. Proceed to Section III,E,1.

D. Cell Permeabilisation Assay

This assay allows evaluation of the efficiency of the SLO treatment. It is recommended to carry out this assay using four additional coverslips supporting 293T cells as described in Section III,A, in addition to those used for the reprogramming reaction. This assay is based on the uptake of the fluorescent DNA stain, propidium iodide, by permeabilised cells, but not by intact cells.

1. Permeabilise cells on two coverslips with SLO as described under Section III,C,2. However, in the first step, add propidium iodide to 0.1 μg/ml to the SLO dilution in HBSS on one of the coverslips. Propidium iodide will be taken up as cells are being permeabilised. The other coverslip receives 250 μl SLO dilution in HBSS without propidium iodide.
2. Two additional coverslips should also be used as controls for the absence of SLO. Add 250 μl HBSS containing propidium iodide to one of the control coverslips as in step 1. The other coverslip receives 250 μl HBSS without propidium iodide.
3. Incubate at 37°C in regular atmosphere for 50 min.
4. For SLO-treated and control coverslips *not* containing propidium iodide, remove HBSS and immediately add 1.5 ml of preheated (37°C) complete RPMI 1640 containing 2 mM Ca^{2+} added from a 1 M stock (see Section II,A). Incubate at 37°C for 2 h to allow resealing of the plasma membranes.
5. For SLO-treated and control coverslips labelled with propidium iodide, remove HBSS, rinse with PBS, and add 250 μl PBS.
6. Assess propidium iodide labelling of the nuclei by epifluorescence microscopy.
7. After the 2-h membrane resealing step described in step 4, remove the culture medium, rinse with PBS, and add 250 μl PBS containing 0.1 μg/ml propidium iodide; incubate for 10 min.
8. Assess propidium iodide uptake, or lack thereof, in the resealed cells as in step 6.

E. Reprogramming Reaction

1. Extract Preparation

During SLO treatment, the extract should be prepared for reprogramming.

1. Prepare the ATP-regenerating system: mix on ice ATP:GTP:creatine kinase:phosphocreatine in a 1:1:1:1 ratio from each separate stock (described in Section II,A) and keep on ice.
2. Add 5 μl of the ATP-regenerating system mix to 100 μl of extract on ice.
3. Add 4 μl of the 25 mM NTP mix (see Section II,A) to 100 μl of extract on ice.
4. Vortex briefly and replace the extract on ice.

2. Reprogramming Reaction

1. Remove SLO from the cells by careful aspiration.
2. Quickly add PBS to prevent drying.
3. Immediately transfer each coverslip into a new dry well of a 24-well plate and carefully lay 65 µl of extract (prepared as described in Section III,B,3) onto each coverslip. Be careful that cells do not dry out upon transfer of the coverslip(s) to the new wells and prior to addition of the extract. It is important that the coverslip be covered by the extract during the entire incubation time. Should the extract spread out of the coverslip, transfer the coverslip into a new well and pipette the extract back onto the cells.
4. Incubate at 37°C for 1 h in regular atmosphere. Note that reprogramming has also proven to be successful upon incubation in a 5% CO_2 incubator at 37°C.

F. Resealing Reprogrammed Cells

1. At the end of incubation, directly add to each well 1.5 ml of preheated (37°C) complete RPMI 1640 containing 2 mM Ca^{2+} added from the 1 M stock (see Section II,A). Do not remove the extract before adding the Ca^{2+}-containing medium.
2. Incubate for 2 h in a 5% CO_2 incubator at 37°C.
3. Remove the Ca^{2+}-containing medium by gentle aspiration and replace with 250 µl of complete RPMI 1640 (Jurkat TAg cell culture medium).
4. Place the cells back into the 5% CO_2 incubator and culture until reprogramming assessments are performed.

G. Assessment of Nuclear Reprogramming

Various assessments of nuclear and cell reprogramming can be performed, depending on the purpose of the experiment. Using the method described here, we have reported changes in the gene expression profile of the reprogrammed 293T cells using cDNA macroarrays from R&D Systems (Abington, UK) (Håkelien et al., 2002). Expression of new proteins can also be monitored at regular intervals after the reprogramming reaction by immunofluorescence or flow cytometry using standard protocols. We have shown the expression of several antigens on the surface of the reprogrammed cells, which are specific for hematopoietic cells. These include CD3, CD4, CD8, CD45, and components of the T-cell receptor (TCR) complex (Håkelien et al., 2002). A variety of functional assays can also be carried out, such as cytokine secretion in response to stimulation of the T-cell receptor/CD3 complex in the reprogrammed cells, or expression of additional cytokine receptors on the cell surface

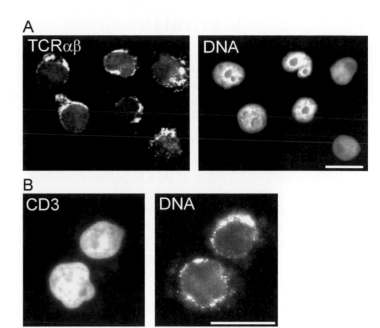

FIGURE 3 Immunofluorescence detection of (A) T-cell receptor (TCR) α and β chain expression and (B) CD3 expression on the surface of 293T cells reprogrammed in a Jurkat TAg cell extract. DNA is labelled with 0.1 µg/ml propidium iodide. Analysis was performed ~14 days after the reprogramming reaction. Bars: 20 µm.

(Håkelien et al., 2002). **Fig. 3** illustrates the expression of the TCR α and β chains and of CD3 molecules on the surface of 293T cells reprogrammed in a Jurkat TAg cell extract.

IV. COMMENTS

1. Commercially available SLO batches vary greatly in activity. Thus, it is recommended to test a range of SLO concentration on the cell type to be reprogrammed prior to initiating reprogramming reactions. The efficiency of SLO-mediated permeabilisation also varies for various cell types.

2. pH of the extract. We usually observe a drop of 1-1.5 pH unit upon extract preparation, which explains the pH 8.2 of the CLB. Notably, raising the pH of CLB to 8.7 with a HEPES buffer does not increase the pH of the final extract. Other buffers with greater buffering capacity have not been tested.

3. The method described here can be used easily with either purified cell nuclei (Landsverk et al., 2002) or permeabilised cells (Håkelien et al., 2002). Procedures for purifying intact (membrane-enclosed) nuclei from interphase cultured cells have been reported earlier (Collas et al., 1999; Landsverk et al., 2002; Martins et al., 2000; Steen et al., 2000).

V. PITFALLS

1. It is currently difficult to objectively assess the extent of sonication of Jurkat TAg cells or any other cell type. It is important to sonicate until all cells and nuclei are completely lysed. Whether extended sonication after cell lysis is complete is detrimental or beneficial is at present unknown.

2. Variability in batches of reprogramming extracts is seen, even among extracts that have been rated as "nontoxic" in the toxicity assay described in Section III,B,4. Variability is evident by the absence of markers of cell reprogramming approximately 1 week after the reprogramming reaction.

3. Currently, the duration of expression of a reprogrammed phenotype is limited to at least 2 months for 293T cells reprogramming in Jurkat TAg extract (Håkelien *et al.*, 2002). The reprogrammed phenotype may also last for shorter periods depending on the marker analyzed.

Acknowledgments

Our work was supported by Nucleotech, LLC and grants from the Research Council of Norway, the Norwegian Cancer Society, and the Human Frontiers Science Program.

References

Blau, H. M., and Blakely, B. T. (1999). Plasticity of cell fate: Insights from heterokaryons. *Semin. Cell D* **10**, 267–272.

Cibelli, J. B., Stice, S. L., Golueke, P. J., Kane, J. J., Jerry, J., Blackwell, C., Ponce, D. L. F., and Robl, J. M. (1998). Cloned transgenic calves produced from nonquiescent fetal fibroblasts. *Science* **280**, 1256–1258.

Collas, P., Le Guellec, K., and Tasken, K. (1999). The A-kinase anchoring protein, AKAP95, is a multivalent protein with a key role in chromatin condensation at mitosis. *J. Cell Biol.* **147**, 1167–1180.

Gurdon, J. B., Laskey, R. A., De Robertis, E. M., and Partington, G. A. (1979). Reprogramming of transplanted nuclei in amphibia. *Int. Rev. Cytol. Suppl.* 161–178.

Håkelien, A. M., Landsverk, H. B., Robl, J. M., Skålhegg, B. S., and Collas, P. (2002). Reprogramming fibroblasts to express T-cell functions using cell extracts. *Nature Biotechnol.* **20**, 460–466.

Landsverk, H. B., Håkelien, A. M., Küntziger, T., Robl, J. M., Skålhegg, B. S., and Collas, P. (2002). Reprogrammed gene expression in a somatic cell-free extract. *EMBO Rep.* **3**, 384–389.

Martins, S. B., Eide, T., Steen, R. L., Jahnsen, T., Skålhegg, B. S., and Collas, P. (2000). HA95 is a protein of the chromatin and nuclear matrix regulating nuclear envelope dynamics. *J. Cell Sci.* **113**, 3703–3713.

Morrison, S. J. (2001). Stem cell potential: Can anything make anything? *Curr. Biol.* **11**, R7-R9.

Munsie, M. J., Michalska, A. E., O'Brien, C. M., Trounson, A. O., Pera, M. F., and Mountford, P. S. (2000). Isolation of pluripotent embryonic stem cells from reprogrammed adult mouse somatic cell nuclei. *Curr. Biol.* **10**, 989–992.

Steen, R. L., Martins, S. B., Tasken, K., and Collas, P. (2000). Recruitment of protein phosphatase 1 to the nuclear envelope by A-kinase anchoring protein AKAP149 is a prerequisite for nuclear lamina assembly. *J. Cell Biol.* **150**, 1251–1262.

Tada, M., Takahama, Y., Abe, K., Nakastuji, N., and Tada, T. (2001). Nuclear reprogramming of somatic cells by *in vitro* hybridization with ES cells. *Curr. Biol.* **11**, 1553–1558.

Walev, I., Hombach, M., Bobkiewicz, W., Fenske, D., Bhakdi, S., and Husmann, M. (2002). Resealing of large transmembrane pores produced by streptolysin O in nucleated cells is accompanied by NF-kappaB activation and downstream events. *FASEB J.* **16**, 237–239.

Western, P. S., and Surani, M. A. (2002). Nuclear reprogramming. Alchemy or analysis? *Nature Biotechnol.* **20**, 445–446.

Wilmut, I., Schnieke, A. E., McWhir, J., Kind, A. J., and Campbell, K. H. S. (1997). Viable offspring derived from fetal and adult mammalian cells. *Nature* **385**, 810–813.

Immortalization

27

Immortalization of Primary Human Cells with Telomerase

Kwangmoon Lee, Robert L. Kortum, and Michel M. Ouellette

I. INTRODUCTION

One major obstacle to the immortalization of primary human cells is telomere-controlled senescence. Telomere-controlled senescence is caused by the shortening of the telomeres that occurs each time somatic human cells divide. The enzyme telomerase can prevent the erosion of telomeres and block the onset of telomere-controlled senescence, but its expression is restricted to the human embryo and, in the adult, to rare cells of the blood, skin, and digestive track. However, we and others have shown that the transfer of an exogenous hTERT cDNA, encoding the catalytic subunit of human telomerase, can be used to prevent telomere shortening, overcome telomere-controlled senescence, and immortalize primary human cells (Bodnar *et al.*, 1998). Most importantly, hTERT alone can immortalize primary human cells without causing cancer-associated changes or altering phenotypic properties (Jiang *et al.*, 1999; Morales *et al.*, 1999; Ouellette *et al.*, 2000). Primary human cells that have been immortalized successfully with hTERT alone include fibroblasts, retinal pigmented epithelial cells, endothelial cells, myometrial cells, esophageal squamous cells, mammary epithelial cells, keratinocytes, osteoblasts, and Nestin-positive cells of the pancreas (Bodnar *et al.*, 1998; Yang *et al.*, 1999; Ramirez *et al.*, 2001; Yudoh *et al.*, 2001; Condon *et al.*, 2002; Lee *et al.*, 2003; Morales *et al.*, 2003). This article describes the use of hTERT for the purpose of immortalizing primary human cells using the primary human fibroblast as an example.

II. MATERIALS AND INSTRUMENTATION

Packaging cell line Phoenix-ampho (φNX-A) for the production of amphotropic viruses is available from Dr. Gary Nolan (Stanford University, Stanford, CA; http://www.stanford.edu/group/nolan/). Retroviral vectors pBabePuro and pBabePuro-hTERT are from Geron Corp. (Menlo Park, CA; Ouellette *et al.*, 1999). Dulbecco's modified Eagle's medium (DMEM; Cat. No. 11995), medium 199 (Cat. No. 11150), gentamicin (Cat. No. 15710), and trypsin–EDTA (0.05% trypsin, 0.53 mM EDTA; Cat. No. 25300) are from Invitrogen Corp. (Carlsbad, CA). Electrophoresis apparatus (Cat. Series 21070), including glass plates, spacers, and combs, are from Life Technologies (Rockville, MD). Cosmic calf serum (Cat. No. SH30087) and fetal bovine serum (Cat. No. SH30070) are from HyClone (Logan, UT). The polysulfone filter (0.45 μm, Cat. No. DD50402550) is from Life Science Products Inc. (Frederick, CO). Dimethyl sulfoxide (DMSO; Cat. No. D 2650), polybrene (Cat. No. H 9268), puromycin (Cat. No. P 7255), and bovine serum albumin (BSA; Cat. No. A 2153) are from Sigma-Aldrich (St. Louis, MO). The MBS mammalian $CaPO_4$ transfection kit is from Stratagene (Cat. No. 200388; La Jolla, CA). The TrapEZE kit is from Serologicals Corp. (Cat. No. S7700; Norcross, GA). *N,N,N',N'*-Tetramethylenediamine (TEMED; Cat. No. 161–0800) and 40% acrylamide:bisacrylamide [19:1] solution (Cat. No. 161-0146) are from Bio-Rad (Hercules, CA). The PhosphorImager is from Molecular Dynamics (Model No. 810; Sunnyvale, CA). Nalgene

215

Cryo 1°C freezing containers are from Nalge Nunc International (Cat. No. 5100-0001; Rochester, NY). T4 polynucleotide kinase is from New-England BioLabs (Cat. No. M0201S; Beverly, MA). [γ-^{32}P]ATP is from New England Nuclear (Cat. No. 502A12070; Boston, MA). The polymerase chain reaction (PCR) is performed using a PCRExpress thermocycler from Hybaid (Ashford, Middlesex, UK). All other chemicals are from Fisher Biotechnology (Fait Lawn, NJ), Fisher Scientific Co. (Pittsburgh, PA), or Sigma-Aldrich (St. Louis, MO).

III. PROCEDURES

A. Production of Replication-Defective Retroviruses Carrying the hTERT cDNA

Primary human cells tend to transfect very poorly. Consequently, the preferred method of transferring exogenous telomerase to such cells is their transduction with replication-defective retroviruses carrying an hTERT cDNA. The following procedure, modified from Pear *et al.* (1993), yields high-titer viruses following the transient transfection of ϕNX-A cells, a retroviral packaging cell line.

Solutions

1. *293T medium*: To 500 ml of DMEM, add 50 ml of fetal bovine serum and 500 μl of 10 mg/ml gentamicin. Store at 4°C and protect from light.
2. *Complete medium X*: To a clean sterile 500-ml bottle, combine 400 ml of DMEM, 100 ml of medium 199, 50 ml of cosmic calf serum, and 500 μl of 10 mg/ml gentamicin. Store at 4°C and protect from light.

Steps

1. In a 37°C water bath, thaw a vial of ϕNX-A cells and transfer cells to a culture dish containing 293T medium. Incubate cells at 37°C under 5% CO$_2$. The next day, replace medium with fresh medium.

2. Expand ϕNX-A cells in 293T medium using split ratios of 1:4 to 1:6. Avoid letting the cells grow beyond 90% confluence. To split cells, remove medium by aspiration, gently wash cells once with PBS. Take extra care not to dislodge the cells, as ϕNX-A cells tend to detach very easily. Add 2 ml of trypsin–EDTA per 100-mm dish and let cells detach at 37°C for 1–2 min. Tap dish to help dislodge cells, add 5 ml of 293T medium, and pipette up and down to break cell clusters. Transfer suspension to a sterile tube and pellet cells by low-speed centrifugation (2000 g for 5 min at room temper-

ature). Remove supernatant and resuspend cells in 5 ml of 293T medium.

3. Following the manufacturer's instructions, count cells using a Coulter counter or hemacytometer. In two 100-mm plates containing 293T medium, seed 5.0×10^6 cells/dish. Freeze remaining cells to replenish archival stocks, if needed (refer to Section III,C).

4. On the following day, ϕNX-A cells should be ~80% confluent and ready to be transfected with the retroviral vectors. Transfect each dish with 10–20 μg of plasmid; one with vector pBabePuro-hTERT and the other with the empty vector, pBabePuro. ϕNX-A cells are transfected most easily using the CaPO$_4$ method, but other methods can be used as well. To CaPO$_4$ transfect the ϕNX-A cells, we have used the Stratagene's MBS mammalian transfection kit following the manufacturer's instructions. In a 5-ml Falcon 2054 polystyrene tube, prepare a calcium–DNA precipitate by mixing plasmid DNA (10–20 μg in 450 μl of sterile water) with 50 μl of solution I (2.5 M CaCl$_2$) and then adding 500 μl of solution II [N,N-bis(2-hydroxyethyl)-2-aminoethanesulfonic acid in buffered saline].

5. Incubate the calcium–DNA mixtures at room temperature for 10–20 min. In the meantime, prepare the cells by washing cells once with PBS and feeding them with 10 ml/dish of 293T medium, in which 6% modified bovine serum (provided by the kit) is used in place of the fetal bovine serum.

6. Resuspend the DNA precipitate gently and then apply dropwise in a circular motion to the dishes, as to distribute the DNA evenly. Swirl dishes once.

7. Incubate cells at 37°C under 5% CO$_2$ for 3 h.

8. Remove medium by aspiration.

9. Wash cells once with sterile PBS and feed with 293T medium. Return cells to 37°C under 5% CO$_2$. Cells have now been transfected and should start producing retroviruses within the next 24 h.

10. Viruses can be collected in three consecutive harvests over the next 24–72 h posttransfection. To collect viruses, replace the 293T medium with 4–5 ml of the target cells culture medium, in which the viruses will now be allowed to accumulate. For primary human fibroblasts, collect viruses in complete medium X. Return cells to 37°C under 5% CO$_2$ with the dishes spread on a leveled shelf to ensure good coverage of all surfaces.

11. After 8–16 h of exposure to the cells, harvest supernatants and then force through a 0.45-μm polysulfone filter so that all remaining cells are removed. Bleach and discard the transfected cells once the last batch of viral supernatant has been collected.

12. Keep viral supernatants either on ice and use within the hour or else aliquot, freeze, and store at

−80°C, where they can last for approximately 6 months. Label each aliquot with name of retroviral vector, harvest medium, date, and harvest number (1st, 2nd, etc . . .). Freezing does not appear to cause substantial drops in titer, but cycles of freeze/thaw do.

B. Transduction of Human Primary Cells with Retroviruses Carrying hTERT

Solutions

1. *400 μg/ml polybrene:* Dissolve 8 mg of polybrene in 20 ml of deionized double-distilled water. Sterilize by filtration through a 0.45-μm polysulfone filter. Store at 4°C.
2. *500 μg/ml puromycin:* Dissolve 10 mg of puromycin in 20 ml of deionized double-distilled water. Sterilize by filtration through a 0.45-μm polysulfone filter. Store aliquots at −20°C.

Steps

1. Thaw primary human fibroblasts in complete medium X. Incubate cells at 37°C under 5% CO_2. The next day, replace medium with fresh medium.

2. Expand cells in complete medium X using split ratios of 1:2, 1:4, and 1:8 for late, mid, and early passage cells, respectively. Avoid leaving monolayers at full confluence for extensive periods of time, as cells may then resist trypsinization. To split cells, remove medium and wash cells twice with 2 ml of trypsin–EDTA per 100-mm dish. Incubate dishes at 37°C until cells have detached (1–3 min). Tap dishes to help dislodge cells, add 5 ml of complete medium X, and pipette up and down to dissociate clumps. Transfer suspension to a sterile tube and pellet cells by low-speed centrifugation (2000 g for 5 min at room temperature). Remove supernatant and resuspend cells in 5 ml of complete medium X. While keeping track of population doublings (see later), expand fibroblasts until sufficient cells are available.

3. Seed four wells of a 6-well plate with target cells so that the cells are in log-phase growth on the next day. For primary human fibroblasts, seed cells in complete medium X at 2×10^5 cells per well. Take note of the population doubling of the seeded cells as the initial PD (or PD_i). Freeze remaining cells to replenish archival stocks, if needed. Label frozen vials with name of strain, date, population doubling, and approximate number of cells per vial (refer to Section III,C).

4. Incubate dishes overnight at 37°C under 5% CO_2.

5. Remove the medium from all dishes. Infect one dish with viruses carrying hTERT (sample A; infected with pBabePuro-hTERT) and a second dish with control viruses carrying no insert (sample B; infected with pBabePuro). Feed the remaining two dishes with virus-free medium (samples C and D; uninfected). Perform infections at 37°C under 5% CO_2 for 8–16 h using 1–2 ml of a mixture containing 1 volume of viral supernatant, 1 volume of the target cells culture medium, and 4 μg/ml polybrene. Within 24–48 h, cells can be infected sequentially with all of the different harvests of the same virus.

6. Remove the last viral supernatants and replenish all four dishes with the target cell culture medium, using complete medium X for primary human fibroblasts. Return cells at 37°C under 5% CO_2.

7. Let cells divide once or twice over the next 24–48 h to allow integration of the viral genomes.

8. Select cells for viral integrations using puromycin. The exact dose of puromycin to use should be determined in a preliminary experiment, in which uninfected cells are exposed to increasing doses of puromycin (e.g., 0, 250, 500, 750, 1000 ng/ml) for 7–10 days; use the lowest dose that kills all cells. For primary human fibroblasts, 500–750 ng/ml is a good starting point. Add puromycin to sample A, B, and C. Leave sample D puromycin free.

9. Maintain cells in continuous log-phase growth. Samples that reach confluence should be trypsinized and expanded to a larger dish; do not discard any excess cells yet. Replace puromycin-containing media as needed or twice a week until selection is complete. Selection is complete when all of the uninfected cells of control sample C have died, after typically 7–10 days of selection.

10. Expand cells until samples A and B have reached the size of a 100-mm dish. Count total number of cells in samples A, B, and D. Evaluate the number of population doublings done during the infection/selection phases of the experiment ($\Delta PD_{i/s}$). For this purpose, the simplest approximation is to assume that samples A, B, and D have done an equivalent number of doublings during this same period of time. Using sample D as a reference, calculate $\Delta PD_{i/s}$ knowing that $\Delta PD_{i/s}$ = log[(final number of cells in sample D) ÷ (initial number of cells plated in step 3)] / log[2]. Discard sample D and set PD of samples A and B to $PD_{time0} = PD_i + \Delta PD_{i/s}$.

11. Cells are now ready to be tested for telomerase activity and assessed for life span extension. Put aside 5×10^4 cells of each sample for a telomerase assay (see Section III,C). Start a growth curve by seeding 2×10^5 cells of each sample in a 100-mm dish (see Section III,C). Freeze remaining cells and label vial with name of strain, vector transduced, population doubling (PD_{time0}), and approximate number of cells.

C. Other Procedures Related to the Immortalization of Human Primary Fibroblasts

1. Freezing φNX-A Cells and Primary Human Cells

Solution

Freeze medium: Mix 10 ml of DMSO with 90 ml of fetal bovine serum. Store aliquots at −20°C and keep working solution at 4°C.

Steps

1. Trypsinize cells as described earlier, resuspend in culture medium, and recover by low-speed centrifugation (2000 g for 5 min at room temperature).
2. Remove supernatant by aspiration, and resuspend pellet in freeze medium so that each milliliter contains the cell equivalent of 25–150 cm^2 of confluent dish surface.
3. Aliquot suspension in 2-ml cryogenic vials, at 1 ml per vial.
4. Label vials with name of sample, population doublings, date, and approximate number of cells.
5. Place vials into a Nalgene Cryo 1°C freezing container filled with fresh isopropanol. Transfer container to −80°C.
6. On the next day, transfer the frozen vials to either a liquid nitrogen tank or a −135°C freezer.

2. Measuring Cellular life Span

To verify that exogenous telomerase has an extended cellular life span, a growth curve is generated to measure and compare the life span of the vector- and hTERT-transduced cells.

Steps

1. Maintain the vector- and hTERT-transduced cells in continuous log-phase growth as cells are counted once a week. For this purpose, seed cells at a density that requires a week for early passage cells to reach confluence. For most primary human fibroblasts, a density of 2×10^5 cells per 100-mm dish is adequate. Take note of the population doublings of the two samples, using the value of PD_{time0} for cells that have just been transduced and then selected.
2. Let cells grow for a week at 37°C under 5% CO_2.
3. Trypsinize, wash, recover in culture media, and then count cells using either a Coulter counter or a hematocytometer.
4. Calculate the number of population doublings done by the cells since they were last seeded (or ΔPD), where ΔPD = log[(number of cells recovered) ÷ (number of cell seeded)] / log[2].

5. Calculate the current PD of each of the two cell populations by adding the value of their ΔPD to that of their previous PD (at which cells were when they had been plated a week earlier). If ΔPD is negative, set the value of ΔPD to zero.
6. Replate cells at the exact same density as in step 1 and freeze all remaining cells. Label frozen vials with name of strain, vector transduced, population doubling (current PD), and approximate number of cells.
7. Repeat steps 2 though 6 until all of the vector-transduced cells have become senescent. For both samples, each cycle of growth will yield a ΔPD that is then added to the previous PD to increase the current PD to its present value.
8. For each sample, plot PD_{n+1} as a function of time. Cells transduced with the empty vector should eventually reach a plateau corresponding to telomere-controlled senescence, at which point cells will cease to divide. Pursue the experiment until the vector-transduced cells have reached senescence. Cells can be considered senescent once they perform less than a doubling over a 2-week period. Most strains of primary human cells reach senescence after 15–90 doublings. A complete bypass of senescence by the hTERT-transduced cells would then confirm that the cellular life span has been extended by exogenous telomerase. In this eventuality, grow the hTERT-transduced cells until the life span has been extended by a factor of at least 3, at which point the cells can be considered functionally immortal.

3. Measuring the Activity of Telomerase Using the Telomere Repeat Amplification Protocol (TRAP) Assay

The TRAP assay is based on an improved version of the original method described by Kim *et al.* (1994). It is typically performed using a commercial kit, the TRAPeze telomerase detection kit (Serologicals Corp., Norcross, GA).

Solutions

1. *50 mg/ml BSA:* Dissolve 0.5 g of BSA in 10 ml of deionized water. Aliquot and store at −20°C.
2. *0.5 M EDTA:* To make 500 ml, add 93.1 g of EDTA (disodium salt) to 400 ml of deionized water. Adjust pH to 8.0 using sodium hydroxyde. Complete to 500 ml with deionized water.
3. *6X loading dye:* To make 10 ml, combine 6.4 ml of deionized water, 3 ml of glycerol, 600 μl of 0.5 M EDTA, 25 mg of bromphenol blue, and 25 mg of xylene cyanol.
4. *12.5% acrylamide gel solution:* To make 50 ml, combine 15.6 ml of a 40% acrylamide:bisacrylamide [19:1] solution with 5 ml of 5X TBE. Complete to 50 ml

with deionized water. Just before casting the gel, add 50 µl of TEMED and 500 µl of 10% ammonium persulfate. Mix well and cast immediately.

5. *5X TBE:* To make 1 liter, add 54 g of Tris base, 27.5 g of boric acid, and 20 ml of 0.5 M EDTA, pH 8.0, to 800 ml of deionized water. Mix until solute has dissolved. If needed, readjust pH to 8.1–8.5. Complete to 1 liter with deionized water.

Steps

1. Prepare the following samples for analysis: uninfected cells, hTERT-transduced cells, vector-transduced cells, and a positive control (telomerase-positive cancer cell line, such as 293, H1299, or HeLa cells). Trypsinize and count cells; for each sample, transfer 5×10^4 cells to an Eppendorf tube.

2. Pelet cells by low-speed centrifugation (2000 g for 5 min) and remove supernatant. Spin sample once more for just a few seconds so that the very last drop of remaining medium can be removed with the use of a micropipette. Freeze pellets at –80°C.

3. Thaw cell pellets on ice. To each sample, add 100 µl of ice-cold TRAPeze kit CHAPS lysis buffer (10 mM Tris–HCl, pH 7.5, 1 mM MgCl$_2$, 1 mM EGTA, 0.1 mM benzamidine, 5 mM β-mercaptoethanol, 0.5% CHAPS, 10% glycerol). Disperse cells by forcing the pellets 5–10 times through the tip of a P200.

4. Incubate cell lysates on ice for 30 min.

5. Spin samples in a microcentrifuge at 12,000 g for 20 min at 4°C.

6. Transfer 80 µl of the supernatant into a fresh Eppendorf tube. Cell extracts should now contain 500 cell equivalents per microliter. Keep on ice and use within the hour or else freeze and store at –80°C.

7. End label the Telomerase Substrate (TS) primer (5'-AATCCGTCGAGCAGAGTT-3'). To prepare a sufficient amount of TS primer for six telomerase reactions, combine 6 µl of TS primer, 1.2 µl of 10X kinase buffer (provided with the enzyme), 3 µl water, 1.5 µl of [γ-32P]ATP (3000 Ci/mmol, 10 mCi/ml), and 0.3 µl of T4 polynucleotide kinase (10 units/µl). Incubate at 37°C for 20 min. Kill the enzyme at 85°C for 5 min.

8. Prepare a master mix containing all components of the telomerase assay, minus the extracts to be tested. To prepare sufficient master mix for six assays, combine 30 µl of 10X TRAP reaction buffer (200 mM Tris–HCl, pH 8.3, 15 mM MgCl$_2$, 630 mM KCl, 0.5% Tween-20, 10 mM EGTA), 3 µl of 50 mg/ml BSA, 6 µl of dNTP mix (2.5 mM each dATP, dTTP, dCTP, and dGTP), 6 µl of TRAP primer mix (contains the PCR primers needed for amplifying the telomerase products), 234.6 µl of water, 12 µl of ^{32}P-labeled TS primer (from the previous step), and 2.4 µl of *Taq* DNA polymerase (5 units/µl). Mix well. Aliquot the master mix in four RNase-free PCR tubes at 49 µl/tube.

9. On ice, thaw all of the cell extracts to assay. Samples should include the uninfected cells, vector-transduced cells, hTERT-transduced cells, a negative control (TRAPeze kit 1X CHAPS lysis buffer), and a positive control (telomerase-positive cancer cell line).

10. Add 1 µl of sample per tube (500 cell equivalents), mix, and place in the thermocycler.

11. Incubate at 30°C for 30 min.

12. Perform the following two-step PCR for 27–30 cycles: 94°C for 30 s, followed by 59°C for 30 s. While these conditions work on most thermocyclers, optimization of the annealing temperature and addition of a 72°C extension step may be necessary on certain machines.

13. Add 10 µl of 6X loading dye to all samples. Store at –20°C in a β blocker.

14. Prepare a vertical 12.5% polyacrylamide gel. Choose glass plates, spacers, and comb so that the gel will be 1.5 mm in thickness and 10–12 cm in length. Once the mold is ready, add TEMED and ammonium persulfate to the 12.5% acrylamide gel solution and pour the gel immediately.

15. Once the gel has polymerized, load 30 µl of each sample per lane. Run at 400 V for 90 min or until the xylene cyanol has run 70–75% of the gel length.

16. Dispose of the electrophoresis buffer as liquid radioactive waste. Wrap gel in saran wrap.

17. Expose gel to an X-ray film or PhosphoImager cassette.

18. The presence of an active telomerase should yield a ladder of products with 6-bp increments, starting at 50 nucleotides. For the TRAP assay to be valid, the following conditions should first be met: (1) the negative control corresponding to the lysis buffer should display no such ladder; (2) the positive control (e.g., 293, H1299, or HeLa cells) should yield an intense ladder; and (3) all samples should display a 36-bp band corresponding to the internal TRAP assay standard (ITAS), a control template included in the TRAP primer mix whose amplification serves to document the efficiency the PCR step of the protocol. The lack of ITAS amplification would suggest that inhibitors of *Taq* DNA polymerase may have been present in some of the samples. If the assay has met these conditions, analysis of the experimental samples can now proceed. A successful reconstitution of telomerase activity by exogenous hTERT should produce a telomerase ladder similar in intensity to that of the positive control, with no such activity detected in either vector-transduced cells or uninfected cells.

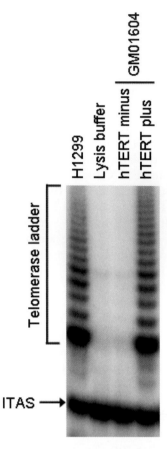

FIGURE 1 Telomerase activity in hTERT-transduced fibroblasts. GM01604 fibroblasts engineered to express exogenous hTERT or no hTERT were assayed for telomerase activity using the TRAPeze telomerase detection kit. H1299 and lysis buffer, respectively, serve as positive and negative controls. ITAS, internal TRAP assay standard.

D. Example of the Use of hTERT to Extend Cellular Life Span

1. Primary Human Fibroblasts

Primary human fibroblasts (GM01604; Coriell Institute, Camden, NJ) were infected with pBabePuro-hTERT retroviruses under the conditions described in this article. Infected cells were selected for 2 weeks with 750 ng/ml puromycin and were then tested for telomerase activity using the TRAPeze telomerase detection kit. Figure 1 shows a strong telomerase ladder in both hTERT-transduced cells and positive control H1299. The 36-bp ITAS is visible in all lanes, indicating that inhibitors of *Taq* DNA polymerase, which could have resulted in false negatives, were absent. The life span of hTERT-transduced cells was then compared to that of uninfected cells. Cells were maintained in continuous log-phase growth and counted once a week. Figure 2 is a graphic represen-

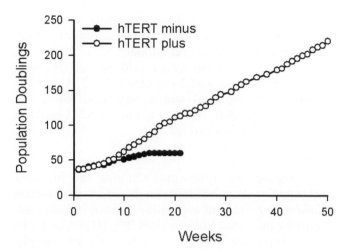

FIGURE 2 Extension of cellular life span by exogenous hTERT. GM01604 fibroblasts engineered to express exogenous hTERT or no hTERT were maintained in continuous log-phase growth and counted once a week. Population doublings executed expressed as a function of time.

tation of the cumulative number of PD executed by the cells as a function of time. Note that hTERT minus cells ceased dividing at PD 60 whereas hTERT-transduced cells grew beyond PD 180, at which point these cells were considered functionally immortal.

IV. COMMENTS

1. Other retroviral vectors carrying hTERT are available from Dr. Robert A. Weinberg. (Whitehead Institute for Biomedical Research, Cambridge, MA). These additional vectors include an alternate version of vector pBabePuro-hTERT, as well as plasmid pCIneo-hTERT, a retroviral construct that confers resistance to G418. Be sure to use a vector carrying an hTERT cDNA that has not been epitope tagged at the C terminus, as these modifications can block the access of telomerase to the telomeres (Ouellette *et al.*, 1999).

2. To adapt the protocol described for other primary human cells, replace complete medium X with a culture medium that is compatible with the long-term growth and survival of your target cells. Transfected φNX-A cells can be made to produce viruses in a large variety of culture media.

3. It should be noted that hTERT alone may not be sufficient to immortalize all types of primary human cells. First, the enzyme telomerase does not appear to alter the phenotypic properties, such that postmitotic terminally differentiated cells are unlikely to be rescued by the enzyme. Second, certain primary

human cells experience additional forms of senescence that are independent of telomere size (Kiyono *et al.*, 1998). It has been suggested that these extra barriers represent a stress response to inadequate culture conditions, in some cases elicited by the loss of mesenchymal–epithelial interactions (Ramirez *et al.*, 2001; Shay and Wright, 2001). In addition to exogenous hTERT, the immortalization of such primary cells might also require a reoptimization of the culture conditions for long-term growth, the cultivation of the cells over feeder layers, or, alternatively, the use of oncogenes that can block pRB function, such as the SV40 large T antigen, HPV type 16 E7, or adenovirus type 5 E1A.

V. PITFALLS

1. When working with amphotropic retroviruses, due caution must be exercised in the production, use, and storage of recombinant viruses. Transfected φNX-A cells, viral supernatants, and all plasticwares that have been in contact with these reagents should be bleached and treated as biohazard.

2. When using ^{32}P, appropriate measures must be taken to ensure that the user remains shielded from radiation and that radioactive by-products and wastes are being contained appropriately. Working areas should also be monitored for radioactive contamination.

3. When performing the TRAP assay, precautions should be taken to limit PCR contamination. To limit such contamination, electrophoresis analysis of the samples should be run in a separate area away from the bench where samples are prepared and TRAP reactions performed.

References

Bodnar, A. G., Ouellette, M., Frolkis, M., Holt, S. E., Chiu, C. P., Morin, G. B., Harley, C. B., Shay, J. W., Lichtsteiner, S., and Wright, W. E. (1998). Extension of life-span by introduction of telomerase into normal human cells. *Science* **279**, 349–352.

Condon, J., Yin, S., Mayhew, B., Word, R. A., Wright, W. E., Shay, J. W., and Rainey, W. E. (2002). Telomerase immortalization of human myometrial cells. *Biol. Reprod.* **67**, 506–514.

Jiang, X. R., Jimenez, G., Chang, E., Frolkis, M., Kusler, B., Sage, M., Beeche, M., Bodnar, A. G., Wahl, G. M., Tlsty, T. D., and Chiu, C. P. (1999). Telomerase expression in human somatic cells does not induce changes associated with a transformed phenotype. *Nature Genet.* **21**, 111–114.

Kim, N. W., Piatyszek, M. A., Prowse, K. R., Harley, C. B., West, M. D., Ho, P. L., Coviello, G. M., Wright, W. E., Weinrich, S. L., and Shay, J. W. (1994). Specific association of human telomerase activity with immortal cells and cancer. *Science* **266**, 2011–2015.

Kiyono, T., Foster, S. A., Koop, J. I., McDougall, J. K., Galloway, D. A., and Klingelhutz, A. J. (1998). Both Rb/p16INK4a inactivation and telomerase activity are required to immortalize human epithelial cells. *Nature* **396**, 84–88.

Morales, C. P., Gandia, K. G., Ramirez, R. D., Wright, W. E., Shay, J. W., and Spechler, S. J. (2003). Characterization of telomerase immortalized normal human oesophageal squamous cells. *Gut* **52**, 327–333.

Morales, C. P., Holt, S. E., Ouellette, M., Kaur, K. J., Yan, Y., Wilson, K. S., White, M. A., Wright, W. E., and Shay, J. W. (1999). Absence of cancer-associated changes in human fibroblasts immortalized with telomerase. *Nature Genet.* **21**, 115–118.

Ouellette, M. M., Aisner, D. L., Savre-Train, I., Wright, W. E., and Shay, J. W. (1999). Telomerase activity does not always imply telomere maintenance. *Biochem. Biophys. Res. Commun.* **254**, 795–803.

Ouellette, M. M., McDaniel, L. D., Wright, W. E., Shay, J.W., and Schultz, R. A. (2000). The establishment of telomerase-immortalized cell lines representing human chromosome instability syndromes. *Hum. Mol. Genet.* **9**, 403–411.

Pear, W. S., Nolan, G. P., Scott, M. L., and Baltimore, D. (1993). Production of high-titer helper-free retroviruses by transient transfection. *Proc. Natl. Acad. Sci. USA* **90**, 8392–8396.

Ramirez, R. D., Morales, C. P., Herbert, B. S., Rohde, J. M., Passons, C., Shay, J. W., and Wright, W. E. (2001). Putative telomere-independent mechanisms of replicative aging reflect inadequate growth conditions. *Genes Dev.* **15**, 398–403.

Shay, J. W., and Wright, W. E. (2001). Aging. When do telomeres matter? *Science* **291**, 839–840.

Yang, J., Chang, E., Cherry, A. M., Bangs, C. D., Oei, Y., Bodnar, A., Bronstein, A., Chiu, C. P., and Herron, G. S. (1999). Human endothelial cell life extension by telomerase expression. *J. Biol. Chem.* **274**, 26141–26148.

Yudoh, K., Matsuno, H., Nakazawa, F., Katayama, R., and Kimura, T. (2001). Reconstituting telomerase activity using the telomerase catalytic subunit prevents the telomere shortening and replicative senescence in human osteoblasts. *J. Bone Miner. Res.* **16**, 1453–1464.

28

Preparation and Immortalization of Primary Murine Cells

Norman E. Sharpless

I. INTRODUCTION

The generation of immortal cell lines from genetically defined mice has proven useful in the understanding of molecular pathways in mammalian biology. This article describes the preparation and immortalization of murine embryo fibroblasts (MEFs), the workhorse cell type of genetically defined mice. These cells are used most frequently because of their ease of preparation, relative homogeneity, and high frequency of immortalization with serial passage in culture. These techniques can be used, with modification, to immortalize rat embryo fibroblasts as well. Furthermore, variations of these techniques can be applied to other murine cell types (e.g., astrocytes or keratinocytes) for the study of cell-type specific pathways (see Section V).

The genetics of immortalization in murine cells are now largely understood [Fig. 1, reviewed in Sharpless and DePinho (1999)]. As opposed to human cells, the spontaneous immortalization of cultured rodent cells occurs with high frequency resulting from a stochastic genetic event. This difference in frequency is attributable to the additional requirement in human cells for telomerase expression as detailed in the previous article. In murine cells, the act of cell culture, with a few exceptions, potently induces both antiproliferative products (p16^{INK4a} and p19^{ARF}) of the *Ink4a/Arf* locus, which regulate the Rb and p53 pathways, respectively (Fig. 1). Immortalization of murine cells requires circumventing at least p19^{ARF}–p53 or, in some cell types, both of the pathways regulated by the products of the *Ink4a/Arf* locus (for examples, see Bachoo *et al.*, 2002; Kamijo *et al.*, 1997; Randle *et al.*, 2001). This can be done in one of three ways: spontaneous, stochastic genetic deletion through serial culture; the use of immortalizing oncogenes; or the use of genetically defined mice. This article deals predominantly with the former method, the latter two are described in Section V.

II. MATERIALS

Sterile 10-cm dishes (Falcon Cat. No. **353003**)
Sterile 6-cm dishes (Falcon Cat. No. **353002**)
Sterile phosphate-buffered saline (PBS, GIBCO Cat. No. **14190-144**)
100 and 70% ethanol
0.25% trypsin–EDTA (GIBCO Cat. No. **25200-056**).
DMEM (with glucose and L-glutamine, GIBCO Cat. No. **11995-065**)
Heat-inactivated fetal calf serum (FCS, Sigma Cat. No. **F-2442**; to heat inactivate, thaw and then place in water bath at 56°C for 30 min)
100× penicillin/streptomycin (P/S 10,000 units/ml, GIBCO Cat. No. **15140-122**)
Optional: β-mercapoethanol (β-ME, Sigma Cat. No. **M-7522**)
Liquid nitrogen-safe cryotybes (Corning Cat. No. **2028**)
Dimethyl Sulfoxide (DMSO, Mallinkrodt AR Cat. No. **4948**)

III. INSTRUMENTATION (ALL STERILE)

6 3/4 in. Mayo scissors (Fisher Cat. No. **13-804-8**)
Large forceps (Fisher Cat. No. **13-812-36**)
Curved fine scissors (Fisher Cat. No. **08-951-10**)

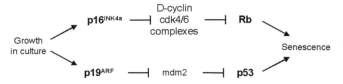

FIGURE 1 The genetics of senescence. In murine cells, the act of culture [("culture shock" (Sherr and DePinho, 2000)] induces both products of the *Ink4a/Arf* locus. p16^{INK4a} inhibits cyclin-dependent kinases 4 and 6, leading to Rb hyphophorylation and growth arrest. p19^{ARF} stabilizes p53 by inhibiting its mdm2-mediated degradation; p53 also induces growth arrest. In MEFs, the p19^{ARF}–p53 axis dominates the growth phenotype (Kamijo *et al.*, 1997; Sharpless *et al.*, 2001), although both p16^{INK4a} and p19^{ARF} contribute to senescence in other cell types (Bachoo *et al.*, 2002; Randle *et al.*, 2001).

Two curved small forceps (Fisher Cat. No. **08-953D**)
Two 5 1/2 in. Kelly clamps (Fisher Cat. No. **08-907**)
9-in. cotton-plugged Pasteur pipettes (Fisher Cat. No. **13-678-8B**).
Razor blades (place in 100% ethanol in 15-m dish prior to use)

IV. PROCEDURES

A. Murine Embryo Fibroblast Production

Day 1

1. The embryo dissection need not be performed in the hood but rather can be done on a clean bench covered with a diaper. Use sterile reagents (e.g., autoclave or flame dissecting equipment).

2. Sacrifice the pregnant mother at approximately noon on postcoital day 13.5 (see Section VI) through CO_2 inhalation. If the embryos are of potentially different genotypes (e.g., the progeny of a heterozygous intercross), regenotype the mother at this stage.

3. Place the mother on a dissection stand and clean by spraying generously with 70% ethanol.

4. Using Mayo scissors and large forceps, carefully open the abdomen, locate the embryos, and remove the bicornuate uterus with embryos. The embryos will appear as ~1-cm "beads on a string," in general there will be 6–12 embryos in a normal litter. If a mating produces an embryo with a genotype associated with developmental lethality, it is often possible to see partially resorbed embryos at this time.

5. Place uterus plus embryos in sterile PBS in 10-cm dish. Using fine forceps and/or scissors, gently dissect each balloon-like embryo with fetal membranes from the thicker uterus and place individually into a 6-cm dish with sterile PBS. If there are a large (>12) number

of embryos, place the 6-cm dishes on ice. If the fetal membranes rupture during dissection, separate the embryos from the placenta and membranes and put in a 6-cm dish in PBS. Some investigators use a dissecting microscope for this step, although we do not find this necessary.

6. Perform the rest of the dissection in a tissue culture hood.

7. Dissect away placenta and fetal membranes. Of note, the placenta is partly of maternal origin and can therefore contaminate genotyping if the mother's genotype differs from those of the embryos. To remove placenta, use a pair of fine forceps. Grasp umbilical vessels from the placenta to the embryo with one set of forceps and grasp the placenta with the other and pull in opposite directions. Pulling the placenta directly can cause the abdominal viscera to herniate and detach. The fetal membranes may still be wrapped around the embryo after placental removal, but these can be gently pulled away from the embryo after placental detachment. The fetal membranes are of embryonic origin and can be used for genotyping, although we prefer to use nonadherent cells obtained on day 2 (see later).

8. Place the entire embryo without membranes or placenta in a 10-cm dish with 1 ml 0.25% trypsin–EDTA (this is preferable to the 0.05% trypsin–EDTA commonly used to passage cells).

9. Hold a razor blade in each of two Kelly clamps and flame after dipping in 100% ethanol. Briefly cool razor blades in sterile PBS in a 10-cm dish and then use to mince the embryos. The pieces of tissue need not be too fine; chopping approximately 30 times per embryo is generally adequate. Change razor blades between embryos if the embryos are of differing genotypes.

10. Allow minced tissue to sit in trypsin–EDTA for 10 min at 37°C, 5% CO_2.

11. Add 2 ml of growth media (DMEM + 10% FCS + pen/strep with 50 μM β-ME; the β-ME is optional) and disaggregate by repeated pipetting: 10 times with a 5-ml pipette and 10 times with a plugged Pasteur pipette. After disaggregation, there should be no large (<1 mm in largest dimension) chunks of tissue remaining.

12. Add 8 ml of growth media and swirl plates for even seeding. We grow cells at this stage in an incubator dedicated for primary cells in the rare event of bacterial contamination.

Day 2

1. The 10-cm dish should be 100% confluent after attachment of MEFs overnight. If subconfluent, this generally indicates poor disaggregation of the embryo,

improper media, or bacterial contamination (see Section VI).

2. Gently remove the media and save nonadherent cells for genotyping (see later). Take care as the cells are poorly adherent at this stage and large chunks of the monolayer can detach with poor handling.

3. Add 10 ml of fresh growth media.

4. If genotyping is necessary (i.e., the embryos are of different genetic composition), this can be done on DNA prepared from nonadherent cells collected on day 2. This is done by pelleting these cells in 15-cm conical tubes (1000 rpm × 5 min), detergent lysis, and precipitation of DNA. The exact method used depends on the manner of genotyping (e.g., polymerase chain reaction vs Southern blot). One can easily obtain >100 μg of good-quality DNA from these pellets if needed. We prepare using a commercially available DNA extraction kit (Promega #A1125).

Day 3

1. The culture should be superconfluent on day 3 and ready to freeze down or passage for further experiments. If cells are less than 100% confluent by day 3, freeze down on day 4.

2. To freeze, wash cells once with sterile PBS and add 1 ml of 0.25% trypsin–EDTA. Cells will detach in 5–10 min and then add to a 15-ml conical tube with 10 ml of growth media.

3. Spin cells at 1000 rpm for 5 min.

4. Resuspend cell pellet in 3 ml of ice-cold freezing media (90% FCS + 10% DMSO) and put in three well-labeled, chilled cryovials. We label with date prepared, the line number, and the passage number upon thawing. The lines are numbered as the mother's number first, followed by the embryo number (e.g., 115-7 would be the seventh embryo of mother 115). The cells will be passage 2 upon thawing if not passaged further before freezing.

5. Keep cells on ice for 10–15 minutes and then put in a −80°C freezer in a sealed styrofoam block overnight. Alternatively, the cells can be placed in a cryo-freezing container (Nalgene Cat. No. **5100-001**) and then placed in a −80°C freezer overnight.

6. The day after freezing, cells should be moved to a liquid nitrogen tank for long-term storage.

B. Thawing MEFs

1. Keep cells on dry ice until absolutely ready to thaw.

2. Place the vial of cells in a 37°C water bath and watch carefully. As soon as the cells begin to thaw, pour the entire pellet into a 15-ml conical tube.

3. Add 10 ml of growth media dropwise over about 30 s with gentle mixing while adding. Mix well after all 10 ml of media is added.

4. Spin cells down (1000 rpm for 5 min), resuspend in 3 ml media, and plate in a 10-cm dish. One vial of frozen cells can go into one 10-cm dish; add another 7 ml media.

5. The cells should be 90–100% confluent the day after thawing. If confluence is less with large numbers of floating (dead) cells noted, this suggests improper freezing, improper storage (e.g., freezer malfunction), or improper thawing (see Section VI).

C. Immortalizing MEFs

This can be done using a rigorously defined passaging protocol (e.g., 3T3 or 3T9 assay) or, more simply, by serial replating of cells. The advantage of the passaging protocol is that one learns about the growth kinetics and immortalization frequency of the cell (see Section V). The immortalization of wild-type MEFs requires a stochastic genetic event: generally p53 loss or, less frequently, loss of p19^{ARF} (Kamijo et al., 1997). Other immortalizing events, such as mdm2 amplification (Olson et al., 1993), have also been noted, but these are considerably rarer. In our experience, this genetic event is most likely to be loss of p53 (~60–70% of immortalized lines) or p19^{ARF} (~20–30%) regardless of growth conditions (passage at high or low density). Therefore, serial passaging without counting cells at every passage is an appropriate way to derive immortalized cells. In fact, the "3T3 cell" has become a nondescript term referring to immortal murine embryo fibroblasts obtained through serial passaging, but few laboratories except those interested in growth and senescence do rigorous 3T3 assays to obtain them.

D. Immortalization by Serial Passage

1. Split MEFs 1:3 twice per week during the rapid growth phase of the culture (the first 6–10 passages).

2. As cells enter senescence, split 1:2 or merely replate (aim to keep cells near 100% confluency on the day of passaging). If the cultures are seeded too sparsely, this will decrease the immortalization frequency.

3. Between passages 10 and 20 (4 to 9 weeks in culture), immortalized lines will begin to overgrow the culture. These cells are smaller and spindle shaped, initially appearing as small nests of cells obvious on the day of passaging. These cells can be subcloned, but are used more frequently as pooled immortalized cells.

E. 3T9 Assay

This assay was originally described by Todaro and Green (1963). The "3" refers to passaging every 3 days, and the "9" refers to 9×10^5 cells plated per passage. 3T3 (3×10^5 cells) and 3T12 (1.2×10^6 cells) assays can also be done to study growth and immortalization at varying densities. In our experience, immortalization frequency is highest for the 3T3 assay (and therefore this assay has lent its name to all immortalized lines of murine fibroblasts), whereas 3T9 is more useful for measuring immortalization frequency (Sharpless *et al.*, 2001). As stated earlier, serial replating of murine fibroblasts at almost any density will eventually yield immortalized lines as long as the cells are not seeded too sparsely.

1. For the first passage, wash cells growing in a 10-cm dish (usually passage 2 at this stage) with PBS, trypsinize (0.05% trypsin–EDTA is adequate for this purpose), dilute into 10 ml of growth media, and count.

2. Spin down and resuspend cells at a concentration of 9×10^5 cells/ml. Seed 1 ml of cells (9×10^5 cells) into a 6-cm dish and add 2.5 ml of media. Swirl the cells to ensure uniform seeding. Label the dish with the MEF line number and the passage number.

3. For subsequent passages, trypsinize and recount cells every 3 days in a manner identical to steps 1 and 2. Record the number of cells counted before each passage. Growth curves and immortalization frequencies can be determined from analysis of these data as described in Section V.

4. Immortalized lines will emerge between passages 10 and 20; virtually all lines proliferating after 20 passages will be immortal. If necessary, these lines can be subcloned by limiting dilution, but this is not generally done. The majority of immortalized lines will have lost either p53 or p19^ARF function (Kamijo *et al.*, 1997), although various other genetic events can increase or decrease the frequency of immortalization (Frank *et al.*, 2000; Jacobs *et al.*, 1999; Kamijo *et al.*, 1999; Sharpless *et al.*, 2001).

V. COMMENTS

A. Immortalization through the Use of Oncogenes or Genetic Background

A problem with immortalization by serial passage is that the nature of the immortalizing genetic event is not known, although predominantly p53 or p19^ARF is inactivated as MEFs escape senescence. As the behavior of p19^ARF-deficient cells may be quite different from p53-deficient cells, however, the stochastic inactivation of these pathways may produce significant line-to-line variability and therefore be undesirable. For example, *p53* null MEFs are more aneuploid and more resistant to DNA damage than *p19^ARF* null MEFs (Kamijo *et al.*, 1997; Pomerantz *et al.*, 1998; Serrano *et al.*, 1996; Stott *et al.*, 1998). To assure that all lines will evade senescence via a similar mechanism, cells can either be immortalized through the use of an immortalizing oncogene or by using cells derived from animals of a genetic background that resists senescence.

Classically, the most commonly used oncoprotein to immortalize cells is the SV40 large T antigen (Colby and Shenk, 1982; Jat and Sharp, 1986; Todaro and Green, 1966). This molecule inactivates the p53 and Rb pathway, and the majority of murine cell types can be immortalized by transfection or retroviral transduction of T Ag. A disadvantage of T Ag, however, is that cells expressing this molecule are unstable and are prone to clonal *in vitro* evolution. Furthermore, as Rb pocket proteins are required for the differentiation of many cells types (Dannenberg *et al.*, 2000), T Ag generally impairs or precludes the study of differentiation. Alternatively, MEFs and several other murine cell types can be immortalized with a dominant-negative form of p53 [e.g., p53-DD (Bowman *et al.*, 1996)], which preserves the Rb pathway, although these cells are still more prone to aneuploidy than p19^ARF-deficient lines.

The most elegant method of obtaining immortal cell lines is by deriving cells from animals resistant to senescence. The most commonly used strains for this purpose are *Ink4a/Arf*-deficient (Serrano *et al.*, 1996), *p19^ARF*-deficient (Kamijo *et al.*, 1997), or *p53*-deficient (Donehower *et al.*, 1992) mice. These strains are widely available and can be obtained from the mouse models of the human cancer consortium (MMHCC http://web.ncifcrf.gov/researchresources/mmhcc/). For studies of genetically defined animals, the genetic background of interest can be crossed two generations to mice of these backgrounds and then MEFs (or other cell types) derived as described earlier, which will be immortal in most cases if derived from p53- or p19^ARF-deficient embryos. This method can be employed to derive immortal cells from difficult genetic backgrounds that undergo premature senescence in culture (Frank *et al.*, 2000; Jacobs *et al.*, 1999). This strategy can also be employed to obtain cells of nonfibroblast lineages. For example, immortal melanocytes (Chin *et al.*, 1997), skin keratinocytes (unpublished observations), glia (Bachoo *et al.*, 2002), lymphocytes (unpublished observations; Randle *et al.*, 2001), and macrophages (Randle *et al.*, 2001) have been derived successfully from *p19^ARF*- or *Ink4a/Arf*-deficient mice using standard culture methods for these cell types. To immortalize with high efficiency, cells must be homozygous null for p53, *p19^ARF*, or *Ink4a/Arf*; therefore, the principal disadvantage of this approach is the extra time needed to backcross to these defined genetic backgrounds.

B. Data Analysis of 3T9 or 3T3 Assay

These data can be plotted as cell number per passage or population doublings (PDs) per passage (Fig. 2). PDs for any given passage = $\log_2(N_f/N_0)$ = $1.44 * \ln(N_f/N_0)$, where N_f = number of cells counted at the end of the passage and N_0 = number of cells seeded at the beginning of the passage (i.e., 9×10^5 for a 3T9 assay). For the purpose of immortalization frequency, "senescence" occurs if less than 9×10^5 cells are recovered for two consecutive passages. The immortalization frequency = 1-number of senescent lines/total number of lines analyzed. Measured in this way, the immortalization frequency of wild-type MEFs can vary significantly depending on culture conditions, method of embryo preparation, and so on and therefore is only meaningful when compared to proper littermate control embryos analyzed concurrently.

VI. PITFALLS

1. *Embryo dates.* By convention, the day a coital plug is detected is day 0.5 for timed matings, and embryos should be made at midday 13 days later. While it is difficult to be sure of correct plugging dates at the time of dissection, in general day 13.5 embryos have paddle-like front paws, whereas day 14.5 embryos have more fully formed individual digits. Embryos that are too large and well developed or too small suggest an incorrect date of plugging. Cells of different embryonic ages do differ in several *in vitro* growth properties; therefore, littermate controls are always preferable in MEF experiments. In our experience, MEFs from embryos older than 13.5 grow less well and immortalize less frequently than 13.5 embryos.

2. *MEFs are not confluent the day after dissection.* In general, cells from a single embryo should cover a 10-cm plate fully the morning after plating. Poor coverage of the dish can result from inadequate embryo disaggregation, tissue mincing with razor blades that were not cooled properly after flaming, improper media, or bacterial contamination. In particular, one should be vigilant for bacterial contamination as this can be difficult to note given that there is copious debris in the culture 1 day after plating.

3. *MEFs are not confluent the day after thawing.* This results most often from freezer or liquid nitrogen tank malfunction, but can also be due to improper technique when the cells were frozen, malfunction of the cryo-freezing container, or improper thawing. DMSO is toxic to these primary cells so it is important to resuspend the frozen cell pellet well in growth media, spin the cells down, and then respsund in fresh growth media prior to plating. Thawed vials of cells should

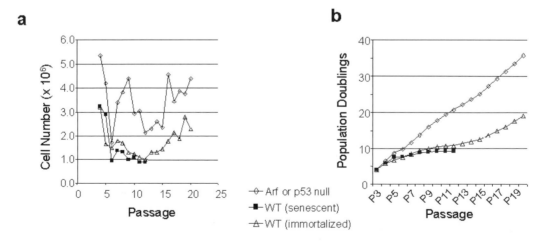

FIGURE 2 The 3T9 assay can be used to quantify both growth and immortalization frequency. Data from a 3T9 (or 3T3) assay can be graphed in either of two ways: (a) as cell number per passage or (b) as population doublings (PDs) per passage (PDs defined in text). The same data are graphed by either method showing a senescent line, an immortalized line, or a line lacking *Ink4a/Arf*. The p19^ARF–dependent slow growth period seen in wild-type MEFs between passages 5–15 is called "senescence," although in actuality it represents a culture-induced phenomena.

not be left in freezing media for a significant period of time prior to replating.

4. *Cells grow poorly and/or fail to immortalize.* This can result from poor growth media (e.g., the fetal calf serum is too old), the use of embryos significantly later than E13.5, or occult pathogen contamination. Embryos of certain genetic backgrounds grow poorly in culture ("premature senescence"), which can sometimes be obviated by backcrossing to *Ink4a/Arf-*, *p19^{ARF}-*, or *p53*-deficient animals (Frank *et al.*, 2000; Jacobs *et al.*, 1999; Kamijo *et al.*, 1999).

References

Bachoo, R. M., Maher, E. A., Ligon, K. L., Sharpless, N. E., Chan, S. S., You, M. J., Tang, Y., DeFrances, J., Stover, E., Weissleder, R., *et al.* (2002). Epidermal growth factor receptor and Ink4a/Arf: Convergent mechanisms governing terminal differentiation and transformation along the neural stem cell to astrocyte axis. *Cancer Cell* **1**, 269–277.

Bowman, T., Symonds, H., Gu, L., Yin, C., Oren, M., and Van Dyke, T. (1996). Tissue-specific inactivation of p53 tumor suppression in the mouse. *Genes Dev.* **10**, 826–835.

Chin, L., Pomerantz, J., Polsky, D., Jacobson, M., Cohen, C., Cordon-Cardo, C., Horner, J. W., II, and DePinho, R. A. (1997). Cooperative effects of INK4a and ras in melanoma susceptibility *in vivo*. *Genes Dev.* **11**, 2822–2834.

Colby, W. W., and Shenk, T. (1982). Fragments of the simian virus 40 transforming gene facilitate transformation of rat embryo cells. *Proc. Natl. Acad. Sci. USA* **79**, 5189–5193.

Dannenberg, J. H., van Rossum, A., Schuijff, L., and te Riele, H. (2000). Ablation of the retinoblastoma gene family deregulates G(1) control causing immortalization and increased cell turnover under growth- restricting conditions. *Genes Dev.* **14**, 3051–3064.

Donehower, L. A., Harvey, M., Slagle, B. L., McArthur, M. J., Montgomery, C. A., Jr., Butel, J. S., and Bradley, A. (1992). Mice deficient for p53 are developmentally normal but susceptible to spontaneous tumours. *Nature* **356**, 215–221.

Frank, K. M., Sharpless, N. E., Gao, Y., Sekiguchi, J. M., Ferguson, D. O., Zhu, C., Manis, J. P., Horner, J., DePinho, R. A., and Alt, F. W. (2000). DNA ligase IV deficiency in mice leads to defective neurogenesis and embryonic lethality via the p53 pathway. *Mol. Cell* **5**, 993–1002.

Jacobs, J. J., Kieboom, K., Marino, S., DePinho, R. A., and van Lohuizen, M. (1999). The oncogene and Polycomb-group gene bmi-1 regulates cell proliferation and senescence through the ink4a locus. *Nature* **397**, 164–168.

Jat, P. S., and Sharp, P. A. (1986). Large T antigens of simian virus 40 and polyomavirus efficiently establish primary fibroblasts. *J. Virol.* **59**, 746–750.

Kamijo, T., van de Kamp, E., Chong, M. J., Zindy, F., Diehl, J. A., Sherr, C. J., and McKinnon, P. J. (1999). Loss of the ARF tumor suppressor reverses premature replicative arrest but not radiation hypersensitivity arising from disabled atm function. *Cancer Res.* **59**, 2464–2469.

Kamijo, T., Zindy, F., Roussel, M. F., Quelle, D. E., Downing, J. R., Ashmun, R. A., Grosveld, G., and Sherr, C. J. (1997). Tumor suppression at the mouse INK4a locus mediated by the alternative reading frame product p19ARF. *Cell* **91**, 649–659.

Olson, D. C., Marechal, V., Momand, J., Chen, J., Romocki, C., and Levine, A. J. (1993). Identification and characterization of multiple mdm-2 proteins and mdm-2-p53 protein complexes. *Oncogene* **8**, 2353–2360.

Pomerantz, J., Schreiber-Agus, N., Liegeois, N. J., Silverman, A., Alland, L., Chin, L., Potes, J., Chen, K., Orlow, I., Lee, H. W., *et al.* (1998). The Ink4a tumor suppressor gene product, p19Arf, interacts with MDM2 and neutralizes MDM2's inhibition of p53. *Cell* **92**, 713–723.

Randle, D. H., Zindy, F., Sherr, C. J., and Roussel, M. F. (2001). Differential effects of p19(Arf) and p16(Ink4a) loss on senescence of murine bone marrow-derived preB cells and macrophages. *Proc. Natl. Acad. Sci. USA* **98**, 9654–9659.

Serrano, M., Lee, H., Chin, L., Cordon-Cardo, C., Beach, D., and DePinho, R. A. (1996). Role of the INK4a locus in tumor suppression and cell mortality. *Cell* **85**, 27–37.

Sharpless, N. E., Bardeesy, N., Lee, K. H., Carrasco, D., Castrillon, D. H., Aguirre, A. J., Wu, E. A., Horner, J. W., and DePinho, R. A. (2001). Loss of p16Ink4a with retention of p19Arf predisposes mice to tumorigenesis. *Nature* **413**, 86–91.

Sharpless, N. E., and DePinho, R. A. (1999). The INK4A/ARF locus and its two gene products. *Curr. Opin. Genet. Dev.* **9**, 22–30.

Sherr, C. J., and DePinho, R. A. (2000). Cellular senescence: Mitotic clock or culture shock? *Cell* **102**, 407–410.

Stott, F. J., Bates, S., James, M. C., McConnell, B. B., Starborg, M., Brookes, S., Palmero, I., Ryan, K., Hara, E., Vousden, K. H., and Peters, G. (1998). The alternative product from the human CDKN2A locus, p14(ARF), participates in a regulatory feedback loop with p53 and MDM2. *EMBO J.* **17**, 5001–5014.

Todaro, G. J., and Green, H. (1963). Quantitative Studies of mouse embyo cells in cutture and their development into established linos. *J. Cell Biol.* **17**, 299–313.

Todaro, G. J., and Green, H. (1966). High frequency of SV40 transformation of mouse cell line 3T3. *Virology* **28**, 756–759.

Somatic Cell Hybrids

29

Viable Hybrids between Adherent Cells: Generation, Yield Improvement, and Analysis

Doris Cassio

I. INTRODUCTION

Somatic cell hybridization was discovered and introduced by Barski *et al.* (1960) and Sorieul and Ephrussi (1961). This technique allows one to examine the result of introducing various genomes in different functional states and from different species into the *same* cell. Hybrid cells have been widely used in various fields (genetics, cell biology, tumour biology, virology) and the most famous hybrids are hybridomas.

One important application of somatic cell hybridization is chromosomal gene assignment. Breeding analysis, which is effective for this purpose in lower animals and plants, is too slow in mammals (even in mice the generation time is about 3 months) and is impossible in humans. Gene mapping techniques based on somatic cell genetics have been central to the study of human genetics. In 1968, only two genes had been mapped to specific autosomes, and a decade later this number had risen to 300, mostly using human–rodent somatic cell hybrids. Such hybrids present the advantage to retain only a few human chromosomes and they are now currently used as donor cells in irradiation and fusion gene transfer (IFGT) experiments for constructing detailed genetic maps (Walter and Goodfellow, 1993).

Cell fusion was also used to analyse how specialized cells acquire and maintain their differentiation. The activities of somatic cells can be divided into two main categories: essential or ubiquitous functions that are indispensable for cell survival and growth and

"luxury" or differentiated functions. Essential functions continue to be expressed in hybrids, whereas differentiated functions are subject to different regulations (expression, extinction, activation) depending on the histogenetic nature of the parental cells that have been fused (see examples in Cassio and Weiss, 1979; Hamon-Benais *et al.*, 1994; Killary and Fournier, 1984; Mevel-Ninio and Weiss, 1981). Cell determination is, however, not modified in hybrids, as extinction or activation requires retention of the chromosomes coding for the appropriate regulatory factors.

Somatic cell hybridization has not only shown that tissue-specific genes are regulated by *trans*-acting factors, but has provided strong evidence for the existence of tumour suppressor genes (Anderson and Stanbridge, 1993). Cell fusion experiments have also demonstrated that cellular senescence is a dominant active process and that several genes or genes pathways are implicated in the senescence program (Goletz *et al.*, 1994).

Spontaneous fusion of cells in culture occurs at a very low frequency. To obtain hybrid cells, inactivated Sendai virus or, more commonly, polyethylene glycol (PEG), which was introduced by Pontecorvo (1976), is used as the fusogen. Cell hybridization has also been performed by electrofusion on filters (Ramos *et al.*, 2002). The inital products of fusion contain within a common cytoplasm two or more distinct nuclei from one single parent (homokaryons) or from both (heterokaryons). Only a very small proportion of these polykaryons will progress to nuclear fusion and then through mitosis. Moreover, the first divisions of the heterokaryons and of their daughter cells often fail because of abnormalities of the mitotic spindle and abnormal chromosome movements. The formation of viable hybrids from heterokaryons is thus a rare event, and the use of selective methods that favour the survival of the hybrids at the expense of the parental cells is often a requisite. These selective methods are also necessary because hybrid cells often grow more slowly than parental cells and are rapidly overgrown by parental cells.

A. Selective Methods

The best known of such methods is the application of hypoxanthine + aminopterin + thymidine (HAT) selection (Littlefield, 1964) for the fusion of cells deficient in hypoxanthine guanosine phosphoryl transferase (HGPRT⁻) with cells deficient in thymidine kinase (TK⁻), but different combinations of selectable markers can be used (Hooper, 1985), provided that the two selective systems do not interfere. If the lines that are fused have no selective markers, a good strategy is to select sequentially for HGPRT deficiency (thioguanine resistance) and ouabain resistance in one parental cell line. Then this marked cell line may be fused with any unmarked cell line and hybrids selected in HAT + ouabain (Jha and Ozer, 1976). Other couples of selective markers, such as TK deficiency (5–bromo-2′deoxyuridine resistance) and neomycine resistance, can also be used. For producing primate–rodent hybrids, the selection of an HGPRT⁻ rodent parent is sufficient because rodent cells are more resistant to ouabain than primate cells. Moreover, hybrid cells can also be isolated on the basis of their size, morphology, growth parameters, and DNA content.

B. Yield of Viable Hybrids

Whatever the method used to isolate hybrids, the most important is to optimize the fusion conditions in order to obtain a number of viable hybrids as high as possible. In the best cases the fusion of several millions of parental cells leads to the formation of only a few hundred hybrids and often the yield of viable hybrids is much lower, as illustrated in Table I for hepatoma-derived hybrids. The protocol described here has been used routinely to produce large amounts of hybrid clones between differentiated rat hepatoma cells and various cells of different histogenetic origin and of different species, particularly mouse and human fibroblasts (Mevel-Ninio and Weiss, 1981; Sellem et al., 1981). Moreover, some of the hybrids obtained have been used themselves as partners of fusion and new hybrids were generated successfully using exactly the same method (Bender et al., 1999; Hamon-Benais et al., 1994). The most important parameters in fusion experiments are the yield of viable and growing hybrids and their stability. Thus it is recommended to define optimal fusion conditions and to vary different parameters, particularly the ratio of parental cells, for improving the yield. It is also recommended to analyze regularly hybrid clones for their phenotype and chromosomal content.

II. MATERIALS

PEG 1000 ultrapure (Merck, Cat. No. 9729)
Trypsin (pig pancreas; United States Bioch. Corp., Cleveland, OH, Cat. No. 22715)

TABLE I   Production of Rat Hepatoma-Derived Hybrids[a]

Partner of fusion	Hybrid yield (10^{-6})	Reference
Normal diploid human fibroblast	<1	Sellem et al. (1981)
Normal diploid mouse fibroblast	5–8	Mevel-Ninio and Weiss (1981) Killary and Fournier (1984)
Normal rat hepatocytes	6	Polokoff and Everson (1986)
Mouse fibroblastic line cl1-D	1000	Mevel-Ninio and Weiss 1981
Mouse hepatoma BW1-J	100	Cassio and Weiss (1979)
Rat hepatoma–mouse fibroblast hybrid	400	Hamon-Benais et al. (1994)
Rat hepatoma–mouse fibroblast monochromosomal hybrid	100	Hamon-Benais et al. (1994)
Rat hepatoma–human fibroblast segregated hybrid	30–300	Bender et al. (1999)

[a] In all cases, parental cells were fused at pH 7.2 in a ratio 1:1, except for the fusion with hepatocytes (hepatoma:hepatocytes = 1:2). The same rat hepatoma parental cells were used in all experiments.

Complete growth medium (available from local suppliers)

Serum-free growth medium (available from local suppliers)

Selective complete growth medium (available from local suppliers)

35- and 50-mm tissue culture dishes (Falcon, Cat. No. 3001, 3002)

15-ml tube (Falcon, Cat. No. 352099)

22 × 22-mm sterile glass coverslips

III. PROCEDURES

A. Before Fusion

Solutions

1. *50% PEG (for fusion)*: Autoclave PEG 1000. This both liquifies and sterilizes the PEG. Cool it to 50°C and mix with an equal volume of sterile serum-free medium prewarmed for a short period at 50°C. Adjust, if necessary, to the desired pH with $1.0 M$ NaOH (range of pH generally used is 7.2–7.9). This solution can be stored at 4°C for up to 2 weeks.

2. *0.25 and 0.05% trypsin (to detach fusion products)*: To make 100 ml of solution, solubilize 0.8 g of NaCl, 0.04 g of KCl, 0.058 g of $NaHCO_3$, 0.1 g dextrose, and 0.25 g (0.25%) or 0.05 g (0.05%) of trypsin. Complete to 100 ml distilled water. Incubate at 37°C for 1–2 h. Sterilize by filtration and store at 4°C (rapid use) or –20°C.

Step

Grow parental cells in *nonselective medium* for a short period.

B. Fusion

Steps

1. Inoculate the mixture of parental cells to be fused into several 50-mm tissue culture dishes containing complete (nonselective) growth medium. The total number of cells has to be adjusted to occupy all the dish surface, such that the fusion will be done on cells that are in close contact. For 50-mm petri dishes, the total cell number could vary from 5×10^5 to 4×10^6 depending on the density at confluence of the cell lines used. Although equal numbers of parental cells are generally recommended, use different ratios of parental cells (see Table II and Section IV).

2. Incubate the mixed cultures a few hours (4 h to overnight). This allows the cells to adhere to the support and to establish contacts with neighbors.

3. Warm the serum-free medium and the 50% PEG solution to 37°C.

4. Remove the medium thoroughly from the culture and wash once with 5 ml of serum-free medium.

5. Add gently 3.0 ml of PEG 50% all over the cell layer.

6. After 45 s aspirate the PEG.

7. Exactly 1 min after PEG treatment, add 5 ml of serum-free medium.

8. Aspirate half the medium and add 2.5 ml of serum-free medium.

9. Repeat step 8 four times.

10. Aspirate all the medium.

11. Add 5 ml of complete growth medium and let the cells recover for at least 2 h and no more than 12 h.

Notes: Steps 4 to 11 must be done *dish per dish*. Some control dishes must be included. They will be treated

TABLE II   Production of Rat Hepatoma × Human Fibroblast Hybrids[a]

Ratio of parental cells[b] (hepatoma:fibroblast)	Fusion conditions	Selection conditions	Hybrid yield[c] (10^{-6})
1:1	PEG, pH 7.2	HAT, ouabain, pH 7.2	<1
4:1			5
10:1			20
1:1	PEG, pH 7.8	HAT, ouabain, pH 7.8[d]	15
4:1			30
10:1			150
10:1	Sendai virus, pH 7.8	HAT, ouabain, pH 7.8[d]	100
1:1	No fusogen, pH 7.8	HAT, ouabain, pH 7.8[d]	<1
	No fusogen, pH 7.2	HAT, ouabain, pH 7.2	<0.5

[a] For details of the parental cell lines, see Sellem *et al.* (1981).

[b] 2×10^6 cells were plated per 50-mm petri dish.

[c] Total number of hybrid clones/total cell number of minority parent.

[d] Cells were maintained at pH 7.8 for 6 days after fusion and then cultured at pH 7.2, as usual.

as the others except that the PEG solution will be replaced by serum-free medium.

C. After Fusion

Steps

1. Aspirate the medium and add 3 ml of 0.25% trypsin per dish.

2. As soon as the cell layer begins to detach, add 3 ml of 0.05% trypsin and detach the cells by repeated pipetting.

3. Add the cell suspension in a 15-ml tube containing 2 ml of complete growth medium (to arrest the trypsin action). If necessary, rinse the dish with 2 ml medium and add this medium to the tube containing the cells.

4. Centrifuge the cells at $500g$ (1500 rpm) for 5 min at room temperature.

5. Resuspend thoroughly the cell pellet in complete growth medium and count the cell number in a hemacytometer. Using the described protocol, generally at least 80% of the cells are recovered after fusion.

6. Pool, if necessary, cells recovered from identical dishes and inoculate different numbers of cells (10^3–10^6) in either culture dishes or on 22-mm glass coverslips (in 35-mm dishes).

7. Incubate the cells at least overnight (eventually a few days) in complete growth medium before adding selective medium that will kill the parental cells and let the hybrid cells survive.

8. At regular intervals, watch for the appearance of growing hybrid colonies. Count their number in dishes or coverslips that contain well-isolated colonies. From this number the yield of growing hybrids can be calculated. For one fusion, this yield is equal to the total number of hybrid clones obtained divided by the total cell number of the minority parent engaged in the fusion.

9. Use dishes that contain a small number of colonies to isolate independent hybrid clones (one per dish will be scraped, subcultured, characterized, as soon as possible, frozen, and recharacterized). From dishes that contain a lot of colonies, if this situation arises, hybrid cell populations can be obtained in mass and studied rapidly.

10. Use glass coverslips to control by cytogenetic methods the hybrid nature of the clones and to test if their chromosomal content is stable with time in culture. The karyotyping *in situ* method, described by Worton and Duff (1979), is highly recommended. This method is easy, of general application for adherent cells, and can be performed on small colonies even a few generations after fusion (Fig. 1). Glass coverslips

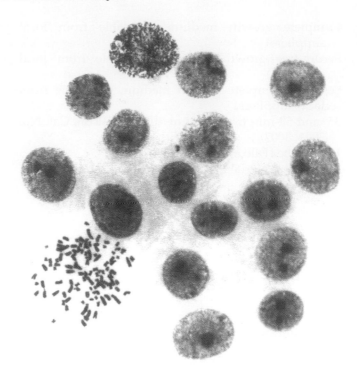

FIGURE 1 Karyotypic analysis of emerging hybrids. The *in situ* karyotyping method (Worton and Duff, 1979) was performed on rat hepatoma-derived hybrid colonies 8 days after fusion. The colony shown was composed of 17 cells, 1 of which was in metaphase. This metaphase contains 100 chromosomes, corresponding to the expected sum of the mean chromosome number of each parent (46 and 52, respectively).

can also be used for phenotypic characterization of hybrid clones.

IV. COMMENTS

The production of hybrid clones in large amounts depends greatly on the parental cells (cell type and growth capacity) and on the fusion conditions. These two points are illustrated in Tables I and II for rat hepatoma-derived hybrids. The frequency of occurrence of hybrids between rat hepatoma and normal fibroblast was particularly low compared to other hepatoma-derived hybrids (Table I). Consequently, various fusion conditions were tested. The use of unbalanced ratios of parental cells is one of the most important parameters to improve the hybrid yield (Table II). Therefore, to save time and to obtain the highest number of hybrid cells, it is recommended to fuse parental cells in different ratios.

Mixed parental cell populations that have not been treated by PEG can give rise to colonies that grow in selective medium at low frequency. These colonies could be either spontaneous hybrids or revertants from parental cells. The isolation of revertants is one of the most common difficulties that may arise in selecting hybrids. Therefore the hybrid nature of the cells selected has to be verified by checking their chromosomal content.

V. PITFALLS

1. Some PEG preparations produce enormous lethality, whereas ultrapure PEG from Merck results in acceptable levels of lethality (generally below 20%).

2. Detach and dissociate very carefully cells after fusion (avoid aggregates).

3. Because it is impossible to predict if hybrid clones will be produced with a high yield or not, once fusion has been performed and the products of fusion detached, inoculate them at different concentrations (that could cover a 1000× range) such that well-isolated hybrids could be obtained even if the yield varies from 10^{-6} to 10^{-3}.

4. A lower number of hybrid clones is obtained if the cells are not detached after fusion.

5. A lower number of hybrid clones is obtained if the mixture of parental cells is fused in suspension. This result is due to the fact that the relative proportion of binucleate heterokaryons is higher in monolayer fusion, whereas suspension fusion favours the formation of giant heterokaryons that die soon after fusion. Hence, there is no reason and no advantage to fuse adherent cells in suspension as often recommended (except if one of the parent grows in suspension).

Acknowledgments

I thank M. C. Weiss for training in cell culture and C. H. Sellem and C. Hamon-Benais for the illustrations.

References

Anderson, M. J., and Stanbridge, E. J. (1993). Tumor suppressor genes studied by cell hybridization and chromosome transfer. *FASEB J.* **7**, 826–833.

Barski, G., Sorieul, S., and Cornefert, F. (1960). Production dans des cultures *in vitro* de deux souches cellulaires en association, de cellules de caractère "hybride". *C. R. Acad. Sci. Paris* **251**, 1825–1827.

Bender, V., Bravo, P., Decaens, C., and Cassio, D. (1999). The structural and functional polarity of the hepatic human-rat hybrid WIF-B is a stable and dominant trait. *Hepatology* **30**, 1002–1010.

Cassio, D., and Weiss, M. C. (1979). Expression of fetal and neonatal hepatic functions by mouse hepatoma-rat hepatoma hybrids. *Som. Cell Genet.* **5**, 719–738.

Goletz, T. J., Smith, J. R., and Pereira-Smith, O. M. (1994). Molecular genetic approaches to the study of cellular senescence. *Cold Spring Harb. Symp. Quant. Biol.* **LIX**, 59–66.

Hamon-Benais, C., Delagebeaudeuf, C., Jeremiah, S., Lecoq, O., and Cassio, D. (1994). Efficiency of a specific albumin extinguisher locus in monochromosomal hepatoma hybrids. *Exp. Cell Res.* **213**, 295–304.

Hooper, M. (1985). In *"Mammalian Cell Genetics"* (E. Bittar, ed.), pp. 77–81. Wiley-InterScience, New York.

Jha, K. K., and Ozer, H. L. (1976). Expression of transformation in cell hybrids. I. Isolation and application of density-inhibited Balb/3T3 cells deficient in hypoxanthine phosphoribosyl transferase and resistant to ouabain. *Som. Cell Genet.* **2**, 215–233.

Killary, A. M., and Fournier, R. E. K. (1984). A genetic analysis of extinction: Trans-dominant loci regulate expression of liver-specific traits in hepatoma-hybrid cells. *Cell* **38**, 523–534.

Littlefield, J. W. (1964). Selection of hybrid from matings of fibroblasts *in vitro* and their presumed recombinants. *Science* **145**, 709–710.

Mevel-Ninio, M., and Weiss, M. C. (1981). Immunofluorescence analysis of the time-course of extinction, reexpression and activation of albumin production in rat hepatoma-mouse fibroblast heterokaryons and hybrids. *J. Cell Biol.* **90**, 339–350.

Polokoff, M. A., and Everson, G. T. (1986). Hepatocyte-hepatoma cell hybrids: Characterization and demonstration of bile acid synthesis. *J. Biol. Chem.* **261**, 4085–4089.

Pontecorvo, G. (1976). Production of indefinitely multiplying mammalian somatic cell hybrids by polyethylene glycol (PEG) treatment. *Somat. Cell Genet.* **1**, 397–400.

Ramos, C., Bonenfant, D., and Teissie, J. (2002). Cell hybridization by electrofusion on filters. *Anal. Biochem.* **302**, 213–219.

Sellem, C. H., Cassio, D., and Weiss, M. C. (1981). No extinction of tyrosine aminotransferase inducibility in rat hepatoma-human fibroblast hybrids containing the human × chromosome. *Cytogenet. Cell Genet.* **30**, 47–49.

Sorieul, S., and Ephrussi, B. (1961). Karyological demonstration of hybridization of mammalian cells *in vitro*. *Nature (Lond.)* **190**, 653–654.

Walter, M. A., and Goodfellow, P. N. (1993). Radiation hybrids: Irradiation and fusion gene transfer. *Trends Genet.* **9**, 352–356.

Worton, R. G., and Duff, C. (1979). Karyotyping. *Methods Enzymol.* **58**, 322–344.

Suggested Reading

Harris, H. (1995). "The Cells of the Body: A History of Somatic Cell Genetics." Cold Spring Harbor Laboratory Press, Cold Spring Harbor, NY.

Ringertz, N. R., and Savage, R. E., (1976). "Cell Hybrids." Academic Press, London.

Zallen, D. T., and Burian, R. M. (1992). On the beginnings of somatic cell hybridization: Boris EPHRUSSI and chromosome transplantation. *Genetics* **132**, 1–8.

Cell Separation Techniques

30

Separation and Expansion of Human T Cells

Axl Alois Neurauter, Tanja Aarvak, Lars Norderhaug, Øystein Åmellem, and Anne-Marie Rasmussen

I. INTRODUCTION

For the characterization of specific cell types and the investigation of their functions, it is essential that the cells can be purified. Cell separation techniques based on the use of antibody-coated magnetic beads, e.g., Dynabeads (Ugelstad *et al.*, 1980, 1994), are now widely used in research and clinical laboratories. Specific cells can, after binding to the magnetic beads, be selected by the use of a magnet and, following brief washing, high cell purity can be achieved. This technique continues to encompass new fields for the selective isolation of eukaryotic cells (Funderud *et al.*, 1987; Luxembourg *et al.*, 1998; Marquez *et al.*, 1998; Soltys *et al.*, 1999; Chang *et al.*, 2002). The use of pure cell populations has also reached the field of therapy. *Ex vivo* expansion and manipulation of isolated cells have given promising possibilities in therapy, especially in immunotherapy. Dynabeads, having approximately the same size as eukaryotic cells, have proven to be very efficient in the *ex vivo* activation of T cells, the prime effectors of the acquired immune system (Garlie *et al.*, 1999; Lum *et al.*, 2001). An *ex vivo*-expanded population of T cells may be administrated to the patient, thereby helping to fight diseases such as cancer, HIV, and autoimmune disorders (Liebowitz *et al.*, 1998; Levine *et al.*, 1998, 2002; Thomas and June, 2001).

There are two main strategies for isolating a specific cell type: a "positive selection" of cells of interest or a "negative selection" where unwanted cells are depleted (Fig. 1). By positive selection, a specific cellular subset is isolated directly from a complex mixture of cells based on the expression of a distinct surface antigen. The resulting immune complexes of beads and target cells are collected using a magnet. By negative selection, all unwanted cell types are removed from the sample by the magnetic beads. Cells isolated by negative selection have not been bound to antibodies at any time. Surface antigen-bound antibodies may elicit the transmission of signals across the cell membrane. However, in general the purity obtained by negative selection is lower than for positive selection.

It is important to note that Dynabeads can be used directly in complex samples such as whole blood, which offers rapid and direct access to the maximum number of target cells that have undergone the minimum amount of interference. Usually, Dynabeads are precoated with target-specific antibodies (direct technique). Alternatively, antibodies are first added to the cell suspension and thereafter the labeled cells are immobilized to Dynabeads through secondary antibodies, e.g., pan antimouse IgG (indirect technique).

The particular properties of the magnetic beads make it possible to use positively selected cells directly for cell stimulation or molecular studies, such as PCR or RT-PCR, without the necessity of removing the beads. However, in functional studies the beads should be removed. For several cell types, Dynabeads can be detached from the cells after isolation using a polyclonal antibody (DETACHaBEAD) that binds to the Fab-region of the cell-specific monoclonal antibody (Rasmussen *et al.*, 1992). No antibodies remain on the isolated cells after detachment when using DETACHaBEAD (Fig. 2). CELLection is another system for the removal of beads from isolated cells. DNase is used to cut a DNA linker between the beads and the antibodies, removing the beads and leaving only the

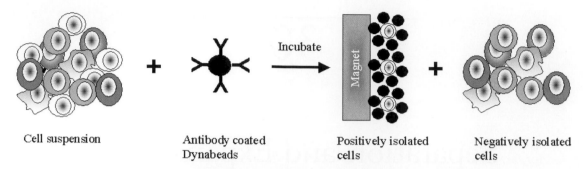

FIGURE 1   Positive and negative isolation of cells using antibody-coated Dynabeads. By positive selection, a specific cellular subset is isolated directly from a complex mixture of cells based on the expression of a distinct surface antigen. The resulting immune complexes of beads and target cells are collected using a magnet. By negative selection, all unwanted cell types are removed from the sample by the magnetic beads.

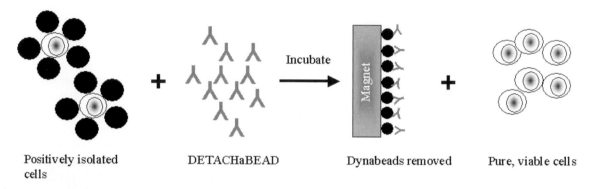

FIGURE 2   Detachment of cells from Dynabeads using DETACHaBEAD. Cells are detached from the beads by a polyclonal antibody (DETACHaBEAD) that binds the Fab-region of the cell-specific monoclonal antibody, thereby altering its affinity for the antigen. The cells are left without antibody on their surface.

antibodies on the cell surface (Soltys *et al.*, 1999; Werther *et al.*, 2000).

Factors such as incubation time, temperature, and concentration of reactants have a measurable effect on the efficiency of cell isolation using magnetic beads. Furthermore, the process is also affected by specific key parameters, such as the nature and state of the target cell, characteristics of the antigen/antibody binding, sample type, concentration, and ratio of beads and cells. Successful cell isolation with Dynabeads, which implies high yield and purity, is dependent on the concentration of the magnetic beads, the ratio of beads to target cells, and the choice of antibody. Monoclonal antibodies are generally recommended due to their high specificity toward the target antigen.

II. MATERIALS AND INSTRUMENTATION

Lymphoprep 6 × 500 ml (Prod. No.1114547), density 1.077 ± 0.001 g/ml, osmolality 290 ± 15 mOsm from

Axis-Shield Poc is stable for 3 years if stored at room temperature in the dark. Human serum, AB 100 ml (Part Code US14–490E) from BioWhittaker is stable for 3 years if stored at −20°C. Human serum, AB should be heat inactivated (56°C, 30 min) before use. All human blood-based products should be handled in accordance with currently acceptable biosafety practices and guidelines for the prevention of blood-borne viral infections. Fetal bovine serum, premium, US origin 500 ml (Part Code US14-501F) from BioWhittaker is stable for 3 years if stored at −20°C. Fetal bovine serum should be heat inactivated (56°C, 30 min) before use. Bovine serum albumin (BSA) >96% by electrophoresis, 500 g (Prod. No. A-4503) is from Sigma. DPBS, without Ca^{2+} or Mg^{2+} 500 ml (Part Code BE17-512F) from BioWhittaker is stable for 2 years if stored at 15–30°C. RPMI 1640, with L-glutamine 500 ml (Part Code BE12-702F) from BioWhittaker is stable for 2 years if stored at 15–30°C. Proleukin interleukin-2 (recombinant IL-2) 22 × 10⁶ IU from Chiron B.V. was reconstituted in sterile H_2O at $6 × 10^5$ IU/ml and stored at −20°C for up to 1 year. CD4-FITC 100 tests/2 ml

(Prod. No. 3021) from Diatec is a mouse monoclonal IgG2a/κ, clone EDU-2 recommended for flow cytometry. CD4-FITC should be stored at 2–8°C (short term) or −20°C (long term). CD25-PE 100 tests (Cat. No. 341010) from BD Biosciences Pharmingen is a mouse monoclonal IgG1, clone 2A3 recommended for flow cytometry. CD25-PE should be stored at 2–8°C. CD8-PC5 100 tests/2 ml (Part No. IM2638) from Beckman Coulter is a mouse monoclonal IgG1, clone B9.11 recommended for flow cytometry. CD8-PC5 should be stored at 2–8°C. Dynal CD4 positive isolation kit, 2 ml (Prod. No. 113.03), Dynabeads CD25, 2 ml (Prod. No. 111.33), DETACHaBEAD CD4/CD8, 5 ml (Prod. No. 125.04), Dynal CD4 negative isolation kit, 5 ml (Prod. No. 113.17), Dynal CD8 negative isolation kit, 5 ml (Prod. No. 113.19), Dynabeads Tcap™ EBV/BMLF-1 (Prod. No. 103.01), and Dynabeads CD3/CD28 T-cell expander, 2 ml (Prod. No. 111.31) are from Dynal Biotech. These products, which contain magnetic beads (2.8–4.5 μm) and/or antibody cocktails (mouse monoclonal antibodies) and/or release agents (polyclonal sheep antimouse antibodies), are stable for 12–36 months when stored at 2–8°C. Magnets: Dynal MPC (Prod. No. 120.01, 120.20, and 120.21) and from Dynal Biotech. Sodium citrate dihydrate pro analysis, >99% pure, 1 kg (Cat. No. 1.06448.1000) is produced by Merck and sold by VWR International. rHLA-A2/GLC-PE MHC tetramers 50 tests (code T2A-G) from ProImmune are stable for >6 months when stored at 2–8°C. Centrifuge: Rotanta 460 R from Hettich was delivered by Nerliens. Sample mixer: Dynal MX1 (Prod. No. 159.07) with 12-tube mixing wheel (Prod. No. 159.03) is from Dynal Biotech ASA. Flow cytometer: BD LSR II with 488- and 633-nm lasers from BD Biosciences was delivered by Laborel. Laminar flow bench: Biowizard Kojair KR 200 from Kojair Tech Oy was delivered by Houm AS. 37°C CO_2 incubator: Forma Scientific Model 3548 (water-jacked incubator) was delivered by Houm AS. Pipettes: Finnpipette 0.5–10 μl, 5–50 μl, 20–200 μl, and 200–1000 μl from Thermo Labsystem were delivered by VWR International.

III. PROCEDURES

This article covers procedures for some basic principles of cell isolation and cell stimulation using Dynabeads.

The isolation of CD4+CD25+ regulatory T cells (Shevach, 2002) demonstrates the different isolation techniques. The isolation is performed in two steps; the first step involves isolation of CD4+ cells by positive selection (protocol A) or negative selection (protocol B). The second step involves isolation of CD25+ cells from the CD4+ cell population (protocol C).

Isolation of antigen-specific CD8+ T cells using recombinant HLA molecules on Dynabeads demonstrates enrichment of rare cells from a complex cell sample (protocol D). Expansion of these antigen-specific CD8+ T cells demonstrates *ex vivo* cell stimulation and manipulation (protocol E). For further questions, contact the supplier at: techcentre@dynalbiotech.com.

A. Positive Selection of CD4+ T Cells (Direct Technique)

CD4+ T cells are isolated from buffy coat by positive selection using the Dynal CD4 positive isolation kit. The isolated CD4+ cells are then ready for further studies, e.g., isolation of the CD4+CD25+ regulatory T-cell subpopulation (protocol C).

Solutions

1. *Phosphate-buffered saline (PBS) pH 7.4*: Dulbecco's PBS without Ca^{2+} and Mg^{2+}
2. *PBS/citrate*: PBS with 0.6% (w/v) sodium citrate (to prevent microcoagulation)
3. *Heat-inactivated fetal calf serum (FCS)*
4. *PBS/BSA or PBS/FCS*: PBS with 0.1% (w/v) bovine serum albumin or PBS with 2% (v/v) FCS.
5. *Culture medium*: RPMI 1640 with 1% (v/v) FCS.

Steps for Cell Isolation

1. Dynabeads required: 2×10^7 beads/ml $\times$ 40 ml = 8×10^8 beads (2.0 ml).
2. Wash Dynabeads; mix vial, transfer beads to a 50-ml tube, add 10 ml PBS/BSA, collect beads on a magnet for 1 min, remove supernatant, and replace with 2 ml PBS/BSA; cool to 2–8°C.
3. Dilute 15 ml buffy coat in 25 ml PBS/citrate; cool to 2–8°C.
4. Add 40 ml of cell suspension to the Dynabeads (2 ml) directly into the 50-ml tube. Mix *gently* by tilting and rotation for 30 min at 2–8°C.
5. Isolate the cells that are attached to beads (rosetted cells) by placing the tube in the magnet for 2 min. Discard the supernatant while the rosetted cells are held at the tube wall by the magnet.
6. Wash cells; remove the tube from the magnet and resuspend the rosetted cells gently in 10 ml PBS/citrate. Repeat steps 5 and 6 four to five times.
7. Resuspend the rosetted cells in 1 ml of culture medium in a 5-ml tube. The cells are now ready for the removal of beads.

Steps for Removal of Dynabeads from Cells (DETACHaBEAD)

1. Add 10 μl DETACHaBEAD per 10^7–10^8 beads used for cell capture.
2. Mix *gently* by tilting and rotation for 45–60 min at room temperature.
3. Pipette gently five to six times to resuspend the cell/beads suspension and place the tube in a magnet for 2 min.
4. Collect the supernatant while the beads are held at the tube wall by the magnet.
5. To obtain residual cells, resuspend the beads in culture medium and repeat steps 3 and 4 twice.
6. Combine the supernatants, and wash the cells twice with PBS/BSA to remove DETACHaBEAD.
7. The cells are now ready for further analysis or isolation of subpopulations (protocol C).

B. Negative Selection of CD4+ T Cells (Indirect Technique)

CD4+ T cells are isolated from peripheral blood mononuclear cell (PBMC) by negative selection using the Dynal CD4 negative isolation kit. Unwanted cells are removed, and the CD4+ T cells are then ready for further studies, e.g., isolation of the CD4+CD25+ regulatory T-cell subpopulation (protocol C).

Solution

PBS/citrate, PBS/BSA, FCS, and culture medium are prepared as in protocol A.

Steps for Cell Isolation

1. Dynabeads required: 4 beads/PBMC × 10^8 PBMC = $4 × 10^8$ beads (1.0 ml).
2. Wash Dynabeads as described in protocol A, resuspend in 1 ml.
3. Isolate PBMC with low platelet content; add 35 ml solution (10 ml buffy coat +25 ml of PBS/citrate) on top of 15 ml of Lymphoprep at room temperature. Centrifuge for 20 min at 160 g at 20°C. Remove 20 ml of supernatant by suction to eliminate platelets. Centrifuge for 20 min at 350 g at 20°C. Recover PBMC from the plasma/Lymphoprep interface. Wash PBMC three times in PBS/BSA (centrifuge for 8 min at 500 g the first time and then at 225 g) and resuspend the PBMC at 10^7 cells per 100–200 μl in PBS/BSA. Keep cells at 2–8°C.
4. Incubate 10^8 PBMC with the 200-μl antibody mix and 200 μl FCS for 10 min at 2–8°C.
5. Wash cells; add 5–10 ml PBS/BSA and centrifuge for 8 min at 500 g at 2–8°C. Resuspend cells in 9 ml PBS/BSA at room temperature.

6. Add 9 ml of PBMC to the depletion Dynabeads (1 ml) directly into the 50-ml tube. Mix *gently* by tilting and rotation for 15 min at room temperature.
7. Pipette gently five to six times to resuspend the rosetted cells and double the sample volume with PBS/BSA. Place the tube in a magnet for 2 min. Collect the supernatant containing CD4+ T cells. The cells are now ready for further analysis or isolation of subpopulations (protocol C).

C. Positive Selection of CD4+CD25+ Regulatory T Cells (Fig. 3)

CD4+ T cells have been isolated by either positive (protocol A) or negative (protocol B) selection. From these cells the CD4+CD25+ regulatory T-cell subpopulation can be isolated using Dynabeads CD25, and the beads are removed with DETACHaBEAD.

Solutions

1. PBS, PBS/BSA, FCS, and culture medium are prepared as in protocol A
2. *PBS/10% FCS*: PBS with 10% (v/v) FCS

Steps for Cell Isolation

1. Dynabeads required: 4 beads/CD4+ cell × $5 × 10^7$ CD4+ cells = $2 × 10^8$ beads (500 μl).
2. Wash Dynabeads as described in protocol A, but in a 15-ml tube, resuspend in 500 μl.
3. Prepare cells by protocol A or B, resuspend at 10^7 cells/ml in PBS/BSA, and cool to 2–8°C.
4. Add 5 ml of cells to the Dynabeads (500 μl) directly into the 15-ml tube. Mix *gently* by tilting and rotation for 20 min at 2–8°C.

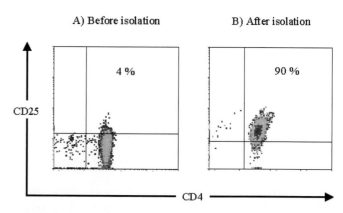

FIGURE 3 Isolation of CD4+CD25+ regulatory T cells. CD4+CD25+ regulatory T cells were isolated according to protocol C. Diatec CD4-FITC and Pharmingen CD25-PE (clone 2A3) were used for cell staining. (A) Staining of CD25+ T-cells in the CD4+ T-cell population. (B) Staining of CD25+ T-cells after positive isolation from the CD4+ T-cell population.

5. Isolate and wash the rosetted cells as described in protocol A (steps 5 and 6), but use 2 ml PBS/BSA for washing.

6. *Use precoated tubes* (10 min at room temperature using PBS/10% FCS) from this step to avoid cell loss. Resuspend the rosetted cells in 200 μl of culture medium in a 1.5-ml tube. The cells are now ready for removal of beads.

Steps for Removal of Dynabeads from cells (DETACHaBEAD)

1. Add 10 μl DETACHaBEAD per 10^7 beads used for cell capture.

2. Incubate, resuspend, collect, and wash cells as described in protocol A; use precoated tubes. The cells are now ready for further analysis and functional studies.

D. Isolation of Antigen-Specific CD8+ T Cells (Fig. 4)

Dynabeads coupled with recombinant HLA (rHLA) class I molecules loaded with the relevant peptide can be used to enrich antigen-specific CD8+ T cells (Garboczi *et al.*, 1992; Luxembourg *et al.*, 1998; Ostergaard Pedersen *et al.*, 2001). The starting material is PBMC or preferably negatively isolated CD8+ T cells (protocol B, using Dynal CD8 negative isolation kit). Antigen-specific CD8+ T cells are isolated using Dynabeads Tcap™ EBV/BMLF-1.

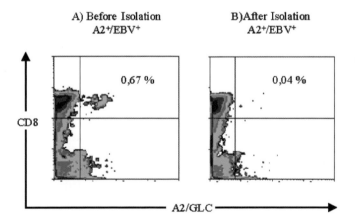

A) Before Isolation A2+/EBV+   **B) After Isolation A2+/EBV+**

0,67 %   0,04 %

CD8   A2/GLC

FIGURE 4 Isolation of rHLA-A2/GLC positive T cells from PBMC. (A) PBMC from a HLA-A2/GLC positive donor stained with rHLA-A2/GLC tetramers. (B) PBMC after removal of rHLA-A2/GLC tetramer positive T cells by rHLA-A2/GLC coated Dynabeads, i.e., Dynabeads HLA-A2 EBV/BMLF-1 (see protocol D). The rHLA-A2 was produced and loaded with the peptide GLCTLVAML according to Ostergaard Pedersen *et al.* (2001).

Solutions

1. PBS/BSA and FCS are prepared as in protocol A.
2. *Culture medium*: RPMI 1640 with 5% (v/v) FCS or 5% (v/v) AB serum.

Steps for Cell Isolation

1. Dynabeads required: 10^7 beads/ml × 1 ml = 10^7 beads (25 μl).

2. Wash Dynabeads; mix vial, transfer beads to a 1.5-ml tube, add 0.5 ml PBS/BSA, collect beads on a magnet for 1 min, remove supernatant, and replace with 100 μl PBS/BSA; cool to 2–8°C.

3. Prepare cells by protocol B (preferably negatively isolated CD8+ T cells), resuspend at $5 × 10^7$ cells/ml in PBS/BSA, and cool at 2–8°C.

4. Add 1 ml of cells to the Dynabeads directly into the 1.5-ml tube. Mix *gently* by tilting and rotation for 30 min at 2–8°C.

5. Isolate and wash the rosetted cells at 2–8°C as described in protocol A (steps 5 and 6), but use 1 ml PBS/BSA for washing. Resuspend the rosetted cells in 100 μl of culture medium (2–8°C). The cells are now ready for further analysis or expansion (protocol E).

E. Activation and Expansion of T Cells Using Dynabeads

T-cell activation *in vivo* is initiated by the binding of T-cell receptors on its surface to appropriate peptide-HLA molecules on the surface of antigen-presenting cells (APC). However, activation of highly pure T cells *in vitro* is difficult to perform without the presence of APC. The Dynabeads CD3/CD28 T-cell expander can mimic the APC and therefore be used to activate pure T cells. Some T cells are CD28 negative and will require other costimulatory signals than through CD28, e.g., through CD137 (Maus *et al.*, 2002).

Solutions

1. FCS is prepared as in protocol A.
2. *Culture medium*: RPMI 1640 with 5% (v/v) FCS or 5% (v/v) AB serum
3. *Culture medium/rIL-2*: culture medium with 20–100 IU/ml recombinant IL-2

Steps for in Vitro Cell Expansion

1. Dynabeads required: 4 beads/T cell × 10^7 cells = $4 × 10^7$ beads (1.0 ml).

2. Starting material may be monocyte-depleted PBMC, pure CD3+, CD4+, or CD8+ T cells (e.g., prepared according to protocol A or B). Resuspend at 10^6 cells/ml in culture medium and add 1 ml of this cell

suspension to each well in a 24-well culture plate. If rare cells are isolated and beads are still attached to the cells (e.g., protocol D), each well should contain a maximum of 10^6 beads (used for isolation), regardless of the number of cells isolated.

3. Add the Dynabeads CD3/CD28 T-cell expander directly to the cells at a ratio of four beads per cell. Mix *gently* by pipetting.

4. Incubate cells for 3 days at 37°C, resuspend, and count the cells. Dilute to 0.5×10^6 cells/ml with culture medium/rIL-2. Split cells every second to third day with culture medium/rIL-2 to keep the cell concentration at $0.5-1 \times 10^6$ cells/ml.

5. Cells can be expanded for 14 days without adding extra beads, and cell expansion rates are usually between 100- and 1000-fold, depending on the donor.

IV. COMMENTS

Immunomagnetic cell isolation offers rapid and direct access to target cells from whole blood and bone marrow without cell loss or damage. If necessary, unwanted elements such as erythrocytes, free DNA, fat, or serum proteins may be removed before cell isolation to improve the performance of the beads. Buffy coat, a concentrate of white blood cells, has the advantage of high target cell concentration. However, the quality of buffy coat preparations may vary considerably. Density gradient isolation of cells provides removal of possible interfering elements and the ability to manipulate target cell concentration to perform cost-efficient cell isolation. Drawbacks include cell losses during centrifugations and negative effects on the cells due to contact with the density gradient medium. Generally the concentration of nucleated cells should be 10^8 cells/ml when performing immunomagnetic cell isolation.

By positive selection the cells of interest are isolated for analysis. General isolation parameters are $\geq 1 \times 10^7$ beads/ml, bead:cell ratio 4:1–10:1, and 10–30 min of incubation. Typically, 95–100% purity and viability are achieved with 60–95% yield. For some downstream applications the beads can remain attached to the cells (e.g., mRNA or DNA isolation). By negative selection (= *depletion*), unwanted cells are removed prior to analysis of the remaining population. General isolation parameters are $\geq 2 \times 10^7$ beads/ml, bead:cell ratio $\geq 4:1$, and 20–60 min of incubation. Typically, 95–99% depletion of unwanted cells is achieved. Two successive depletion cycles may result in higher purity for small cell populations. The direct technique offers fast cell isolation with antibody-coated beads. Cell handling is minimized, reducing the risk for cell damage and loss. The indirect technique is especially useful when the affinity/avidity of the primary antibody is low or when the epitope density on the target cell is limited. The disadvantage of the indirect technique is cell handling (centrifugation).

Cross-reactivity and Fc binding can be blocked by the addition of free proteins (e.g., Fc receptor blocking with γ-globulin). Nonspecific binding of "sticky" cells to beads can be avoided by predepletion of these cells with protein-coated beads (e.g., secondary coated or BSA-coated beads). Free DNA will contribute to binding of unwanted cells to the beads. Freezing/thawing may damage cells and release free DNA. DNase treatment prior to cell selection will abolish this problem. Platelets may also induce nonspecific binding of cells to Dynabeads. Preparation of PBMC with a low platelet content (see protocol B) will reduce this problem.

Primary-coated Dynabeads are ready-to-use products for a wide variety of cell surface markers. In addition, secondary-coated Dynabeads offer an excellent possibility to make beads with the reactivity of choice using mouse, rat, or rabbit antibodies directly from culture supernatant, ascites, or polyclonal sera (without the need of purification). This is especially useful when only small amounts of nonpurified antibody are available. However, affinity-purified antibodies are preferred.

V. PITFALLS

1. Prolonged incubation (>60 min) and increased bead concentration (>1×10^8 beads/ml) will rarely improve the cell selection efficiency. However, nonspecific binding may increase, damage of cells from sheer forces of the beads may occur, and risk of cell trapping increases.

2. A soluble form of cell surface antigens or other serum components can reduce the efficiency of immunomagnetic cell isolation. One or two washing steps will overcome this problem.

3. Nonspecific binding. Genomic DNA from lysed cells (e.g., present in buffy coat, PBMC, or after freezing/thawing of cells) will induce non-specific binding of cells to beads. DNase treatment of the cell suspension prior to cell selection will prevent this problem without harming intact cells. Some sample tubes (e.g., glass or polystyrene) tend to bind cells nonspecifically, which can be a major problem when working with minor cell populations (e.g., rare, circulating tumour

cells). Precoating of sample tubes with a protein solution before use or the use of low-binding plastic tubes is recommended.

4. Phagocyte cells (e.g., monocytes) will bind and engulf beads if incubation is performed at temperatures above 2–8°C.

References

Chang, C. C., Ciubotariu, R., Manavalan, J. S., Yuan, J., Colovai, A. I., Piazza, F., Lederman, S., Colonna, M., Cortesini, R., Dalla-Favera, R., and Suciu-Foca, N. (2002). Tolerization of dendritic cells by T(S) cells: The crucial role of inhibitory receptors ILT3 and ILT4. *Nature Immunol.* **3**, 237–243.

Funderud, S., Nustad, K., Lea, T., Vartdal, F., Gaudernack, G., Stenstad, P., and Ugelstad, J. (1987). Fractionation of lymphocytes by immunomagnetic beads. *In "Lymphocytes: A Practical Approach"* (G. G. B. Klaus, ed.), pp. 55–65. IRL Press, Oxford.

Garboczi, D. N., Hung, D. T., and Wiley, D. C. (1992). HLA-A2-peptide complexes: Refolding and crystallization of molecules expressed in *Escherichia coli* and complexed with single antigenic peptides. *Proc. Natl. Acad. Sci. USA* **89**, 3429–3433.

Garlie, N. K., LeFever, A. V., Siebenlist, R. E., Levine, B. L., June, C. H., and Lum, L. G. (1999). T cells coactivated with immobilized anti-CD3 and anti-CD28 as potential immunotherapy for cancer. *J. Immunother.* **22**, 336–345.

Levine, B. L., Bernstein, W. B., Aronson, N. E., Schlienger, K., Cotte, J., Perfetto, S., Humphries, M. J., Ratto-Kim, S., Birx, D. L., Steffens, C., Landay, A., Carroll, R. G., and June, C. H. (2002). Adoptive transfer of costimulated CD4+ T cells induces expansion of peripheral T cells and decreased CCR5 expression in HIV infection. *Nature Med.* **8**, 47–53.

Levine, B. L., Cotte, J., Small, C. C., Carroll, R. G., Riley, J. L., Bernstein, W. B., Van Epps, D. E., Hardwick, R. A., and June, C. H. (1998). Large-scale production of CD4+ T cells from HIV-1-infected donors after CD3/CD28 costimulation. *J. Hematother.* **7**, 437–448.

Liebowitz, D. N., Lee, K. P., and June, C. H. (1998). Costimulatory approaches to adoptive immunotherapy. *Curr. Opin. Oncol.* **10**, 533–541.

Lum, L. G., LeFever, A. V., Treisman, J. S., Garlie, N. K., and Hanson, J. P. Jr. (2001). Immune modulation in cancer patients after adoptive transfer of anti-CD3/anti-CD28-costimulated T cells-phase I clinical trial. *J. Immunother.* **24**, 408–419.

Luxembourg, A. T., Borrow, P., Teyton, L., Brunmark, A. B., Peterson, P. A., and Jackson, M. R. (1998). Biomagnetic isolation of antigen-specific CD8+ T cells usable in immunotherapy. *Nature Biotech.* **16**, 281–285.

Marquez, C., Trigueros, C., Franco, J. M., Ramiro, A. R., Carrasco, Y. R., Lopez-Botet, M., and Toribio, M. L. (1998). Identification of a common developmental pathway for thymic natural killer cells and dendritic cells. *Blood* **91**, 2760–2771.

Maus, M. V., Thomas, A. K., Leonard, D. G., Allman, D., Addya, K., Schlienger, K., Riley, J. L., and June, C. H. (2002). Ex vivo expansion of polyclonal and antigen-specific cytotoxic T lymphocytes by artificial APCs expressing ligands for the T-cell receptor, CD28 and 4-1BB. *Nature Biotechnol.* **20**, 143–148.

Ostergaard Pedersen, L., Nissen, M. H., Hansen, N. J., Nielsen, L. L., Lauenmoller, S. L., Blicher, T., Nansen, A., Sylvester-Hvid, C., Thromsen, A. R., and Buus, S. (2001). Efficient assembly of recombinant major histocompatibility complex class I molecules with preformed disulfide bonds. *Eur. J. Immunol.* **31**, 2986–2996.

Rasmussen, A.-M., Smeland, E., Eriksten, B. K., Calgnault, L., and Funderud, S. (1992). A new method for detachment of Dynabeads from positively selected B lymphocytes. *J. Immunol. Methods* **146**, 195–202.

Shevach, E. M. (2002). CD4+CD25+ suppressor T cells: More questions than answers. *Nature Rev.* **2**, 389–400.

Soltys, J., Swain, S. D., Sipes, K. M., Nelson, L. K., Hanson, A. J., Kantele, J. M., Jutila M. A., and Quinn, M. T. (1999). Isolation of bovine neutrophils with biomagnetic beads: Comparison with standard Percoll density gradient isolation methods. *J. Immunol. Methods* **226**, 71–84.

Thomas, A. K., and June, C. H. (2001). The promise of T-lymphocyte immunotherapy for the treatment of malignant disease. *Cancer J.* **7**, S67–S75.

Ugelstad, J., Kilaas, L., Aune, O., Bjørgum, J., Herje, R., Schmid, R., Stenstad, P., and Berge, A. (1994). Monodisperse polymer particles. *In "Advances of Biomagnetic Separation"* (M. Uhlén, E. Hornes, and Ø. Olsvik, eds.), pp 1–20, Eaton Publ. Comp., Natick, MA.

Ugelstad, J., Mørk, P. C., Herder Kaggerud, K., Ellingsen, T., and Berge, A. (1980). Swelling of oligomer particles: New methods of preparation of emulsions and polymer dispersions. *Adv. Colloid Interface Sci.* **13**, 101.

Werther, K., Normark, M., Hansen, B. F., Brunner, N., and Nielsen H. J. (2000). The use of the CELLection kit in the isolation of carcinoma cells from mononuclear cell suspensions. *J. Immunol. Methods* **238**, 133–141.

31

Separation of Cell Populations Synchronized in Cell Cycle Phase by Centrifugal Elutriation

R. Curtis Bird

I. INTRODUCTION

Centrifugal elutriation is the only method whereby large numbers of cells can be separated rapidly on the basis of size (Diamond, 1991; Merrill, 1998; Davis *et al.*, 2001). The capability of discriminating between very small differences in cell size provides the ability to separate cells into sequential cell cycle phase populations of relatively high purity without the use of drugs or inhibitors (Bludau *et al.*, 1986; Braunstein *et al.*, 1982; Hann *et al.*, 1985; Iqbal *et al.*, 1984; Wu *et al.*, 1993; Brown and Schildkraut, 1979; Bialkowski and Kasprzak, 2000; Datta and Long, 2002; Deacon *et al.*, 2002; Hengstschlager *et al.*, 1999; Houser *et al.*, 2001; Karas *et al.*, 1999; Rehak *et al.*, 2000; Sugikawa *et al.*, 1999; Syljuasen and McBride, 1999; Van Leeuwen-Stok *et al.*, 1998). It has been used successfully to separate a wide variety of cell types from suspension and substrate-dependent cultures and to separate mixed cell populations liberated directly from tissues or body fluids (Boerma *et al.*, 2002; Dagher *et al.*, 2002; Wong *et al.*, 2001, 2002). The purity of the samples is relatively high and the cells proceed to grow, following separation, without a detectable lag period. Thus, centrifugal elutriation combines speed of separation of large numbers of cells with little or no perturbation of the cell growth cycle and avoids the use of agents that might induce artifact. As additional advantages, centrifugal elutriation overcomes the limits on cell number imposed by fluorescence-activated cell sorting and the long separation times required for unit gravity sedimentation, as well as problems associated with osmotic stress in centrifugation media. The only real compromise is that the purity of the samples is somewhat lower than commonly achieved with alternative methods. The developmental history and theory of centrifugal elutriation are reviewed elsewhere (Conkie, 1985; Beckman Instruments, 1990).

Centrifugal elutriation was developed by Lindahl (1948) for the separation of cells and particulate fractions of cells (Lindahl and Nyberg, 1955; Lindahl, 1986) based on the original work of Lindbergh (1932). Counterstreaming centrifugation was used by these investigators to separate a variety of cell types but was not generally available to other laboratories until the introduction of a commercial centrifugal elutriator by Beckman Instruments in 1973 (reviewed in Beckman Instruments, 1990). Once available commercially, centrifugal elutriation was applied to the separation of many kinds of cells, including bacteria, yeast, and mammalian cells, grown in culture or liberated directly from tissues and solid tumors (reviewed in Beckman Instruments, 1990). In addition, elutriation has been used to separate cells in different phases of the cell cycle based on the small differences in size as cells gradually grow between divisions (Bird *et al.*, 1996a,b). In all cases, successful separation is dependent on complete dissociation of the cells to a single cell suspension. Failure to accomplish this affects the quality of separation as well as cell yield and thus cell clumps must be removed or dissociated prior to elutriation.

Centrifugal elutriation imposes two opposing forces on mixed cell populations to facilitate their fractionation into subpopulations (Fig. 1 and see Section IV).

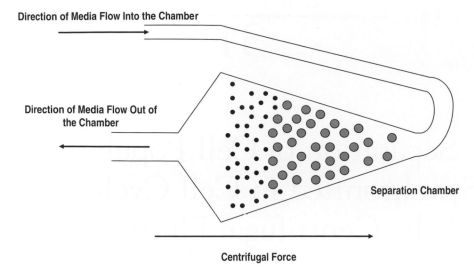

Direction of Media Flow Into the Chamber

Direction of Media Flow Out of the Chamber

Separation Chamber

Centrifugal Force

FIGURE 1 Cell separation dynamics and opposing forces in the elutriation chamber. The flow of media is opposed by centrifugal force in a balance that holds particles in equilibrium in the separation chamber. Due to different levels of force applied proportional to surface area presented, smallest particles sort to the inside of the chamber relative to larger particles. Separation is achieved by increasing the flow rate incrementally to push the smallest population of particles past the widest part of the chamber and into the inner narrowing section where media flow accelerates affecting elutriation of the population.

This technology has proven to be effective in fractionating cells, based on very small differences in cell size, with nominal cross contamination, and in numbers unmatched by other methods of cell separation. In addition, centrifugal elutriation can be performed rapidly, requiring only a few minutes (usually 20–120 min) to affect separation and this occurs in media containing no special additives that might affect the osmolarity or viscosity of the medium. Thus, with very little physiological change perceived by the cells, separation can be affected rapidly. Some shear force is exerted on the cells during separation, but this is not sufficient to appreciably affect viability or behavior of the cells in most cases. Compared to other methods of separation, such as fluorescence-activated cell sorting or unit gravity separation, centrifugal elutriation is by far the most gentle and the most rapid in manipulation and separation of the cell populations.

In this example application, synchronous fractions of HeLa S3 cells were analyzed, following centrifugal elutriation, by flow cytometry and [^{3}H]thymidine incorporation into acid-precipitable materials (Wu *et al.*, 1993; Pai and Bird, 1994). Both means of analysis demonstrated that sequential fractions of elutriated cells represent sequential cell cycle phases as determined by the analysis of cell volume, DNA content, and ability to incorporate thymidine during five sequential 1-h periods following return to culture. From this analysis, cells collected at flow rates of 21–25 ml/min were designated the G_1-phase population,

cells collected at flow rates of 29–35 ml/min were designated the S-phase population, and cells collected at flow rates of 43 ml/min were designated the G_2/M-phase population. Flow cytometric analysis, based on measurements of DNA content and cell volume, were also used to determine the level of contamination of S-phase cells in the G_1-phase fractions (approximately 3%) and G_1-phase cells in the S-phase cell population (approximately 10%) (Hann *et al.*, 1985). Thus, large populations of cells were separated rapidly into seven to eight synchronous fractions without the use of drugs, with little evidence of perturbation, and with low levels of contaminating cells.

II. MATERIALS AND INSTRUMENTATION

The centrifugal elutriator is from Beckman Instruments (elutriator rotor assembly Cat. No. JE-6B was run in a Model J2-21 elutriation centrifuge). Accessories are used as specified by the manufacturer throughout, and the rotor is equipped with a standard separation chamber. A Masterflex digital peristaltic pump (0–100 rpm Model 07523-70 fitted with a model 7014-21 pump head, Cole-Parmer) is used to pump cells and media through the rotor.

Materials for cell culture include α-modification of Eagle's minimal essential growth medium (α-MEM,

Invitrogen, Cat. No. 11900-073), fetal bovine serum (FBS, Hyclone, Cat. No. SH30070.03), donor horse serum (DHS, ICN Biomedicals, Inc., Cat. No. 29-211-49), 100× antibiotic/antimycotic solution (Invitrogen, Cat. No. 15240-062), 10× trypsin solution (Invitrogen, Cat. No. 15090-046), and Hanks' balanced salt solution (Sigma, Cat. No. H9394). All plasticware is tissue culture grade (Corning Plasticware, Fisher Scientific). All other reagents are standard reagent grade and available from numerous sources. The water used throughout this procedure is ultrapure in quality and is prepared by ion-exchange chromatography (Barnstead Nanopure) to 18-MΩ resistance and then glass distilled to remove residual endotoxin and RNase activity. Solutions are sterilized by autoclaving or ultrafiltration (0.2 μm).

III. PROCEDURES

A. Cell Culture

Growth medium and elutriation medium are used per standard protocols for the growth of HeLa cells (Pai and Bird, 1992). If different cell lines are to be elutriated, appropriate media should be substituted (see Section V).

Solutions

1. *Growth and elutriation media*: Dissolve powdered α-MEM (10-liter pack) in ultrapure water containing a final concentration of 1× antibiotics and 22.2 g NaHCO$_3$ and make up to a total volume of 10 liters. Place the medium in a pressurizable vessel (Millipore), filter through a 0.2-μm filter (Corning) by positive pressure into sterile 0.5-liter bottles, and store at 4°C. Prior to use, add FBS (10%, v/v) to make growth medium or add DHS (5%, v/v) to make elutriation medium. Use of DHS reduces the cost of elutriation greatly without detectable effects on cell growth or quality of the fractionation (see Section V).

2. *0.5 M HEPES*: Dissolve 14.17 g HEPES buffer (Fisher, Cat. No. BP310-100) in ~80 ml water and adjust pH to 7.2. Adjust volume to 100 ml, filter sterilize, and store in the cold at 4°C.

3. *0.5 M Na$_2$EDTA*: Begin to dissolve 14.61 g EDTA-free acid (Fisher, Cat. No. BP118-500) in ~40 ml water with a magnetic stir bar. Monitor pH of the solution continuously while slowly adding 1 M NaOH (40 g/liter for 1 M stock) dropwise. Continue to adjust the pH up to ~8.0. As the EDTA dissolves, the pH will continue to fall. Carefully adjust the final pH to 8.0 without exceeding this value. Adjust volume to 100 ml, filter sterilize, and store at 4°C.

4. *Trypsin solution*: Add 8 ml of 2.5% trypsin stock solution (10 ×), 2 ml 0.5 M HEPES, pH 7.2, and 2 ml 0.5 M Na$_2$EDTA, pH 8.0, to 100 ml of Hanks' salt solution. Make additions from sterile stock solutions, maintaining sterility of the final solution. Store at 4°C.

5. *70% ethanol*: Combine 700 ml absolute (not denatured) ethanol with 300 ml water. Store in 0.5-liter bottles at room temperature.

Steps

1. Grow cultured HeLa S3 cells in 20 ml of modified Eagle's α-MEM medium (Invitrogen/Gibco) with 10% FBS and antibiotics in a 15-cm diameter culture plates at 37°C with 5% CO$_2$ and 100% humidity (Pai and Bird, 1992).

2. Collect cells at 60–70% confluence by trypsin digestion. Harvest just before the rotor is filled with medium.

3. Concentrate cells by low-speed centrifugation (3000 g for 5 min) at 4°C and resuspend in 5 ml of ice cold α-MEM with 5% DHS for every three plates (15 cm). Use this medium throughout the elutriation procedure. Maintain the cells on ice until they are loaded into the elutriator.

B. Elutriation

Steps

1. Arrange the elutriation system as described in the schematic diagram (Fig. 2A). Assemble the elutriator rotor according to the manufacturer's directions (Fig. 2B). Lightly lubricate each O ring, which seals the rotating assembly on top of the rotor, with silicone grease, taking care to wipe off any excess. Place the lower washer and spring on top of the rotor followed by the outer ring and rotating seal. Note that the screw threads on the top assembly are reversed. Lightly tighten the top and check that the outer ring spins freely. Tighten the side set screw and repeat the check for a freely spinning assembly. Loosen the lower washer inside the outer ring half a turn. Check that the assembly spins freely, retighten the lower washer, and check that it spins freely again. Care should be taken to ensure that the rotating seal connecting the rotor to the fluid lines is freshly cleaned and polished with the scintered glass plate and polishing paper provided. Only a lint-free tissue with an appropriate solvent (e.g., CHCl$_3$) should be used as even a small speck of lint can cause leakage. A *very thin* layer of silicone grease can be applied to the upper edge of the rotating seal to help create and maintain a good seal but all excess silicone must be removed. Stick the seal to the polypropylene disk on the bottom surface of the top of

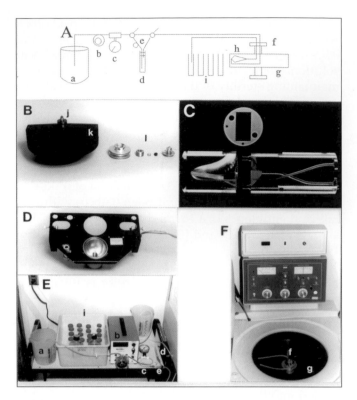

FIGURE 2 Assembly and setup of centrifugal elutriator. (A) Schematic of the centrifugal elutriator. Elutriation medium is pumped from the reservoir beaker (a) by the peristaltic pump (b) through the pressure gauge (c) and the sample tube (d) or the bypass harness (e) through the rotating seal (f) and into the rotor (g). The sample is separated in the elutriation chamber (h) and is pumped back to the sample collection tubes on ice (i). (B) Rotor and rotating seal assembly, including O rings on top of the rotor (j), which seal the rotating assembly to the rotor, the elutriation chamber (k), and, from left to right, the outer ring, lower washer, the spring that is placed inside the lower washer, rotating seal, and top of the seal assembly (l). (C) Elutriation chamber showing left (outer) and right (inner) halves separated by the gasket (above). Note the actual separation chamber within the left side and the set screws and alignment pins extending beyond both edges of the right side of the chamber. (D) Strobe assembly located below the rotor. (E) Elutriation setup showing the reservoir beaker (a), the peristaltic pump (b), pressure gauge (c), sample tube (d), bypass harness (e), and sample collection tubes (i). Note the positions of beakers to supply/accept media. (F) Elutriation centrifuge showing the complete rotating seal assembly (f), elutriation rotor (g), with strobe wires/waste tubing extending through the right centrifuge wall, and the inlet/outlet tubing extending through the left centrifuge wall.

the seal assembly by gently seating the silicone on the seal with a half turn. Carefully screw the top of the seal assembly onto the outer ring (note reverse threads) and hand tighten. This connection should not leak more than 1–3 ml during a normal elutriation run of <1 h. Even very small leaks encountered at low initial flow rates can result in significant loss of fluid and cells toward the end of the run as the flow rate rises. If per-

sistent leaks are encountered, reexamine the seal and check for a perfectly clean and even surface on both edges. Repolish the rotating seal if necessary and then repeat the assembly procedure. If none of these measures seals the rotor connections adequately, the seal should be replaced.

2. Assemble the elutriator chamber according to the manufacturer's directions (Fig. 2B). Ensure that both halves of the chamber are perfectly clean and that the polypropylene gasket is positioned to allow alignment of the sample tube (see Section V). Tighten the screws to assemble but do not overtighten. Lightly lubricate both O rings, on the base of the chamber assembly, with silicone grease, taking care to wipe off any excess. Insert the chamber and align the base pin and sample tube connections. Secure it in place with the metal plug. The screw threads on the plug should be lubricated with Spincoat (Beckman) provided with the elutriator. Tighten with the tool provided but do not overtighten as the O rings can be crushed.

3. Assemble the elutriator centrifuge according to the manufacturers directions (Fig. 2C). Remove the high-speed rotor (if present) from the centrifuge and wipe out any moisture in the centrifuge chamber. Install the strobe assembly in the centrifuge chamber and secure with thumb screws while ensuring that it is centered over the rotor spindle in the center (Fig. 2D). Feed both wire connections through the ports on the right side of the centrifuge chamber and secure them with the metal plate. Carefully place the rotor on the spindle in the center of the centrifuge and ensure that it spins freely. Connect the three pieces of silicone tubing to the rotor (inlet, outlet, and overflow) and feed them through the appropriate port (inlet and outlet to the left, overflow to the right with the wires). Ensure that all are pulled tight enough so that none have any slack in them but not so tight as to pinch or pull off any of the connections (Fig. 2F). The overflow tube should wrap around the top of the seal assembly and pass under the inlet port (the upper of the two ports on the sides of the top of the seal assembly). Seal each of the ports around each tube and wire with a slit stopper (provided with the centrifuge), where they pass through the centrifuge chamber wall, to ensure a good vacuum.

4. Prepare a large ice bucket containing two 0.5-liter bottles of elutriation medium, 30 sterile-capped tubes (50 ml), and the cell suspension (Fig. 2E). Place the bucket on a cart or table next to the centrifuge with the pump and three 2-liter beakers. One beaker contains 1 liter of sterile water. Include a 20-ml syringe with an 18-gauge needle and the sample injection harness with a pressure gauge (Cole-Parmer, Cat. No. EW-07380-75). If sterile samples are to be collected, an additional

beaker of 70% ethanol and a bottle of sterile water must also be included. If elutriated cells are to be cultured for more than a few hours after separation, a sterile hood must be positioned next to the centrifuge to allow both reservoir beakers and samples to be collected under sterile conditions.

5. Clean the exterior of the silicone tubing with ethanol and place the inlet in the beaker containing the sterile water and the outlet in an empty beaker. The inlet tubing should also be installed into the pump head and attached to the sample application and bypass harness, with an inline pressure gauge, so that they are between the pump and the rotor (Fig. 2E). Begin pumping water through the rotor (which is stationary and the centrifuge is off, see Section V) at 45 ml/min (>200 ml). Carefully observe the water as it fills the equipment (~100 ml) and dislodge any bubbles with gentle pressure on the tubing or tapping of other components. Release the air from dead spaces within the pressure gauge by pinching the tubing just after the gauge and releasing it rapidly. Do not let the pressure rise above 15 psi or the tubing may burst. Adjust the stopcocks on the sample/bypass harness to allow the harness, stopcocks, and sample tube to fill completely. Continue to pump water through the equipment until all the bubbles have been cleared. Observe all connections, particularly at the rotor, for leakage.

6. Close the centrifuge door and start the rotor. Allow the speed to gradually rise and stabilize at 2000 rpm (±1 rpm is acceptable) at 4°C. Be sure to increase the speed slowly as the fine speed control can easily overshoot the set point on acceleration. Check each of the stoppers to ensure that the seals are adequate to allow development of a vacuum in the centrifuge. Observe the chamber through the periscope assembly in the door and adjust the strobe firing timer to center the image. If the chamber appears to have a rod running along the center rather than two screws running near each edge, then the strobe is 180° out of phase with the rotor and is allowing visualization of the balance chamber, not the elutriation chamber. Continue to adjust the strobe until the elutriation chamber is visible. Check for bubbles at the outlet using the squeeze-and-release technique described earlier. Turn off the pump and observe the outlet tubing as you raise it out of the beaker and suspend it free in the air. If a slow leak occurs, the fluid and air will be drawn back up into the tube as the fluid leaks out of the system. The system is sealed if the fluid level in the outlet tube does not change.

7. If sterile collection is required, switch to 70% ethanol and pump 200 ml through the rotor followed by 200 ml of sterile water from a bottle in the laminar flow hood. Switch to elutriation medium (without DHS) on ice and pump 200 ml (if sterile collection is not required, omit ethanol rinse). Be sure to include all sections of the sample injection harness, including the sample tube and the stopcocks, during the rinse with each of these solutions. Turn off the pump before switching the stopcocks. Turn the pump off and change to elutriation medium with 5% DHS to prevent the cells from sticking. Pump 100 ml at 45 ml/min through the sample tube. At this point, two people are required to operate the system: one collects samples and one monitors rotor and pump speeds as well as managing sample injection.

8. Disperse the trypsinized cells (~ 1–2×10^8), which have been held on ice in 5 ml of elutriation medium (containing 5% DHS), with seven *very gentle* passes through an 18-gauge needle on a 20-ml syringe. Be careful not to introduce bubbles. Turn off the pump. Gently inject the cells into the sample tube (positioned with the stopper up), allowing the cells to settle to the bottom of the tube. Carefully turn the tube over (stopper down) and *gently* inject about 2 ml of air into the sample tube to act as a shock absorber against the pulsing peristaltic action of the pump. Adjust the pump, which is still off, to zero and then turn the pump on. Gradually adjust the pump up to 10 ml/min, being careful not to overshoot this value. Observe the sample tube and watch the cells enter the system. Take care to avoid pulsing of the medium or leaving a residual pools of cells in the sample tube. Collect three tubes of 50 ml each in the sample collection tubes on ice. Once the cells are loaded, the harness can be set on the bench in a stable position that maintains the inverted orientation of the sample tube (stopper down). Continue to monitor the entry of the cells into the elutriation chamber through the periscope. Carefully increase the pump speed to 15 ml/min. Do not overshoot. We collect G_1-phase cells between 21 and 25 ml/min, S-phase cells at 29–35 ml/min, and G_2/M-phase cells at 43 ml/min. Be prepared to work quickly at the end of an elutriation run as the flow rate becomes very fast and tubes containing elutriated cells fill at the rate of approximately one every minute. Following elutriation, pellet the cells by centrifugation (3000 g 5 min) and resuspend in growth medium. Adjust cell density with additional growth medium, after determination of cell concentration in each sample by counting in a hemacytometer, and transfer to tissue culture dishes.

C. Cell Cycle Analysis

Part of the synchronous cell fractions are fixed and analyzed by flow cytometry and the remainder of the cells are immediately prepared for RNA isolation or placed back into culture for further manipulation (Fig. 3). Cultured cells are analyzed for their ability to synchronously enter S phase by determining the kinetics of [³H]thymidine incorporation (Fig. 4).

Solutions

1. *Phosphate-buffered saline (PBS)*: Dissolve 0.71 g of Na_2HPO_4 (0.01 M final, Sigma, Cat. No. S-1934) and 4.5 g NaCl (0.9%, w/v final) in ~400 ml water and adjust pH to 7.6. Adjust volume to 500 ml, filter sterilize, and store at 4°C.

2. *Staining solution*: Make 50 μg/ml propidium iodide (PI) and 40 μg/ml RNase A by dilution of stock solute/ons. Add 111 μl PI stock (4.5 mg/ml, Sigma, Cat. No. P-1764) and 20 μl RNase A stock (20 mg/ml, Sigma, Cat. No. R-1859) to 10 ml water. Do not attempt to weigh RNase as any contamination will make future isolation of RNA very difficult. Open a 250-mg bottle in the fume hood and add 12.5 ml water. Cap, dissolve, and boil for 15 min to inactivate DNase (Sambrook *et al.*, 1989). Aliquot with a plugged disposable pipette (tissue culture type) in 1-ml lots in microcentrifuge tubes. Store at approximately –20°C. Use only disposable tubes and pipettes with RNase solutions and avoid any contamination or aerosols. Dissolve 100 mg of PI in 22.22 ml of water directly in the bottle. Do not weigh out. PI is extremely toxic and is a carcinogen as well as a potential mutagen and should be handled with care, including the use of gloves. Be careful not to create aerosols or liberate dust from the granular reagent. Dispose of all PI solutions and contaminated materials as hazardous waste. Use only disposable tubes and pipettes. Store PI at –20°C in the dark as it is light sensitive.

3. *100% trichloroacetic acid*: To a 500 g bottle of trichloroacetic acid (TCA, Sigma, Cat. No. T-2069), add sufficient water to bring the volume in the bottle to approximately the shoulder. Add a stir bar, cap the bottle, and stir to dissolve the contents. Carefully adjust the volume to 500 ml. TCA is extremely caustic. Do not attempt to measure or weigh TCA granules. Use caution when pipetting the solution. TCA 100% solution is very stable when stored at approximately ~4°C. Dilutions should be freshly prepared from the stock daily. Store dilutions on ice while in use and dispose of unused portions.

4. *[³H]Thymidine growth medium*: Add 10 μCi/ml [³H]thymidine (1 μCi/μl stock, PerkinElmer Life and Analytical Sciences, Inc., Cat. No. NET-027Z) directly to growth medium under sterile conditions at the rate of 10 μl/ml of medium. Prepare fresh on a daily basis.

5. *1.0 M Tris buffer*: Dissolve 60.55 g Tris buffer (Fisher, Cat. No. BP152-1) in ~400 ml water and adjust pH to 7.6. Adjust volume to 500 ml, filter sterilize, and store at 4°C.

6. *20% sodium dodecyl sulfate (SDS)*: Dissolve 20 g SDS (Sigma, Cat. No. L-1926) in sterile water using a sterile beaker and a stir bar rinsed in ethanol. Adjust

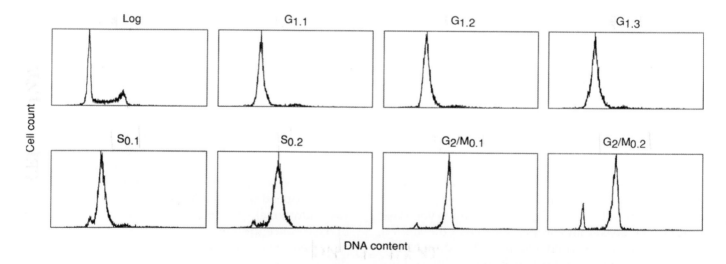

FIGURE 3 Flow cytometric analysis of cell cycle synchrony in sequential centrifugal elutriator fractions of HeLa S3 cells. Synchronous populations of HeLa S3 cells were selected by centrifugal elutriation of exponential cultures (log), and the degree of synchrony was analyzed by flow cytometry. Cell number was plotted against DNA content based on propidium iodide fluorescence for each cell cycle fraction. Flow cytometric analysis of sequential G_1-phase fractions collected at flow rates of 21 ml/min ($G_{1.1}$), 23 ml/min ($G_{1.2}$), and 25 ml/min ($G_{1.3}$). S-phase cells were selected by centrifugal elutriation at flow rates of 29 ml/min ($S_{0.1}$) and 35 ml/min ($S_{0.2}$), and G_2/M-phase cells were collected at 43 ml/min.

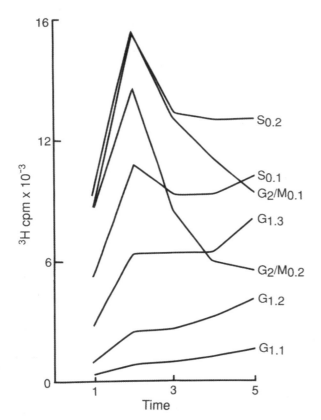

FIGURE 4 Analysis of synchrony of entry into DNA synthesis in centrifugal elutriation fractions of HeLa S3 cells. HeLa cells were separated into synchronous fractions by centrifugal elutriation, and kinetics of entry into DNA synthesis phase (S phase) was determined by measuring mean acid-precipitable [³H]thymidine incorporation (10 μCi/ml in growth medium) for five sequential 1-h incubations, in duplicate, for each cell cycle fraction identified in Fig. 3. Note that each cell fraction reaches S phase at sequentially later times after return to culture.

final volume to 100 ml. SDS cannot be autoclaved or filtered. Store at room temperature.

7. *TES buffer*: Add 10 mM Tris–HCl, pH 7.6 (1 ml of 1 M stock), 1 mM EDTA, pH 8.0 (0.5 ml of 0.5 M stock), and 1% SDS (5 ml of 20% stock) to 93.5 ml water, filter, and store at room temperature.

8. *4% paraformaldehyde*: Dissolve 4 g paraformaldehyde (Electron Microscopy Sciences, Cat. No. 15710) in 100 ml water and stir to dissolve by heating gently to 60–65°C in a fume hood. This can take an extended period and the solution will still appear cloudy. Clarify by the addition of a few drops of 1 M NaOH (up to ~20 drops). Cool before use and store at 4°C. Fixative fumes are extremely toxic. Always use in a fume hood and avoid any contact.

Steps

1. Analyze cell cycle fractions by flow cytometric analysis (Fig. 3). Wash approximately 5 × 10⁵ cells by centrifugation and resuspend in ice-cold Hanks' balanced salt solution. Fix cells by addition of an equal volume of ice-cold 4% paraformaldehyde. After a 24-h incubation at 4°C, collect the cells by centrifugation and resuspend in 2 ml of ice-cold PBS. Alternatively, fix cells by slow dropwise addition of 3 volumes of 70% ethanol (–20°C) while applying continuous gentle agitation with a vortex mixer. Allow cells to fix for at least 1 h at 4°C and then wash as described earlier. Approximately 30 min prior to flow cytometric analysis, add 3 ml of staining solution to each aliquot of 300 μl of cell suspension and incubate at room temperature. Analyze fluorescence on a flow cytometer (Elite Flow Cytometer, Coulter Electronics).

2. Analyze acid precipitable [³H]thymidine incorporation by synchronous cell populations by pulse labeling each separated fraction of cells with 10 μCi/ml [³H]thymidine in complete growth medium (Fig. 4) for 1 h at hourly intervals after return of the cells to culture (Wu *et al.*, 1993; Bird *et al.*, 1988). Place cells in 96-well plates (2 × 10⁴/well) in 100 μl growth medium and incubate under normal growth conditions. Add 1 μCi [³H]thymidine to each well at the appropriate time and incubate for 1 h at 37°C. Wash each cell fraction twice with Hanks' balanced salt solution, drain fluid, lyse in 100 μl TES, load 100 μl of lysate onto a 2.4-cm filter paper circle (Whatman 540, Cat. No. 1540-324) labeled with a No. 2 pencil, and allow to air dry. Precipitate samples with excess solutions of 200 ml for up to 50 filters for 20 min (do not exceed this time or the filters may be damaged): 20% TCA, 10% TCA, absolute ethanol, ether, absolute ethanol. Then, air dry and determine the radioactivity in each sample by liquid scintillation counting in a 5-ml ScintiSafe Econ 1 scintillation fluor (Fisher, Cat. No. SX20-5) as described previously (Wu *et al.*, 1993; Bird *et al.*, 1988).

IV. COMMENTS

A. Theoretical Basis of Separation

Cells are separated in the centrifugal elutriator by a process that was originally called counterstream centrifugation. This process is based on the opposition of two forces: media flow and angular acceleration due to an applied centripetal force. All of the cells in the suspension to be separated by centrifugal elutriation are pumped under gentle pressure into the separation chamber, entering at the extreme outside end of the chamber where the radius of rotation is maximal (r_{max}) to begin accumulating in the chamber (Fig. 1). As the cells fill the chamber, they change direction (as the loading tube turns) from flowing out to the perimeter of the rotor to flowing toward the center of the rotor. As they do, the particles enter a chamber that is designed with a conical profile. As the particles enter the chamber, the cross-sectional area increases rapidly, dramatically decreasing inward movement of the cells under a constant flow rate. Thus, the cells begin to decelerate as the cross-sectional area of the chamber increases. At some point, determined by cell size, media flow rate and viscosity, and centripetal force, the cells cease to move inward and become suspended in the chamber at a position where cell movement inward due to flow rate is balanced by apparent centrifugal force outward. At this point, medium is flowing by the cells but inward movement of the cells is offset by the apparent outward centrifugal force applied due to the angular momentum of the rotor. The population of cells has thus reached equilibrium and their position remains unchanged, although the cells tend to sort themselves under this combination of opposing forces with the smallest particles sorting closest to the center of the rotor.

There are now two choices for the operator with respect to fractionation of the cells: rotor speed can be decreased incrementally or the flow rate can be increased incrementally. In practice, the latter choice is the more practical due to the ease and precision with which the pump can be adjusted compared to the centrifuge. In either case, adjustments are made such that the equilibrium position of the smallest population of cells can move inward into the chamber until it reaches the constriction in the chamber where flow rate accelerates due to a rapid decrease in cross-sectional area. Thus, by adjusting the flow rate up in small increments (size of the increments must be determined empirically), cells can be fractionally separated from the population.

The actual elutriation characteristics depend on the shape and size of the cells in the population. For trypsinized cells, whose specific gravity is not much greater than the buffer in which they are suspended, shape generally approaches spherical unless cell processes or applied forces alter this. Thus, separation is usually the product of apparent cell diameter. Centrifugal elutriation has been shown to be unusually subtle in its ability to discriminate subfractions based on very small differences in cell size. For example, in separation of cells in different phases of the cell cycle, cells approximately double in volume as they pass through the cell cycle from G_1 phase to mitosis (Mitchison, 1971). This translates into an increase of only about 26% in the radius of an average premitotic cell compared to an average freshly divided G_1 phase cell. This relationship can be demonstrated by the

following equations where V is cell volume and r is cell radius, assuming a spherical shape. The G_1 phase cell is represented by a volume of V_{G1} and an apparent radius of r_{G1}, the premitotic cell is represented by a volume of V_M and an apparent radius of r_M, and the assumption is made that cell volume exactly doubles on average during the intervening period.

Thus

$$4/3\,\pi\,(r_{G1})^3 = V_{G1}$$
1 (volume of a spherical G_1 phase cell) (1)

$$4/3\,\pi\,(r_{G1})^3 = V_M$$
2 (volume of a spherical premitotic cell) (2)

$$2\,V_{G1} = V_M$$
3 (volume doubles during the cell cycle) (3)

Substituting Eqs. (1) and (2) into Eq. (3) for each volume and then simplifying,

$$2[4/3\,\pi\,(r_{G1})^3] = [4/3\,\pi\,(r_M)^3]$$

$$2\,(r_{G1})^3 = (r_M)^3$$

Taking the cube root of each side,

$$1.26\,r_{G1} = r_M$$

Thus, the radius of the largest cell in the cell cycle, on average, has an apparent radius (and diameter) only 26% larger than the smallest cell on average. We have successfully separated cells in this narrow continuum of cell sizes into seven distinct fractions with little overlap. This demonstrates that the subtle differences in cell size that occur as cells proceed through the cell cycle can be detected and efficiently separated using this technique. Successful separation was based on only about a 3–5% difference in radius (or diameter) between succeeding fractions that could be discriminated clearly by centrifugal elutriation.

The theoretical basis on which centrifugal elutriation is capable of separating cells, by balancing changes in medium flow balanced against angular momentum, has revealed that sequential fractions were frequently differentiated by as little as a 3% difference in cell diameter. In most cases, apart from the expense of purchase of the elutriation centrifuge, the only compromise consists of a somewhat lower level of purity in some samples, although purities approaching 100% have been reported. To balance this drawback, great improvements in yield, speed, and gentleness of handling of the cells have been achieved. Due to these obvious benefits, centrifugal elutriation has been applied to a wide variety of cells and cell types to successfully affect separation based on cell size.

The theoretical considerations described earlier have been applied to allow separation of a great variety of cell types and separation of cultured cells based on cell cycle phase (reviewed extensively in Conkie, 1985; Beckman Instruments, 1990). Using this protocol, it is possible to separate approximately 1×10^8 to 1×10^9 cells into up to eight sequential cell cycle phase fractions in about 40–60 min (sequential fractions designated: $G_{1.1}$, $G_{1.2}$, $G_{1.3}$, $S_{0.1}$, $S_{0.2}$, $G_2/M_{0.1}$, $G_2/M_{0.2}$, see Figs. 3 and 4). Later cell cycle fractions contained increasing numbers of contaminating cells from the phase preceding it. The first fractions were of the highest purity (approximately 97%, see Fig. 3) while the purity of the fractions dropped (to approximately 90%) in samples representing later times during the cell cycle. In G_2/M fractions, significant numbers of cells synthesizing DNA are recovered, although this activity declines rapidly once the cells are placed back in culture (Fig. 4). Only centrifugal elutriation can produce so many fractions of relatively high purity and of such large size so quickly with little detectable lag in cell growth and without drug/inhibitor-induced artifacts.

V. PITFALLS

1. In the time since we acquired our centrifugal elutriator, Beckman Instruments has released newer models designed to separate, among other qualities, larger numbers of cells. While the principles governing separation remain the same, details of procedures necessary in setting up the rotor and the seals may differ, depending on the model, and thus the manufacturer's directions should be adhered to carefully to ensure correct assembly and operation.

2. It is critical that separations be attempted only with single cell populations. Steps should be taken to ensure complete dissociation of cells liberated from tissues or culture vessels. When in doubt, dispersion of samples should be monitored microscopically.

3. Only Beckman neutral pH rotor detergent should be used to clean the rotor and separation chamber. It is particularly important to ensure that the sample tube at the outer edge of the separation chamber is soaked in detergent overnight to remove any cell debris that has accumulated as this will affect flow rate greatly as well as the fluid dynamics of sample loading in the chamber. This aperture is too small to be cleaned with a tool, and no instrument should be applied that could scratch the interior surface of the chamber. The operator must be extremely careful with the separation chamber, as overtightening of the screws or scratches

to the surface will damage the chamber, affecting performance significantly.

4. If alterations to the growth medium or elutriation medium are contemplated, the new medium must be tested to empirically determine at what flow rate cell cycle fractions elute. Even very small changes in media formulation (e.g., as little as 0.5% change in DHS concentration) will change the fluid dynamics of the elutriation system dramatically. Changes in cell-loading density and temperature can also create detectable effects on elutriation profiles. A simple pilot experiment is usually sufficient to determine what effects such alterations have on elution flow rate if it is followed by flow cytometric analysis.

5. All O rings and gaskets should be inspected before each run to ensure that they are in good condition. All worn seals should be replaced.

6. Failure to secure the wires and tubing connecting the rotor and strobe light to the exterior of the centrifuge will result in them becoming wrapped around the rotor, resulting in serious damage to the equipment.

7. We have replaced the Oakridge-style sample application tube supplied by the manufacturer with a straight glass test tube (13×100 mm) with the same size aperture at the top as the tube supplied. This eliminates the shoulder at the top of the Oakridge tube, which can trap cells, resulting in a continuous residual loading of cells throughout the elutriation procedure. The tube should be siliconized to within approximately 2 cm from the top with Gel Slick (FMC Bio-Products). Gel Slick should not be allowed to contact the glass surface above this point as the stopper will no longer hold securely under pressure.

8. Rotor speed should be monitored frequently during the run as fluctuations of only a small amount can affect purity of the samples greatly. It is particularly important to check rotor speed each time that the refrigeration system in the centrifuge cycles on.

9. If sterilization of the assembly is required, be sure that the centrifuge is *turned off* while the ethanol is in the system. Failure to observe this safety measure could result in a fire hazard. Alternatively, the system can be sterilized with 6% hydrogen peroxide while the centrifuge is running (Conkie, 1985).

Acknowledgments

The author thanks Dr. Gin Wu, Dr. Suresh Pai, and Dr. Shiawhwa Su for consultation on this protocol and Patricia DeInnocentes and Randy Young-White for valuable technical support. The author also thanks Dr. Lauren G. Wolfe for critical reading of the manuscript.

References

Beckman Instruments (1990). Centrifugal Elutriation of Living Cells: An Annotated Bibliography. In Applications Data, Number DS-534, pp. 1–41, Beckman Instruments Inc., Palo Alto, CA.

Bialkowski, K., and Kasprzak, K. S. (2000). Activity of the anti-mutagenic enzyme 8-oxo-2'-deoxyguanosine 5'-triphosphate pyrophosphohydrolase (8-oxo-dGTPase) in cultured chinese hamster ovary cells: Effects of cell cycle, proliferation rate, and population density. *Free Radic. Biol. Med.* **28**, 337–344.

Bird, R. C. (1996a). Cell separation by centrifugal elutriation. *In* "Cell Biology: A Laboratory Handbook" (J. E. Celis, ed.), 2nd Ed., pp. 205–208. Academic Press, New York.

Bird, R. C. (1996b). Synchronous populations of cells in specific phases of the cell cycle prepared by centrifugal elutriation. *In* "Cell Biology: A Laboratory Handbook" (J. E. Celis, ed.), 2nd Ed., pp. 209–217. Academic Press, New York.

Bird, R. C., Bartol, F. F., Daron, H., Stringfellow, D. A., and Ridell, G. M. (1988). Mitogenic activity in ovine uterine fluids: Characterization of a growth factor which specifically stimulates myoblast proliferation. *Biochem. Biophys. Res. Commun.* **156**, 108–115.

Bludau, M., Kopun, M., and Werner, D. (1986). Cell cycle-dependent expression of nuclear matrix proteins of Ehrlich ascites cells studied by *in vitro* translation. *Exp. Cell Res.* **165**, 269–282.

Boerma, M., van der Wees, C. G., Wondergem, J., van der Laarse, A., Persoon, M., van Zeeland, A. A., and Mullenders, L. H. (2002). Separation of neonatal rat ventricular myocytes and non-myocytes by centrifugal elutriation. *Pflug. Arch.* **444**, 452–456.

Braunstein, J. D., Schulze, D., DelGiudice, T., Furst, A., and Schildkraut, C. L. (1982). The temporal order of replication of murine immunoglobulin heavy chain constant region sequences corresponds to their linear order in the genome. *Nucleic. Acids Res.* **10**, 6887–6902.

Brown, E. H., and Schildkraut, C. L. (1979). Perturbation of growth and differentiation of Friend murine erythroleukemia cells by 5-bromodeoxyuridine incorporation in early S phase. *J. Cell. Physiol.* **99**, 261–277.

Conkie, D. (1985). Separation of viable cells by centrifugal elutriation. *In* "Animal Cell Culture: A Practical Approach" (R. I. Freshney, ed.), pp. 113–124. IRL Press, Oxford.

Dagher, R., Long, L. M., Read, E. J., Leitman, S. F., Carter, C. S., Tsokos, M., Goletz, T. J., Avila, N., Berzofsky, J. A., Helman, L. J., and Mackall, C. L. (2002). Pilot trial of tumor-specific peptide vaccination and continuous infusion interleukin-2 in patients with recurrent Ewing sarcoma and alveolar rhabdomyosarcoma: An inter-institute NIH study. *Med. Pediatr. Oncol.* **38**, 158–164.

Datta, N. S., and Long, M. W. (2002). Modulation of MDM2/p53 and cyclin-activating kinase during the megakaryocyte differentiation of human erythroleukemia cells. *Exp. Hematol.* **2**, 158–165.

Davis, P. K., Ho, A., and Dowdy, S. F. (2001). Biological methods for cell-cycle synchronization of mammalian cells. *Biotechniques* **30**, 1322–1331.

Deacon, E. M., Pettitt, T. R., Webb, P., Cross, T., Chahal, H., Wakelam, M. J., and Lord, J. M. (2002). Generation of diacylglycerol molecular species through the cell cycle: A role for 1-stearoyl, 2-arachidonyl glycerol in the activation of nuclear protein kinase C-betaII at G2/M. *J. Cell Sci.* **115**, 983–989.

Diamond, R. A. (1991) Separation and enrichment of cell populations by centrifugal elutriation. *Methods* **2**, 173–182.

Hann, S. R., Thompson, C. B., and Eisenman, R. N. (1985) c-*myc* oncogene protein synthesis is independent of the cell cycle in human and avian cells. *Nature* **314**, 366–369.

Hengstschlager, M., Holzl, G., and Hengstschlager-Ottnad, E. (1999). Different regulation of c-Myc- and E2F-1-induced apoptosis during the ongoing cell cycle. *Oncogene* **18**, 843–848.

Houser, S., Koshlatyi, S., Lu, T., Gopen, T., and Bargonetti, J. (2001). Camptothecin and zeocin can increase p53 levels during all cell cycle stages. *Biochem. Biophys. Res. Commun.* **289**, 998–1009.

Iqbal, M. A., Plumb, M., Stein, J., Stein, G., and Schildkraut, C. L. (1984). Coordinate replication of members of the multigene family of core and H1 human histone genes. *Proc. Natl. Acad. Sci. USA* **81**, 7723–7727.

Karas, M., Zaks, T. Z., Liu, J. L., and LeRoith, D. (1999). T cell receptor-induced activation and apoptosis in cycling human T cells occur throughout the cell cycle. *Mol. Biol. Cell* **10**, 4441–4450.

Lindahl, P. E. (1948). Principle of counterstreaming centrifuge for the separation of particles of different sizes. *Nature* **161**, 648–649.

Lindahl, P. E. (1986). On counterstreaming centrifugation in the separation of cells and cell fragments. *Biochim. Biophys. Acta* **21**, 411–415.

Lindahl, P. E., and Nyberg, E. (1955). Counterstreaming centrifuge for the separation of cells and cell fragments. *IVA* **26**, 309–318.

Lindberg, C. A. (1932). A method for washing corpuscles in suspension. *Science* **75**, 415–416.

Merrill, G. F. (1998). Cell synchronization. *Methods Cell Biol.* **57**, 229–249.

Mitchison, J. M. (1971). "The Biology of the Cell Cycle." Cambridge Univ. Press, Cambridge.

Pai, S. R., and Bird, R. C. (1992). Growth of HeLa S3 cells cotransfected with plasmids containing a c-fos gene under the control of the SV40 promoter complex, pRSVcat, and G418 resistance. *Biochem. Cell Biol.* **70**, 316–323.

Pai, S. R., and Bird, R. C. (1994). Overexpression of c-*fos* induces expression of the retinoblastoma tumor suppressor gene in transfected cells. *Anticancer Res.* **14**, 2501–2508.

Rehak, M., Csuka, I., Szepessy, E., and Banfalvi. G. (2000). Subphases of DNA replication in Drosophila cells. *DNA Cell Biol.* **19**, 607–612.

Sambrook, J., Fritsch, E. F., and Maniatis, T. (1989). "Molecular Cloning: A Laboratory Manual" (J. Sambrook, E. F. Fritsch, and T. Maniatis, eds.), 2nd Ed., p B.17. Cold Spring Harbor Laboratory Press, Cold Spring Harbor, NY.

Sugikawa, E., Hosoi, T., Yazaki, N., Gamanuma, M., Nakanishi, N., and Ohashi, M. (1999). Mutant p53 mediated induction of cell cycle arrest and apoptosis at G1 phase by 9-hydroxyellipticine. *Anticancer Res.* **19**, 3099–3108.

Syljuasen, R. G., and McBride, W. H. (1999). Radiation-induced apoptosis and cell cycle progression in Jurkat T cells. *Radiat. Res.* **152**, 328–331.

Van Leeuwen-Stok, E. A., Jonkhoff, A. R., Visser-Platier, A. W., Drager, L. M., Teule, G. J., Huijgens, P. C., and Schuurhuis, G. J. (1998). Cell cycle dependency of 67gallium uptake and cytotoxicity in human cell lines of hematological malignancies. *Leuk. Lymphoma* **31**, 533–544.

Wong, E. C., Lee, S. M., Hines, K., Lee, J., Carter, C. S., Kopp, W., Bender, J., and Read, E. J. (2002). Development of a closed-system process for clinical-scale generation of DCs: Evaluation of two monocyte-enrichment methods and two culture containers. *Cytotherapy* **4**, 65–76.

Wong, E. C., Maher, V. E., Hines, K., Lee, J., Carter, C. S., Goletz, T., Kopp, W., Mackall, C. L., Berzofsky, J., and Read, E. J. (2001). Development of a clinical-scale method for generation of dendritic cells from PBMC for use in cancer immunotherapy. *Cytotherapy* **3**, 19–29.

Wu, G., Su, S., Kung, T.-Y. T., and Bird, R.C. (1993). Molecular cloning of G₁-phase mRNAs from a subtractive G₁-phase cDNA library. *Biochem. Cell Biol.* **71**, 372–380.

Polychromatic Flow Cytometry[1]

Stephen P. Perfetto, Stephen C. De Rosa, and Mario Roederer

I. INTRODUCTION

The flow cytometer has proven to be one of the most powerful scientific techniques for the analysis of immunobiology in the past 35 years. Recent advances allowing the detection of 14 distinct cell parameters on each cell have revealed the immense heterogeneity of the immune system; e.g., the identification of more than 100 functionally distinct cell phenotypes in the peripheral blood of humans. In addition, the evolution in computer technology brings to bear the ability to analyze very large sample sets using sophisticated algorithms, increasing the power of analysis of low-frequency cell populations.

Current state-of-the-art cytometric evaluation allows for distinct cell population delineation using two physical parameters—side scattered light (SSC) and forward scattered light (FS)—and 12 separate fluorescent parameters. Each fluorescence parameter can be used to independently measure the expression of a protein or function; combined, these measurements may predict disease progression, vaccine responses, or other immunological parameters.

Still, the use of high-end multicolor flow cytometry is in its infancy; many challenges must still be overcome before this technology will become routinely available in research laboratories. The most difficult obstacle at this time is reagent availability; many laboratories are forced to conjugate fluorochromes not yet available commercially. Many dyes are now available

that can be conjugated easily to antibodies for use in polychromatic flow cytometry (PFC) to supplement the commercially available conjugates. With regard to instrument development, by far the most important requirement is automation. Instrument setup and calibration are still far more complex with the use of multiple lasers and detectors for PFC; computer-aided validation of instrument performance is necessary. In addition, automated compensation (fluorescence spillover) is necessary; however, this is adequately handled by most contemporary software packages.

Data analysis is by far the most time-consuming aspect of PFC experiments. The complexity of data is such that, with current software tools, analysis of individual samples requires inordinate amounts of time. There is a considerable demand for tools that can organize the analyses into databases, as well as assist in the exploration of the complex data sets. To help researchers analyze such data, automated multivariate techniques have been developed by this laboratory. These algorithms are designed to compare multidimensional distributions to identify and quantify the degree of difference between data sets (De Rosa *et al.*, 2001; Roederer and Hardy, 2001). In addition, these algorithms can rapidly identify regions of multivariate distributions that differ, thereby providing a mechanism for identifying those cells that are different between two samples. Tools such as these help researchers explore the complex data sets and identify interesting aspects that no combination of two-dimensional graphs could have revealed. The probability binning (PB) algorithm can be used to quantify the degree of similarity or disparity of highly complex distributions for the purposes of ranking these distributions (Roederer *et al.*, 2001a,b). This may be useful for quantifying the number of cells that respond to a particular stimulus (perhaps having responded in complex multivariate patterns) or for identifying variations in immunophenotyping patterns that correlate

---

[1] The National Institutes of Health does not endorse or recommend any commercial products, processes, or services. The views and opinions of authors expressed in this manuscript do not necessarily state or reflect those of the U.S. Government and may not be used for advertising or product endorsement purposes. The U.S. Government does not warrant or assume any legal liability or responsibility for the accuracy, completeness, or usefulness of any information, apparatus, product, or process disclosed.

with pathogenic states (Baggerly, 2001). These strategies allow for the most useful and comprehensive display of complex data in the field of flow cytometry today.

This article presents some of the latest techniques used in our implementation of our 12-color flow cytometric technology. It discusses the reagents, calibration and setup, and analysis of the resulting data, including some of the hurdles and pitfalls encountered. This guide will aid research laboratories wishing to implement flow cytometric technology capable of more than 5 or 6 colors.

II. MATERIALS AND INSTRUMENTATION

A. Monoclonal Antibodies

Purified monoclonal antibodies are available from manufacturers as bulk concentrated protein in the absence of any exogenous protein. All conjugations are performed as detailed (http://www.drmr.com/abcon/). Large quantities of reactive fluorochromes are prepared and stored (and are stable for many months at 4C); the conjugation to antibodies is a fairly rapid procedure that can then be accomplished in 2–3 h.

Figure 1 shows the common fluorochromes currently used in our laboratory, which can all be directly conjugated to monoclonal antibodies. Also shown are the excitation and emission spectra and suggested bandpass filters for each fluorochrome. Ideally, each fluorochrome should be conjugated to a wide variety of monoclonal specificities in order to provide a wide range of possible panels; of course, this requires a fairly substantial investment of both fluorochromes and monoclonal antibodies. New fluorochromes are becoming available that are more photo stable and easier to conjugate, such as the Alexa family of fluorochomes (i.e., Alexa, 488, 532, 660, 633, 647, and 680). Commercial manufacturers currently offer only a limited range of fluorochromes conjugated to specific

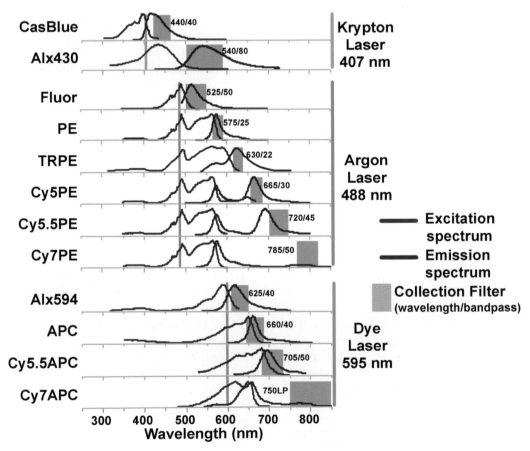

FIGURE 1 Conjugate spectrum chart of excitation and emission curves for common fluorochomes used in PFC. Laser lines show which conjugates are excited by the individual laser lines and the selected bandpass filters used to measure the specific emission.

monoclonal antibodies. Table I shows that only 7 of the 12 fluorochromes used in our laboratory are available commercially; this range is currently increasing but is still a limiting factor. In addition, newer conjugates have far more limited reagent combinations available until manufacturers can build up a significant inventory. Hence, it is likely that researchers who wish to perform more than 6-color flow cytometry will likely need to invest in the ability to manufacture reagents in-house. In general, this is incorporated most efficiently into a core facility that can supply reagents to multiple research laboratories.

B. Qualification of Fluorescently Conjugated Monoclonal Antibodies (mAbs)

Regardless of the conjugate used (i.e., commercial or in-house), all mAbs require titration against a target cell population to determine the optimal concentration for staining. Note that staining conditions (tempera-ture, time, and volume) impact the titration; thus, qualification should be performed under the same conditions as experimental staining. By plotting the median of the positive cell population against the serial dilution of the mAb as illustrated in Fig. 2, the lowest concentration at which the maximum separa-tion can be discerned; in general, this is the optimal concentration to use (in this example, $10 \mu g/ml$). Lower concentrations of mAb result in loss of resolu-tion (note, however, that for some mAbs the separa-tion is still highly adequate at very low concentrations, allowing the use of the reagent at a more economical rate). It cannot be overemphasized that proper titration and selection of mAb concentration are paramount for successful multicolor analysis. Adding too much anti-body conjugate can have as much effect on sample analysis as the addition of too little antibody conju-gate. In this situation, backgrounds increase due to nonspecific binding of the antibody conjugate, thus reducing the signal-to-background ratio dramatically.

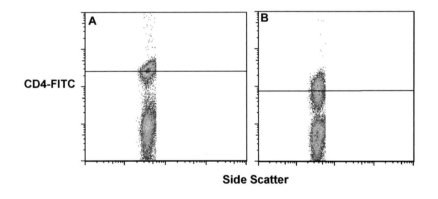

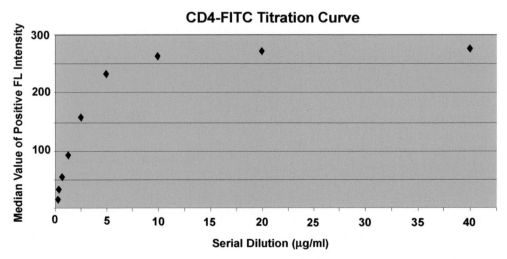

FIGURE 2 Serial dilution of anti-CD4-FITC-labeled PBMCs and the median value of the positive signal (black line). (A) The point on the titration curve ($10 \mu g/ml$) yielding the greatest separation (highest median value with the lower auto fluorescence) as compared to poor separation (B) at a lower titer (<$2 \mu g/ml$).

TABLE I Current Breadth of Availability of Antibody Conjugates[a]

Fluorochrome	VRC	Caltag	Coulter	BDIS
FITC	>100	>100	>100	>100
PE	>100	>100	>100	>100
TRPE	26	5	20	1
Cy5PE	96	38	30	76
Cy55PE	43	—	—	—
Cy7PE	54	6	15	18
Ax594	66	—	—	—
APC	>100	28	15	137
Cy55APC	46	—	—	—
Cy7APC	44	7	—	14
CasBlue	58	—	—	—
Ax430	12	—	—	—

[a] The approximate number of distinct specificities available from three major manufacturers compared to what we have conjugated in our laboratory ("VRC") is shown. A wide range of conjugates is necessary to have the greatest flexibility in creating multicolor antibody staining panels.

C. Instrumentation

Samples are acquired on a modified FACSDiVa flow cytometer (BDIS, San Jose, CA), which measures 12 fluorescent parameters and 2 physical parameters (FS and SSC). Figure 3 illustrates the optical configuration and filter selection of this instrument. The geometry of this instrument is based on the traditional collection optics and is far more complex than more recent instrumentation using optical fibers such as the LSR II (BDIS). In these instruments, photons of light from the laser-excited conjugates are transmitted through optical fibers and measured in unique optical arrays. Figure 4 shows an example of an argon laser array containing eight detectors called an Octagon. The advantage of these systems includes increased efficiency of photon transfer, thus lower-energy lasers can be used with good signal reproducibility as compared to higher-end systems equipped with powerful lasers. Figure 5 shows the laser configuration of older traditional high-end systems as compared to newer instruments using low-powered diode lasers (5–10 mW). Clearly another advantage of lower-power lasers and optical fibers is the smaller footprint, yielding a more compact system with the measurement capability of a high-end research instrument.

D. Alignment and Calibration Beads

Alignment beads containing a single peak bead (single peak Rainbow beads) are from Spherotech Inc. (Rainbow beads, Cat. No. RFP-30-5A). Calibration beads containing eight separate peaks (eight peak Rainbow beads) are from Spherotech Inc. (Rainbow beads, Cat. No. RFP-30-5A) and Blank Beads are from Becton Dickinson Immunocytometry Systems (Cat. No. RFP-30-5A). All beads are diluted by the addition of one drop (approximately 20 µl) per milliliter of phosphate-buffered saline (PBS) containing 1% HIFCS (Quality Biological—PAA labs, Cat. No. 110-001-101) and 1 mg/ml of sodium azide (Sigma Chemical, Cat. No. S202).

E. Monoclonal Antibody Selection and Combinations

A laboratory Web-based database containing all of the antibody conjugate reagents is used to select the antibody conjugate combinations. This database lists the correct mAb concentration (determined from a titration curve; see the titration procedure in Section III) and displays the concentration, specificity, lot number, and location on a laboratory worksheet. Such databases become necessary—laboratories doing 6–12 color flow cytometry will inevitably have a very large storehouse of reagents; it is necessary to have a centralized repository of the qualification and validation data for each reagent in an easily accessible location. Researchers planning experiments need access to this information in order to know how much of each reagent to use, as well as some idea of the quality of the staining that can be expected with that reagent.

F. Compensation Beads

Latex beads coated with anti-mouse κ antibody are from Becton Dickinson Immunocytometry Systems (Cat. No. 557640). After incubating with mAb conjugate, beads are fixed in a final concentration of 0.5% paraformaldehyde (PFA). These "capture" beads are used with each antibody conjugate tested to set up the compensation matrix.

III. PROCEDURES

A. Alignment and Instrument Calibration

All flow cytometers, regardless of engineered advances in alignment techniques, require alignment and calibration quality control to determine reproducibility and sensitivity. For this purpose, alignment beads and calibration standards must be stable and reproducible from day to day. It is important to point out that such quality control measurements are not a substitute for proper cell controls as outlined in this

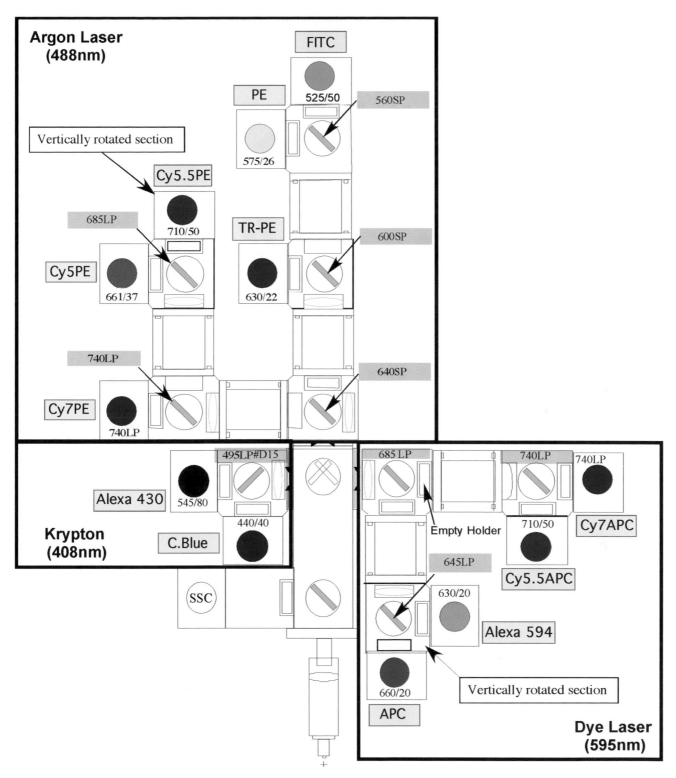

FIGURE 3   Digital Vantage optical configuration used to detect 12 colors using three lasers: 488-nm argon, 408-nm krypton, and 595-nm dye lasers. Each section of detectors is outlined (black boxes) to indicate the laser associated with the emission. Also illustrated are the dichroic mirrors and bandpass filter combinations for each PMT.

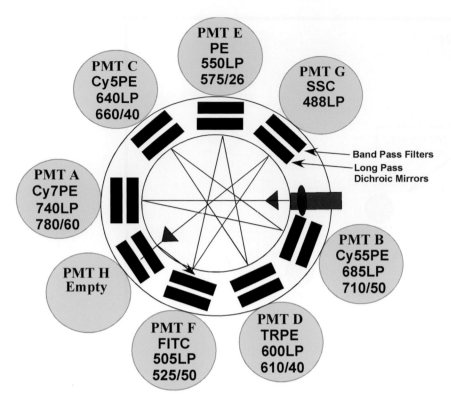

FIGURE 4 LSR II optical configuration of detectors for only the 488-nm diode laser. This configuration shows the light path with the typical dichroic mirrors and bandpass filters used to detect six different conjugates excited by this laser plus side scattered light.

article to assure testing quality. Alignment beads are used to determine good instrument performance and, if successful, should determine proper light collection in all detectors as measured by fluorescence intensity and fluorescence CV at a consistent voltage. Once the instrument lasers are aligned properly, calibration beads can be used to determine the correct tolerance range of fluorescent intensity by adjusting the detector sensitivity (e.g., photomultiplier voltage). These tolerance ranges are determined by unstained cell analysis set to a predetermined value. In addition, calibration standards determine the signal-to-background ratio, which must remain consistent within at most a 5% variation.

Steps

1. Place diluted single peak Rainbow beads onto the sample insertion tube and adjust the sample pressure to approximately 600 beads per second.

2. While observing dual-parameter histograms of FS vs SSC and all other fluorescence parameter combinations, adjust instrument to achieve the narrowest CV and highest intensity possible according to the manufacturer's instructions. Figure 6 shows the incorrect (A) and the correct (B) display for two fluorescence parameters. It is important to note that while slight misalignments may have only a small impact on the measurement of any given single parameter, they can have a serious impact on the visualization of data after compensation (Roederer, 2001a,b). This is because the spread in compensated data is directly related to the efficiency of light collection; larger CVs, as illustrated in Fig. 6, result from decreased light efficiency and result in much poorer data quality after compensation.

3. Adjust the voltages for each PMT to result in predetermined intensity levels for the bead population. Such levels are defined previously as optimal for the particular types of samples that are being analyzed; this calibration procedure ensures that the instrument sensitivity is comparable across experiments, allowing for the best comparison of data. As an example, for the analysis of lymphocytes, we typically analyze completely unstained lymphocytes and adjust the PMT voltages such that the upper end of the unstained cells is at the top of the first (of four) decades of fluorescence. Then the beads are reanalyzed at those voltages

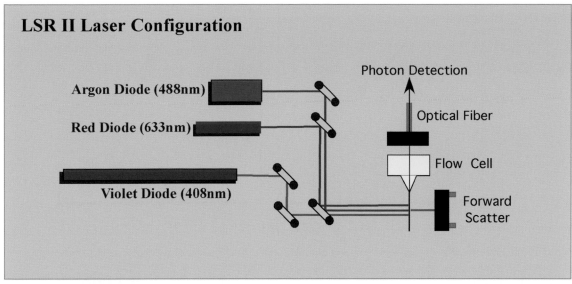

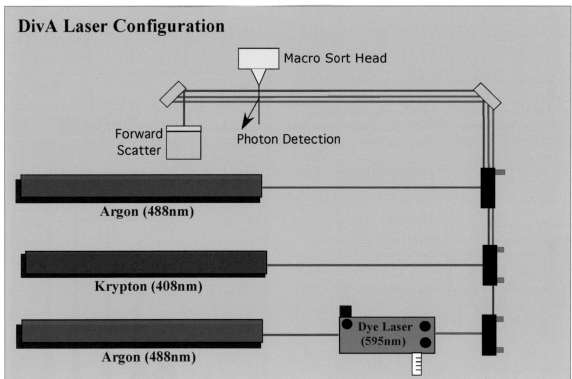

FIGURE 5 Comparison between the Digital Vantage and the LSR II laser configuration system. Due to the use of the low-powered diode laser in the LSR II system, this instrument can be very compact and inexpensive to operate.

and the intensities are recorded for future target settings. These same voltage settings are used for sample acquisition and instrument calibration.

4. Place diluted eight peak Rainbow beads onto the sample insertion tube and adjust the sample pressure to approximately 600 beads per second.

5. Collect and save all single parameter histograms for future analysis.

6. Place diluted Blank beads onto the sample insertion tube and adjust the sample pressure to approximately 600 beads per second.

7. Collect and save all single parameter histograms for future analysis.

8. From data collected on the single peak beads, chart voltage versus time. Correct tolerances for each PMT should be ±5% variance.

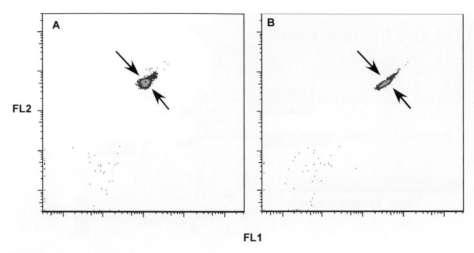

FIGURE 6 Rainbow beads with incorrect (A) and correct (B) alignment of two fluorescent parameters. Tolerance ranges for the coefficient of variance and fluorescent intensity are established for all instrument parameters. Setting quality control of all parameters within these tolerance ranges can avoid incorrect alignment and compensation error.

9. Data collected on the eight peak beads (median channel of the eight peak) divided by the median channel of the blank bead yield the signal-to-background ratio (see Fig. 7). Each PMT will have a characteristic S/B ratio; however, the correct tolerance plotted against time should produce a range of ±10%.

B. Compensation

Compensation and analysis of samples can be done either online or after data collection with appropriate software. In general, if cell samples are sorted, the user must perform compensation online; however, in most cases, sample compensation and analysis are performed off-line. Regardless of when compensation is performed, the same rules apply to correctly compensate the sample. It is important to note that each experiment must have matching compensation controls. These controls must produce signals, which are of as high or higher fluorescence intensity than the test sample. In many cases the use of compensation beads will satisfy this condition; however, if any of the experimental samples are more than severalfold brighter than the beads, then single-stained cells of the appropriate reagents must be used as compensation controls. In addition to matched compensation controls, a negative (unstained beads) must be collected. This control and individually labeled compensation tubes are used in the compensation algorithm to calculate compensation. Finally, to verify cell autofluoresence relative to the unstained bead location, an unstained cell sample control must be collected. Online compen-

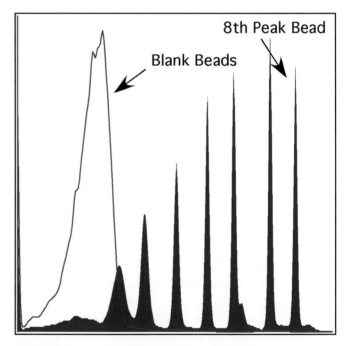

FIGURE 7 An overlay of eight peak rainbow beads with blank beads of 1 of the 12 parameters. The signal/background ratio is calculated using the median channel of the eighth peak bead and dividing by the median channel of the blank bead.

sation for the FACSDiVa is described later. Figure 8 shows an example of compensation beads labeled with anti-CD8 APC. Figure 8A shows the location of unstained beads as compared to the stained beads in Fig. 8B. As expected, the degree of spillover for the excitation of APC from the argon into the Cy5PE channel is negligible (Fig. 8C). However, the dye laser (595 nm) excites APC and the spillover is seen in the Cy55APC channel. This will require compensation correction of the Cy55APC channels due to the spillover of APC (Fig. 8D). As shown in Fig. 9, the degree of compensation (percentage of spillover) or light contamination can be considerable, and the complexity of this issue is only magnified by the addition of multiple parameters. A detailed explanation of compensation and controls is beyond the scope of this article; however, additional information can be found at http://www.drmr.com/compensation/index.html.

Steps

1. Into a 12 × 75 test tube add 40 µl of compensation beads (mouse anti-κ beads) and the volume of a pre-viously tittered antibody conjugate. Dilute volume to 100 µl with PBS containing 1 mg/ml of sodium azide and 1% fetal calf serum (FCS). Note that each antibody conjugate tested will have a single stained tube for the compensation control used in the experiment. Each compensation control tube will be used to set the compensation matrix.

2. Incubate in the dark for 15 min at room temperature.

3. Wash once in PBS containing 1 mg/ml of sodium azide and 1% FCS.

4. Remove supernatant and resuspend in 250 µl of PBS containing 1 mg/ml of sodium azide and 1% FCS.

5. Vortex and add 250 µl of 0.5% PFA.

6. Acquire each compensation control tube and the unstained bead control on the flow cytometer using previously defined voltage settings.

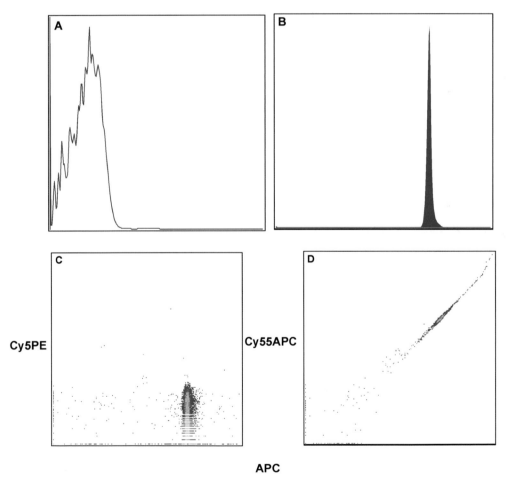

FIGURE 8 Fluorescent intensity of unstained (A) compensation beads (anti-mouse κ) and stained beads with anti-CD8 APC (B). (C and D) The effect of contaminated light from the excitation of CD8-APC. In C, no compensation is required because no contaminating light or spillover occurs into the Cy5PE channel. However, D shows that compensation correction is required because of the large spillover of contaminating light into the Cy55APC channel. After applying compensation correction, D will appear like C.

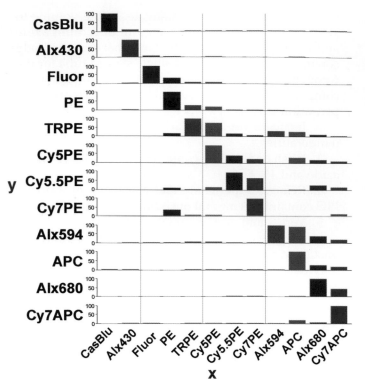

FIGURE 9 Compensation percentage of contaminated light, which must be subtracted from all of the other detectors for accurate analysis. The *x* axis shows the signal, and the *y* axis indicates the percentage of contaminated light removed from the signal. In general the largest contamination occurs within the same laser excitation group.

7. Set automated compensation matrix either on the instrument or by off-line software. Once completed, the test sample is correctly compensated and is ready for acquisition.

8. Compensation should be checked frequently by acquiring cell samples stained with combinations of antibody conjugates. After applying the compensation matrix, median values of the unstained cells should match the median values of the positive cells.

C. Sample Acquisition for Immunophenotyping

Steps and Considerations

1. Altering sample pressure and sheath velocity should be avoided. When possible, cells should be analyzed at low sheath pressure to ensure greatest sensitivity; increasing sample pressure increases the CV of the measurement, thus affecting alignment and compensation negatively. Therefore, it is recommended to maintain the lowest useful sample pressure and sheath velocity.

2. Collecting enough events for statistical analysis is key. However, the number of events needed for 12-color analysis is no different than for 2-color analysis. What is important is the size of the population of cells of interest. For example, if the subset of interest represents 0.1% of the input population, then collecting 1 million events yields 1000 events of the subset of interest. In general, 1000 events is more than enough for phenotypic analysis; for quantitative enumeration, a count of 1000 has an associated precision of 3% (i.e., the square root of 1000 divided by 1000). If a 1% precision was required, then 10 million events of the original sample need to be acquired.

3. Particle size and cell aggregates should be considered before sample collection. In addition to sample clogging the macro-sort tip, small changes in alignment can alter fluorescent intensity, compensation, and forward scatter detection. In general, if the particle size is greater than one-fourth the size of the macro-sort tip, the user should consider a larger macro-sort tip. Alternatively, cell aggregates can be avoided by prefiltering samples through a 100-μm filter cap tube prior to sample acquisition.

D. Sample Analysis

Steps and Considerations

1. Fluorescence minus one (FMO) control refers to a staining strategy, which uses all mAbs in the staining mix as in the test sample except for one mAb. This method allows for the correct determination of gate selection and verification of cell percentages. Figure 10 shows an example of the FMO strategy. In this example, cells were stained with four colors: anti-CD3-FITC, anti-CD4-PE, anti-CD8-Cy5PE, and anti-CD45RO-Cy7PE. The FMO control lacked anti-CD4-PE. In Fig. 10A, the sample is compensated correctly and shows that the FMO control is a better indicator of negative cell control cursor position than the unstained control sample (compare line 1 and line 2). Figure 10B demonstrates that even poorly compensated samples can benefit from the FMO control sample (compare line 1 and line 2). After setting the positive gate cutoff the test sample percentages can be determined more accurately. Thus this control reduces false negatives (best specificity) and better identifies positive cell populations (lower sensitivity). In addition, the FMO control can help identify compensation issues within the test sample. Therefore, FMO controls should be used whenever accurate discrimination is essential or when antigen expression is relatively low. While theoretically there could be as many FMO controls for each staining combination as

	Unstained Control		FMO Control	Test Stain
FITC	–		CD3	CD3
PE	–		–	CD4
Cy5PE	–		CD8	CD8
Cy7PE	–		CD45RO	CD45RO

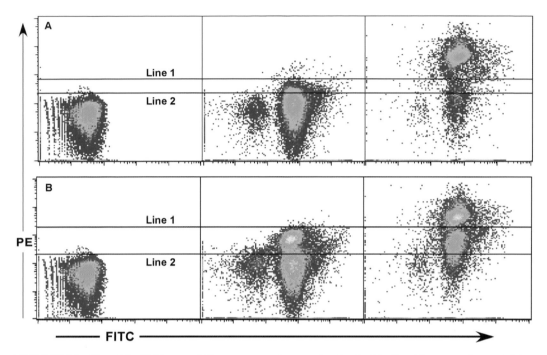

FIGURE 10 Use of the FMO control (fluorescence minus one) stained with all mAbs except for anti-CD4-PE. (A and B) The unstained control, the FMO control, and the fully stained cell sample (anti-CD3-FITC, anti-CD4-PE, anti-CD8-Cy5PE, and anti-CD45RO-Cy7PE) are compared. (A) The sample is compensated correctly and shows that the FMO control is a better indicator of negative cell control cursor position than the unstained control sample (compare line 1 and line 2). (B) Even samples that are poorly compensated can benefit from the FMO control sample (compare line 1 and line 2).

there are colors, in reality most of these are not necessary. For example, in many cases (e.g., CD3 or CD8 staining), the distinction of positive and negative cells is made easily enough based on visual inspection. In no case, however, is a completely unstained (or completely isotype-stained) sample the appropriate control for setting discriminatory gates.

2. Use only titered antibody conjugates prior to use in combinations as described earlier.

IV. COMMENTS

A. Sample Viability

Dead cells will bind many antibody conjugates nonspecifically and erroneously count these as a positively labeled cell. Therefore, gating strategies must be employed to properly gate out these cells. Intercalating dyes such as ethidium monoazide (EMA) or propidium iodide (PI) can be useful in gating out these nonspecifically labeled cells. One advantage of the use of PI over EMA is the ability to use the same channel (Cy5PE channel) for both PI and another mAb stained with Cy5PE. This can be done due to the high intensity of PI over most mAbs sharing this channel. Figure 11 shows an example of EMA used as a dead cell discriminator. In this example, unfixed cells were stained with 0.5 µg/ml of EMA (Molecular Probes Inc, Cat. No. E1374) for 10 min on ice covered with aluminum foil followed by 15 min under a bright fluorescent light. Samples can be fixed with 2% PFA and run within the same day. Dead cells are labeled positive with EMA, as seen in Fig. 11B, and can be removed from the gated live cells (live cell gate, B). Without the use of EMA, Fig. 11A shows the total number of dead cells and live cells combined. Cells within a standard

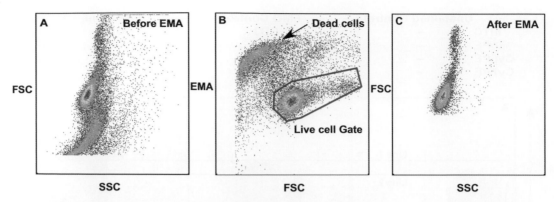

FIGURE 11   Use of ethidium monazide (EMA) as a method to discriminate live cells from dead cells. Dead cells are labeled positive with EMA as seen in B and can be removed from gated live cells (live cell gate). Without the use of EMA, A shows the total number of dead cells and live cells combined. Cells within a standard light scatter gate could contain dead cells as shown in A; however, after the dead cells were gated out using the live cell gate in B, dead cells were removed from the analysis (C).

light scatter gate could contain dead cells as shown in Fig. 11A; however, after the dead cells were gated out using the live cell gate in Fig. 11B, the dead cells were removed from the analysis (Fig. 11C).

B. Antibody Aggregation

Cyanine conjugates can potentially form immune complexes or aggregates during storage. Typically, cyanine tandem dyes form these aggregates and must be removed before using in the staining procedure. Ultracentrifugation of the antibody conjugate mixture at 13,000 g for 3 min will remove these aggregates.

V. CONCLUSION

Routine polychromatic flow cytometry is now closer to reality than ever before. Recent advances in instrumentation, computer technology, and biochemistry will prove to be the ingredients necessary to successfully understand the human immune system (Eckstein *et al.*, 2001; Roederer *et al.*, 1997). As engineering goals meet science objectives, the last frontier to cross will be analysis and comprehension of data never seen before or not very well understood. It will be this area where intense effort is needed to understand the massive amount of information collected and interpreted. Nonetheless, significant hurdles remain to be crossed by all laboratories wishing to implement this technology, and significant education of all immunologists regarding the interpretation of data generated by this technology is crucial to the understanding of its vagaries.

References

Baggerly, K. A. (2001). Probability binning and testing agreement between multivariate immunofluorescence histograms: Extending the chi-squared test. *Cytometry* **45**, 141–150.

De Rosa, S. C., Herzenberg, L. A., and Roederer, M. (2001). 11-color, 13-parameter flow cytometry: Identification of human naive T cells by phenotype, function, and T-cell receptor diversity. *Nature Med.* **7**, 245–248.

Eckstein, D. A., Penn, M. L., Korin, Y. D., Scripture-Adams, D. D., Zack, J. A., Kreisberg, J. F., Roederer, M., Sherman, M. P., Chin, P. S., and Goldsmith, M. A. (2001). HIV-1 actively replicates in naive CD4(+) T cells residing within human lymphoid tissues. *Immunity* **15**, 671–682.

Roederer, M. (2001a). Compensation is not dependent on signal intensity or on number of parameters. *Cytometry* **46**, 357–359.

Roederer, M. (2001b). Spectral compensation for flow cytometry: Visualization artifacts, limitations, and caveats. *Cytometry* **45**, 194–205.

Roederer, M., De Rosa, S., Gerstein, R., Anderson, M., Bigos, M., Stovel, R., Nozaki, T., Parks, D., and Herzenberg, L. (1997). 8 color, 10-parameter flow cytometry to elucidate complex leukocyte heterogeneity. *Cytometry* **29**, 328–339.

Roederer, M., and Hardy, R. R. (2001). Frequency difference gating: A multivariate method for identifying subsets that differ between samples. *Cytometry* **45**, 56–64.

Roederer, M., Moore, W., Treister, A., Hardy, R. R., and Herzenberg, L. A. (2001a). Probability binning comparison: A metric for quantitating multivariate distribution differences. *Cytometry* **45**, 47–55.

Roederer, M., Treister, A., Moore, W., and Herzenberg, L. A. (2001b). Probability binning comparison: A metric for quantitating univariate distribution differences. *Cytometry* **45**, 37–46.

33

High-Speed Cell Sorting

Sherrif F. Ibrahim, Timothy W. Petersen, Juno Choe, and Ger van den Engh

I. INTRODUCTION

Flow cytometric sorting is an extremely versatile technology that has established itself as a cornerstone in biological research for the foreseeable future. The strength and long-term success of this approach to cell purification can be attributed to its highly quantitative, rapid, and serial nature of analysis. Each passing particle is individually interrogated for the presence or absence of potentially limitless combinations of light scattering and fluorescence parameters. If user-defined threshold values are met, desired cells are then isolated from a heterogeneous sample. As such, the cell sorter becomes a launching point for many downstream cellular and molecular investigations, as well as a powerful tool for the analysis of complex mixtures.

The modern age of discovery-based, system-wide approaches to research is contingent upon high throughput analysis, and cell sorters have had to evolve accordingly to keep pace with these changes. Aptly, the most significant improvements in sorting technology were catalyzed by large-scale DNA sequencing efforts such as the Human Genome Project (Van Dilla and Deaven, 1990). It was obvious that by providing chromosome-specific DNA libraries, cell sorters could accelerate as well as improve sequencing results. The problem, however, was that literally days of sorting would be necessary to isolate sufficient genetic material for sequencing [modern polymerase chain reaction (PCR) techniques had not yet been reported]. Ensuing advances in the fluidics, illumination, optics, computers, and electronics of sorters led to the development of several experimental instruments that could sort at rates nearly two orders of magnitude above existing machines, and

eventually laid the groundwork for modern, high-end, high-speed sorters (Peters *et al.*, 1985; Gray *et al.*, 1987).

It is important to stress that the term "high speed" does not refer to simply speeding up the sort process, but is the product of careful engineering of each individual component in order to optimize overall performance. Coupled with these developments, the newest instruments have also become modular, with openly accessible and changeable parts, resulting in greater flexibility and sophistication with simpler and more stable operation.

The continuing evolution of high-speed cell sorters illustrates how advances in scientific knowledge and laboratory techniques allow for novel applications of existing methodologies, whereas improvements in instrument design facilitate previously unapproachable experiments. Accordingly, we have divided this article into two main parts. The first presents ideas that will yield better results in any experimental setting, with focus geared toward cytometer setup and general instrumentation. The following sections discuss biological aspects of study design, particularly pertaining to newer applications that rely on the added features and throughput of high-speed machines. The latter protocols do not include the traditional bulk separation of biological material, but instead take advantage of the capacity to screen rapidly through extremely complex cell and molecular populations for the isolation of highly defined, rare events. Specifically, fluorescent reporter proteins have provided the cytometrist with a real-time window into gene expression and have become the focus of many of these experimental approaches. It is these newest applications of high-speed cell sorting coupled with our increasing ability to study fewer, more characterized cells that will have the greatest impact on basic science research and our knowledge of human biology.

II. HIGH-SPEED CELL SORTERS: TECHNIQUES FOR RELIABLE SORTING

High-speed cell sorters are currently available from a handful of manufacturers with varying specifications, capabilities, facility requirements, and user interfaces. The purchase of any one particular instrument will depend on the desires of individual laboratories and should come after careful consideration of these differences. Although there are a number of mechanisms for sorting cells into discrete populations, the most widely used method for high-speed cell sorting involves using a jet in air, where cells contained in charged droplets are deflected using a static electric field. While the discussion that follows is most relevant to these sorters, the majority of protocols will have broader application. This section outlines some suggestions for consistent, reliable cell sorting that can be instituted at any sorting laboratory.

A. Cell Sorter Maintenance

Successful high-speed cell sorting requires both careful preparation of the sample to be studied and maintenance of the instrument being used. In most cases, the upkeep of electronics, lasers, deflection plates, software, and other hardware is left to service professionals, while alignment of optics and maintenance of fluidics and associated systems remain the purview of the user. In practice, however, basic knowledge of all aspects of the sort process is needed to ensure optimal and consistent operation.

In cell sorters, the fluidics layout is composed of two subsystems: the sample injection system and the sheath fluid system that surrounds the sample core. Together, these two subsystems contain several feet of tubing, and it is into the interstices and cracks of the tubing and valves that contaminants collect and can affect sort purity. As a result, the tubing material should be chosen to be as inert as possible and changed on a regular basis. Likewise, valves and corners should be kept to a minimum. Typically, the sheath tubing, sample tubing, and associated in-line filters should be changed at least monthly. New tubing should be washed and flushed for at least 10 min prior to use with a 2% bleach solution, followed by sterile sheath fluid. For cell sorting where even slight contamination can skew results greatly, such as cases in

which the sorted fraction will be used for quantitative PCR, it is not farfetched for the sample tubing to be replaced at the start of each sort. For its combination of inertness and cost, materials such as FEP (Teflon) are a good choice for the sheath tubing. FEP is chemically inert to a wide range of materials, holds up well to both acids and alkalis, and is autoclavable. An alternative to FEP that has better mechanical properties, particularly with respect to plastic deformation, is Tygon S-50-HL surgical grade tubing. Like FEP, this material can be sterilized by autoclaving (30 min, 15 psi, 250 F). Unfortunately, sterilization does not always free tubing of inert inorganic contaminants, and thorough flushing prior to each use is recommended. For sample injection tubing, an ideal choice is PEEK tubing. Like FEP, it is chemically inert and has good mechanical properties necessary for rigidity of the smaller diameter tubing.

At the start of each day, the sorter should be flushed for 10 min with a 2% bleach solution through both the sample and sheath lines, again followed by sterile sheath fluid to remove the bleach. In a similar manner, the sample injection should be flushed between samples when cross contamination is an issue. Finally, at the end of each day of sorting, it is recommended to flush the sheath and sample lines with a 2% bleach solution for 10 min followed by deionized water to reduce salt crystal accumulation and then air-dried. Attaching an empty sample tube and sheath container to the sorter and pressurizing each for a brief period can accomplish air drying of the sample and sheath lines. The tubing and the sheath containers should be stored dry in a clean place.

B. Sheath Fluid

It is often overlooked that cell samples can be sensitive to even minor changes in the composition of the sheath fluid. One may erroneously assume that the sample does not interact significantly with the sheath, as there is no mixing between the two (the flow at the sample injection is laminar) and ions and solutes can move only by diffusion. This is not always the case, and the sample often interacts significantly with the sheath fluid, as shown in Fig. 1. In Fig. 1, zymosan particles have been labeled with FITC, a fluorophore whose absorption cross section is pH sensitive at 488 nm, but not at 458 nm. In Fig. 1a, fluorescence efficiency at 488 nm (normalized by the fluorescence at 458 nm) changes when the pH of the sample is altered, causing an apparent change in measured fluorescence and potentially altering experimental results. However, these changes are observed only when the pH of the sheath is altered in the same manner as the

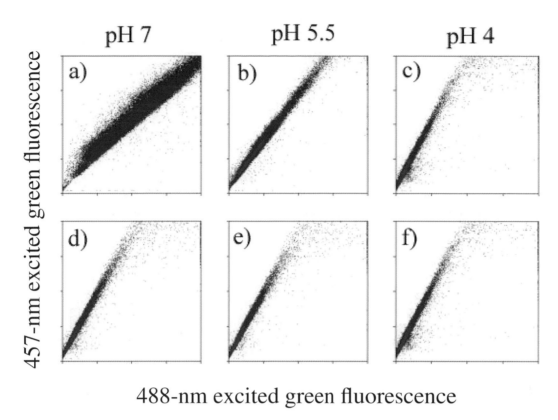

pH 7   **pH 5.5**   **pH 4**

457-nm excited green fluorescence

a)   b)   c)

d)   e)   f)

488-nm excited green fluorescence

FIGURE 1   FITC-labeled zymosan particles excited at two wavelengths. When excited with 488-nm light, FITC shows a pH-dependent fluorescence emission that is not seen when 457-nm light is used. Columns show the pH of the *sample*. In the top row, the pH of the sheath was adjusted to match that of the sample, but in the bottom row, it was maintained at pH 4. One can see that the pH of the sheath is a determinant for fluorescence efficiency.

sample fluid and are no longer apparent without changes in sheath fluid pH (Fig. 1b). Commercial sheath fluid can be obtained with various chemical properties and constituents that can potentially affect the characteristics of the sample being studied. It is necessary to be familiar with the exact makeup of any sheath selection and to determine its likelihood of either chemical or biological interaction with the experiment. A simpler and less expensive alternative is to prepare a 0.9% NaCl solution and filter sterilize it or to filter the solution and then sterilize by autoclaving. Filtration of self-prepared sheath will reduce the number of nozzle clogs dramatically.

C. Sample Preparation and Determination of Sort Rate

High-speed cell sorters separate complex cell mixtures into constituent fractions so they can be studied in isolation. In the sorter, a jet of sheath fluid emanating from a nozzle surrounds the cell suspension coaxially. In order to separate events from one another, an acoustic vibration is coupled to the tip of the nozzle, and the vibrating nozzle leaves a trail of cyclical imprints onto the surface of the liquid. Because of surface tension energy, the cylinder of fluid is inherently less stable than a sphere of fluid, and soon the jet separates into regularly spaced droplets. Therefore, the cells, first carried by a cylinder of liquid, soon find themselves distributed randomly among the droplets. To separate the desired fraction of cells, an electrical charge may be applied to the droplets containing cells of interest as they separate from the main jet. In this manner, droplets with different cell types are directed toward collection vials by a static electrical field. On average, there are several empty droplets between each passing cell. As a result, the event rate (number of cells analyzed per second) should not be greater than the droplet formation rate. A typical droplet formation rate for a high-speed machine will be on the order of 50,000 drops/s, and the maximal throughput will be obtained if, on average, there is one cell per droplet. For this situation, however, statistics regarding the distribution of the cells among the droplets dictate that only 37% of the droplets will have one cell. The remainder of the droplets will be empty or contain more than one cell. Thus, the maximal sort rate becomes roughly 37% of the droplet formation rate; higher rates will cause too many cells to be lost due to coincident drop occupation (and hence, equivocal signal measurement), and lower rates will result in unnecessarily slow sorts. Without this information, standard coincidence rejection protocols require a 1.5 droplet interval between sortable events. As a result, we recommend limiting sort rates to roughly one-third of the droplet formation rate. Again, each sorting facility will have to perform a series of basic experiments to determine what is most appropriate for a given application. A more thorough derivation of the physics of drop formation and determinants of sort rate can be found in van den Engh (2000).

As a consequence of the desire to sort cells with maximal throughput, one must also determine the density of cells in the sample suspension. This consideration arises from needing to control the droplet occupancy rate (as described earlier) independent of the droplet formation rate and the sheath fluid pressure. Control of the sample injection rate is achieved by independently pressurizing the sample and the sheath fluid and altering the differential pressure between the two fluidic systems. There is a limit, however, to the degree one can increase the sample injection rate by increasing the sample pressure. Jet velocity is proportional to the square root of jet pressure, so large increases in pressure will result in relatively smaller changes in jet velocity. Furthermore, as the sample pressure continues to increase, a critical point will be reached that will disrupt the laminar injection of the sample into the core of sheath fluid. In this situation, the cells will not be hydrodynamically focused by the sheath fluid and will disperse throughout the fluid column. Cells would then encounter different laser light intensity as a function of their position in the sheath fluid, compromising measurements of fluorescence and scattered light from particles and resulting in decreased sort purity. To preserve fluidics integrity, the sample injection pressure should be within 1–3 psi of the sheath pressure. Additionally, it is difficult to control the event rate of a sample that has far too many cells per unit volume, as small changes in the differential pressure will result in uncontrollably large changes in the event rate. Although the desired density of the sample depends critically on the diameter of the sample injection tube (most are of the order of 125 μm), the majority of situations dictate sample densities of roughly $1–10 \times 10^6 \, ml^{-1}$ with higher densities appropriate for smaller diameter sample injection tubes. Other factors to con-

sider will depend on the inherent biological characteristics of the sample itself. For instance, certain cell lines are adhesive and may clump easier, making control of the sort rate difficult.

III. TROUBLESHOOTING

Cell sorters are complex machines and, as such, seemingly minor alterations or inconsistencies in set up and/or calibration can have an additive effect on sort purity. This section lists some common problems and suggested solutions.

A. Droplet Formation

Poor droplet formation, as evidenced by movement of the break-off point (the point at which the stream becomes discrete drops) with respect to the laser, will result in both poor yield and contamination. If the break-off point is not well stabilized, check to ensure that the nozzle is free of trapped air or debris and that the sheath fluid has equilibrated to the temperature of the room in which sorting is conducted. Other causes of poor droplet formation can be the presence of detergent in the system (which changes the surface tension energy of the fluid column) or a piezoelectric element that is poorly coupled to the nozzle.

B. High Background Event Rates

High background count rates (triggering of sort electronics without sample injection and/or event rates disproportionately above expected rates for a given sample) are generally the result of poorly maintained tubing or improperly prepared sheath fluid. These considerations are of particular importance when sorting or analyzing samples that are relatively dilute. We have found that a well-maintained sorter should have a background count of less than three particles per second. A good way to measure this is by using 1-μm beads and adjusting the forward and perpendicular scatter detector gains to one-fourth of their full range. After the sample injection valve is closed, one should then be able to differentiate background count rates from particles that are larger than 0.5 μm. Should the background be unacceptably high, filtering of sheath fluid through a 0.22-μm filter or changing the sample and/or sheath fluid lines may be necessary.

C. Nozzle Clogs

Frequent nozzle clogs are usually the result of poor cleanup from a prior sort, suboptimal sample preparation, or inappropriate injection rate. Most commonly, sheath fluid left throughout the fluidics system from a previous sort dries and salt crystals are formed that clog the nozzle at the start of the following sort. Following the suggestions for flushing at the conclusion of each sort will remedy this issue. Additionally, one should examine the nozzle tip with a microscope to ensure that it is free of contamination and is uniformly smooth and patent. Certain samples will be more prone to causing clogs and, as such, the maximum injection pressure that will avoid clogs will have to be determined for each case.

The preceding points of discussion will vary to some degree among individual machines, laboratory settings, and instrument operators. In practice, machine start up, shut down, maintenance, sample manipulation, and so on, will all have an impact on overall, effective sort rate and general practice habits, being somewhat unique to each application. With consistent and thorough protocols for maintaining the sorting machinery, familiarization with the components of the sort process, and a systematic, skilled approach to troubleshooting, one can focus on the more exciting and challenging experimental aspects of high-speed cell sorting.

IV. STRATEGIES FOR EXPERIMENTAL DESIGN

While the original applications of high-speed cell sorting focused on the bulk purification of cellular and molecular material, modern biology has become increasingly focused on the minor nuances that distinguish one cell type from another. Although many biological experiments can still be approached through the analysis of cell populations, there is a growing need and improving ability to attain information from highly specified phenotypes, or even single cells. High-speed cell sorters are ideally suited for these applications because, by definition, a more highly classified cell will be present in less numbers, and it is often necessary to serially screen millions of individual events in a relatively short time period to find those particles of interest. Recent examples in the literature include the analysis of B cells from multiple myeloma patients to isolate subpopulations of cells with specific mutations in pathogenic genes (Kalakonda et al., 2001) or the haplotyping of individual sperm cells to look for trinucleotide repeat expansion in the Huntington's disease locus (Chong et al., 1997).

Beyond the study of naturally occurring variations and mutations, cell sorters have emerged as tools to screen engineered cell-based or solid support molecular libraries to select genetic clones and/or molecules of interest. Ingenious fluorescent-labeling schemes are employed to screen immensely complex libraries of particles to detect rare yet significant molecular processes such as enzyme activity or antibody binding and to monitor gene expression (Boder et al., 2000; Olsen et al., 2000). A revolution in molecular biology that is ideally suited to flow cytometry is the cloning of fluorescent reporter genes such as green fluorescent protein (GFP), discosoma Red (dsRed), and their variants (Chalfie et al., 1994; Matz et al., 1999). When fused to coding sequences of interest, these proteins generate fluorescent molecules from cells that express the cloned sequences in a dose-dependent fashion. Simple excitation of fluorescent proteins with a laser of appropriate wavelength results in the emission of a highly quantitative signal easily detectable by a flow cytometer. Combinations of these proteins allow for the design of gene expression studies under a variety of experimental conditions. Cells expressing the desired proteins (and hence cloned sequences) can be separated in a cell sorter and interrogated in further studies.

V. PROCEDURES

A. Library Construction and Clone Selection

The ability to isolate useful DNA fragments from larger complex mixtures is critical for all of molecular biology. Because single molecules of DNA cannot be readily seen or handled, target DNA is ligated into a larger vector that can be manipulated and propagated in a host bacterial strain, most commonly E. scherichia coli. This construct usually consists of a plasmid or phage, but can be larger, such as with bacterial artificial chromosomes (BACs), or the entire bacterial chromosome. Well-described molecular biology techniques can be applied in this fashion to create highly diverse libraries rapidly with complexities greater than 10^9 and lie outside the scope of this article. With traditional methods, E. coli cells are grown on agar plates under antibiotic selection, with each colony representing a construct that has entered a bacterial cell successfully. Colonies with the desired genotype are then selected based on a colormetric phenotype and transferred to liquid media or streaked out on another agar plate for growth. Selected species can then be isolated efficiently with commercially available plasmid prep kits (Qiagen, Inc.) or by other established DNA extraction and amplification methods.

B. Cell Sorters as a Tool for Clone Selection and Genetic Screening

While the aforementioned approach is accurate and has been successfully employed for decades, cell sorters are ideally equipped to perform the same tasks in less time with added flexibility. At the most basic level, markers can be devised to simply indicate the presence or absence of a cloned insert. With more sophisticated study design and genetic manipulation, sorters become tools to potentially screen genomes of entire organisms in rapid fashion, providing a window into genetic regulation. Libraries can be screened and clones displaying a desired phenotype can be isolated, their DNA amplified, and eventually sequenced. As an example, an experiment could be fashioned to screen the entire *Saccharomyces cerevisiae* genome (~14 Mbp) by generating 100-bp fragments placed strategically in relation to a reporter construct. High-end sorters would be capable of scanning the genome to 10X coverage for specific expression events in a matter of minutes. A similar approach was reported in a study by Barker *et al.* (1998) that generated a library of 200- to 1000-bp fragments of the *Mycobacterium marinum* genome fused to a promoterless copy of GFP. Like its counterpart *Mycobacterium tuberculosis*, this bacterium is able to avert the normal function of the immune system and survive within macrophages. By isolating phagosomes emitting green fluorescence and sequencing the regions of DNA upstream of GFP, they were able to determine which regions of the genome may be responsible for circumventing normal bacterial killing.

C. Fluorescence Encoding Strategies

The major challenge in experimental design is then to devise a strategy for expressing some characteristic of the cloned and/or expressed DNA as fluorescence. Because fluorescent proteins such as GFP and its variants can be coupled to genetic sequences and produced endogenously within bacteria without the need for added substrates, they become the obvious starting point. Basic approaches to clone selection indicate the presence of foreign DNA with the loss of a fluorescent signal. Inouye *et al.* (1997) developed a plasmid vector that directs the production of GFP under a constitutive promoter. When an insert is cloned into the vector, stop codons within the insert prevent the successful translation of GFP. Therefore, nonfluorescent colonies can be picked to obtain individual clones. In our experience, negative selection (based on the absence of fluorescence) can be a poor criterion for the analysis of individual cells in a cell sorter. Background particles are usually weakly fluorescent and, as such, it becomes difficult to segregate them from nonfluorescent (i.e., insert containing) bacteria. Additionally, a large degree of cell-to-cell variability exists that is widely determined by different sizes of bacterial cells and varying levels of protein expression. The combination of these factors makes it difficult to be certain that a nonfluorescent cell is not simply expressing a fluorescent marker at low levels and will result in sort contamination.

More complex schemes can be derived to represent unique characteristics of individual clones as positive fluorescence signals that result in sorts of higher purity. For instance, Olsen *et al.* (2000) created a scheme in which protease variants from a mutant library were expressed on the surface of *E. coli*. Functional enzyme mutants successfully cut a synthetic substrate bound to the *E. coli* membrane containing BODIPY and tetramethylrhodamine fluorophores connected by a linker with a specific target site. The tetramethylrhodamine acts as a quencher for the BODIPY dye by fluorescence resonance energy transfer. If the OmpT mutant on a particular cell cleaved the substrate, then the tetramethylrhodamine diffused away and BODIPY was no longer quenched. Sorting the fluorescent particles resulted in a 5000-fold enrichment for clones containing active variants of the protease.

As an additional example of immense potential for such techniques, Koo *et al.* (2004) devised a means to encode the activity of mRNA processing in *E. coli* as a fluorescent signal. A GFP construct was created as a fusion to a mRNA processing region. If functional mRNA processing was present, many of the GFP transcripts were degraded, leading to a low level of GFP fluorescence. If the ability to perform mRNA processing was disrupted, cells had high levels of GFP and were isolated. Using this method, over 60% of bacteria were highly fluorescent after four rounds of sorting.

VI. PRACTICAL CONSIDERATIONS

Once an algorithm for encoding biological function as a fluorescent signal has been conceived, it is vital to test the system under a variety of culture and cellular conditions. Often times, fluorescence from GFP and other fluorescent proteins is not visible in actively growing *E. coli* cultures. Because the rate of replication of *E. coli* is so high, any fluorescent protein that is produced is diluted continuously and, as such, fluorescence will not usually begin to appear until late log phase or stationary phase. To obtain higher cell-to-cell consistency of fluorescence, it is often helpful to grow the cells under conditions that repress fluorescent protein production during active growth. This can be achieved by cloning the fluorescent protein downstream from a tightly regulated promoter. If a *lac* promoter is used, growth in 0.5% glucose results in almost complete repression. In late log or stationary phase, cells can be resuspended in an equal volume of culture medium that does not repress protein production. With this approach, all cells begin producing fluorescent protein at approximately the same time and the rate of division has slowed, resulting in more consistent levels of fluorescent protein expression.

During testing, it is often convenient to use a fluorescence microscope with appropriate excitation and emission filters, as well as high-power objectives to image individual bacterial cells. Direct observation of cultured cells can be a quick method for assessing the levels of various fluorescent markers, as well as looking for undue elongation of cells that can indicate unhealthy cellular states. Only after thorough testing should bacterial cultures be inspected by flow cytometry. Typically, forward scatter is not an informative measurement to record but is an excellent parameter to trigger on as it can easily detect bacterial cells in a flow stream regardless of their orientation. This is a better method than triggering on fluorescence to prevent nonfluorescent cells from altering sort outcome.

A. Optical Setup and Instrument Settings

Fluorescence measurements can be performed with a variety of optical configurations. If a 488-nm laser is used to detect forward scatter, it is convenient to use this same laser for GFP measurements. Additionally, some variants of YFP (yellow mutant of GFP) and DsRed can be excited readily with a 488-nm beam. Other laser lines such as multiline UV or 514.5 nm can be used as needed. If multiple fluorescent proteins or markers are being utilized it is often efficient to use a dichroic beam splitter to efficiently separate fluorescence into short and long wavelength pathways. Further filtering with long-pass, band-pass, or short-pass filters can fine tune wavelength ranges to increase the number of measured parameters and to minimize cross talk between signals.

Bacterial cultures should be diluted approximately 200:1 in saline before injection into the flow cytometer. This can be variable depending on densities of cultures and cytometer configuration. The sample injection pressure on the flow cytometer should be adjusted for optimal event rate on each individual sorter as discussed previously. Test sorts should be performed using control cells first until the researcher attains

some level of confidence that desired cell populations can be isolated successfully. In addition, these initial tests can often provide some estimate as to the enrichment that can be achieved with a particular fluorescence-encoding scheme. Tests should also be performed to assess the viability of bacteria after culturing and sorting. This can be achieved easily by sorting a particular number of cells into LB medium or saline and then serial dilutions of these sorted cells can be plated. After overnight growth, colony counts will provide a measure of viability. Optimally, 50–70% of injected cells should be viable. However, viability can drop below 1% depending on factors such as cell strain and culture conditions.

When initial testing has been completed, the actual experiment can proceed. Cells can be batch sorted with hundreds to thousands of cells per tube or individual cells can be sorted into 96-well plates for clone isolation. If an inadequate level of enrichment is achieved after the first round of sorting, cells can be cultured and then resorted. This process can be repeated until the desired level of enrichment is achieved, and final processing of the cells will be highly dependent on experimental design. Cells can be further cultured for DNA isolation, or PCR amplification of genetic elements can be performed from bacterial cultures and/or single cells. Other protocols may culture sorted cells before isolating proteins for various *in vitro* assays.

VII. CONCLUSIONS

The sections presented herein serve to provide the sorter with a foundation in appropriate sorting techniques, as well as to provoke thought toward elegant experimental design. These approaches take advantage of the latest advances in molecular biology as well as the full range of capabilities inherent to high-end cell sorters. As sorting technology has improved, these instruments have become more stable, easier to operate, and amenable to almost any laboratory setting as opposed to dedicated, core-sorting facilities. The high-speed cell sorter has become a tool for discovery —diversity of heterogeneous populations can be assessed, entire genomes and molecular libraries can be screened in short periods of time, pure populations can be purified, and their constituents can be amplified and studied further with minimal confounding variables. No other technology has the ability to purify cells based on so many parameters with such high speed and accuracy as the high-speed cell sorter. The tight interplay between biology and technology has resulted in an improved ability to highlight increasingly specified nuances between cells. In combination with the latest tools, such as mass spectrometry and expression arrays, these differences will unravel the earliest molecular underpinnings in processes such as disease pathogenesis and developmental biology.

References

Barker, L. P., Brooks, D. M., and Small, P. L. (1998). The identification of Mycobacterium marinum genes differentially expressed in macrophage phagosomes using promoter fusions to green fluorescent protein. *Mol. Microbiol.* **5**, 1167–1177.

Boder, E. T., Midelfort, K. S., and Wittrup, K. D. (2000). Directed evolution of antibody fragments with monovalent femtomolar antigen-binding affinity. *Proc. Natl. Acad. Sci. USA* **97**(20), 10701–10705.

Chalfie, M., Tu, Y., Euskirchen, G., Ward, W. W., and Prasher, D. C. (1994). Green fluorescent protein as a marker for gene expression. *Science* **263**(5148), 802–805.

Chong, S. S., Almqvist, E., *et al.* (1997). Contribution of DNA sequence and CAG size to mutation frequencies of intermediate alleles for Huntington disease: Evidence from single sperm analyses. *Hum. Mol. Gene.* **6**, 301–309.

Gray, J. W., Dean, P. N., Fuscoe, J. C., Peters, D. C., Trask, B. J., van den Engh, G. J., and Van Dilla, M. A. (1987). High-speed chromosome sorting. *Science* **238**(4825), 323–329.

Inouye, S., Ogawa, H., *et al.* (1997). A bacterial cloning vector using a mutated Aequorea green fluorescent protein as an indicator. *Gene* **189**, 159–162.

Kalakonda, N., Rothwell, D. G., *et al.* (2001). Detection of N-Ras codon 61 mutations in subpopulations of tumor cells in multiple myeloma at presentation. *Blood* **98**, 1555–1560.

Koo, J. T., Choe, J., and Moseley, S. L. (2004). HrpA, a DEAH-box RNA helicase, is involved in mRNA processing of a fimbrial operon in *Escherichia coli*. *Mol. Microbio.*

Matz, M. V., Fradkov, A. F., Labas, Y. A., Savitsky, A. P., Zaraisky, A. G., Markelov, M. L., and Lukyanov, S. A. (1999). Fluorescent proteins from nonbioluminescent Anthozoa species. *Nature Biotechnol.* **10**, 969–973.

Olsen, M. J., Stephens, D., Griffiths, *et al.* (2000). Function-based isolation of novel enzymes from a large library. *Nature Biotechnol.* **18**, 1071–1074.

Peters, D., Branscomb, E., Dean, P., Merrill, T., Pinkel, D., Van Dilla, M., and Gray, J. W. (1985). The LLNL high-speed sorter: Design features, operational characteristics, and biological utility. *Cytometry* **6**(4), 290–301.

van den Engh, G. (2000). High speed cell sorting. *In* "Emerging Tools for Single Cell Analysis: Advances in Optical Measurement Technologies" (G. Durack, and J. P. Robinson, eds.). Wiley-Liss, New York.

Van Dilla, M. A., and Deaven, L. L. (1990). Construction of gene libraries for each human chromosome. *Cytometry* **11**(1), 208–218.

Cell Cycle Analysis

34

Cell Cycle Analysis by Flow and Laser-Scanning Cytometry

Zbigniew Darzynkiewicz, Piotr Pozarowski, and Gloria Juan

I. INTRODUCTION

The cytometric methods for cell cycle analysis can be grouped into three categories. The first comprises methods that rely on a single time point ("snapshot") measurement of the cell population. This analysis may be either univariate, based on the measurement of cellular DNA content alone (Crissman and Hirons, 1994), or multivariate (multiparameter), when in addition to DNA content another cell attribute is measured (Darzynkiewicz *et al.*, 1996; Endl *et al.*, 2001; Larsen *et al.*, 2001). The measured attribute is expected to provide information about a particular metabolic or molecular feature(s) of the cell that correlates with a rate of cell progression through the cycle or is a marker the cell proliferative potential or quiescence. While the single time measurement reveals the proportions of cells in G_1 *vs* S *vs* G_2/M, it provides no direct information on cell cycle kinetics. However, if duration of the cell cycle (or time of doubling of cells in the culture) is known, the length of G_1, S, or G_2/M phase can be estimated from the percentage of cells in the respective phase.

In the second category are methods that combine time-lapse measurements of populations of cells synchronized in the cycle or whose progression through the cycle was perturbed, e.g., halted by the agent arresting them at a specific point of the cycle. These methods reveal kinetics of cell progression through the cycle. A classical example of this group is the stathmokinetic approach where cells are arrested in mitosis, e.g., by vinblastine or colcemide, and the rate of cell entrance to mitosis ("cell birth" rate) is estimated from the slope representing a cumulative increase in the percentage of mitotic cells as a function of time of the arrest (Darzynkiewicz *et al.*, 1987).

Methods of the third category rely on analysis of DNA replication concurrent with measurement of DNA content. They may be either single time point measurements or use the time-lapse strategy to measure cell cycle kinetics. Incorporation of the thymidine analogue and the S phase marker 5'-bromo-2'-deoxyuridine (BrdUrd) is detected either cytochemically, based on the use of DNA dyes such as Hoechst 33258, whose fluorescence is quenched by BrdUrd (Poot *et al.*, 2002), or immunocytochemically using fluoresceinated BrdUrd antibodies (Dolbeare and Selden, 1994). Still another method of this category detects incorporated BrdUrd by the increased sensitivity of DNA to photolysis: utilizing terminal deoxynucleotidyl transferase, the photolytically generated strand breaks are then labeled with fluorochrome-tagged deoxynucleotides (Li *et al.*, 1996). Because the latter method escapes the harsh conditions used to induce DNA denaturation (heat or acid), it is applicable in conjunction with immunocytochemical detection of intracellular proteins. The time-lapse measurements of the cohort of BrdUrd-labeled cells allows one to estimate their rate of progression through different points of the cell cycle (Terry and White, 2001).

The methods described in this article, representative of each of the three categories, can be applied both to cells measured by flow cytometry and to cells mounted on slides. The latter can be analyzed by laser-scanning cytometer (LSC), the microscope-based cytofluorimeter that measures the fluorescence of individual cells deposited on slides rapidly, with sensitivity and accuracy comparable to that of flow cytometry (Kamentsky, 2001). One advantage of LSC is that the cells are stained and measured while attached to microscope slides. This eliminates cell loss that inevitably occurs due to repeated centrifugations during sample preparation for flow cytometry. Another advantage stems from the possibility of relocation of particular cells on slides for their visual inspection or morphometry, following the initial measurement of a large cell population and electronic selection (gating) of cells of interest. The instrument thus combines advantages of both flow and image cytometry.

Only a few selected methods are presented in this article. More detailed descriptions of these and other methods, their applicability to different cell systems, and advantages and limitations are provided elsewhere in books devoted specifically to the cell cycle (Gray and Darzynkiewicz, 1987; Fantes and Brooks, 1993; Studzinski, 1995, 1999) or flow cytometry (Darzynkiewicz *et al.*, 1994, 2001; Gray *et al.*, 1990).

II. MATERIALS AND INSTRUMENTATION

The materials listed for each of the different procedures can be purchased from the following sources: Triton X-100 (Cat. No. T 9284), Pipes (Cat. No. P 3768), RNase A (Cat. No. R 5000), and 5′-bromo-2′-deoxyuridine (Cat. No. B 5002) are from Sigma Chemical Co.; DAPI (4′,6′-diamidino-2-phenylindole; Cat. No. D 1306), propidium iodide (PI, Cat. No. P-1304), and high-purity acridine orange (AO, Cat. No. A-1301) are from Molecular Probes; and methanol-free formaldehyde (Cat. No. 4018) is from Polysciences Inc.

The greatest selection of monoclonal and polyclonal antibodies applicable to cell cycle analysis is offered by DACO Corporation, Sigma Chemical Co., Upstate Biotechnology Incorporated, B.D. Biosciences PharMingen, and Santa Cruz Biotechnology, Inc.

A variety of models of flow cytometers of different makers can be used to measure cell fluorescence following staining according to the procedures listed in this article. The manufacturers of the most common flow cytometers are Becton Dickinson Immunocytometry Systems, Beckman/Coulter Inc., DACO/Cytomation, and PARTEC GmbH. The multiparameter laser-scanning cytometer is manufactured by CompuCyte Corp.

The software used to deconvolute DNA content frequency histograms, to estimate the proportions of cells in the respective phases of the cycle, is available from Phoenix Flow Systems and Verity Software House.

III. PROCEDURES

A. Univariate Analysis of Cellular DNA Content

Progression through S phase and completion of mitosis (cytokinesis) result in changes in cellular DNA content. The cells position in the major phases ($G_{0/1}$ *vs* S *vs* G_2/M) of the cycle, therefore, can be estimated based on DNA content measurement. A variety of fluorochromes and numerous methods are used for DNA content analysis. A simple protocol, which can be modified to accommodate different dyes, has been developed and applied to numerous cell types.

1. Cell Staining with DAPI

Solution

Staining solution: Phosphate-buffered saline (PBS) containing 0.1% (v/v) Triton X-100 and 1 μg/ml DAPI (final concentrations)

Steps

1. Suspend approximately 10^6 cells in 0.5 ml of PBS. Vortex gently (~3 s) or gently aspirate several times with a Pasteur pipette to obtain a mono-dispersed cell suspension, with minimal cell aggregation.

2. Fix cells by transferring this suspension, with a Pasteur pipette, into 12×75-mm centrifuge tubes containing 4.5 ml of 70% ethanol on ice. Keep cells in ethanol for at least 2 h at 4°C (cells may be stored in 70% ethanol at 4°C for weeks).

3. Centrifuge the ethanol-suspended cells 3 min at 200 g. Decant thoroughly ethanol.

4. Suspend the cell pellet in 5 ml of PBS, wait ~30 s, and centrifuge at 300 g for 3 min.

5. Suspend the cell pellet in 1 ml of DAPI staining solution. Keep in the dark, at room temperature, for 10 min.

6. Transfer sample to the flow cytometer and measure cell fluorescence. Maximum excitation of DAPI, bound to DNA, is at 359 nm and emission is at 461 nm. For fluorescence excitation, use the available ultraviolet (UV) light laser line at the wavelength nearest to 359 nm. When a mercury arc lamp serves as the excitation source, use a UG1 excitation filter. A combination of appropriate dichroic mirrors and emission filters should be used to measure cell fluorescence at wavelengths between 450 and 500 nm.

7. The data acquisition software of most flow cytometers/sorters allows one to record the fluorescence intensities (the integrated area of the electronic pulse signal) of 10^4 or more cells per sample. Data are presented as cellular DNA content frequency his-

tograms (Fig. 1). The data analysis software packages (e.g., Rabinovitch, 1994) that deconvolute the frequency histograms to obtain percentage of cells in $G_{0/1}$, S, and $G_2 + M$ are either included with the purchase of the flow cytometer or are available commercially from other vendors, as listed under Section II. For details of the methods of deconvolution of DNA content frequency histograms, see Rabinovitch (1994).

2. Staining with Propidium Iodide

If excitation with UV light is not possible, the procedure given earlier for DAPI can be modified to apply PI as the DNA fluorochrome. Thus, instead of DAPI, PI is included into the staining solution at a concentration of 10 μg/ml. Because PI also stains double-stranded RNA, RNA is removed enzymatically during the staining reaction. This is accomplished by the addition of RNase A into the staining solution.

Solution

Staining solution: PBS containing 0.1% (v/v) Triton X-100, 10 μg/ml of PI, and 100 μg/ml of DNase-free RNase A.

Steps

1–4. Follow steps 1 to 4 as described earlier for cell staining with DAPI.

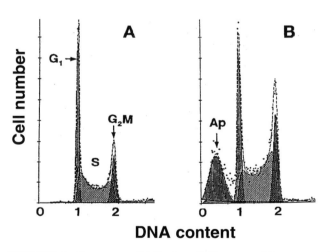

FIGURE 1 Frequency distribution DNA content histograms of human leukemic HL-60 cells untreated (A) or treated with DNA topoisomerase II inhibitor fostriecin (B) and stained with PI as described in the text. The "multicycle" deconvolution program (Phoenix Flow Systems) has been used to identify and calculate percentage of cells with fractional DNA content (apoptotic cells; Ap) and cells in G_1, S, and G_2/M phases of the cycle content, as shown. The drug treatment caused an increase in the proportion of S and G_2M cells and induced apoptosis.

5. Suspend the cell pellet in 1 ml of PI staining solution. Keep in the dark, at room temperature, for 30 min or at 37°C for 10 min.

6. Transfer sample to the flow cytometer and measure cell fluorescence. Maximum excitation of PI, bound to DNA, is at 536 nm and emission is at 617 nm. Blue (488 nm) or green light lines of lasers are optimal for excitation of PI fluorescence. Emission is measured using the long-pass 600- or 610-nm filter. Data acquisition and analysis are as described earlier for DAPI-stained cells.

Comments

All cells in G_1 have a uniform DNA content, as do cells in G_2 and M; the latter have twice as much DNA as G_1 cells. Under ideal conditions of DNA staining, the fluorescence intensities of all G_1 or G_2/M cells are expected to be uniform and, after analog to digital conversion of the electronic signal from the photomultiplier (representing their fluorescence intensity), to have uniform numerical values, respectively. In practice, however, G_1 and G_2 cell populations are represented on frequency histograms by peaks of various width. The coefficient of variation (*cv*) of the mean value of DNA-associated fluorescence of the G_1 population (width of the peak) is a reflection of an accuracy of DNA content measurement and should not exceed 8%. Improper staining conditions, instrument missadjustment, and the presence of a large number of dead or broken cells all result in high *cv* of the G_1 cell populations.

Apoptotic cells often have fractional DNA content due to the fact that the fragmented (low MW) DNA undergoes extraction during the staining procedure. Some cells may also lose DNA (chromatin) by shedding apoptotic bodies. Only a fraction of DNA thus remains within apoptotic cells. They are represented then on the DNA content frequency histograms by the "sub-G_1" peak (Fig.1). Commercially available software packages to deconvolute DNA histograms are able to identify and quantity the "sub-G_1" cell population.

If the length of the cell cycle (or cell doubling time) is known, the duration of each of the phases can be estimated from the percentage (fraction) of cells in that phase. For example, during the exponential phase of cell growth, the duration of G_1 (T_{G1}) can be calculated from

$$T_{G1} = \frac{T_C \times \ln(f_{G1}+1)}{\ln 2}$$

where T_C is duration of the cell cycle and f_{G1} is a fraction of cells residing in G_1. T_C, with generally acceptable approximation, equals the cell doubling time in cultures. The latter can be estimated from the growth curve.

B. Multiparameter Analysis

Multiparameter analysis of other attributes of the cell, in addition to DNA content, allows one not only to distinguish cells in G_1 *vs* S *vs* G_2/M, but also to identify quiescent (G_0) or mitotic cells. Thus, bivariate analysis of cell population with respect to their cellular DNA and RNA content discriminates between G_0 and G_1 cells. Bivariate analysis of DNA content and proliferation-associated proteins, particularly cyclins, provides another means to distinguish between proliferating and quiescent cells and yields additional information about the proliferative potential of cell populations (Darzynkiewicz *et al.*, 1996). Immunocytochemical detection of histone H3 phosphorylation combined with DNA content analysis offers a convenient approach to distinguish M from G_2 cells and to quantify the mitotic index in the cell population (Juan *et al.*, 2001). This section presents examples of these methods.

1. Differential Staining of Cellular DNA and RNA

Quiescent (G_0) cells are characterized by a many-fold lower RNA content compared to their cycling G_1 counterparts (Darzynkiewicz *et al.*,1976). Simultaneous staining of RNA and DNA, therefore, allows one to distinguish G_0 from G_1 cells (based on differences in RNA content), as well as to identify cells in S and G_2/M. Differential staining of cellular DNA and RNA can be accomplished with the metachromatic fluorochrome acridine orange (AO). At appropriate concentrations and ionic conditions AO intercalates into dsDNA and fluoresces green, while its interactions with RNA result in red fluorescence (Darzynkiewicz *et al.*, 1994). Prior to staining with AO the cells are permeabilized with Triton X-100 in the presence of $0.08M$ HCl and serum proteins. Such treatment makes cells permeable to AO, yet they are not lysed and their DNA and RNA content is preserved. Alternatively, the cells may be prefixed in 70% ethanol, as described previously for univariate DNA content analysis. Apoptotic cells stained under these conditions are characterized by markedly diminished DNA associated (green) AO fluorescence.

Solutions

1–4. *AO stock solution*: Dissolve 1 mg AO in 1 ml of distilled water. This solution of AO is stable for several months when kept at 4°C in the dark (foil wrapped).

2. *First-step solution (solution A)*: Dissolve 0.1 ml of Triton X-100 and 0.87 g of NaCl in 92 ml of distilled water. Add 8 ml of $1M$ HCl solution.

3. *Second-step solution (solution B)*: Prepare 100 ml of the phosphate – citric acid buffer at pH 6.0 by combining 37 ml of $0.1M$ citric acid with 63 ml of Na_2HPO_4. Add 0.87 g of NaCl and 34 mg EDTA, dissodium salt. Stir until dissolved. Add 0.6 ml of AO stock solution. This solution is stable for several months when kept at 4°C in the dark.

Steps

1. Transfer a 0.2-ml aliquot of the cell suspension ($<5 \times 10^5$ cells) directly from tissue culture to a 2- or 5-ml tube.
2. *Gently* add 0.4 ml of ice-cold solution A. Wait 15 s, keeping sample on ice.
3. *Gently* add 1.2 ml of ice-cold solution B. Keep sample on ice prior the measurement.

NOTE 1 *Vortexing, syringing, and vigorous mixing of cells when suspended in solution A or B break down the cells and should be avoided.*

NOTE 2 *As an alternative to detergent (Triton X -100) treatment, cells may be fixed in 70% ethanol as described for staining with DAPI (Section III, A,1). Carry out steps 1–4 of the DAPI staining protocol, suspend $<10^5$ cells in 0.2 ml of PBS, and follow with steps 2 and 3 as just described.*

4. Measure cell fluorescence during the next 2–10 min after the addition of solution B. Excite AO fluorescence with blue light (use 488- or 457-nm laser lines or BG12 excitation filter in the case of mercury lamp illumination). Measure the green fluorescence of AO bound to DNA at 530 ± 20 nm and red fluorescence of AO bound to RNA at >640 nm (long-pass filter).

Comments

A characteristic distribution of G_0, G_1, S, G_2/M, and apoptotic cells, differing in RNA and DNA content, is shown in Fig. 2. The differences in RNA content enable G_0 cells to be discriminated from G_1 cells. However, the differences in DNA content provide the basis to identify apoptotic and nonapoptotic cells and, among the latter, to distinguish $G_{0/1}$, S, and G_2/M cell subpopulations.

The major advantages of this assay are its simplicity, applicability to different instruments that use either laser or mercury lamp as a light source for fluorescence excitation, and the possibility it offers to distinguish G_0 from G_1 cells. Differential stainability of DNA *vs* RNA, however, requires very stringent conditions of cell staining in terms of dye concentration and

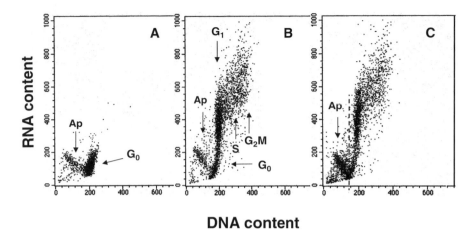

DNA content

FIGURE 2   Bivariate distributions (scatter plots) demonstrating differential staining of DNA and RNA with acridine orange (AO). Nonstimulated (A) and stimulated mitogenically with phytohemagglutinin (PHA) (B and C) lymphocytes were cultured for 48 h and then stained with AO according to the presented protocol. Nonstimulated G_0 cells have minimal RNA content and uniform DNA content. During stimulation, a subset of lymphocytes undergoes apoptosis (Ap; activation-induced apoptosis), another subset enters cell cycle, while some cells remain in G_0. All these subpopulations can be identified after differential staining of DNA and RNA: cells in G_1 *vs* S *vs* G_2/M differ in DNA content, whereas G_0 cells are distinct from G_1 cells due to differences in RNA content. The antitumor drug onconase was included in one of the PHA-treated cultures (C) to enhance "activation-induced apoptosis." The dashed line in C separates apoptotic from nonapoptotic cells.

ionic composition of the medium. AO also has a propensity to attach to sample flow lines of flow cytometers and therefore requires careful rinsing of the instrument with a bleaching solution (~15 min) to lower the background fluorescence for the subsequent analysis of weakly fluorescent samples. Further details of the AO methodology are presented elsewhere (Darzynkiewicz *et al.*, 1994).

2. Cellular DNA Content and Expression of Proliferation-Associated Proteins

The expression of proliferation-associated proteins often varies during the cell cycle, as well as is different in cycling and quiescent cells. Their immunocytochemical detection, therefore, provides information on the proliferative status of the cell. The most common markers of proliferating cells are the proliferating cell nuclear antigen (PCNA) (Larsen *et al.*, 2001), the antigen detected by the Ki-67 antibody (Endl *et al.*, 2001) and certain cyclins (Darzynkiewicz *et al.*, 1996).

Methods for detection of the proliferation associated proteins, particularly the choice of optimal fixative, may vary depending on the particular antigen (Jacobberger, 2001). The following method is applicable not only to cyclins (Darzynkiewicz *et al.*, 1994, 1996), but also other intracellular antigens.

Solutions

1. Fixatives: Methanol (see Comments)
2. *Cell permeabilizing solution*: 0.25% Triton X-100, 0.1% sodium azide in PBS
3. *Rinsing solution*: 1% bovine serum albumin (BSA), 0.1% sodium azide in PBS

Steps

1. Prepare the fixative by filling 5-ml polypropylene tubes with 4.5 ml of methanol (or 70% ethanol; see Comments). Keep tubes on ice.
2. Suspend $1–2 \times 10^6$ cells in 0.5 ml of PBS. Fix the cells by transferring this suspension with a Pasteur pipette into an ice-cold methanol tube. Keep cells in the fixative at $-20°C$ at least overnight (cells can be stored in the fixative at $-20°C$ for days).
3. Centrifuge at 300 g for 3 min. Resuspend the cell pellet in 5 ml PBS. Keep at room temperature for 5 min. Spin at 200 g for 5 min.
4. Resuspend cells in 0.5 ml of the permeabilizing solution. Keep at room temperature for 5 min. Centrifuge as in step 3.
5. Resuspend cell pellet in 100 μl of the rinsing solution that contains the primary Ab. Follow instructions supplied by the vendor regarding the final titer of the

supplied antibody (0.5–1.0 µg of the Ab per 10^6 cells suspended in 100 µl is generally optimal). Incubate for 60 min at room temperature with gentle agitation or overnight at 4°C.

6. Add 5 ml of the rinsing solution. Centrifuge at 300 g for 5 min.

7. *Use the isotype immunoglobulin as a negative control. Process as in steps 5 and 6.*

8. Resuspend cells in 100 µl of rinsing solution that contains the fluoresceinated secondary Ab, generally at a final 1:20 to 1:40 dilution. Incubate at room temperature for 30–60 min, agitating gently.

9. Add 5 ml of the rinsing solution and centrifuge at 300 g for 5 min.

10. Suspend the cell pellet in 1 ml PBS containing 5 µg/ml of PI and 100 µg of DNase-free RNase A. Keep in the dark at room temperature for 1 h.

11. Transfer cells to flow cytometer. Use blue light (488-nm laser line) for fluorescence excitation. Measure cell fluorescence in green (FITC, 530 ± 20 nm) and red (PI, >620 nm) light wavelengths.

NOTE If the primary Ab is fluorochrome tagged, skip step 8.

Comments

The critical steps for immunocytochemical detection of intracellular proteins are cell fixation and permeabilization. The fixative is expected to stabilize the antigen *in situ* and preserve its epitope in a state where it remains reactive with the Ab. The cells have to be permeable to allow the access of the Ab to the epitope.

The choice of optimal fixative and permeabilizing agent varies, primarily depending on the intracellular antigen, less on the cell type. General strategies of cell fixation, permeabilization, and stoichiometry of antigen detection are discussed by Jacobberger (2001). Cold methanol appears to be optimal for the detection of D-type cyclins. For cyclins E, A, and B1, 70% cold ethanol is equally good.

Also critical is choice of a proper Ab. Often, the Ab applicable to immunoblotting fails in immunocytochemical application, and *vice versa*. This may be due to differences in the *in situ* accessibility of the epitope or differences in degree of denaturation of the antigen on the immunoblots compared to that within the cell. Some epitopes may not be accessible *in situ* at all, whereas the accessibility of others may vary depending on their functional state, e.g., due to phosphorylation or steric hindrance. Because there is strong homology between different cyclin types, cross-reactivity may also be a problem. Because commercially available Abs may differ in specificity, degree of cross-reactivity, and so on, it is essential to provide detailed information (vendor and the hybridoma clone number) of the reagent used in the study.

Cyclins are key components of the cell cycle progression machinery (Sheer, 2000). During unperturbed growth of normal cells the timing of expression of several cyclins, particularly cyclins D, E, A, and B, is discontinuous, occurring as discrete and well-defined periods of the cell cycle (Fig. 3). This periodicity in cyclins expression provides new cell cycle landmarks that can be used to subdivide the cell cycle into several

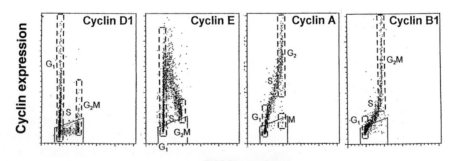

DNA content

FIGURE 3 Typical expression of cyclins D1, E, A, and B1 *vs.* DNA content as seen in normal, nontumor cells, processed as described in the text. Expression of cyclin D1 is shown in exponentially growing human normal fibroblasts. Expression of cyclins E, A, and B1 in PHA-stimulated human lymphocytes 48 h after administration of PHA. Boundaries of G_1 and G_2/M populations are marked by dashed lines. Trapezoid windows show the level of the unspecific, background, fluorescence measured separately using the isotypic irrelevant Ab. It is evident that cyclin D1 is expressed by a fraction of G_1 cells; cells entering and progressing through S and most in G_2/M cells are cyclin D1 negative. Cyclin E is maximally expressed by cells entering S phase and its level drops during progression through S. Cyclin A is expressed by S phase and maximally by G_2 cells; mitotic cells (postprometaphase) are cyclin A negative. Cyclin B1 is expressed by late S cells, maximally in G_2 and M.

subcompartments, additional to the subdivision of into four major phases (Darzynkiewicz *et al.*, 1996). Furthermore, bivariate analysis of cyclins expression *vs* DNA content makes it possible to discriminate between cells having the same DNA content but residing in different phases of the cycle, such as between G_2 and M cells (based on differences in cyclin A content), or between G_2 diploid and G_1 tetraploid cells (based on differences in expression of cyclins E and/or B1). Likewise, G_0 cells lacking expression of D-type cyclins or cyclin E can be distinguished from cells that entered cell cycle and become cyclins D, and subsequently cyclin E, positive. Strategies for the use of cyclins as additional markers of the cell cycle position are discussed elsewhere (Darzynkiewicz *et al.*, 1996). It should be noted, however, that some tumor cell lines, or normal cells when their cell cycle progression is perturbed, show unscheduled expression of cyclins D, E, and B1; namely G_1 cyclins (e.g., cyclin E) are expressed during G_2/M and the G_2/M cyclins (cyclin B1) during G_1 (Darzynkiewicz *et al.*, 1996).

3. Identification of Mitotic Cells by Cytometry

There is often a need to estimate mitotic index, e.g., to assess effectiveness of the drugs that disrupt microtubules or in stathmokinetic experiment (Darzynkiewicz *et al.*, 1987) to reveal the rate of cell entrance to mitosis. The cytometric methods used to identify mitotic cells are reviewed by Juan *et al.* (2001). The most convenient immunocytochemical method appears to be the one that utilizes Ab that is specific to histone H3 phosphorylated on *Ser*-10 (H3-P), the event that occurs during mitosis (Juan *et al.*, 2001). Because histone H3 is phosphorylated during prophase and is dephosphorylated late in telophase, the "time

window" of detection of mitosis by this Ab spans these two mitotic stages. Histone H3-P-specific Abs are offered by Sigma Chemical Co. (monoclonal) and Upstate Biotechnology, Inc. (polyclonal). The methodology of cell staining and fluorescence measurement is similar to that described earlier for analysis of DNA content and proliferation-associated proteins. Optimal cell fixation, however, requires a brief (15 min) pretreatment with 1% formaldehyde (in PBS, on ice) followed by postfixation in 70% ethanol. The results of fluorescence measurement are shown in Fig. 4.

C. Analysis of DNA Replication and Cell Cycle Kinetics

1. Stathmokinetic Approach

In a classical stathmokinetic experiment, the agent arresting cells in mitosis (e.g., colcemide or vinblastine) is added into the culture during the exponential phase of cell growth and the proportion of cells in mitosis is estimated as a function of the time of arrest. The slope of the plot representing an increase in the percentage of M cells during stathmokinesis reveals the *rate* of cell entry to M ("mitotic rate"; "cell birth rate").

Flow cytometric analysis of the stathmokinetic experiment can be based either on quantification of the increased proportions of cells in $G_2 + M$, represented by the G_2/M peak on the DNA content frequency histograms (by DNA content measurement followed by univariate data analysis), or by enumeration of cells in M (by selective staining of M cells, e.g., as shown in Fig. 4, followed by multivariate analysis of such data). Depletion of cells from the G_1 compartment (G_1 exit rate), as well as the rate of cell progression through S

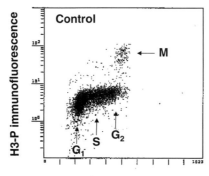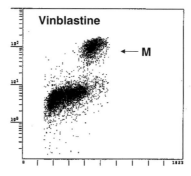

FIGURE 4 Immunocytochemical identification of mitotic cells (M) using Ab that reacts with histone H3 phosphorylated on *Ser*-10. To increase the proportion of mitotic cells, the culture shown at the right was treated for 4 h with the metaphase-arresting agent vinblastine.

phase can also be measured during stathmokinesis (Darzynkiewicz *et al.*, 1987). This section describes the scheme of a simple stathmokinetic *in vitro* experiment.

Solutions

Depending on the method used to stain DNA (Fig. 1) or detect mitotic cells (Fig. 4), appropriate solutions, as described earlier in the article, should be applied.

Steps

1. To the exponentially and asynchronously growing cell culture add the stathmokinetic agent (e.g., colcemid, vinblastine, or nocodazole) at the concentration that arrests all cells entering mitosis and yet does not perturb the progression through other phases. Different cell types show different sensitivities to particular agents and pilot experiments testing various concentrations of the agents are often needed to estimate efficiency of the cell arrest. Vinblastine, at a final concentration of 50 ng/ml, is quite effective in arresting most cell types of hematopoietic lineage in mitosis.

2. Collect cells hourly, during a time interval equivalent to approximately one-third of the cell doubling time, and fix them in suspension.

3. Use the flow cytometric staining techniques that allow either identification of cells in G_1, S, and G_2 + M (e.g., as in Fig. 1) or multiparameter analysis, which allows one to distinguish M cells (e.g., as in Fig. 4).

4. Analyze data to obtain the percentage of cells in the respective phases of the cycle per each sample.

5. Plot data as in Fig. 5. From the graphic display estimate the kinetic parameters, as shown in Fig. 5 and described in the legend. A more detailed analysis of the stathmokinetic experiment was presented elsewhere (Darzynkiewicz *et al.*, 1987).

Comments

A major drawback of the methods based on single time point measurement is lack of kinetic information. These methods, thus, cannot distinguish whether cell cycle progression is accelerated, slowed down, or even halted, e.g., during drug treatment, if G_1, S, and G_2/M phases are affected proportionally to each other. The stathmokinetic approach can be used in such instances to reveal cell kinetics. The alternative method, namely cell synchronization followed by observation of the cycle progression of the synchronized cells cohort, is more complex and time-consuming.

2. BrdUrd Incorporation

Incubation of cells in medium containing BrdUrd results in its incorporation during DNA replication (S phase). The incorporated BrdUrd can be detected

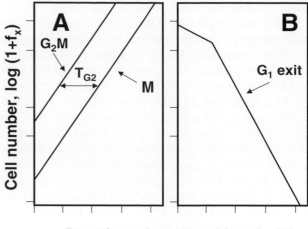

Duration of stathmokinesis (h)

FIGURE 5 The scheme of analysis of the stathmokinetic experiment. Asynchronously and exponentially growing cell cultures were treated with the metaphase-arresting stathmokinetic agent and subsequently sampled to obtain percentage of cells in M or G_2/M and in G_1 (e.g., as in Fig. 1 or 4). The percentage of cells in these phases (expressed as fraction of total; f_x) at a given time point of mitotic arrest is then plotted as log (1 + f_x). The slope representing the rate of entrance to M (or G_2 + M) reveals duration of the cell cycle (T_c). The duration of G_2 (T_{G2}) is estimated as the time–distance of the G_2 + M vs M slopes (A). Because of block of the cell cycle progression in M, the rates of emptying the G_1 (G_1 exit) can also be estimated (B). The stochastic component of the rate of cell exit from G_1 manifests as the straight-line slope that reveals the half-time of cell residence in G_1 (Darzynkiewicz *et al.*, 1987).

either cytochemically, by virtue of its propensity to quench the fluorescence of several DNA fluorochromes such as Hoechst 33358 or AO, or immunocytochemically using poly- or monoclonal Abs against this precursor.

Continuous or pulse-chase cell labeling with BrdUrd, followed by detection of BrdUrd simultaneously with measurement of cellular DNA content and bivariate data analysis (Dolbeare *et al.*, 1983; Terry and White, 2001), allows one to estimate a variety of cell cycle parameters. The protocol of Dolbeare *et al.*, (1983), with more recent modifications (Gray *et al.*, 1990), is given here. DNA denaturation by acid (HCl) gives more satisfactory results in some cell types.

a. Thermal Denaturation of DNA

Solutions

1. DNA denaturation buffer: 0.1 mM Na–EDTA in 1 mM Na–cacodylate; final pH 6.0
2. *Diluting buffer*: PBS containing 0.1% Triton X-100 and 0.5% bovine serum albumin (BSA)

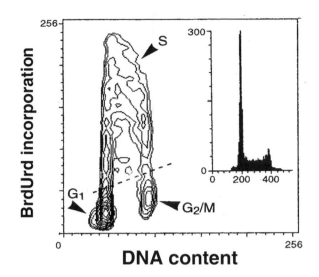

FIGURE 6 Bivariate distribution of cellular DNA content and BrdUrd incorporation. HL-60 cells were incubated with BrdUrd for 30 min, fixed, DNA was denatured by 2 M HCl, the incorporated BrdUrd was detected by the monoclonal antibody, and DNA was counterstained with PI, as described in the text.

Steps

1. Incubate cells with 10–30 μg/ml of BrdUrd under light-proof conditions.
2. Fix cells in suspension in 70% ethanol.
3. Centrifuge cells (1–2×10^6) at 200 g for 3 min, resuspend cell pellet in 1 ml of diluting buffer containing 100 μg/ml of RNase A, and incubate at 37°C for 30 min.
4. Centrifuge cells (300 g, 3 min) and suspend cell pellet in 1 ml of ice-cold 0.1 M HCl containing 0.1% Triton X-100. After 1 min, centrifuge cells again. Drain thoroughly and resuspend in 5 ml of DNA denaturation buffer.
5. Centrifuge cells again and resuspend cell pellet in 1 ml of DNA denaturation buffer.
6. Heat cells at 90 or 95°C for 5 min and then place on ice for 5 min.
7. Add 5 ml of diluting buffer. Centrifuge (300 g).
8. Drain well and suspend cells in 100 μl of BrdUrd Ab, dissolved in diluting buffer, for 30 min at room temperature (follow the instructions provided by the supplier regarding the dilution, time, and temperature of incubation with anti-BrdUrd).
9. Add 5 ml of dilution buffer and centrifuge.
10. Suspend cells in 100 μl of FITC-labeled goat antimouse IgG (dissolved in diluting buffer) and incubate for 30 min at room temperature.
11. Add 5 ml of diluting buffer, centrifuge, drain, and resuspend in 1 ml of this buffer containing 5 μg/ml of PI.

12. Measure the BrdUrd-associated green fluorescence and DNA-associated red fluorescence as described earlier in procedure B2.

NOTE: If the primary Ab is fluorochrome-tagged, skip step 10.

b. Denaturation of DNA by HCl

Solution

Diluting buffer: same as for thermal denaturation of DNA

Steps

1. Follow steps 1–4 as described earlier for thermal denaturation of DNA.
2. Centrifuge cells (300 g, 3 min) and resuspend cell pellet in 1 ml of 2 M HCl. After 20 min at room temperature, add 5 ml of HBSS, centrifuge, and drain well. Resuspend cells in 5 ml of 0.2 M phosphate buffer at pH 7.4 to neutralize traces of the remaining HCl.
3. Centrifuge cells at 300 g
4. Follow steps 8–12 as described earlier for thermal denaturation of DNA.

Comments

The critical step in this procedure is induction of partial DNA denaturation. This step often results in cell damage and leads to significant cell loss. Use of silinized tubes during centrifugations may decrease cell loss. Also, there are differences in sensitivity of DNA to denaturation between cell types, depending on their chromatin structure. Thus, while induction of DNA denaturation by acid may prove to be satisfactory with one cell type, it may fail with another. Some cell types require higher acid concentration (4 M) for optimal results.

The alternative approach is based on selective photolysis of DNA that contains the incorporated BrdUrd followed by DNA strand break labeling (Li *et al.*, 1996). Because no heat or acid treatment is required, the latter procedure is applicable in combination with immunocytochemical analysis.

The scope of this article makes it impossible to describe all possibilities of analysis of the cell cycle based on BrdUrd incorporation, either after the pulse-chase or continuous cell labeling. Readers are advised to consult Dolbeare and Selden (1994) and Terry and White (2001) for a more detailed description of these methods.

IV. CELL ANALYSIS BY LASER-SCANNING CYTOMETER

All the methods described earlier can be adapted to stain cells mounted on microscope slides, to be analyzed by the multiparameter (three-laser excitation, four-color fluorescence detection) LSC (Darzynkiewicz *et al.*, 1999; Kamentsky, 2001). To be analyzed by LSC the cells are attached to the slides electrostatically or by cytospinning, fixed, rinsed and then subjected to the procedures as presented earlier. To attach cells by cytospinning, 300 μl of cell suspension in tissue culture medium (with serum) containing approximately 20,000 is added to a cytospin chamber. The cells are then cytocentrifuged at 1000 rpm for 6 min and are submerged in the respective fixative in Coplin jars.

Small volumes of the respective buffers, rinses, or staining solutions as described for each of the methods in this article, are carefully layered on the cytospin area of the horizontally placed slides. At appropriate times these solutions are removed with a Pasteur pipette (or vacuum suction pipette). Small pieces (1 × 1 cm) of thin polyethylene or Parafilm foil may be layered on slides atop of the drops of the solutions used for cell incubations, over the cytospins, to prevent drying. The incubations should be carried out in a moist atmosphere.

At the final step of each particular staining procedure the cells are mounted in a drop of the respective staining solution, made identical as for flow cytometry, under the coverslip. The coverslips may be sealed with melted paraffin or a gelatin-based sealer. The cell fluorescence is measured by LSC, and the choice of the fluorescence excitation wavelength and emission filters is the same as described earlier for flow cytometers.

V. LIMITATIONS AND PITFALLS

1. Each approach has different type of limitations and possible pitfalls. The most frequent pitfall in univariate cell cycle analysis is lack in accuracy of DNA content measurement. As noted earlier, the accuracy is expressed as cv of the mean DNA content of G_1 cells. The results are unacceptable if the cv is larger than 8%; optimally, cv should be below 2%. Inappropriate adjustment of optics and fluidics of the flow cytometer are the most common causes of inaccuracy of DNA content analysis.

2. Common pitfalls in analysis of proliferation-associated antigens are due to inappropriate cell fixation and inappropriate choice of the antibody. Pilot experiments, testing different fixatives, different means of cell permeabilization, and different titer of the antibody, should be performed in the case of each new antigen. Many monoclonal antibodies that are available commercially have been developed to denatured proteins may not be applicable for immunocytochemistry. Although the isotypic IgG is commonly used as a negative control, some IgG preparations may show reactivity to various cell constituents. The optimal negative control is the same cell line having gene coding for the studied protein deleted, if available.

3. In the case of BrdUrd techniques, the unpredictable variable that affects cell stainability is variation in chromatin structure between different cell types. Hence, the methods should often be optimized for a particular cell type by testing different temperatures of DNA denaturation (80–100°C) or different strength of HCl used to induce DNA denaturation (1–4 M).

Acknowledgment

Supported by NCI RO1 96 704. Dr. Gloria Juan is currently at the Sloan-Kettering Cancer Institute, New York. Dr. P. Pozarowski is on leave from the Department of Immunology, School of Medicine, Lublin, Poland.

References

Bauer, K. D., and Jacobberger, J. (1994). Analysis of intracellular proteins. *Methods Cell Biol.* **41**, 352–376.

Crissman, H. A., and Hirons, G. T. (1994). Staining of DNA in live and fixed cells. *Methods Cell Biol.* **41**, 196–209.

Darzynkiewicz, Z., Crissman, H. A., and Robinson, J. P. (eds.) (2001). *Methods Cell Biol.* **63** and **64**.

Darzynkiewicz, Z., Gong, J., Juan, G., Ardelt, B., and Traganos, F. (1996). Cytometry of cyclin proteins. *Cytometry* **25**, 1–13.

Darzynkiewicz, Z., Gong, J., and Traganos, F. (1994). Analysis of DNA content and cyclin protein expression in studies of DNA ploidy, growth fraction, lymphocyte stimulation and cell cycle. *Methods Cell Biol.* **41**, 422–436.

Darzynkiewicz, Z., Robinson, J. P., and Crissman, H. A. (eds.) (1994). *Methods Cell Biol.* **41** and **42**.

Darzynkiewicz, Z., Traganos, F., and Kimmel, M. (1987). Assay of cell cycle kinetics by multivariate flow cytometry using the principle of stathmokinesis. In "Techniques in Cell Cycle Analysis" (J. W. Gray and Z. Darzynkiewicz, eds.), pp. 291–336. Humana Press, Clifton, NJ.

Darzynkiewicz, Z., Traganos, F., Sharpless, T., and Melamed, M. R. (1976). Lymphocyte stimulation: A rapid multiparameter analysis. *Proc. Natl. Acad. Sci. USA* **73**, 2881–2884.

Dolbeare, F., and Selden, J. L. (1994). Immunochemical quantitation of bromodeoxyuridine: Application to cell kinetics. *Methods Cell Biol.* **41**, 298–316.

Endl, E., Hollmann, C., and Gerdes, J. (2001). Antibodies against the Ki-67 protein: Assessment of the growth fraction and tools for cell cycle analysis. *Methods Cell Biol.* **63**, 399–418.

Fantes, P., and Brooks, R. (eds.) (1993). *"The Cell Cycle: A Practical Approach."* Oxford Univ. Press, Oxford.

Gray, J. W., and Darzynkiewicz, Z. (eds.) (1987). *"Techniques in Cell Cycle Analysis."* Humana Press, Clifton, NJ.

Gray, J. W., Dolbeare, F., and Pallavicini, M. G. (1990). Quantitative cell cycle analysis. *In "Flow Cytometry and Sorting"* (M. R. Melamed., T. Lindmo, and M. L. Mendelsohn, eds.), pp. 445–467. Wiley-Liss, New York.

Jacobberger, J. M. (2001). Stoichiometry of immunocytochemical staining reactions. *Methods Cell Biol.* **63**, 271–298.

Juan, G., Traganos, F., and Darzynkiewicz, Z. (2001). Methods to identify mitotic cells by flow cytometry. *Methods Cell Biol.* **63**, 343–354.

Kamentsky, L. A. (2001). Laser scanning cytometry. *Methods Cell Biol.* **41**, 51–88.

Larsen , J. K., Landberg, G., and Roos, G. (2001). Detection of proliferating cell nuclear antigen. *Methods Cell Biol.* **63**, 419–431.

Li, X., Melamed, M. R., and Darzynkiewicz, Z. (1996). Detection of apoptosis and DNA replication by differential labeling of DNA strand breaks with fluorochromes of different color. *Exp. Cell Res.* **222**, 28–37.

Poot, M., Silber, J. R., and Rabinovitch, P. S. (2002). A novel flow cytometric technique for drug cytotoxicity gives results comparable to colony-forming assays. *Cytometry* **48**, 1–5.

Rabinovitch, P. S. (1994). DNA content histogram and cell cycle analysis. *Methods Cell Biol.* **41**, 364–387.

Sheer, C. J. (2001). The Pezcoller lecture: Cancer cell cycles revisited. *Cancer Res.* **60**, 3689–3695.

Studzinski, G. P. (1995). *"Cell Growth and Apoptosis: A Practical Approach."* Oxford Univ. Press, Oxford.

Studzinski, G. P. (1999). *"Cell Growth, Differentiation and Senescence: A Practical Approach."* Oxford Univ. Press, Oxford.

Terry, N. H. A., and White, R. A. (2001). Cell cycle kinetics estimated by analysis of bromodeoxyuridine incorporation. *Methods Cell Biol.* **63**, 355–374.

35

Detection of Cell Cycle Stages *in situ* in Growing Cell Populations

Irina Solovei, Lothar Schermelleh, Heiner Albiez, and Thomas Cremer

I. INTRODUCTION

Microscopic *in situ* detection of the cell cycle stages is based on immunocytochemical techniques and allows one to determine the stage of the cell cycle of individual cells. Of numerous antigens specific for a certain cell cycle stage, the most often used one is nuclear protein Ki-67 (pKi-67). This protein is expressed only in cycling cells and is, therefore, used routinely in histopathological analyses to assess proliferating activity in suspected neoplastic tissues. Despite many investigations, its function remains not wholly clear (Endl and Gerdes, 2000). Importantly, pKi-67 changes its intranuclear distribution during progression of the cell cycle (Bridger *et al.*, 1998; Kill, 1996; Starborg *et al.*, 1996). **Figure 1** shows the distribution of pKi-67 at the main cell cycle stages of human primary fibroblasts. In early G1, pKi-67 accumulates in the assembling nucleoli and in sites enriched in heterochromatin; it forms numerous granules scattered over the nucleoplasm ("jaguar" pattern). Then—in late G1, during the entire S phase, and in G2—pKi-67 is localized predominantly in the nucleoli, mostly at their periphery. During late G2 and early prophase, when the nucleoli disassemble, the protein is diffusely distributed in the nucleoplasm. In later prophase, pKi-67 coats the surface of condensing chromosomes (forming "perichromosomal layer") and persists on the surface of mitotic chromosomes until late telophase. The protein especially strongly decorates heterochromatin regions of mitotic chromosomes (with exception to the centromeric heterochromatin), while NORs are free from pKi-67 (Traut *et al.*, 2002). Biochemical studies indicate that pKi-67 resides in the regions of the nucleus containing densely packed chromatin, most probably, heterochromatin (Kreitz *et al.*, 2000), and presumably, is involved in chromatin compaction. Indeed, the C-terminal domain of pKi-67 binds mammalian heterochromatin protein 1 (HP1) (Scholzen *et al.*, 2002) and sites where pKi-67 accumulates in early G1 become later HP1-binding foci (Kametaka *et al.*, 2002).

Based on pKi-67 immunostaining (see protocol **A**), one can distinguish between proliferating and quiescent cells and identify cells in early G1. Additional markers are still required to distinguish among late G1, S, and G2. Labeling with halogenated thymidine analogues (replication labeling) allows the recognition of S-phase cells (Aten *et al.*, 1992; Gratzner, 1982). The simple addition of bromdeoxyuridine (BrdU), chlorodeoxyuridine (CldU), or iododeoxyuridine (IdU) to growth medium leads to uptake and incorporation of these substances into nascent DNA. Halogenated nucleotides can be visualized by immunostaining after denaturation of DNA, which makes incorporated halogenated dUTPs accessible for antibodies. Denaturation is a critical step of detection because it can negatively affect preservation of cell morphology. In particular, incubation of cells in $2N$ HCl for 30–60 min (often used for denaturation) strongly impairs cell and nuclear morphology and, therefore, should be avoided in assays where the conservation of cell morphology is important. Heat denaturation in 70% formamide (**protocol B**) is a more protective method (Solovei *et al.*, 2002a,b), although it also damages cell morphology to a certain degree and can destroy useful antigens. Noteworthy, this treatment causes significant losses in GFP fluorescence (Solovei *et al.*, 2002a). The recently developed protocol, based on enzymatic digestion of DNA (Tashiro *et al.*, 2000), combines good preservation of cell morphology

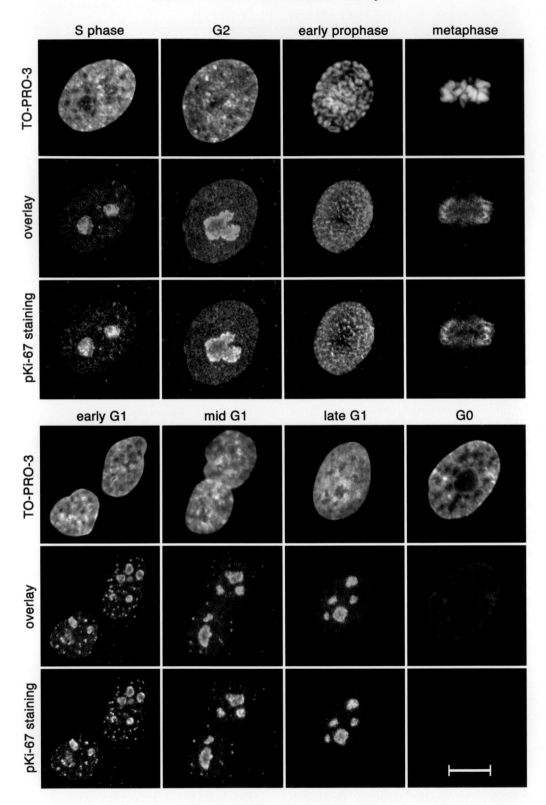

◀ **FIGURE 1** **Immunostaining of pKi-67 (protocol A).** The cells stained are human primary fibroblasts. For each stage of the cell cycle, three images represent DNA staining with TO-PRO-3 (top), pKi-67 immunostaining (bottom), and an overlay of the two with pseudocolors (mid): red for TO-PRO-3 and green for pKi-67. Each image is a maximum intensity projection of four midconfocal sections (1 μm) through a nucleus. **S phase** (S-phase cells were identified by BrdU labeling, data not shown): only nucleoli are stained; a weak diffuse staining is observed in the nucleoplasm. **G2:** nucleoli often fuse, forming one huge nucleolus of a lobulated shape with an irregular edge; staining of the nucleoplasm is more intense than in S phase. **Early prophase:** pKi-67 is distributed over the entire nucleus and occupies all the space between condensing chromosomes. **Metaphase:** protein covers condensed chromosomes. The same distribution is observed in later prophase and in anaphase (not shown). **Early G1:** characteristic "jaguar" staining pattern; pKi-67 is located in reassembling nucleoli and in small multiple foci in the nucleoplasm. **Mid G1:** the number of foci in the nucleoplasm decreases. **Late G1:** staining is limited to the nucleoli. **G0:** in contrast to cycling cells, quiescent cells do not express pKi-67. Note that in cycling cells (but not in G0 cells), small foci of pKi-67 are also present in the cytoplasm; apparently, they are sites of pKi-67 synthesis. Scale bar: 10 μm.

and antigenes with effective disclosure of incorporated halogenated dUTPs (**protocol C**).

BrdU labeling also allows one to differentiate among early, mid, and late S-phase stages (**Fig. 2**). Pulse labeling with two halogenated nucleotides during two different time points results in differential staining of chromatin replicating at different periods of S phase in the same nuclei (Aten *et al.*, 1992, 1993, 1994) (**protocol D**). Double replication labeling is an indispensable method used to establish the time sequence of replication events and is widely used in the studies of DNA replication in different cells and organisms (e.g., Ma *et al.*, 1998; Visser *et al.*, 1998; Zink *et al.*, 1999; Habermann *et al.*, 2001; Alexandrova *et al.*, 2004). To exemplify the application of this method, **Fig. 3** shows different nuclear location of early (green) and mid (red) replicating chromatin in human fibroblast nucleus visualized by double replication labeling.

A combination of pKi-67 immunostaining and BrdU immunodetection (**protocol E; Fig. 4**) allows one to differentiate between all cell cycle stages with exception to late G1 and G2. It may be noted though that G2 nuclei have a more diffuse distribution of pKi-67 and fusing nucleoli; they are also noticeably larger than G1 nuclei. Though these characters are not clearly manifested in all cell types, with some experience they often help distinguish between G1 and G2 cells.

It is noteworthy, for a growing cell population with a known doubling time, that combined pKi-67 staining and BrdU labeling are sufficient to estimate duration of the main cell cycle stages. Duration of the cell cycle (or doubling time), T_d, is calculated from the formula:

$$T_d = \frac{\log 2 \times \Delta t}{\log N_2 - \log N_1}$$

where N_1 and N_2 are the number of counted cells per given area of a coverslip at time points 1 and 2, corre-

spondingly, and Δt is the duration between two observations. The doubling time is equivalent to the cell cycle length when 100% of the cells proliferate, as can be estimated by immunodetection of pKi-67.

The proportion of quiescent (G0) cells, p_{G0}, is simply the proportion of pKi-67 negative cells. The length of S phase (T_S) is proportional to the proportion of S-phase (p_S, BrdU positive) cells:

$$T_S = T_d \cdot p_S / (1 - p_{G0}).$$

Similarly, the lengths of each substage of S phase can be estimated from proportions of cells showing early, mid, and late replication patterns, correspondingly. The proportion of early G1 cells, p_{G1e}, can be estimated by counting nuclei with specific "jaguar" pattern and duration of early G1:

$$T_{G1e} = T_d \cdot p_{G1e} / (1 - p_{G0}).$$

The length of mitosis and G2 do not vary significantly in mammalian cells and it is therefore assumed that they take about 1 and 2–3 h, respectively. On this basis, one can also assess the total duration of G1, T_{G1}, as difference between doubling time and duration of other cell cycle stages:

$$T_{G1} = T_d - (T_S + T_M + T_{G2}).$$

These calculations are simplified for immortalized cell lines in which there are no quiescent cells, i.e., essentially all cells are cycling, as, e.g., HeLa cells. Such populations indeed show a nearly 100% positive staining with pKi-67. In such cases,

$$T_S = T_d \cdot p_S \quad \text{and} \quad T_{G1e} = T_d \cdot p_{G1e}.$$

Table I exemplifies estimation of cell cycle stages lengths for two immortalized cell lines: neuroblastoma cells SH-EP N14 and HeLa.

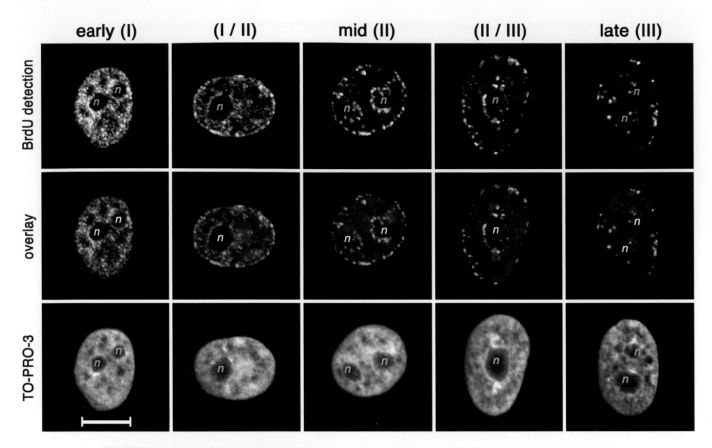

FIGURE 2   Replication labeling with BrdU (protocol C). Neuroblastoma SH-EP N14 cells were pulse labeled with BrdU for 40 min; BrdU immunodetection was performed using DNase digestion, counterstaining with TO-PRO-3. For each stage of the S phase, three images represent BrdU labeling (top), DNA staining with TO-PRO-3 (bottom), and overlay of the two with pseudocolors (mid): red for TO-PRO-3 and green for BrdU. Each image is a maximum intensity projection of four midconfocal sections (1 μm) through a nucleus. Three characteristic patterns change one another as the S phase proceeds. **Early S phase (pattern I):** all nucleoplasm is labeled with exception to the nucleoli (*n*) and speckles (dark areas in counterstaining). **Mid S phase (pattern II):** characteristic labeling is seen along the nuclear periphery and the border of the nucleolus (*n*); between these sites, signals are rare or lacking. **Late S phase (pattern III):** replication foci are significantly larger than in early and mid S phase; signals are observed at the periphery of the nucleus, at the border of nucleoli (*n*), and in nucleoplasm between them. Two transient patterns (I/II and II/III) are sometimes considered as separate S-phase stages (O'Keefe *et al.*, 1992). Scale bar: 10 μm.

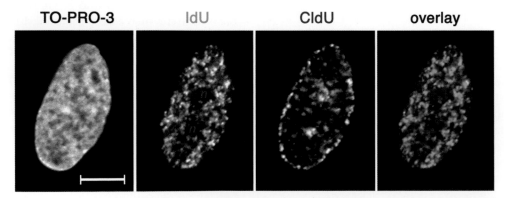

FIGURE 3   Double replication labeling with IdU and CldU (protocol D). Primary human fibroblasts were pulse labeled with IdU (green) in early S phase and with CldU (red) in mid S phase after a 4-h chase. Counterstaining with TO-PRO-3. Images represent the same midoptical section through the nucleus. Note that the midreplicating chromatin is mainly located at the very nuclear periphery and around nucleoli (*n*), whereas early replicating chromatin is distributed throughout the nuclear interior with exception of the nucleoli. Scale bar: 10 μm.

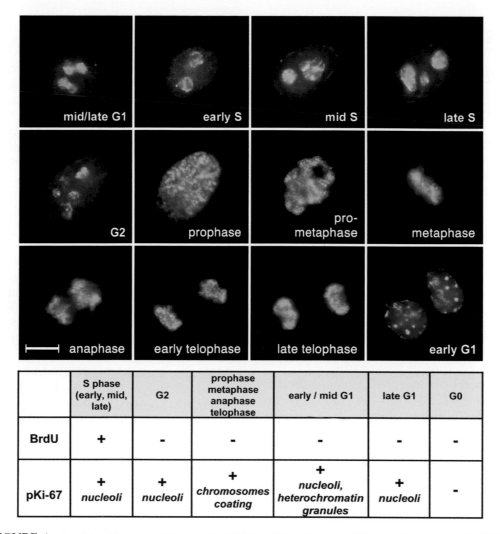

	S phase (early, mid, late)	G2	prophase metaphase anaphase telophase	early / mid G1	late G1	G0
BrdU	+	-	-	-	-	-
pKi-67	+ *nucleoli*	+ *nucleoli*	+ *chromosomes coating*	+ *nucleoli, heterochromatin granules*	+ *nucleoli*	-

FIGURE 4   Combined immunodetection of pKi-67 and BrdU (protocol E). Neuroblastoma cells SH-EP N14 were pulse labeled with BrdU for 30 min. After simultaneous detection of pKi-67 (green) and BrdU (red), nuclei were counterstained with DAPI (blue). Images are overlays of three gray scale pictures (one for each fluorochrome) collected using an epifluorescence microscope. Differences among stages of the cell cycle are summarized in Table I. Scale bar: 10 μm.

TABLE I   Estimated Lengths of Main Cell Cycle Stages in Two Immortalised Nonsynchronized Cell Lines, Neuroblastoma Cells SH-EP N14, and HeLa Cells

	SH-EP N14		HeLa	
Doubling time = cell cycle length	~13 h		~18 h	
Proportion of cycling cells (pKi67 positive)	100%		100%	
Cell cycle stages	% of cells	Length (h)	% of cells	Length (h)
S phase (BrdU positive)	47	**~6.1**	36	**~6.5**
Early S	*25*	*~3.2*	*21.1*	*~3.8*
Mid	*17.5*	*~2.3*	*12.2*	*~2.2*
Late	*4.5*	*~0.6*	*2.7*	*~0.5*
G2 (constant)		**~2.5**		**~2.5**
Mitosis (constant)		**~1**		**~1**
Early G1 (specif. pKi67 pattern)	15	**~2**	15	**~2**
Mid–late G1 (remaining)		**~1.4**		**~6**

II. MATERIAL AND INSTRUMENTATION

1. Grow cell on coverslips of desirable size and thickness of 0.17 ± 0.01 mm. In case of cells growing in suspension, we recommend attaching living cells to coverslips coated with polylysine as described in detail elsewhere (Solovei *et al.*, 2002b).

2. We recommend performing fixation and washings in 6-well plates; each well accommodates one coverslip of size ranging from 15×15 to 22×22.

3. To handle coverslips, use fine forceps (e.g., Dumont No. 4 or 5).

4. Special care must be taken not to dry cells at any steps of immunostaining. Make sure that when solutions are changed, coverslips always remain covered with liquid.

5. For incubation with antibody, use humid dark chambers with Parafilm on the bottom. Place drops of antibody solutions on Parafilm, and lay coverslips with cells on these drops with the cell side down. Carry out incubations at room temperature.

6. We recommend using Vectashield (Vector Laboratories; Cat. No. H-1000) as the antifade mounting medium. To mount cells, place a drop of Vectashield on a clean microscopic slide and quickly put coverslip with cells upside down on the antifade. Gently remove excess antifade medium with soft paper and seal with colourless nail polish (preferably, so called, base coat).

7. Mounted and sealed preparations can be stored in the dark at 4°C for months.

III. PROCEDURES

A. Immunostaining with Anti-Ki67 Antibodies

Solutions

1. Cell culture mediums used for respective cells

2. *Phosphate-buffered saline (1× PBS):* 140 mM NaCl, 2.7 mM KCl, 6.5 mM Na$_2$HPO4, 1.5 mM KH2PO4, pH 7.2

3. *Fixative:* 3.7% formaldehyde made before use from 37% stock solution (Sigma, Cat. No. F-1268) by dilution with 1× PBS

4. *PBST:* PBS with 0.05% Tween 20 (Calbiochem, Cat. No. 655204)

5. 1× PBS with 0.04% sodium azide (Merck, Cat. No. 106688)

6. 0.5% Triton X-100 (Sigma, Cat. No. T-8787) in 1× PBS

7. *Blocking solution:* 4% bovine serum albumin (BSA) in PBST. Store the stock solution of 12% BSA in PBS with 0.04% sodium azide at 4°C

8. Mouse anti-pKi-67 (DakoCytomation, Cat. No. M7240); working dilution = 1:100 in blocking solution

9. Any antimouse antibodies conjugated to a desirable fluorochrome. As two examples, we suggest goat antimouse-Cy3 (orange–red fluorescence, Jackson ImmunoResearch Laboratories, Cat. No. 115-165-072; working dilution = 1:200) or goat antimouse-Alexa 488 (green fluorescence; Molecular Probes, Cat. No. A-11017; working dilution = 1:500); antibodies are diluted in blocking solution

10. DNA-specific dye. The choice depends on the available microscopic filters or confocal laser lines and on fluorochrome chosen for secondary antibodies. The most popular counterstaining dyes are 4′,6-diamidino-2-phenylindole (DAPI, Sigma, Cat. No. D-9564) with blue fluorescence; propidium iodide (PI, Sigma, Cat. No. P-4170) with red fluoroscence; and TO-PRO-3 (Molecular Probes, Cat. No. T-3605) with emission in the far-red part of the spectrum. DAPI stock = 5 μg/ml in H$_2$O; working solution = 0.05 μg/ml. PI stock = 500 μg/ml in H$_2$O; working solution = 25 μg/ml. TO-PRO-3 stock = 1 mM in dimethyl sulfoxide (DMSO); working solution = 1 μM. All stock solutions are kept frozen; working dilutions in PBST are done before use

11. RNase (Roche, Cat. No. 109169); stock solution = 20 mg/ml in 10 mM Tris–HCl + 15 mM NaCl; working solution = 0.2 mg/ml in PBS

Steps

1. Fix cells in 3.7% formaldehyde/PBS at room temperature for 10 min; wash in PBST, 3×5 min. If needed, fixed cells can be kept in PBS with 0.04% sodium azide at 4°C overnight or longer.

2. Permeabilize with 0.5% Triton X-100 in PBS for 5 min.

3. Incubate in blocking solution for 10 min.

4. Incubate with primary antibodies, mouse anti-pKi-67 for 30 min and wash with PBST 3×5 min.

5. Incubate with secondary antibodies, e.g., goat antimouse Alexa488 or goat antimouse-Cy3, for 30 min and wash with PBST 3×5 min.

6. Counterstain nuclei with DNA-specific dye (DAPI, PI, or TO-PRO-3) for 5 min and then rinse briefly in PBST. *Optional:* Because PI and TO-PRO-3 also stain RNA, slides can be preincubated in RNase working solution at 37°C for 30 min.

7. Mount cells in antifade medium and seal with nail polish.

B. Replication Labeling with BrdU and Detection with Heat Denaturation

Solutions

1. 5-bromo-2'-deoxyuridine (Sigma, Cat. No. B-9285). Make a 1 mM stock solution in H_2O, store aliquots at −20°C
2. 2 × SSC, pH 7.0; 1 × SSC is 0.15 M NaCl and 0.015 M sodium Na-citrate
3. 0.1 N HCl
4. 50% formamide (Merck, Cat. No. 1.09684) in 2 × SSC. Formamide is toxic, so all steps that involve the use of this reagent should be performed in a hood and gloves should be worn.
5. 70% formamide in 2 × SSC.
6. Mouse anti-BrdU (Roche, Cat. No. 7580), working dilution = 1:200 in blocking solution.
7. Any antimouse antibodies coupled to desirable fluorochrome. As two examples, we suggest the same secondary antibodies as for detection in protocol **A**.

The rest of the solutions are the same as for protocol **A**.

Steps

1. Grow cells to ≥50% confluency.
2. Add BrdU to the cell culture medium to a final concentration of 10–20 μM and incubate for 15–60 min at 37°C in a CO_2 incubator.
3. Fix and permeabilize cells as described in protocol **A**.
4. Incubate in 0.1 N HCl for 10 min.
5. Wash with 2 × SSC for 5 min and equilibrate in 50% formamide/2 × SSC at room temperature for a minimum of 15 min. At this step, slides may be kept in 50% formamide at 4°C overnight or longer.
6. For cell DNA denaturation, warm up 70% formamide/2 × SSC to 70°C, immerse coverslips with cells for 2 min, and then immediately move them to ice-cold PBST and wash with PBST 3 × 5 min.
7. Incubate in blocking solution for 10 min.
8. Incubate with primary antibodies, mouse anti-BrdU for 30 min and wash with PBST 3 × 5 min.
9. Incubate with secondary antibodies, e.g., goat antimouse Alexa488 or sheep antimouse-Cy3, for 30 min and wash with PBST 3 × 5 min.
10. Mount cells in antifade medium and seal with nail polish.

Comments

BrdU pulse length can vary: only 5 min may be sufficient to detect the typical replication pattern. However, extended incubation time results in brighter signals, as more BrdU per replication focus is incorporated. At the same time, the replication pattern gradually becomes less distinct as pulse length exceeds 1 h.

C. Replication Labeling with BrdU and Detection with DNase Digestion

Roche Applied Science sells a kit for BrdU detection for immunofluorescence microscopy (Cat. No. 1296736), which is based on enzymatic digestion of DNA. The antibody to BrdU supplied with the kit contains nucleases generating single-stranded DNA fragments in the nucleus. As an alternative to the kit, this section describes a simple and inexpensive method that allows incorporated molecules of BrdU to be exposed to antibodies after DNase digestion.

Solutions

1. DNase I (Roche, Cat. No. 104 159); to make a stock solution, dilute to 1 mg/ml in 0.15 M NaCl in 50% glycerol, make aliquots, and store them at −20°C.
2. *DNase incubation buffer:* 0.5 × PBS, 30 mM Tris, 0.3 mM $MgCl_2$, 0.5 mM 2-mercaptoethanol, 0.5% bovine serum albumin, and 10 μg/ml DNase I.

The rest of the solutions are the same as for protocol **A**.

Steps

Steps 1–3 are the same as in protocol **A**.

4. Incubate in blocking solution for 10 min.
5. Dissolve primary antibodies against BrdU in the DNase incubation buffer and incubate as specified in protocol **B**.

The rest of the steps are the same as steps 9 and 10 in protocol **B**.

Pitfalls

The most probable cause for weak staining is bad DNase. In addition, an extra step with incubation in 0.1 N HCl for 5–10 min (see step 4 in protocol **B**) is recommended.

D. Double Labeling with IdU and CldU

Solutions

1. 5-Iodo-2'-deoxyuridine (IdU) (Sigma, Cat. No. I-7125). Make a 1 mM stock solution in 40% DMSO, store aliquots at −20°C
2. 5-Chloro-2'-deoxyuridine (CldU) (Sigma, Cat. No. C-6891). Make a 1 mM stock solution in H_2O, store aliquots at −20°C

3. Mouse anti-IdU and -BrdU [Becton Dickinson; Cat. No. 347580 (7580), clone B44], working dilution = 1:50

4. Goat antimouse-Alexa488 highly cross-adsorbed (Molecular Probes; Cat. No. A-11029), working dilution = 1:500

5. Rat anti-CldU and -BrdU [Harlan Sera-Labs, Cat. No. MAS 250, clone BU1/75 (ICR1)], working dilution = 1:200

6. Donkey antirat-Cy3 (Jackson ImmunoResearch, Cat. No. 712-165-153), working dilution = 1:200. All antibodies are diluted in blocking solution. Of course one can use a different pair of fluorochromes conjugated to secondary antibodies than Alexa488 and Cy3. Secondary antibodies themselves can also be changed without serious risk to spoil staining, whereas the choice of primary antibodies is crucial for successful staining.

7. *High stringency buffer*: 0.5 M NaCl, 36 mM Tris, 0.5% Tween 20, pH 8.0

The rest of the solutions are the same as for protocol **A**.

Steps

1. Grow cells to ≥50% confluency.
2. Add the required amount of IdU to the cell culture medium at a final concentration of 1–10 μM; incubate for 30–45 min.
3. Remove incubation medium, rinse cells twice in PBS, and replace with fresh medium. Chase: incubate cells in fresh medium for the period equal to one-half to three-fourth of the S-phase duration. A chase time of 4–6 h is recommended for, e.g., human and mouse fibroblasts, HeLa cells, stimulated human lymphocytes, and lymphoblastoid cells.
4. Add the required amount of CldU to the cell culture at a final concentration of 1–10 μM; incubate for 15 to 45 min.
5. Fix cells in 3.7% formaldehyde/PBS at room temperature for 10 min; wash in PBST, 3 × 5 min. If necessary, fixed cells could be kept in PBS with 0.04% sodium azide at 4°C overnight or longer.
6. Permeabilize cells with 0.5% Triton X-100 in PBS for 5 min.
7. The denaturation of DNA can be done either by heating in formamide (steps 5–7 in protocol **B**) or by DNase digestion (as described in protocol **C**). In the latter case, antibodies diluted in DNase incubation buffer should be rat anti-CldU/-BrdU (see later).
8. The recommended rat anti-BrdU antibodies detect both BrdU and CldU, but not IdU, and should be applied first. The recommended mouse anti-BrdU antibodies (Becton Dickinson), which are used for the

detection of IdU, also have some affinity to CldU and must therefore be used after the detection of CldU. Apply antibodies against CldU (rat anti-CldU and -BrdU), incubate for 30 min, and wash in PBST 3 × 5 min.

9. Apply secondary antibodies for CldU detection, donkey antirat-Cy3; incubate for 30 min and wash in PBST 3 × 5 min.

10. Apply antibodies against IdU (mouse anti-IdU/-BrdU), incubate for 30 min, and wash in PBT 3 × 5 min.

11. Wash in the stringency buffer for 10 min to remove nonspecifically bound antibodies. This step is important, as some binding of mouse anti-IdU antibodies to CldU epitops still takes place, even after primary rat anti-CldU and secondary goat antirat antibodies were applied.

12. Apply secondary antibodies for IdU detection, goat antimouse-Alexa488, incubate in a dark humid chamber for 30 min, and wash in PBST 3 × 5 min.

13. Counterstain nuclei (with DAPI and/or TO-PRO-3), mount cells in antifade medium, and seal with nail polish.

Pitfalls

When the mouse anti-IdU/-BrdU antibody cross-reacts with CldU, the time of washing in the stringency buffer (step 11) should be increased.

E. Combination of Both Methods with Anti-Ki-67 Staining

Epitops of Ki-67 protein are stable and the protein can be detected after 0.1 N HCl and denaturation steps (with heat denaturation or DNase). Therefore, pKi-67 immunostaining can be performed either simultaneously or after detection of incorporated halogenated thymidine analogues. For simultaneous detection, primary antibodies against pKi-67 and BrdU are mixed; both secondary antibodies may also be applied as one solution.

Solutions

1. Mouse anti-BrdU (Roche, Cat. No. 7580); working dilution = 1:200
2. Rabbit anti-Ki-67 (DakoCytomation, Cat. No. A0047); working dilution = 1:50
3. Goat antimouse-Cy3 (Jackson ImmunoResearch Laboratories, Cat. No. 115-165-072); working dilution = 1:200
4. Goat antirabbit-FITC (BioSource, Cat. No. ALI 4408); working dilution = 1:200

All antibody dilutions are done in blocking solution.

Steps

Steps 1–7 are the same as in protocol **B**.

8. Incubate in mixture of mouse anti-BrdU and rabbit anti-Ki-67 antibodies at room temperature and wash with PBST 3×5 min.
9. Incubate in a mixture of goat antimouse-Cy3 and goat antirabbit-FITC and wash with PBST 3×5 min.
10. Counterstain nuclei, mount cells in antifade medium, and seal with nail polish.

Pitfalls

Special care should be taken to prevent cross-reaction of antibodies: use preadsorbed secondary antibodies and avoid using in the same experiment primary antibodies raised in closely related animals, e.g., mouse and rat or sheep and goat.

References

Alexandrova, O., Solovei, I., Cremer, T., and David, C. (2003). Replication labeling patterns and chromosome territories typical of mammalian nuclei are conserved in the early metazoan Hydra. *Chromosoma* **112**, 190–200.

Aten, J. A., Bakker, P. J., Stap, J., Boschman, G. A., and Veenhof, C. H. (1992). DNA double labelling with IdUrd and CldUrd for spatial and temporal analysis of cell proliferation and DNA replication. *Histochem. J.* **24**, 251–259.

Aten, J. A., Stap, J., Hoebe, R., and Bakker, P. J. (1994). Application and detection of IdUrd and CldUrd as two independent cell-cycle markers. *Methods Cell Biol.* **41**, 317–326.

Aten, J. A., Stap, J., Manders, E. M., and Bakker, P. J. (1993). Progression of DNA replication in cell nuclei and changes in cell proliferation investigated by DNA double-labelling with IdUrd and CldUrd. *Eur. J. Histochem.* **37**(Suppl. 4), 65–71.

Bridger, J. M., Kill, I. R., and Lichter, P. (1998). Association of pKi-67 with satellite DNA of the human genome in early G1 cells. *Chromosome Res.* **6**, 13–24.

Endl, E., and Gerdes, J. (2000). The Ki-67 protein: Fascinating forms and an unknown function. *Exp. Cell Res.* **257**, 231–237.

Gratzner, H. G. (1982). Monoclonal antibody to 5-bromo- and 5-iododeoxyuridine: A new reagent for detection of DNA replication. *Science* **218**, 474–475.

Habermann, F., Cremer, M., Walter, J., Hase, J., Bauer, K., Wienberg, J., Cremer, C., Cremer, T., and Solovei, I. (2001). Arrangements of macro- and microchromosomes in chicken cells. *Chromosome Res.* **9**, 569–584.

Kametaka, A., Takagi, M., Hayakawa, T., Haraguchi, T., Hiraoka, Y., and Yoneda, Y. (2002). Interaction of the chromatin compaction-inducing domain (LR domain) of Ki-67 antigen with HP1 proteins. *Genes Cells* **7**, 1231–1242.

Kill, I. R. (1996). Localisation of the Ki-67 antigen within the nucleolus: Evidence for a fibrillarin-deficient region of the dense fibrillar component. *J. Cell Sci.* **109**, 1253–1263.

Kreitz, S., Fackelmayer, F. O., Gerdes, J., and Knippers, R. (2000). The proliferation-specific human Ki-67 protein is a constituent of compact chromatin. *Exp. Cell Res.* **261**, 284–292.

Ma, H., Samarabandu, J., Devdhar, R. S., Acharya, R., Cheng, P. C., Meng, C., and Berezney, R. (1998). Spatial and temporal dynamics of DNA replication sites in mammalian cells. *J. Cell Biol.* **143**, 1415–1425.

O'Keefe, R. T., Henderson, S. C., and Spector, D. L. (1992). Dynamic organization of DNA replication in mammalian cell nuclei: Spatially and temporally defined replication of chromosome-specific alpha-satellite DNA sequences. *J. Cell Biol.* **116**, 1095–1110.

Schmidt, M. H., Broll R., Bruch H. P., Bogler O., and Duchrow, M. (2003). The proliferation marker pKi-67 organizes the nucleolus during the cell cycle depending on Ran and cyclin B. *J. Pathol.* **199**, 18–27.

Schmidt, M. H., Broll, R., Bruch, H. P., and Duchrow, M. (2002). Proliferation marker pKi-67 affects the cell cycle in a self-regulated manner. *J. Cell Biochem.* **87**, 334–341.

Scholzen, T., Endl, E., Wohlenberg, C., van der Sar, S., Cowell, I. G., Gerdes, J., and Singh, P. B. (2002). The Ki-67 protein interacts with members of the heterochromatin protein 1 (HP1) family: A potential role in the regulation of higher-order chromatin structure. *J. Pathol.* **196**, 135–144.

Solovei, I., Cavallo, A., Schermelleh, L., Jaunin, F., Scasselati, C., Cmarko, D., Cremer, C., Fakan, S., and Cremer, T. (2002a). Spatial preservation of nuclear chromatin architecture during three-dimensional fluorescence in situ hybridization (3D-FISH). *Exp. Cell Res.* **276**, 10–23.

Solovei, I., Walter, J., Cremer, M., Habermann, F., Schermelleh, L., and Cremer, T. (2002b). FISH on three-dimensionally preserved nuclei. *In* "FISH: A Practical Approach" (B. Beatty, S. Mai, and J. Squire, eds.), pp. 119–157. Oxford Univ. Press, Oxford.

Starborg, M., Gell, K., Brundell, E., and Hoog, C. (1996). The murine Ki-67 cell proliferation antigen accumulates in the nucleolar and heterochromatic regions of interphase cells and at the periphery of the mitotic chromosomes in a process essential for cell cycle progression. *J. Cell Sci.* **109**, 143–153.

Tashiro, S., Walter, J., Shinohara, A., Kamada, N., and Cremer, T. (2000). Rad51 accumulation at sites of DNA damage and in postreplicative chromatin. *J. Cell Biol.* **150**, 283–291.

Traut, W., Endl, E., Garagna, S., Scholzen, T., Schwinger, E., Gerdes, J., and Winking, H. (2002). Chromatin preferences of the perichromosomal layer constituent pKi-67. *Chromosome Res.* **10**, 685–694.

Visser, A. E., Eils, R., Jauch, A., Little, G., Bakker, P. J., Cremer, T., and Aten, J. A. (1998). Spatial distributions of early and late replicating chromatin in interphase chromosome territories. *Exp. Cell Res.* **243**, 398–407.

Zink, D., Bornfleth, H., Visser, A., Cremer, C., and Cremer, T. (1999). Organization of early and late replicating DNA in human chromosome territories. *Exp. Cell Res.* **247**, 176–188.

36

In vivo DNA Replication Labeling

Lothar Schermelleh

I. INTRODUCTION

DNA replication in higher eukaryotes takes place in a well-defined spatial and temporal manner. Large numbers of replication sites are simultaneously active, creating characteristic replication patterns during progression through S phase (Nakamura *et al.*, 1986; O'Keefe *et al.*, 1992). Each site is typically active for ~45 min, forming a replication focus of some 100 kb up to several Mb in size, consisting of a cluster of 1–10 simultanously firing replicons (Berezney *et al.*, 2000). Because the higher order chromatin structures revealed by replication foci persist through all stages of the cell cycle, they are referred to as ~1-Mb chromatin domains. Importantly, replication timing reflects important functional chromatin features, as the gene-rich, transcriptionally active euchromatin replicates during the first half of S phase, whereas gene-poor, heterochromatic sequences are later replicating (Cremer and Cremer, 2001).

In contrast to DNA labeling with halogenated thymidine analogues, which can be detected immunocytochemically after cell fixation and DNA denaturation, usage of fluorescent precursors allows the investigation of dynamic properties of ~1-Mb chromatin domains in living cells. Depending on the time period in S phase when the labeling is accomplished, early and mid to late replication foci can be specifically marked (Schermelleh *et al.*, 2001). Moreover, further proliferation of directly labeled cells results in the random distribution of sister chromatids with labeled and unlabeled chromatin domains during the second mitosis. Subsequent cell generations result in nuclei with few labeled chromosome territories (within a bulk of unlabeled chromatin). These cells are ideally suited to study chromatin domain/chromosome territory movements *in vivo* in relation to other nuclear components tagged, e.g., with green fluorescent protein (Walter *et al.*, 2003).

Several fluorochrome-coupled deoxynucleotides (dNTPs) are available commercially. The charged phosphate group and the attached fluorophore prevent the uptake of these nucleotides across the cell membrane. Hence, a procedure is required for dNTPs to pass the cell membrane barrier. A well-established and reliable method is microinjection of a nucleotide solution through a thin glass capillary directly into the cytoplasm or nuclei of adherently growing cells. (Ansorge and Pepperkok, 1988; Pepperkok and Ansorge, 1995; Zink *et al.*, 1998). However, this procedure is tedious, requires costly equipment, and allows only labeling of a rather small number of cells in each experiment.

A transient permeabilization that allows the uptake of macromolecules from the surrounding medium can be achieved by a number of methods (for review, see McNeil, 1989), yet not all of them are useful for *in vivo* replication labeling. Detergents (e.g., Triton X-100, saponin, or digitonin) permeabilize cells effectively, but have detrimental effects on cell viability, whereas more gentle "noninvasive" methods, such as lipofection and osmotic shift, are not efficient in our experience. More suitable regarding loading efficiency and viability are methods creating slight mechanical damage to the cell membrane in the presence of nucleotides. This can be achieved by shaking a labeling solution with small glass beads over a cell layer or by detaching a cell layer with a cell scraper.

The best results in terms of labeling efficiency, reproducibility, and amount of nucleotides needed were obtained by a modified scratch-loading protocol, termed scratch replication labeling. This protocol can be applied on any adherently growing cell line. It allows the labeling of a high number of cells within one experiment with little effort. Most importantly, it does not impair further growth of a large fraction of affected cells (Schermelleh *et al.*, 2001). This protocol has been applied for live cell studies of chromatin domains and chromosome territories (Walter *et al.*, 2003).

II. MATERIALS AND INSTRUMENTATION

Cy3-dUTP (Amersham Bioscience, Cat. No. PA53022), Cy5-dUTP (Amersham Bioscience, Cat. No PA55022), fluorescein-12-dUTP (Roche Cat. No. 1373242), disposable hypodermic needles (e.g., 0.45 mm × 25 mm), 15 × 15-mm square coverslips, 60/15-mm tissue culture dishes, phase-contrast microscope, appropriate cell culture medium (with HEPES), paper wipes (e.g., Kimwipes lite precision wipes, Kimberly-Clark), fine forceps, live cell chamber with fitting coverslips (e.g., Bioptechs FCS2).

III. PROCEDURE

Solutions

Appropriate cell culture medium (with 25 mM HEPES and supplemented with 10% fetal calf serum and antibiotics); labeling solution: 20 μM (Cy3-dUTP) or 50 μM (Cy5-dUTP, fluorescein-12-dUTP) in medium.

Steps

1. Seed cells on small coverslips (15 × 15 mm) and grow them until they reach near confluence.

2. Pick up the coverslip with fine forceps, drain excess medium, dry the bottom side of the coverslip briefly with a wipe, and place it into an empty tissue culture dish. This will prevent the coverslip from sliding during the subsequent scratching procedure.

3. Add 8–10 μl of the labeling solution onto the coverslip and distribute it evenly over the cells by gently tilting the dish. Surface tension will prevent the solution from running off the coverslip.

4. With the tip of a hypodermic needle, apply parallel scratches into the cell layer. For a high fraction of labeled cells, scratches should be performed a few cell diameters apart from each other and cover the complete coverslip (Fig. 1). For optimal coverage, the procedure can be performed under a low magnification phase-contrast microscope (e.g., 5× objective lens). The procedure should not take longer than a few minutes to avoid drying of the cells.

5. Add 5 ml prewarmed medium and incubate further. Exchange medium after 30–60 min to remove nonincorporated nucleotides.

6. To obtain nuclei with segregated labeled and unlabeled chromosome territories, the cells are harvested by trypsinization some hours later and cultivated further for two or more cell cycles prior to live cell observation. The day before live cell observations are carried out, seed cells on coverslips fitting the live cell chamber (Fig. 2).

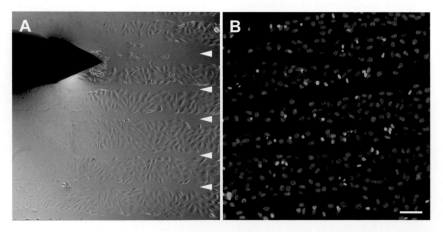

FIGURE 1   (A) Scratch replication labeling of human neuroblastoma cells. Arrowheads indicate scratches in the monolayer applied with the tip of a hypodermic needle (top left) in the presence of Cy3-dUTP. (B) Cells were fixed with 4% formaldehyde two hours after the scratching procedure and counterstained with 4′, 6-diamidino-2-phenylindole (DAPI, blue). Numerous cells display Cy3-dUTP labeled nuclei (red) along the scratch lines. Bar: 100 μm.

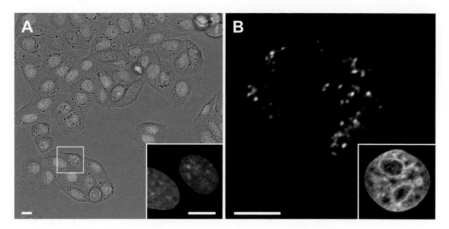

FIGURE 2 (A) Living HeLa cells with green fluorescent protein (GFP)-tagged histone H2B reveal green fluorescent nuclear chromatin. The cell culture was scratch labeled with Cy3-dUTP 5 days prior to observation in the live cell chamber. Numerous cells display nuclei with a few Cy3-labeled chromosome territories/~1-Mb chromatin domains. The framed cells are shown at higher resolution in the inset. Bars: 10 μm. (B) Confocal midsection of a HeLa cell expressing histone H2B-GFP fixed 3 days after labeling with Cy3-dUTP shows segregated chromosome territories/~1-Mb chromatin domains at high resolution. (Inset) GFP-labeled chromatin of the same nucleus. Bar: 5 μm.

IV. COMMENTS

The scratch procedure creates damage in the cell membrane that may last for only a few seconds ("transient holes"). This allows the uptake of charged macromolecules (such as fluorochrome-coupled dUTPs) from the surrounding medium, to which cells would normally be impermeable. Accordingly, most S-phase cells damaged alongside the scratch line or lifted off from the surface by the needle will incorporate the modified nucleotides.

The fraction of replication-labeled cells depends on the (1) density of scratches applied, i.e., how many cells are affected, and (2) number of cells that are in S phase. Synchronization of cells at the G1/S transition (e.g., by aphidicolin) is recommended to obtain a high yield of labeled cells.

A pool of fluorescent nucleotides is available for DNA replication over a time scale of roughly 1 h. The labeling pattern therefore resembles a BrdU-labeling experiment with ≤1-h pulse length.

Cy3-dUTP is significantly brighter and more photostable than Cy5- and fluorescein-dUTP and thus is best suited for *in vivo* observation. In cases where especially high fluorescence intensities are desirable (e.g., for long-term *in vivo* observations after segregation), a higher concentration of nucleotides (100 μM) is advantageous. We did not note any improvement of label intensities when adding nonfluorescent dNTPs to the labeling solution or by using different labeling buffers (phosphate-buffered saline or Tris-buffered saline instead of medium).

For studies of nascent RNA formation, the scratch protocol can be applied accordingly using BrUTP (5 mM in medium) followed by immunofluorescent detection.

References

Ansorge, W., and Pepperkok, R. (1988). Performance of an automated system for capillary microinjection into living cells. *J. Biochem. Biophys. Methods* **16**, 283–292.

Berezney, R., Dubey, D. D., and Huberman, J. A. (2000). Heterogeneity of eukaryotic replicons, replicon clusters, and replication foci. *Chromosoma* **108**, 471–484.

Cremer, T., and Cremer, C. (2001). Chromosome territories, nuclear architecture and gene regulation in mammalian cells. *Nature Rev. Genet.* **2**, 292–301.

McNeil, P. L. (1989). Incorporation of macromolecules into living cells. *Methods Cell Biol.* **29**, 153–173.

Nakamura, H., Morita, T., and Sato, C. (1986). Structural organizations of replicon domains during DNA synthetic phase in the mammalian nucleus. *Exp. Cell Res.* **165**, 291–297.

O'Keefe, R. T., Henderson, S. C., and Spector, D. L. (1992). Dynamic organization of DNA replication in mammalian cell nuclei: Spatially and temporally defined replication of chromosome-specific alpha-satellite DNA sequences. *J. Cell Biol.* **116**, 1095–1110.

Pepperkok, R., and Ansorge, W. (1995). Direct visualization of DNA replication sites in living cells by microinjection of fluorescein-conjugated dUTPs. *Methods Mol. Cell Biol.* **5**, 112–117.

Schermelleh, L., Solovei, I., Zink, D., and Cremer, T. (2001). Two-color fluorescence labeling of early and mid-to-late replicating chromatin in living cells. *Chromosome Res.* **9**, 77–80.

Walter, J., Schermelleh, L., Cremer, M., Tashiro, S., and Cremer, T. (2003). Chromosome order in HeLa cells changes during mitosis and early G1, but is stably maintained during subsequent interphase stages. *J. Cell Biol.* **160**, 685–697.

Zink, D., Cremer, T., Saffrich, R., Fischer, R., Trendelenburg, M. F., Ansorge, W., and Stelzer, E. H. (1998). Structure and dynamics of human interphase chromosome territories *in vivo*. *Hum. Genet.* **102**, 241–251.

37

Live Cell DNA Labeling and Multiphoton/Confocal Microscopy

Paul J. Smith and Rachel J. Errington

I. INTRODUCTION

Techniques for the discrimination and location of DNA, chromatin architecture, chromosomes, nuclear superstructures, and cell nuclei in their various forms through the cell cycle are of increasing interest in the biosciences and the generation of automated screening systems. The appropriate selection of a DNA-labeling dye may also be important for tracking events in microscale devices (e.g., lab on a chip). Here the critical issue is the transmittance advantage of longer wavelengths using external laser excitation or an incorporated source (e.g., a red laser diode within an optical biochip).

Advanced microscopy methods such as multiphoton excitation laser scanning (MPLSM) are now becoming more generally accessible, as they provide significant advantages in live cell studies. Although it is apparent that MPLSM technology will evolve, providing more convenient and adaptable systems, the principles discussed here will still apply. Currently, MPLSM instrumentation usually consists of a tuneable femtosecond pulsed infrared (IR) laser attached to a scanning microscope with IR-compatible optics to ensure efficient transfer of long wavelength light (690–1000 nm). Such a laser configuration provides virtually simultaneous delivery of two photons to a focal point of the microscope objective lens, achieving volume limited excitation by two or more photons of any fluorophor molecule with appropriate absorption characteristics. The advantages of this approach compared to confocal laser scanning microscopy (CLSM) are numerous. Because photons are only emitted from the "in focus" plane, all the photons can be collected via a nondescanning route, including those photons scattered by the sample and usually rejected at the confocal aperture. This provides a massively improved efficiency of emission light collection, an important consideration in photon-limited systems (White and Errington, 2002). Importantly, infrared light penetrates biological material further than shorter wavelengths and the potential damaging effects of ultraviolet (UV)-visible wavelengths are avoided. Reviews of the subject and information on two-photon fluorescence excitation cross sections are available (e.g., Denk *et al.*, 1995; Williams *et al.*, 2001; Albota *et al.*, 1998). However, a pragmatic outcome of the combined advantages of MPLSM is a resurgence of the use of UV-excited fluorescent probes, although these have pitfalls for use with GFP-based fluorophors (see later). Two-photon fluorescence absorption and emission spectra have been obtained for DNA probes relevant for live cell work. The technique of multi-photon excitation of a fluorophore, using a pulsed laser light source, provides solutions to several problems associated with continuous wave excitation via single photon absorption. Typically for multi-photon excitation the sample is illuminated with light of a wavelength which is approximately twice (or three times) the wavelenghth of the absorption peak of the fluorophore in use (Errington *et al.*, 2005) (Bestvater *et al.*, 2002; Smith *et al.*, 2000; van Zandvoort *et al.*, 2002) and a selection is shown in Table I.

DNA offers a highly attractive substrate for chemical and biophysical probe interactions and there are excellent references for the chemical descriptions and cytometric applications of a wide range of DNA dyes (Darzynkiewicz *et al.*, 1990; Latt and Langlois, 1990; Waggoner, 1990). The forms of interaction range from covalent binding, inter- and intrastrand cross-linking, adduct formation, to ternary complex formation with

TABLE I   Cell-Permeant Nucleic Acid Probes Characterized for Both Single-Photon and Multiphoton Studies

Probe (product code)	SPE Ex$_\lambda$max^a (nm)	MPE Em$_\lambda$max^b (nm)	Em max (nm)	Reference for more MPE information
Acridine orange (A-3568)	500 (DNA)	837 > 882 >> 981	526 (DNA)	Bestvater et al. (2002)
	460 (RNA)		650 (RNA)	
DAPI (D-1036)	358	685 > 697	461 (DNA)	Bestvater et al. (2002)
		970^c	500 (RNA)	
DRAQ5	647	1064	670	Smith et al. (2000)
Hoechst 33342 (H-3570)	350	660 > 715	461	Bestvater et al. (2002)
SYTO 13 (S-7575)	488	800	509	van Zandvoort et al. (2002)

^{a} Single photon excitation wavelength maximum.
^{b} Multiphoton excitation emission wavelength maximum.
^{c} Three-photon excitation (Lakowicz et al., 1997).

DNA-binding proteins. Additionally, major or minor groove residence are potentially stabilized through hydrogen bonding and van der Waals forces, phosphate group interactions, and various levels of intercalation between the base pairs. A restricted number of agents are now available for DNA labeling in live cell systems and MPLSM. The most frequently used are UV-excitable fluorochromes such as the bisbenzimidazole dyes Hoechst 33258 and Hoechst 33342, and the agent DAPI (4′, 6-diamidino-2-phenylindole). Cell-permeant, acridine orange displays metachromatic staining of nucleic acids that precludes its convenient use as an exclusively DNA-discriminating probe. The introduction of cell-permeant cyanine SYTO nucleic acid stains (Frey, 1995; Frey et al., 1995) has provided many different reagents with a wide range of spectral characteristics. These agents passively diffuse through the membrane of most cells and can be excited by UV or visible light but stain RNA and DNA in both live and dead eukaryotic cells. Such dyes would be accessible to two-photon excitation (Table I). Their extremely low intrinsic fluorescence, with quantum yields typically less than 0.01 unbound (>0.4 when bound to nucleic acids), reduces the need to remove extracellular dye prior to imaging. It has been stressed that the highly versatile SYTO dyes do not act exclusively as nuclear stains in live cells and that they should not be equated in this regard with compounds such as DAPI or the Hoechst 33258 and Hoechst 33342 dyes.

At the other end of the fluorescence spectrum, DRAQ5, a DNA-binding anthraquinone derivative, can provide sufficient discrimination of cellular DNA content in multiparameter fluorescence studies. Anthraquinones are a group of synthetic DNA-intercalating agents (Lown et al., 1985) but are only weakly fluorescence (Bell, 1988). DRAQ5 is a modified anthraquinone with enhanced DNA affinity and intracellular selectivity for nuclear DNA (Smith et al., 1999, 2000). DRAQ5 appears to achieve its live cell nuclear discrimination by its high affinity for DNA. Excitation at the 647-nm wavelength, close to the Ex$_{\lambda max}$, produces a fluorescence spectrum extending from 665-nm out to beyond 780-nm wavelengths. Thus the emission spectrum is beyond that of fluorescein, phycoerythrin, Texas Red, Cy 3, and, perhaps most importantly, EGFP. (Waggoner, 1990) DRAQ5 enters cells and nuclei rapidly, the broad excitation spectrum of DRAQ5 means that flow cytometric applications can utilize excitation wavelengths down to 488 nm (Smith et al., 2000); two-photon excitation is also permissible at wavelengths >1047 nm. The two-photon dark excitation region of DRAQ5 (720–860 nm) permits detection of low-intensity intracellular fluorescence of other probes by MPLSM and then definition of nuclear location by CLSM. This article provides simple procedures for two DNA-specific probes for living cells and their use in MPLSM studies. These dyes sit at either end of the excitation spectrum, Hoechst 33342 (UV excitable) and DRAQ5 (far red excitable), and therefore offer maximum flexibility in multiprobe analyses.

II. MATERIALS AND INSTRUMENTATION

Materials

DRAQ5 (molecular weight 412.54) dye is supplied as an aqueous stock solution of 5mM (Biostatus Ltd) and can be diluted in aqueous buffers or added directly to full culture medium. DRAQ5 is stable at room temperature as a stock solution but can be stored routinely at 4°C, although freezing of the stock solution should be avoided. 33342 Hoechst (H-3570; termed here Hoechst 33342) has a molecular weight of 615.99 in the trihydrochloride/trihydrate form and is supplied by Molecular Probes Inc. as a 10-mg/ml stock solution in water in a light-excluding container. It is routinely stored at 4°C, without freezing. With both DNA dyes, appropriate safety information is available from the suppliers. Because both agents have the potential to damage DNA, they should be considered both cytotoxic and potentially mutagenic. Currently available information indicates that although they are not listed as carcinogens by the National Toxicology Program (NTP), the International Agency for Research on Cancer (IARC), or the Occupational Safety and Health Administration (OSHA), the dyes should be treated as such. Mitotracker Orange CMTMRos (M7510) is supplied by Molecular Probes in 20×50-µg aliquots, stored at $-20°C$. Before use a stock solution is made by the addition of 50 µl dry dimethyl sulfoticle, which can be subsequently frozen; however, repeated freeze–thawing is not recommended. Zinquin ethyl ester [Zinquin E, [2-methyl-8-(4-methylphenylsulfonylamino) quinolinyl] oxyacetic acid ethyl ester] (Alexis Corporation, Nottingham, UK) is stored as a 5 mM stock solution in ethanol at $4°C$. The HEPES buffer (1 M) (Sigma; H-0887) filter-sterilized solution is stored at $4°C$. Stock preparations of 8% paraformaldehyde (Sigma; P 6148) are made up in phosphate-buffered saline (PBS) lacking calcium and magnesium and are kept in aliquots of 20 ml at $-20°C$. All live cell work is conducted with cells seeded onto Nunc Lab-tek (#1 coverglass) chambered wells, appropriate for high-resolution imaging and recommended when handling cytotoxic agents.

Cytometry

The flow cytometry system is a FACS Vantage cell sorter (Becton Dickinson Inc., Cowley, UK) incorporating two lasers: (1) a Coherent Enterprise II laser simultaneously emitting at multiline UV (350- to 360-nm range) and 488-nm wavelengths and (2) a Spectra Physics 127–35 helium–neon laser (maximum 35-mW output) emitting at 633 nm with a temporal separation of about 25 µ from that of the primary 488-nm beam for the UV or red lines. Light scatter signals are collected as standard. The analysis optics are (i) primary beam-originating signals analyzed at FL3 (barrier filter of LP715 nm) after reflection at a SP610 dichroic and (ii) delayed beam-originating signals analyzed at FL4 (barrier filter of LP695 nm for DRAQ5 emission or DF660/20 nm for Hoechst 33342 emission) or at FL5 (barrier filter of DF424/44 nm for Hoechst 33342 emission). Forward and 90° light scatter are analyzed to exclude any cell debris. Optical filters are originally sourced from Becton Dickinson Inc. or Melles Griot Inc.). All parameters are analyzed using CellQuest software (Becton Dickinson Inc.).

Imaging

Multiphoton excitation at wavelengths between 720 and 980 nm uses a laser-scanning microscope comprising a 1024 MP scanning unit controlled by LaserSharp software (Bio-Rad Cell Science Division) attached to a Zeiss Axiovert 135 (Carl Zeiss Ltd.). MPLSM mode is achieved with a Verdi- Mira excitation source (Coherent UK Ltd.). The Mira Optima 900-F (Coherent UK Ltd.) is a simple, stable titanium:sapphire, KLM mode-locking system with X-wave optics, a broadband, single optics set tuning 700–1000 nm, and a purged enclosure. The continuous wave diode pump laser (Verdi) operates at 5 W at 532 nm. CLSM mode is achieved using the same scanning unit and 488-, 568-, and 647-nm lines of a krypton–argon laser. All images are collected with either a ×63, 1.4 NA oil immersion lens or a ×40 1.3 NA oil immersion lens. Typically, each optical slice consists of 512 × 512 (x,y resolution = 0.35 µm, z resolution is approximately 1 µm). Multiphoton excitation at 1047 nm uses a similar BioRad 1024 MP system incorporating an all solid-state excitation source Nd:YLF laser (Coherent UK Ltd.) and mode-locked femtosecond pulsed laser providing two-photon excitation of DRAQ5 at a 1047-nm wavelength at 15 mW with detection of fluorescence in the far red.

III. PROCEDURES

A. General Considerations

Hoechst 33342 is more lipophilic than Hoechst 33258 and is the form recommended for live cell studies showing high affinity for A-T base pairs through noncovalent binding in the minor groove of DNA. The ligand shows fluorescence enhancement upon binding. DRAQ5 is weakly fluorescent but has high DNA affinity and intracellular selectivity for nuclear DNA due to increased AT base pair preference but with no apparent fluorescence enhancement upon binding. When considering live cell labeling, the performance status of the cell is the most critical issue for achieving optimal staining. There is a subtle shift in the violet to red bias in the emission spectrum of Hoechst 33342 upon DNA binding. A small (approximately 10 nm) red shift in peak wavelength also occurs upon DRAQ5 binding. Nuclear staining is straightforward and most cultured cells label well in full culture medium (containing 10% fetal calf serum) supplemented with 10 mM HEPES, reaching equilibrium over some 5–60 min for concentrations in the 1–20 µM range. Staining in phosphate buffers is less efficient. Staining rate is enhanced at $37°C$. The following procedures are for generic uses of the dyes.

B. Two-Photon Excitation of Hoechst 33342

Steps

1. Seed cells onto a sterile glass coverslip in a culture dish (approximately 1×10^5 cells/25 cm^2) or

seed into a Nunc chamber coverslip. Incubate attached cultures under standard conditions until analysis is required. Add Hoechst 33342 directly to culture medium at a final concentration of $5 \mu M$ ($3.1 \mu g/ml$).

2. For greater pH control conditions, especially when manipulating cells out of 5% CO_2 incubation, supplement the culture medium with HEPES buffer (final concentration $5–25 mM$).

3. After a 60-min incubation (minimum of 15 min) at 37°C, wash coverslips gently in PBS to preserve mitotic cell attachment. Mount coverslips in an inverted position onto a microscope slide in PBS. To avoid cell crushing, the edges of the coverslip can be supported and sealed (e.g., using a piping of petroleum jelly). To clean the noncell side of the coverslip, the surface can be wiped (lens tissue).

4. Imaging conditions and typical results are shown in Fig. 1.

C. Two-Photon Coexcitation of Different Fluorophors

Steps

1. Generate cultures and label Hoechst 33342 as in Section III,B.

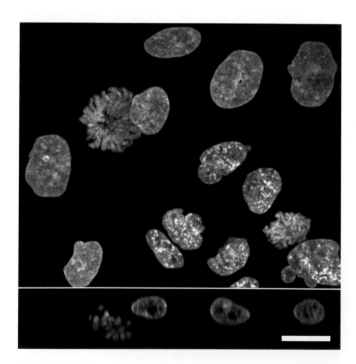

FIGURE 1 Two-photon excitation (Ex_λ 750 nm) of nuclear-located Hoechst 33342 in live HeLa cells ($5 \mu M \times 60$ min). (Top) Main image is a z-axis projection of a series of optical slices ($Em_\lambda > 460$ nm) revealing intranuclear chromatin architecture and a display of condensed metaphase chromosomes in a mitotic cell. (Bottom) The corresponding z-axis slice for an x axis intercepting the mitotic and three adjacent cells. Size bar: 10 μm.

2. A second cell-permeant probe may be introduced concurrently with Hoechst 33342 staining or after culture washing. Typically, Hoechst 33342-labeled cells will lose approximately 25% nuclear fluorescence over a 15- to 30-min postlabeling period in fresh medium. Here we exemplify the signal separation of Hoechst 33342 and the probe Mitotracker Orange CMTMRos, the latter introduced into the culture medium at 5 ng/ml for the final 5-min period of Hoechst 33342 labeling.

3. Sample preparation is as in Section III,B, whereas dual-imaging conditions and typical results are shown in Fig. 2.

D. Hoechst 33342 Staining Kinetics and Population Spectral Shift Analysis

Steps

1. Detach cultured cells by a standard trypsin/EDTA method and resuspend at 2×10^5 cell/ml full culture medium supplemented with 10 mM HEPES and allow to equilibrate at 37°C prior to Hoechst 33342 addition (see Section III,B).

2. Analyze single cell suspensions, in the presence of the dye, by one-photon UV excitation flow cytometry and acquire data for different dye uptake periods.

3. The change of spectral emission with dye uptake by mouse L cells is shown for populations of 10^4 cells in Fig. 3. Cells showing a rapid, essentially immediate, red shift are damaged and can be excluded by additional electronic gating using light scatter changes.

E. Two-Photon Excitation of Hoechst 33342 and Spectral Shift Analysis

Steps

1. Generate cultures and label Hoechst 33342 as in Section III,B.

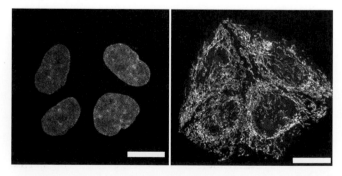

FIGURE 2 Dual images of four live HeLa cells using two-photon coexcitation (Ex_λ 800 nm) of nuclear-located Hoechst 33342 ($5 \mu M \times 60$ min; left side Em_λ 455/25 nm) and cytoplasmic Mitotracker orange CMTMRos (5 ng/ml × 5 min; right side Em_λ 585/32 nm). Size bar: 10 μm.

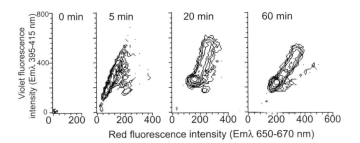

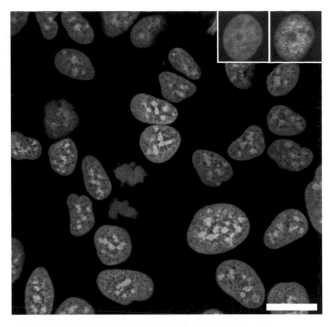

FIGURE 3 Flow cytometric contour plots of mouse L-cell populations undergoing a violet-to-red spectral shift in fluorescence during nuclear localization of Hoechst 33342 ($10\mu M$; 4×10^5 cells/ml) using one-photon excitation by multiline UV. The broad fluorescence emission spectrum was monitored simultaneously in the red region (Em_λ 660/20nm) and the violet region (Em_λ 405/20nm) for 10^4 cells at the specified time points. Data show the rapid development of violet fluorescence and the later red shift. Residual levels of damaged cells not excluded by light scatter gating are evident by their rapid red-shift pattern.

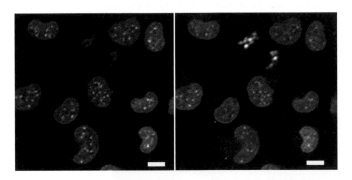

FIGURE 4 Dual images of live mouse L cells using two-photon excitation (Ex_λ 750nm) of nuclear-located Hoechst 33342 ($5\mu M \times$ 60min) showing a red-biased fluoresence emission in metaphase versus interphase cells (right side shows Em_λ 585/32nm and the left side shows Em_λ 405/35nm). Size bar: $10\mu m$.

2. Sample preparation is as in Section III,B with dual-imaging conditions and typical results shown in Fig. 4. The dual images of live mouse L cells show the differences in emission by condensed chromatin in metaphase cells lacking a nuclear envelope and interphase nuclei.

F. One-Photon Excitation of DRAQ5 for Live and Persistence in Fixed Cells

Steps

1. Generate cultures and mount coverslips as in Section III,B.

2. Cultures can be exposed to DRAQ5 by direct addition to full culture medium ($10\mu M \times 10min$) prior to coverslip washing (optional) and mounting.

FIGURE 5 Confocal image of live MCF-7 breast tumor cells using one-photon excitation (Ex_λ 647nm; Em_λ 680/32nm) of nuclear-located DRAQ5 ($10\mu M \times 10min$) revealing nuclear architecture and the presence of an anaphase cell. (Left insert) Nuclear retention of the dye following preparation washing and conventional paraformaldehyde fixation for the purposes of subsequent immunofluorescence colocalization analysis. (Right insert) An example of immunolocalization is shown for the expression of nuclear-located protein detected using a secondary antibody conjugated to Alexa488 (Ex_λ 488nm; Em_λ 522/35nm). Size bar: $10\mu m$.

However, the rapid staining kinetics of DRAQ5 (see Section III,G) permit direct introduction of the dye to already mounted samples and imaging immediately for nuclear tracing.

3. Figure 5 shows one-photon excitation images of nuclear-located DRAQ5 and dye persistence following standard 4% paraformaldehye fixation and the sample subsequently processed for immunofluorescence, which permits colocalization of an immunofluorescence signal (Alexa 488) with nuclear structures.

G. DRAQ5 Staining Kinetics and DNA Content Analysis

Steps

1. Prepare cells as in Section III,D and the conditions given in Fig. 6.

2. DRAQ5 far-red fluorescence is detectable within seconds of dye addition to cell suspensions and can be generated by both 488- and 633-nm wavelength one-photon excitation with similar outcomes. The optical configurations are shown in Fig. 6. The flow cytometry protocol acquires fluorescence emissions with time

Latt, S. A., and Langlois, R. G. (1990). Fluorescent probes of DNA microstructure and DNA synthesis. *In "Flow Cytometry and Sorting"* (M. R. Melamed, T. Lindmo, and M. L. Mendelsohn, eds.), pp. 249–290. Wiley-Liss, New York.

Lown, J. W., Morgan, A. R., Yen, S.-F., Wang, Y. H., and Wilson, W. D. (1985). Characteristics of the binding of the anticancer agents mitoxantrone and ametantrone and related structures to deoxyribonucleic acids. *Biochem. J.* **24**, 4028–4035.

Morgan, S. A., Watson, J. V., Twentyman, P. R., and Smith, P. J. (1990). Reduced nuclear binding of a DNA minor groove ligand (Hoechst 33342) and its impact on cytotoxicity in drug resistant murine cell lines. *Br. J. Cancer* **62**, 959–965.

Smith, P. J., Blunt, N., Wiltshire, M., Hoy, T., Teesdale-Spittle, P., Craven, M. R., Watson, J. V., Amos, W. B., Errington, R. J., and Patterson, L. H. (2000). Characteristics of a novel deep red/infrared fluorescent cell-permeant DNA probe, DRAQ5, in intact human cells analyzed by flow cytometry, confocal and multiphoton microscopy. *Cytometry* **40**, 280–291.

Smith, P. J., Lacy, M., Debenham, P. G., and Watson, J. V. (1988). A mammalian cell mutant with enhanced capacity to dissociate a bis-benzimidazole dye-DNA complex. *Carcinogenesis* **9**, 485–490.

Smith, P. J., Wiltshire, M., Davies, S., Patterson, L. H., and Hoy, T. (1999). A novel cell permeant and far red-fluorescing DNA probe, DRAQ5, for blood cell discrimination by flow cytometry. *J. Immunol. Methods* **229**, 131–139.

Van Zandvoort, M., de Grauw, C., Gerritsen, H., Broers, J., oude Egbrink, M., Ramaekers, F., and Slaaf, D. (2002). Discrimination of DNA and RNA in cells by a vital fluorescent probe: Lifetime imaging of SYTO13 in healthy and apoptotic cells. *Cytometry* **47**(4), 226–235.

Waggoner, A. S. (1990). Fluorescent probes for cytometry. *In "Flow Cytometry and Sorting"* (M. R. Melamed, T. Lindmo, M. L. Mendelsohn, eds.), pp. 209–225. Wiley-Liss, New York.

White, N. S., and Errington, R. J. (2002). Multi-photon microscopy: Seeing more by imaging less. *Biotechniques* **33**(2), 298–305.

Williams, R. M., Zipfel, W. R., and Watt, W. W. (2001). Multi-photon microscopy in biological research. *Curr. Opin. Chem. Biol.* **5**, 603–608.

Zalewski, P. D., Forbes, I. J., and Betts, W. H. (1993). Correlation of apoptosis with change in intracellular labile Zn(II) using zinquin [(2-methyl-8-p-toluenesulphonamido-6-quinolyloxy)acetic acid], a new specific fluorescent probe for Zn(II). *Biochem J.* **296**, 403–408.

Cytotoxic and Cell Growth Assays

allowing the cells to settle or adhere to the flask before transfering to the improved Neubauer chamber.

2. Mix thoroughly 50 μl of cell suspension with 50 μl of trypan blue in a 500-μl Eppendorf tube. Leave the mixture no more than 1–2 min because longer incubation with the dye may be toxic to viable cells and will result in overestimating the number of dead cells.

3. Place a coverslip over the hemocytometer so that it covers the central 1 mm^2 of the semisilvered counting area.

4. With a micropipette, collect the mixture of 1:1 cell suspension and trypan blue and transfer it to the edge of the hemocytometer chamber.

5. Let the mixture flow under the coverslip by capillary action, being careful not to overfill or underfill the chamber, as it will affect the counting.

6. If any surplus fluid is present over the edges, use absorbing paper to remove it.

7. Place the Neubauer chamber under the microscope and select the 10× objective. Focus on the center of the semisilvered counting area where the grid lines are evident by contrast.

8. Triple parallel grid lines surround a 1-mm^2 area divided in 25 smaller squares further subdivided in 16 smaller squares that are used for counting.

9. Unstained cells with a refractile ring around them are the viable cells, whereas dark blue colored cells that do not have refractile ring around them are nonviable cells.

10. Count the total amount of cells, stained and unstained. The percentage of unstained cells gives you the percentage of viable cells with this method. For routine culture, count 100 cells/mm^2. Counting more cells makes the test more accurate.

B. Cell Viability Assays

1. XTT/PMS Assay

This procedure exploits the fact that the internal environment of proliferating cells is more reduced than one of nonviable cells. Tetrazolium salts are used to measure this reduced state. Among them, XTT is preferred to MTT because it is more soluble. However, there are some disadvantages with this method. XTT is generally cytotoxic and destroys the cells under investigation, allowing only a single evaluation. It requires the presence of phenazine methosulfate for efficient reduction.

Materials and Instrumentation

Falcon Microtest tissue culture plates (96 wells) (Cat. No. 35-3072) and Falcon polystirene pipette (Cat.

No. 35-7551) are from Becton-Dickinson (Franklin Lakes, NJ); RPMI 1640 medium with L-glutamine (Cat. No. 11875-093), fetal bovine serum (FBS) (Cat. No. 10437-028), phosphate-buffered saline (PBS), 7.4 (Cat. No. 10010-023), and trypsin–EDTA (Cat. No. 25200-056) are from GIBCO BRL (Rockville, MD); multichannel pipette (50–300 μl) (Cat. No. P3970-18), micropipettes (200 μl) (Cat. No. P-3950-200), and pipette tips (Cat. No. P3020-CPS) are from Deville Scientific (Metuchen, NJ); XTT sodium salt [2,3-bis(2-methoxy-4-nitro-5-sulfophenyl)-2H-tetrazolium-5-carboxanilide inner salt] (Cat. No. X 4626) and phenazine methosulfate (N-methylphenazonium methyl sulfate salt) (Cat. No. P 9625) are from Sigma-Aldrich (St. Louis, MO); and microplate reader SpectraMax Plus is from Molecular Devices (Sunnyvale, CA).

Two 96-wells plates are required: one for cells and one for drug dilutions.

a. Preparation of Cells in Plate A

Steps

1. The method may be used on cells that are adherent or growing in suspension.

2. Culture cell lines in RPMI 1640 media with 10% FBS, 1% glutamine, and 1% pen/strep or other appropriate media.

3. For harvesting, the cells should be in log-phase growth (300–500 × 10^3 cells/ml) or, if dealing with adherent cells, trypsinization must be done before cells reach 80% confluence.

4. Harvest 100,000 cells per each 96-well plate and resuspend in a total volume of 10 ml/medium with 20% FBS, 1% glutamine, and 1% pen/strep.

5. From 100,000 cells in 10 ml medium, pipette 100 μl of medium+cells in each well to have 1000 cells/well.

6. Leave the first row for the blank. The second row is a control (cells without drug).

7. For adherent cells, allow 1 h for cells to reattach before adding the drug under study.

The amount of medium per well in each experiment may change depending on the amount of drug that is added after the cells are plated. The following example uses 100 μl of medium containing 1000 cells and 100 μl of drug, resulting in 200 μl of medium in each well.

b. Drug Preparation in Plate B

Steps

1. Pipette 125 μl of RPMI 1640 medium in each well of a 96-well plate.

2. Add 125 μl of the drug in each well in the first row. Then, after mixing, transfer 125 μl of the mix to

the following row and repeat the procedure up to the last row. In this way X concentration of the drug will be present in the first row, a $X/2$ concentration in the second row, and so on.

3. Pipette 100 μl from the 10th row of the plate (plate B) with the drug dilutions to the last row of the plate with the cells (plate A) so that the lower concentrations do not affect subsequent transfers.

4. When all the transfers are completed, add 100 μl of plain RPMI 1640 medium to the control row. You will have 200 μl of medium with 10% FBS in each well.

5. Add 200 μl of medium to the blank row. Start the incubation.

c. Assay Procedure

Steps

1. Warm 5 ml of plain RPMI 1640 medium at 50°C for each plate tested. This temperature allows the XTT salt to dissolve better.
2. Add 5 mg of XTT powder to the 5 ml of RPMI 1640 (it is important that no more than 1 mg of XTT/ml of medium is used).
3. Prepare a stock solution of 5 mM PMS. Add 326 μl of warm PBS to the vial containing the 0.5 mg of PMS (FW 306.3).
4. Add 25 μl of the stock 5 mM PMS to the solution containing 5 ml of medium + 5 mg of XTT.
5. Pour the solution in a reservoir and, with a multichannel pipette, transfer 50 μl of it per each well. The ratio of 0.25 ml of the XTT/PMS solution/ml of cell culture must be maintained. In the procedure described earlier there is 200 μl of medium/well × 96 wells, or a total of 1820 μl, and 50 μl of the XTT/PMS solution should be added to each well.
6. Incubate at 37°C for 2–4 h.
7. Measure absorbance with a microplate reader at the wavelength of 450 and 630 nm as a reference wavelength.

Pitfalls

Warming up the RPMI 1640 media is critical for total XTT solubilization. When dissolved incompletely, XTT salt will affect the results. The incubation time after XTT is added may vary and could be longer than the 2–4 h as suggested earlier.

2. MTS/PMS Assay

The tetrazolium compound [3-(4,5-dimethylthiazol-2-yl)-5-(3-carboxymethoxyphenyl)-2-(4-sulfophenyl)-2H-tetrazolium, inner salt MTS, in the presence of the electron coupling reagent phenazine methosulfate (PMS)] is bioreduced by viable cells into a formazan product that is soluble in culture media. The advan-

tage of MTS over XTT is that it is more soluble and nontoxic, allowing the cells to be returned to culture for further evaluation. The disadvantage is that like XTT it requires the presence of PMS for efficient reduction.

Materials and Instrumentation

Falcon Microtest tissue culture plates (96 wells) (Cat. No. 35-3072) and Falcon polystirene pipette (Cat. No. 35-7551) are from Becton-Dickinson (Franklin Lakes, NJ); RPMI 1640 medium with L-glutamine (Cat. No. 11875-093), fetal bovine serum (Cat. No. 10437-028), Dulbecco's PBS (Cat. No. 14190-136), and trypsin–EDTA (Cat. No. 25200-056) are from Gibco BRL (Rockville, MD); multichannel pipette (50–300 μl) (Cat. No. P3970-18), micropipettes (200 μl) (Cat. No. P-3950-200), and pipette tips (Cat. No. P3020-CPS) are from Deville Scientific (Metuchen, NJ); phenazine methosulfate (N-methylphenazonium methyl sulfate salt) (Cat. No. P 9625) is from Sigma-Aldrich (St. Louis, MO); CellTiter 96 AQ$_{ueous}$ MTS reagent powder (Cat. No. G1111) is from Promega Co. (Madison, WI); and microplate reader Spectra®Max Plus[384] is from Molecular Devices (Sunnyvale, CA).

The **preparation of cells and drug preparation** are similar to the XTT/PMS assay.

Assay Procedure

Steps

1. Add 2 mg of MTS powder to each 1 ml of Dulbecco's PBS. Per each 96-well plate add 4 mg of MTS to 2 ml of DPBS.
2. Prepare a stock solution of PMS at a concentration of 0.92 mg/ml.
3. Add 100 μl of PMS to the MTS solution immediately before addition to the cultured cells.
4. Pour the solution in a reservoir and, with a multichannel pipette, transfer 20 μl of it per each well.
5. Incubate the plate for 1–4 h at 37°C in a humidified, 5% CO_2 atmosphere.
6. Measure absorbance with a microplate reader at a wavelength of 490 and 630 nm as a reference wavelength.

Comments

The incubation time after MTS/PMS is added may vary and could be longer than the 1–4 h suggested earlier.

3. Sulforhodamine B Assay (SRB)

The SRB assay is based on binding of the dye to basic amino acids of cellular proteins, and colorimetric evaluation provides an estimate of total protein mass, which is related to cell number. This assay has been widely used for the *in vitro* measurement of cellular protein content of both adherent and suspension cultures. The advantages of this test as compared to other tests include better linearity, higher sensitivity, a stable end point that does not require time-sensitive measurement, and lower cost. The disadvantage lies in the need for the addition of TCA for cell fixation. This step is critical because, if not added gently, TCA could dislodge cells before they become fixed, generating possible artifacts that will affect the results.

Materials and Instrumentation

Falcon microtest tissue culture plates (96 wells) (Cat. No. 35-3072) and Falcon polystirene pipette (Cat. No. 35-7551) are from Becton-Dickinson (Franklin Lakes, NJ); RPMI 1640 medium with L-glutamine (Cat. No. 11875-093), FBS (Cat. No. 10437-028), PBS, 7.4 (Cat. No. 10010-023), and trypsin–EDTA (Cat. No. 25200-056) are from GIBCO BRL (Rockville, MD); multichannel pipette (50–300 µl) (Cat. No. P3970-18), micropipettes (200 µl) (Cat. No. P-3950-200), and pipette tips (Cat. No. P3020-CPS) are from Deville Scientific (Metuchen, NJ); trichloroacetic acid (TCA) (Cat. No. T9159), Trizma base [tris (hydroximethyl)aminomethane] (Cat. No. 25-285-9), acetic acid (Cat. No. A6283), and sulforhodamine B sodium salt (Cat. No. S 9012) are from Sigma-Aldrich (St. Louis, MO). Microplate reader SpectraMax®Plus is from Molecular Devices (Sunnyvale, CA).

The **preparation of cells and drug preparation** are similar to the XTT/PMS assay.

Assay Procedure

Steps

1. Prepare a stock solution of 50% TCA and add 50 µl of this cold solution (4°C) to each well containing 200 µl of medium + cells so that a final concentration of 10% TCA is reached in each well.
2. Place the 96-well plate for 1 h at 4°C to allow cell fixation.
3. Prepare a 0.4% SRB (w/v) solution in 1% acetic acid and add 70 µl of this solution to each well and leave at room temperature for 30 min.
4. Wash the plate with 1% acetic acid five times in order to remove unbound SRB.
5. Prepare a stock solution of 10 mM Trizma base and add 200 µl of this solution to each well in order to solubilize bound SRB. Place the 96-well plate on a plate shaker for at least 10 min.
6. Read abosrbance with a microplate reader at 492 nm, subtracting the background measurement at 620 nm.

Pitfalls

The addition of TCA for fixation is critical and if it is not done with caution can cause dislodgement of the cells before fixation and subsequent alteration of the results.

4. Alamar BlueAssay

AlamarBlue is used to monitor the reducing enviroment of proliferating cells. Because it is not toxic, cells exposed to it can be returned to culture or used for other purposes. AlamarBlue takes advantage of mitochondrial reductase to convert nonfluorescent resazurin to fluorescent resorufin.

Proliferation measurements with alamarBlue may be monitored using a standard spectrophotometer, a standard spectrofluorometer, or a spectrophotometric microtiter well plate reader.

Materials and Instrumentation

AlamarBlue (Cat. No. DAL1100) is from Biosource (Camarillo, CA); Falcon microtest tissue culture plates (96 wells) (Cat. No. 35-3072) and Falcon polystirene pipettes (Cat. No. 35-7551) are from Becton-Dickinson (Franklin Lakes, NJ); RPMI 1640 medium with L-glutamine (Cat. No. 11875-093), FBS (Cat. No. 10437-028), PBS, 7.4 (Cat. No. 10010-023), and trypsin–EDTA (Cat. No. 25200-056) are from GIBCO BRL (Rockville, MD); multichannel pipettes (50–300 µl) (Cat. No. P3970-18), micropipettes (200 µl) (Cat. No. P-3950-200), and pipette tips (Cat. No. P3020-CPS) are from Deville Scientific (Metuchen, NJ); and microplate reader SpectraMax Plus[384] is from Molecular Devices (Sunnyvale, CA).

The **preparation of cells** is similar to XTT/PMS assay, except for the following.

Steps

1. The assay can be performed on adherent cells or suspension culture.
2. Culture cells in RPMI 1640 medium with 10% FBS, 1% glutamine, and 1% pen/strep and amphotericin to avoid microbial contaminants that may reduce AlamarBlue.
3. For harvesting, the cells in suspension must be in log-phase growth (300–500 × 10^3 cells/ml) or, if dealing with adherent cells, trypsinization must be done before the cells reach 80% confluence.
4. Harvest 100,000 cells for each 96-well plate re-suspended in a total volume of 10 ml/medium with 20% FBS, 1% glutamine, 1% pen/strep, and amphotericin.
5. Pipette 100 µl of medium+cells in each well to have 1000 cells/well.

Assay Procedure

Steps

1. Leave the first row for the blank. The second row is used as a control (cells without drug)

2. For adherent cells, allow 1 h for cells to reattach before adding the drug in study.

3. Add 25 µl of alamarBlue *is then added* to a resulting final volume of 250 µl of media+cells.

4. Measure viability after a 1-h incubation at 37°C in humidified 5% CO_2 when the medium in the control row turns from blue to pink. If the reduction observed is insufficient, you may allow the incubation to proceed for a longer period of time.

5. Place the 96-well plate in a automated plate-reading spectrofluorophotometer, with excitation at 530 nm and emission at 590 nm. Fluorescence is expressed as a percentage of control (cells with no drug) after reading the substraction of background fluorescence (blank without cells). AlamarBlue reduction can also be measured spectrophotometrically at two wavelengths, 570 and 600 nm, which are the wavelengths where the reduced and oxidized forms of AlamarBlue absorb maximally.

Pitfalls

The whole procedure has to be performed under aseptic conditions because proliferating bacterial and fungal cells are able to reduce AlamarBlue and may affect the results.

5. ATP Cell Viability Assay

ATP is the most important source of energy for the living cells and can be quantitated in a luminometer by measuring the light generated using the luciferase-luciferin reagent. Typically, apoptotic cells exhibit a significant decrease in ATP levels due to loss of cell integrity.

The ATP cell viability assay is based on two steps. In the first step, ADP is added as a substrate for adenylate kinase and, in the presence of this enzyme, ADP is converted to ATP. In the second step, the enzyme luciferase catalyzes the formation of light from ATP and luciferin. The intensity of the light emitted is measured using a luminometer or a β counter. When the measurement is done on cells in culture using microtiter plates, it is necessary to perform this procedure using white-walled microtiter plates suitable for measuring luminescence.

Materials and Instrumentation

White-walled tissue culture plates (96 wells) (Cat. No. LT07-102) and ToxiLight nondestructive cytotoxicity assay (Cat. No. LT07-117) are from BioWhittaker-Cambrex (Rutherford, NJ); Falcon polystyrene pipette (Cat. No. 35-7551) is from Becton-Dickinson (Franklin Lakes, NJ); RPMI 1640 medium with L-glutamine (Cat. No. 11875-093), FBS (Cat. No. 10437-028); PBS, 7.4 (Cat. No. 10010-023), and trypsin–EDTA (Cat. No. 25200-056) were from GIBCO BRL (Rockville, MD); multichannel pipette (50–300 µl) (Cat. No. P3970-18), micropipettes (200 µl) (Cat. No. P-3950-200), and pipette tips (Cat. No. P3020-CPS) are from Deville Scientific (Metuchen, NJ); and the Reporter microplate luminometer (Cat. No. 9600-001) is from Turner BioSystem (Sunnyvale, CA).

The **preparation of cells** is similar to the XTT/PMS assay except that white-walled microtiter plates are used.

Drug Preparation

Steps

1. Pipette 75 µl of RPMI 1640 medium in each well of a 96-well plate.

2. Add 75 µl of the drug in each well in the first row. Then, after mixing, transfer 75 µl of the mix to the following row and repeat the procedure up to the last row. In this way you will have 1X concentration of the drug in the first row, a X/2 concentration in the second row, and so on.

3. Pipette 50 µl from the 10th row of the plate with the drug dilutions to the last row of the plate with the cells so that the lower concentrations will not affect the subsequent transfers.

4. When all the transfers are completed, add 50 µl of plain RPMI 1640 medium to the control row. Each well will contain 100 µl of medium with 10% FBS.

5. Add 100 µl of medium to the blank row. Start the incubation.

Assay Procedure

Steps

1. Reconstitute the AK detection reagent by adding 10 ml of Tris–AC buffer and, after mixing it, gently allow the reagent to equilibrate at room temperature for 15 min.

2. After the planned period of incubation for cells and drug in the white-walled microplates, remove the plate from the incubator and allow the plate to equilibrate to room temperature prior to measurement.

3. Add 100 µl of the AK detection reagent to all the wells with a multichannel pipette.

4. Wait 5 min before reading to allow for detectable ADP conversion to ATP. Measurement of the light emission should be performed within 30 min from the addition of the AK detection reagent.

5. Place the plate into a luminometer or a β counter. Measure the light emission. Results are expressed as relative light units (luminometer) or counts per second (β counter).

6. [³H]-Thymidine Incorporation Assay

This assay is based on the ability of proliferating cells to incorporate [³H]-thymidine into replicating DNA. Despite its precision to produce accurate data on DNA synthesis, this assay has some disadvantages. It uses radioactivity, requires extensive sample preparation, and the method is sample destructive as compared to a clonogenic assay. The assay described is for human cells grown on agar; the assay may also be used for cells grown in suspension or attached to glass or plastic.

Materials and Instrumentation

Falcon Microtest tissue culture plates (24 wells) (Cat. No. 35-3047), Falcon polystirene pipette (Cat. No. 35-7551), and BlueMax Falcon 15-ml tubes (Cat. No. 35-2097) are from Becton-Dickinson (Franklin Lakes, NJ); RPMI 1640 medium with L-glutamine (Cat. No. 11875-093), FBS (Cat. No. 10437-028), and PBS, 7.4 (Cat. No. 10010-023) are from GIBCO BRL (Rockville, MD); micropipettes (200 μl) (Cat. No. P-3950-200) and pipette tips (Cat. No. P3020-CPS) are from Deville Scientific (Metuchen, NJ); agar (Cat. No. A 7002), sodium azide (Cat. No. S 2002), and TCA (Cat. No. T 9159) are from Sigma-Aldrich (St. Louis, MO); KHO (Cat. No. SP208-500), 20-ml Wheaton glass liquid scintillation vials (Cat. No. 03-341-25G), and ScintiVerse scintillation liquid (Cat. No. SX18-4) are from Fisher Scientific Co. (Suwanee, GA); thymidine [6-³H] specific activity 10 Ci (370 GBq/mmol) at a concentration of 1 m Ci/ml (Cat. No. 355001MC) is from Perkin Elmer Life Science (Boston, MA); and LS 6500 liquid scintillation counter is from Beckman Coulter Inc. (Fullerton, CA).

a. Preparation of Cells

Steps

1. Prepare 0.5% agar by mixing 3.5 ml of 3% Noble agar with 16.5 ml of RPMI 1640 containing 20% fetal bovine serum.

2. Add 500 μl of this agar mixture to each well of a 24-well plate. Then refrigerate the plate at 4°C for 10 min.

3. Resuspend cells in a mixture containing 0.4% agar in RPMI 1640 containing 20% fetal bovine serum at a final concentration of 10^4 cells/ml.

4. Add 1 ml of the cell suspension to each well containing the hardened underlayer.

5. Incubate the plate at 37°C in a 5% CO_2 atmosphere for 24 h.

Assay Procedure

Steps

1. Add sodium azide at a final concentration of 4×10^3 μg/ml to the contol wells (cells without drug).

2. Add the drug to all the remaining wells at the concentrations planned.

3. Incubate cells + drug for 72 h.

4. At the end of this incubation, layer 5 μ Ci of [³H]-thymidine over each well.

5. Incubate the plate for an additional 24 h.

6. Transfer the agar layers from each well to 15-ml centrifuge tubes and bring the volume to 13 ml adding PBS to each tube.

7. Boil the tubes for 30 min and then centrifuge at 1000 rpm for 5 min.

8. Aspirate the supernatants and wash each pellet two times with cold PBS.

9. Centrifuge the tubes and collect the precipitates. Then wash each pellet with 5% TCA.

10. Dissolve each pellet by adding 0.3 ml of 0.075 N KOH and pipetting up and down to completely solubilize cells.

11. Transfer each solubilized cell solution into a scintillation vial containing 5 ml scintillation liquid.

12. Count the radioactivity of each vial in a LS 6500 Beckman liquid scintillation counter.

Comments

When used for cells in suspension the assay may be modified to obtain several time points, e.g., 5, 10, 20, 40, and 60 min, thus generating a rate of thymidine incorporation into DNA and more quantitative data.

III. CLONOGENIC ASSAYS

One of the most important methods for the assessment of survival is the measurement of the ability of a single cell to form colonies. This is usually done by simple dilution after generating a single cell suspension and counting the colonies that arise from single cells. For effective and correct counting, a lower threshold, such as five or six doublings (32 or 64 cells/colony), is quantitated, taking into account the doubling time. Thus the effect of a concentration of a drug on cell survival may be measured with this assay.

In addition to counting colonies, as some drugs may have a delayed effect on cell proliferation, it might be necessary to do colony size analysis. This can be done by counting the cells per colony, by measuring the diameter, or by measuring the absorbance of colonies stained with 1% crystal violet.

The clonogenic assay for tumor colony-forming cells has applicability to a broad spectrum of cell lines and fresh cells obtained from human tumors and has provided information on the biology, clinical course, and chemosensitivity of human cancers

A. Monolayer Cloning

In this method, adherent cells are plated onto a plastic or glass surface and colonies formed are stained and counted. The method is straightforward and useful for cell lines that grow on plastic if a reasonable percentage of cells generate colonies.

Materials and Instrumentation

25-cm² flasks (Corning, Cat. No. 430639)

Phosphate-buffered saline (GIBCO, Cat. No. 10010-023)

Trypsin (GIBCO, Cat. No. 25200-056)

Petri dishes (Falcon, Cat. No. 1007, 60 × 15 mm)

Methanol (J. T. Baker, Cat. No. 9069-03)

1% crystal violet

Hemocytometer (Reichert-Improved Neubauer, Cat. No. 132501)

Steps

1. Prepare replicate 25-cm² flasks, two for each concentration of drug and two for controls.
2. Add the drug to the test flask and solvent to the control flask when the cells reach the required growth phase (usually 24 h after plating) and incubate for 1 h at 37°C.
3. Remove the drug, rinse the monolayer with PBS, and prepare a single cell suspension by trypsinization (desirable).
4. Count the cells and dilute the cell concentration to give 100–200 colonies per 6-cm petri dish. The cell number used per dish depends on the efficiency of plating and the effect of the drug. Plate setup should contain at least two different cell concentrations: one for lower concentrations of the drug and one for higher drug concentrations.
5. Plate out the appropriate number of the cells and incubate at 37°C with 5% CO_2 until colonies grow. This time varies according to the doubling time of the cells, but generally ranges from 10 to 21 days. The colonies should grow to 1000 cells or more on average for the survival assays.
6. Rinse dishes with PBS, fix in 1% methanol or 0.5% glutaraldehyde, and stain with 1% crystal violet. Rinse in running tap water, distilled water, and dry. Count colonies above threshold and calculate as a fraction of control. Plot on a log scale against drug concentration.

Comments

Longer drug exposures can also be assessed by incubating cells with drug for different times, removing media and adding fresh media.

B. Cloning by Limiting Dilution

Puck and Marcus (1955) first established this method. It is more useful for suspension cultures. To improve plating efficacy, modifications for improving the yield of harvested cells, such as using a rich medium that has been optimized for the cell type in use, may be necessary. Cells in log phase should be selected for this method. Also, where serum is required, fetal bovine serum is generally better than calf or horse serum. Sometimes changing the conditions may be useful for obtaining high colony efficiency such as filtering the media or incubating cells for a further 48 h.

Materials and Instrumentation

DMEM, high glucose (Life Technologies, Inc., Cat. No. 10313-021 or equivalent)

Fetal bovine serum (Gibco-Invitrogen, Co. Cat. No. 10437-028 or equivalent)

L-Glutamine (Gibco-Invitrogen, Co. Cat. No. 25030-081 or equivalent)

Hybridoma cloning factor (Fisher, Cat. No. IG50-0615)

50-ml sterile centrifuge tubes (Falcon, Cat. No. 2070)

15-ml sterile centrifuge tubes (Falcon, Cat. No. 2099)

24- and 96-well culture plate (Falcon 353047-0413 and Falcon 353072-0664)

Hemocytometer (Reichert-Improved Neubauer, Cat. No. 132501)

Trypan blue, 0.4% (Sigma Chemicals, Co. Cat. No. 72K2328)

Multichannel pipetter (Thermo Labsystem, Cat. No. 4610050) and sterile tips (Denville Scientific, Cat. No. P-3950-200)

Reagent reservoir (Labcor, Inc., Cat. No. 730-004)

HT (Life Technologies, Inc., Cat. No. 11067-30)

Steps

1. Refeed cells in 24-well plates or flasks with fresh medium 24 h before cloning.
2. Prepare the cloning media by using 10% hybridoma cloning factor, 20% FBS, 4 mM L-glutamine, and DMEM.
3. Resuspend the cells to be cloned in 15-ml sterile tubes; use the trypan blue dye exclusion method to determine viability. Viability should be greater than 80%.

4. For each cell line calculate the dilutions to give 4, 2, and 1 cell/ml in cloning medium. Using 50-ml tubes, serially dilute to contain 4, 2, and 1 cell/ml. The final dilution tube should contain 50 ml of cloning medium at 1 cell/ml.

5. Pour each of the dilutions into a sterile reservoir. Plate 250 µl/well into 96-well plates (one plate with 4 cells/ml, one plate with 2 cells/ml, and two plates with 1 cell/ml). Complete dilutions and plating for each cell line.

6. Incubate all plates at 37°C with 8–10% CO_2 for 5–7 days. At the end of this time, examine all plates microscopically to ensure cloning and plating efficiency before refeeding the plates.

7. Count colonies

C. Soft Agar Clonogenic Assay

Another useful method for cytotoxicity studies is the soft agar technique. It is particularly useful for cells that grow in suspension, but may also be used for cells that attach to glass or plastic. The cells are treated with drug, washed, instead of creating a growth curve as in the outgrowth method, the cells are cloned in soft agar as described next. Agar solution, medium, and cell suspensions are the three basic components in the cloning technique.

Materials and Instrumentation

Noble agar (Agar-Noble Difco Lab)
Fetal bovine serum (Gibco-Invitrogen, Co. Cat. No. 10437-028 or equivalent)
RPMI 1640 medium (GIBCO, Cat. No. 11875-093)
Large culture tubes (Daigger, Cat. No. EF4003)
24-well plate (Falcon, Cat. No. 3487)

Use the following steps for preparing the agar.

Steps

1. Weigh 0.11 g Noble agar and put into a dry flat-bottom bottle that can hold 50 ml.
2. To the 0.11 g agar, add 5.2 ml distilled water. In adding the water, be sure that the water runs in gently so that the agar does not explode.
3. Autoclave 15 min, slow exhaust, and remove immediately upon completion of sterilization.
Use the following steps for preparing the medium.

Steps

1. Measure 50 ml of medium plus serum into a bottle and store at 37°C. (The medium should contain serum in excess of the normal amount used for liquid cultures, such as 15–20%.)

2. For each condition being tested, prepare a culture tube 125 × 20 mm containing 9.0 ml of medium.

3. Treat the cells with drug and resuspend in 15–20% serum-supplemented medium. Cells will need to be diluted so that no more than 1.0 ml (tube cloning) and 0.5 ml (double-layer cloning) of cell suspension will contain the desired number of cells.

Example: For L5178Y cells, a mouse leukemia cell line, the cloning efficiency is 88%. In order to get a cloning tube with 20 clones per tube, 10 ml of cell suspension is made having 120 cells in 10 ml. This is done in the tube containing 9.0 ml of medium, as described previously.

Example: Cell stock after centrifuging is 2×10^4 cells/ml. Dilute 1 : 100; take 0.6 ml of the 1 : 100 dilution, and add to the 9.0 ml of medium. Bring the volume to 10 ml by adding 0.3 ml of medium and 0.1 ml of appropriate drug solution. Each condition tested will require a separate cell suspension, i.e., each cell suspension tube supplies cells for a maximum of five cloning tubes. Only four are generally used.

For Tube Cloning Procedure

1. Add the previously measured 50 ml of medium to the bottle of liquified agar solution. It should be cooled enough so that it can be held by hand comfortably.
2. Distribute 3 ml of agar-medium mixture to each cloning tube.
3. Add 2 ml of cell suspension in the large culture tubes, being sure they are well suspended.
4. Tighten the cap, and mix in the following manner: Hold the tube horizontally and rotate. At the same time, rock the tube to mix. Do this gently to avoid bubbles.
5. Place the tube upright in ice for 2 min.
6. Remove from ice and place in culture tube rack.
7. Keep at room temperature for 15 min.
8. Incubate in an upright position at 37°C.
9. Clones of fast growing lines, such as L5178Y and L1210, are counted on the 10th day. Others take longer, depending on the generation of time of the cancer cell line.

Double-Layer Soft Agar Clonogenic Assay Procedure

This method has some additional benefits compared to the monolayer agar method (Runge et al., 1985). It is very useful for cell cultures whose cloning capacity is low and for fresh cells obtained from tumor biopsy samples.

Materials and Instrumentation

The materials are the same used in other agar-clonogenic assays mentioned earlier.

Steps

Plate 1 ml of underlyer (feeder layer) consists of 15–20% serum-supplemented RPMI 1640 medium and 0.5% Noble agar in 24-well culture plates. The underlayers have to be gelled at least 1 h prior to plating the 1 ml cell and drug(s) (based on design) containing upper layer. For each condition, a suspension is prepared with 4 ml of 20% supplemented RPMI 1640 medium, 0.5 ml of 3% agar solution (final concentration 0.3%), and 0.5 ml of cell suspension containing 5000 cells and appropriate concentration(s) of drug(s). Plate 1 ml of such suspension in each well with a gelled underlayer. Each condition is in quadruplicate. Place all double-layered plates at room temperature for 20 min and then incubate for 10–14 days at 37°C in 100% humidity in 5% CO_2 of atmosphere. For continuous exposure, leave the drug(s) in culture for the entire period of incubation. For time point exposures, such as 4 h, 24 h, 48 h, and even 7 days, incubate cells with drug in suspension culture, wash cells twice with PBS, harvest, resuspend, and finally clone per the procedure described earlier. After 10–14 days of incubation, count clones greater than 50 cells in each well under an inverted microscope (×40). Results are expressed as the mean of 4 well as the percentage of untreated control colony counts.

D. Use of Image Analysis System to Count Colonies

The clonogenic assay for tumor colony-forming cells has applicability on a broad scope of human tumors and has proved valuable in studies of biology, clinical course, and chemosensitivity of human cancers. However, visual counting of colonies has several problems: it is time-consuming and therefore very expensive, the size of colonies changes very rapidly, and there is variability in counting from one researcher to another, partly because of differences in criteria for what constitutes a colony and fatigue.

Bausch and Lomb Omnicon FAS-II image analysis system provides sufficient reliability to be used for counting human tumor colonies grown *in vitro* (Kressner *et al.*, 1980). In addition, the colony counter performed the petri dish counts 10 times faster than experienced technicians did and without associated operator fatigue (Salmon *et al.*, 1984)

References

Cory, A. H., Owen, T. C., Barltrop, J. A., and Cory, J. G. (1991). Use of an aqueous tetrazolium/formazan assay for cell growth assays in culture. *Cancer Commun.* **3**, 207–212.

Denizot, F., and Lang, R. (1986). Rapid colorimetric assay for cell growth and survival: Modification to the tetrazolium dye procedure giving improved sensitivity and reliability. *J. Immunol. Method* **89**, 271–277.

Donacki, N. http://www.protocol-online.org/protocols/cloning_by_limiting_dilution.htm

Freshney, R. I. (1994). Cloning and selection of specific cell types. In "Culture of Animal Cells" (I. R. Freshney, ed.), pp. 161–178. Wiley-Liss, New York.

Garewal, H. S., Ahmann, F. R., Schifman, R. B., and Celniker, A (1986). ATP assay: Ability to distinguish cytostatic from cytocidal anticancer drug effect. *J. Natl. Cancer Inst.* **77**, 1039–1045.

Goegan, P., Johnson, G., and Vincent, R. (1995). Effects of serum protein and colloid on the almarBlue assay in cultures. *Toxicol. in Vitro* **9**, 257–266.

Hamburger, A. W. (1987). The human tumor clonogenic assay as a model system in cell biology. *Int. J. Cell Cloning* **5**(2), 89–107.

H-Zanki, S. U., and Kern, D. H. (1987). *In vitro* assay for new drug screening: Comparison of a thymidine incorporation assay with the human tumor colony-forming assay. *Int. J. Cell Cloning* **5**, 421–431.

Kaltenbach, J. P., Kaltenbach, M. H., and Lyons, W. B. (1958). Nigrosin as a dye for differentiating live and dead ascites cells. *Exp. Cell Res.* **15**, 112–117.

Kangas, L., Gronroos, M., and Nieminen, A. L. (1984). Bioluminescence of cellular ATP: A new method for evaluating cytotoxicity agents *in vitro*. *Med. Biol.* **62**, 338–343.

Kressner, B. E., Morton, R. R. A., Martens, A. E., Salmon, S. E., Von Hoff, D. D., and Soehlen, B. (1980). Use of image analysis system to count colonies in stem cell assays of human tumors. In "Cloning of Human Tumor Cells," pp. 179–193. A. R. Liss, New York.

Mollgard, L., Tidefelt, U., Sundman-Engberg, B., Lofgren, C., and Paul, C. (2000). *In vitro* chemosensitivity testing in acute non lymphocytic leukemia using the bioluminescence ATP assay. *Leuk. Res.* **24**, 445–452.

Mosmann, T. (1983). Rapid colorimetric assay for cellular growth and survival: Application to proliferation and cytotoxicity assays. *J. Immunol. Methods* **65**, 55–63.

Papazisis, K. T., Geromichalos, G. D., Dimitriadis, K. A., and Kortsaris, A. H. (1997). Optimization of the sulforhodamine B colorimetric assay. *J. Immunol. Methods* **208**, 151–158.

Puck, T. T., and Marcus, P. I. (1955). A rapid method for viable cell titration and clone production with HeLa cells in tissue culture: The use of X-radiated cells to supply conditioning factors. *Proc. Natl. Acad. Sci. USA* **41**, 432–437.

Roehm, N. W., Rodgers, G. H., Hatfield, S. M., and Glasebrook, A. L. (1991). An improved colorimetric assay for cell proliferation and viability utilizing the tetrazolium salt XTT. *J. Immunol. Methods* **142**, 257–265.

Rynge, H. M., Neuman, H. A., Bucke, W., and Pfleiderer, A. (1985). Cloning ovarian carcinoma cells in an agar double layer versus a methylcellulose monolayer system: A comparison of two method. *J. Cancer Res. Clin. Oncol.* **110**(1), 51–55.

Salmon, S. E., Young, L., Lebowitz, J., Thompson, S., Einsphar, J., Tong, T., and Moon, T. E. (1984). Evaluation of an automated image analysis system for counting human tumor colonies. *Int. J. Cell Cloning* **2**, 142–160.

Scudiero, D. A., Shoemaker, R. H., Paull, K. D., Monks, A., Tierney, S., Nofziger, T. H., Currens, M. J., Seniff, D., and Boyd, M. R. (1988). Evaluation of a soluble tetarzolium/formazan assay for cell growth and drug sensitivity in culture using human and other tumor cell line. *Cancer Res.* **48**, 4827–4833.

Sevin, B. U., Peng, Z. L., Perras, J. P., Ganjei, P., Penalver, M., and
Averette, H. E. (1988). Application of an ATP-bioluminescence
assay in human tumor chemosensitivity testing. *Gynecol. Oncol.*
31, 191–204.

Skehan, P., Storeng, R., Scudiero, D., Monks, A., McMahon, J.,
Vistica, D., Warren, J. T., Bokesh, H., Kenney, S., and Boyd, M. R.
(1990). New colorimetric cytotoxicity assay for anticancer-drug
screening. *J. Natl. Cancer Inst.* **82**, 1107–1112.

Soehneln, B., Young, L., and Liu, R. (1980). Cloning of human tumor
stem cells. *In "Standard Laboratory Procedures for in Vitro Assay of
Human Tumor Stem Cells,"* pp. 331–338.

Westermark, B. (1974). The deficient density dependent growth
control of human malignant glioma cells and virus-transformed
glia-like cells in culture. *Int. J. Cancer* **12**, 438–451.

White, M. J., Di Caprio, M. J., and Greenberg, D. A. (1996) Assessment of neuronal viability with Alamar blue in cortical and
granule cell cultures. *J. Neurosci. Methods* **70**, 195–200.

39

Micronuclei and Comet Assay

Ilona Wolff and Peggy Müller

I. INTRODUCTION

To evaluate genotoxic effects of chemical substances *in vitro*, various methods are available differing in their sensitivity, their practicability, and, finally, the genetic end points considered. The micronucleus assay is one of the methods used to detect chromosomal aberrations in proliferating cell culture systems (Fenech and Morley, 1985). Its advantages are sensitivity as well as uncomplicated realisation and evaluation of results. Because of its relevance and applicability for human cell systems, the cytokinese-block Micronucleus assay with human lymphocytes is described in detail (Fenech, 2000). Micronuclei represent chromosome fragments or whole chromosomes which are not incorporated into the main nuclei at mitosis and they consequently appear only in dividing eukaryotic cells. In order to score micronuclei exclusively in cells that have completed one nuclear division only, the cytokinese-block method is applied in this test version, which prevents the cytoplasmic division after nuclear division by use of cytochalasin B.

The comet assay, or single cell gel electrophoresis, is another widely used assay for identifying genetic damage such as DNA strand breaks. The alkaline version of the assay introduced by Singh *et al.* (1988) allows the identification of DNA single strand breaks in a very sensitive manner. In addition, there are further advantages, such as the relatively low cell number (<10,000 cells per slide), its ease of application, the applicability of nearly all human and other eukaryotic cells proliferating *in vitro*, and its time- and cost-saving performance, which argue for the use of this assay. To summarize main test principles: agarose-embedded cells are lysed by the help of high salts and detergents; DNA unwinding is promoted by incubat-ing in alkaline electrophoresis buffer (pH > 13); and single-stranded DNA is electrophoresed under alkaline conditions, allowing the DNA fragments to migrate to the anode. After neutralization of slides, the amount of DNA damage may be detected by use of a fluorescent dye such as ethidium bromide visualizing the migrated DNA as the so-called comet tail. The migration length (i.e., the tail length) depending on the DNA fragment size is the most commonly used parameter for quantifying DNA damage. Other metrics considered only by use of image analysis systems are the percentage of migrated DNA and the tail moment (Olive *et al.*, 1990).

Exemplarily, three cell types are presented considering different species (human and rat cells), different culture types (suspension or monolayer culture), and diversity in metabolic activation capacity (normal human bronchial epithelial cells and human lymphocytes versus rat hepatocytes). Selected cell types are capable for genotoxicity testing of single substances and complex mixtures (Müller *et al.*, 2002), and the presented test protocols may be transferred to any other eukaryotic cell type proliferating *in vitro*.

II. MATERIALS AND INSTRUMENTATION

RPMI 1640 instamed (T 121-01) is from Biochrom AG Seromed, as well as the trypan blue solution (L-6323), phosphate-buffered saline (PBS, L-182-01), collagenase CLS II (C 2-22), fibronectin from human plasma (L 7117), and gentamycine (A-271-23).

Epith-o-ser (FM-56-L) and Leibovitz's L-15 (PM-23-S) are from cc-pro, as well as fetal calf serum (FCS, S-10-L), trypsin–EDTA (Z-26-M), and penicilline–streptomycine solution (Z-13-M). Minimum essential medium Eagle (MEM, M-1018) is from Sigma-Aldrich Chemie, as well as phytohaemagglutinine (PHA, L-9132), cytochalasine B (C-6762), $CaCl_2$ anhydride (C-1016), $MgSO_4 \times 7H_2O$ (M-1880), acridine orange (A-6014), phenol red (P 3532), insulin from bovine pancreas (I-6634), N-[tris(hydroxymethyl)methyl]-2-aminoetharesuforic acid (TES, T 5691), HEPES (H 4034), tricine (T 5816), $MgCl_2 \times 6H_2O$ (24,696-4), $Na_2HPO_4 \times 2H_2O$ (21,988-6), and $Na_2HPO_4 \times 7H_2O$ (22,199-6) and Na_2HPO_4 anhydride (S-9763). Normal melting temperative agarose (NMA, 11400) and collagen R (47254.02) are from Serva, as well as ethidium bromide (EtBr, 21251), glucose-6-phospate (22775.01), and Na-pyruvic acid (15220). Low melting temperature agarose (LMA, 35-2010) is from peqLab. Triton X-100 (6683.1), NaCl (3957.2), Na-EDTA (8043.1), Tris (4855.2), dimethyl sulfoxide (DMSO, 7029.1), ethanol (P.076.1), and isopropanole (6752.1) are from Roth. Lymphocyte separation medium (Ficoll-Paque, 17 1440-03) is from Amersham Biosciences. NADP (93208) and formic acid (33015) are from Fluka. Aroclor 1254 (RPC-1254) is provided by Ultra Scientific. All other chemicals are from VWR International: glucose anhydride (1.08337.1000), Na_2HCO_3 (1.063290.500), methanol (1.06009.1000), glacial acetic acid (1.00056.1000), NaOH (1.06462.1000), KH_2PO_4 (1.04873.0250), methoxymagresium–methylcarborate (MMC, 8.18156.0100), KCl (1.04933.0500), Na_2SO_4 (106647), H_2O_2 (1.08597.1000), 2-mercaptoethanol (1.15433.0050), and L-glutamine (1.00289.0100). Heparin sodium salt (101931) is from ICN Biomedicals Inc. Bovine serum albumin fraction V (11018-025) is from Invitrogen. Peanut oil is from the pharmacy of the clinic of Martin Luther University Halle-Wittenberg.

For preparation of lymphocytes, Leucosep vials (227.290) are used, which are from Greiner. For Aroclor treatment of rats, a gauge needle (Sterican, 4667123) from B. Braun Melsungen AG is used. Surgery instruments are from Sigma Chemie: curved microdissecting forceps (F 4142), dressing tissue forceps (F 4267), straight microdissecting forceps (F 4017), curved microdissecting scissors with a sharp point (S 3271), straight microdissecting scissors with a sharp point (S 3146), scalpel blade (S 2771), and handle (S 2896). Cell culture supplies are from TPP: culture dishes ($60 \, cm^2$, 93100), cell culture vials ($20 \, cm^2$, 91106), cryo vials (2.0 ml, 89020), 15-ml centrifuge vials (91015), 50-ml centrifuge vials (91050), serological pipettes (10 ml, 94010), and culture flasks ($25 \, cm^2$, 90026). Automatic pipettes (Eppendorf research variable: 0.5–10, 10–100, 20–200, 100–1000, and 500–5000 μl) are from Eppendorf. The pipette aid (Drummond pipette aid, 28081410) is from Heinemann Labortechnik.

Heating plate (453N1120), water bath (Memmert WB 7, 4623520), swap thermostat (MP 5, 4613411), membrane pump for liquids (ND 100 KT.18, 2245110), electrophoresis chamber (Phero-Sub 2, 344797), power supply (Consors E835, 5822150), and the gel carrier (5817129) are from VWR International. The cell counting chamber (Fuchs-Rosenthal, T731.1), minicentrifuge (X409.1), staining tanks (Hellendahl, H549.1), slides ($76 \times 26 \times 1 \, mm$, 0656.1), and coverslips ($24 \times 40 \, mm$, 1870) are from Roth. For the comet assay, fully frosted slides (61224) from Menzel are used. The CO_2 incubation chamber CB 150 (9040-0001) is from wtb-Binder. Clean bench Uniflow (KR-125-GS) is from UniEquip. The table centrifuge (Megafuge, 75003060) is from Kendro. For microscopic analysis, a Nikon fluorescence microscope Eclipse E 600 (MBA 70400) is used. Data processing is performed with the Komet analysis system, including software Komet 4.0 from Kinetik Imaging Ltd., purchased by BFI Optials.

III. PROCEDURES

A. Preparation and Cultivation of Cells

All cell types are cultivated in a CO_2-incubation chamber under the same conditions: 37°C, 5% CO_2, and 95% relative humidity.

1. Human Lymphocytes

Solutions:

Heat-inactivated FCS: Fill the necessary volume of FCS in a centrifuge vial, deposit the vial in a water bath, and heat the water continuously up to a temperature of 56°C. Hold this temperature for 30 min.

RPMI 1640 cell culture medium: To prepare 1 litre of cell culture medium, add the specified volume of powdered RPMI 1640, 2 g $NaHCO_3$, and 750 ml dH_2O while stirring. Complete to 890 ml with dH_2O, adjust pH to 7.4, and store at 4°C. Before use, add 100 ml of heat-inactivated FCS and 10 ml of penicilline–streptomycine and proof the pH to be 7.4.

Phosphate-buffered saline (PBS): To make 1 litre of PBS, add the specified volume of powdered PBS (Ca^{2+}/Mg^{2+} free) and 750 ml dH_2O while stirring. Complete to 1 litre with dH_2O and store at 4°C.

Steps

Before starting cell preparation, warm up all media, buffer, and supplements in a water bath (37°C). Use certified buffy coat or other human blood preparation.

1. For setup of the vials, pour 15 ml Ficoll-Paque in 50 ml Leucosep tubes and centrifuge at 900 g for 2 min.
2. Mix cold buffy coat with warm PBS in 1:2 proportion.
3. Layer 30 ml of this buffy coat PBS mixture onto a separation ring of the centrifuged Leucosep tubes and centrifuge at 800 g for 25 min, not using the brake.
4. After centrifugation, remove the lymphocytes carefully from the white lymphocyte ring, deliver them in a new tube, wash with a double quantity of PBS, and then centrifuge again at 800 g for 10 min.
5. Perform a second washing with 40 ml PBS and repeat centrifugation (800 g for 10 min).
6. Finally, resuspend lymphocytes in warm RPMI 1640 culture medium and cultivate the cells overnight in cell culture vials to allow sedimentation of the mononuclear cells, B lymphocytes, and regeneration of lymphocytes from the isolation procedure.

2. Hepatocytes of Rat

Solutions

Presuspension buffer: To make 2 litres, add 1500 ml dH$_2$O, 16.6 g NaCl, 1 g KCl, and 4.8 g HEPES while stirring. After dissolving, complete to 2 litres with dH$_2$O and adjust pH to 7.4. Keep at 4°C.

Collagenase buffer: To make 1 litre of stock solution, add 750 ml dH$_2$O, 3.9 g NaCl, 0.5 g KCl, 0.7 g CaCl$_2$ × 2 H$_2$O, and 2.4 g HEPES while stirring. After dissolving, complete to 1 litre with dH$_2$O and adjust pH to 7.6. Keep at 4°C. To make 200 ml of working solution, add 27,300 U of collagenase.

Suspension buffer: To make 1 litre, add 750 ml dH$_2$O, 4 g NaCl, 0.4 g KCl, 0.13 g MgCl$_2$ × 6 H$_2$O, 0.15 g KH$_2$PO$_4$, 0.1 g Na$_2$SO4, 7.2 g HEPES, 6.9 g TES, and 6.5 g Tricine while stirring. After dissolving, adjust pH to 7.6. Furthermore, add 1 g glucose, 30 g BSA, and 5 mg phenol red and sterilize by filtration. Keep at 4°C.

Pentobarbital solution: To get 1 ml, weigh 50 mg of pentobarbital and complete to 1 ml with dH$_2$O.

Heparin solution: To make 2.5 ml, weigh 1 mg heparin sodium salt and complete to 2.5 ml with 0.9% NaCl.

MEM culture medium with Hank's salts, L-gluthamine, and nonessential amino acids: To prepare 1 litre, add the specified volume of powdered MEM, 320 mg NaHCO$_3$, and 750 ml dH$_2$O while stirring, complete to 1 litre, adjust pH to 7.6, and aliquot. Keep at 4°C. Before use, weigh 0.5 µg/ml insulin from bovine pancreas and 50 µg/ml gentamicin and add to medium.

Steps

Hepatocytes may be isolated from untreated young laboratory rats by the *in situ* perfusion method with collagenase buffer following the modified methods by Berry and Friend (1969) and Seglen (1976). *Note*: It needs a special authorisation and large experience to perform manipulations with laboratory animals. Therefore, the method described in the following is only a summary.

1. Sterilise all surgical instruments, several glass beakers, and tubes. Warm up presuspension buffer and collagenase buffer (water bath, 40°C).
2. Anaesthetize rat with ip injection of 5 mg/100 g body mass pentobarbital (0.1 ml/100 g body mass) and inject 1 ml heparin solution per 250 g body mass through the penis vein.
3. Open the rat, dissect vena portae, and induct a gauge needle for injection the buffer solutions. Fix the needle by ligature.
4. Open the thorax, cut the superior caval vein, and start perfusion with warm presuspension buffer for 15–20 min at a constant flow rate of 10 ml/min.
5. Continue perfusion with warm collagenase buffer for about 10–20 min until the liver loses its typical elasticity and colour and becomes soft and pale. (Among other things, perfusion time depends on the activity of collagenase, flow rate, and constitution of liver.)
6. Gently dissect liver lobes and transfer to a culture dish with cold suspension buffer.
7. Carefully remove liver capsule and mechanically disaggregate cells using two forceps. Filtrate the cell suspension through different layers of gauze, transfer suspension to centrifuge vials, and centrifuge at 50 g for 6 min (4°C).
8. Discard supernatant and resuspend pellet in MEM medium. Transfer cell suspension to culture vials. Adjust cell number to $1 × 10^7$ cells/ml after counting and estimation of viability by the trypan blue method (see later).
9. Use freshly prepared hepatocytes in tests without cultivation.

3. Normal Human Bronchial Epithelial Cells (NHBEC)

NHBEC may be isolated from explants from the bronchial airways of patients after lob- or pneumectomia (i.e., the resection of one of the lobes or of the lung at a hemithorax because of bronchial carcinoma) following the method of Lechner and LaVeck (1989) modified by Stock (2002).

Solutions

Epith-o-ser culture medium with supplements (serum free): Thaw the frozen supplements delivered and add to 500 ml of basal medium. Prepare the medium before use.

L-15 medium: To make 1 litre, dissolve the specified volume of powdered L-15 in 750 ml dH$_2$O and complete to 1 litre with dH$_2$O.

Coating solution: To make 100 ml solution, dissolve 1 mg fibronectine, 3 mg collagen R, and 1 mg BSA in 90 ml L-15 medium while stirring. Complete to 100 ml with L-15 and filtrate. Keep at 4°C. *Trypsin–EDTA solution*: Aliquot the delivered solution (0.05% trypsin/0.02% EDTA in Ca^{2+}- and Mg^{2+}-free PBS).

Steps

1. Coat culture dishes: Transfer 0.2 ml/cm^2 coating solution into culture dishes (60 mm diameter) while working under a clean bench. Incubate at 37°C, 5% CO2, and 95% relative humidity for 24 h. Aspirate remaining solution and let the layer dry. Dishes may be stored enveloped in folio at 4°C for about 6 weeks.

2. Fragment the bronchial tissue into pieces of 5 to 10 mm^2 and wash in PBS tree times.

3. Place six to eight tissue pieces into a coated culture dish (see earlier discussion), close, and allow adhesion for 5 min.

4. Add 4 ml of epith-o-ser medium and cultivate cells while changing culture medium every 2–3 days until a subconfluent status (about 2 weeks).

5. Before use of cells in the comet assay, carefully remove culture medium from the well, add warm 1.5 ml trypsin–EDTA solution, and allow digestion for about 5 min (37°C).

6. To stop digestion of cells, add 4.5 ml cold PBS and gently move the well in order to get all cells free from the well surface.

7. Transfer cell suspension into a centrifuge vial, centrifuge at 125 g for 10 min, remove the supernatant, and resuspend the pellet in fresh medium.

8. Calculate cell number and test viability by the trypan blue method (see later).

Viability Test by Trypan Blue Method

Dilute 1 ml of cell suspension with 9 ml culture medium or buffer, mix gently, extract 1 ml of this mixture, and add 1 ml trypan blue solution. Identify living (pale) and nonliving (blue) cells.

B. Micronucleus Assay

Peripheral lymphocytes are one of the most applied cell types for the identification of micronuclei in human tissue. Therefore, the method of the cytokinese-block micronucleus assay with human lymphocytes is described (Fenech, 2000).

Solutions

RPMI 1640 cell culture medium: To prepare 1 litre of cell culture medium, add the specified volume of powdered RPMI 1640 as described in the catalogue to 750 ml dH$_2$O while stirring. Complete to 890 ml with dH$_2$O, adjust pH to 7.4, and store at 4°C. Before use, add 100 ml of FCS and 10 ml of penicilline–streptomycine and proof the pH to be 7.4.

Fixative: To make 100 ml, add 75 ml of methanol and 25 ml of glacial acetic acid and keep at 4°C.

Staining solution: To make 100 ml of stock solution (0.24 mM), take 100 μl acridine orange and complete to 100 ml with dH$_2$O. Store at 4°C. To make 100 ml of working solution, add 1 ml of stock solution and 15 ml of Soerensen buffer.

Soerensen buffer: 60 mM Na$_2$HPO$_4$:60 mM KH$_2$PO$_4$ = 1:1, pH 6.8.

Solution 1: To make 1 litre, weigh 11.876 g of Na$_2$HPO$_4$ and complete to 1 litre with dH$_2$O.

Solution 2: To make 1 litre, weigh 9.078 g of KH$_2$PO$_4$ and complete to 1 litre with dH$_2$O.

To get 100 ml of buffer with pH of 6.8, add 50 ml of Solution 1 and 50 ml of Solution 2.

Steps

Perform the micronucleus test following the modified methods of Fenech and Morley (1985) and Sgura et al. (1997).

1. Resuspend cultivated lymphocytes, centrifuge at 800 g for 5 min, and remove the medium above the pellet.

2. Resuspend the cells in fresh RPMI medium supplemented with 10% inactivated FCS and 1% penicilline–streptomycine-solution at a concentration of 1 × 10^6 cells ml, add PHA in a final concentration of 2 mg/ml, and cultivate for 20 h.

3. After this time, add the test substance in the selected test concentrations. Cultivate cells for 24 h.

4. Add cytochalasin B in the final concentration of 3 μg/ml and continue cultivation for another 24 h. (Cytochalasin B prevents cells from completing cytokinesis, resulting in the formation of multinucleated cells.)

5. Centrifuge this suspension at 425 g for 10 min, remove the supernatant, wash the pellet with 2 ml PBS, and centrifuge again at 425 g for 10 min.

6. After removing the supernatant, resuspend cells in 1 ml of fixative, centrifuge at 200 g for 5 min, aspirate

the supernatant, and dissolve the pellet in few drops of fixative.

7. Transfer the fixed cells to normal, precleaned air-dried slides. Stain the cells with acridine orange solution for 3 min, rinse three times with Sörensen buffer, and finally cover with coverslips.

8. Score the slides (two slides per concentration and control) for micronucleated lymphocytes: 1000 binucleate lymphocytes have to be scored for the number of micronuclei using a fluorescent microscope (400× magnification) fitted with a epifluorescent condensor and filter set (excitation filter 510–590 nm). Perform two replicate experiments for each test.

9. Statistical differences between controls and each treated sample may be identified with the one-tailed χ^2 test following Lovell *et al.* (1989).

C. Alkaline Single Cell Gel Electrophoresis (Comet Assay)

Solutions

Trypsin–EDTA solution

NMA in aqueous solution: To make 100 ml 1% agarose solution, add 1 g NMA to 100 ml distilled water, heat carefully until the agarose is dissolved, and store at 4°C.

NMA in PBS: To make 100 ml of a 0.6% solution, add 0.6 g NMA to 100 ml PBS (without Ca^{2+} and Mg^{2+}), heat carefully until agarose is dissolved completely, and store at 4°C.

LMA in PBS: To make 100 ml of a 0.5% solution, add 0.5 g LMA to 100 ml PBS (without Ca^{2+} and Mg^{2+}), heat carefully until agarose is dissolved completely, and store at 4°C.

Lysing solution: To make 1 litre, add 1.21 g Tris (10 mM), 146. 1 g NaCl (2.5 M), and 37 g Na–EDTA (100 mM) to about 700 ml dH_2O while stirring. After dissolving, complete to 890 ml with dH_2O, adjust pH to 10, and keep at room temperature. Immediately before use, add 1% Triton-X and 10% DMSO.

Electrophoresis buffer: To make 1 litre, add 0.37 g Na–EDTA (1 mM) and 12 g NaOH (300 mM) to about 700 ml dH_2O under stirring. After dissolving, complete to 1 litre with dH_2O and adjust pH to 13. Prepare fresh.

Neutralisation buffer: Per litre, add 48 g Tris to 750 ml dH_2O and complete to 1 litre with dH_2O (400 mM, pH 7.4).

Ethidium bromide solution: To get 10 ml, dissolve 200 μg EtBr in distilled water and complete to 10 ml. Aliquot in 1-ml portions and keep at 4°C.

Steps

1. Preparation and Treatment Schedule of Cells

1. *Human lymphocytes and rat hepatocytes (or other cells in suspension culture)*: Decant medium from lymphocyte overnight culture or from hepatocyte suspension, respectively. Centrifuge cells for 5 min at 200 g. Resuspend cells in fresh culture medium. Calculate cell number (Fuchs–Rosenthal counting chamber) and test viability by the trypan blue method (see earlier discussion).

2. *In case of culture treatment with test substance* (in vitro *experiments*): Decant medium from deposited cells, add mixture of culture medium, test substance in the required test concentration, resuspend cells in this mixture, and cultivate while stirring vials gently.

3. After treatment, centrifuge cells for 5 min at 200 g, discard supernatant, resuspend pellet in fresh culture medium, decant medium from deposited cells, and resuspend cells gently in 85 μl melted LMA per slide.

NHBEC (or Other Monolayer Cultures)

Short-term Treatment (1–2 hs)

1. Carefully remove media from subconfluent cell culture, add 1.5 ml of warm trypsin–EDTA solution, and incubate for 5 min.
2. Follow steps 6–8 as described in Section III,A,3.
3. Centrifuge cells for 5 min at 200 g, discard supernatant, add mixture of culture medium and test substance in the required test concentration, resuspend cells in this mixture, and cultivate while stirring vials gently.
4. After treatment, centrifuge cells for 5 min at 200 g, discard supernatant, resuspend pellet in fresh culture medium, decant medium from deposited cells, and resuspend cells gently in 85 μl melted LMA per slide.

Long-term Treatment (>2 hs)

1. Carefully remove media from subconfluent cell culture, add mixture of culture medium, and test substance in the required test concentration. Cultivate cells over the required treatment period.
2. After treatment, remove the treatment solution carefully, rinse with fresh medium, remove it, and add 1.5 ml of warm trypsin–EDTA solution. Incubate for 5 min.
3. Follow steps 6–8 as described in Section III,A,3.
4. Centrifuge cells for 5 min at 200 g, discard supernatant, resuspend pellet in fresh culture medium, decant medium from deposited cells, and resuspend cells gently in 85 μl melted LMA per slide.

2. Preparation of Cells from in Vivo Experiments

Follow the procedures described earlier except for the treatment of cells.

Cells from Suspension Culture. Decant medium from cultured cells and resuspend cells gently in 85 µl melted LMA per slide.

Cells from Monolayer Culture

1. Carefully remove media from subconfluent cell culture, add 1.5 ml of warm trypsin–EDTA solution, and incubate for 5 min.
2. Follow steps 6–8 as described in Section III,A,3.
3. Centrifuge cells for 5 min at 200 g, discard supernatant, resuspend pellet in fresh culture medium, decant medium from deposited cells, and resuspend cells gently in 85 µl melted LMA per slide.

Note: For administration of LMA, do not exceed temperature of 37°C. Adjust cell number to 10,000 per slide. In human cell culture systems (lymphocytes and NHBEC), S9 mix as an external metabolising system should be involved (described in Section III,D) in parallel experiments. Do not treat cells longer than 3 h with S9 mix.

3. Slide Preparation

The slide preparation follows the basic method of Singh *et al.* (1988). All steps following step 3 are performed in a dark room under yellow light.

1. Melt NMA (1% in dH$_2$O) in a microwave (1 ml per slide). Cover a labelled fully frosted microscopic slide with 1 ml of NMA, spread it evenly, and remove the layer after drying by scratching.
2. Overload the slide with 300 µl of melted NMA (0.6% in PBS), cover with a large coverslip, and store at room temperature until the agarose is solidified.
3. Meanwhile, prepare cell suspension (see earlier discussion).
4. Remove coverslip gently and transfer the cell agarose suspension onto the prepared slide.
5. Cover with a fresh coverslip and incubate for 10 min on ice. After the agarose is solidified, slide off coverslip gently and apply a layer of 85 µl melted LMA.
6. Cover with a fresh coverslip and incubate for 10 min on ice.
7. Slide off coverslip gently and incubate slide in freshly prepared chilled lysing solution for 1 h at 4°C in the dark using a staining tank.

Note: Steps 5 and 6: For incubation of slides on ice, prepare an ice bath, cover it with a piece of blotting paper, and apply an appropriate pane of glass.

4. Electrophoresis

1. Remove slide from the lysing solution and wash gently with fresh electrophoresis buffer (pH > 13).
2. Apply slides side by side (without distances) on the horizontal gel box of the electrophoresis chamber, fill the chamber with freshly made electrophoresis buffer, avoiding bubbles, and incubate slides for 1 h in order to allow DNA unwinding. The buffer just has to cover the slides!
3. Switch on power supply: Conduct electrophoresis at 25 V and 300 mA (0.66 V/cm) for 30 min and cover the chamber with aluminium foil.
4. Switch off power supply, remove slides from the chamber gently, and rinse carefully with neutralisation buffer three times to remove alkali and detergents. It is possible to store slides in neutralisation buffer until examination in the dark at 4°C, but do not exceed 24 h.

5. Examination

1. Gently remove slides from neutralisation buffer, place on a slide drainer, and stain every slide after draining with 60 µl EtBr. Cover with a coverslip and allow staining for 10 min.
2. For DNA analysis, use an epifluorescence microscope with a 510- to 590-nm excitation filter and a short arc mercury lamp. Examine at 400-fold magnification. Per test concentration not less than 50 cells should be scored; two replicate experiments should be performed per test. It is possible to determine comet tail length as a result of DNA damage by use of a micrometer in the microscope eyepiece. However, this method is time-consuming and is subject to various mistakes. A better method of processing data is to use an imaging analysis system in connection with special software for the comet assay, such as described earlier. In this case, follow the manual.
3. To determine the lowest test concentration at which a significant increase in DNA damage (given as comet tail length or tail moment) has occurred, multiple pairwise comparisons have to be conducted between control data and each dose using the Student's *t* test.

Note: To avoid confusion of results by concomitant processes leading to apoptosis, do not observe highly damaged cells. To ensure reproducibility of data between various experiments, do not use short arc mercury lamps with different wattage.

D. Further Procedures

1. Preparation of S9 Mix from S9 Fraction

Most of the cell culture systems, apart from hepatocytes, should be exposed to test substance both in the presence and in the absence of an external metabolising system, i.e., the postmitochondrial supernatant of the rat hepatocyte preparation of Aroclor 1254 pretreated male rats (S9 mix) as described by Czygan *et al.* (1973), Ames *et al.* (1975), and Natarajan *et al.* (1976), to consider the effectivity of test substance in dependency on its metabolisation. Usually, the final concentration of S9 mix ranges from 1 to 10%, depending on the classification of test substance.

2. Preparation of S9 Fraction

Solutions

1. *Aroclor 1254*: Suspend 200 mg Aroclor in 1 ml peanut oil.
2. *0.15 M KCl*: Dissolve 5.59 g KCl in destilled water and fill up to 500 ml; sterilize this solution by filtration. Store at 4°C.

Steps

1. Aroclor is administered to rat by ip injection using a final dose of 500 mg/kg. Because of the high viscosity of the Aroclor suspension, use a gauge needle not smaller than 18 gauge.
2. Five days after injection, anaesthetize the rat, open the abdomen, and bleed and remove the liver immediately while working under sterile conditions at 4°C.
3. Weigh the liver, cut it into small pieces, and mix with cold sterilized KCl (1 g per 3 ml KCl).
4. Homogenize this mixture and centrifuge at 6000 g at 4°C for 10 min using sterile centrifuge vials.
5. Aliquot the supernatant to sterile cryo vials and store frozen at −80°C.

3. Preparation of S9 Mix

Solutions

Stock phosphate solution: (a) Dissolve 4.804 g KH_2PO_4 in dH_2O and complete to 100 ml. (b) Dissolve 6.286 g Na_2HPO4 in dH_2O and complete to 100 ml. Mix 19.6 ml of solution a with 80.4 ml of solution b, adjust pH to 7.4, and sterilize by heat (20 min at 121°C in autoclave). Store at 4°C.

$MgCl_2$: Dissolve 406.51 mg $MgCl_2 \times 6\ H_2O$ in dH_2O and complete to 50 ml, aliquot, and sterilize by microfiltration. Store at 4°C.

KCl: Dissolve 615 mg KCl in dH_2O and complete to 50 ml and sterilize by microfiltration. Store at 4°C.

NADP: Dissolve 726 mg NADP in distilled water and complete to 20 ml and sterilize by microfiltration. Aliquot and store frozen at −30°C.

Glucose-6-phosphate: Dissolve 304 mg glucose-6-phosphate in dH_2O and complete to 20 ml, aliquot, and sterilize by microfiltration. Store at −30°C.

Steps

1. To thaw components slowly, put all vials and bottles in an ice bath.
2. To make 10 ml of S9 mix, take 3 ml of stock phosphate solution, add 2 ml of KCl solution, 2 ml of MgCl, and 1 ml (10%) of S9 fraction. Complete this mixture with 1 ml NADP solution and finally add 1 ml of glucose-6-phosphate solution. It is very important to follow this sequence exactly and to take care that all components are mixed completely before adding the next one.

IV. COMMENTS

In each experiment, in addition to the test concentrations, a negative control (culture medium) and a positive control should be included for validation of results [micronucleus test: 0.5 M MMC, following Surralés and Natarajan (1997), comet assay: 300 μM H_2O_2]. If a test substance has to be dissolved or suspended in any vehicle, a vehicle control has to be included too.

It is important to determine the concentration range as well as the treatment period in prestudies for every test substance and for various cell systems. The number of different test concentrations should be three or more. For more detailed description of test parameters, see Tice *et al.* (2000).

V. PITFALLS

1. In the case of elevated room temperature (about 30°C or more) it is possible that LMP agarose will not solidify; therefore, keep the slides together with the ice bath into the refrigerator.
2. To assign the slides, use a pen capable for cryopreservation or a pen for scratching on glass.

References

Ames, B. N., McCann, J., and Yamasaki, E. (1975). Methods for detecting carcinogens and mutagens with Salmonella/mammalian microsome mutagenicity test. *Mutat. Res.* **31**, 347–364.

Berry, M. N., and Friend, D. S. (1969). High-yield preparation of isolated rat liver parenchymal cells. *J. Cell Biol.* **43**, 506–520.

Czygan, P., Greim, H., Garro, J., Hutterer, F., Schaffner, F., Popper, H., Rosenthal, P., and Cooper, D. Y. (1973). Microsomal metabolism of dimethylnitrosamine and the cytochrom P-450 dependency of its activition to a mutagen. *Cancer Res.* **33**, 2983–2986.

Fenech, M. (2000). The in vitro micronucleus technique. *Mutat. Res.* **455**, 81–95.

Fenech, M., and Morley, A. A. (1985). Measurement of micronuclei in lymphocytes. *Mutat. Res.* **147**, 29–36.

Lechner, J. F., LaVeck, M. A., Gerwin, B. I., and Matis, E. A. (1989). Differential responses to growth factors by normal human mesothelial cultures from individual donors. *J. Cell Physiol.* **139**(2), 295–300.

Lovell, D. P., Albanese, R., Clare, G., Richold, M., Savage, J. R. K., Anderson, D., Amphlett, G. E., Ferguson, R., and Papworth, D. G. (1989). Statistical analysis of *in vivo* cytogenetic assays. *In* "*Statistical Evaluation of Mutagenicity Test Data*" (D. J. Kirkland, ed.). Cambridge Univ. Press, Cambridge.

Müller, P., Stock, T., Bauer, S., and Wolff, I. (2002). Genotoxicological characterisation of complex mixtures: Genotoxic effects of a complex mixture of perhalogenated hydrocarbons. *Mutat. Res.* **515**, 99–109.

Natarajan, A. T., Tates, A. D., van Buul, P. P. W., Meijers, M., and de Vogel, N. (1976). Cytogenetic effects of mutagens / carcinogens after activation in a microsomal system *in vitro*. I. Induction of chromosome aberrations and sister chromatid exchanges by diethylnitrosamine (DEN) and dimethylnitrosamin (DMN) in CHO cells in the presence of rat-liver microsomes. *Mutat. Res.* **37**, 83–90.

Olive, P. L., Banath, J. P., and Durand, R. E. (1990). Heterogeneity in radiation-induced DNA damage and repair in tumor and normal cells using the "comet" assay. *Radiat. Res.* **122**, 86–94.

Seglen, P. O. (1976). Preparation of isolated rat liver cells. *Methods Cell Biol.* **XIII**, 29–83.

Sgura, A., Antoccia, A., Ramirez, M. J., Macros, R., Tanzerella, C., and Degrassi, F. (1997). Micronuclei, centromere-positive micronuclei and chromosome nondisjunction in cytogenesis blocked human lymphocytes following mitomycin C or vincristine treatment. *Mutat. Res.* **392**, 97–107.

Singh, N. P., McCoy, M. T., Tice, R. R., and Schneider, A. L. (1988). A simple technique for quantitation of low levels of DNA damage in individual cells. *Exp. Cell Res.* **175**, 184–191.

Stock, T. (2002). Culture of normal human bronchial epithelial cells and its test for toxicological application. Martin Luther University Halle-Wittenberg, medical thesis. (http://sundoc.bibliothek.uni-halle.de/diss-online/02/02H153/index.htm)

Surrallés, J., and Natarajan, A. T. (1997). Human lymphocytes micronucleus assay in Europe. An international survey. *Mutat. Res.* **392**, 165–174.

Tice, R. R., Agurell, E., Anderson, D., Burlinson, B., Hartmann, A., Kobayashi, H., Miyamae, Y., Rojas, E., Ryu, J.-C., and Sasaki, Y. F. (2000). Single cell gel/comet assay: Guidelines for *in vitro* and *in vivo* genetic toxicology testing. *Environ. Mol. Mutagen.* **35**, 206–221.

Apoptosis

40

Methods in Apoptosis

Lorraine O'Driscoll, Robert O'Connor, and Martin Clynes

I. INTRODUCTION

Apoptosis and necrosis are two mechanisms of cell death, each with its own distinguishing morphological and biochemical features. Necrosis, which occurs within seconds of cell insult (Majno and Joris, 1995), may be described as "cell murder" resulting from external damage to the cell membrane, loss of homeostasis, water and extracellular ion influx, intracellular organelle swelling, cell rupture (lysis), and so inflammatory cell attraction. Initially described by Kerr *et al.* (1972), apoptosis is a much slower process of events than necrosis, requiring from a few hours to several days (depending on the initiator) and resulting from molecular signals initiated within individual cells (see Nagata, 1997; Barinaga, 1998; Van Cruchten and Van Den Broeck, 2002). The initiators of apoptosis that instigate the cascade of events leading to activation of a series of cytoplasmic proteases, termed caspases (cysteinyl-asparatate-specific proteinases), are multiple. Two such pathways involve (i) activation of cell surface death receptors, resulting in direct activation of caspases, and (ii) cytochrome *c* release from the mitochondria into the cytoplasm following induction of leakiness in its membrane. The terminal caspases downstream from these initiator mechanisms lead to the morphological and biochemical events of apoptosis.

The mechanisms of apoptosis, which is analogous to "cell suicide," are essentially the same, whether induced by genetic signals or through external initiators. The events involved include cell membrane blebbing; chromatin aggregation; nuclear and cytoplasmic condensation leading to cell shrinkage; and partitioning of cytoplasm and nucleus into membrane-bound apoptotic bodies, which contain ribosomes, morpho-

logically intact mitochondria, and nuclear material. As a result of the efficient mechanism for the removal of apoptotic cells by phagocytic cells *in vivo*, an inflammatory response is not stimulated. *In vitro* apoptotic bodies swell and finally lyse (Darzynkiewicz and Traganos, 1998; Kiechle and Zhang, 2002).

Apoptosis is key to many fundamental aspects of biology, including embryonic development and normal tissue homeostasis, as well as in many pathological events, such as loss of regulated cell death in cancer, response of cancer cells to chemo- and radiotherapy (Clynes *et al.*, 1998), and death of cells in diabetes (Sesti, 2002) and neurodegenerative diseases (Vila and Przedborski, 2003). Accurate detection of apoptosis is of great importance to increase our understanding of biological events that may allow us to understand and to manipulate these events as a form of therapy.

Major elements involved in the apoptotic pathway, which should be considered when selecting suitable methods for apoptosis detection, include the following.

a. Death receptor activation: Following receptor cross-linking by ligand (e.g., the Fas receptor by the CD95 (APO-1 Fas) or Tumor Necrosis Factor (TNF) receptor type 1 by TNF), signal transduction leads to caspase activation (see (f) below).

b. Changes in cellular morphology: As described earlier.

c. Membrane alterations: Translocation of phosphatidylserine (PS) from the cytoplasmic to the extracellular side of the cell membrane is an early event in apoptosis.

d. DNA fragmentation: Prior to the induction of cell membrane permeability, fragmentation of genomic DNA at sites located between nucleosomal units, generating mono- and oligonucleosomal DNA fragments, irreversibly commits the cell to die.

e. Disruption of mitochondria: As described, disruption of the mitochondrial membrane results in cytochrome *c* (Apaf-2) release. This subsequently promotes caspase activation by binding to Apaf-1 and inducing activation of Apaf-3 (caspase-9). Similarly, release of apoptosis-inducing factor (AIF) induces apoptosis.

f. Activation of caspases: At least 11 different caspases have been identified in mammalian cells. Activation of this protease cascade *via* a range of stimuli is central to the execution of apoptosis.

The range of techniques and methods for analysis of apoptosis is extensive. Due to space limitations in this article, we do not propose to describe all methods comprehensively. Table I lists a number of techniques for analysis of the cellular events described earlier. A selection of techniques used for studying these events is described in further detail.

II. MATERIALS AND INSTRUMENTATION

A. Light Microscopy (LM) and Fluorescence Microscopy (FM)

Frost-ended slides and coverslips (Chance Propper); ice-cold methanol; Coplin jars; forceps; micropipettes; grease pen (DAKO S2002); and mounting medium (Vectashield mounting solution with antifade additive [Vector Labs.; H-1000]) suitable for fluorescence slides and may also be used for LM slides. Alternatively, 20% glycerol prepared in H_2O is also suitable for mounting slides for LM.

If analysing suspension cells, a cytospin (e.g., Heraeus Labofuge 400) and cytospin cups are required.

For LM only: haematoxylin, aluminium potassium sulphate, citric acid, and chloral hydrate.

For FM only: Stains include 4′,6-diamidino-2-phenylindole (DAPI, Sigma D-9542), propidium iodide (PI, Sigma P-4170), Hoechst 33258 (Sigma B-2883), Hoechst 33342 (Sigma B-2261), and acridine orange (AO, Sigma A-6014) in phosphate-buffered saline (PBS), pH 7.4.

B. Gel Electrophoresis for DNA Ladder

Horizontal agarose gel electrophoresis chamber and combs (Bio-Rad); electric power supply; UV transilluminator or gel analyser (e.g., EpiChemi II Darkroom, UVP Laboratory Products); PBS (Oxoid BR14a); ethid-

TABLE I   **Methods Used for Detecting Cellular Changes That Occur during Apoptosis**

Cellular morphology
  Cellular features by light microscopy and time-lapse video microscopy
  Fluorescence microscopy and laser-scanning confocal microscopy, e.g., using DNA stains for nuclear morphology

Membrane alterations
  Annexin V binding for phospholipid externalisation
  Impermeable dyes (PI) and permeable DNA stains (DAPI, Hoechst) for membrane permeability

Biochemical activation events
  Caspase cleavage products
  Caspase activity
  PARP activity
  Transglutaminase activity

Nuclear events and DNA fragmentation
  Gel electrophoresis for DNA ladder (internucleosomal cleavage) detection
  in situ nick translation
  Comet assay
  Immunohistochemistry for single-stranded DNA
  FACS analysis for cell cycle (pre-G1 peak) dissolution
  TUNEL

Mitochondrial permeability
  Metabolic activity
  Accessibility of mitochondrial antigens
  Permeability of vital dyes
  Detection of cytochrome *c* release

Detection of apoptosis-related genes (e.g., bcl-2 family members, survivin, caspases) and death antigens
  For gene transcript analysis
    RT-PCR
    Northern blotting
    RNase protection assay
    Microarrays
  For protein analysis
    Immunocytochemistry/immunofluorescence
    Western blotting
    ELISA
    Two-dimensional gel electrophoresis surface enhanced laser desorption Ionisation (SELDI)

ium bromide (Sigma E-8751); agarose (Sigma A-9539); Tris (Sigma T-8524); EDTA (Tris E-5134); NaCl (Sigma S-9899); sodium dodecylsulfate (SDS) (BDH 442152V); RNase A (Sigma R-5250); proteinase K (Sigma P-2308); boric acid (Sigma B-7901); bromphenol blue (Sigma B-5525); glycerol (Sigma G-2025); molecular markers, e.g., Phi X174 DNA *Hae*III digest (Sigma D-0672); micropipettes.

C. Terminal Deoxynucleotidyl Transferase-Mediated Deoxyuridine Triphosphate Nick End-Labelling Assay (TUNEL)

Apoptosis detection system: (1) fluorescein-containing equilibrating buffer, (2) nucleotide mix, (3)

TdT enzyme, (4) 20X SSC solution, (5) proteinase K, (6) protocol (Promega; G3250); plastic coverslips, glass slides and coverslips (Chance Propper); propidium iodide (Sigma P-4170); Coplin jars; forceps; humidifying chamber; 37°C incubator; Triton X-100 (Sigma T-8787), PBS (Oxoid BR14a); 4% paraformaldehyde (Sigma P-6148) in PBS (pH 7.4) (*freshly prepared*); Vectashield mounting solution with antifade additive (Vector Labs.; H-1000); 70% ethanol [prepare from absolute ethanol (Sigma E-7037)]; and micropipettes.

Note: Items 1–6 are included in the apoptosis detection system available commercially from Promega (G3250). There are, however, other detection kits available commercially that may be equally suitable.

D. Reverse Transcriptase–Polymerase Chain Reaction

As all general laboratory glassware, spatulas, etc., are often contaminated by RNases, these items should be treated by baking at 180°C for a minimum of 8 h. Sterile, disposable plasticware is essentially free from RNases and so generally does not require pretreatment. All solutions/buffers used should be prepared in baked glassware using sterile ultrapure water treated by the addition of diethylpyrocarbonate (DEPC) [Sigma D-5758, (0.1%, v/v)] and autoclaved. As for all laboratory procedures described in this article, gloves should be worn at all times to protect both the operator and the experiment. This, too, prevents the introduction of RNases and foreign RNA or DNA in the reverse transcriptase (RT) and polymerase chain reaction (PCR).

1. For RNA Isolation and Quantification

TRI Reagent (Sigma T-9424), chloroform (Sigma C-2432), isopropanol (Sigma I-9516), ethanol [Sigma E-7037; prepare as 75% (v/v) in H_2O], DEPC, micropipettors, tips, Eppendorf tubes, etc., spectrophotometer (e.g., SpectraMax Plus plate reader, Molecular Devices), and quartz cuvettes or Nanodrop (ND-1000; Labtech Int. Ltd.)

2. For RT and PCR Reactions

DEPC-treated H_2O; oligo(dT)$_{12-18mer}$ (Oswel, Southampton, UK); MMLV-RT enzyme (200 U/μl) (Sigma M-1302); 5X buffer (Sigma B-0175); dithiothreitol (DTT, 100 mM) (Sigma D-6059); RNasin (40 U/ml) (Sigma R-2520); dNTPs (10 mM each of dATP, dCTP, dGTP, and dTTP for RT; 1.25 mM each for PCR) (Sigma DNTP-100); MgCl$_2$ (25 mM) (Sigma M-8787), *Taq* DNA polymerase enzyme (5 U/μl) (Sigma D-6677); 10X buffer (Sigma P-2317); target primers and internal control primer (see Table II) (for further details on primer selection criteria, see O'Driscoll *et al.*, 1993); thermocycler.

3. Gel Electrophoresis

Amplified products are analysed by gel electrophoresis (see Section IIIB).

Note: Several of the stains and other reagents used (e.g., DEPC) should be handled with caution as they are known or thought to be toxic, carcinogenic, and/or mutagenic.

III. PROCEDURES

A. Light and Fluorescence Microscopy Detection of Apoptotic Cell Morphology

As described previously, a cell undergoing apoptosis proceeds through various stages of morphological change (Wilson and Potten, 1999). Light microscopy and fluorescence microscopy are probably the simplest and most basic techniques by which such apoptotic cell death can be investigated. A broad range of stains and dyes are available to assist in the assessment of nuclear morphology. For light microscopy, the nuclear stain haematoxylin is used frequently (often with eosin as a counterstain). The most commonly used DNA nucleic acid-reactive fluorochromes for UV light fluorescence microscopic analysis of fixed (porated using, e.g., methanol or ethanol) cells include DAPI, PI, Hoechst 33258 and 33342, and AO. (As mentioned previously, PI, DAPI, and Hoechst are also very useful for assessing the membrane permeability of cells).

Using LM, cells dying by apoptosis are identified by their reduced size, cell membrane "blebbing/budding," and loss of normal nuclear structural features—nuclear fragmentation and chromatin condensation. In contrast, characteristic features of necrotic cell death include cell and nuclear swelling, cytoplasmic vacuolisation, patchy chromatin condensation, and plasma membrane rupture.

If assessing adherent cell types, the cells may be grown directly on coverslips (plated in Petri dishes). For suspension cells, cytospins are generally prepared. For both monolayer and suspension cultures, it is important to consider cell concentration. Fifty to 70% confluency of the area being analysed is generally considered optimal—if the cells are more confluent, overlapping may occur, which may hinder analysis; if cells are too sparse, an accurate examination may not be achievable. Typically, $1–2 \times 10^5$ cells suspended in 100 μl PBS containing 1% (w/v) fetal calf serum (FCS) should be cytocentrifuged onto microscope slides

41

Cellular Assays of Oncogene Transformation

Michelle A. Booden, Aylin S. Ulku, and Channing J. Der

I. INTRODUCTION

A critical step in the validation of the role of candidate oncogenes in oncogenesis involves a demonstration of their ability to endow nonneoplastic cells with growth properties characteristics of tumor cells. The most widely utilized model system for the analyses of oncogene function involves the use of NIH 3T3 mouse fibroblasts. NIH 3T3 mouse fibroblasts have been a useful model cell system to evaluate transformation caused by aberrant activation of many different functional classes of oncogenes, including serine/threonine (e.g., Raf) and tyrosine (e.g., Src) protein kinases, nuclear transcription factors (e.g., Fos), G-protein-coupled receptors (GPCRs; e.g., Mas, PAR1), and small (e.g., Ras) and large (e.g, $G\alpha_{12}$) GTPases. This article describes the application of various cellular assays for the study of growth transformation caused by GPCR oncogenes that promote the aberrant activation of Rho family small GTPases.

Mutationally activated and transforming forms of the three human Ras proteins (H-, K-, and N-Ras) are found in 30% of human cancers. In contrast, mutations in the Ras-related Rho GTPases are not found in human cancers (Sahai and Marshall, 2002). Instead, Rho GTPases are activated indirectly by alterations in expression or activity. In particular, Rho GTPases are activated by persistent upstream signaling, e.g., by Dbl family guanine nucleotide exchange factors (GEFs) or by activated GPCRs (Schmidt and Hall, 2002; Whitehead *et al.*, 2001). The first section of this article describes methods that assess the transforming capabilities of proteins that cause transformation, in part, by activation of Rho family proteins. The assays described provide both qualitative and quantitative information of transforming potential.

In addition to their abilities to induce tumorigenesis, Ras and Rho family proteins have attracted considerable research attention as important mediators of tumor cell invasion and metastasis. For example, previous studies from our laboratory determined that Rac1, Cdc42, and RhoA can promote breast cancer cell invasion (Keely *et al.*, 1997), while Hynes and colleagues showed that upregulation of RhoC promoted metastatic growth of melanoma cells (Clark *et al.*, 2000). Therefore, the second section of this article describes assays to examine human tumor cell invasion *in vitro* induced by the activation of Rho family proteins. The applications described can be used to implicate GPCRs, direct activators of Rho family members, and Rho proteins themselves as important for promoting cellular invasion.

II. NIH 3T3 MOUSE FIBROBLAST TRANSFORMATION ASSAYS

Three types of focus formation assays can be used to evaluate the transforming potential of oncoproteins that cause persistent activation of Rho family proteins (Clark *et al.*, 1995; Solski and Der, 2000). The most commonly used assay is the primary focus formation assay, which provides a straightforward method for quantitation of transforming potential. A second assay is a primary focus formation cooperation assay, where activated Raf is coexpressed along with the oncogene under evaluation. Our laboratory and others showed that Rho GTPase activators cooperate with activated

(Hanahan and Weinberg, 2000). Colony formation in soft agar is the most widely used assay to evaluate anchorage-independent growth *in vitro* and correlates strongly with *in vivo* tumorigenic growth potential. Untransformed NIH 3T3 cells need to adhere to a solid matrix in order to remain viable and proliferate. NIH 3T3 cells transformed by activated Ras and Rho GTPases lose this requirement, which results in their ability to form proliferating colonies of cells when suspended in a semisolid agar.

Briefly, transfect NIH 3T3 cells as described earlier for focus-formation assays. Three days after transfection, replate the transfected cells onto three individual 100-mm dishes containing growth medium supplemented with the appropriate drug for selection (400 µg/ml G418, 400 µg/ml hygromycin, or 1 mg/ml puromycin). After approximately 2 weeks, trypsinize and pool together the multiple drug-resistant colonies (>100) that have arisen to establish mass populations of stably transfected cells. Perform Western blot analysis to confirm ectopic overexpression of the introduced gene; these stably transfected cells are then used to assay for the oncoprotein-mediated acquisition of various transformation phenotypes described later.

Materials and Reagents

1.8% Bacto-Agar (Difco 214050): Prepare a 1.8% stock (w/v, in distilled water) by boiling the solution in a microwave to dissolve the agar. Place 50-ml aliquots in 100-ml bottles and autoclave to sterilize. Store at room temperature. Microwave to melt.

2× DMEM: 2× DMEM and other culture medium components can be purchased from GIBCO (Invitrogen Life Technologies).

C. Soft Agar Assay

This modified protocol is based on a similar protocol described previously (Clark *et al.*, 1995). Aliquot a layer of bottom agar (0.6% in growth medium) and allow to solidify. Trypsinize the cell lines of interest to generate a single-cell suspension, count, and resuspend in a layer of top agar (0.4% in growth medium). For weakly transforming Dbl family proteins or GPCRs, it is best to resuspend 10^4 to 10^5 cells per plate in the top agar layer given that they cause a low efficiency of colony formation. Further, if these analyses require stimulation with a ligand or pharmacological inhibition of a specific signaling pathway, the ligand or drug should be added to both the top and the bottom layers of agar.

Compared to Ras-transformed cells that develop colonies within 10 days, Dbl and GPCR-induced colonies are only visible after 21 to 30 days (Fig. 1C).

Therefore, it is important to include both a positive (Ras-transformed cells) and negative (empty vector-transfected) control cell population. These controls are also important given that it is very easy to kill the cells if the agar solution is too hot or is made incorrectly. The following section describes the procedure to prepare reagents for a 14 dish assay where duplicates are used for each condition; one set would be used for a negative control, one set for a positive control, and five sets for five cell lines or experimental conditions.

Melt the sterile 1.8% Bacto-agar stock solution in a microwave and place in a 55°C water bath to cool. Place the complete growth medium, 2× DMEM, and calf serum in a 37°C water bath to prewarm.

Preparation of the bottom agar layer (0.6% agar in medium) is as follows (Table I). For 14- to 60-mm dishes, add 33 ml of 2 × DMEM, 27 ml of regular growth medium, 7.3 ml serum, and 73 µl of 1000× penicillin/streptomycin to a sterile bottle and place in a 37°C water bath to prewarm. Add 33 ml of the melted agar that was cooled to 55°C. Mix the solution and quickly aliquot 5 ml onto 60-mm dishes, allow to solidify for 10 to 15 min at room temperature, and transfer to a 37°C incubator until you are ready to pour the top agar layer. Avoid introducing bubbles when pouring the bottom agar.

Trypsinize each set of cells and transfer to a sterile 15-ml conical tube. Spin the cells out of the trypsin and resuspend in complete growth medium. Count the cells and transfer anywhere from $5 × 10^3$ to 10^5 cells per 0.5 ml to a 5-ml Falcon tube. The appropriate cell densities should be optimized for each cell line. Place the cell suspension in a 37°C water bath to prewarm while preparing the top agar layer.

TABLE I   Preparation of Solutions for Soft Agar Assay

	Bottom layer (0.6% agar)		Top layer (0.4% agar)	
	Per dish	14 dish assay	Per dish	14 dish assay
1.8% Bacto-agar	1.65 ml	33 ml	0.33 ml	1.65 ml
2× DMEM	1.65 ml	33 ml	0.33 ml	1.65 ml
Complete growth medium[a]	1.35 ml	27 ml	0.27 ml	1.35 ml
Calf serum	365 µl	7.3 ml	73 µl	365 µl
1000× pen/strep	3.65 µl	73 µl	0.73 µl	3.65 µl
Cell suspension[b]	N/A	N/A	0.50 ml	N/A
Total volume	5.00 ml	100.3 ml	1.50 ml	5.0 ml

[a] DMEM supplemented with final concentrations of 10% calf serum and 1× penicillin/streptomycin.

[b] In complete growth medium.

To prepare 5 ml of the top agar layer, combine 1.65 ml of 2× DMEM, 1.35 ml growth medium, 365 μl serum, and 3.65 μl of 1000× penicillin/streptomycin in a 15-ml conical tube and place in a 37°C water bath to prewarm. If the experimental conditions require the addition of a ligand or inhibitor, those compounds should be added before the solution is prewarmed. Add 1.65 ml of the melted 1.8% Bacto-agar cooled to 55°C and mix. Quickly add 1 ml of the top agar-medium solution to the 0.5-ml cell suspension and gently pipette the cell agar/solution onto the solid bottom agar layer. Allow the top agar to solidify for 30 min at room temperature and then transfer to a 37°C humidified 10% CO_2 incubator. Feed each dish with 2 to 3 drops of complete growth medium every 2 to 3 days.

III. IN VITRO CELLULAR INVASION ASSAYS

Three general types of *in vitro* assays can be used to assess the invasive potential of oncogene proteins that cause activation of Rho GTPases. The first and most widely used assay is the Matrigel transwell invasion assay, which provides quantitation of invasive potential. A second set of assays involves qualitative assessment invasive capabilities. These assays evaluate basal cellular invasion and correlate with *in vivo* metastatic potential. This section describes the analyses of human tumor cells where invasion is mediated, in part, by the persistent activation of Rho GTPase function by upstream activation of the PAR1 GPCR.

Reagents for Invasion Assays

Human breast carcinoma cell lines: Obtain from ATCC and culture according to their recommended media. Maintain MDA-MB-231 cells in DMEM supplemented with 10% fetal bovine serum. Maintaine MCF-7 cells in α-MEM supplemented with 10% FBS and 10 μg/ml insulin.

Matrigel invasion chamber: Purchase BD BioCoat growth factor reduced (GFR) Matrigel invasion chambers with a 8-μm pore size PET membrane (Cat. No. 354483) from BD Biosciences.

G8 Myoblasts: Maintain mouse fetal G8 myoblasts (American Type Culture Collection, Rockville, MD) in DMEM supplemented with 10% fetal calf serum and 10% horse serum and grow at 37°C with 10% CO_2. Cells should be subcultured when they reach 70–80% confluency to avoid alterations in the growth properties of the cells.

Matrigel: Purchase Matrigel from Collaborative Biomedical Products (Cat. No. 356234)

Cell dissociation media: Enzyme-free cell dissociation solution PBS based (1× liquid) can be purchased from Specialty Media (Cat. No. S-014-B)

A. Transwell Matrigel Invasion Assay

This modified assay is based on a similar protocol described previously (Albini *et al.*, 1987) and it should be emphasized that it is extremely important to optimize the starvation, cell density, and chemoattractant conditions for each cell line used in this assay. For example, we have determined that to assay the invasive potential of MDA-MB-231 cells, the cells must be starved in serum-free medium for a minimum of 24 h and no more than 5×10^3 cells per well should be added to the upper chamber of the Matrigel transwell chamber. If these conditions are not used, we have found that too many cells invade into the Matrigel to accurately quantitate. Therefore, we have described the procedures and amounts of MDA-MB-231 cells required for a 12 transwell chamber assay, where duplicate dishes are used for each experimental condition. This would involve one set of chambers for a positive control, one set for a negative control, and four sets for experimental conditions. We usually use MDA-MB-231 as a positive control and MCF-7 cells as a negative control (Booden *et al.*, 2003) (Fig. 2A). For the negative control, we generally seed 2×10^5 cells per 0.5 ml in the upper chamber of each transwell.

Plate the cells of interest on 100-mm dishes 2 days before the invasion assay at a density of 10^6 cells per dish. After an overnight incubation, rinse the cells with PBS and replace the normal grown media with starvation media [base medium supplemented with 1% bovine serum albumin (BSA) and 10 mM HEPES]. Starve the cells for at least 24 h.

Matrigel is a solubilized basement membrane preparation extracted from the Engelbreth–Holm–Swarm (EHS) mouse sarcoma cell line, which gels quickly at room temperature to form a reconstituted basement membrane. The growth factor-reduced (GFR) Matrigel is a more defined and characterized reconstituted basement membrane than Matrigel. Remove GFR Matrigel invasion chambers from −20°C and gently add 0.5 ml of starvation medium to the inner chamber. Allow the Matrigel to rehydrate for 2 h at 37°C. Once the Matrigel is rehydrated, it is important not to let it dry out. Therefore, do not remove the rehydration medium until it is time to add the cell suspension to the upper chamber.

While the Matrigel is rehydrating, dissociate the serum-starved cells of interest with enzyme-free cell

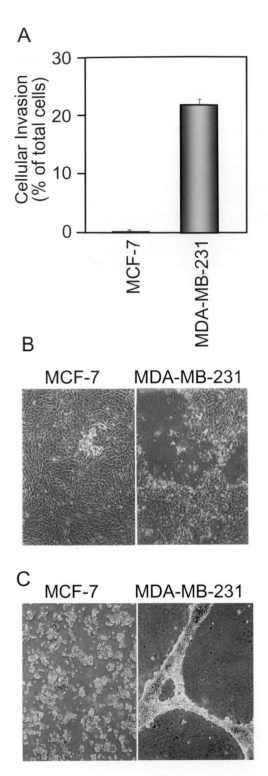

FIGURE 2 *In vitro* invasion assays. The behavior of noninvasive MCF-7 (low PAR1 expression) and highly invasive MDA-MB-231 (high PAR1 expression) human breast carcinoma cells was evaluated in the Matigel transwell invasion assay (A), the G8 myoblast invasion assay (B), or the two-dimensional Matrigel invasion assay (C).

dissociation media. We use the enzyme-free dissociation medium instead of trypsin, as trypsin can act as a mitogen that stimulates multiple signaling pathways, including GPCRs, that can cause Ras and Rho GTPase activation. Make sure to rinse the cells thoroughly with calcium-free PBS and, if need be, incubate them in calcium-free PBS for 5 to 10 min at 37°C before stimulating with enzyme-free dissociation medium. This incubation will facilitate a more rapid dissociation of the cells and will decrease the sheeting of epithelial cells, making them easier to count and plate. Spin the cells out of the dissociation medium and prepare cell suspensions in starvation medium containing 1% BSA and 10 mM HEPES at 10^4 cells/ml. For each transwell chamber, add 750 µl of medium to the bottom well of the 24-well companion plate. Remove the base medium from the inner chamber with a pipette, one chamber at a time, and immediately add 0.5 ml of cell suspension (5×10^3 cells) to the inner chamber. Transfer the Matrigel chambers to the wells containing the starvation medium. If these assays require receptor stimulation, allow the cells to incubate for 30 min and then stimulate with the ligand of interest. Be sure that no air bubbles are trapped beneath the membrane. Air bubbles will impede cells from invading through the Matrigel. Incubate the Matrigel transwell chambers for 24 h at 37°C and 5% CO_2.

After incubation, remove the noninvading cells by pipetting off the medium in the upper chamber of the transwell insert. Remove the Matrigel from the surface of the filter in the inner chamber by swabbing the insert with a cotton tip. These steps should be done quickly and just prior to fixing and staining the cells to avoid drying of the cells adhering to the bottom surface of the membrane.

Cells on the lower surface of the membrane should be fixed and stained. We use a Diff-Quick Kit (Dade Behring Inc.) to stain the cells on the lower surface of filter. With this kit, the cell nucleus stains purple and the cytoplasm stains pink. Alternative staining methods include fixation followed by hematoxylin and eosin or crystal violet. To fix and stain the cells using the Diff-Quick kit, add 750 ml of each Diff-Quick solution to three individual wells in a 24-well plate and add distilled water to two beakers. Immediately after removal of noninvading cells and Matrigel, sequentially transfer the transwell chambers, every 45 s, through each stain solution and the two beakers of water. Remove excess solution by gently swabbing the inner chamber with a cotton tip and allow the inserts to air dry for a minimum of 60 min.

Depending on the proportion of cells that invade, quantitation can be done by directly counting the stained cells by visual inspection using a phase-contrast microscope. We generally place the chamber on a grid and count the stained cells on one-half of the filter of duplicate membranes at 10× magnification. It is also important to examine the bottom well for cells that have invaded through the Matrigel but did not adhere to the filter. This phenomenon is quite common for highly invasive breast carcinoma cell lines such as MDA-MB-231, BT549, and Hs578T. Quantitate these cells using a hemacytometer. We do not recommend removing the membrane from the insert housing as the filter generally tears.

B. G8 Myoblast Invasion Assay

G8 monolayer invasion provides a simple and highly reproducible assay to test the ability of tumor cells to invade a monolayer that is enriched in a variety of extracellular matrix components. This provides a qualitative assay that monitors activity that correlates well with *in vivo* metastasis. The G8 myoblast assay was determined to correlate with the invasive and metastatic ability of TA3/St mouse mammary carcinoma cells *in vivo* (Yu and Stamenkovic, 1999; Yu *et al.*, 1997). While they cannot be employed to directly quantitate *in vitro* invasion mediated by a specific protein or signaling pathway, the G8 myoblast and two-dimensional Matrigel assays report the ability of cells to invade and degrade extracellular matrix, as well as the cell monolayer that has deposited the matrix proteins, and, more importantly, they report the ability of the cells to organize, move through an unnatural environment, and proliferate at a distant site (Fig. 2B).

The G8 invasion assay is relatively simple. Plate G8 myoblasts at a density of 2×10^5 cells per well in 6-well dishes 3 days before the assay begins. This time in culture allows the myoblasts to deposit extracellular matrix proteins on the cell surface. When the cells reach confluence, wash the monolayer once with PBS and fix with dimethyl sulfoxide for 2 h at room temperature, which results in a monolayer of both extracellular matrix protein and myoblast cells. After fixation, rinse the monolayer gently four times with PBS and incubate with 1 ml of culture medium.

Dissociate the cells to be evaluated with the enzyme-free dissociation medium as described earlier. Then collect the cell suspension by centrifugation to remove the dissociation medium, resuspend in growth medium, and count. Add a 1-ml suspension of the cells to the top of the G8 myoblast monolayer and observe daily. Invasion generally occurs within 1 to 3 days of seeding. The appropriate cell density and growth medium for each cell line used in this assay should be determined empirically. The percentage of fetal bovine serum used in the culture medium is a critical consideration; different cells require different amounts of FBS for the assay to be interpretable. In our experience with the MDA-MD-231, MCF-7, OVCA-420, -429, 432, 433, OVCAR-3, and -5 human tumor cell lines, a good starting point for the optimization of this assay is a cell density of 5×10^3 cells per well in growth medium supplemented with 1% FBS. The ability of cells to invade and degrade the myoblast monolayer and begin to proliferate can be visualized using an inverted phase-contrast microscope. A good positive control for this assay is the OVCA-432 cell line, and a good negative control is the NIH 3T3 cell line.

C. Two-Dimensional Matrigel Assay

The two-dimensional Matrigel assay provides a qualitative assay that correlates well with *in vivo* invasion and metatasis (Petersen *et al.*, 1992) (Fig. 2C). Thaw Matrigel overnight on ice at 4°C. Matrigel will begin to solidify at temperatures above 4°C. Therefore, it is important to thaw the Matrigel slowly and not allow it to warm before it is aliquoted. Each 12-well dish should be placed on ice for 10 to 15 min in order to cool the dish to 4°C. After chilling each dish, add 700 µl of Martigel to each well. Transfer the Matrigel-coated dishes to 37°C and incubate for a minimum of 30 min. While the Matrigel is solidifying, dissociate the cells of interest with enzyme-free cell dissociation medium. Spin the cells out of the dissociation medium and plate 2×10^5 per well in 1 ml of growth media. The ability of the cells to invade the Matrigel and organize into honeycomb-like structures can be assessed 24 h after incubation using an inverted phase-contrast microscope. A good positive control for this assay is the highly invasive MDA-MB-468 cell line, and a good negative control is the noninvasive MCF-7 cell line (Eckert et al. 2004).

References

Albini, A., Iwamoto, Y., Kleinman, H. K., Martin, G. R., Aaronson, S. A., Kozlowski, J. M., and McEwan, R. N. (1987). A rapid *in vitro* assay for quantitating the invasive potential of tumor cells. *Cancer Res.* **47**, 3239–3245.

Booden, M. A., Campbell, S. L., and Der, C. J. (2002). Critical but distinct roles for the pleckstrin homology and cysteine-rich domains as positive modulators of Vav2 signaling and transformation. *Mol. Cell. Biol.* **22**, 2487–2497.

Booden, M. A., Eckert, L. B., Der, C. J., and Trejo, J. (2004). Persistent signaling by dysregulated thrombin receptor trafficking promotes breast carconoma cell invasion. *Mol. Cell. Biol.* **24**, 1990–1999.

Clark, E. A., Golub, T. R., Lander, E. S., and Hynes, R. O. (2000). Genomic analysis of metastasis reveals an essential role for RhoC. *Nature* **406**, 532–535.

Clark, G. J., Cox, A. D., Graham, S. M., and Der, C. J. (1995). Biological assays for Ras transformation. *Methods Enzymol.* **255**, 395–412.

Eckert, L. B., Repasky, G. A., Ülkü, A. S., McFall, A., Zhou, H., Sartor, C. I., and Der, C. J. (2004). Involvement of Ras activation in human breast cancer cell signaling, invasion, and anoikis. *Cancer Res.* **64**, 4585–4592.

Fabian, J. R., Daar, I. O., and Morrison, D. K. (1993). Critical tyrosine residues regulate the enzymatic and biological activity of Raf-1 kinase. *Mol. Cell. Biol.* **13**, 7170–7179.

Hanahan, D., and Weinberg, R. A. (2000). The hallmarks of cancer. *Cell* **100**, 57–70.

Keely, P. J., Westwick, J. K., Whitehead, I. P., Der, C. J., and Parise, L. V. (1997). Cdc42 and Rac1 induce integrin-mediated cell motility and invasiveness through PI(3)K. *Nature* **390**, 632–636.

Khosravi-Far, R., Solski, P. A., Kinch, M. S., Burridge, K., and Der, C. J. (1995). Activation of Rac and Rho, and mitogen activated protein kinases, are required for Ras transformation. *Mol. Cell. Biol.* **15**, 6443–6453.

Martin, C. B., Mahon, G. M., Klinger, M. B., Kay, R. J., Symons, M., Der, C. J., and Whitehead, I. P. (2001). The thrombin receptor, PAR-1, causes transformation by activation of Rho-mediated signaling pathways. *Oncogene* **20**, 1953–1963.

Petersen, O. W., Ronnov-Jessen, L., Howlett, A. R., and Bissell, M. J. (1992). Interaction with basement membrane serves to rapidly distinguish growth and differentiation pattern of normal and malignant human breast epithelial cells. *Proc. Natl. Acad. Sci. USA* **89**, 9064–9068.

Qiu, R.-G., Chen, J., Kirn, D., McCormick, F., and Symons, M. (1995). An essential role for Rac in Ras transformation. *Nature* **374**, 457–459.

Sahai, E., and Marshall, C. J. (2002). RHO-GTPases and cancer. *Nature Rev. Cancer* **2**, 133–142.

Schmidt, A., and Hall, A. (2002). Guanine nucleotide exchange factors for Rho GTPases: Turning on the switch. *Genes Dev.* **16**, 1587–1609.

Solski, P. A., and Der, C. J. (2000). Analyses of transforming activity of Rho family activators. *Methods Enzymol.* **325**, 425–441.

Whitehead, I. P., Zohn, I. E., and Der, C. J. (2001). Rho GTPase-dependent transformation by G protein-coupled receptors. *Oncogene* **20**, 1547–1555.

Yu, Q., and Stamenkovic, I. (1999). Localization of matrix metalloproteinase 9 to the cell surface provides a mechanism for CD44-mediated tumor invasion. *Genes Dev.* **13**, 35–48.

Yu, Q., Toole, B. P., and Stamenkovic, I. (1997). Induction of apoptosis of metastatic mammary carcinoma cells *in vivo* by disruption of tumor cell surface CD44 function. *J. Exp. Med.* **186**, 1985–1996.

42

Assay of Tumorigenicity in Nude Mice

Anne-Marie Engel and Morten Schou

I. INTRODUCTION

Tumorigenicity is defined as the ability of viable cultured cells to give rise to progressively growing tumor nodules, showing viable and mitotically active cells, in immunologically nonresponsive animals over a limited observation period (WHO, 1987). It is, however, no absolute property in a cell line, but will always be defined by the assay in which it is tested. In order to ensure reproducibility, it is therefore very important that the cells tested are from the same batch and that the animals used are syngenic. The only way of testing whether a cell line has achieved all the characteristics needed for solid tumor growth is by testing its ability to form such tumors *in vivo*. In order to rule out the interference of the immune defense of the recipient, testing must take place in a host organism that will not reject the transplant. If the cells being tested are originally from mice or rats, syngenic animals can be used as recipients of cell inocula. Otherwise, immune-deficient animals that are incapable of recognizing and rejecting nonself tissue are used. The nude mouse is ideal in this respect in that its immune defect affects thymic development, resulting in a lack of functional T lymphocytes in the animals (Rygaard, 1973). Therefore, a large proportion of mammalian tumors can be transplanted to nude mice without being rejected. If a cell line will not produce tumors in nude mice, it will sometimes help to mix irradiated mouse or human fibroblasts 1:1 with the tumor cell inoculum (Wilson *et al.*, 1984). The severe combined immune deficiency (SCID) mouse, in which both T and B lymphocyte maturation are impaired due to a deficiency in a gene coding for recombinase, which participates in the recombination of genes coding for T

and B lymphocyte antigen receptors, can also be used in this kind of assay (Bosma *et al.*, 1983). Some reports say that SCID mice will let some cell lines form tumors even though they are not tumorigenic in nude mice (Xie *et al.*, 1992).

Testing the tumorigenicity of cells propagated and perhaps transformed *in vitro* is relevant in a number of different contexts, such as transfection with antisense DNA in order to decrease tumor formation (Cheng *et al.*, 1996) or investigation of the role of specific genes in increased or decreased tumorigenicity after *in vitro* manipulation (Sun *et al.*, 1996). In the production of vaccines, it is important to know that the cells used for propagating a virus are not tumorigenic *in vivo* (WHO, 1987). The potential tumorigenic effect of different genes can be tested by transfecting immortalized cell lines with oncogenes and subsequently inoculating them on nude mice (Chisholm and Symonds, 1992) or by assaying the baseline expression of a gene in a cell line and relating this to the tumorigenicity of the cells. Immortalized cell lines are also used for determining the carcinogenic effect of chemical compounds by *in vitro* treatment and subsequent *in vivo* tumorigenicity testing in nude mice (Iizasa *et al.*, 1993). A number of phenotypical features are characteristic of malignantly transformed cells. As opposed to normal cells, they are immortalized, i.e., they can grow in culture for more than 100 population doublings. They lose contact inhibition, resulting in piling up of cultured cells, and they exhibit high plating efficiency (i.e., a high number of colonies obtained per 100 cells plated at 2–50 cells/cm^2). The growth rate of such cells is normally higher than for normal cells, and they can exhibit chromosomal abnormalities such as aneuploidy or heteroploidy (Shimizu *et al.*, 1995). These are all characteristics that can be observed in *in vitro* experimental setups. The ability to form colonies in soft agar is another characteristic closely connected to the malignant phenotype. During the multistep process of malignant transformation, cells become anchorage independent, probably as a result of cell surface modifications (Nicolson, 1976). This enables them to grow in suspension or in semisolid media such as soft agar. Testing for this characteristic is widely used to evaluate the possible tumorigenic effect of different manipulations performed on cells *in vitro* (Hamburger and Salmon, 1977).

Angiogenesis, the ability to induce blood vessel formation, is characteristic of cancer cells. The angiogenic potential of a cell line can be assayed by implanting tumor cells at the edge of rabbit cornea or by incubating it with tumor cell extracts and observing the formation of blood vessels. Implantation into the chorioallantroic membrane of chicken embryos followed by observation of the formation of blood vessels is also used for assaying angiogenesis (Folkman, 1985). Invasiveness is another characteristic of tumor cells, which can be assayed in *in vitro* setups.

The *in vivo* protocol suggested here is adapted from the guidelines of WHO given for tumorigenicity testing of tumor cells used for propagation of the poliomyelitis virus used in the manufacturing of polio vaccine (WHO, 1987).

II. MATERIALS AND INSTRUMENTATION

Mice used are 6- to 8-week-old Balb/cA nu/nu females. They are from Taconic M&B, Ry, Denmark, and are kept at our institute for 1 week before experiments are initiated. They are fed autoclaved water and sterilized food pellets and kept in Macrolon II cages under SPF conditions.

All plasticware used for tissue culture and harvest of cells is from Greiner GmbH. Tissue culture medium is RPMI 1640 with UltraGlutamin (BioWhittaker, Cat. No. 12-702F/U1). The trypsin–EDTA solution (Cat. No. 17-161E) and phosphate-buffered saline (PBS) without Ca^{2+} and Mg^{2+} (Cat. No. 17-516F) are also from BioWhittaker. Syringes for injection of tumor cells are 1-ml Luer tuberculin from Once (Cat. No. 1202) fitted with a 0.5 × 16 Terumo Neolus needle (Cat. No. NN-2516R).

For fixation of excised tissue, buffered formalin (Lillies fixative) is prepared: 163.1 g 24.5% formaldehyde solution, 4.5 g natriumdihydrogenphosphatdihydrate (Merck 6345), and 8.2 g dinatriumphosphatdihydrate (Merck 6580). Add sterile water until the volume reaches 1000 ml.

III. PROCEDURES

A positive and a negative control cell line should always be included in an assay of tumorigenicity in order to ensure that the procedure has been carried out correctly. HeLa cells (American Type Culture Collection CLL-2) can be used as positive controls, and MRC-5 cells (American Type Culture Collection CLL-171) can be used as negative controls.

A. Preparation of Cells

Solutions

1. *Tissue culture medium:* Prepare the same cell culture medium for each of the cell lines tested that you would normally use for propagation *in vitro*. For this experiment, we use RPMI 1640 with 50 ml/500 ml of heat-inactivated calf serum (heat inactivate in a water bath at 56°C for 30 min) and penicillin/streptomycin to a final concentration of 100 IU/ml penicillin and 10 mg/ml streptomycin.
2. *PBS without Ca^{2+} and Mg^{2+}*
3. *Trypsin–EDTA solution*

Steps

1. Expand the cell line(s) to be tested for tumorigenicity and the two control cell lines in culture until an appropriate number of cells has been obtained. At least 10 mice should be used for each cell line. Each mouse must receive 10^7 cells. Harvest the cells when they are subconfluent and thus still in the logarithmic growth phase (Fig. 1A).
2. The harvest of cells from culture flasks depends on whether they grow in suspension or adhere to the flask. If cells grow adherently, they can be removed either by scraping with a cell scraper or by trypsination for 5–10 min at 37°C (in an incubator). Prior to adding trypsin, remove culture medium and wash cells with 10 ml PBS without calcium and magnesium in order to remove any traces of serum that will otherwise inhibit the action of trypsin.
3. After removal from the flask, centrifuge the cells for 5 min at 200 g, discard the supernatant, and resuspend the cells in 10 ml PBS or culture medium without antibiotics and serum. Centrifuge again at 200 g for 5 min. Discard supernatant and resuspend cells in 10 ml PBS or medium as described earlier.
4. Dilute a small sample (e.g., 0.5 ml) of the cell suspension 1:10 in PBS or medium. Dilute this solution 1:1 with 0.4% trypan blue. Let the suspension stand for 2–5 min and then count cells in a hemocytometer. Keep the rest of the cells on ice while preparing and counting the sample. Calculate the number of viable

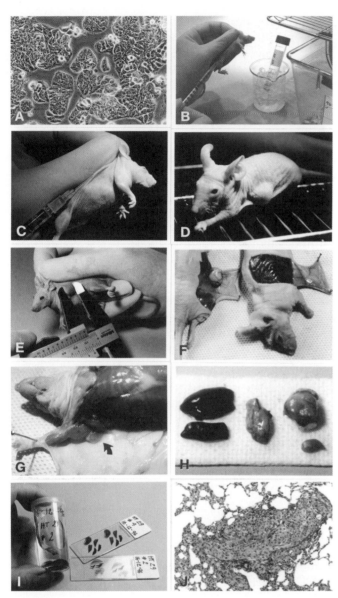

FIGURE 1   (A) Subconfluent cells (colon adenocarcinoma HT29) ready to be harvested. (B) Cells kept on ice, ready to be injected. (C) Subcutaneous injection of 0.2 ml cell suspension. (D) Mouse with tumor growth. (E) Measurement of tumor. (F) Autopsy after 21 days. Mouse with and without tumor growth. (G) Axillary lymph node (arrow). (H) Organs removed for fixation in formalin. (I) Fixed tissue and histological preparations. (J) Lung metastasis (objective x).

suspension as some of it will be trapped in the syringe. An extra 0.5 ml for 10 mice to be injected is sufficient.

6. Keep cells on ice until use and be sure to inoculate as soon as possible after the adjustment of concentration (Fig. 1B).

B. Inoculation and Inspection

Steps

1. Inject 0.2 ml of the cell suspension subcutaneously in the right lateral aspect of the thoracic wall (Fig. 1C).
2. Mice should be observed daily and inspected for tumor growth at least three times a week (Fig. 1D).

C. Examination of Recipients and Tumors

Solution

Prepare Lillies fixative (see Section II) and store at room temperature.

Steps

1. Tumors are measured in two perpendicular dimensions (Fig. 1E). After 21 days of observation, sacrifice mice and autopsy immediately. Inspect the inoculation site from the deep aspect and excise (Fig. 1F). In order to describe the metastatic potential of the cell line tested, excise the ipsilateral axillary lymph node (Fig. 1G), part of the spleen, one-third of the liver, and the inferior lobe of the left lung for histological examination (Fig. 1H).

2. Fix tissues in buffered formalin (Lillies fixative), embed in paraffin, and section in 5-μm sections. Stain with hematoxylin and eosin and study under a light microscope (Fig. 1I). Histological examination enables the investigator to see if the tumor has retained the morphological characteristics of the primary tumor from which the cell line was established and whether it gives rise to metastases (Fig. 1J).

D. Reporting of Findings

Reports of macroscopic and microscopic findings should be given for each cell line. Macroscopic findings include the conditions of mice during the test period and macroscopic tumor growth, if any. The macroscopic appearance of the tumor-inoculation site at autopsy should also be reported. Microscopic findings should include a description of the tumor tissue or the inoculation site if no tumor appears. Metastases, location, and frequency should be reported.

An example of an observation table is given in Table I for HeLa cells used as a positive control in a tumorigenicity assay carried out in nude Balb/cA mice.

cells per milliliter of the original suspension as follows: viable cells in larges squares counted/number of large squares counted × dilution of cells × 10^4.

5. Adjust the concentration of the cell suspension to be inoculated to 5×10^7/ml viable cells in serum-free culture medium or PBS. This may include one more step of centrifugation and resuspension in a suitable volume. Always make sure to have an excess of cell

TABLE I   HeLa Cells Used as Positive Controls in a Tumorigenicity Assay

Mouse No.	Site of inoculation/ no cells	Lymph node	Liver	Spleen	Lung
1	Subcutaneous in the right flank / 10^7 cells	0	0	0	0
2	Subcutaneous in the right flank / 10^7 cells	0	0	0	0
3	Subcutaneous in the right flank / 10^7 cells	0	0	0	0
4	Subcutaneous in the right flank / 10^7 cells	0	0	0	0
5	Subcutaneous in the right flank / 10^7 cells	0	0	0	0
6	Subcutaneous in the right flank / 10^7 cells	0	0	0	0
7	Subcutaneous in the right flank / 10^7 cells	0	0	0	0
8	Subcutaneous in the right flank / 10^7 cells	0	0	0	0
9	Subcutaneous in the right flank / 10^7 cells	0	0	0	0
10	Subcutaneous in the right flank / 10^7 cells	0	0	0	0

E. Conclusions

According to the World Health Organization, the positive control cell (HeLa) should give rise to progressively growing tumors in 9 out of 10 mice if the assay is carried out correctly. Furthermore, when a histopathological examination of tumor nodules is performed, mitotic activity must be detected in the tumor tissue.

For the cell line tested for tumorigenicity, at least 9 out of 10 mice should develop a tumor if the cell line is to be designated tumorigenic. If only some mice in a group develop tumors, the assay should be repeated with the same cell dose and with an increased cell dose, e.g., 5×10^7. As mentioned in Section I, it is a prerequisite that the cells used are from the same batch and that the recipient mice are syngenic. The cell line used as a negative control should not, of course, give rise to tumor growth in any of the 10 animals. Some

authors use a more differentiated assay, administering cells in differentiated doses and categorizing cell lines as high, moderate, low, or nontumorigenic (Gurtsevitch and Lenoir, 1985).

IV. PITFALLS

It is of the utmost importance for the interpretation of results that the cell lines tested be free of mycoplasma infection, as mycoplasma can alter their ability to form tumors (Fogh, 1973). If there is not a routine of regular mycoplasma testing in the cell culture laboratory, a mycoplasma test should be performed and a negative outcome ensured before cell lines are used in a tumorigenicity assay.

References

Bosma, G. C., Custer, R. P., and Bosma, M. J. (1983). A severe combined immunodeficiency in the mouse. *Nature* **301**, 527–530.

Cheng, J. Q., Ruggeri, B., Klein, W. M., Sonoda, G., Altomare, D. A., Watson, D. K., and Testa, J. R. (1996). Amplification of AKT2 in human pancreatic cells and inhibition of AKT2 expression and tumorigenicity by antisense RNA. *Proc. Natl. Acad. Sci. USA S* **93**, 3636–3641.

Chisholm, O., and Symonds, G. (1992). An *in vitro* fibroblast model system to study myc-driven tumour progression. *Int. J. Cancer* **51**, 149–158.

Fogh, J. (ed.) (1973). "Contamination in Tissue Culture." Academic Press, New York.

Folkman, J. T. (1985). Tumor angiogenesis. *Adv. Cancer Res.* **43**, 175–203.

Gurtsevitch, V. E, and Lenoir, G. M. (1985). Tumorigenic potential of various Burkitt's lymphoma cell lines in nude mice. *In* "Immune-Deficient Animals in Biomedical Research" (J. Rygaard *et al.*, eds.), pp. 199–203. Karger, Basel.

Hamburger, A. W., and Salmon, S. E. (1977). Primary bioassay of human tumor stem cells. *Science* **197**, 461–463.

Iiaza, T., Momiki, S., Bauer, B., Caamano, J., Metcalf, R., Lechner, J., Harris, C. C., and Klein-Szanto, A. J. (1993). Invasive tumors derived from xenotransplanted, immortalized human cells after *in vivo* exposure to chemical carcinogens. *Carcinogenesis* **14**, 1789–1794.

Nicolson, G. L. (1976). Trans-membrane control of the receptors on normal and tumor cells. II. Surface changes associated with transformation and malignancy. *Biochim. Biophys. Acta* **458**, 1–72.

Rygaard, J. (1973). "Thymus and Self: *Immunobiology of the Mouse Mutant Nude*." Wiley, London.

Shimizu, T., Kato, M. V., Nikaido, O., and Suzuki, F. (1995). A specific chromosome change and distinctive transforming genes are necessary for malignant progression of spontantous transformation in cultured Chinese hamster embryo cells. *Jpn. J. Cancer Res.* **86**, 546–554.

Sun, Y., Kim, H., Parker, M., Stetler-Stevenson, W. G., and Colburn, N. H. (1996). Lack of suppression of tumor cell phenotype by overexpression of TIMP-3 in mouse JB6 tumor cells: Identification of a transfectant with increased tumorigenicity and invasiveness. *Anticancer Res.* **16**, 1–17.

Wilson, E. L., Gärtner, M., Campbell, J. A. H., and Dowdle, E. B. (1984). Growth and behaviour of human melanomas in nude mice: Effect of fibroblasts. *In "Immune–Deficient Animals"* (B. Sordat, ed.), pp. 357–361. Karger, Basel.

World Health Organization (WHO) (1987). "Requirements for Continuous Cell Lines Used for Biologicals Production." *WHO Technical Report Series*, No. 745, Annex 3.

Xie, X., Brünner, N., Jensen, G., Albrectsen, J., Gotthardsen, B., and Rygaard, J. (1992). Comparative studies between nude and scid mice on the growth and metastatic behavior of xenografted human tumors. *Clin. Exp. Metast.* **10**, 201–210.

43

Transfilter Cell Invasion Assays

Garth L. Nicolson

I. INTRODUCTION

The ability to invade surrounding extracellular matrices and tissues is an important phenotype of malignant tumor cells. To determine the invasive properties of malignant cells, several invasion assays have been developed. For example, organ fragments (Nicolson *et al.*, 1985), reconstituted tissue spheroids (Mareel *et al.*, 1988), membranous tissues (Nabeshima *et al.*, 1988; Yagel *et al.*, 1989), cultured cell monolayers (Kramer and Nicolson, 1979; Waller *et al.*, 1986), or extracellular matrices (Albini *et al.*, 1987; Schor *et al.*, 1980) have been used as tissue or matrix for cell invasion studies. Commercially available Matrigel, a mouse EHS tumor extract consisting of major basement membrane components polymerized into a gel, has been the most commonly used material to determine the invasiveness of various types of cells (Albini *et al.*, 1987). The apparatus of choice for measuring cell invasion has been a modified Boyden chamber or Transwell. Briefly, the apparatus contains two chambers (upper and lower) that are separated by a microporous polycarbonate filter, the upper surface of which is coated with a thin layer of Matrigel. Tumor cells are placed into the upper chamber where they settle by gravity onto the Matrigel layer. In the lower chamber a chemotactic agent can be placed to stimulate directional cell migration of cells that invade the Matrigel layer. The invasive ability of these cells can be expressed as the number of cells invading through the Matrigel layer and filter with time. For the most part, the invading cells are found on the lower surface of the filter and are not released into the fluid of the lower chamber. Although there are some exceptions to this (Simon *et al.*, 1992), the invasive abilities of tumor cells in the transfilter invasion assay are usually related to their *in vivo* invasion behavior (Albini *et al.*, 1987; Repesh, 1989; Hendrix *et al.*, 1987). This article describes the basic protocol of the Matrigel transfilter invasion assay.

II. MATERIALS AND INSTRUMENTATION

Culture medium DME/FI2 (Cat. No. 11330-032), fetal bovine serum (FBS), phosphate-buffered saline (PBS) (Cat. No. 14040-133), and trypsin (2.5%, Cat No. 15090-046) are from GIBCO-BRL. Matrigel (Cat. No. 40234) and fibronectin (Cat. No. 40008) are from Collaborative Biomedical Products. Bovine serum albumin (BSA) (Cat. No. 810661, fraction V) is from ICN. EDTA (Cat. No. 423-384) is from CMS. Hematoxylin (Cat. No. GHS-1-16), eosin (Cat. No. HTIIO-3-16), and 10% neutralized formaldehyde solution (Cat. No. HT50-1-128) are from Sigma. Transwells (Cat. No. 3421, 6.5-mm diameter, 5-μm pore size) and 24-well culture plates (Cat. No. 25820) are from Corning Costar Corporation. Additional equipment includes general tissue culture supply and equipment, forceps, cotton swabs, a reticle (1/10-mm measurement), and a cell-counting device (Coulter counter, hemocytometer, or equivalent equipment).

III. PROCEDURES

A. Matrigel Coating on the Microporous Transwell Filter

Solutions

1:30 diluted Matrigel solution (sterile): To make 900 μl Matrigel solution for 12 Transwells (50 μl X the number of Transwells), thaw Matrigel at 4°C overnight and mix 30 μl with 870 μl of ice-cold PBS. Keep ice cold until the solution is applied to the Transwell. Aliquot the remaining Matrigel (e.g., 1 ml each), freeze, and store at −80°C.

1 : 100 diluted Matrigel solution (sterile): To make 2 ml, mix 20 μl of Matrigel with 2 ml of ice-cold PBS. Keep solution ice-cold.

Steps

1. Place 500 μl of 1 : 100 diluted Matrigel solution in a small culture dish (i.e., 30 mm diameter) on an ice-cold plate.

2. Use a forceps for the handling of Transwells. The Transwell invasion chamber system is shown in Fig. 1. Soak the lower surface of the polycarbonate filter of each Transwell in the Matrigel solution. After briefly removing any excessive amount of the solution, place the Transwell into a 24-well culture plate and allow it to dry in a hood overnight at room temperature.

3. Pour 50 μl of the 1 : 30 diluted Matrigel solution into the upper chamber of the Transwell.

4. Carefully overlay 200 μl of sterilized, double-distilled H₂O to each filter and allow it to dry completely in a hood at room temperature under occasional ultraviolet light. This usually takes about 2 or 3 days.

5. Proceed to the next step or store the coated Transwell sets in a scaled 24-well culture plate at 4°C.

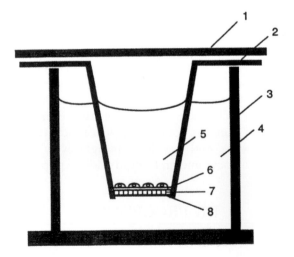

FIGURE 1 Transwell invasion chamber system: (1) lid, (2) Transwell, (3) 24-well culture plate, (4) lower chamber (800 μl of 10 μg/ml fibronectin in invasion buffer), (5) upper chamber (200 μl), (6) Matrigel layer, (7) microporous polycarbonate filter (5-μm pore size), and (8) lower surface of the filter coated with a thin layer of Matrigel.

B. Setting up the Invasion Assay

Solutions

2% BSA stock solution: To make 100 ml, slowly dissolve 2 g of BSA in ice-cold PBS for 20 min and sterilize by filtration through a 0.22-μm filter.

Invasion buffer, 0.1% BSA in DME/F12: To make 40 ml, mix 2 ml of 2% BSA stock solution with 38 ml of DME/FI2 medium.

EDTA stock solution (200 mM): To make 100 ml, dissolve 7.44 g EDTA with double-distilled H₂O and adjust the pH to 7.4.

Trypsin–EDTA (0.25%): To make 100 ml, add 10 ml of 2.5% trypsin solution and 1 ml of EDTA stock solution into 89 ml of PBS.

10% FBS-DME: To make 20 ml, mix 2 ml of FBS with 18 ml of DME/FI2 medium.

Fibronectin solution (10 μg/ml): To make 10 ml, mix 100 μl of fibronectin stock solution (1 mg/ml) with 10 ml of invasion buffer. In the example here, fibronectin is used as a chemoattractant.

Steps

1. These procedures should be performed under sterile conditions. Culture (4 × 100-mm-diameter dishes) B16-F10 cells in DME/FI2 medium supplemented with 5% FBS at 37°C in an atmosphere of 5% CO₂–95% air and grow to about 80% confluency.

2. Prewarm the Matrigel-coated Transwells in the 24-well culture plate to room temperature, and rehydrate the Matrigel with 200 μl of invasion buffer for 1 h.

3. Wash the B16-F10 culture dishes with 10 ml of DME/FI2 medium and incubate with 3 ml each of trypsin–EDTA for 5 min. Place in a 50-ml tube and sediment cells by centifugation at 1000 rpm for 5 min.

4. Wash the cells with 10 ml of 10% FBS-DME medium by centrifugation at 1000 rpm for 5 min and resuspend the cells in 10 ml of 10% FBS-DME medium. Let the cells incubate for 15 min at room temperature.

5. Wash the cells with 40 ml of DME/FI2 medium twice by centrifugation at 1000 rpm for 10 min. Resuspend the cells in 5 ml of invasion buffer.

6. Prepare 2.5 × 10⁵ cells/ml in invasion buffer by counting the cells with a Coulter counter, hemocytometer, or equivalent equipment.

7. Carefully discard the buffer in the upper chambers of the Transwells, wash the chambers again with 200 μl of invasion buffer, and carefully pour 800 μl of the fibronectin solution (chemoattractant solution) through the slit of the Transwell into the lower chamber.

8. Add 200 μl of the cell suspension into the upper chamber. Avoid making bubbles in the upper chamber

so that the cells can settle down evenly on the filter. Carefully put the Transwell invasion chamber system into an incubator and incubate for 48 h at 37°C in 5% CO_2–95% air.

C. Staining Invading Cells with Hematoxylin–Eosin

Steps

1. Prepare a 24-well culture plate with 400 µl of 10% formaldehyde, hematoxylin, or eosin in each well and fill four jars (about 500 ml each) with distilled water.

2. Place the Transwell in the 10% formadehyde solution for 10 min and then rinse the Transwell by submerging it in a jar containing distilled water.

3. Transfer the Transwell to the hematoxylin solution for 10 min and then rinse with water. Place the Transwell into the well with warm distilled water (about 40°C) for 10 min.

4. Transfer the Transwell to the eosin well for 5 min and then place it into a jar containing distilled water. Gently rub the cells off the upper side of the filter using a cotton swab and rinse the Transwell with distilled water.

D. Counting Invading Cells Using a Phase-Contrast Microscope (or Coulter Counter)

Steps

1. Place the Transwell into a 24-well culture place. Make sure that there is moisture on the inside of the culture plate. Count the cells in the 16 fields indicated in Fig. 2 at 200 × magnification.

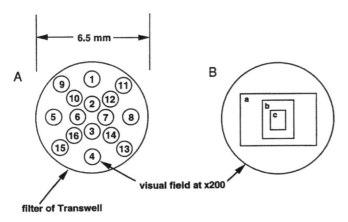

FIGURE 2 A Transwell filter in a microscopic field. (A) Spots used to count the cells on the filter are shown as numbered circles. The numbers of the spots are an example of the sequence recommended for cell counting. (B) Photo frames of the Diaphot (Nikon). Relative sizes of the frames to the field are 47 (a), 17 (b), and 4.3% (c).

2. Using a reticle with 1/10-mm guides (Bausch & Comb Inc.), place the cell suspension onto the stage of a phase-contrast microscope, and measure the diameter of the field (r; mm) with the magnification at 40×. The total number of invading cells (1) can be calculated with the formula

$$I = c \times (5 \times R/r)^2,$$

where c is the average number of invading cells per ×200 magnification field and R is the diameter of the filter (6.5 mm). For example, the value of r of a Diaphot (Nikon) is 4.45. Therefore, $(5 \times R/r)^2 = 53.3$. If the cells are too dense to count the entire field, a photo as shown in Fig. 2 can be used. Using a Diaphot (Nikon), the areas of a, b, and c in Fig. 2 are 47, 17, and 4.3% of the total area of the field, respectively.

E. Reusing Transwells

Steps

1. After counting invading cells, wipe the cells off the lower surface of the filter using a cotton swab and soak the Transwells in a detergent solution (e.g., 1% Contrad 70) for at least 1 day.

2. Wipe both surfaces of the filter gently and thoroughly using a cotton swab and place the Transwells into the 24-well culture plate.

3. Fill the entire Transwell plate with distilled water, cover the plate with a lid, shake it several times, and discard the water.

4. Repeat step 3 at least three times. After the final wash, let the Transwell plate sit for 1 h filled with distilled water.

5. Wash the plate with double-distilled water several times and rinse with 70% ethanol as in step 3. Dry the Transwells in a tissue culture hood overnight under ultraviolet light.

IV. COMMENTS

Although tumor cells *in vivo* usually invade connective or parenchymal tissue, a number of reports indicate that the invasive and metastatic phenotypes of cells correlate with their *in vitro* Matrigel invasion capacity (Albini *et al.*, 1987; Repesh, 1989; Hendrix *et al.*, 1987). This may be due to the fact that invasion *in vivo* into connective or parenchymal tissue is limited by the capability of tumor cells to undergo basement membrane invasion, a key event in malignant progression. The invasive process involves at least three events (adhesion, degradation, and migration) (Liotta,

Matrigel. In 1–2 weeks the segments are monitored for the outgrowth of endothelial cells and for the formation of three-dimensional tube-like structures. The main drawback of this system is using adult aorta endothelial cells in the case of the aortic ring assay or microvascular endothelial cells obtained from embryonic arch, which are composed of cells dividing before exposure to angiogenic factors. Nevertheless the ability of endothelial cells to form tube-like structures in Matrigel makes this *in vitro* cell culture model the most faithful assay system.

This article describes the modification of a tube forming assay in Matrigel in which a dense clump of cultured endothelial cells is formed by cultivation of the cells in a hanging drop of growth media. During incubation, cells form dense "clump" at the bottom of the drop. The sediment of the cells is transferred onto a layer of Matrigel and is covered with a thin layer of Matrigel in which an interconnecting network of endothelial capillary tubes is formed rapidly. Tube formation occurs through an ordered sequence of events. Endothelial cells localized on the surface of the "cellular island" first develop large, dynamic cellular protrusions and then form small aggregates and cord-like structures. These early cord structures are dense and do not have lumens. Cells start to migrate to form a complex network of tube-like structures. The advantages of the assay are the ability to directly visualize the changes in cell morphology and the ability of endothelial cells to generate tube-like structures. Endothelial cells in the "clump" can survive for more than a few days, leaving adequate time for assessing angiogenic reactions. Introduction of angiogenesis-inducing factors or cells into Matrigel at some distance from the "island" allows one to test **chemotactic** activity. The assay is readily quantifiable but utilizes already mentioned problems characteristic of the *in vitro* angiogenesis assay.

Another advantage is that in this assay, cultivated endothelial cell lines can be used, which can give more standard results than the use of primary cultures, or organ cultures.

II. MATERIALS AND INSTRUMENTATION

Growth medium: Dulbecco's modified Eagle's medium (DMEM) with a high concentration of glucose (Cat. No. 31966-021, GIBCO) supplemented with fetal bovine serum (FBS, Cat. No. F 7524, Sigma) at 10%, 100× penicillin/streptomycinx (Cat. No. 15140-148, GIBCO).

PBS(-) Ca^{2+} and Mg^{2+}-free Dulbecco's phosphate-buffered saline (PBS, 2.7 mM KH$_2$PO$_4$, 8.1 mM Na$_2$HPO$_4$, 137 mM NaCl), sterile.
Trypsin–EDTA solution (Cat. No. 25200-056, GIBCO)
HEPES buffer solution 1 M (Cat. No. 15630-049, GIBCO)
Stock of Matrigel – basement membrane matrix (BD Biosciences, Cat. No. 354234). Matrigel is supplied frozen and is stored at −70°C.
25-cm^2 tissue culture flasks, 48-well tissue culture plates, Pasteur pipettes, 10-ml pipettes (sterile), and syringe needles (23 gauge).

III. PROCEDURES

1. For culture in 75-cm^2 flasks, suspend murine microvascular endothelial SVEC4-10 cells in culture medium at a concentration of 2×10^5 cells/ml and plate out 2–4 ml of cell suspension/flask. Place the flasks in a 37°C tissue culture incubator and incubate in an atmosphere of 5% CO$_2$ overnight.

2. The next day, examine cells under an inverted microscope fitted with phase-contrast objectives. Choose the flasks with the cell density >80% of confluence.

3. Remove media and wash the cells with 10 ml PBS(−).

4. Replace the PBS with 1 ml of trypsin–EDTA and incubate at 37°C for approximately 1 min. Cell rounding should be observed in the inverted microscope. When the cells are rounded, detach them by strong agitation. Add 10 ml of culture medium supplemented with 10% FBS to the flask, pipette the cells up and down five times, and transfer contents to a 15-ml centrifuge tube.

5. Centrifuge the cells at 800 g for 5 min at room temperature. Resuspend the cells in 10 ml DMEM/FBS. Count an aliquot of the cell suspension with the Coulter counter or in a hemocytometer.

6. Centrifuge the cells at 800 g for 5 min at room temperature.

7. Resuspend the cells in DMEM supplemented with 20% FBS, 1 M HEPES, pH 7.5, at a concentration of 3×10^6 cells/ml.

8. Dispense 1.0 ml of DMEM/FBS into each well of a 48-well tissue culture plate.

9. Turn the lid of the plate upside down. Plate out 0.02 ml of the cell suspension in the middle of the inner side of the lid. The cell suspension should form a drop (Fig. 1A).

10. Carefully turn over the lid and place it on the plate. The drops of the cell suspension will hang over the media in the wells (Fig. 1B).

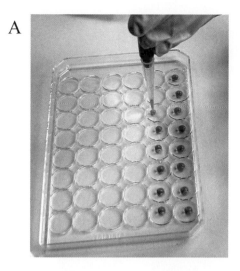

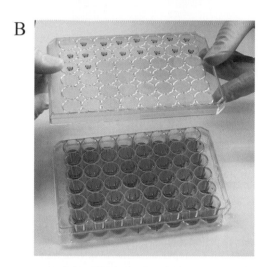

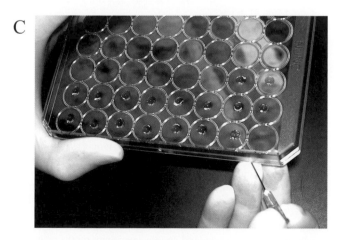

FIGURE 1 Culturing cells in the hanging drops. (A) Plating of the cell suspension in the middle of the inner side of the lid. (B) Turning over the lid with the hanging drops. (C) Transferring the cell "clump" with a syringe needle.

11. Incubate plates at 37°C in a tissue culture incubator. During overnight incubation, cells form a dense "clump" at the bottom of the drop.

12. Defreeze the necessary amount of Matrigel overnight at 4°C with rotation.

13. The next day, examine all the wells under a stereomicroscope and select those that have the most compact, well-formed "clump" of cells.

14. Place Matrigel on ice. Add 5 × DMEM and FBS to obtain a final concentration of 1× DMEM and 10% FBS. Keep the mixture on ice.

15. Place a fresh 48-well plate on ice. Dispense 0.15 ml/well of Matrigel/mix into the plate. Incubate the plate at 37°C in an incubator for 3–4 hours. The Matrigel should solidify.

16. Carefully lift the lid of the plate with hanging drops, stick the cell "clump" to the tip of a syringe needle, and transfer the "clump" on the surface of the solidified Matrigel in the 48-well tissue culture plate (Fig. 1C).

17. Cover the cell "clump" with 0.01 ml of Matrigel/mix and incubate the plate for 15 min at 37°C in an incubator.

18. Add 0.3 ml of DMEM/FBS in each well and return the plate to the incubator. The cells will remain viable for several days.

IV. COMMENTS

Using the protocol described in this assay, it is possible to demonstrate the effects of compounds on the ability of endothelial cell to form capillary-like tubes *in vitro*. We have also used this protocol to cocultivate endothelial cells with other cell types and to study the influence of the molecules produced by these cells on the ability of endothelial cells to form capillaries (Fig. 2D).

V. PITFALLS

1. To obtain maximum viability of cells growing in the hanging drops, avoid long exposure of the cells to trypsin during harvesting cells from the 75-cm² flasks.

2. Choose only well-formed, round-shaped, dense cellular "clumps."

3. When lifting the lid with the hanging drops, caution must be exercised to avoid disrupting the cell "clumps."

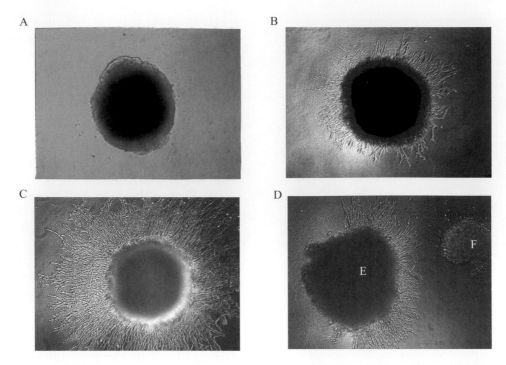

FIGURE 2 Mouse endothelial cells SVEC4-10 migrating from a dense "clump" of the cells into Matrigel containing an angiogenic compound: (A) 0 h of cultivation, (B) 24 h, (C) 48 h, and (D) chemotactic activity of an angiogenic compound. E, endothelial cells; F, tumor cells.

References

Folkman, J. (1986). How is blood vessel growth regulated in normal and neoplastic tissue? G.H.A. Clowes memorial award lecture. *Cancer Res.* **46**, 467–473.

Folkman, J. (1992). The role of angiogenesis in tumor growth. *Semin. Cancer Biol.* **3**, 65–71.

Folkman, J., and Shing (1992). Angiogenesis. *J.Biol.Chem.* **267**, 10931–1093.

Gale, N., and Yancopoulos, G. D. (1999). Growth factors acting via endothelial cell-specific receptor tyrosine kinases: VEGFs, angiopoietins and ephrins in vascular development. *Genes Dev.* **13**, 1055–1066.

Grant, D. S., Kleinman, H. K., Leblong, C. P., Inoue, S., Chung, A. E., and Martin, G. R. (1985). The basement-membrane-like matrix of the mouse EHS tumor. II. Immunochemical quantitation of six of its components. *Am.J.Anat.* **174**, 387–398.

Gumkowski, F., Kaminska, G., Kaminski, M., Morrissey, L. M., and Auerbach R. (1991). Heterogeneity of mouse vascular endothelium: *In vitro* studies of lymphatic, large blood vessels and microvascular endothelial cells. *Blood Vessels* **24**, 11–23.

Kråling, B. M., Jimenez, S. A., Sorger, T., and Maul, G. G. (1994). Isolation and characterization of microvascular endothelial cells from the adult human dermis and from skin biopsies of patients with systemic sclerosis. *Lab. Invest.* **71**, 745–754.

Lamszus, K., Schmidt, N. O., Ergun, S., and Westphal, M. (1999). Isolation and culture of human neuromicrovascular endothelial cells for the study of angiogenesis in vitro. *J. Nerosci. Res.* **55**, 370–381

Madri, J. A., Pratt, B. M., and Tucker, A. M. (1988). Phenotipic modulation of endothelial cells by transforming growth factor-β depends upon the composition and organization of the extracellular matrix. *J. Cell Biol.* **106**, 1375–1384.

Muthukkaruppan, V. R., Shinners, B. L., Lewis, R., Park, S-J., Baechler, B. J., and Auerbach, R. (2000). The chick embryo aortic arch assay: A new, rapid, quantifiable in vitro method for testing the efficacy of angiogenic and anti-angiogenic factors in a three-dimensional, serum-free organ culture system. *Proc. Am. Assoc. Cancer Res.* **41**, 65.

Nicosia, R. F., and Ottinett, A. (1990). Growth of microvessels in serum-free matrix culture of rat aorta. A quantitative assay of angiogenesis in vitro. *Lab Invest.* **63**, 115–122.

Soker, S., Takashima, S., Miao, H. Q., Neufeld, G., and Klagsbrun, M. (1998). Neuropilin1 is expressed by endothelial and tumor cells as an isoform-specific receptor for vascular endothelial growth factor. *Cell* **92**, 735–745.

Springhorn, J. P., Madri, J. A., and Squinto, S. P. (1995). Human capillary endothelial cells from abdominal wall adipose tissue: Isolation using an anti-PECAM antibody. *In Vitro Cell Dev. Biol* **31**, 473–481.

Wang, H. U., Chen, Z. F., and Anderson, D. J. (1998). Molecular distinction and angiogenic interaction between embryonic arteries and viens revealed by ephrin B2 and its receptor EphB4. *Cell* **93**, 741–753.

Zetter, B. R. (1988). Endothelial heterogeneity: Influence of vessel size, organ localization, and species specificity on the properties of cultured endothelial cells. *In "Endothelial Cells"* (U.S. Ryan, ed.), vol. 2, pp. 63–79. V. 2. CRC, Boca Raton, FL.

45

Analysis of Tumor Cell Invasion in Organotypic Brain Slices Using Confocal Laser-Scanning Microscopy

Takanori Ohnishi and Hironobu Harada

I. INTRODUCTION

Organotypic cultures of nervous tissue, including those of the hippocampal and cortical regions, have been produced successfully with a simple method in which brain slices are maintained in a culture at the interface between air and culture medium (Yamamoto *et al.*, 1989, 1992; Stoppini *et al.*, 1991; Tanaka *et al.*, 1994). In these organotypic brain slice cultures, not only is the normal cytoarchitecture such as cortical lamination and pyramidal cells preserved, but the biochemical and electrophysiological properties of neuronal cells are also maintained for 2 or 3 months. By modifying this organotypic culture of nervous tissues, we established a model for glial tumor cell invasion with conditions analogous to those of normal brains *in situ* (Ohnishi *et al.*, 1998; Matsumura *et al.*, 2000). This model enables not only to quantitatively analyze the tumor cell invasion in brain tissues, but also to investigate molecular events *in vitro* (events actually occur between transplanted cells and brains *in vivo*).

II. MATERIALS

Hanks' balanced salt solution (HBSS) (Cat. No. H9269), Eagle's minimum essential medium (MEM) with HEPES (Cat. No. M7278), D-glucose (Cat. No. G7021), penicillin–streptomycin solution (Cat. No. P0781), amphotericin B (Cat. No. A2942), propidium iodide (PI) (Cat. No. P4170), L-glutamine (Cat. No.

G5763), *N*-methyl-D-aspartate (NMDA) (Cat. No. M3262), and agar (Cat. No. A5431) are from Sigma. Dulbecco's phosphate-buffered saline, calcium-magnesium free [PBS(-), pH 7.4] (Cat. No. 14190-250), horse serum (Cat. No. 16050-122), and fetal bovine serum (FBS) (Cat. No. 16000-044) are from Invitrogen Corp. Culture plate inserts with a 0.4-μm-pore membrane, 30mm (Millicell-CM) (Cat. No. PICM 030 50) are from Millipore Corp. Six-well culture plates (Cat. No. 3506), 60-mm culture dishes (Cat. No. 430166), and 100-mm culture dishes (Cat. No. 430167) are from Corning. The PKH2 fluorescent cell staining kit is from ZYNAXIS Cell Science. C6 rat glioma cells and T98G human glioma cells are from American Type Culture Collection. For these cell cultures, Ham's F10 powder (Cat. No. N6635) and MEM (Cat. No. M4655) are from Sigma.

III. PROCEDURES

A. Preparation of Brain Slice and the Oraganotypic Culture

This procedure is modified from the method of Stoppini *et al.* (1991).

Solutions and Instruments

Scissors (large one for decapitation and small one for dissection of brains)

Microforceps with fine chips

10% povidone–iodine solution

Phoshate-buffered saline without calcium and magnesium (pH 7.4)

Microslicer with a sliding cut mode (possible to cut nonfrozen fresh brains with a range of 50 to 1000 μm thick)

Culture plate inserts with 0.4 μm-pore membranes (30 mm in diameter) (Millicell-CM)

Six-well culture plates (35 mm in diameter/well)

60-mm culture dishes

CO₂ incubator

Culture medium: 50% Eagle's MEM (Earle salt with L-glutamine, 25 μM HEPES, and NaHCO3), 25% HBSS, 25% heat-inactivated horse serum, 6.5 mg/ml D-glucose, 100 U/ml penicillin, 100 μg/ml streptomycin, and 2.5 μg/ml amphotericin B. To make 200 ml, add 96 ml of a Eagle's MEM solution, 50 ml of HBSS, 50 ml of horse serum, 1.3 g of D-glucose, 2 ml of a penicillin (10,000 U/ml)–streptomycin (10 mg/ml) solution, and 2 ml of an amphotericin B solution (250 μ/ml). Keep at 4°C.

Steps

1. Anesthetize a 2-day-old neonatal rat with diethyl ether and plunge into a 10% povidone–iodine solution.

2. Cut off the head with large scissors, remove the skin and the skull with a small scissors, and take out the whole brain quickly and place in a 60-mm culture dish with HBSS.

3. Cut the brain vertically to the base, 1 mm inward from both rostral and caudal ends of the cerebrum with a blade, and mount on the stage of a microslicer, which is sterilized with 70% ethyl alcohol.

4. 300-μm-thick cut brain slices and transfer each slice onto a porous (0.4 μm pore size) membrane of a culture plate insert, which is placed in a well of a six-well culture plate filled with PBS.

5. After aspiration of PBS from the outer well of the six-well culture plate, add 1 ml of culture medium to the outer well but without covering the brain slice placed on the membrane.

6. Incubate the brain slice at 37°C under standard conditions of 100% humidity, 95% air, and 5% CO₂.

7. After 3 days of the culture, replace half of the medium with fresh medium twice a week. Reduce the volume of the medium after the second change to 0.8 ml so that the slices remain well exposed to the air. (This is critical for long-term survival of the neuronal cells.) (Fig. 1).

B. Assessment of Viability of Brain Slices

The viability of cultured brain slices can be assessed by morphological observation, neuronal activity, electrophysiological features, and production of bioactive substances such as γ-aminobutyric acid and neuropeptides. Normal cytoarchitecture such as cortical lamination and hippocampal structure is clearly

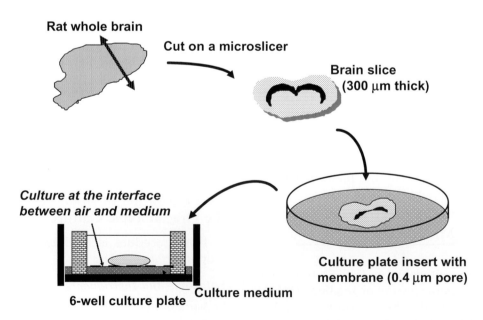

FIGURE 1 Illustrative procedures of brain slice culture. A rat whole brain slice 300 μm thick is placed on a porous membrane affixed to the culture plate insert and cultured at the interface between air and culture medium.

observed for about 2 months after the slice culture if the culture condition is kept properly (Fig. 2). This section describes the method used to assess the neuronal viability of brain slices by NMDA insult that can induce early and delayed neuronal cell death (Sakaguchi *et al.*, 1996).

Solutions and Instruments

100 μM N-methyl-D-aspartate solution: Dissolve 1.47 mg of NMDA in 100 ml of artificial cerebrospinal fluid (CSF). Prepare artificial CSF from three stock solutions, A, B, and C, before use. Stock solution A consists of 18.12 g of NaCl, 0.488 g of $NaH_2PO_4 \cdot 2H_2O$, 0.932 g of KCl, and 1.232 g of $MgSO_4 \cdot 7H_2O$ in 100 ml of H_2O. Stock solution B contains 2.22 g of $CaCl_2$ in 100 ml of H_2O, and stock solution C contains 4.62 g of $NaHCO_3$ in 100 ml of H_2O. Store at 4°C. To make 100 ml of artificial CSF, add 4 ml of stock solution A, 4 ml of stock solution C, and 91 ml of H_2O and then place the mixed solution under the current of 95% air and 5% CO_2 to lower the pH of the solution. Then, add 1 ml of solution B and 0.18 g of D-glucose to the mixed solution (the final pH is 7.4).

4.6 μg/ml propidium iodide solution. Dissolve PI in a serum-free solution containing 75% MEM, 25% HBSS, 2 mM L-glutamine, and 6.5 mg/ml D-glucose to a final concentration of 4.6 μg/ml.

Fluorescence microscope with a tetramethylrhodamine isothiocyanate (TRITC) filter

Steps

1. Incubate brain slices in 1 ml of the artificial CSF solution containing 100 μM NMDA, which is placed in the bathing well of a six-well culture plate for 15 min. Incubate the slices in PI solution for 1 h to measure early neuronal death or for 24 h to measure delayed neuronal cell death.

2. View PI signals under a fluorescence microscope with a TRITC filter.

3. As a control, incubate brain slices in PI solution for 1 h or 24 h following incubation in the CSF solution without NMDA for 15 min. (Fig. 3).

C. Preparation of Tumor Cell Spheroids

Tumor Cells, Solutions, and Instruments

Tumor cells in culture and their culture medium (Ham's F10 medium containing 10% FBS is used for C6 rat glioma cells, and MEM supplemented with 1% nonessential amino acid, 1% sodium pyruvate, and 10% FBS is used for T98G human glioma cells)

PKH2 fluorescent cell-staining kit

1.25% agar-coated culture dish (100 mm in diameter)

Place 5 ml of 1.25% agar solution on a culture dish and dry under air.

Reciprocating shaker (usable in a CO_2 incubator)

CO_2 incubator

Steps

1. Grow tumor cells as a monolayer culture under standard conditions.

2. Harvest the tumor cells by trypsinization, wash twice, and resuspend in labeling diluent "A" (provided with the PKH2 staining kit) at a concentration of 2×10^7 cells/ml (cell/diluent suspension).

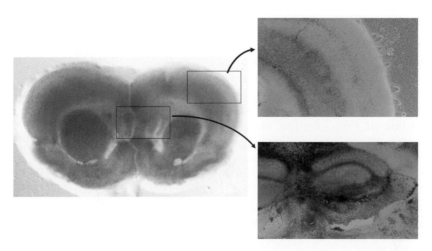

FIGURE 2 Morphological pictures of a rat brain slice after 8 days of culture. Normal cytoarchitecture, including cortical lamination and hippocampal structure, is clearly observed (left: macroscopic picture, ×0.5; right: phase contrast, ×40).

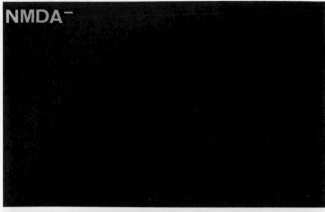

FIGURE 3 Neuronal viability of rat brain slices assessed by cellular uptake of propidium iodide (PI) without (upper) and with (lower) treatment of *N*-methyl-D-aspartate (NMDA). Normally functioned neurons can exclude PI and show early or delayed neuronal cell death by NMDA insult, thus permitting entry of the PI into the cells.

3. Add PKH2 dye to an equal volume of diluent "A" to make a $4\mu M$ solution. Add this solution to the cell/diluent suspension and mix by gentle agitation.

4. After incubating the cells at room temperature for 5 min, stop the labeling reaction by adding a double volume of the culture medium containing 10% FBS and four times the volume of FBS into the sample tubes.

5. Wash the cells and resuspend in the culture medium with 10% FBS.

6. Seed the labeled tumor cells (5×10^6) into a 1.25% agar-coated culture dish and incubate under continuous agitation at a speed of 40 rpm on a reciprocating shaker at 37°C in a humidified atmosphere of 5% CO_2 and 95% air for 2 to 3 days.

7. For the experiments, select cell aggregates with a size of 150 to 200 μm.

D. Migration Assay of Tumor Cells on Brain Slices (Fig. 4)

Instuments and Molecules

Barin slices after 7 days culture
Tumor cell spheroids
Micropipette with a volume of 10 μm
Molecules or agents affecting cell migration
Fluorescence microscope with a FITC filter
Color-chilled 3-CCD camera
Personal computer

Steps

1. Using a micropipette, take one spheroid of tumor cells place on the surface of brain slice, and coculture at 37°C under standard conditions.

2. Four hours later, apply 2 µl of molecules in investigation directly to the tumor spheroid. Carry out the application of the molecule once a day for 3 to 6 days.

3. To estimate the extent of cell migration, calculate the distance between the margin of the initially placed spheroids and the population of the migrating cells showing half of the density (area) of the maximum density of migrating cells from the tumor spheroid by using computer images for which the original fluorescent pictures of the slices are taken with a color-chilled 3-CCD camera.

4. For this calculation, draw concentric circles 10 µm apart around the margin of the spheroid and measure cell density (area of fluorescence-stained cells) within each ring by an NIH image. Then, do the summation of area of the cells contained in each ring and plot as a function of the distance from the margin of the tumor spheroid. Thus the distribution curve of migrating cells outside the spheroid is constructed for each brain slice (Fig. 5). The migratory strength of the cells on the slice is defined as the distance (µm) that shows half of the value of the maximum density (area) of migrating cells on the distribution curve.

E. Invasion Assay in Brain Slices (Fig. 4)

Instruments

Barin slices after 7 days culture
Tumor cell spheroids
Micropipette with a volume of 10 µm
Inverted confocal laser-scanning microscope with FITC filter optics
Personal computer

Steps

1. With a micropipette, take one spheroid of tumor cells tagged with the PKH2-fluoresent dye, place on

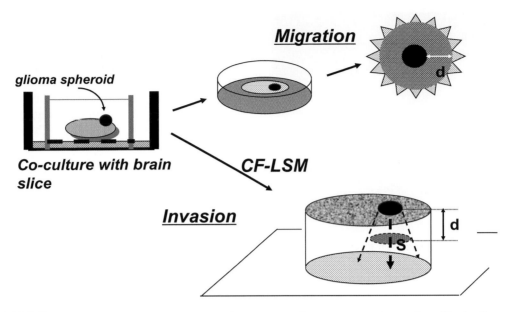

FIGURE 4 Illustrative procedures of tumor cell migration and invasion assay in cocultured brain slices. A tumor (glioma) spheroid is placed on the brain slice and is cocultured at the interface between air and culture medium. For tumor cell migration, the extent of the spread of fluorescent dye-stained tumor cells on the surface of the slice is measured. For tumor cell invasion, the spatial extent of the tumor cell infiltration in the slice is analyzed by confocal laser-scanning microscopy (CF-LSM). d, distance; S, area.

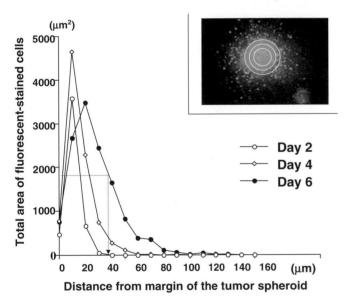

FIGURE 5 Construction of distribution curves of migrating cells for a quantitative analysis of tumor cell migration in brain slices. The total area of the labeled cells contained in each ring is plotted as a function of the distance from the margin of the tumor spheroid. (Inset) Concentric circles 10 μm apart are drawn around the margin of the tumor spheroid, and cell density (area of fluorescent-stained cells) within each ring is measured by an NIH image. Distribution curves represent L1-stimulated C6 glioma migration at day 2, day 4, and day 6 after the coculture with brain slices. The migratory strength of the cells is determined as the distance (μm) that shows half of the value of the maximum density (area) of migrating cells on the distribution curve (see the distribution curve on day 6).

the surface of brain slice, and coculture at 37°C under standard conditions.

2. To detect PKH2-stained tumor cells in brain slices, use an inverted confocal laser-scanning microscope with FITC (520 nm) filter optics.

3. At the first observation, determine the level of the basal plane (0 μm) in accordance with the upper surface of the brain slice.

4. Obtain serial sections every 20 μm downward from the basal plane to the bottom of the slice (Fig. 6).

F. Quantitative Analysis of Tumor Invasion

The total area of PKH2-stained cells in each section is calculated with NIH image software. The area is plotted as a function of the distance from the basal plane of the brain slice and the distribution curve is constructed for each experiment. The extent of tumor cell invasion in the slice is defined as the depth (μm) that shows half of the maximum density (area) of invasive cells on the distribution curve.

IV. COMMENTS

1. As a source of brain slices, brain tissues from mice and humans (obtained from epilepsy surgery) are also applicable. In the case of rats, 2- to 5-day-old

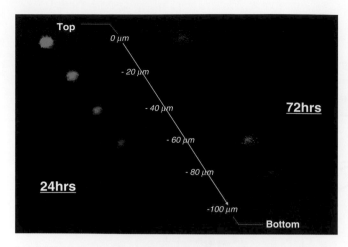

FIGURE 6 Confocal laser-scanning microscopic pictures of invading glioma cells in rat brain slices. At 24 h after coculture of T98G glioma spheroid and the brain slice, most glioma cells remain at the top of the slice (0 µm), while the glioma cells migrate extensively within the brain slice (show the maximum spread at –40 µm from the top of the slice) at 72 h after coculture.

neonatal brains are best. As brains from much younger rats are smaller and more soft, it is difficult to manuplate the whole brain as intact slices. Brains prepared from rats of an older age have a tendency to be resistant to tumor invasion into the slices.

2. For about 4–5 days after initiating the brain slice culture, several neuronal death and glial cell migration to the bottom of the slices occur. After 5 days in culture, the exposure of slices to PI alone without NMDA insult does not elicit a detectable fluorescence signal. Therefore, for tumor invasion experiments, brain slices after 5 days in culture should be used. Usually, brain slices after 7 days in culture are used for any kind of studies. At this time, the thickness of the brain slices is reduced to about 200 µm from the original thickness of 300 µm.

Brain slices maintain their normal structures, such as cortical lamination, and are functionally viable for about 2 months after the culture, but it seems that the best time for experiments is from 7 to 30 days after culture.

3. To detect migrating tumor cells in the brain slice, a tumor-labeling method using green fluorescent protein (GFP) is also applicable. Once tumor cell clones with persistent expression of GFP are established, their use is of great advantage in analyzing the behavior of the tumor cells because the GFP is transmitted to the tumor cells after cell division. An EGFP vector (pEGFP-C1, obtained from Clonetics Corp., San Diego, CA) can be used to transfect and label malignant glioma cells.

References

Matsumura, H., Ohnishi, T., Kanemura, Y., Maruno, M., and Yoshimine, T. (2000). Quantitative analysis of glioma cell invasion by confocal laser scanning microscopy in a novel brain slice model. *Biochem. Biophys. Res. Commun.* **269**, 513–520.

Ohnishi, T., Matsumura, H., Izumoto, S., Hiraga, S., and Hayakawa, T. (1998). A novel model of glioma cell invasion using organotypic brain slice culture. *Cancer Res.* **58**, 2935–2940.

Sakaguchi, T., Okada, M., Kuno M., and Kawasaki, K. (1996). Dual mode of N-methyl-D-aspartate-induced neuronal death in hippocampal slice cultures in relation to N-methyl-D-aspartate receptor properties. *Neuroscience* **76**, 411–423.

Stoppini, L., Buchs, P.-A., and Muller, D. (1991). A simple method for organotypic cultures of nervous tissue. *J. Neurosci. Methods* **37**, 173–327.

Tanaka, M., Tomita, A., Yoshida, S., Yano, M., and Shimuzu, H. (1994). Observation of the highly organized development of granule cells in rat cerebellar organotypic cultures. *Brain Res.* **641**, 319–327.

Yamamoto, N., Kurotani, T., and Toyama, K. (1989). Neural connections between the lateral geniculate nucleus and visual cortex *in vitro*. *Science* **245**, 192–194.

Yamamoto, N., Yamada, K., Kurotani, T., and Toyama, K. (1992). Laminar specificity of extrinsic cortical connections studied in coculture preparations. *Neuron* **9**, 217–228.

46

Angiogenesis Assays

Yihai Cao

I. INTRODUCTION

All healthy and pathological tissue growth requires the formation of functional blood vessels (Hanahan and Folkman, 1996). Angiogenesis, sprouting of new capillaries from the existing vessels, is the key process that contributes to both physiological and pathological neovascularization. Angiogenesis has paradoxical implications in the treatment of the most severe human diseases. In disorders such as atherosclerosis/ infarction of the heart, stroke, and peripheral occlusive limb disease, the growth of new vessels in the affected tissue is obviously beneficial for the improvement of blood circulation (Cao *et al.*, 2003). In addition, organ transplantation and tissue regeneration need new blood vessels. However, inhibition of angiogenesis has become an important therapeutic approach for the treatment of diseases such as cancer, diabetic retinopathy, and arthritis in which angiogenesis plays an important role (Cao, 2001; Folkman, 1995). It has been reported that therapeutic angiogenesis or antiangiogenesis could delay the onset and progression of obesity (Rupnick *et al.*, 2002). Based on their broad therapeutic implications and potential economic values, discovery and development of novel angiogenic and antiangiogenic agents have become a competitive business for pharmaceutical companies. As a result, a novel angiogenesis regulator is identified almost every week. There are at present many assay systems to be used for testing the angiogenic or antiangiogenic activity of compounds (Jain *et al.*, 1997; Kenyon *et al.*, 1996). The process of angiogenesis includes several critical steps: endothelial cell morphological changes, endothelial migration, endothelial proliferation, endothelial cell reorganization, and lumen formation, formation of new branches, followed by reconstitution of the basement membrane and vas-cular remodeling. Each of these steps can be monitored using appropriate *in vitro* and *in vivo* assay systems. This article describes *in vivo* and *in vitro* assays that are reliable, reproducible, and convenient and that are important for the characterization of compounds for use in angiogenesis–related disorders. Migration assays are described elsewhere in this volume, while this article includes the capillary endothelial cell proliferation assay and functional vessel formation assays, including the shell-less *in vivo* chick chorioallantoic membrane (CAM) assay and the mouse corneal angiogenesis assay.

II. MATERIALS AND INSTRUMENTS

A. Endothelial Proliferation Assay

Primary bovine capillary endothelial (BCE) cells are from Dr. Judah Folkman's laboratory at the Children's Hospital, Boston. All tissue culture plastic bottles and discs are from Falcon, Becton Dickinson. DME (low glucose) medium is from JRH Biosciences Limited (Cat. No. 2D0113). Gelatin is from Difco (Becton Dickinson, Cat. No. 214340). Recombinant human fibroblast growth factor (FGF-2) is from Scios Nova (Mountain View, CA). Bovine calf serum (BCS) is from Boule Nordic AB (Cat. No. SH300072.03). Trypsin solution (0.05%) is from Sigma (Cat. No. T9906). The Isoton phosphate-buffered saline (PBS) solution is from KEBO (Stockholm, Sweden). The Coulter counter is by Coulter, KEBO, Sweden.

B. Chick Chorioallantoic Membrane Assay

Fertilized white leghorn eggs are from OVA Production, Sörgåden, Sweden. Epigallocatechin-3-gallate is from Sigma (Cat. No. E4143). Methylcellulose is from Sigma (Cat. No.M-0262). Tissue culture plates (100 × 20 mm) are from Falcon, Becton Dickinson. The chick embryo incubator is made in Germany. The stereomicroscope is from Nikon, Japan.

C. Mouse Corneal Angiogenesis Assay

C57Bl6/J male and female mice are from the Microbiology and Tumor Biology Center, Karolinska Institute, Stockholm, Sweden. Vascular endothelial growth factor (VEGF) is from R&D System Inc. (Cat. No. 293-VE-050). FGF-2 is from Scios Nova. Sulcralfate is from Bukh Meditec, Vaerlose, Denmark. Hydron (type NCC) is from Interferon Sciences, Inc. (New Brunswick, NJ) (Cat. No. NCC-97001). Nylon mesh (15 × 15 mm) is from Tetko (Lancaster, NY). Methoxyflurane and proparacaine are from Ophthetic (Alcon, TX). The operation microscope is made by Zeiss, Oberkochen, Germany. Surgical blades are from Bard-Parker No. 15; Becton Dickinson (Franklin Lakes, NY). The eye examination microscope is from Nikon, Sf-2, Tokyo, Japan.

III. PROCEDURES

A. Endothelial Cell Proliferation Assay

For additional information, see Cao *et al.* (1996, 1997, 1999, 2001).

1. Stimulation of Endothelial Cell Proliferation
Steps

1. Set a CO_2 incubator to 10% CO_2.
2. Set up tissue culture hood for at least 15 min before use.
3. Coat 24-well tissue culture plates with 0.5 ml 1.5% gelatin in PBS solution (previously sterilized by autoclaving)/well for at least 1 h at 37°C or overnight at room temperature.
4. Prewarm the following solution to room temperature: 1x PBS, 0.05% trypsin solution, and 10% BCS-DMEM medium.
5. Thaw the FGF-2 solution in ice and dilute with sterile 10% BCS-DMEM to appropriate concentrations.
6. Add FGF-2 to 10% BCS-DMEM to a final concentration of 3 ng/ml.
7. Remove culture medium from BCE cells (cultured in 6-well plates) and wash cells with PBS twice.

8. Add 0.5 ml trypsin solution/well and expose cells to trypsin solution until they detach from the bottom.
9. Once cells are detached, add immediately 5 ml/well of 10% BCS-DMEM to cells.
10. Transfer cell solution into a centrifuge tube and centrifuge cells at 1500 rpm for 5 min.
11. Remove supernatant and resuspend cells in 10 ml 10% BCS-DMEM containing 3 ng/ml FGF-2.
12. Count cell numbers and seed cells at a density of 10,000 cells/well. Make sure that each sample is at least in triplicate.
13. Incubate cells at 37°C and 10% CO_2 for 72 h.
14. After 72 h, remove the culture medium and wash cells twice with PBS.
15. Add 0.5 ml of 0.05% trypsin solution to each well.
16. When cells are detached completely, resuspend cells into a single cell solution by tituration.
17. Transfer cell solution to a Coulter cup containing 10 ml Isoton solution and counter cell numbers.

2. Inhibition of Endothelial Cell Proliferation

1. Repeat steps 1–10 described in Section III,A,1.
2. Resuspend cells in 5% BCS-DMEM in an appropriate volume.
3. Seed BCE cells at a density of 10,000 cells/well in a volume of 0.5 ml of 5% BCS-DMEM.
4. Add tested angiogenesis inhibitors (e.g., angiostatin or EGCG) at various concentrations and incubate the cells for 1 h at 37°C with 10% CO_2.
5. Add FGF-2 to a final concentration of 1 ng/ml and incubate the cells for 72 h.
6. Repeat steps 14–17 described in Section III,A,1.

B. Chick Chorioallantoic Membrane Assay

For additional information, see Cao *et al.* (1998, 1999, 2001).

Steps

1. All procedures should be carried out in a 37°C warm room equipped with a sterile ventilation hood and a humidifier.
2. Set up the chick incubator at 37°C with 75% humidity.
3. Place freshly fertilized eggs onto the selves of the incubator and incubate the eggs for 72 h.
4. Switch on the humidifier to make sure that the room air has a similar humidity as that in the egg incubator.
5. Prewarm plastic tissue dishes (100 × 20 mm) at 37°C.

6. In the sterile hood, hold the egg in one hand and gently make a small crack using a metal or a glass stick. Open the egg using both thumbs placed on each side of the crack and gently place the contents into a prewarmed tissue culture dish. It is important that the yolk sac remains intact.
7. Immediately place the dish into a tissue culture incubator supplied with 3 or 4% CO_2 and 75% humidity at 37°C.
8. Incubate the embryos for 48h in case of testing inhibitors of angiogenesis. Use 6 days for testing angiogenesis stimulators.
9. During the incubation, prepare 0.9% methylcellulose in H_2O as autoclaved sterile solution and tested angiogenic or antiangiogenic compounds at appropriate concentrations (e.g., 2mg/ml) with H_2O.
10. Place a piece of nylon mesh onto a glass beaker in a sterile ventilation hood.
11. Mix 10µl of 0.9% methylcellulose with 10µl of tested compounds and carefully transfer the mixture onto the nylon mesh (the drop will hang on the mesh).
12. When the drop has dried out, carefully cut out the area as a square (about 3 × 3mm).
13. For antiangiogenesis, after a 48-h incubation carefully place the nylon mesh containing the test compounds onto the newly formed chorioallantoic membrane.
14. Return the chick embryos to the incubator and incubate the embryos for 1–2 days.
15. Detect the antiangiogenic effect of the tested compound under a stereomicroscope.
16. For a typical potent angiogenesis inhibitor, the formation of avascular zones around the implant can be detected readily after 24–48h of implantation.
17. Testing of angiogenesis stimulators takes place after 6 days of incubation. Carefully place the nylon mesh on the almost fully developed chorioallantoic membrane, preferably at a site with sparse vessel formation.
18. Return the chick embryos to the incubator and incubate the embryos for 3–5 days. For a typical potent angiogenic factor, the formation of new microvessel sprouts and branches can now be detected (see Fig. 1).

C. Mouse Corneal Angiogenesis Assay

For additional information, see Cao *et al.* (2003), Kenyon *et al.* (1996); and Cao and Cao (1999).

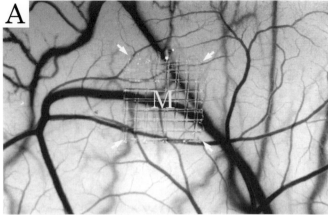
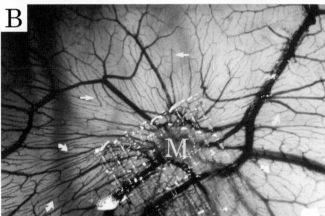
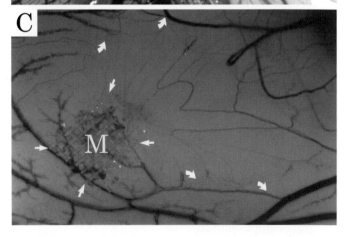

FIGURE 1 (A) Normal chick chorioallantoic membrane after 8-day incubation. (B) A typical example of the CAM stimulated by FGF-2. New vessel sprouting stimulated by 2.5µg of FGF-2 after a 5-day implantation (arrows). (C) A typical example of inhibition of CAM angiogenesis by 50µg of epigallocatechin-3-gallate. Formation of avascular zones is marked by arrows.

1. Stimulation of Angiogenesis

Steps

1. Order male or female mice at the age of between 6 and 9 weeks. For routine angiogenesis analysis, the C57Bl/6 strain is commonly used.
2. Suspensions of sterile H_2O containing the appropriate amount of angiogenic factors such as FGF-2 or VEGF and 10 mg of sucralfate.
3. Vacuum the mixture solution for 5 min in a speed centrifuge.
4. Add 10 μl of 12% hydron in ethanol.
5. Deposit the suspension onto a nylon mesh (pore size 0.4 × 0.4 mm) and embed between the fibers (the total grid area: 15 × 15 mm).
6. Cover both sides with a thin layer of hydron and allow the materials to dry at room temperature.
7. When the immobilized angiogenic factors are dried on the mesh, pull apart the grid fibers and uniformed pellets (0.4 × 0.4) are released.
8. Choose equal sized pellets; each pellet contains equal amounts of angiogenic factors.
9. Anesthetize mice with methoxyflurane and anesthetize topically with 0.5% proparacaine.
10. Proptose the eye globes with a jeweler's forceps.
11. Under an ophthalmological operation microscope, perform a central intrastromal linear keratotomy with a surgical blade.
12. Using a modified von Graefe knife (2 × 3 mm), dissect a lamellar micropocket toward the temporal limbus.
13. The micropocket is usually extended to 1.0–1.2 mm of the temporal limbus.
14. Place a single pellet on the corneal surface at the base of the pocket with jeweler's forceps.
15. Using one arm of the forceps, push the pellet to the end of the pocket.
16. Apply erythromycin ointment immediately to the operated eyes.
17. Examine the implanted eyes routinely under a slit lamp biomicroscope between days 4 and 7 after pellet implantation.
18. Anesthetize mice with methoxyflurane and position the eyes properly.
19. Measure the maximal vessel length of neovascularization zone from the limbal vascular plexus toward the pellet with a linear reticule through the microscope (equipped with the microscope).
20. Measure the contiguous circumferential zone of neovascularization as clock hours (if the eye is visualized as a clock).

2. Inhibition of Angiogenesis

Steps

1. Repeat steps 1–16 described in Section III,C,1.
2. Systemically inject potential angiogenesis inhibitors into mice with angiogenic factor-implanted eyes. The injection routes include intravenous, intraperitoneal, subcutaneous, and intramuscular injections. The injected volume usually should not exceed 0.1 ml/10 g mouse.
3. The duration and frequency of treatment are dependent on the potency and the half-lives of tested angiogenesis inhibitors.
4. In a control group of mice, inject the relevant buffer that is used to suspend the angiogenesis inhibitors.
5. The examination of corneal neovascularization is described in steps 17–20 in Section III,C,1 (see Fig. 2).

IV. COMMENTS

Endothelial cell division is an essential process for the growth of new blood vessels. Therefore, the endothelial cell proliferation assay is the most relevant and reliable assay system used to detect endothelial cell growth. As new blood vessels usually bud from microvessels, endothelial cells isolated from capillaries are most optimal for endothelial cell growth assay. This article described a method that employs BCE cells as primary endothelial cells for the *in vitro* proliferation assay. Among *in vivo* angiogenesis models, the mouse corneal angiogenesis model is the most rigorous and clear system to detect angiogenic responses. As the corneal organ is naturally avascular, corneal neovascularization indicates that all blood vessels are newly formed vessels. In contrast to the CAM assay, corneal angiogenesis is devoid of preexisting blood vessels. Thus, this system is convenient for studying new blood vessel formation, new blood vessel stability, and structures of newly formed vessels. For example, a recent study of blood vessel stability in the cornea has found that not all angiogenic factors can stabilize the newly formed vasculature (Cao *et al.*, 2003). As most transgenic or knockout animal models are performed in mice, the mouse corneal model is very valuable in detecting the impact of overproduction or loss of a particular gene on new blood vessel formation. With exceptional skills, the corneal angiogenic responses in knockout mice can be detected in newborns (Zhou *et al.*, 2000). Although the CAM assay is not an ideal quantitative angiogenesis assay, this assay is a fast screening system that allows the determination of angiogenic or antiangiogenic responses within

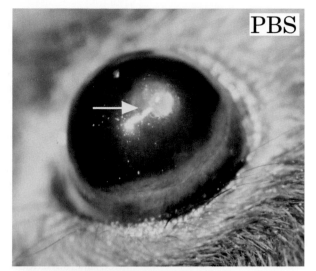

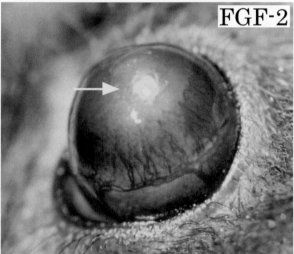

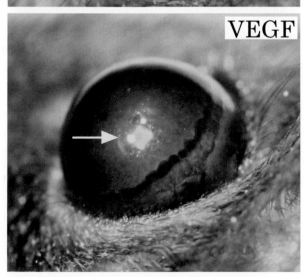

FIGURE 2 Typical examples of FGF-2- and VEGF-induced corneal neovascularization on day 5 after implantation. (Top). PBS buffer and sucralfalte polymer alone without angiogenic factors. (Center) FGF-2 at a 80-ng dose induces intense neovascularization originating from the limbal vessels toward the implanted pellet. (Bottom) VEGF at a 160-ng dose induces a robust neovascularization response.

a relatively short time. In addition, the CAM assay requires less sophisticated equipment and less cost.

V. PITFALLS

A. Endothelial Cell Proliferation Assay

1. The success of *in vitro* endothelial cell proliferation assay is entirely dependent on the number of population doublings (PDLs) of the BCE cells. In principle, low PDLs of BCE cells will increase successful rates. Primary BCE cells will stop their responses to angiogenic stimuli around the age of 40 PDLs. Ultimately, these cells enter into a senescent state at the average of 45 PDLs. Therefore, it is important to avoid using presenescent BCE cells in endothelial cell proliferation assay.

2. Primary endothelial cells isolated from different tissues or organs may react to different mitogens. Indeed, a tissue-specific endothelial growth factor has been identified (LeCouter *et al.*, 2001). Thus, it is important to determine the stimulatory/inhibitory activity of a compound using primary endothelial cells isolated from different tissues.

3. Not all angiogenic factors are able to induce endothelial cell proliferation or migration due to the diverse biological functions of various angiogenic factors. For example, FGF-2 is a potent angiogenic factor that preferentially stimulates endothelial cell proliferation but not cell migration. In contrast, VEGF is a potent endothelial chemotactic factor but a poor endothelial mitogen (Yoshida *et al.*, 1996). Thus, the stimulatory effect of an angiogenic factor should be tested in various endothelial cell assays.

4. In endothelial cell inhibition assays, it is critically important that low concentrations of angiogenic factors be used. For example, in the BCE cell assay, usually 1 ng/ml of FGF-2 is used to stimulate cell growth. If higher concentrations of FGF-2 were used, angiogenesis inhibitors would not be able to counteract the stimulatory effect.

B. CAM Assay

1. Make sure that the entire experimentation is performed with optimal temperature and humidity. The chick embryos are extremely vulnerable to pathogen infections. Thus, a strict sterile condition is required, including washing hands and wearing gloves when handing eggs and embryos.

2. Use fresh fertilized eggs, which should not be more than 3 days old, as the survival rate of embryos will be reduced dramatically.

3. Do not shake or crack eggs by hitting them against other solid objects.

4. Avoid cutting the embryonic vessels by the sharp edge of the egg shell.

5. Do not keep embryos with nonintact yolk.

6. Avoid hemorrhages of the CAM during the implantation of meshes.

7. The release half-life of small chemical compounds immobilized on the nylon mesh can be different. Thus, it is important to analyze the results at different time points.

C. Mouse Corneal Assay

1. Do not choose aged mice that usually produce delayed angiogenic responses.
2. Avoid using Balb/C or other white background mice. Their red-color eyes may disturb the quantification of corneal neovascularization.
3. Avoid accommodating many mice, especially male mice, in one case. Their unusual behaviors may interrupt corneal neovascularization.
4. Prepare corneal implants under sterile conditions, as corneal inflammation could cause neovascularization.

References

Cao, R., Brakenhielm, E., Pawliuk, R., Wariaro, D., Post, M. J., Wahlberg, E., Leboulch, P., and Cao, Y. (2003). Angiogenic synergism, vascular stability and improvement of hind-limb ischemia by a combination of PDGF-BB and FGF-2. *Nature Med.* **31**, 31.

Cao, R., Brakenhielm, E., Wahlestedt, C., Thyberg, J., and Cao, Y. (2001). Leptin induces vascular permeability and synergistically stimulates angiogenesis with FGF-2 and VEGF. *Proc. Natl. Acad. Sci. USA* **98**, 6390–6395.

Cao, R., Wu, H. L., Veitonmaki, N., Linden, P., Farnebo, J., Shi, G. Y., and Cao, Y. (1999). Suppression of angiogenesis and tumor growth by the inhibitor K1–5 generated by plasmin-mediated proteolysis. *Proc. Natl. Acad. Sci. USA* **96**, 5728–5733.

Cao, Y. (2001). Endogenous angiogenesis inhibitors and their therapeutic implications. *Int. J. Biochem. Cell. Biol.* **33**, 357–369.

Cao, Y., and Cao, R. (1999). Angiogenesis inhibited by drinking tea. *Nature* **398**, 381.

Cao, Y., Chen, A., An, S. S., Ji, R. W., Davidson, D., and Llinas, M. (1997). Kringle 5 of plasminogen is a novel inhibitor of endothelial cell growth. *J. Biol. Chem.* **272**, 22924–22928.

Cao, Y., Ji, R. W., Davidson, D., Schaller, J., Marti, D., Sohndel, S., McCance, S. G., O'Reilly, M. S., Llinas, M., and Folkman, J. (1996). Kringle domains of human angiostatin. Characterization of the anti-proliferative activity on endothelial cells. *J. Biol. Chem.* **271**, 29461–29467.

Cao, Y., Linden, P., Farnebo, J., Cao, R., Eriksson, A., Kumar, V., Qi, J. H., Claesson-Welsh, L., and Alitalo, K. (1998). Vascular endothelial growth factor C induces angiogenesis *in vivo*. *Proc. Natl. Acad. Sci. USA* **95**, 14389–14394.

Folkman, J. (1995). Angiogenesis in cancer, vascular, rheumatoid and other disease. *Nature Med.* **1**, 27–31.

Hanahan, D., and Folkman, J. (1996). Patterns and emerging mechanisms of the angiogenic switch during tumorigenesis. *Cell* **86**, 353–364.

Jain, R. K., Schlenger, K., Hockel, M., and Yuan, F. (1997). Quantitative angiogenesis assays: progress and problems. *Nature Med.* **3**, 1203–1208.

Kenyon, B. M., Voest, E. E., Chen, C. C., Flynn, E., Folkman, J., and D'Amato, R. J. (1996). A model of angiogenesis in the mouse cornea. *Invest. Ophthalmol. Vis. Sci.* **37**, 1625–1632.

LeCouter, J., Kowalski, J., Foster, J., Hass, P., Zhang, Z., Dillard-Telm, L., Frantz, G., Rangell, L., DeGuzman, L., Keller, G. A., Peale, F., Gurney, A., Hillan, K. J., and Ferrara, N. (2001). Identification of an angiogenic mitogen selective for endocrine gland endothelium. *Nature* **412**, 877–884.

Rupnick, M. A., Panigrahy, D., Zhang, C. Y., Dallabrida, S. M., Lowell, B. B., Langer, R., and Folkman, M. J. (2002). Adipose tissue mass can be regulated through the vasculature. *Proc. Natl. Acad. Sci. USA* **99**, 10730–10735.

Yoshida, A., Anand-Apte, B., and Zetter, B. R. (1996). Differential endothelial migration and proliferation to basic fibroblast growth factor and vascular endothelial growth factor. *Growth Factors* **13**, 57–64.

Zhou, Z., Apte, S. S., Soininen, R., Cao, R., Baaklini, G. Y., Rauser, R. W., Wang, J., Cao, Y., and Tryggvason, K. (2000). Impaired endochondral ossification and angiogenesis in mice deficient in membrane-type matrix metalloproteinase I. *Proc. Natl. Acad. Sci. USA* **97**, 4052–4057.

47

Three-Dimensional, Quantitative *in vitro* Assays of Wound Healing Behavior

David I. Shreiber and Robert T. Tranquillo

I. INTRODUCTION

Wound healing *in vivo* is a dynamic process involving the coordinated regulation of cell proliferation, cell migration, cell traction, and apoptosis (Clark, 1996). For instance, during dermal wound healing, inflammatory cells are induced to infiltrate a wound site primarily by factors released from platelets. Fibroblasts are stimulated to migrate up a chemotactic gradient of soluble factors, and possibly a haptotactic gradient of matrix-bound factors, released by the inflammatory cells and platelets into a provisional matrix composed primarily of fibrin and fibronectin. These fibroblasts proliferate and secrete collagen and other extracellular matrix molecules to form granulation tissue. The cells often contract this granulation tissue while continuing to secrete collagen. Ultimately, the cells die through apoptosis and leave a dense, collagenous, acellular scar as a reparative patch. The progression of fibroblast behavior is dictated in part by cues from the wound healing environment, such as soluble growth factors, integrin binding to network proteins, and mechanical stress associated with wound contraction. Therefore, it becomes a great challenge to design and implement bioassays that capture quantitatively the key features of wound healing in a controlled, but physiologically relevant manner. This article describes several assays that allow quantitative evaluation of fundamental aspects of cell behavior involved in the wound healing response—cell migration, chemotaxis, cell traction, and cell proliferation—in controlled environments with improved physiological relevance. The relevance of the assays is improved by examining cellular phenomena within three-dimensional (3D) hydrogels of biopolymers involved in wound healing, namely type I collagen and fibrin. Many studies have demonstrated dramatic differences in tissue cell behavior when cultured in a 3D gel rather than on a 2D substrate (Bell *et al.*, 1979; Nusgens *et al.*, 1984).

II. MATERIALS

Trypsin (Product No. T6763), paraformaldehyde (Product No. 158127), ethylenediaminetetraacetic acid (EDTA) (Product No. E26282), $CaCl_2$ (Product No. 21075), NaOH (Product No. 72079), bovine fibrinogen (Product No. 46312), bovine thrombin (Product No. T4265), and agarose (Product No. A2790) are from Sigma Chemical Company (St. Louis, MO). Vitrogen 100 bovine type I collagen (Product No. FXP-019) is from Cohesion Technologies, Inc. (Palo Alto, CA). Tissue culture medium, penicillin/streptomycin (pen-strep; Cat. No. 15070063, fungizone (Cat. No. 15240062), HEPES buffer (Cat. No. 15630080), phosphate-buffered saline (PBS; Cat. No. 10010023), and L-glutamine (Cat. No. 21051024) are from GIBCO Laboratories (Grand Island, NY). Fetal bovine serum (FBS; Product No. SH30073.02) is from HyClone Laboratories (Logan, UT). Polystyrene beads (Product No. 64130) are from Polysciences, Inc. (Warrington, PA). Stock Teflon (Product No. B-ZRT-2), stock polycarbonate (Product No. B-211040), and stock hydrophilic porous polyethylene disk (Product No. B-PEH-060/50) are from Small Parts (Miami Lakes, FL).

III. EQUIPMENT

Inverted microscope with computer-controlled stage and on-stage incubation system, biological hood, air or CO_2 incubator.

IV. METHODS

A. Biopolymer Gel Solution Preparations

Type I Collagen Gel (2.0 mg/ml)

Collagen gels are prepared by neutralizing stock type I collagen solution and raising the temperature to facilitate self-assembly of monomeric collagen into fibrils and forming an entangled network of fibrils with interstitial medium (Knapp et al., 1997).

Steps

1. To make 1 ml of collagen, add the following reagents to a 15-ml conical tube in order in a biological safety cabinet/laminar flow hood under sterile conditions: 20 μl 1 M HEPES buffer, 132 μl 0.1 N NaOH, 100 μl 10X MEM, 60 μl FBS, 1 μl pen/strep, 10 μl L-glutamine, and 677 μl Vitrogen 100.
2. Mix gently by pipetting.
3. Keep the solution on ice until ready to prepare the assay.

Note: The final solution should be a light pink color/red color indicating neutral pH.

B. Fibrin Gels (3.3 mg/ml)

Fibrin gels are prepared by enzymatically cleaving fibrinogen with thrombin in the presence of Ca^{2+} ions (Knapp et al., 1999).

Solutions

1. *Fibrinogen solution* A: Dissolve fibrinogen powder in a 20 mM HEPES-buffered saline solution to a concentration of 30 mg/ml. Pass the solution through a 0.20-μm filter. Store in 1-ml aliquots at −80°C.
2. *Thrombin solution* B: Dissolve thrombin (250 units) in 1 ml sterile water and 9 ml of PBS. Pass the solution through a 0.20-μm filter. Store in 100-μl aliquots at −80°C.

Gel Preparation

1. Add 1 aliquot of fibrinogen solution A to 5 ml 20 mM HEPES-buffered saline to make a 5-mg/ml fibrinogen solution.
2. In a separate vessel, add one aliquot of thrombin solution B to 1 ml unsupplemented M-199 no serum and 15 μl of 2 M $CaCl_2$ solution to make solution C.
3. Prepare cell suspension D in cell culture medium with the cell concentration six time the desired final concentration. Solution D will be diluted 1:6.
4. Keep the solutions separated and on ice until you are ready to prepare the assay.
5. To make the fibrin gel, mix one part of thrombin/Ca^{2+} solution C, one part of cell suspension D, and four parts of fibrinogen solution A. Mix gently by pipetting and fill the assay chamber quickly.

Pitfalls

Frequently, mixing of fibrin and collagen solutions generates many bubbles, which can affect the geometry and rheology of gels and blur microscopy images. To limit bubbles, apply a vacuum to conical tubes holding the solutions (for collagen, degas the solution after mixing but before gelation, for fibrin, degas the fibrinogen solution, thrombin solution, and cell culture medium before mixing) to draw dissolved air out of solution and into the vacuum. Make sure that the solution is not too close to the top of the conical tube (10 ml or less).

Because the fibrin gel can form quickly, add the fibrinogen solution last and pipette the mixed solution into the assay chamber quickly.

Disrupting forming gels can affect structural and rheological properties that are important for cell behavior assays. Take care when handling the gels during and after formation.

C. Cells

The following assays were developed to examine phenomena associated with dermal wound healing and therefore incorporate dermal fibroblasts as the cell of choice. However, the assays can be adapted easily for other cell types (e.g., smooth muscle cells or corneal fibroblasts.) The key is to maintain tight control over cell populations and the cell density used in the assays, as cell behavior can vary widely from passage to passage, and cell density dramatically affects the potential for cell–cell signaling, which, if not accounted for, can cloud results. Cell densities delineated below are recommended but should be optimized according to the cell type and the phenomena to be studied.

V. CHEMOTAXIS ASSAY

Chemotaxis experiments are performed in a conjoined 3D gel system (Knapp *et al.*, 1999; Moghe *et al.*, 1995). Manufacture of chemotaxis chambers and image analysis are involved. The experiments require machining of chemotaxis chambers (see later). The chambers allow the generation of a gradient of a protein/growth factor-sized diffusible species (Fig. 1). Briefly, one-half of the chamber is filled with collagen or fibrin solution and a defined concentration of chemotactic species, and the other, initially separated by a thin divider, is filled with an equal volume of biopolymer solution with a defined density of cells of interest. After removing the divider, a gradient of the species is formed in the gel with cells.

1. Have chemotaxis chambers (at least 6–10) machined according to Fig. 2.
   a. Chambers are designed to fit on top of a standard microscope slide (7.5 × 2.5 cm).
   b. Chambers have a groove at the midline for a thin, Teflon divider.
   c. Thoroughly clean chambers with soapy water and autoclave prior to each use

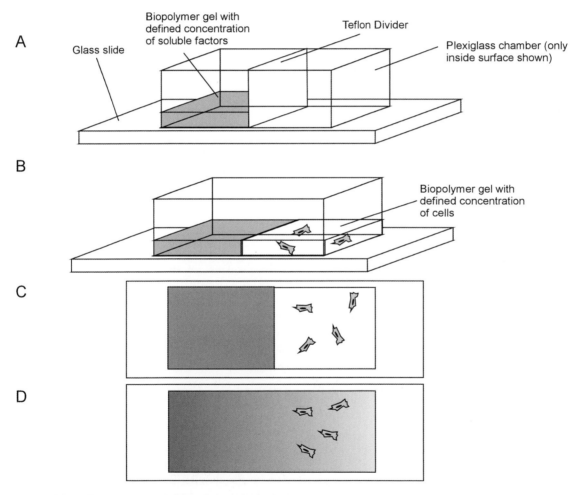

FIGURE 1 Schematic of linear chemotaxis chamber preparation. The chamber consists of a hollow, plexiglass box that is sealed to a standard glass slide with vacuum grease. A Teflon plate divides the chamber into two sections. (A) One side of the chamber is filled with biopolymer gel with a defined concentration of chemotactic factor. For collagen assays, the chamber is placed in the incubator, and the biopolymer solution is allowed to gel. (B) The Teflon divider is removed, and the other half of the chamber is filled with biopolymer solution with a defined cell concentration. The chamber is returned to the incubator to facilitate gelation. (C) Initially, all of the chemotactic factor is in the left half of the chamber, and the cells in the right half are oriented randomly. (D) Over time, the soluble factor diffuses into the right half, and the cells (if responsive to the soluble factor) reorient and migrate in the direction of the gradient.

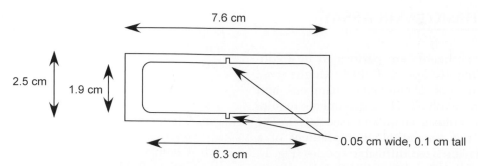

FIGURE 2 Mechanical drawing of linear chemotaxis chamber. The chamber is machined from plexiglass (polycarbonate) and is 1.3 cm high. A Teflon plate ($1.2 \times 0.05 \times 2$ cm) is also required to fit into the notched area and divide the chamber in half.

2. Sterilize all chambers and glass slides.

3. Under sterile conditions, secure bottom of chamber to glass slide with vacuum grease.

4. Place Teflon dividers into grooves.

5. Prepare solutions for assay (either fibrin or collagen).

   a. Each chamber requires ~3.5 ml biopolymer solution, which should be divided into two equal volumes of 1.75 ml/chamber.

   b. Add enough chemotactic factor to one of the volumes of the solution to generate the desired final concentration. Generally for growth factors, the working range is 0.1–100 ng/ml. This is now called "solution A."

   c. Add cells to the other half of the solution to a final concentration of 10,000 cells/ml. This is now called "solution B."

6. Add 1.75 ml of solution A to one-half of each chamber. This should fill the half approximately 3 mm.

7. Secure another glass slide to the top of each chamber and place the chambers in a humidified incubator until gelation/self-assembly is complete.

8. After gelation, remove chambers from the incubator and, again under sterile conditions, remove the Teflon divider.

9. Mix solution B by pipetting to ensure uniform distribution of cells and fill the empty half of the chamber with 1.75 ml of solution B.

10. Replace top glass slide, ensuring a good seal, and place back in humidified incubator for 24–36 h

11. At desired time points (typically 12–36 h after gelation), place chamber under microscope and capture images of all cells through the thickness of the gel. This should be done with sufficient objective power ($4\times$ or greater) to observe the orientation of cells. This is facilitated greatly using automated microscopy/confocal microscopy with a motorized stage to build a mosaic, but can be done by hand if necessary. With automation, a projection mosaic of the gel containing cells is generated incorporating all planes throughout the thickness of the gel.

12. For each cell, using standard image analysis packages (such as NIH Image), draw a line segment representing the prevailing orientation of the cell (Fig. 3) and calculate and record the angle, θ (from 0° to 90°), the line segment makes with the horizontal axis, which represents the direction of the chemotactic gradient.

13. For each cell, calculate $\sin^2(\theta)$

14. Determine the average value of $\sin^2(\theta)$ for all of the cells in a given chamber. In other words,

$$\Phi = \sum_{i=1}^{n} \frac{\sin^2 \theta_i}{i} \qquad (1)$$

where n represents the number of cells (Fig. 3).

In the case of no chemotactic factor, cells should be oriented randomly. Therefore, the average angle should be 45° and $\Phi = 0.5$. For a pure chemotactic response, $\theta = 0°$ and $\Phi = 0$. Generally, values of $\Phi < 0.5$ signify a chemotactic response, but results should be analyzed statistically after measuring the response in multiple chambers.

Helpful Hints and Pitfalls

1. Control experiments should include no chemotactic factor, as well as uniform chemotactic factor (half-loading in both solution A and solution B).

2. If imaging requires a large block of time, gels can be fixed with 2–4% paraformaldehyde to ensure equal exposure times to chemotactic gradients across experiments.

3. To avoid settling of cells to the bottom of the gel due to gravity, prewarm solution B to ~28–30°C prior to filling the second half of the chamber. This is more crucial for experiments with collagen.

4. Alignment of cells in the direction of the gradient may indicate a negative chemotaxis response. If the

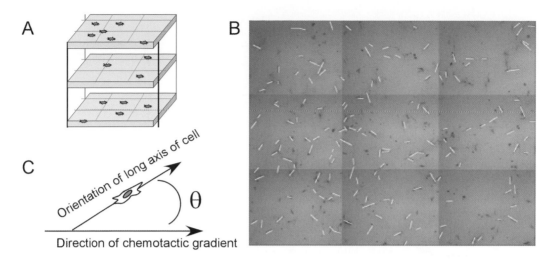

FIGURE 3 Chemotaxis analysis. (A) After a period of time (or at regular intervals, if desired), a composite, mosaic image of the cell-populated half of the chamber is captured and projected into one image. (B) The orientation of each cell is defined by tracing the long axis of the cell with a line. (C) The orientation of each line is compared to the orientation of the gradient via the angle θ. The response of the population is then evaluated using Eq. (1).

polarity of the cells cannot be determined easily, it is necessary to record the migration of the cells to confirm that they migrate toward the source of the factor (positive chemotaxis or chemoattraction) or away from the source (negative chemotaxis or chemorepulsion).

A similar assay can be developed to study chemotaxis toward peptide fragments much smaller than a growth factor (e.g., RGD) (Fig. 4) (Knapp *et al.*, 1999). Because these species are much smaller, they diffuse more rapidly and would equilibrate in concentration if the linear chambers described earlier were used. By maintaining a divider between the two halves with a small notch in the divider, diffusion across the two halves is restricted and gradients can be maintained for at least 24 h.

1. Machine chambers identical to those described earlier, but add a small notch to the Teflon divider.
2. When preparing the chamber, include a glass coverslip alongside the Teflon divider.
3. Fill one-half of the chamber with biopolymer solution with a defined concentration of the peptide sequence and allow to gel.
4. Fill the other half with biopolymer solution with defined cell concentration and allow to gel.
5. Remove coverslip, but leave notched Teflon divider in place. This should create radial gradients (see Fig. 4).
6. Transfer to microscope and quantify as described previously, with the exception that the direction of the gradient is now radially outward from the notch.

VI. CELL TRACTION ASSAY

Simple cell traction assays can be performed with either mechanically constrained, stressed gels or unconstrained, unstressed gels in various geometries (Neidert *et al.*, 2002; Ehrlich and Rajaratnam, 1990). This section describes the simplest one to implement— cylindrical disks (Neidert *et al.*, 2002; Tuan *et al.*, 1994). The geometry is a hemisphere initially, but evolves to a cylindrical disk shape as the cell traction proceeds.

1. With a sterile scribe, score a 1-cm-diameter circle in each well of a six-well plate.
2. Prepare collagen or fibrin solution with a final cell concentration between 10,000 and 500,000 cells/ml.
3. Carefully pipette 0.5 ml collagen or fibrin solution into the scored region. The scratch in the culture plate should force the solution to maintain shape.
4. Carefully place the plate in an incubator and allow solution to gel.
5. After gelation, fill each well with 2.5 ml of culture medium with defined concentrations of the desired soluble factors.
6. If the assay is for unstressed gels, with a sterile, flat spatula, gently pry the gels off of the bottom surface so that they are "free floating."
7. Transfer the plate to a microscope and measure the thickness of the gel. When using a microscope equipped with motorized focus, this is done most easily by focusing on the top of the gel, recording the motor position, and then focusing on the bottom of the

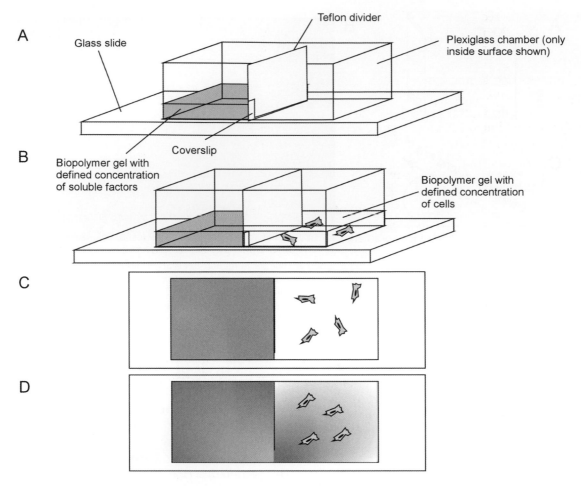

FIGURE 4 Schematic of radial chemotaxis assay for small chemotactic molecules. The assay is similar to the linear chambers, except that the Teflon divider remains in the chamber and has a small notch to restrict diffusion between the two halves. A glass coverslip prevents premature diffusion.

gel and determining the change in motor position, and therefore the thickness. For unstressed gels, also measure the diameter of the gel. (The diameter should remain unchanged for stressed gels.)

8. Return the gels to the incubator.

9. Repeat measurement performed in step 7 at regular intervals of your discretion. Once every 1–2 h is generally sufficient for a short duration (12–24 h) experiment. Once every 4–6 h is sufficient for longer duration experiments. Return the samples to the incubator immediately after recording the thickness.

10. After completion, normalize results by the original dimension and plot percentage compaction vs time for the various conditions.

Remarks

1. Gel thickness measurements reflect the gel rheology (and possibly cell proliferation) as much as cell traction. An intrinsic measure of the latter can be

obtained by analyzing data using a mechanical model for cell–matrix interactions (Barocas *et al.*, 1995; Barocas and Tranquillo, 1997a,b).

2. If only measurements are desired for the constrained case, the sample can be attached to a force transducer (Eastwood *et al.*, 1996; Kolodney and Wysolmerski, 1992).

3. Cell alignment that typically results during the contraction of mechanically constrained gels can complicate the interpretation of data (Barocas and Tranquillo, 1997a,b).

VII. CELL MIGRATION

Time-lapse cell migration assays in a 3D matrix can be performed to evaluate cell migration in two or three dimensions (Knapp *et al.*, 2000; Shreiber *et al.*, 2001,

2003). Both techniques require an automated image analysis system and an XY motorized stage. Motorized focus is advantageous for evaluation in 2D and is required for 3D.

Generally, it is advised to perform each experimental condition in triplicate. For instance, if the assay was to determine the effects of PDGF-BB on fibroblast migration and the experiment called for testing cells in the presence of 0, 10, 20, and 50 ng/ml, then 12 wells would be needed.

The migration assay can be performed in practically any assay chamber. It is written here for a 96-well plate arrangement.

In these assays, data are recorded to fit to the persistent random walk model of cell migration (Dunn and Brown, 1987). This model implies that cell migration is purely random over long periods of time, but can have persistent direction over short period of times. To assess cell migration, there are three parameters, two of which are independent: persistence time, P; cell speed, S; and cell motility, μ. These three parameters are related via calculation of the mean-squared displacement (MSD) of the cells:

$$\text{MSD} = \langle d^2(t) \rangle = 2 n_d \mu [t - P(1 - e^{-1/p})]$$
$$S = \sqrt{\mu n_d / P} \tag{2}$$

where n_d = number of dimensions tracked (two for X–Y tracking, three for X–Y–Z tracking).

Assays can be designed to improve measurement accuracy of any of these three parameters (Dickinson and Tranquillo, 1993); it is left to the researcher to decide which is most appropriate and design accordingly. Briefly, in order to evaluate the persistence time accurately, cell positions must be recorded at a time lapse significantly less than the actual persistence time in order to observe the directed motion of the cell. For example, if the persistence time of a cell is estimated to be 30 min (on average it changes direction in its motion every half-hour), then to observe this phenomenon, cell position must be monitored at a minimum every 5–10 min, as the cell will appear to be moving randomly without any directional persistence. However, if the position of each cell needs to be recorded every 5 min, then the total number of cells that can be monitored is decreased. (More images of wells of a 96-well plate or individual cells can be captured in 10 min than in 5 min.) Thus, the amount of data that is averaged to determine S, P, and μ is decreased. The timing of the imaging sequence must be determined by the individual laboratories and can be influenced by the imaging hardware and software and the inherent motility of the cells. Generally, this timing issue is negligible for 2D analysis and only

becomes a problem in 3D, where each cell is generally monitored individually with high magnification.

Both 2D and 3D analyses require complex image analysis codes that can perform object identification and/or correlation. Many software packages now include such algorithms, and they can also be programmed by the individual laboratories. The actual codes will differ according to the software packages, and presentation of a code is beyond the scope of this article. A brief outline for 2D and 3D is presented.

The general steps are the same for 2D and 3D analysis for preparing the assays.

1. Prepare collagen or fibrin solution with a final cell concentration between 7500 and 30,000 cells/ml and 5000–10,000 10-µm polystyrene after beads/ml. Beads serve as fiduciary markers to allow measurement of any drift in the stage or movement of the gel. They are especially crucial for 3D, high-magnification tracking and analysis.

2. Pipette 100 ml of collagen or fibrin solution into a well of the 96-well plate. Fill as many wells as warranted to complete the experimental test matrix.

3. Carefully place the plate in an incubator and allow solution to gel.

4. After gelation, fill each well with 100 ml of cell culture medium with twice the defined concentration of the desired soluble factor(s) to yield the correct final concentration.

5. Transfer the plate to an inverted microscope that includes an automated image analysis system and motorized stage, potentially motorized focus, and an environmental chamber to maintain proper humidity and CO_2 concentration. Air-buffered media can be used (e.g., M199) if on-stage CO_2 regulation is unavailable.

The followings steps are used for analysis of migration in 2D (without automated focus).

1. Using a 10x objective, select a focal plane that is consistent among all of the wells.

2. Move the stage from desired well to well and record images of each well. Build a mosaic image if necessary. Develop a numbering scheme to save the images (e.g., [date]_[condition]_[sample]_[image number]). Be sure to record the centroid position of each well so that the computer can move the microscope stage to those same positions at future time points.

3. Determine the total duration of an individual interval, and the total number of intervals, which define the overall length of the time-lapse experiment. Be certain that your system is capable of capturing the desired number of images required for one complete

interval in the time allocated. Generally, intervals should be as short as possible, so it frequently helps to work backward and determine how quickly the desired number of images can be captured. In this case, be sure to include a safety factor (usually a minute or two is sufficient to ensure that all required images are captured).

4. Instruct the computer to return to each position, in sequence, record an image(s) of the well, move to the next position, etc. After the last picture in one interval has been recorded, the computer should instruct the stage to wait until the beginning of the next time interval to return to the first well. For instance, if a time interval is 10 min and all pictures from an interval are captured in 8 min, the computer should wait 2 min to begin the next interval instead of immediately beginning the next time interval. Maintaining a consistent time interval simplifies data analysis greatly.

5. Object identification can be performed during an experiment or off-line. In either case, the general scheme is the same: (a) capture image and (b) filter image to accentuate "objects," i.e., cells and beads.

Filtering usually involves

i.   Equalizing contrast.
ii. Low-pass filter.
iii. Edge detection filter (e.g., Sobel).
iv. Generating a binary image based on an appropriate gray-scale level.
v.   Rejecting objects that are too big and/or too small.
vi. Recording the centroid position of objects, and correlating those positions, and possible shapes to the previous interval to track individual cells properly. Shape matching is not necessarily advised for tracking cells, as they change morphology during migration, but will certainly work for beads.
vii. Tabulate the cell/bead positions in a text file in a rational sequence. Include a flag to represent whether the object is a bead (1) or a cell (0). For instance.

Interval 1, Well 1, Object (cell or bead) 1, Type of object (Cell = 0, Bead = 1), Xposition 1, Yposition 1
Interval 1, Well 1, Object 2, Type, Xposition 1, Yposition 1 . . .
Interval 1, Well 2, Object 1, Type, X1, Y1
Interval 1, Well 2, Object 2, Type X1, Y1 . . .
Interval 2, Well 1, Object 1, Type X2, Y2
Interval 2, Well 1, Object 2, Type X2, Y2 . . .
Proceed to step 10.

Use the following steps for analysis of migration in 3D (must have automated focus).

6. Use the lowest power objective that generally allows you to focus on any object so that it is the only object in the volumetric field of view (FOV). To do this:
   a. Scan through the sample until you find an object.
   b. Maneuver the stage so that the object is in the center of the FOV and in focus.
   c. Make sure that the object is the only one in the volume immediately surrounding that object in all directions (including the focal plane, Z).
   d. Recenter and focus the object and record the position of the stage and focus on the computer

7. Select the cells and beads to track. Optimally, search for a bead with many (at least three to four; the more the better) cells within 200–400 μm of the bead, but not within the FOV. The motion of that bead would then represent the local absolute motion of the stage and gel to allow the subtraction of any convective effects that occur due to gel contraction.
   a. Find a bead that is the only object in its volumetric FOV.
   b. Scan the regions immediately outside that FOV for cells. Find beads with multiple cells in the immediately adjacent FOV.
   c. Return to the bead. Focus and center the bead and then record the X, Y, Z position with the computer
   d. Move to the cells in the adjacent volumes, focus and center each cell, and record the X, Y, Z position of each cell.
   e. Proceed to another bead and repeat

As with 2D tracking and analysis, there are opportunity costs that allow the optimization of the number of assay conditions, the number of beads/cells tracked per condition, and the duration of a single interval. This optimization is even more crucial for 3D tracking. Ideally, at least 20–30 cells are tracked for each condition.

8. When finished identifying the final object to be tracked, program the computer to return the stage to the first cell.
   a. Execute an autofocus routine to focus the object. If it has moved from the center, recenter.
   b. Run the object identification algorithm (see earlier discussion) to locate the centroid of the object.
   c. Record the position of the object and write the position of the object to a database. Include a flag to represent whether the object is a bead (1) or a cell (0). For instance: Interval #, Well #, Object #(cell or bead), Type (Cell = 0; Bead = 1), Xposition, Yposition, Zposition
   d. Automatically move to the next object and repeat until all objects are recentered.

9. Initiate the time-lapse loop. The time-lapse loop essentially performs the same procedures as step 8, but

the computer recenters the object. The main steps that need to be programmed are as follows.

  a. Move to the recorded XYZ position of an object, $X_iY_iZ_i$.
  b. Search in the Z direction for that object using an autofocus routine, returning the objective to the position of best focus. This is the new Z position of the cell, Z_{i+1}.
  c. Identify the centroid of the object with the object identification algorithm.
  d. Calculate the X and Y distance between the position of the centroid and the center of the FOV in pixels. Convert this to micrometers or stage position units and add this to the current stage position. This is the new XY position of the object, $X_{i+1}Y_{i+1}$.
  e. Move the stage to the new XY position, $X_{i+1}Y_{i+1}$, which you just calculated.
  f. Proceed to the next object and complete all objects for a given interval.
  g. Wait until the duration of the interval is over before returning to the first object. See 2D tracking and analysis for an explanation.
  h. Following completion of the time lapse, proceed to step 10.

The reasoning behind the high magnification cell-tracking algorithm is that in a given time interval, a cell cannot migrate fast enough to exit the FOV scanned during the autofocus. By recentering the object at each time interval, the likelihood of finding that object successfully in the next interval is maximized (see Fig. 5 for a schematic of the high magnification tracking algorithm).

10. Cell track analysis. Analysis requires three general steps: (A) correcting cell tracks for stage error and gel compaction (cell convection), (B) generating mean-squared displacement data, and (C) fitting data to the persistent random walk model.

A. Correcting cell tracks

a. Subtract initial position for each object from subsequent positions for that object. Each object should now begin at 0,0,0.
b. Subtract the XYZ position of the bead from XYZ colocalized cells for each time interval.
c. Write a new file of corrected object positions.

B. Calculating MSD

$$MSD = \Delta x^2 + \Delta y^2 + \Delta z^2 \qquad (3)$$

Mean-squared displacement data can be generated in two manners: overlapping intervals and nonoverlapping intervals (Fig. 6). Overlapping intervals generates more data but introduces covariance into the

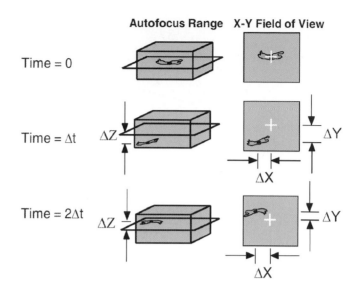

Autofocus Range   X-Y Field of View

FIGURE 5 Schematic of autofocus and cell-tracking scheme for high magnification tracking in 3D. Initially, cells are focused and positioned in the center of the field of view. At the next time interval, the computer-controlled stage and focus move to the previously recorded position and execute an autofocus routine to locate the cell and determine the distance moved in the Z direction. A snapshot is taken, the centroid of the cell is located, and the X and Y distances moved by the cell are recorded. The process is repeated for the next and subsequent intervals, always returning the stage to the centroid recorded at the previous interval for each cell.

error. Analyses have been formulated to account for these (and other) complex errors (Dickinson and Tranquillo, 1993), but a full discussion is beyond the scope of this article. The two techniques are discussed briefly, and then we proceed assuming we are using overlapping intervals.

In an MSD calculation, we generate X-Y data that address the question: How far did an object, on average, migrate over a time interval of a given duration? In analyzing cell migration, we assume that cells have no inertia. That is, the fact that a cell is moving at a particular speed in a particular direction at time 1 has no assumed influence on the speed or direction at time 2, or any future time. Therefore, to determine the average distance traveled by a cell over one time interval, we average all displacements that occurred over 1 time interval. Thus, if the time lapse were four intervals long (five time points), we would calculate:

From $t = 1$ to $t = 2$:   MSD1
From $t = 2$ to $t = 3$:   MSD2
From $t = 3$ to $t = 4$:   MSD3
From $t = 4$ to $t = 5$:   MSD4

MSD for 1 time interval = average (MSD1–4)

We then average these over all of the cells in a given condition to arrive at the MSD value for that number of time intervals.

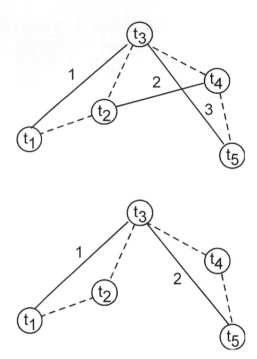

FIGURE 6 Overlapping vs nonoverlapping intervals for calculating mean-squared displacement (MSD). In this example, an object is imaged five times (t1–t5) over four time intervals. Suppose we are determining the MSD over two time intervals. With the overlapping interval procedure (top), the distance traveled by the object during two time intervals can be measured three times (solid lines), whereas with the nonoverlapping technique, the same measurement can only be made twice.

The use of overlapping vs nonoverlapping intervals arises when calculating the MSD over durations greater than 1 time interval (Fig. 6). For instance, in the example just given, if we want to calculate the MSD that occurs over 2 time intervals, we have two choices:

Option 1: Overlapping intervals
   From $t = 1$ to $t = 3$:   MSD1
   From $t = 2$ to $t = 4$:   MSD2
   From $t = 3$ to $t = 5$:   MSD3

Option 2: Nonoverlapping intervals
   From $t = 1$ to $t = 3$:   MSD1
   From $t = 3$ to $t = 5$:   MSD2

As can be seen, overlapping intervals result in 3 MSD measurements to average, while nonoverlapping results in only 2. However, the overlapping intervals are interdependent; any error that occurs from interval 2 to 3 will be included in the calculation from interval 1 to 3 and from interval 2 to 4. Generally, overlapping intervals are preferred as the improvement in signal-noise outweighs the cost of error covariance. Also, as mentioned, there are statistical means of accounting for this error in determining cell traction parameters.

The MSD should be calculated for all possible interval durations. Again, using our 5 time point, 4 interval example and assuming overlapping intervals we would have

Over 1 time interval: MSD1 = average of 4 MSD measurements (1–2, 2–3, 3–4, 4–5)

Over 2 time intervals: MSD2 = average of 3 MSD measurements (1–3, 2–4, 3–5)

Over 3 time intervals: MSD3 = average of 2 MSD measurements (1–4, 2–5)

Over 4 time intervals: MSD4 = 1 MSD measurement (1–5)

X-Y data are then generated as follows

X = duration	Y = average MSD over that number of intervals
$1 \Delta t$	MSD1
$2 \Delta t$	MSD2
$3 \Delta t$	MSD3
$4 \Delta t$	MSD4

A full time-lapse experiment would include many more intervals and therefore a much greater amount of data.

C. Fitting data to the random walk model. The easiest way to fit data is to use a program (or write one yourself) capable of nonlinear regression. Fit X–Y data to Eq. (2) to identify μ and P (or S and P). If a nonlinear regression package is not available, the parameters can be estimated by recognizing that the MSD vs duration plot can be separated into two distinct regions: at short durations, the curve goes as time^2, with slope $\sim n_d S^2$, and at long durations the curve goes as time, with slope $\sim 2 n_d \mu$. An investigator can split MSD curves into these two regions and then use standard linear regression to identify cell traction parameters.

Finally, Shreiber *et al.* (2003) have detailed a technique to temporally resolve mean-squared displacement data.

Potential Pitfalls and Helpful Hints

1. Do not trust that all objects are tracked appropriately. Review time-lapse movies of each experiment and note which objects are lost or switched to a different object and disregard these cells in data analysis.

2. Equation (2) assumes that cells move randomly. If it appears that cells are not moving randomly, examine the cell tracks of individual cells (look at *all*, not just a few) to visually inspect for any directional bias. Also, mean-squared displacement data can be generated for one direction at a time and the MSD in

each direction independently fit to Eq. (2). If the migration is indeed random, then cell migration parameters should be (within error) the same for analysis of the X direction, the Y direction, and the Z direction.

3. Note that in calculating values for MSD, more measurements go into calculating the average MSD over one interval than over two intervals, more over two than over three intervals. Thus, there is generally more error in the calculation of MSD over long durations than short durations. This can be accounted for in the analysis by repeating data for a given duration to represent the number of samples that were included in the average calculation.

4. The increase in error in the MSD calculation over long durations can also produce aberrant behavior in the MSD plot. Specifically, the MSD plot may deviate from linear behavior as the error in the MSD calculation increases. These points can be ignored in processing data.

VIII. MIGRATION/TRACTION

The combined migration/traction assay allows assessment of the contractile ability of cells and their motility (Knapp *et al.*, 2000; Shreiber *et al.*, 2001, 2003). It is more technically challenging than either assay is individually and requires machining of special chambers. A schematic of the chamber is shown in Fig. 7, but the dimensions are somewhat arbitrary and can be tuned to the specific need of the investigator. Our assay chamber consists of a stainless-steel annulus (2.5 cm o.d., 1.5 cm i.d., 0.6 cm thick) with 3-mm holes located 1 mm from the bottom surface, bored through the side of the annulus. For stressed gel assays, two stainless-steel posts (3 mm diameter, 1.2 and 0.8 cm in length) with flared ends fit securely into opposing holes. On the inside ends of the two posts is glued a 3-mm-diameter, 1-mm-thick disk of porous polyethylene. A polycarbonate tube with an inner diameter matching the diameter of the posts/disks is supported by the two posts and serves as a mold for the assay. When the tube is filled with a collagen or fibrin solution, the solution penetrates the pores to form a fixed boundary condition upon gel formation. For unstressed, free-floating gels, the stainless-steel posts are replaced with polycarbonate posts that are each 1 mm longer to account for the lack of a porous polyethylene disk. In these cases, the gel forms without a rigid attachment, and the gel is free to compact uniformly in all directions.

All steps are performed under sterile conditions.

1. Prepare collagen or fibrin solution as described earlier. Be sure to degas the collagen/fibrin.

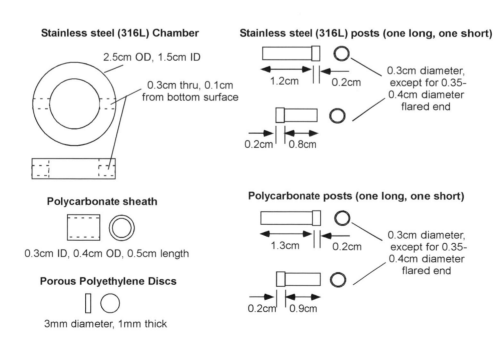

Stainless steel (316L) Chamber
2.5cm OD, 1.5cm ID
0.3cm thru, 0.1cm from bottom surface

Polycarbonate sheath
0.3cm ID, 0.4cm OD, 0.5cm length

Porous Polyethylene Discs
3mm diameter, 1mm thick

Stainless steel (316L) posts (one long, one short)
1.2cm   0.2cm
0.3cm diameter, except for 0.35-0.4cm diameter flared end
0.2cm   0.8cm

Polycarbonate posts (one long, one short)
1.3cm   0.2cm
0.3cm diameter, except for 0.35-0.4cm diameter flared end
0.2cm   0.9cm

FIGURE 7 Individual components of a single traction/migration chamber. The stainless-steel posts and porous polyethylene discs are used in the stressed, constrained assay. The polycarbonate posts are used in the unstressed, free floating assay and are slightly longer to accommodate the length lost by not including the porous polyethylene.

2. Add cells to desired concentration (7500–30,000 cells/ml).
3. Add beads (7500–10,000 beads/ml).
4. Prepare culture medium with defined soluble factors of interest (e.g., 50 ng/ml PDGF).
5. Under sterile conditions, hold the assay chamber in one hand with the sheath supported by one post. Fill the sheath with the gel solution from steps 1–3.
6. Gently push the opposing post into the mold.
7. With vacuum grease, secure a coverslip to the top and bottom of the chamber and place in an incubator.
8. Gently flip the chamber every 5 min until the solution is gelled to prevent cells and beads from settling to the bottom.
9. Remove chamber from incubator, place right side up, and carefully remove the top coverslip.
10. Fill the chamber with medium from step 4.
11. With sterile tweezers, gently slide the sheath over the long post to expose gel to culture medium.
12. Replace the coverslip, again securing with vacuum grease.
13. Transfer chambers to microscope stage and prepare tracking algorithm (section VIII,A).

A schematic of the final steps is provided in Fig. 8.

A. Microscopy

The migration/traction assay requires at least an *XY* motorized stage. Automated focus is preferred and rotating objectives are ideal. The routines for tracking cell migration are identical to those described previously. To record cell traction, follow initial selection of cell/bead positions, move the stage to the center of each chamber, and instruct the computer to build a mosaic image of the gel at the midplane. This is done most easily with a low power objective; if available, use the motorized objective feature to switch objectives to 2 or 4X. This process should be repeated automatically at the end of each time interval. A typical time lapse is run for 12–48 h. Typical "before" and "after" images of the stressed and unstressed assays are shown in Fig. 9.

B. Analysis

Cell migration and traction analyses are performed as before. For traction, measure the diameter at the midplane of the gel at each time point. The fixed, stressed gel should form an hourglass during cell-mediated gel compaction, so the diameter measurement should essentially be the middle of the hourglass. The free floating, unstressed gels should compact uni-

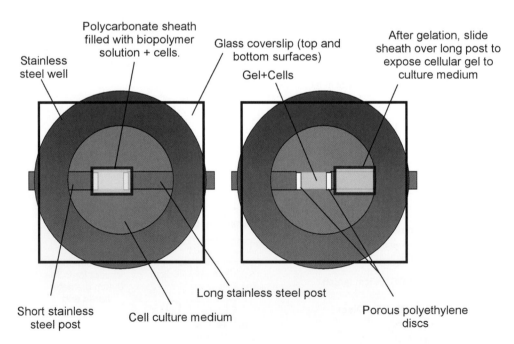

FIGURE 8 Schematic of cell traction/cell migration assay preparation. Fill the polycarbonate sheath with the biopolymer solution (+cells and beads) and support with stainless-steel posts, use vacuum grease to cover top and bottom surfaces with glass coverslips, and transfer to the incubator to facilitate gelation. After gelation, remove the top coverslip and fill the chamber with culture medium with defined concentrations of soluble factors of interest. Slide the sheath over the long post to expose the gel to the culture medium. Replace the top coverslip and transfer to the microscope stage for traction and migration tracking.

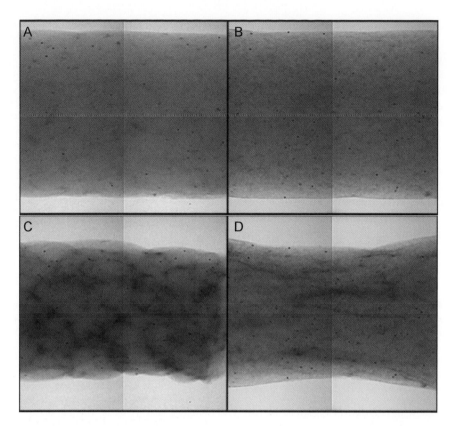

FIGURE 9 Examples of unstressed, free-floating assay (A,C) and stressed, constrained assay (B,D). Migration–traction assays are prepared as described and are nearly identical after initial preparation. (A) Free-floating cylindrical gel at time = 0. The gel was formed by supporting a sheath with smooth-ended polycarbonate posts. (B) Contstrained cylindrical gel at time = 0. The gel was prepared by filling an identical sheath that is supported by stainless-steel posts with porous polyethylene discs glued to the ends (outside field of view). The biopolymer solution penetrates the pores to form a fixed boundary condition upon gelation. Following incubation, cells that are entrapped in the gel exert traction and compact the gel. (C) Without physical constraints to compaction, the free-floating, cylindrical gel compacts (roughly) uniformly. (D) In contrast, the physical connection to the posts via the porous polyethylene prevents the constrained gel from compacting in the axial direction. The result is pure radial compaction and subsequent fiber and cell alignment. The characteristic "hourglass" shape results. In both cases, the degree of cell traction is related to the amount of gel compaction, measured as the decrease in radius at the midplane of the sample.

formly, so the diameter measurement can be taken from any cross-section. The result is X–Y data of percentage compaction vs time, which can be examined for temporal trends or quantified on the basis of final midplane compaction and compared to results of the migration analysis.

C. Pitfalls

1. Be careful not to introduce bubbles into the assay that can affect gel compaction and obscure imaging.

2. Similarly, do not smear vacuum grease over the coverslip in an area that will obscure imaging.

3. Use hydrophilic porous polyethylene (Small Parts, Miami Lakes, FL) to ensure good penetration of the gel-forming solution for the fixed case.

4. In the free-floating case, cell traction and/or convective motion of the culture medium may lead to rigid body motion of the cylindrical gel. Because cell migration may be tracked at high magnification, a small translocation of the gel can result in "losing" all of the cells during the traction algorithm, as it is only prepared to follow movements of 0.5 xFOV per interval. Use a low percentage, sterile, agarose solution (1%) to increase the viscosity of the culture medium and to reduce the ability for gel movement.

5. Similarly, under some conditions (high cell concentration, certain concentrations of soluble factors, etc.), gels may compact too quickly to monitor cell position. Some trial and error is likely necessary to converge on appropriate conditions.

6. Bubbles in the medium can also diminish imaging capabilities. Degas the culture medium and be careful when filling and sealing the chambers to avoid introducing air.

7. Strong compaction in the stressed case leads to fiber alignment in the axial direction, which can lead to anisotropic migration. It may be necessary to quantify migration in the individual directions (X, Y, and Z) for these cases.

8. Generally, because of the high variability in cell lines from culture to culture and passage to passage, it is recommended to run all experimental conditions in one set of experiments rather than running all controls one day, all at one condition the next, and so on.

References

Barocas, V. H., Moon, A. G., and Tranquillo, R. T. (1995). *The fibroblast-populated collagen microsphere assay of cell traction force 2. Measurement of the cell traction parameter. J. Biomech. Engin.* **117**(2), 161–170.

Barocas, V. H., and Tranquillo, R. T. (1997a). *A finite element solution for the anisotropic biphasic theory of tissue-equivalent mechanics: The effect of contact guidance on isometric cell traction measurement. J. Biomech. Engin.* **119**(3), 261–268.

Barocas, V. H., and Tranquillo, R. T. (1997b). *An anisotropic biphasic theory of tissue-equivalent mechanics: the interplay among cell traction, fibrillar network deformation, fibril alignment, and cell contact guidance. J. Biomech. Engin.* **119**(2), 137–145.

Bell, E., Ivarsson, B., and Merrill, C. (1979). *Production of a tissue-like structure by contraction of collagen lattices by human fibroblasts of different proliferative potential in vitro. Proc. Natl. Acad. Sci. USA* **76**(3), 1274–1278.

Clark, R. A. F. (ed.) (1996). *"The Molecular and Cellular Biology of Wound Repair."* Plenum Press, New York.

Dickinson, R. B., and Tranquillo, R. T. (1993). *Optimal estimation of cell movement indices from the statistical analysis of cell tracking data. AIChE J.* **39**(12), 1995–2010.

Dunn, G. A., and Brown, A. F. (1987). *A unified approach to analysing cell motility. J. Cell Sci. Suppl.* **8**, 81–102.

Eastwood, M., et al. (1996). *Quantitative analysis of collagen gel contractile forces generated by dermal fibroblasts and the relationship to cell morphology. J. Cell. Physiol.* **166**, 33–42.

Ehrlich, H. P., and Rajaratnam, J. B. (1990). *Cell locomotion forces versus cell contraction forces for collagen lattice contraction: An in vitro model of wound contraction. Tissue Cell* **22**(4), 407–417.

Knapp, D. M., et al. (1997). *Rheology of reconstituted type I collagen gel in confined compression. J. Rheol.* **41**(5), 971–993.

Knapp, D. M., Helou, E. F., and Tranquilllo, R. T. (1999). *A fibrin or collagen assay for tissue cell chemotaxis: Assessment of fibroblast chemotaxis to GRGDSP. Exp. Cell Res.* **247**, 543–553.

Knapp, D. M., et al. (2000). *Estimation of cell traction and migration in an isometric cell traction assay. AIChE J.* **45**(12), 2628–2640.

Kolodney, M. S., and Wysolmerski, R. B. (1992). *Isometric contraction by fibroblasts and endothelial cells in tissue culture: A quantitative study. J. Cell Biol.* **117**(1), 73–82.

Moghe, P. V., Nelson, R. D., and Tranquillo, R. T. (1995). *Cytokine-stimulated chemotaxis of human neutrophils in a 3-D conjoined fibrin gel assay. J. Immunol. Methods* **180**(2), 193–211.

Neidert, M. R., et al. (2002). *Enhanced fibrin remodeling in vitro with TGF-beta1, insulin and plasmin for improved tissue-equivalents. Biomaterials* **23**(17), 3717–3731.

Nusgens, B., et al. (1984). *Collagen biosynthesis by cells in a tissue equivalent matrix in vitro. Coll Relat Res.* **4**(5), 351–363.

Shreiber, D. I., Barocas, V. H., and Tranquillo, R. T. (2003). *Temporal variations in cell migration and traction during fibroblast-mediated gel compaction. Biophys J.* **84**(6), 4102–4114.

Shreiber, D. I., Enever, P. A., and Tranquillo, R. T. (2001). *Effects of pdgf-bb on rat dermal fibroblast behavior in mechanically stressed and unstressed collagen and fibrin gels. Exp. Cell Res.* **266**(1), 155–166.

Tuan, T. L., et al. (1996). *In vitro fibroplasia: Matrix contraction, cell growth, and collagen production of fibroblasts cultured in fibrin gels. Exp Cell Res.* **223**(1), 127–134.

Electrophysiological Methods

48

Patch Clamping

Beth Rycroft, Fiona C. Halliday, and Alasdair Gibb

I. INTRODUCTION

The patch-clamp technique was first utilized in 1976 with the exclusive intention of recording single channel currents from acetylcholine receptor ion channels in frog skeletal muscle fibres (Neher and Sakmann, 1976). The following 25 years have witnessed refinements of the technique such that it is now applicable to a diversity of biological preparations, including animal and plant cells, intracellular organelles, yeast, fungi, and bacteria. Furthermore, the versatility of the technique permits many questions to be addressed. Undoubtedly, the most common application of the patch-clamp technique is to manipulate the cell membrane voltage and measure the electrical currents that flow across the membrane through ionic channels. These channels include voltage-operated ion channels (e.g., Na^+, K^+, Ca^{2+}, and Cl^- channels), channels regulated by intracellular second messengers (cAMP, cGMP, Ca^{2+}, G-proteins as well as numerous kinases and phosphatases), and neurotransmitter-activated receptor-operated channels. Currents can be recorded either from individual channels or from an entire cellular population of channels. The patch pipette can be used in whole-cell mode to introduce Ca^{2+}, second messengers, or ion-sensitive dyes into the cell (Park *et al.*, 2002). Combining patch clamp with optical imaging (Park *et al.*, 2002) has become a very powerful two-dimensional technique used to investigate the concentration or movement of labeled cellular biomolecules and their effect on the cell (Voipio *et al.*, 1994). Furthermore, the versatility of the patch-clamp technique extends beyond simply measuring ionic currents. It can be used to measure changes in cell membrane area caused by vesicular secretion. Using flash photolysis, rapid concentration jumps of caged intracellular messengers (e.g., Ca^{2+} or inositol trisphosphate) introduced through the patch pipette can be achieved (Gurney, 1990; 1994). This article provides a general introduction to the principles as well as an overview to the practical side of patch clamping. More detailed descriptions of patch clamping and other electrophysiological techniques can be found elsewhere (Ogden and Stanfield, 1994; Sakmann and Neher, 1995; Levis and Rae, 1998).

Although the choice of preparation used depends on the questions being asked, several limitations must be considered before embarking on an experiment. First, the technique relies on formation of a tight seal between the cell membrane and the patch pipette (see later). For this reason, cells with clean membranes must be used. Examples include tumour-derived cell lines and cells in primary culture, which tend to have a less extracellular matrix. Adult cells from complex tissues require enzymatic treatment to clean the membranes of connective tissue and extracellular matrix. Enzymes used most commonly include collagenase and proteases (Sigma-Aldrich). Enzymatic methods have the potential disadvantage that the enzyme treatment may damage or alter the properties of the channels or receptors of interest. However, a major drawback of cultured cells is that depending on the growth phase, different subpopulations of cells can be expressed that have properties different to freshly isolated cells.

In addition, some cell types are not an ideal shape for voltage clamping. If a cell has long processes, such as dendrites or axons, then the membrane voltage in these regions may not be controlled, leading to inaccuracies in current measurement.

II. MATERIALS AND INSTRUMENTATION

In addition to equipment required for fabricating patch pipettes (Section VIII), the following are required for a patch-clamp setup.

A. Microscope

A dissection microscope may be adequate in order to make patch-clamp recordings from *Xenopus* oocytes and other large cells. For anything smaller, i.e., cultured or freshly isolated cells, inverted microscopes (Narishige) are used commonly for visualizing the patch pipette and cell.

B. Micromanipulators

Precise movement of the patch pipette in three axes is required at the submicrometer level prior to giga seal formation. For very small cells, fine movement can be obtained using remotely controlled manipulators, such as piezoelectric (Newport, Burleigh intracell) or hydraulic (Narishige) manipulators. Once the giga seal has been formed, the pipette must not drift but should remain in a fixed position so that whole-cell and cell-attached recordings can be sustained.

C. Flotation Table

These are required to dampen out both horizontal and vertical vibrations from the surrounding environment down to a few hertz (Section VIIIB).

D. Faraday Cage

To shield the recording electronics from electromagnetic radiation, primarily line-frequency pick-up (50 Hz), the patch-clamp setup is enclosed in an earthed cage consisting of a metal framework with three of its sides and top covered with wire mesh. The fourth side is open for access to the microscope and recording electronics. Faraday cages are available commercially (Newport, Intracel). They can be made easily, however, and the parts for the cage can be obtained from most hardware shops.

E. Oscilloscope

Software packages are available (Axotape, Axon Instruments) that enable the computer to emulate the oscilloscope. This may be a less expensive alternative than buying both an oscilloscope and a computer, but should be avoided because the display on the computer monitor may be slow and it may also be difficult to scale appropriately. It is also common practice to tape record raw data (see later) as it is produced and store it on the tape for later "off-line" analysis on the computer.

F. Amplifiers

Commercially available patch-clamp amplifiers provide compensation for both pipette capacitance and cell capacitance associated with whole-cell recordings (Axon, Heka, Cairn). They also enable series resistance to be compensated for, thus eliminating voltage error that arises from the voltage drop across the access resistance of the electrode. Low-pass filters are incorporated in the amplifier so that the recording bandwidth can be adjusted in order to be compatible with the acquisition rate. Amplifiers also permit the membrane potential to be altered by the user. Several different types are available that can be used to record from a diversity of preparations, including oocytes and bilayers. For patch-clamp recordings, Axopatch-1D and Axopatch 200B amplifiers are available, which permit both whole-cell and single-channel recordings. The advantage of the latter model is that the internal circuitry within the amplifier headstage is cooled to $-15°C$ in order to reduce thermal noise, and thus the contribution of noise from the amplifier is minimised.

G. Tape Recorder

It is often necessary to store patch-clamp recordings, particularly of single-channel openings, on tape for "off-line" analysis. This provides a permanent record of data, which can be replayed under different gain and filter settings. Digital audio tape recorders (e.g., Biologic DTR 1200 or DTR 1600) are probably the most commonly used.

H. Equipment for Online Data Acquisition, Data Storage, and Analysis

Computers are increasingly used "online" to generate voltage protocols and store data as it occurs. An interface (e.g., Axon Instruments Digidata 1200, Cambridge Electronic Design 1401) between the computer and the recording setup provides both analogue-to-digital (ADC) and digital-to-analogue (DAC) conversion. ADC conversion enables the analogue signal from the cell, which is a continuous time-varying voltage, to be converted into a digitized record. DAC conversion allows voltage signals generated by the computer in data form (binary digits) to be applied to the cell via the amplifier voltage command input. Software used to generate voltage protocols and analyse acquired data can either be bought individually or along with the interface. pCLAMP (Axon) is a commonly used software package for PC users, which can be bought with the Digidata 1200 interface, as is Axograph for Macintosh users. Academic users can also use free packages such as the Strathclyde Electrophysiology Software (www.strath.ac.uk/Departments/PhysPharm).

I. Filters

Although most commercially available patch-clamp amplifiers have built-in filters, it is useful to have an additional filter/amplifier system with different characteristics from those of the patch-clamp amplifier. This provides a more finely tuned filtering system for single channel data or noise analysis. The most suitable type of filter for single-channel recordings is the Bessel filter, which does not produce an oscillation in response to a rectangular input. Because single-channel currents are essentially rectangular, filters lacking in this property will cause the currents to become distorted. Filter and digitizing frequencies must be chosen carefully in order to prevent aliasing. To prevent this from occurring, the sample rate should be at least five times the cutoff (–3 dB) frequency of the Bessel filter.

III. PRINCIPLES OF PATCH-CLAMP RECORDING

Electrical or chemical stimulation of the cell membrane causes a change in membrane potential. Normally the change is counteracted quickly by the activation of voltage-dependent ion channels, which helps restore the membrane potential back to resting levels. The flux of the different ionic species responsible for shaping the depolarization/repolarization is too brief to enable the experimenter to identify the ions involved or the change in membrane conductance. The voltage-clamp technique permits the experimenter to clamp the membrane voltage at a fixed level so that the ion channels responsible for the current at that particular voltage can be identified.

The key element in the patch clamp is the feedback amplifier in the headstage (Fig. 1). The feedback amplifier controls the membrane potential. Current flows from the output of the amplifier when the voltage of its two inputs is not equal. The amplifier receives two inputs: a positive input from a command potential

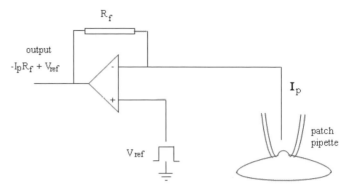

FIGURE 1 A headstage current/voltage amplifier.

source of variable setting (V_{ref}), which is determined by the user, and a negative input from the pipette potential (V_p). When both inputs are at the same potential, the output will be zero. When a discrepancy arises between the two inputs, the amplifier strives to null this discrepancy and force V_p to equal V_{ref}. This is achieved by the amplifier passing current across the feedback resistor, R_f, to drive the inside of the cell to the reference potential. This current supplied by the feedback amplifier is equal and opposite to the current carried by ions flowing across the membrane. The current flowing through the pipette (I_p) to clamp the cell membrane is proportional to the voltage drop ($V_{out} - V_p$) across the resistor, R_f, according to Ohm's law

$$I_p = V_{out} - V_p / R_f \qquad (1)$$

Because the feedback amplifier clamps V_p at V_{ref}, then in addition,

$$i_p = V_{out} - V_{ref} / R_f \qquad (2)$$

In practice, therefore, pipette current is monitored by a differential amplifier that constantly measures the difference between V_{ref} and V_{out}. From Eq. (2) it follows that the sensitivity of current measurement is inversely proportional to the size of the feedback resistor, with a large resistor enabling measurement of smaller current amplitudes. Because the gain of the amplifier is set by the resistor connecting V_o to the (–) input, the larger the value of R_f, the larger the gain and the more closely V_p approaches V_{ref}.

The feedback resistor also contributes thermal noise, which can contribute to noise in the current recording. The variance of the current noise (s_i^2) through the feedback resistor is related to Johnson noise due to the resistance (R_f), being given by

$$s_i^2 = 4kTf_c / R_f \qquad (3)$$

where k is Boltzmann's constant (1.381×10^{-23} VCK^{-1}), T is the absolute temperature (° Kelvin), and f_c is the bandwidth (Hz), i.e., the low-pass filter setting. It follows that for a high-resolution, low-noise recording, R_f should be high, and is usually around 50 GΩ. However, the amplifier cannot put out more voltage than is provided by its power supply, which is approximately ±12 V, indicating from Eq. (1) that the headstage output will be saturated if i_p exceeds 240 pA. For this reason, large whole cell currents are measured with a lower feedback resistor, and the value of R_f must be chosen to suit the experiment, typically 500 MΩ.

IV. PATCH-CLAMP CONFIGURATIONS

The principle of the technique is to electrically isolate a patch of membrane from the external solution and record current flowing into the patch. This is achieved by pressing the tip of a heat-polished pipette onto a clean membrane. The resulting seal between pipette tip and membrane is very tight, with a resistance greater than 10 GΩ, hence the term "giga seal." This is obtained using the following procedure.

1. After lowering the pipette into the solution in the recording chamber, zero the voltage output to subtract the junction potential (see later).

2. Apply a regular rectangular test voltage pulse of 1–10 mV for 5 ms every 20 ms to the pipette to measure its resistance. This should result in the appearance of a rectangular waveform on the current trace, which can be monitored on the oscilloscope. The pipette resistance can be worked out from Ohm's law by dividing the test voltage by the resulting current amplitude.

3. Position the pipette so that it is just above the cell. Apply positive pressure to prevent the pipette tip from becoming clogged with dirt and gently blowing away any debris near the cell. This procedure will also indicate when the pipette tip is very near the cell, as the positive pressure will cause a small dimple to appear on the cell surface.

4. When a small dimple is seen, a small decrease in the size of the test voltage current pulse should occur, indicating an increase in pipette resistance following contact between membrane and pipette. At this point, stop moving the pipette, release the positive pressure, and apply gentle suction, e.g., with a 1-ml syringe or by mouth. Progress of seal formation is indicated by the rectangular waveform on the current trace becoming smaller. Application of negative voltage (−50 to −60 mV) in the pipette at this stage encourages seal formation. When the pipette resistance is greater than 1 GΩ, a giga seal has been formed.

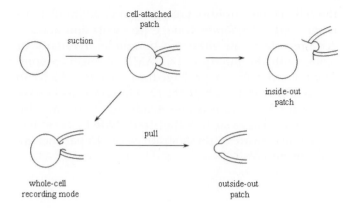

FIGURE 2 Giga-seal recording configurations.

After giga-seal formation, it is now possible to record either currents from the entire cell using the whole-cell recording mode or single channel activity with cell-attached, inside-out or outside-out modes (Fig. 2).

A. Cell-Attached Recording

This configuration is not only the simplest to achieve, as it is formed by obtaining a giga seal, but it is also the most physiological. The major disadvantage of this configuration is that the resting potential of the cell is unknown, which is added to the applied pipette potential.

B. Inside-out Recording

Upon formation of a giga seal, withdrawing the pipette from the cell usually results in an excised membrane patch with its cytoplasmic side exposed to the bath solution. Inside-out patches can be used to study the influence of cytoplasmic constituents or second messengers on channel activity. A disadvantage of this configuration is that cytoplasmic constituents are lost, which may be important in modulating the behavior of ion channel proteins.

C. Whole-Cell Recording

Instead of withdrawing the patch pipette from the membrane after seal formation, application of gentle suction will disrupt the membrane patch directly under the pipette, leading to the formation of a low resistance pathway between the cell interior and the solution in the pipette. Because the cell interior is being perfused with pipette solution, this configuration has the disadvantage that certain cytosolic factors important for cellular function may be washed out. Formation of the whole-cell configuration becomes immediately apparent by the sudden appearance of large capacity transients at the leading and trailing ends of the test pulse, which reflect the charging and discharging of the capacitance of the cell membrane. These transients can be minimized using the whole-cell capacitance cancellation and series resistance (Section IXA) compensation dials on the patch-clamp amplifier. This allows a crude estimation of the cell capacitance (and hence cell size) to be made because cell membranes have a fairly constant specific capacitance of $1 \mu F/cm^2$.

D. Outside-out Recording

Upon obtaining the whole-cell configuration, slow withdrawal of the pipette from the cell causes the membrane to stretch until it finally breaks. The membrane should reseal to form a patch with its intracellular face in contact with the pipette solution. This configuration can be used to study the effects of extracellular agents on single-channel activity.

E. Perforated Patch Recording

It is possible to prevent important cytosolic components from being washed out of the cell interior, whilst also allowing the electrical contact between cell interior and pipette to be sustained. Two polyene antibiotics are available, nystatin and amphotericin B, which, when present in the pipette solution, generate ionic channels within the cell membrane (Korn *et al.*, 1991; Rae *et al.*, 1991). These channels are small enough to permit the passage of monovalent cations and anions across the membrane, but will impede the passage of larger molecules (MW > 300). Thus, it is possible to control the concentrations of internal Na^+, K^+, and Cl^- (in contrast, gramicidin allows only positively charged ions to cross the cell membrane, hence avoiding disruption of the membrane chloride gradient) whilst ensuring that cytosolic constituents will remain trapped within the cell. It is important that the very tip of the pipette contains only nystatin-free solution because the antibiotic impedes the formation of giga seals. To fill the pipette, the tip must first be dipped into nystatin-free solution for a few seconds and then filled with nystatin-containing solution in the usual way. Once a giga seal has formed between pipette and cell, the development of ionic channels in the membrane can be monitored by applying a $-10 mV$ hyperpolarising pulse. As more and more channels are formed, capacity transients at the leading and trailing ends of the pulse should get larger and faster, as the series resistance of the membrane patch decreases. This process can take between 5 and 30 min. The path that is eventually formed between pipette and cell interior should be of low enough resistance to permit recordings in the whole-cell configuration. Furthermore, channels formed by the antibiotics are virtually voltage independent, enabling studies on voltage-dependent channels to be performed.

F. Planar Electrode Array and Automated Patch-Clamp Recording

In addition to the conventional patch-clamp techniques described earlier, advancements have been made towards amplification and automation of this time-consuming process (Sigworth and Klemic, 2002). At present, these techniques can only be performed on cultured or dissociated cells and in the whole-cell configuration. One method employs a planar chip made from quartz in which a submicron aperture has been made (Fertig *et al.*, 2002). The cell suspension solution is poured onto the chip and suction is applied to the apertures to secure a single cell to the aperture. Although giga-ohm resistance seals are hard to obtain with this material, advances have been made using hydrophilic-oxidised Sylgard [polydimethylsiloxane (PDMS)] instead of quartz. Advantages of the planar electrode include high-resolution, low-noise recordings due to the small diameter of the aperture; with the development of several apertures on the same chip, multiple parallel experiments can be performed simultaneously. Other automated methods, such as the PatchXpress 7000A (Axon), employ conventional glass micropipettes but use a new protocol that involves filling a glass pipette with the cell suspension solution and flushing it towards the pipette tip to create a seal with a cell (Lepple-Wienhues *et al.*, 2003).

V. RECORDING SOLUTIONS FOR PATCH CLAMPING

The composition of solutions used to fill the pipette or bathe the cells will depend on the nature of the ion channels being investigated. A standard extracellular solution might have the following composition (mM): KCl 5, NaCl 140, $CaCl_2$ 1, $MgCl_2$ 1, and HEPES 10; while the pipette solution may be KCl 140, HEPES 10, and EGTA 10, with pH adjusted to 7.3 with NaOH in both cases. EGTA is used to buffer the intracellular calcium concentration to very low levels. Specific EGTA/Ca^{2+} mixtures can be used to buffer intracellular calcium at a particular value, e.g., $100 nM$, and ATP, Mg^{2+}, kinases, or other intracellular enzymes may also be added.

VI. ELECTRICAL CONTINUITY BETWEEN MEMBRANE PATCH AND RECORDING CIRCUITRY

A problem that can arise at the interface between the metal recording electrode and the pipette solution is the development of a junction potential. This potential difference arises due to the diffusional flux of anions and cations between mediums of different ionic composition or concentration. In order to overcome this problem, connections made to the recording circuitry are via nonpolarisable reversible Ag/AgCl electrodes. These electrodes consist of a silver wire with a chlorided tip produced by dipping in bleach (20% sodium hypochlorite solution) for about 1 min to make a nonpolarisable electrode. The electrode will turn grey/black with the formation of AgCl on the surface. The AgCl coating is fairly robust, but needs to be renewed every few days. It is also good practice to heat polish the back end of the patch pipettes in the flame of a spirit burner (Merck) before pulling so that sliding the pipette onto the silver wire electrode prevents the AgCl from being scratched off. A second AgCl wire, the reference electrode, is present in the recording chamber, which is connected to the headstage ground socket. For either of these electrodes, a good indication of the AgCl coating beginning to deteriorate can be seen from the current trace on the oscilloscope, which may display erratic 50-Hz interference or develop a considerable offset potential (50–100 mV).

VII. PATCH PIPETTES

Many of the electrophysiological properties of patch pipettes depend on the type of glass and the size and shape of the pipette tip. Pipette glass (Harvard) has an outer diameter of 1.5–2 mm and is available with an internal filament running the whole length of the tube to ease filling of the pipette. The glass of choice for single-channel recording is thick-walled glass, such as borosilicate glass, as pipette capacitance and hence background noise levels are low. Noise can be reduced further by using quartz electrodes for single-channel recordings, although this incurs additional costs for materials and a specialized puller (Levis and Rae, 1998). Thin-walled borosilicate glasses are preferable for whole-cell recordings, as these provide a lower access resistance. The access resistance can also be minimized by producing pipettes with a steep angle of taper and a tip diameter of 1 μm after fire polishing to yield a low-resistance pipette (<5 MΩ). In general, for small cells, relatively high pipette resistances are necessary to prevent the patched cell from being sucked up the pipette during formation of the giga seal. Low-resistance, wide-bore pipettes tend to cover a greater surface area of membrane and hence a high number of channels, and so for this reason higher pipette resistances are preferable for single-channel recordings.

A. Making Patch Pipettes

There are three stages involved in making patch pipettes: pulling the pipette, coating it with Sylgard, and fire polishing the pipette tip.

Several types of pipette pullers are available commercially (Sutter, Narishige) that pull the glass either horizontally or vertically, by mechanical or gravitational force, respectively. The pipettes are pulled over two stages: during the first, the centre of the tubing is heated in a coil of nichrome wire until it stretches over a length of around 10 mm to form an hour-glass shape; at its thinnest part, the pipette should measure approximately 400–500 μm in diameter. The length of the first pull determines the taper of the pipette, with a higher heat causing a steeper taper, thus reducing electrode resistance and capacitance. Once the glass has cooled, the coil is recentred around the thinnest part of the tube, heated at a lower temperature, and the two ends pulled apart to form two similar pipettes. The temperature of the second pull determines the final tip diameter, with high temperatures producing pipettes with narrow tips.

B. Coating Pipettes with Sylgard

Coating the pipette shank with an inert, hydrophobic material such as Sylgard (Merck) resin or beeswax (Sigma-Aldrich, Merck) is necessary for two reasons: (1), the pipette wall is an insulator separating two electrolyte solutions, which means that it acts as a capacitor and is thus a source of noise. This may be of particular significance for single-channel recording, where currents as small as a fraction of a pA are being measured. Because capacitance is proportional to 1/thickness, the capacitance can be reduced by coating the pipette with a thick, nonconducting layer. The hydrophobic nature of either Sylgard or beeswax also prevents the bath solution from creeping up the sides of the pipette. This will decrease the area of electrical contact between pipette and bath solution and also therefore the pipette capacitance. Sylgard is available as a resin and a curing agent, which must be mixed together in a 10:1 ratio. A good idea is to mix 10 ml of resin with 1 ml of curing agent and dispense the resulting mixture into 1 ml Eppendorf tubes. The aliquots can then be stored in the freezer until required, where the mixture remains in a fluid state. Because polymerization is temperature dependent, an aliquot will remain fluid for several hours at room temperature. Sylgard should be applied as near the pipette tip as possible without causing it to become blocked, i.e., ~100 µm, and extend along the pipette to just beyond the shoulder of the pipette. It is cured by placing the pipette tip in a heated wire coil for a few seconds.

Note: Sylgard residue on the pipette tip is a potential source of consistent failure to obtain giga-ohm seals. If this happens, it is worth trying to seal with a few pipettes that have not been coated.

Beeswax is almost as effective as Sylgard in reducing noise associated with pipette capacitance, but has the advantage that its application is simpler in that wax only requires gentle heating in order to melt and dries almost instantaneously at room temperature when applied to the pipette tip.

C. Fire Polishing

Fire polishing produces a clean and smooth pipette tip with which tight seals can be obtained. In general it is best to make patch pipettes on the day that they will be used. The basic fire-polishing apparatus consists of a micromanipulator to which the pipette is attached, which is brought into close proximity to a heated platinum wire. The wire is mounted onto the stage of a microscope and is bent into a V shape in order to focus the heat onto the pipette tip. A small blob of glass is positioned onto the tip by melting the end of a pipette onto the wire. The glass blob prevents tiny fragments of the heated platinum from spluttering onto the pipette tip and making it dirty. To polish the pipette tip, the wire is heated until it glows a dull red, and the pipette is advanced towards the wire until the tip is seen to darken slightly and shrink back (by ~2 µm).

VIII. PITFALLS

A. Series Resistance

Because the membrane possesses channels that have finite resistance, the cell membrane can be modeled by a circuit comprising a variable resistor, R_m, in series with a parallel plate capacitor, C_m (Fig. 3). The series resistance (R_s) is introduced into the model to represent the pipette and access resistance, which is in series with the cell and the pipette during whole-cell recordings of currents. The rate at which the mem-

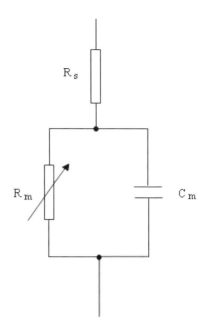

FIGURE 3 A simple electrical analog of the cell.

brane potential is changed is restricted by the charging of the membrane capacitance. This capacity current decays exponentially with a time constant (τ) equivalent to the product of the capacitance and resistance R, where

$$\tau = RC \qquad (4)$$

where $R = R_m R_s/(R_m + R_s)$. This means that the cell capacitance can limit the speed of the voltage clamp and the ability to observe rapidly activating ionic currents. It is also clear that the membrane resistance and series resistance influence the speed of the membrane charging. Keeping R_s low by using large-tipped pipettes with low resistance is essential for rapid clamping of the membrane voltage. Minimising R_s is also important for reducing errors in the measurement of membrane voltage and current. Several resistances contribute to R_s, with the major one being the access resistance of the pipette, usually in the order of a few MΩ. When a current flows across the membrane, the presence of R_s leads to a discrepancy between the clamped pipette potential V and the true value of the cell membrane potential, V_m. The size of the error is $I \times R_s$ and is therefore large if the current being recorded is large. It is possible to compensate for R_s electronically by adding to the command voltage a voltage signal proportional to the membrane current. In commercial amplifiers this is achieved by multiplying the series resistance value from the whole-cell transient cancellation by I_p and adding a proportion, between 80 and 90%, to V_{ref}. Overcompensation of R_s causes the

Organ Cultures

49

Preparation of Organotypic Hippocampal Slice Cultures

Scott M. Thompson and Susanne E. Mason

I. INTRODUCTION

Brain slice cultures are prepared by cutting thin sections of brain tissue from neonatal animals and culturing them as intact slices of tissue rather than as dissociated cells. Like all culture methods, these preparations offer the advantages of (1) long-term survival; (2) precise control of the experimental conditions; (3) excellent accessibility for viral vectors, biolistic transfection, and other means of gene transfection; (4) survival of tissue from neonatal-lethal transgenic animals; and (5) excellent visibility of cells and subcellular structures for morphological and electrophysiological studies. Unlike cell cultures, however, these tissue slices retain many features of their organotypic organization, as has been described extensively (e.g., Zimmer and Gähwiler, 1984; Gähwiler, 1984; Gähwiler *et al.*, 1997). This permits the identification of defined cell groups, the stimulation or lesioning of specific axonal pathways, and the formation of relatively normally sized synaptic connections (Debanne *et al.*, 1995). Other advantages of the slice culture technique include the ability to coculture slices from different brain regions, thus facilitating the experimental manipulation of long-distance connections *in vitro* (e.g., Gähwiler *et al.*, 1987), and the ability to induce conventional long-term potentiation (e.g., Debanne *et al.*, 1994).

There are two variations of the technique that are in use currently. In the roller tube technique, pioneered by Gähwiler (1981), the slices are attached to glass coverslips and placed in sealed test tubes on a roller drum in a dry air incubator. In the membrane or interface technique, pioneered by Stoppini *et al.* (1991), the slices are placed on semipermeable membranes and grown statically in CO_2 incubators. Primary differences between the two techniques are that the roller tube cultures generally become thinner than the membrane cultures, but may be slightly more demanding and time-consuming to prepare.

This article provides a concise description of the steps involved in preparing and maintaining hippocampal slice cultures. More details can be obtained in Gähwiler *et al.* (1998). Of course, many other brain structures can be cultured readily using these techniques. It is recommended that beginners start with the hippocampus, as it is large, easy to dissect, has a readily visible cell body layer, and has proven to be robust when cultured with these methods.

II. SOLUTIONS

Keep all solutions refrigerated until use. Maintain sterility.

1. *Roller tube culture medium*:

100 ml	basal medium Eagle (GIBCO, product # 21010)
50 ml	Hank's balanced salt solution (HBSS) with Earle's salts (GIBCO, product # 24020) or Hanks' balanced salt solution without phenol red for fluorescence applications (GIBCO product # 14025 or Cellgro, product # MT21-023-CV)
50 ml	horse serum (GIBCO, heat inactivated previously at 56°C for 30 min in a water bath, product # 16050)
4 ml	50% glucose solution, and
1 ml	200 mM glutamine (from frozen aliquots)

2. *Membrane culture medium*:

100 ml	MEM with Hank's salts and glutamine (GIBCO product # 11575)
50 ml	HBSS as described earlier
50 ml	horse serum as described earlier
1 ml	penicillin/streptomycin solution (Sigma product # P-4333)
1 g	HEPES
1 ml	50% glucose solution

3. *HBSS + glucose*:

500 ml	Hanks' balanced salt solution and
6 ml	50% glucose solution

4. *HBSS + glucose + kynurenate*:

50 ml	*HBSS + glucose* and
0.5 ml	300 μM kynurenate stock solution

5. *Chicken plasma*: There is variability in the amount and type of anticoagulants contained in commercial plasmas. We recommend lyophilized chicken plasma from Cocalico Biologicals (product # 30-0390L). Reconstitute to appropriate volume with tissue culture water. Centrifuge for 18–20 min at 2500 rpm.

6. *Thrombin*: Prepare aliquots at 150 units/ml. Store frozen. Add 0.75 ml *HBSS + glucose* to 1 ml aliquot of thrombin. This dilution can be adjusted to modify the firmness of the plasma clot (see later).

7. *Antimitotics*: 3 mg each of cytosine-β-D-arabinofuranoside, uridine, and 5-fluoro-2'-deoxyuridine in 100 ml HBSS. Aliquot and freeze.

III. MATERIALS

A. Coverslips

Purchase 12 × 24-mm coverslips of 0/1 thickness. Place coverslips individually in the bottom of a large glass dish. Fill the dish with enough 95% ethanol to cover the coverslips. Soak them overnight. Replace the 95% ethanol with 100% ethanol and soak overnight. Let the coverslips dry overnight with the dish covered with a paper towel. Transfer the coverslips to a petri dish, wrap in aluminum foil, and bake at 200°C for 4–8 h.

B. Membrane Culture Dishes

Corning Costar Transwell polyester membrane inserts and multiwell dishes (product # 3460) (also available with various substrate coatings). Millipore Millicell membrane inserts can also be used.

C. Instruments

Small (ca. 3-cm blades) surgical scissors (1×)
Large (ca. 5-cm blade) surgical scissors (1×)

Scalpel or holder for razor blade shards (1×)
Small (ca. 3 × 20-mm) flat spatulas (6×)
Curved surgical forceps (1×)
Alcohol lamp (1×)

D. Plasticware

Aclar plastic (Ted Pella, Inc., product # 10501-25) (Cut into 4 × 4-cm squares.)
Culture tubes (Nunclon Flat sided TC tubes, 110 × 16 mm)
Petri dishes (60 × 15 and 35 × 10 mm)

E. Roller Drum and Drive Unit

Available from Bellco Glass (product # 7736-10164 and -20351). Set tilt angle to ca. 12° and rotation to ca. 10 rpm.

F. Tissue Chopper

McIlwain Mechanical Tissue Chopper (Brinkmann Instr., product # 023401002).

IV. PROCEDURES

A. Prior to Dissection

Day before Culturing

1. Using sterile forceps, place five coverslips in the upside down lid of a 60 × 15-mm petri dish.
2. Place 10 μl of poly-D-lysine (MW 30,000–70,000) on each coverslip and spread thoroughly with a small sterile spatula.
3. Allow to dry overnight before culturing.

Day of Culturing

1. Set out three 150-ml glass beakers with distilled water, 70% ethanol, and 95% ethanol. Turn on hood and light alcohol lamp.
2. Sterilize instruments by dunking in 70% ethanol and then in 95% ethanol. Large scissors and spatulas should be flamed in the alcohol lamp after removing from 95% ethanol.
3. Fill culture tubes with 750 μl of medium. Seal and store refrigerated.
4. Sterilize one aclar sheet for each animal by dunking in 70% ethanol and then in 95% ethanol. Allow to dry on sterile gauze pads in the hood.
5. Break a double-edged razor blade in half, wipe with 95% ethanol, and insert into the tissue chopper.

Swab the stage and mounted blade with 95% ethanol.

6. Set the micrometer on the chopper for the desired slice thickness (start with 400 µm). Set blade force (start at the "9 o'clock" position).

7. Thaw and prepare thrombin and chicken plasma.

8. Fill one 60 × 15-mm petri dish with *HBSS + glucose* solution to cover the bottom of the dish. These dishes are used for the dissected hippocampi. One dish is needed for each animal to be dissected. Store in a refrigerator.

9. Fill one 35 × 10-mm petri dish with *HBSS + glucose + kynurenate* to cover the bottom of the dish. These dishes are for the cut slices. One dish is needed for each animal to be dissected. Store in a refrigerator.

B. Tissue Dissection

1. Mount an aclar sheet on the chopper stage.

2. Place a 60 mm × 15-mm petri dish containing chilled *HBSS + glucose* in the hood.

3. It is recommended that you start with rat or mouse pups that are 5–7 days old. Younger animals can also be used, but the dissection will be more challenging. Animals older than 10 days rarely survive more than a few days *in vitro*.

4. Place one pup in a closed beaker with a small piece of dry ice for anesthesia.

5. When anesthetized, hold animal gently by head, rinse neck area with 70% ethanol, and decapitate with large scissors.

6. Hold head right side up with the nose pointing away. Insert tip of small scissors into the foramen magnum toward the nose with the flat of the blades in the horizontal plane. Cut by moving primarily the blade on the outside of the skull. Repeat on the other side. Discard bottom of head.

7. Hold top of head nose down over the petri dish. Using a spatula dipped in HBSS, push the brain stem, cerebellum, and midbrain down gently, leaving the cortex and hippocampus inside the dorsal skull. While sliding the wet spatula along the sides of the cortex, slide the cortex and hippocampus gently into the petri dish.

8. Under a dissecting microscope, position the brain dorsal side down. Use one spatula in your left hand to hold the brain in place by impaling the anterior brain. Use the razor blade chip to free one hippocampus at a time (Fig. 1). First, cut the lateral end of the hippocampus and then cut posterior to the hippocampal fissure, using the prominent blood vessel as a guide. Repeat for the other hippocampus, making one cut along the midline and another between septal nuclei and the hippocampus.

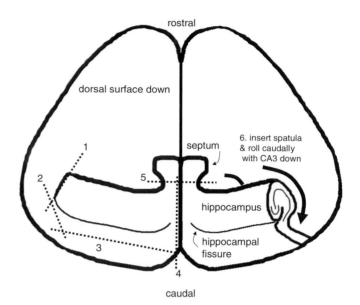

FIGURE 1 Dissection of the hippocampus. The diagram shows a view of the brain after peeling away the brain stem, midbrain, and thalamus and removing it from the skull as described in the text. It is pictured laying on its dorsal (i.e., cortical) surface and is viewed from the ventral aspect. The hippocampus is dissected free by making a series of sequential cuts with a razor blade shard as indicated by the numbered dashed lines. After cutting, a spatula is placed under the hippocampus and it is rolled caudally so that it becomes free of the cortex and rests on its CA3/DG surface.

9. Using a spatula dipped in HBSS, gently lift and roll the hippocampus away from the cortex in the rostral to caudal direction so that it rests with area CA1 up. **It is important that this be done without bending the hippocampus.**

10. Transfer the hippocampus to the aclar sheet on the chopper stage with CA1 up. The long axis of the hippocampus should be perpendicular to the length of the razor blade. Use a spatula to wipe any excess HBSS away from the hippocampus on the aclar. Failure to do so will result in the hippocampus being lifted by the blade while chopping. Repeat for the other hippocampus.

11. Chop slices.

12. Remove the aclar sheet from the chopper and, holding one corner, trim the aclar sheet into a "tongue" shape around the sliced hippocampi. Insert the aclar tongue into the small petri dish with *HBSS + glucose + kynurenate* and slide the slices off.

13. Under a dissecting microscope, separate the slices using two sterile spatulas. It is important that this be done gently, without bending the slices.

14. With proficiency, repeat steps 1–12 with a second animal.

15. If desired, X-irradiate the dishes containing the slices (1500 rad over ca. 1.5 min).

C. Mounting Slices for Roller Drum Cultures

1. Place one 20-μl drop of *chicken plasma* in the center of each coverslip in one 60-mm dish.

2. Under the dissecting scope, choose a healthy slice and transfer to the plasma droplet with a spatula. Healthy slices have a clear, well-defined continuous cell body layer and no obvious signs of damage. Repeat for all coverslips in dish.

3. Spread the plasma around and around the entire surface of the first coverslip around the slice (~3 s). Add a 20-μl drop of *thrombin* to the coverslip and mix thoroughly with the plasma over the entire coverslip (~3 s). Position the slice in the center of the coverslip and wipe excess plasma/thrombin off with a spatula. Repeat with the other coverslips and then for the second dish. After 5 min, the clot should have the consistency of a fairly liquid gelatin and should retain an indentation produced with a spatula.

4. Lift coverslips from dishes and slide into tissue culture tubes so that the bottom surface of the coverslip rests on the flat surface of the tube. Cap the tube tightly and gently tap the coverslip down to the bottom of the tube, if necessary. If slices fall off coverslips, increase the concentration of the thrombin to produce a firmer clot.

5. Place tubes in incubator.

D. Mounting Slices in Interface Culture Wells

The tissue dissection procedures are identical for the roller tube and membrane culture methods. Procedures for the latter preparations are different primarily in the substrate upon which the slices are placed. For membrane cultures, transfer slices with a wide-bore, fire-polished glass Pasteur pipette into the wells. To facilitate removal of slices from wells, small membrane pieces (e.g., Whatman Nucleopore membranes # 112107) can be cut and placed at the bottom of the insert before plating the slices.

E. Antimitotics

1. Add antimitotics after the first 4 or 5 days in culture.
2. Add 20 μl to each tube.
3. Exchange culture medium after 24 h.

F. Feeding the Cultures

1. Feed the cultures once per week.
2. Pour off medium into a beaker in the hood. About 250 μl will remain in tube.
3. Replace with 500 μl fresh medium.

G. Assessing the Health of the Cultures

1. Cultures should remain adhered firmly to the plasma and coverslip. If they fall off despite a firm clot at the time of culturing, suspect either insufficiently cleaned coverslips or a problem with the poly-L-lysine.

2. After 24–48 h *in vitro*, the slices will appear more opaque as the macrophages and other cells proliferate; remove damaged tissue. The cell body layer should remain relatively clear and visible. Glial cells migrating out of the edges of the slice should be apparent.

3. After 14 days *in vitro*, most of the macrophages should have disappeared and the cell body layer should be continuous and more transparent than the dendritic layers. Our studies indicate that the cultures are not completely mature before this time.

4. In our experience, unhealthy cultures can usually be attributed to mishandling of the tissue during dissection and mounting. If you have had good cultures previously and an entire batch goes bad, suspect a bad ingredient common to all of the cultures. If some cultures are good and some are bad, suspect a problem of dissection (it is helpful to note which slices came from which pups).

References

Debanne, D., Gähwiler, B. H., and Thompson, S. M. (1994). Asynchronous pre- and postsynaptic activity induces associative long-term depression in area CA1 of the rat hippocampus *in vitro*. *Proc. Natl. Acad. Sci. USA* **91**, 1148–1152.

Debanne, D., Guerineau, N. C., Gähwiler, B. H., and Thompson, S. M. (1995). Physiology and pharmacology of unitary synaptic connections between pairs of cells in areas CA3 and CA1 of rat hippocampal slice cultures. *J. Neurophysiol.* **3**, 1282–1294.

Gähwiler, B. H. (1981). Organotypic monolayer cultures of nervous tissue. *J. Neurosci. Methods*. **4**, 329–342.

Gähwiler, B. H. (1984). Development of the hippocampus *in vitro*: Cell types, synapses, and receptors. *Neuroscience* **11**, 751–760.

Gähwiler, B. H., Capogna, M., Debanne, D., McKinney, R. A., and Thompson, S. M. (1997). Organotypic slice cultures: A technique has come of age. *Trends Neurosci.* **20**, 471–477.

Gähwiler, B. H., Enz, A., and Hefti, F. (1987). Nerve growth factor promotes development of the rat septo-hippocampal cholinergic projection *in vitro*. *Neurosci. Lett.* **75**, 6–10.

Gähwiler, B. H., Thompson, S. M., McKinney, R. A., Debanne, D., and Robertson, R. T. (1998). Organotypic slice cultures of neural tissue. *In* "Culturing Nerve Cells" (G. Banker and K. Goslin, eds.), 2nd Ed., pp. 461–498. MIT Press, Cambridge, MA.

Stoppini, L., Buchs, P.-A., and Muller, D. (1991). A simple method for organotypic cultures of nervous tissue. *J. Neurosci. Methods* **37**, 173–182.

Zimmer, J., and Gähwiler, B. H. (1984). Cellular and connective organization of slice cultures of the rat hippocampus and fascia dentate. *J. Comp. Neurol.* **228**, 432–446.

50

Thyroid Tissue-Organotypic Culture Using a New Approach for Overcoming the Disadvantages of Conventional Organ Culture

Shuji Toda, Akifumi Ootani, Shigehisa Aoki, and Hajime Sugihara

I. INTRODUCTION

The organ culture of thyroid tissue has been applied to the studies of thyroid biology (Bussolati *et al.*, 1969; Cau *et al.*, 1976; Young and Baker, 1982). However, the conventional organ culture system can not retain viable three-dimensional (3D) thyroid follicles containing both thyrocytes and C cells for a term long enough to investigate their biological behavior, as the tissue becomes necrotic progressively. Although the conventional method allows thyrocytes to grow out only in a monolayer from the tissue periphery placed on culture dishes, it cannot enable them to organize and maintain 3D follicles due to the lack of a 3D microenvironment of extracellular matrix (ECM) (Toda *et al.*, 1996, 2001). Thyroid follicles *in vivo* are embedded in an interfollicular ECM, supported by a dense network of fenestrated capillaries (Fujita and Murakami, 1974). This suggests that both ECM and sufficient oxygen supply are important for the maintenance of follicular structure and function. By simulating this *in vivo* microenvironment of follicles, we have established a new organotypic culture using a 3D collagen gel culture of thyroid tissue fragments with improved oxygenation through air exposure (Toda *et al.*, 2002). This system maintains very well the 3D follicle structures containing both thyrocytes and C cells for more than 1 month. Furthermore, the new follicle formation from preexisting follicles (mother follicle-derived folliculogenesis) takes place actively in the peripheral zones of each tissue fragments in our system (Toda *et al.*, 2003). We herein describe a useful tool for the long-term organ culture of thyroid tissue. In relation to our method, we propose a new approach to cell type-specific culture systems on the basis of *in vivo* microenvironments of various cell types.

II. MATERIALS AND INSTRUMENTATION

Materials, reagents, and equipment are as follows: (1) 6-month-old porcine or human thyroid, (2) Eagle's MEM (EMEM, Cat. No. 05900, Nissui Pharmaceutical Co., Ltd., Tokyo, Japan), (3) dispase I solution (bacterial neutral protease, 1000 protease U/ml in EMEM, Cat. No. GD 81020, Goudoh-Shusei, Tokyo, Japan), (4) Ham's F12 (Cat. No. 05910, Nissui Pharmaceutical Co., Ltd.), (5) fetal bovine serum (FBS, Cat. No. F9423, Lot No. 92K2301, Sigma Chemical Co., MO), (6) gentamicin (Gentamicin, Shering-Plough Co., Ltd., Osaka, Japan), (7) complete medium (Ham's F-12 medium supplemented with 10% FBS and 50 μg/ml gentamicin), (8) acid-soluble type I collagen (Cellmatrix type I-A, Nitta Gelatin Co. Ltd., Osaka, Japan), (9) reconstructive buffer (2.2 g NaHCO$_3$ and 4.77 g HEPES in 100 ml 0.05 M NaOH), (10) a special culture dish of which the bottom is made with nitrocellulose membrane (30 mm diameter, Millicell-CM, Cat. No. PICAM 3050, Millipore, Bedford, MA), (11) 90-mm-diameter bacterial dish (Cat. No. SH-2OS, Terumo Co., Ltd., Tokyo, Japan), and (12) stainless-steel mesh (840 μm, Cat. No. Testing Sieve 840, Ikemoto Rikakogyo Co., Ltd., Tokyo, Japan). All materials, reagents, and equipment used in culture must be sterile.

III. PROCEDURES

A. Initial Preparation of Tissue for the Organotypic Culture of the Thyroid

1. For porcine thyroid, tissues must be kept in ice-cold EMEM from the slaughterhouse to laboratory. Wash the tissue three times with ice-cold EMEM and remove connective and adipose tissues.

2. For human thyroid, obtain tissues from surgical materials with permission and follow the same procedures as in step 1.

3. For obtaining C cell-rich follicles, use tissue derived from only the middle to upper third of the lateral lobe of porcine and human thyroids, as C cells are mainly restricted to this area.

Steps

1. Mince thyroid tissue into small fragments (~2 mm) with sterile scissors.
2. Transfer the tissue fragments (5 g) into a 100-ml beaker containing 50 ml dispase I solution and incubate at 37°C for 1 h.
3. Remove the dispase I solution by filtrating the fluid through mesh (840 μm).
4. Further mince the tissue remaining on the mesh into smaller pieces (~0.5 mm) with sterile scissors.
5. Transfer the fragments into a 50 ml-test tube, wash the tissue fragments with ice-cold EMEM by pipetting, and then centrifuge the tissue suspension at 186 g for 5 min at room temperature. Repeat these procedures twice.
6. The thyroid tissue fragments obtained as a pellet after the final centrifugation are subjected to the following organotypic culture.

B. Preparation of Culture Assembly for Organotypic Culture

In comparison with the conventional organ culture, our culture system is characterized by the following items. (1) Minced tissues are placed in a 3D collagen gel. (2) They are supplied sufficient oxygen through air exposure. The two conditions result in allowing the tissue fragments to situate under a 3D air–liquid (A-L) interface, but not under a submerged state. In this system, the tissues are kept moist and fed with the culture medium that percolated by capillary action from the medium-containing outer dish, through the acellular layer, and into the cellular layer (Toda *et al.*, 2000, 2002, 2003). Likewise, the culture cells can be stimulated by various reagents added to the culture medium of the outer dish. In addition, exposing the cellular layer to various concentrations of oxygen permits the embedded cells to be supplied those of oxygen. Figure 1 illustrates our organotypic culture system.

Steps

1. Prepare the collagen gel solution as follows (Elsdale and Bard, 1972). First, mix 8 volumes of acid-soluble type I collagen with 1 volume of 10× concentrated Ham's F-12 medium by gently pipetting in a test tube. Second, add 1 volume of reconstructive buffer to the mixture and pipette it gently. Keep this mixture on ice.

2. Place 2 ml cold collagen gel solution in a special dish containing the nitrocellulose membrane and immediately warm the dish to 37°C for at least 30 min in a 5% CO_2 incubator for gel formation. The collagen gel layer is called the acellular layer. Preparation of this layer should be completed before the beginning the steps in Section III,A.

3. Pore 1 ml cold collagen solution into a test tube containing the tissue fragments obtained as a pellet. Then, gently and fully mix 1 ml cold collagen gel solution with the tissue fragments (a total of 0.5 g). The initial amount of 5 g tissue results in the preparation of 10 culture dishes.

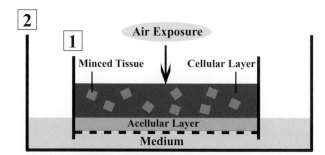

FIGURE 1 Scheme of thyroid tissue-organotypic culture system. Minced tissues embedded in type I collagen gel (cellular layer) are placed on the acellular gel (acellular layer) in the inner dish (1). The inner dish (1) is put in the outer dish (2) with culture medium. In this way, tissues in the cellular layer are localized under air exposure-induced oxygenation. Tissues are kept moist and are fed by culture medium that percolates by capillary action from the medium-containing outer dish through the acellular layer and into the cellular layer.

4. Place 1 ml collagen gel solution containing the fragments on the acellular layer and immediately warm the dish to 37°C for at least 30 min in a 5% CO_2 incubator to allow gel formation. The resultant overlayer is the cellular layer. The culture dish prepared in this way is referred to as the "inner" dish.

5. After at least 30 min, when the gel is fully firm, place the inner dish in a larger "outer" dish (90 mm in diameter) containing 10 ml complete medium.

6. Place this culture assembly in a conventional culture incubator, thereby exposing the cellular layer to a humidified air atmosphere supplemented with 5% CO_2 at 37°C. In this way, the tissue fragments are situated under a microenvironment that consists of both type I collagen and air exposure-induced oxygenation. We call this culture condition a 3D A-L interface.

C. Analyses of Organotypic Culture Cells

Culture cells can be observed by phase-contrast microscopy. The collagen gel layer containing viable cells is scraped easily from the culture assembly. The layer can be treated similar to the various tissues resected from the body and used for analyzing the cellular behavior as follows. (1) The cellular layer gels are fixed in 4% formalin, processed routinely, and embedded in paraffin. The deparaffinized and frozen sections are applied easily to histochemistry, immunohistochemistry, and *in situ* hybridization. (2) To examine the fine structure of the cells, transmission electron microscopy is carried out using cellular layer gels fixed in 2.5% glutaraldehyde and prepared by a standard method. (3) Biochemical and genetic analyses of the cells can be carried out by the various methods described in this volume.

D. Examples of Thyroid Tissue-Organotypic Cultures

In this system, viable 3D follicles within thyroid tissue fragments are maintained for more than 1 month. These follicle structures consist of thyrocytes and C cells with their specific differentiation (Fig. 2). In the tissue periphery, thyrocytes undergo actively growth and mother follicle-derived thyroid folliculogenesis. Likewise, isolated or clustered thyrocytes, which were localized in the tissue periphery at the starting time of the culture, reconstruct follicles. C cells show no proliferative ability and cannot grow even with the stimulation of various concentrations of free calcium. Most endothelial cells of capillaries disappear until 7 days in culture (for further details, see Toda *et al.*, 1990, 1992, 1993, 1997, 2002, 2003).

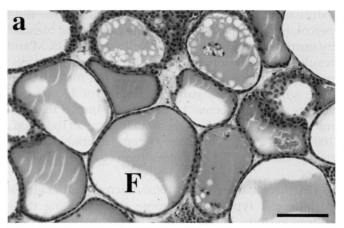

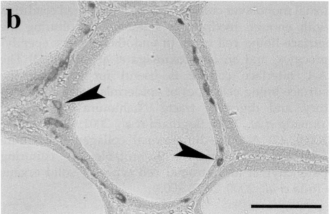

FIGURE 2 Histology of thyroid tissues (a) and immunohistochemistry for calcitonin (b) in the organotypic culture. (a) At 40 days in culture, viable thyroid follicles enclosed by thyrocytes contain colloid substance in their lumens (F). (b) At 30 days in culture, thyroid follicles consisting of both thyrocytes and C cells (arrowheads) are clearly maintained. Scale bar: 100 μm.

IV. COMMENTS

In relation to our culture system, it seems that the time has come to reconsider conventional culture methods in a cell type-specific way. The microenvironments of many cell types of the body are subdivided mainly into the following three types. The first is that of parenchymal and stromal cell types of solid organs, e.g., thyroid, adrenal, and liver. The second is that of surface-lining cell types on which sufficient liquid is not overlayed, e.g., those of the skin, cornea, respiratory, and digestive tracts. The third is the microenvironment of surface-lining cell types on which enough fluid is overlayed, e.g., those of the cardiovascular system and cerebral ventricle. With respect to the first microenvironment, the following is our notion regarding that of the *in vivo* extravascular stroma by which various cell types are supported. The

P A R T

B

VIRUSES

Growth and Purification of Viruses

14

Growth and Purification of Viruses

51

Growth of Semliki Forest Virus

Mathilda Sjöberg and Henrik Garoff

I. INTRODUCTION

Semliki Forest virus (SFV) is an enveloped RNA virus, with a genome of positive polarity that belongs to the Alphavirus group of the family Togaviridae. It readily infects a variety of mammalian and insect cells and can be grown to high titres in tissue culture. Upon infection, host cell-specific synthesis of macromolecules is suppressed within a few hours, structural viral components are made, and new virus particles bud out from the plasma membrane of the infected cell. The SFV particle is spherical with a diameter of approximately 65 nm (molecular mass ≈ 42×10^3 kDa). It consists of a nucleocapsid (NC), a single copy of the RNA genome packed together with 240 copies of a capsid (C) protein (33 kDa), that is surrounded by a lipid membrane in which 80 glycoprotein complexes, the viral spikes, are anchored. The viral spikes are trimeric associations of a protein complex: two membrane-spanning proteins, E1 and E2 (49 and 52 kDa), and a peripheral protein, E3 (10 kDa) (Garoff *et al.*, 1982; Strauss and Strauss, 1994). This article provides protocols to grow SFV in intermediate scale (up to ≈1.5 mg; protocol A) and small scale (^{35}S-methionine labelled; protocol B).

II. MATERIALS AND INSTRUMENTATION

Culture medium Glasgow minimum essential medium (MEM) (BHK-21) (Cat. No. 21710), foetal bovine serum (FBS) (Cat. No. 10106), tryptose phosphate broth (Cat. No. 18050), 1 *M* HEPES (Cat. No. 15630), L-glutamine 200 m*M* (100X) (Cat. No. 25030), penicillin–streptomycin (Cat. No. 15140), MEM (Cat. No. 21090-022), bovine albumin fraction V solution 7.5% (BSA; Cat. No. 15260-037), and phosphate-buffered saline (PBS) Dulbecco's with Ca^{2+} and Mg^{2+} (Cat. No. 14040) are from GIBCO BRL. Sea-plaque agarose (Cat. No. 50100) is from FMC Bio Products. Redivue-[^{35}S]methionine (Cat. No. AG 1094) is from Amersham Biosciences. Sucrose (Cat. No. 0335) is from Amresco. Tris (Cat. No. 146861) is from Angus. Sodium chloride (NaCl) (Cat. No. 106404), HCl (Cat. No. M317), and Titriplex III (EDTA; Cat. No. 108418) are from Merck. NaOH (Cat. No. 05-400201) is from EKA Nobel AB (Tamro). Cholesterol (Cat. No. C3045), neutral red (Cat. No. N6634), and methionine-free MEM (Cat. No. 31900-012) are from Sigma-Aldrich. The density gradient fractionator (Model 185) is from Instrumentation Specialities Company (ISCO). Filter papers No. 1 (Cat. No. 1001 090) are from Whatman. The 75-cm² flasks (Cat. No. 3375) and 162-cm² flasks (Cat. No. 3150) are from Corning Life Sciences. Sixty-millimeter tissue culture plates (Cat. No. 50288) are from Nunc. BHK-21 cells C-13 (Cat. No. CRL-8544) are from American Type Culture Collection. Cotton-tipped applicators are from Solon manufacturing company. Fifty-milliliter Nalgene tubes (Cat. No. 3139-0050) are from Nalge Incorporated. SW 28 tubes (Cat. No. 344058), SW 40 tubes (Cat. No. 331374), and 6 ml-scintillation vials (Cat. No. 566831) are from Beckman. Eppendorf tubes, 1.5 ml (Cat. No. 0030 102.002) and 2.0 ml (Cat. No. 0030 120.094), are from Eppendorf-Netheler-Hinz GmbH. Emulsifier Safe is from Packard Instrument Co. Inc.

III. PROCEDURES

A. Growth of SFV

Solutions

1. *Complete BHK-21 medium*: To make 585 ml, add 50 ml of tryptose phosphate broth, 25 ml of FBS, 5 ml of 1 M HEPES, and 5 ml of 200 mM glutamine to 500 ml of Glasgow MEM (BHK-21). Store at 4°C.

2. *Complete BHK-21 medium + cholesterol* (optional): To make 250 ml, add 0.5 ml 10 mg/ml cholesterol stock solution (dissolve at 37°C prior to use) to 250 ml complete BHK medium.

3. *10 mg/ml cholesterol stock solution* (optional): Add 50 mg cholesterol to 5 ml 99.5% ethanol. Dissolve. Aliquot and store at −20°C.

4. *Supplemented MEM*: To make 529 ml, add 14 ml of 7.5% BSA, 5 ml of 1 M HEPES, 5 ml of 200 mM glutamine, 5 ml of 10,000 U/ml penicillin/10,000 µg/ml streptomycin to 500 ml of MEM. Store at 4°C.

5. *TN*: 50 mM Tris–HCl, pH 7.4, 100 mM NaCl. To make 500 ml, add 3.0 g of Tris and 2.9 g NaCl to distilled water, adjust pH to 7.4 by adding 1 M HCl, and complete the volume to 500 ml. Autoclave. Store at room temperature.

6. *0.25 M EDTA pH 8.0 stock solution*: To make 50 ml, add 4.65 g of Titriplex III to 30 ml of distilled water. Adjust the pH to 8.0 by adding 1 M NaOH and complete the volume to 50 ml. Autoclave. Store at room temperature.

7. *TNE*: 50 mM Tris–HCl, pH 7.4, 100 mM NaCl, 0.5 mM EDTA. To make 100 ml, add 200 µl 0.25 M EDTA, pH 8.0, stock solution to 100 ml TN. Store at room temperature.

8. *200 g/kg sucrose solution*: To make 100 g, weigh 20 g of sucrose and adjust to 100 g with TNE. Store at −20°C.

Steps

1. Grow BHK-21 cells to 100% confluency in a 162-cm^2 tissue culture bottle in complete BHK medium with or without cholesterol ($\approx$2.2 × 10^7 cells/bottle).

2. Dilute the virus to a concentration of 1.1 × 10^6 pfu/ml in supplemented MEM.

3. Remove the medium and wash the cells with 10 ml of PBS (with Ca^{2+} and Mg^{2+}). Add 2.0 ml of the diluted virus to the cells. Incubate for 60 min at 37°C and 5% CO$_2$. Tilt the bottle every 20 minute to ensure even distribution of virus particles.

4. Remove virus solution and rinse with 10 ml PBS (with Ca^{2+} and Mg^{2+}). Add 30 ml complete BHK medium with or without cholesterol and incubate for 18 h at 37°C and 5% CO$_2$.

5. Transfer the virus containing medium to a 50-ml Nalgene tube (30 ml/tube) and centrifuge at 26,500 g and 4°C for 10 min in a J2-21 Beckman centrifuge equipped with a JS 13.1 rotor (13,000 rpm).

6. Pipette the medium to a fresh tube, without disturbing the pellet, and centrifuge as in step 5.

7. Repeat step 6.

8. Transfer 20 ml of the clarified medium to an SW 28-tube. Layer 4.0 ml 200 g/kg sucrose in TNE under the sample. Add the remaining 8–10 ml of clarified medium to the top of the tube and centrifuge at 112,000 g and 4°C for 90 min in a L8-M Beckman centrifuge equipped with a SW 28 rotor (25,000 rpm).

9. Aspirate off the supernatant. Pour the last 0.5 ml into the pipette by tilting the tube. Wipe off the last drops of supernatant from the inside of the tube with a sterile, cotton-tipped applicator.

10. Add 200 µl TNE, cover the tube with Parafilm, and leave on ice for 15 h.

11. Pass the virus suspension slowly up and down in a Gilson P-200 pipette to resuspend the virus. Transfer the suspension to a 1.5-ml Eppendorf tube, rinse the SW 28 tube with 100 µl TNE, and pool.

12. Mix 10 µl of the virus suspension with 90 µl TNE (dilution factor, $D = 10^{-1}$) and measure the optical density at 260 and 280 nm, e.g., in a Pharmacia Ultrospec plus spectrophotometer equipped with a 50-µl cuvette with a 10-mm path length. Calculate the ratio $R = A_{260}/A_{280}$ ($R = 1.4 \pm 0.1$ for pure SFV particles). Estimate the virus concentration (C_{SFV}) if the preparation is sufficiently pure; $C_{SFV} = A_{260}/(D \times 8)$ [mg/ml].

13. Analyse the virus preparation by SDS–PAGE under nonreducing conditions and visualise the bands by Coomassie brilliant blue staining.

14. Aliquot the virus suspension in smaller portions, quick freeze in dry ice/ethanol, and store at −70°C.

B. Growth of ^{35}S-Methionine-Labelled SFV

Solutions

1. *Starvation medium*: To make 103 ml, add 1 ml of 200 mM glutamine, 1 ml of 1 M HEPES, and 1 ml of 10,000 U/ml penicillin/10,000 µg/ml streptomycin to 100 ml of methionine-free MEM. Store at 4°C.

2. *Labelling medium*: To make 7.5 ml, add 50 µl Redivue [^{35}S]methionine (370 MBq/ml) to 7.5 ml starvation medium.

3. *550 g/kg sucrose*: To make 50 g, weigh 27.5 g sucrose and adjust to 50 g with TNE. Store at −20°C.

4. *Complete BHK medium, supplemented MEM, and 200 g/kg sucrose*: see Section III,A.

Steps

1. Grow BHK-21 cells to 100% confluency in a 75-cm^2 tissue culture bottle in complete BHK medium ($\approx 1 \times 10^7$ cells/bottle).

2. Dilute the virus to a concentration of 1×10^8 plaque-forming units (pfu)/ml in supplemented MEM.

3. Remove the medium and wash the cells with 10 ml of PBS (with Ca^{2+} and Mg^{2+}). Add 1.0 ml of the diluted virus to the cells. Incubate for 60 min at 37°C and 5% CO_2. Tilt the bottle every 20 min to ensure even distribution of virus particles.

4. Remove virus solution, add 15 ml complete BHK medium, and incubate for 3.5 h at 37°C and 5% CO_2.

5. Rinse the cells two times with 10 ml PBS (with Ca^{2+} and Mg^{2+}). Add 15 ml of starvation medium, and incubate for 30 min at 37°C and 5% CO_2.

6. Remove the starvation medium, add 7.5 ml of labelling medium, and continue incubation at 37°C and 5% CO_2 for 15–16 h.

7. Harvest the labelling medium and dispense in four 2-ml Eppendorf tubes.

8. Centrifuge at 12,400 g and 4°C for 5 min in an Eppendorf 5416 centrifuge equipped with a 16 F24-11 rotor (11,000 rpm).

9. Transfer the supernatants into new tubes, without disturbing the pellet, and repeat centrifugation as in step 8.

10. Repeat step 9.

11. Pool the clarified medium in an SW 40 tube. Layer 4.5 ml of 200 g/kg sucrose under the radioactive medium and 1 ml 550 g/kg sucrose under the lighter sucrose.

12. Centrifuge at 143,000 g and 4°C for 2 h in an L8-M Beckman centrifuge equipped with a SW 40 rotor (30,000 rpm).

13. Connect the ISCO density gradient fractionator to a peristaltic pump (Pharmacia Pump P1, inner diameter of tubing = 1.0 mm, speed setting = 2 × 10) and a Gilson FC 203B fraction collector. Clamp the SW 40 tube in the fraction collector, perforate the tube from the bottom, and collect 20 fractions of five drops each.

14. To measure the radioactivity in each fraction, pipette 50 μl H_2O followed by 2 μl of the fraction (delivered into the water droplet) and 3 ml Emulsifier-Safe into twenty 6-ml scintillation vials. Mix and count using the ^{35}S window in a liquid scintillation counter.

15. Pool the peak fractions (usually fractions 5–9) and analyse the virus preparation by SDS–PAGE under nonreducing conditions. Aliquot 25-μl portions in 1.5-ml Eppendorf tubes, quick freeze in dry ice/ethanol, and store at –70°C.

C. Quantitation of Infectious Virus Particles by Plaque Titration

Solutions

1. *Complete BHK medium and supplemented MEM:* see Section III,A.

2. *Agarose stock solution:* To make 100 ml, add 1.9 g of Seaplaque, low melting point agarose to 100 ml MEM. Autoclave and store at 4°C.

3. *BHK-medium + 2x additives:* To make 136 ml, add 20 ml of tryptose phosphate broth, 10 ml of FBS, 2 ml of 1 M HEPES, 2 ml of 200 mM glutamine, and 2 ml of 10,000 U/ml penicillin/10,000 μg/ml streptomycin to 100 ml of Glasgow-MEM (BHK-21). Store at 4°C.

4. *Neutral red (2% stock solution):* To make 50 ml, add 1.0 g neutral red to 50 ml H_2O. Filter through a Whatman No. 1 paper and store at room temperature.

5. *Neutral red stain:* To make 100 ml, add 3 ml of neutral red, 2% stock solution to 99 ml of PBS (with Ca^{2+} and Mg^{2+}). Use fresh.

Steps

1. Grow BHK-21 cells to $\approx$90% confluency on 60-mm tissue culture plates in complete BHK medium ($\approx 3.4 \times 10^6$ cells per plate). Prepare 10 plates per virus preparation to be titrated and 2 extra plates to be used as negative and positive controls.

2. To make a serial dilution of the virus preparation, label ten 2.0-ml Eppendorf tubes (1–10) and pipette 445 μl supplemented MEM to the first tube and 1.35 ml supplemented MEM to the following nine tubes. Add 5 μl of the virus preparation to the first tube, mix thoroughly, and transfer 150 μl of the mixture to the second tube. Mix the contents of the second tube and transfer 150 μl to the third tube using a fresh pipette tip. Continue in the same fashion with the last seven tubes. Make two parallel dilution series for each virus preparation to be titrated.

3. Melt the agarose stock solution in a microwave oven. Mix 35 ml of the agarose stock solution and 35 ml of the BHK-medium + 2× additives to complete the overlay solution. Keep in a 37°C water bath until use.

4. Remove the medium from the cells and wash with 2 ml of PBS (with Ca^{2+} and Mg^{2+}).

5. Add 1.0 ml of diluted virus (use tubes 6–10 from the two dilution series; these correspond to dilution factors 10^{-7} through 10^{-11}). Use supplemented MEM

as a negative control and a suitable dilution of a known virus stock (if available) as a positive control.

6. Incubate for 60 min 37°C and 5% CO_2. Tilt the plates every 20 min to ensure even distribution of virus particles.

7. Remove the virus solution from the cells, rinse with 2 ml of PBS (with Ca^{2+} and Mg^{2+}), and add 4 ml of overlay solution (keep the bottle in a beaker filled with 37°C water). Leave the plates at room temperature until the agarose solidifies.

8. Incubate the plates for 48 h at 37°C and 5% CO_2.

9. Add 3 ml of neutral red stain and incubate for 3 h at 37°C (5% CO_2 is optional).

10. Score the number of plaques (diffuse, clear areas on a dark red background) on each plate. To calculate the virus titre as plaque-forming units per milliliter, divide the number of plaques per plate by the appropriate dilution factor.

IV. COMMENTS

Cells used for SFV infections in this protocol (old BHK cells) are BHK-21 cells that with time in culture have transformed further. In doing so, they have lost the extended form of normal BHK-21 cells (freshly obtained from ATCC) and appear more like penta- or hexagons. The SFV strain used [SFV4 (Liljeström et al., 1991)] had undergone an unknown number of passages in the old BHK cells (Glasgow et al., 1991) before it was cloned. At present the specific titre (i.e., the number of infectious virus particles divided by the total number of virus particles produced) is approximately 10 times higher when a virus preparation is titrated on old BHK cells as compared to normal BHK-21 cells. This is also the case when the virus is produced in normal BHK-21 cells and most likely reflects an adaptive change in SFV4 that facilitates entry into the old BHK cells.

The expected yield of SFV particles is approximately 1 μg/cm^2 of confluent BHK cells. When larger amounts (mg) of SFV are desirable, the amount of complete BHK medium used in the production step (Section III,A, step 4) can be reduced down to 20 ml per 162-cm^2 bottle.

An alternative method to estimate the amount of SFV in a preparation is to use CBB-stained SDS–PAGE gels and compare the intensity of the capsid protein band to that of known amounts of BSA ran under reducing conditions on the same gel (five wells with 0.15, 0.3, 0.6, 1.2, and 2.4 μg, respectively, is sufficient). The amount of C protein (m_C) in a band is

half the amount of BSA in a band of equal intensity. The amount of SFV (m_{SFV}) is calculated as $m_{SFV} = (3 \times m_C)/2$.

The stability of the produced SFV is improved if the producer cells are supplied with cholesterol in the growth media. This procedure is indicated as optional in the protocol and is not necessary for the production of stock virus intended for infection of new cells.

The crude virus preparation obtained in Section III,A, step 11 can be purified by isopycnic tartrate gradient centrifugation as described by Haag and colleagues (2002). To this end, cholesterol should be used during virus production and the TNE used in step 10 should be replaced by TNM (50 mM Tris–HCl, 50 mM NaCl, 10 mM $MgCl_2$, pH 7.4) for improved virus stability.

The E1 and E2 proteins of SFV comigrate upon SDS–PAGE under reducing conditions. Without reduction, E1 and E2 are separated readily, with E2 showing a higher apparent molecular mass than E1.

Intact SFV particles contains 88% (w/w) of protein and 12% (w/w) of RNA (Garoff et al., 1982). This is equivalent to an A_{260}/A_{280} ratio of 1.4, provided that A_{260}/A_{280} of pure RNA equals 2.0 (Glaser, 1995; Manchester, 1995). Deviations from this figure ($A_{260}/A_{280} = 1.4 \pm 0.1$) imply that the SFV preparation contains impurities and/or defective particles.

To maintain high virus quality in successive SFV preparations, it is important to use a low multiplicity of infection (MOI). The use of MOI = 0.1 (i.e., 0.1 infectious particle per cell) or less ensures that the initial infection is caused by a single virus particle. In this case, virus particles that carry deletions or other deleterious mutations in their genomes cannot be rescued by multiple infection with functional virus particles. If a high MOI is used in a series of successive infections, the number of so-called defective interfering (DI) particles will increase dramatically (Stark and Kennedy, 1978). The presence of high numbers of DI particles may express itself by low specific infectivity of the newly produced virus. In single round infections, such as radiolabelling experiments, a MOI of 5 to 10 can be advantageous as this will produce a synchronised burst of SFV production in the shortest possible time.

V. PITFALLS

Avoid repeated freeze/thaw cycles, as this will reduce virus infectivity.

If SFV infection is carried out in serum-containing medium, e.g., complete BHK medium, the infectivity of the particles is reduced dramatically. Without interference, at maximum 30% complete BHK medium may be present during infection.

Efficient clarification of the virus containing medium (Section III,A, steps 5–7) is important. Cell debris present during virus pelletation (step 8) will glue the virus particles together and make resuspension difficult.

To preserve the three-dimensional structure of the virus particles, it is important to allow sufficient time for resuspension. Do not decrease the time that the virus is left on ice (Section III,A, step 10).

When trace amounts of SFV proteins are separated on SDS–PAGE, the C protein tends to smear over the lane. This can be avoided if a small volume of BHK cell lysate is included in the sample buffer prior to heating (add $1\,\mu l$ BHK cell lysate for every $10\,\mu l$ of SDS–PAGE sample buffer). To make BHK cell lysate, grow BHK-21 cells to 100% confluency in a 35-mm tissue culture plate, lyse in $300\,\mu l$ 1× lysis buffer, and remove cell nuclei by low-speed centrifugation. Store the BHK cell lysate at $-20°C$. The addition of cell lysate is not necessary when the amount of virus protein in the gel is sufficient for Coomassie brilliant blue staining. Silver staining is not recommended.

References

Garoff, H., Kondor-Koch, C., and Riedel, H. (1982). Structure and assembly of alphaviruses. *Curr. Top. Microbiol. Immunol.* **99**, 1–50.

Glaser, J. A. (1995). Validity of nucleic acid purities monitored by 260 nm/280 nm absorbance ratios. *Biotechniques* **18**, 62–63.

Glasgow, G. M., Sheahan, B. J., Atkins, G. J., Wahlberg, J. M., Salminen, A., and Liljeström, P. (1991). Two mutations in the envelope glycoprotein E2 of Semliki Forest virus affecting the maturation and entry patterns of the virus alter pathogenicity for mice. *Virology* **185**, 741–748.

Haag, L., Garoff, H., Xing, L., Hammar, L., Kan, S.-T., and Cheng, R. H. (2002). Acid-induced movements in the glycoprotein shell of an alphavirus turn the spikes into membrane fusion mode. *EMBO J.* **21**, 255–264.

Liljeström, P., Lusa, S., Huylebroeck, D., and Garoff, H. (1991). In vitro mutagenesis of a full-length cDNA clone of Semliki Forest virus: The 6000-molecular-weight membrane protein modulates virus release. *J. Virol.* **65**, 4107–4113.

Manchester, K. L. (1995). Value of A_{260}/A_{280} ratios for measurement of purity of nucleic acid. *Biotechniques* **19**, 208–210.

Stark, C., and Kennedy, S. I. T. (1978). The generation and propagation of defective-interfering particles of Semliki Forest virus in different cell types. *Virology* **89**, 285–299.

Strauss, J. H., and Strauss, E. G. (1994). The alpha viruses: Gene expression, replication and evolution. *Microbiol. Rev.* **58**, 491–562.

52

Design and Production of Human Immunodeficiency Virus-Derived Vectors

Patrick Salmon and Didier Trono

I. INTRODUCTION

Lentiviral vectors (LV) can govern the efficient delivery and stable integration of transgenes both *in vitro* and *in vivo*, can transduce a wide range of targets, including stem cells, and can be used for generating transgenic animals from several species. Lentiviral vector-mediated gene transfer results in the ubiquitous, tissue-specific, and/or regulated expression of these transgenes, depending on the promoter contained in the vector. Finally, this gene delivery system can mediate the knockdown of endogenous genes by RNA interference via the polymerase III promoter-driven production of small hairpin RNAs.

II. DESIGN

The potential of lentiviral vectors was first revealed in 1996 through the demonstration that they could transduce neurons *in vivo* (Naldini *et al.*, 1996). Since then, many improvements have been brought to achieve high levels of efficiency and biosafety. The principle, however, remains the same and consists of building replication-defective recombinant chimeric lentiviral particles from three different components: the genomic RNA, the internal structural and enzymatic proteins, and the envelope glycoprotein. A schematic diagram of the evolution of human immunodeficiency virus (HIV)-based vectors is represented in Fig. 1. The genomic RNA contains all *cis*-acting

sequences, whereas packaging plasmids contain all the *trans*-acting proteins necessary for adequate transcription, packaging, reverse transcription, and integration.

A. Envelope

Although various envelope proteins can efficiently pseudotype LV particles (Sandrin *et al.*, 2002), the G protein of the vesicular stomatitis virus (VSV) is the most widely used. The main reasons for this choice are that the VSV envelope (1) allows for high titers achieved in unconcentrated supernatants; (2) provides an extremely wide range for the transduction of target cells (virtually all mammalian cells of any tissue tested so far can be transduced by VSV-G pseudotyped LVs); (3) is very robust, allowing for concentration by ultracentrifugation; and (4) has a good resistance to freeze–thaw cycles.

B. Core and Enzymatic Components ("Packaging System")

The first-generation lentiviral vectors were manufactured using a packaging system that comprised all HIV genes but the envelope (Naldini *et al.*, 1996). In a so-called second-generation system, five of the nine HIV-1 genes were eliminated, leaving the *gag* and *pol* reading frames, which encode for the structural and enzymatic components of the virion, respectively, and the *tat* and *rev* genes, fulfilling transcriptional and posttranscriptional functions (Zufferey *et al.*, 1997). Sensitive tests have so far failed to detect replication-competent recombinants (RCRs) with this system. This

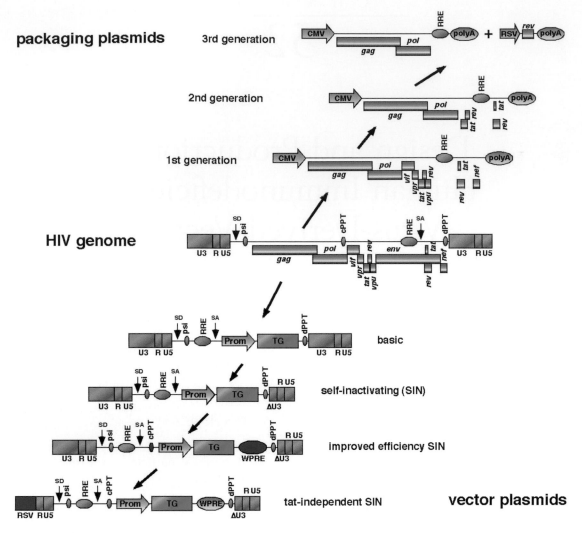

FIGURE 1 **Evolution in the design of HIV-1-based LV vectors.** HIV-1-based LV vectors are derived from wild-type HIV-1 by dissociation of the *trans*-acting components (blue boxes, above HIV genome) coding for structural and accessory proteins and the *cis*-acting sequences required for packaging and reverse transcription of the genomic RNA (golden boxes, below HIV genome). Sequences added between two vector versions are in red. CMV, human cytomegalovirus immediate-early promoter; RRE, rev-responsive element; RSV, Rous sarcoma promoter; polyA, polyadenylation site; U3-R-U5, HIV-1 LTR: SD, major splice donor; psi, HIV-1 packaging signal; cPPT, central polypurine tract; SA, splice acceptor; dPPT, distal (3′) polypurine tract; Prom, promoter of the internal expression cassette; TG, transgene of the internal expression cassette; ΔU3, self-inactivating deletion of the U3 part of the HIV-1 LTR; WPRE, woodchuck hepatitis virus posttranscriptional regulatory element.

good safety record, combined with its high efficiency and ease of use, explains why the second-generation lentiviral vector packaging system is utilized for most experimental purposes. In a third-generation system, geared toward clinical applications, only *gag*, *pol*, and *rev* genes are still present, using a chimeric 5′ long terminal repeat (LTR) to ensure transcription in the absence of Tat (see later).

C. Genomic Vector

The genetic information contained in the vector genome is the only one transferred to the target cells. Early genomic vectors were composed of the following components: the 5′ LTR, the major splice donor, the packaging signal (encompassing the 5′ part of the gag gene), the Rev-responsive element (RRE), the envelope

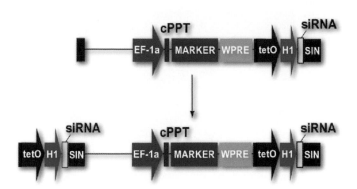

FIGURE 2 Vector for RNA interference. Small hairpin (interfering) RNAs (shRNA or siRNA) are expressed from a polymerase III promoter as described (Brummelkamp *et al.*, 2002). In this example, the expression cassette is placed in the U3 region of the 3' LTR. Because of the modalities of reverse transcription, two copies of the siRNA inducing module will be present in the integrated provirus, facilitating high levels of production. For convenience, a transgene can be placed in the same vector, downstream of an internal promoter.

splice acceptor, the internal expression cassette containing the transgene, and the 3' LTR. In the latest generations, several improvements have been introduced. The woodchuck hepatitis virus posttranscriptional regulatory element (WPRE) has been added to increase the overall levels of transcripts in both producer and target cells, hence increasing titers and transgene expression (Zufferey *et al.*, 1999). The central polypurine tract of HIV has also been added back in the central portion of the genome of the transgene RNA (Follenzi *et al.*, 2000; Zennou *et al.*, 2000). This increases titers in some targets. The 3' LTR has been deleted in the U3 region to remove all transcriptionally active sequences, creating the so-called self-inactivating (SIN) LTR (Zufferey *et al.*, 1998). Finally, chimeric 5' LTRs have been constructed in order to render the LV promoter Tat independent. This has been achieved by replacing the U3 region of the 5' LTR with either the CMV enhancer (CCL LTR) or the corresponding Rous sarcoma virus (RSV) U3 sequence (RRL LTR) (Dull *et al.*, 1998). Vectors containing such promoters can be produced at high titers in the absence of the Tat HIV transactivator. However, the Rev dependence of these third-generation LV has been maintained in order to maximize the number of recombination events that would be necessary to generate an RCR (Fig. 2).

III. MATERIALS AND REAGENTS

293T and HeLa cell lines (ATCC) (www.atcc.org)
Chemicals (Sigma-Fluka) (www.sigmaaldrich.com)

Cell culture media and additives (GIBCO-BRL-Life Technologies) (www.gibcobrl.com) and CellGenix Technologie Transfer GmbH, Germany (www.cellgenix.com)
Plastics for tissue cultures, flow cytometer (BD Biosciences) (www.bdbiosciences.com)
DNA purification kits (Qiagen) (www1.qiagen.com) and Genomed (www.genomed-dna.com)
Centrifuges (Sorvall) (www.sorvall.com)
Filters (Millipore) (www.millipore.com)
Ultracenrifuge tubes (Beckman) (www.beckman.com)
Sequence detector ABI7700 plus ABI SDS software for analysis, optical reaction plates and caps for QPCR (Applied Biosystems) (http://www.appliedbiosystems.com)
QPCR reaction mixes and probes (Eurogentech) (www.eurogentec.com)
HIV-1 p24 antigen capture assay (AIDS Vaccine Program) (Frederick, MD. email: schadent@mail.ncifcrf.gov—http://web.ncifcrf.gov)

IV. PRODUCTION

The production of LV can be achieved by transient transfection of the plasmid set into 293T cells by the calcium phosphate method or from stable producer cell lines. Although proof of principle for the latter approach has been provided (Klages *et al.*, 2000), it still suffers from limitations, e.g., for the production of SIN vectors. Unless very large amounts of a same vector are needed on a regular basis, transient production is still the method of choice for research purposes.

A. Cells

Cells (293T/17 from ATCC Cat. No. SD-3515) are probably the most critical factor for good titers. They need to be passaged every 2–3 days, as they start to form clumps that cannot be dissociated with one round of trypsin.

B. Solutions

1. *0.5 M CaCl₂*: Dissolve 36.75 g of CaCl₂ and 2H₂O (MW 147) (*SigmaUltra* Cat. No. *C5080*) into 500 ml of H₂O (distilled or double distilled). Store at −70°C in 50-ml aliquots. Once thawed, the CaCl₂ solution can be kept at 4°C for several weeks without observing a significant change in the transfection efficiency.

2. *2x HeBS*: Dissolve 16.36 g of NaCl (MW 58.44) (*SigmaUltra* Cat. No. *S7653* (0.28 M final), 11.9 g of

HEPES (MW 238.3) (*SigmaUltra* Cat. No. *H7523*) (0.05 *M* final), and 0.213 g of Na$_2$HPO$_4$, anhydrous (MW 142) (*SigmaUltra* Cat. No. *S7907*) (1.5 m*M* final) into 800 ml of H$_2$O (distilled or double distilled). Adjust pH to 7.00 with 10 *M* NaOH. Be careful, as obtaining a proper pH is *very* important. Below pH 6.95, the precipitate will not form, above pH 7.05, the precipitate will be coarse and transfection efficiency low. Then add H$_2$O to 1000 ml and make the final pH adjustment. Store at –70°C in 50-ml aliquots. Once thawed, the HeBS solution can be kept at 4°C for several weeks without observing a significant change in the transfection efficiency.

3. *Solution for mixing with plasmids*: 50 ml H$_2$O distilled or double-distilled, 1 *M* HEPES, pH 7.3 (*Gibco-BRL*, Cat. No. *15630-056*) 125 μl (2.5 m*M* final). We have observed that the aspect and quality of the precipitates can vary among batches of distilled water. To circumvent this problem, we advise buffering distilled water that is used to dilute the plasmids. A final concentration of 2.5 m*M* HEPES in the water will help maintain a proper pH and will not compete for the final pH with the HeBS, pH 7.00 (HEPES, pH 7.00, provided by HeBS is 25 m*M* final, whereas HEPES, pH 7.3, provided by water is 0.625 *M* final). Store at 4°C.

C. DNA

We use JetStar kits (Genomed GmbH, Germany) or Qiagen kits (Qiagen, GmbH, Germany) to prepare DNAs for transfection. In any case, the last step of the DNA prep should be an additional precipitation with ethanol (EtOH) and resuspension in 10 m*M* Tris (*SigmaUltra* Cat. No. *T6791*)/1 m*M* EDTA (*SigmaUltra* Cat. No. *E6758*) (TE 10/1). Do not treat DNA with phenol/chloroform as it may result in chemical alterations. Also, to avoid salt coprecipitation, we do not precipitate DNA below 20°C.

D. Transfection and Harvesting

The day before transfection, detach the 293T cells using trypsin/EDTA (Cat. No. 25300, GibcoBRL Life Technologies). Seed the 293T cells at 1 to 3 millions (cells must be approximately one-fourth to one-third confluent on the day of transfection) per 10-cm culture dishes (Cat. No. 353003, BD Biosciences) with 10 ml of D10 complete medium composed of DMEM (Cat. No. 41966, GibcoBRL Life Technologies) supplemented with 10% fetal calf serum (FCS, Cat. No. 10099, Gibco-BRL Life Technologies), 1% penicillin–streptomycin (Cat. No. 15140, GibcoBRL Life Technologies), and 1% L-glutamine (Cat. No. 25030, GibcoBRL Life Technologies) (Fig. 3).

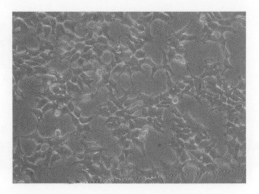

FIGURE 3 Phase-contrast photograph of 293T cells. 293T/17 cells are seeded the day before transfection at 1 to 3 million per 10-cm culture dish. At the time of transfection, cells must have the morphology and density as shown here.

On day 0, in the evening, make the precipitate according to the recipes[1] (cf Table I for one 10-cm plate).

Adjust to 250 μl with buffered water. Add 250 μl of 0.5 *M* CaCl$_2$. Mix well and add this mix, *dropwise, slowly* (one drop every other second), on 500 μl of HeBS 2×, while vortexing at maximum speed. Let stand still for 20 to 30 min minimum (40 min max) on the bench. This time is critical for optimal formation of the precipitate. Excessive incubation times may induce the formation of a coarse precipitate, which is detrimental to transfection efficiency. A too short incubation will not induce the formation of a significant amount of precipitate, which will have no chance to develop further once diluted in the 10 ml of culture medium. Also, it is advisory to incubate the solutions at room temperature or 37°C before mixing to standardize the precipitate characteristics. Add the precipitate slowly, dropwise on the cell monolayer. The plate should then be shaken gently but not stirred or all the precipitate will be in the center. At this point, the precipitate may not be visible, for it is too fine and will take several hours to sediment on the cells. If the precipitate is readily visible, it is too coarse and the transfection will most likely be less efficient.

On day 1 in the morning, check the cells. At this point, a very fine and sandy precipitate all over the plate should be seen, except on the cells and in their vicinity, possibly because cells absorb the CaPO$_4$/DNA precipitate. Discard the medium after gentle, but firm, stirring (to eliminate the maximal amount of precipitate) and replace it with 10–15 ml of fresh medium.

---

[1] These recipes are the result of long optimizations in our laboratory and in the laboratory of Luigi Naldini and discussions with Antonia Follenzi.

TABLE I Recipes

	Second generation		Third generation	
Recipe 1				
Envelope plasmid	pMD2G	5 μg	pMD2G	5 μg
Packaging plasmid	pCMVΔR8.74	15 μg	pMDLg/pRRE	10 μg
Rev-expression plasmid	pRSV-Rev		pRSV-Rev	5 μg
Transfer vector	pHR[a]	20 μg	pRRL[a]	20 μg
Recipe 2				
Envelope plasmid	pMD2G	3 μg	pMD2G	3 μg
Packaging plasmid	pCMVΔR8.74	6.5 μg	pMDLg/pRRE	5 μg
Rev-expression plasmid	pRSV-Rev	2 μg	pRSV-Rev	2.5 μg
Transfer vector	pHR[a]	10 μg	pRRL[a]	10 μg

[a] pHR refers to second-generation vectors with wild-type 5' LTR, and pRRL refers to third-generation vectors with chimeric 5' LTR.

Notes: The medium must be prewarmed at 37°C because the 293T are very sensitive to thermal shock and can shrink and detach. Also, the medium must be added very gently to the cells. Even so, it is difficult not to make a hole in the monolayer.

On day 2, harvest the supernatant and replace with 10–15 ml of medium as the day before. Spin the supernatant at 2500 rpm for 10 min at 4°C [in a tabletop centrifuge such as a Multifuge 3SR (Sorvall)], filter through a 0.45-μm PVDF filter (such as Millex-Durapore, Cat. No. SLHV 033, Millipore), and store at 4°C.

On day 3, harvest the supernatant, spin at 2500 rpm for 10 min at 4°C, filter through a 0.45-μm filter, and pool with the supernatant of day 2. At this point, the pool of supernatants must contain approximately 10^6 transducing units (TU) per milliliter (as titered on HeLa cells, see later).

V. CONCENTRATION AND STORAGE

If a more concentrated lentivector suspension is required, the vector preparation needs to be concentrated. For concentration, we use Beckman Conical tubes (*Cat. No. 358126, Beckman-Coulter*) in a Surespin 630 rotor and a Discovery 90SE ultracentrifuge (Sorvall). Put 4–5 ml of 20% sucrose (SigmaUltra, Cat. No. S7903) in water at the bottom of the tube and fill up to the top with filtered supernatant, spin at 2600 rpm for 90 min and discard gently the supernatant by inversion. Let the tube dry inverted and resuspend the pellet (not always visible) with complete medium or serum-free medium, such as CellGro stem cell growth medium (Cat. No. 2001, CellGenix Technologie

Transfer GmbH, Germany) if the subsequent experiments require the absence of serum.

The use of protein-containing medium is preferable over PBS to resuspend the viral pellet for two reasons: (1) the presence of proteins will stabilize the viral particles and (2) the surfactant effect of proteins will help redissolve the pellet.

VI. TITRATION

Titers of viruses in general and lentivectors in particular critically depend on the method and cells used for titration. The quantification of vector particles capable of achieving every step from cell binding to expression of the transgene depends on both vector and cell characteristics. First, the cell used as the target must be readily permissive to all steps from viral entry to integration of the vector genetic cargo. Second, the expression of the foreign gene must be monitored easily and rapidly reach levels sufficient for reliable quantification. Early vectors had the LacZ bacterial gene as the reporter under the control of the CMV promoter (Naldini *et al.*, 1996). Current vectors now have the green fluorescent protein (GFP) gene as a reporter under the control of promoters that are active in most primary cells (Salmon *et al.*, 2000). Measured titers can also vary with the conditions used for titration, i.e., volume of sample during vector-cell incubation, time of vector-cell incubation, and number of cells used. For several years now, we have been using HeLa cells as target. These cells are stable, easy to grow, and 100% susceptible to transduction by VSV-G-pseudotyped LVs. There are also now used commonly by many laboratories, which helps in comparing titers between laboratories.

(Curran *et al.*, 2000), and bovine (Berkowitz *et al.*, 2001) immunodeficiency viruses and the equine infectious anemia virus (Mitrophanous *et al.*, 1999). These other systems seem to have the same general properties as their HIV counterpart, but in human cells, HIV-derived vectors appear to be more generally efficient, illustrating the fact that a virus adapts to its cognate target. From a biosafety standpoint, HIV-derived vectors may be safer than their nonhuman virus homologues. First, the genomic complexity of HIV is far greater than that of most other lentiviruses, including feline immunodeficiency virus and equine infectious anemia virus, which each have only six genes instead of the nine present in their human counterpart. Because in all cases a minimum of three genes, *gag*, *pol*, and *rev*, will likely be required for generating vector particles efficiently, the multiply attenuated HIV-based packaging system will be the farthest away from its parental virus. In addition, past experience with zoonoses teaches us that the pathogenicity of a given organism is largely unpredictable when it is transferred from its normal animal host into humans. Finally, millions of individuals worldwide have been screened for lentivirus-related diseases. No pathology has been associated with massively deleted forms of HIV-1; however, well-documented cases of long-term clinical nonprogression have occurred in patients infected with HIV-1 strains that carry genetic alterations far more subtle than those introduced in the third-generation HIV-1 packaging system.

Most recent developments of lentivector technology, such as its application for transgenesis (Lois *et al.*, 2002) and RNA interference (Brummelkamp *et al.*, 2002), indicate that the exploitation of this formidable tool is only beginning. Exciting times are ahead, in both experimental and therapeutic arenas.

Acknowledgments

We thank Antonia Follenzi for helpful discussions and Maciej Wiznerowicz for Fig. 2. Additional resources, such as vector sequences, are available at http://www.tronolab.unige.ch/. Downloading of QPCR oligonucleotide sequences and Excel QPCR calculation sheets can be made at http://www.medecine.unige.ch/~salmon/.

References

Berkowitz, R., Ilves, H., Lin, W. Y., Eckert, K., Coward, A., Tamaki, S., Veres, G., and Plavec, I. (2001). Construction and molecular analysis of gene transfer systems derived from bovine immunodeficiency virus. *J Virol.* 75, 3371–3382.

Brummelkamp, T. R., Bernards, R., and Agami, R. (2002). A system for stable expression of short interfering RNAs in mammalian cells. *Science* 296, 550–553.

Curran, M. A., Kaiser, S. M., Achaso, P. L., and Nolan, G. P. (2000). Efficient transduction of nondividing cells by optimized feline immunodeficiency virus vectors. *Mol. Ther.* 1, 31–38.

Dull, T., Zufferey, R., Kelly, M., Mandel, R. J., Nguyen, M., Trono, D., and Naldini, L. (1998). A third-generation lentivirus vector with a conditional packaging system. *J Virol.* 72, 8463–8471.

Follenzi, A., Ailles, L. E., Bakovic, S., Geuna, M., and Naldini, L. (2000). Gene transfer by lentiviral vectors is limited by nuclear translocation and rescued by HIV-1 pol sequences. *Nature Genet.* 25, 217–222.

Klages, N., Zufferey, R., and Trono, D. (2000). A stable system for the high-titer production of multiply attenuated lentiviral vectors. *Mol. Ther.* 2, 170–176.

Lois, C., Hong, E. J., Pease, S., Brown, E. J., and Baltimore, D. (2002). Germline transmission and tissue-specific expression of transgenes delivered by lentiviral vectors. *Science* 295, 868–872.

Mitrophanous, K., Yoon, S., Rohll, J., Patil, D., Wilkes, F., Kim, V., Kingsman, S., Kingsman, A., and Mazarakis, N. (1999). Stable gene transfer to the nervous system using a non-primate lentiviral vector. *Gene Ther.* 6, 1808–1818.

Naldini, L., Blomer, U., Gallay, P., Ory, D., Mulligan, R., Gage, F. H., Verma, I. M., and Trono, D. (1996). *In vivo* gene delivery and stable transduction of nondividing cells by a lentiviral vector. *Science* 272, 263–267.

Negre, D., Mangeot, P. E., Duisit, G., Blanchard, S., Vidalain, P. O., Leissner, P., Winter, A. J., Rabourdin-Combe, C., Mehtali, M., Moullier, P., Darlix, J. L., and Cosset, F. L. (2000). Characterization of novel safe lentiviral vectors derived from simian immunodeficiency virus (SIVmac251) that efficiently transduce mature human dendritic cells. *Gene Ther.* 7, 1613–1623.

Salmon, P., Kindler, V., Ducrey, O., Chapuis, B., Zubler, R. H., and Trono, D. (2000). High-level transgene expression in human hematopoietic progenitors and differentiated blood lineages after transduction with improved lentiviral vectors. *Blood* 96, 3392–3398.

Sandrin, V., Boson, B., Salmon, P., Gay, W., Negre, D., Le Grand, R., Trono, D., and Cosset, F. L. (2002). Lentiviral vectors pseudotyped with a modified RD114 envelope glycoprotein show increased stability in sera and augmented transduction of primary lymphocytes and CD34+ cells derived from human and nonhuman primates. *Blood* 100, 823–832.

Zennou, V., Petit, C., Guetard, D., Nerhbass, U., Montagnier, L., and Charneau, P. (2000). HIV-1 genome nuclear import is mediated by a central DNA flap. *Cell* 101, 173–185.

Zufferey, R., Donello, J. E., Trono, D., and Hope, T. J. (1999). Woodchuck hepatitis virus posttranscriptional regulatory element enhances expression of transgenes delivered by retroviral vectors. *J. Virol.* 73, 2886–2892.

Zufferey, R., Dull, T., Mandel, R. J., Bukovsky, A., Quiroz, D., Naldini, L., and Trono, D. (1998). Self-inactivating lentivirus vector for safe and efficient *in vivo* gene delivery. *J. Virol.* 72, 9873–9880.

Zufferey, R., Nagy, D., Mandel, R. J., Naldini, L., and Trono, D. (1997). Multiply attenuated lentiviral vector achieves efficient gene delivery *in vivo*. *Nature Biotechnol.* 15, 871–875.

53

Construction and Propagation of Human Adenovirus Vectors

Mary M. Hitt, Phillip Ng, and Frank L. Graham

I. INTRODUCTION

Adenoviruses (Ads), which have been used extensively as a model system for molecular studies of mammalian cell DNA replication, transcription, and RNA processing, are now being increasingly investigated as potential mammalian expression vectors for gene therapy and for recombinant vaccines (Berkner, 1988; Graham and Prevec, 1992; Hitt *et al.*, 1999). There are many reasons for this renewed popularity of Ad vectors: the 36,000-bp double-stranded DNA genome of Ad is relatively easy to manipulate by recombinant DNA techniques; Ad infects a wide variety of mammalian cell types, both proliferating and quiescent, with high efficiency; the genome does not undergo rearrangement at a high rate; the viral particle is relatively stable; and the virus replicates to high titer in permissive cells, producing up to 10,000 plaque-forming units (PFU) per infected cell. Late in infection, most of the infected cell protein is virally encoded, potentiating the use of replication-proficient recombinant Ads as short-term high-level expression vectors. In nondividing, nonpermissive cells the viral genome may persist as an episome and continue to express for long periods *in vitro*. This holds true *in vivo* as well in the absence of an immune response against vector-infected cells. This article describes methods for inserting foreign genes into the Ad genome and for purifying, growing, and titrating the recombinant viruses. Our vectors are based on the human Ad5 genome, the structure of which is shown in Fig. 1. In a wild-type infection, early genes (E1a, E1b, E2, E3, and E4) are expressed prior to DNA replication, and late gene expression, driven predominantly by the major late promoter at 16 map units, occurs after the initia-

tion of DNA replication. Deletion of the E1 region renders the virus replication defective, which is desirable for most gene therapy applications. However, such vectors must then be propagated in E1-complementing cells, such as the 293 cell line (Graham *et al.*, 1977). Deletion of E3, which is nonessential for virus growth *in vitro*, together with deletion of E1, allows insertion of foreign genes up to about 8 kb in length. Without deleting E3, the insertion capacity of Ad is about 5 kb.

The vector systems described here rely on site-specific recombination in 293 cells between a shuttle plasmid derived from the E1 region at the "left" end of Ad and a larger plasmid carrying nearly the entire Ad genome in a circular form (Fig. 2). In the Cre/loxP system (Ng *et al.*, 2000a), the genomic plasmid pBH-GloxΔE1,E3Cre carries an expression cassette encoding the Cre recombinase in a region of the plasmid that is excluded from the final vector genome. Cre mediates recombination between a loxP site downstream of the E1 insertion site in the shuttle plasmid and a loxP site in the E1 region of the genomic plasmid. The FLP/frt system (Ng *et al.*, 2000b) is identical except that it uses the FLP recombinase to mediate recombination between frt sites in the shuttle and genomic plasmids. The Ad packaging signal has been deleted from both genomic plasmids, which virtually eliminates the generation of nonrecombinant infectious progeny following cotransfection with an E1 shuttle plasmid.

With either system, vectors can be generated with inserts in place of E1, E3, or both (Fig. 2). E1 replacement vectors, by far the most common, are constructed by insertion of a transgene expression cassette into the E1 shuttle plasmid and subsequent rescue by recombination with the genomic plasmid in 293 cells. Some

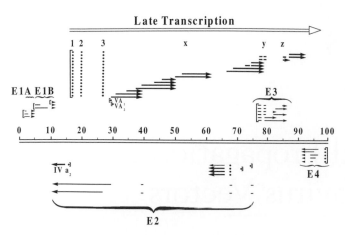

Late Transcription

FIGURE 1 *Transcription map of the human adenovirus type 5.* The approximately 36-kb genome of Ad5 is divided here into 100 map units. Messages from the early regions are indicated as light lines and late messages are indicated in bold. Late transcription originating from the major late promoter at 16 map units and terminating near the right end of the genome is indicated by the open arrow. This transcript is processed into five families of late mRNAs spliced to a common tripartite leader (1, 2, and 3 at map units 16.5, 19.5, and 26.5, respectively), although some mRNA species contain additional leaders. (For more details, see Ginsberg, 1984.)

of the most commonly used shuttle plasmids for this system are illustrated in Fig. 3. E3 replacement vectors are constructed by inserting the transgene expression cassette directly into the genomic plasmid at the *Pac*I site engineered in place of E3 (Bett *et al.*, 1994). The genomic plasmid carrying an insert in E3 can be cotransfected with a shuttle plasmid carrying no insert (i.e., E1 deleted) to generate a replication-defective vector. Double recombinant Ad vectors can be produced by cotransfecting 293 cells with the E3 replacement genome-size plasmid together with an E1 shuttle plasmid encoding a second transgene. This latter strategy has been used successfully to rescue a recombinant Ad vector containing the p35 subunit of interleukin-12 (IL-12) in E1 and the IL-12 p40 subunit in E3 (Bramson *et al.*, 1996).

Foreign coding sequences, including their own or heterologous promoters, can be inserted into the shuttle plasmids or into the E3 region of the pBHG plasmids in an orientation either parallel or antiparallel to the E1 or E3 transcription unit. In general, higher expression levels have been obtained with inserts in the parallel orientation in either E1 or E3 (unpublished results); however, the sequence of the insert itself can affect expression levels, particularly for E3 insertions. Once the desired plasmids have been constructed, the following protocols are used to produce and purify the recombinant Ad viruses.

II. MATERIALS AND INSTRUMENTATION

The Cre/loxP and FLP/frt based AdMax vector rescue systems are available, as Kit D (Cat. No. PD-01-64) and Kit E (Cat. No. PD-01-65) respectively, from Microbix Biosystems, Incorporated. Minimal essential medium (MEM) F11 (Cat. No. 61100-087), L-glutamine (Cat. No. 25030-081), penicillin/streptomycin (Cat. No. 15140-122), horse serum (Cat. No. 16050-159), newborn calf serum (NCS) (Cat. No. 16010-159), agarose (Cat. No. 15510-027), and dithiothreitol (Cat. No. 15508-013) can be obtained from Invitrogen. Bovine serum albumin fraction V (Cat. No. A2153), fetal bovine serum (FBS) (Cat. No. F4135), Joklik's modified MEM (Cat. No. M0518), salmon sperm DNA (Cat. No. D1626), and orcein (Cat. No. O7380) are available from Sigma Chemical Company. All sera are inactivated prior to use by heating to 56°C for 30 min. Fungizone can be purchased from Bristol-Myers-Squibb (Cat. No. 043780). Nunc tissue culture dishes (Cat. No. 1-68381A) can be obtained from VWR. Sterile petri dishes (Cat. No. 08-757-12), Difco agar (Cat. No. 0145-17-0), Difco Bacto Lennox LB broth base (Cat. No. 0402-07-0), Becton Dickinson BBL trypticase peptone (Cat. No. B11921), and yeast extract (Cat. No. B11929) can be obtained from Fisher Scientific. Pronase (Cat. No. 1459643) and bovine pancreatic deoxyribonuclease I (Cat. No. 104-159) can be purchased from Roche Diagnostics. Analytical grade $CaCl_2 \cdot 2H_2O$ (Cat. No. B10070) can be obtained from BDH. All other chemicals can be purchased from standard chemical suppliers (e.g., BDH). Spinner flasks (1969 series) can be obtained from Bellco and Pierce Slide-A-Lyzer 10K dialysis cassettes (Cat. No. 66425) from Chromatographic Specialities. Beckman SW41 Ti and SW 50.1 rotors are also required. Reagents for plasmid DNA isolation, as well as restriction enzymes and reagents and apparatus for horizontal slab gel electrophoresis, are described in a number of cloning manuals (e.g., Sambrook and Russell, 2001).

III. PROCEDURES

A. Preparation of Plasmid DNA for Cotransfections

In order to minimize the generation of bacterial clones containing rearranged plasmid DNA, which is occasionally observed in preparations of very large plasmids such as pBHGlox∆E1,E3Cre and pBHGfrt∆E1,E3FLP, we have adopted the following

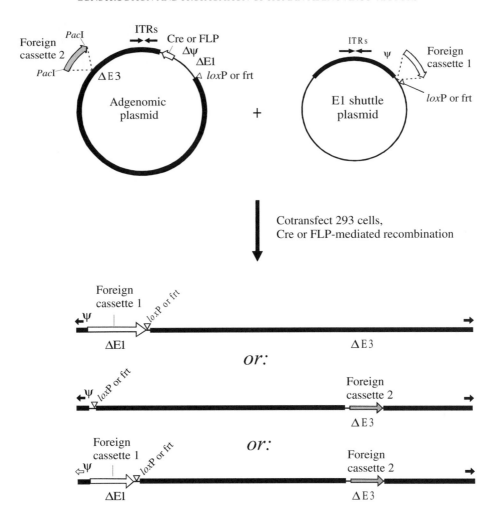

Recombinant Ad vector

FIGURE 2 *Construction of Ad vectors by two-plasmid site-specific recombination in 293 cells.* The AdMax strategy used to introduce foreign DNA inserts into the E1 and/or E3 regions for rescue into virus is illustrated. Expression cassettes can be inserted in place of the E1 region (Ad5 nucleotides 455–3523) by cloning into E1 shuttle plasmids carrying the recognition site (e.g., *lox*P) for a site-specific recombinase. E1 shuttle plasmids are described further in Fig. 3. The 293 cells are then cotransfected with this shuttle plasmid and an Ad genomic plasmid carrying the appropriate recombinase (Cre for *lox*P-containing shuttles or FLP for frt-containing shuttles). Genomic plasmids are available with E3 deleted (Ad5 nucleotides 28138–30818; pBHGloxΔE1,E3Cre or pBHGfrtΔE1,E3FLP) as shown or with a wild-type E3 region (pBHGloxE3Cre and pBHGfrtE3FLP) (Ng and Graham, 2002). Expression of recombinase in 293 cells results in recombination between recognition sites in the shuttle and in the genomic plasmid, generating an E1 replacement Ad vector (top vector in illustration). E3 replacement Ad vectors are constructed by inserting the transgene expression cassette into a unique *Pac*I site that replaces the E3 region in the Ad genomic plasmid. The 293 cells cotransfected with this genomic construct and an "empty" (i.e., no transgene) E1 shuttle plasmid will produce a replication-defective E3 replacement Ad vector (middle vector in illustration). Note that it is also possible to cotransfect with a plasmid containing the intact left end of Ad5 to produce an E1+ nondefective vector by overlap recombination (Bett *et al.*, 1994). Double recombinant vectors are generated by recombination between an E1 shuttle plasmid carrying one expression cassette and a genomic plasmid carrying a second expression cassette (bottom vector in illustration). Ad and bacterial sequences are indicated by thick and thin black bars, respectively, *lox*P and frt sites by open triangles, inverted terminal repeats (ITRs) by black arrows, and the packaging signal by the symbol ψ.

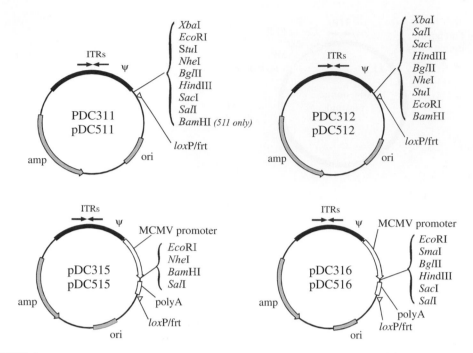

FIGURE 3 *Structure of E1 shuttle plasmids used for vector rescue by* in vivo *site-specific recombination.* The shuttle plasmids pDC311, pDC312, pDC315, and pDC316 are used to rescue vectors by Cre-mediated recombination. The shuttle plasmids pDC511, pDC512, pDC515, and pDC516 are used to rescue vectors by FLP-mediated recombination. Plasmids pDC311, pDC312, pDC511, and pDC512 are designed for insertion of a cassette consisting of a promoter, transgene, and polyadenylation signal sequence. The polycloning sites of plasmids pDC315, pDC316, pDC515, and pDC516 are flanked 5' by the murine CMV promoter and 3' by the SV40 polyadenylation signal sequence. Coding sequences cloned into the latter plasmids generate vectors with high levels of expression in both human and murine cells (Addison *et al.*, 1997).

protocol for bacterial growth prior to plasmid DNA isolation.

Solutions

1. *Super broth (SB)*: Dissolve 5 g NaCl, 32 g trypticase peptone, 20 g yeast extract, and 1 g glucose in 1 liter H_2O. Add 5 ml 1 *N* NaOH. Sterilize by autoclaving.
2. *LB-agar plates*: Dissolve 10 g BBL Lennox LB broth base in 500 ml H_2O. Add 7.5 g agar and sterilize by autoclaving. Cool LB-agar to about 50°C, add antibiotics as required, and pour 25 ml into each of 20 sterile petri dishes. Store at 4°C.
3. *Reagents for isolating plasmid DNA on CsCl gradients*: Not described here.

Steps

1. Streak plasmid-bearing bacteria on an LB-agar plate containing appropriate antibiotics and grow overnight at 37°C.
2. Pick two or more colonies off the plate, resuspend each in 5 ml SB plus antibiotics, and incubate at 37°C on a shaker for several hours.
3. Add each 5 ml culture to 500 ml SB plus antibiotics and continue incubating overnight.

4. Purify the plasmid DNA from each culture separately by alkaline lysis of the bacteria and CsCl banding as described in standard cloning manuals (e.g., Sambrook and Russell, 2001). Plasmid DNA that has not undergone any detectable rearrangement, as indicated from comparison between predicted and observed restriction enzyme cleavage pattens, is suitable for use in cotransfections.

B. DNA Transfection for Rescue of Recombinant Adenovirus Vectors: Calcium Phosphate Coprecipitation

Solutions

1. *Complete MEMF11 (or Joklik's modified MEM)*: Add 5 ml 0.2 *M* L-glutamine, 5 ml penicillin/streptomycin (10,000 U/ml and 10 mg/ml, respectively), and 5 ml 0.25 mg/ml fungizone to 500 ml MEMF11 (or Joklik's modified MEM). Store at 4°C for up to 2 weeks. Add 55 ml heat-inactivated NBS, FBS or HS prior to use in cell culture.
2. *10X citric saline*: Dissolve 50 g KCl and 22 g trisodium citrate dihydrate ($Na_3C_6H_5O_7 \cdot 2H_2O$) in H_2O to

a final volume of 500 ml. Sterilize by autoclaving and store at 4°C. Dilute 1:10 in sterile H_2O to prepare 1X citric saline.

3. *HEPES-buffered saline (HEBS)*: Dissolve 5 g HEPES free acid, 8 g NaCl, 0.37 g KCl, 0.1 g Na_2HPO_4, and 1 g glucose in 900 ml H_2O. Adjust pH to 7.1. Adjust volume to 1 liter with H_2O. Aliquot into small glass bottles, sterilize by autoclaving, and store at 4°C.

4. *10X SSC*: Dissolve 8.7 g NaCl and 4.4 g tri-sodium citrate dihydrate in H_2O to a final volume of 100 ml. Adjust pH to 7.0. Autoclave. Prepare 0.1X SSC by diluting 10X SSC and then autoclaving.

5. *2 mg/ml carrier DNA*: Dissolve 100 mg salmon sperm DNA in 50 ml sterile 0.1X SSC by stirring overnight at room temperature. Determine concentration by reading the OD at 260 nm (one OD unit = 50 μg/ml). Store in small aliquots at −20°C.

6. *2.5 M $CaCl_2$*: Add H_2O to 36.8 g $CaCl_2 \cdot 2H_2O$ (analytical grade) to a final volume of 100 ml. Sterilize by filtration and store in small plastic tubes at 4°C.

7. *MEMF11-agarose overlay*: To make approximately 200 ml, add 10 ml horse serum (HS) and 2 ml each of L-glutamine, penicillin/streptomycin, fungizone (at concentrations given earlier), and autoclaved 5% yeast extract (w/v in H_2O) to 100 ml 2X MEMF11. Autoclave 1 g agarose in 100 ml H_2O. Bring the 2X medium and agarose to 44°C before mixing and use within an hour.

Steps

1. Grow monolayer cultures of 293 cells in 150-mm dishes in complete MEMF11 medium plus 10% FBS (or NBS). At 90% confluence, remove medium, wash each dish twice with 10 ml 1X citric saline, and then incubate for a maximum of 15 min at room temperature in 3 ml 1X citric saline to detach cells. Resuspend cells in medium and divide between two or three 150-mm dishes. 293 cells should be refed with fresh medium every 3 days if not ready to passage.

2. Set up low-passage (<p40) 293 cells in 60-mm dishes to be about 70–80% confluent at the time of use. As a rule of thumb, one 150-mm dish of nearly confluent 293 cells can be split into eight 60-mm dishes each containing 5 ml complete MEMF11 + 10% FBS, which will be ready for transfection the next day.

3. Add 0.005 volume 2 mg/ml carrier DNA to 1X HEBS and shear by vortexing for 1 min.

4. For each virus to be rescued, aliquot 2 ml HEBS + carrier DNA (enough for four dishes) into each of three sterile clear plastic tubes.

5. To these tubes add E1 shuttle plasmid DNA and the appropriate genomic plasmid (e.g., pBH-GloxΔE1,E3Cre) in the following amounts: 20 μg of each plasmid, 8 μg of each plasmid, and 2 μg of each plasmid. If a negative control is desired, set up similar

coprecipitations omitting the first plasmid. A useful positive control is 2 μg of the infectious plasmid pFG140.

6. Gently mix by shaking and then slowly add 0.1 ml 2.5 M $CaCl_2$ to each tube.

7. Gently mix and let stand at room temperature for 15–30 min. (A fine precipitate should form within a few minutes.) Without removing the growth medium, add 0.5 ml DNA suspension to each dish of cells (four dishes for each tube of coprecipitate) and then incubate at 37°C in a CO_2 incubator for at least 5 h, preferably overnight.

8. Remove the medium and add to each dish 10 ml MEMF11-agarose overlay previously equilibrated to 44°C. After the agarose solidifies, incubate at 37°C. Plaques should appear after about 5–14 days. When dishes are examined from below by eye, plaques appear turbid as a consequence of light scattering by dead cells in an otherwise smooth cell monolayer. Microscopic examination reveals plaques as zones of dead or lysed cells surrounded by rounded infected cells.

9. At about 10 days posttransfection, pick well-isolated plaques by punching out agar plugs from the cultures using a sterile Pasteur pipette. Transfer each agar plug to 1 ml sterile PBS^{++} + 10% glycerol in a sterile vial. Store at −70°C until use.

C. Screening Adenovirus Plaque Isolates

The following protocol describes the expansion of plaque isolates by growth in monolayer cultures of 293 cells. A portion of the infected cell material is stored for further purification; the remainder is harvested for analysis of viral DNA.

Solutions

1. *Phosphate-buffered saline^{++} (PBS^{++})*: To make solution A, dissolve 80 g NaCl, 2 g KCl, 11.5 g Na_2HPO_4, and 2 g KH_2PO_4 in H_2O to a final volume of 1 liter. To make solution B, add 1 g $CaCl_2 \cdot 2H_2O$ to 100 ml H_2O. To make solution C, add 1 g $MgCl_2 \cdot 6H_2O$ to 100 ml H_2O. Sterilize solutions separately by autoclaving. For 100 ml PBS^{++}, mix 88 ml sterile H_2O with 10 ml solution A and 1 ml each of solutions B and C.

2. *PBS^{++} + 10% glycerol*: Add 10 ml sterile glycerol to 90 ml PBS^{++}.

3. *Pronase stock solution*: Dissolve 0.5 g pronase in 100 ml 10 mM Tris–HCl, pH 7.5; heat at 56°C for 15 min and then incubate at 37°C for 1 h. Aliquot and store at −20°C. To prepare working solution, thaw stock solution just before use and add 0.1 volume to 10 mM Tris–HCl, pH 7.5, 10 mM EDTA, 0.5% (w/v) sodium dodecyl sulfate (SDS).

4. *Complete MEMF11 + 5% HS*: Add 25 ml HS to 475 ml complete MEMF11.

5. *0.1X SSC*: See Section III.B, solution 4.

6. *Reagents for restriction analysis of viral DNA*: Not described here.

Steps

1. Set up 60-mm dishes of 293 cells to be 80–90% confluent at time of infection. The denser and older the cell monolayer, the longer it takes for the cytopathic effect to reach completion.

2. Remove medium from 293 dishes and add 0.2 ml virus (agar plug suspension). Rock dishes once and adsorb at room temperature for 30 min. Add 5 ml complete MEMF11 + 5% HS and incubate at 37°C.

3. A cytopathic effect should be visible within 1–2 days. Harvest virus and extract infected cell DNA (steps 4 to 6) when all cells are rounded and most have detached from the dish (usually 3–4 days).

4. Release semiadherent cells from the dish by gentle pipetting. Transfer 3.5 ml of the cell suspension to a sterile vial containing 0.5 ml sterile glycerol. Store at –70°C until you wish to amplify the vector.

5. Transfer the remaining 1.5 ml to a microfuge tube and spin 2 min at 7000 rpm. Aspirate all but about 0.1 ml supernatant. Vortex well to suspend infected cells. To extract DNA, add 0.5 ml pronase working solution to the cells in the microfuge tube and incubate at 37°C for 4–18 h.

6. Add 1 ml cold 96% ethanol to precipitate the DNA. Mix well by inverting the tube several times—a fibrous precipitate should be easily visible and the solution should no longer be viscous. Spin 5 min at 14,000 rpm and then aspirate supernatant. Wash pellet twice with 70% ethanol and air dry.

7. Dissolve DNA pellet in 50 μl 0.1X SSC by heating at 65°C with occasional vortexing. Digest 5 μl with *Hind*III (1 unit overnight is usually sufficient for complete digestion).

8. Apply digested samples and appropriate markers (a *Hind*III digest of wild-type Ad5 being one convenient marker) to a 1% agarose gel containing ethidium bromide and subject to electrophoresis until the dye front has migrated at least 10 cm. If the cytopathic effect was complete, viral DNA bands should be easily visible (under ultraviolet light) above a background smear of cellular DNA. Note that in *Hind*III digests of human DNA there will be a band of cellular repetitive DNA at 1.8 kb.

9. Verify candidate recombinants using other diagnostic restriction enzymes. Although generally 100% of viral plaques obtained using AdMax are correct, it is good laboratory practice to carry out one round of plaque purification, as described later, and screening

as described in this section prior to preparation of high-titer stocks.

D. Plaque Assays for Purification and Titration of Adenovirus

Solutions

PBS⁺⁺ and MEMF11-agarose overlay (at 44°C): See Section III.C.

Steps

1. Set up 60-mm dishes of 293 cells to be confluent at time of infection.

2. Remove medium from dishes. Add 0.2 ml virus (dilution of agar plug suspension in PBS⁺⁺ if you wish to plaque purify or dilution of stock for titration). We typically assay dilutions ranging from 10^{-3} to 10^{-6} for plaque purification or 10^{-4} to 10^{-10} for virus titration. Adsorb the virus for 30–60 min in an incubator, occasionally rocking the dishes. Add 10 ml MEMF11-agarose overlay, cool, and then continue incubation at 37°C.

3. Plaques should be visible within 4–5 days and should be counted for titration at 7 days and again at 10 days. For plaque purification, proceed as for isolation of plaques following transfections (Section III.C).

E. Preparation of High-Titer Viral Stocks (Crude Lysates) from Cells in Monolayer

Because most of the virus remains associated with the infected cells until very late in infection, high-titer stocks can be prepared easily by concentrating infected 293 cells as described here.

Solutions

PBS⁺⁺, PBS⁺⁺ + 10% glycerol, and complete MEMF11 + 5% HS: See Section III.C.

Steps

1. Set up 150-mm dishes of 293 cells to be 80–90% confluent at time of infection. We generally use eight or more dishes for each virus.

2. To prepare high-titer stocks, remove medium from the 293 cells and infect at a multiplicity of infection (MOI) of 1–10 PFU per cell (1 ml diluted virus per 150-mm dish). For the initial stock preparation, we dilute virus (from the untitered 4-ml sample stored at –70°C after the last round of viral screening) 1:8 with PBS⁺⁺. To minimize generation and amplification of rearranged forms of the vector, always prepare high-titer stocks from viral screening samples, not from CsCl-banded stocks (Section III.F).

3. Adsorb for 30–60 min and then refeed with complete MEMF11 + 5% HS. Incubate at 37°C and examine daily for signs of a cytopathic effect.

4. When the cytopathic effect is nearly complete, i.e., most cells rounded but not yet detached, harvest by scraping the cells off the dish, combining the cells plus spent medium, and centrifuging at 800 g for 15 min. Aspirate the medium and resuspend the cell pellet in 2 ml PBS++ + 10% glycerol per 150-mm dish. Freeze (−70°C) and thaw (37°C) the crude virus stock two or three times prior to titration. Store aliquots at −70°C.

F. Preparation of High-Titer Viral Stocks (Purified) from Cells in Suspension

Recombinant Ads can be purified from crude lysates of either monolayer or suspension cultures. Due to the greater ease of handling suspension cultures, however, this source is preferable for the preparation of purified high-titer viral stocks as described here. Similar yields can be obtained from thirty to sixty 150-mm dishes of 293 cell monolayers.

Solutions

1. *Complete Joklik's modified MEM + 10% HS*: Add 50 ml HS to 450 ml complete Joklik's modified MEM (described in Section III.B). Store at 4°C.

2. *1% sodium citrate*: Dissolve 1 g trisodium citrate dihydrate in H_2O to a final volume of 100 ml.

3. *Carnoy's fixative*: Add 25 ml glacial acetic acid to 75 ml methanol.

4. *Orcein solution*: Add 1 g orcein dye to 25 ml glacial acetic acid plus 25 ml H_2O. Filter through Whatman No. 1 paper.

5. *0.1 M Tris–HCl, pH 8.0*: Add 1.2 g Tris base to 80 ml H_2O. Adjust pH to 8.0 with HCl. Adjust volume to 100 ml and autoclave.

6. *5% Na deoxycholate*: Add 5 g Na deoxycholate to 100 ml H_2O.

7. *2 M MgCl$_2$*: Add 40.6 g MgCl$_2$·6H$_2$O to 100 ml H_2O and sterile filter.

8. *DNase I solution*: Dissolve 100 mg bovine pancreatic deoxyribonuclease I (DNase I) in 10 ml of 10 mM Tris–HCl, pH 7.4, 50 mM NaCl, 1 mM dithiothreitol, 0.1 mg/ml bovine serum albumin, 50% glycerol. Store in small aliquots at −20°C.

9. *CsCl solutions for banding*: Transfer the indicated amounts of analytical grade CsCl into small beakers to give the desired final densities:

Solution	Density of final solution (g/ml)	Weight of CsCl (g)
1.5 d	1.5	83.1
1.35 d	1.35	54.3
1.25 d	1.25	37.0

Add 100 ml 10 mM Tris–HCl, pH 8, to each beaker and stir to dissolve. Verify density by weighing 1.0 ml of each solution (e.g., 1.0 ml of the 1.35 d solution should weigh 1.35 g). Sterile filter and store at room temperature.

10. *Sterile glycerol*: Prepare by autoclaving.

11. *TE/SDS*: Add 0.5 ml 20% SDS to 100 ml 10 mM Tris–HCl, 1 mM EDTA, pH 8.

Steps

1. Grow 293N3S cells in spinner culture to a density of $2–4 \times 10^5$ cells/ml in 3 liters of complete Joklik's modified MEM + 10% HS. Centrifuge cell suspension at 750 g for 20 min, saving half of the conditioned medium. Resuspend the cell pellet in 0.1 vol fresh medium and transfer to a sterile 500-ml bottle containing a sterile stir bar.

2. Add virus at an MOI of 10–20 PFU/cell and stir gently at 37°C. After 1 h, return culture to the spinner flask and bring to the original volume using 50% conditioned medium and 50% fresh medium. Continue stirring at 37°C.

3. Monitor infection daily by inclusion body staining as follows.
   a. Remove a 5-ml aliquot from the infected spinner culture. Spin for 10 min at 750 g and resuspend the cell pellet in 0.5 ml of 1% sodium citrate.
   b. Incubate at room temperature for 10 min, add 0.5 ml Carnoy's fixative, and fix for 10 min at room temperature.
   c. Add 2 ml Carnoy's fixative, spin for 10 min at 750 g, aspirate, and resuspend the pellet in a few drops of Carnoy's fixative. Add one drop of fixed cells to a slide, let air dry for about 10 min, add one drop orcein solution and a coverslip, and examine in the microscope. Inclusion bodies appear as densely staining nuclear structures resulting from the accumulation of large amounts of virus and viral products at late times in infection. A negative control should be included in initial tests.

4. When inclusion bodies are visible in 80–90% of the cells (1.5 to 3 days), harvest by centrifugation at 750 g for 20 min in sterile 1-liter bottles. Combine pellets in a small volume of medium and spin again. Resuspend pellet in 15 ml 0.1 M Tris–HCl, pH 8.0. Store at −70°C until use.

5. Thaw the frozen crude stock and add 1.5 ml 5% Na deoxycholate. Mix well and incubate at room temperature for 30 min. This disrupts cells without disrupting virions, resulting in a relatively clear, highly viscous suspension.

6. Add 0.15 ml 2 M MgCl$_2$ and 0.075 ml DNase I solution and then mix well. Incubate at 37°C for

54

Production and Quality Control of High-Capacity Adenoviral Vectors

Gudrun Schiedner, Florian Kreppel, and Stefan Kochanek

I. INTRODUCTION

High-capacity adenovirus (HC-Ad) vectors [also called pseudoadenovirus (PAV), helper-dependent (HD-Ad), gutted, or gutless adenovirus vectors] have been developed to address capacity, toxicity, and immunogenicity problems of first- and second-generation adenovirus vectors. As only viral elements this vector type contains the inverted terminal repeats (ITRs), which are essential for replication of the viral DNA, and the packaging signal close to the left terminus that is required for encapsidation of the DNA into the viral capsids. Since the size of the ITRs and the packaging signal together are less than 0.6 kb, up to 37 kb of foreign DNA can be transported.

For practical reasons, most HC-Ad vectors will carry genes or expression cassettes that are smaller than 37 kb. For stability reasons during amplification in most cases additional "stuffer" DNA has to be incorporated into the vector DNA to increase the genome size to at least around 27 kb.

HC-Ad vectors cannot be produced similar to helper-independent vectors, in which most viral functions are provided from the vector. Because adenovirus is a relatively large DNA virus that expresses many different protein and RNA functions, it is unlikely that complementing cell lines can be generated to provide appropriate levels of all viral functions in *trans*. Therefore, a helper virus is used for production that subsequently is eliminated from the end product.

The currently preferred production system is based on excision of the packaging signal of the helper virus by a recombinase expressed in the producer cell line. Most HC-Ad vectors so far have been produced using the Cre-loxP recombination system of the bacteriophage PI. In this system the packaging signal of the helper virus is flanked by two loxP sites (Hardy *et al.*, 1997; Parks *et al.*, 1996). The HC-Ad vector is produced in E1-complementing cells that express the recombinase constitutively. The Cre-mediated excision is surprisingly efficient and the contamination of vector by helper virus is reduced compared to the earlier production system.

The complete characterization of HC-Ad vector preparations comprises three parameters: (1) the number of infectious particles, (2) the number of total particles, and (3) the number helper virus particles remaining in the preparation after purification (Kreppel *et al.*, 2002). Due to the fact that HC-Ad vectors do not possess any viral coding sequences, the number of infectious particles cannot be determined by plaque assay or tissue culture infectivity dose $TCID_{50}$ because these methods rely on vector replication and viral protein expression in E1-transformed cell lines. The number of total particles can be determined by particle lysis and subsequent measuring light absorbance at 260 nm. However, the reliability of this method is usually low and strongly depends on the purity of the vector preparations. The DNA-based method described here allows for fast and reliable determination of all three parameters with standard laboratory equipment independent of viral or reporter gene expression. For quantifying the number of infectious particles, a reference cell line with defined susceptibility for Ad5 is transduced with the HC-Ad vector, cell lysates are prepared, and vector genomes that entered the cells are detected after immobilization on a nylon membrane by hybridization with a radiolabeled vector-specific probe. Total particle numbers can easily be determined with the same probe on the

Cell Biology

445

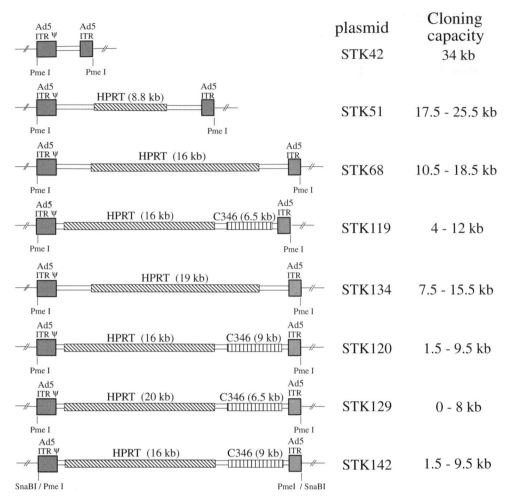

plasmid	Cloning capacity
STK42	34 kb
STK51	17.5 - 25.5 kb
STK68	10.5 - 18.5 kb
STK119	4 - 12 kb
STK134	7.5 - 15.5 kb
STK120	1.5 - 9.5 kb
STK129	0 - 8 kb
STK142	1.5 - 9.5 kb

FIGURE 2 Plasmids for cloning HC-Ad vectors. Plasmids contain Ad5 left (Ad5 sequences nucleotides 1 to 440) and right (Ad5 sequences nucleotides 35818 to 35935) termini. In addition, cloning plasmids contain noncoding stuffer sequences from the HPRT gene locus or the C346 cosmid. Again, to release the pBluescript backbone, both inverted terminal repeats are flanked by *Pme*I sites (or *Sna*BI for pSTK42). In order to incorporate different sizes of transgenes, the plasmids contain several unique cloning sites.

10% SDS: Dissolve 10 g SDS in 100 ml sterile H_2O

Superfect transfection kit (Qiagen Cat. No. 301305)

Tris-buffered saline (TBS): 25 mM Tris, 137 mM NaCl, 2.7 mM KCl, pH 7.4, autoclave

TBS/10% glycerol: TBS containing 10% glycerol

TBS/CsCl: Dissolve 10 g CsCl (Roche, Cat. No. 757306) in 20 ml sterile TBS

Ultracentrifuge, SW41 rotor: e.g., Beckmann

Yeast extract (5%): Dissolve 2 g yeast extract (AppliChem, Cat. No. A1552,0100) in 40 ml H_2O, sterile filter

Production of Helper Virus

Preparation of DNA for Transfection

1. Digest 10 µg of plasmid pGS102#21 with 20–50 units of *Swa*I (New England Biolabs) at 25°C for 2 h.

2. Add TE buffer, pH 8.5, up to 200 µl, add 200 µl phenol, vortex gently, and precipitate for 5 min at 14,000 rpm.

3. Take upper phase, add 200 µl chloroform/ Isoamyl alcohol, vortex gently, and centrifuge for 5 min at 14,000 rpm.

4. Take upper phase, add 20 µl 3 M Na-acetate, mix, add 400 µl ethanol, mix, and precipitate for 30 min at 14,000 rpm and 4°C.

5. Discard supernatant, add 400 µl 70% ethanol, centrifuge for 10 min at 14,000 rpm and 4°C, and discard supernatant.

6. Dissolve pellet in 20 µl TE, pH 7.5 (0.5 µg/µl).

Preparation of N52.E6 Cells for Transfection

N52.E6 cells are used for the production of helper virus and are cultivated in αMEM supplemented with

10% FBS and antibiotics in 5% CO_2 at 37°C. N52.E6 cells are usually passaged twice a week 1:4–5.

7. The day before transfection, wash cells in PBS, detach cells in trypsin, and resuspend in culture medium. Count cells and plate 2×10^6 cells on a 6-cm dish. The cells should be ready for transfection at 60–80% confluency the following day.

Transfection

8. Transfect N52.E6 cells in 6-cm dishes using 5 µg of the SwaI-linearized pGS102#21 and the Superfect transfection kit. Perform transfection according to the manufacturer.

Comments

DNA transfections can also be performed using the calcium–phosphate coprecipitation method as described in a number of different manuals. The commercially available Effectene transfection Kit (Qiagen) also shows very high efficiency in transfection but should not be used for large-sized linearized plasmids.

Plaque Isolation and Preparation of High-Titer Helper Virus Stocks

9. Six to 7 days posttransfection, harvest cells by scratching cells into medium, pellet cells for 10 min at $400 g$, and resuspend in 2 ml Tris-buffered saline (TBS). Lyse cells by three cycles of freeze/thawing in dry ice/ethanol. Lysate can be stored at −80°C.

10. Split 2×10^6 N52.E6 cells into 6-cm dishes the day before. Infect cells by removing medium, adding 3 ml fresh medium plus 0.2 ml of the lysate. After 2 h add 2 ml medium. Five days after infection, harvest cells as described earlier and lyse by freeze/thawing.

11. Split six 6-cm dishes N52.E6 the day before with 2×10^6 cells per dish. Dilute the lysate 1:100 in TBS and infect each two dishes with 1, 5, and 25 µl of the lysate.

12. Three to 4 h later, overlay cells with agarose: remove medium from infected cells and carefully add 10 ml of prewarmed MEM-agarose solution. After the agarose solidifies, incubate at 37°C until plaques appear (usually 7–10 days).

13. Pick well-isolated single plaques by punching out agar using a sterile Pasteur pipette and transfer agar to 0.2 ml TBS/10% glycerol. Plaques can be stored at −80°C.

14. Infect 2×10^6 N52.E6 cells on 6-cm dishes with 100 µl of the plaque lysate. After complete cytopathic effect (CPE) has occurred, harvest cells in 2 ml TBS and lyse by freeze/thawing.

15. Amplify helper virus by first infecting 2×10^6 N52.E6 cells on a 6-cm dish with 1 ml of the lysate.

Complete CPE should be visible 48 to 72 h after infection. Harvest cells and titer helper virus in the lysate by infecting N52.E6 cells on a 6-cm dish (2×10^6 cells per dish, split the day before) with 30, 10, 5, and 1 µl (lysate 1:10 diluted).

16. Infect ten 15-cm dishes N52.E6 cells (2×10^7 per dish) with the amount of lysate titered before and resulting in complete CPE after 48 h.

17. Harvest cells 48 h after infection and centrifuge cells 10 min at $400 g$. Dissolve pellet in 5 ml TBS. Lyse by freeze/thawing. Centrifuge for 15 min at $400 g$ and 4°C.

Gradient Purification of Helper Virus

18. Take supernatant and add TBS up to exactly 10 ml. Add 5 g CsCl and mix gently. Be sure that CsCl is dissolved completely. Transfer lysate to SW41 centrifuge tube.

19. Spin in Beckmann SW41 for 16–22 h at 4°C and 32,000 rpm.

20. Collect the helper virus band by puncturing the tube with a needle and by drawing off the virus band in minimal volume.

21. Add TBS/CsCl solution up to 10 ml. Transfer to centrifuge tube and spin in a Beckmann SW41 as described earlier. Collect virus in minimal volume as before.

22. Desalt virus using the PD-10 desalting column according to the manufacturer.

Titration of Helper Virus

23. Determine particles and infectious units using the slot-blot protocol described later.

24. Alternatively, determine the approximate amount of helper virus resulting in complete cpe of N52.E6 cells after 48 h.

DNA Preparation from CsCl-Purified Helper Virus Particles

In order to exclude helper virus rearrangement, viral DNA should be isolated from CsCl-purified particles. Take 200 µl of gradient-purified and desalted vector. Isolate DNA using the QiaAmp DNA Minikit according to the manufacturer. Subsequent to ethanol precipitation, dissolve the pellet in 20 µl AE buffer. Digest 10 µl with the appropriate enzyme and run in an 0.8% agarose gel containing EtBr. As a positive control, double digest the corresponding helper virus plasmid with SwaI and the aforementioned enzyme.

Production of HC-Ad Vectors

Cloning of HC-Ad Vector Plasmids

General cloning reagents and equipment, as well as procedures, have been described in a number of cloning manuals. Thus, only important steps, observations, and suggestions are discussed in this section.

i. The gene of interest including promoter and polyadenylation signal should be excisable in one piece.

ii. The fragment containing the gene of interest should be purified in an agarose gel using standard protocols (e.g., QIAEXII gel extraction kit, Qiagen Cat. No. 20021 for fragments <5 kb, electroelution for DNA fragments >5 kb). In order to improve the efficiency of cloning, it is important that the agarose gel does not contain EtBr. The DNA in the gel should never be exposed to a UV screen. Instead, the DNA fragments should be stained using Sybr-Gold (Molecular Probes Cat. No. S-11494) and visualized with a Dark Reader (e.g., Molecular Probes).

iii. For cloning, highly competent bacteria should be used (e.g., Stratagene XL2-Blue ultracompetent cells, Cat. No. 200150).

iv. Despite the size of 30–35 kb, the vector plasmid DNA can be isolated in a good amount and quality using standard plasmid DNA isolation protocols or kits.

v. Plasmids should never be stored as glycerol stock but as DNA in TE either at 4°C or in ethanol at –20°C. Repeated freezing and thawing should be avoided. In addition, plasmid colonies on agar plates containing antibiotics should not be stored for more than 3 days.

Preparation of Plasmid DNA for Transfection

In order to release the left and right adenoviral termini (ITR), HC-Ad vector plasmids usually contain restriction sites flanking both ITRs. ITRs in the HC-Ad vector plasmids depicted in Fig. 2 are flanked by unique *Pme*I sites (or *Sna*BI site in pSTK142). The HC-Ad vector plasmid is digested and further purified as described earlier for the helper virus plasmid.

Culture of 73/29 Cells

Cre-expressing 73/29 cells are used for the production of HC-Ad vectors and are cultivated in αMEM containing 10% FBS and antibiotics, supplemented with 200 μg/ml G418. 73/29 cells were usually passaged twice a week and diluted 1 : 4–5.

Amplification of HC-Ad Vectors

First Amplification/Transfection

1. Seed 2×10^6 73/29 cells in one 6-cm dish the day before transfection.
2. Transfect 5 μg of *Pme*I-digested phenol/chloroform-purified and ethanol-precipitated HC-Ad vector plasmid using the Qiagen Superfect trans-

fection kit according to the manufacturer. Incubate at 37°C in a CO_2 incubator.
3. Four to 16 h after transfection, infect cells with 5 MOI of helper virus (multiplicity of infection).
4. Forty-eight hours later, centrifuge infected cells (complete cpe should have occurred) for 10 min at 400 g. Dissolve pellet in 2 ml TBS and lyse by freeze/thawing. Lysate can be stored at –80°C.

Second and Third Amplification

5. Seed 2×10^6 73/29 cells in one 6-cm dish the day before.
6. The next day infect cells with 1 ml of the lysate from the first amplification. At the same time, add 5 MOI of helper virus.
7. Harvest the cells 48 h later and centrifuge for 10 min at 400 g. Dissolve pellet in 2 ml TBS and lyse by freeze/thawing.
8. Infect 73/29 cells in one 6-cm dish (2×10^6 cells seeded the day before) with 1 ml of the lysate from the second amplification and coinfect with 5 MOI of helper virus.
9. Again harvest cells 48 h after infection when complete cpe is visible and spin for 10 min at 400 g. Dissolve the pellet in 2 ml TBS and lyse three times by freeze/thawing.

Fourth and Fifth Amplification

10. Seed $1–2 \times 10^7$ 73/29 cells in one 15-cm dish the day before.
11. Infect cells with half of the lysate from the third amplification and coinfect with 5 MOI of helper virus.
12. Forty-eight hours later harvest cells and pellet for 10 min at 400 g. Dissolve cells in 2 ml TBS and lyse three times by freeze/thawing.
13. Infect two 15-cm dishes of 73/29 cells (2×10^7 cells per dish seeded the day before) with 2 ml of the lysate from the fourth amplification and coinfect with 5 MOI of helper virus.
14. Fourty-eight hours later harvest cells and pellet for 10 min at 400 g. Dissolve cells in 2 ml TBS and lyse three times by freeze/thawing. Centrifuge for 15 min at 400 g and 4°C. Keep the supernatant for further amplifications. Keep the pellet for further testing (see later).

Titering Amplifications

After the fifth amplification it is important to titer the amount of vector present in the lysate. In case HC-Ad vectors express a reporter gene (e.g., β-gal, EGFP), the vector can be titered easily in a reporter gene assay.

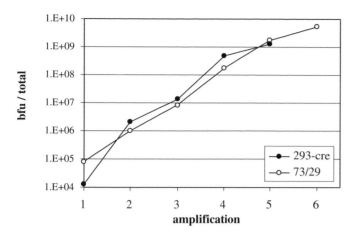

FIGURE 3 Amplification of HC-Ad vector AdGS46 using helper virus AdGS102#21 and a HEK293-based (293-cre) or N52.E6-based (73/29) Cre-expressing producer cell line. Both cell lines were transfected with *Pme*I-digested pGS46 and, at the same time, infected with helper virus (amplification 1). AdGS46 was serially amplified in both cell lines. Titers in each amplification were assayed by infecting N52.E6 cells with aliquots of the lysate followed by staining for β-galactosidase and quantitation as blue forming units (bfu).

Figure 3 shows blue forming unit (bfu) titers in amplifications of AdGS46 (see Fig. 1). However, most HC-Ad vectors do not contain a reporter gene and thus the ratio of helper and vector genomes and possible vector genome rearrangements should be tested using DNA isolated from infected cells.

15. Resuspend the pellet obtained in step 14 in 500 μl TBS.
16. Take 100 μl, add 50 μl TE, 20 μl 10% SDS, 20 μl proteinase K, 10 μl 0.5 M EDTA, vortex, and incubate for 2 h at 37°C.
17. Extract DNA by phenol/chloroform extraction and subsequent precipitation with ethanol.
18. Digest 1 μg DNA with an appropriate enzyme and add RNase to the digest. As a useful positive control, double digest 200 ng of the corresponding HC-Ad vector plasmid with *Pme*I and the aforementioned enzyme. Separate fragments in an 0.8% agarose gel containing EtBr (run gel at low voltage, preferably overnight).

Comments

At this stage of amplification the majority of DNA within the producer cell is vector DNA with only little helper virus contamination. Therefore, if the amplification was successful, only bands corresponding to the vector should be visible with a weak background of the helper virus bands and a faint smear of cellular DNA. In addition, rearranged vector genomes can be detected. If there are no or only weak vector bands visible, one or two additional amplifications can be performed. However, it is not recommended to carry out more than seven amplifications, as the risk of outgrowth of mutated helper virus and vector rearrangements increases significantly.

Preparation of Gradient-Purified HC-Ad Vector

19. Plate ten 15-cm dishes of 73/29 cells with 2×10^7 cells per dish.
20. The next day infect cells with the lysate from the last amplification (supernatant obtained in step 14) and 5 MOI of helper virus. Incubate for 48 h until complete cpe has occurred.
21. Harvest the cells by spinning for 15 min at 400 g. Resuspend cells in 5 ml TBS and lyse by freeze/thawing. Spin for 15 min at 400 g.
22. Split supernatant into two sterile 15-ml tubes, fill up to 10 ml with TBS, and add 5 g CsCl to each tube. Mix carefully until CsCl is dissolved. Transfer to two ultraclear centrifuge tubes.
23. Spin in a Beckmann SW41 for 16–22 h at 4°C and 32,000 rpm.
24. In optimal vector preparations, only one band containing vector particles should be visible. If helper virus contamination is high, a second helper virus particle band is visible in addition to the upper vector particle band separated by only a few millimeters. The upper vector band can be collected by puncturing the tube with a needle just below the vector band and drawing off the vector band in a small volume. Combine vector particles from both tubes. It is also possible to start vector purification by a CsCl step gradient followed by one or two continuous gradients.
25. Add TBS/CsCl solution up to 10 ml. Transfer to a new ultraclear centrifuge tube and spin in a Beckmann SW41 as described previously. Collect the vector in a small volume as described earlier.
26. Desalt vector particles using the PD-10 desalting column according to the manufacturer.

DNA Preparation from CsCl-Purified Vector Particles

In order to finally exclude vector rearrangement vector DNA should be isolated from CsCl-purified particles as described for extraction of helper virus DNA. The DNA should be digested with the appropriate enzyme and run on a 0.8% agarose gel containing EtBr. As a positive control, double digest the corresponding HC-Ad vector plasmid with *Pme*I and the aforementioned enzyme.

Troubleshooting

Problem	Comments and suggestions
Low helper virus titer	Helper virus might be impaired in DNA replication. Test for rearrangements in the helper virus genome. Alternatively, use an E3-containing helper virus
Instability of HC-Ad vector cloning plasmids	pBluescript containing large inserts tend to rearrange. DNA should be isolated from colonies that have been transformed the day before. Increasing the ampicillin concentration to 100 µg/ml is recommended
Poor transfection efficiency	Transfection efficiency should be tested and optimized using a linearized HC-Ad vector plasmid expressing a reporter gene (β gal, EGFP). Transfection can be inhibited by traces of phenol present in the linearized DNA. Purify DNA by an additional ethanol precipitation
Adding 5 MOI of helper virus during amplification does not results in complete cpe after 48 h	Because the amount of vector increases during amplification, it might also be necessary to increase the amount of helper virus. Because amplification efficiency depends on the producer cell and the type of helper virus used, it is highly recommended to standardize amplification using a reporter gene expressing vector for each type of producer cell and helper virus
Vector rearranges during amplification	Make sure that the vector size ranges between 27 and 35 kb. Smaller vectors tend to rearrange
No vector obtained even after sever or eight rounds of amplification	Make sure your that the transgene is not toxic for producer cells and that the vector did not rearrange during amplifications. Another possible explanation for a failed vector amplification is helper virus outgrowth due to deletion of a loxP site
High helper virus content in CsCl-purified vector	There are several possible explanations for high helper virus contamination: (1) The producer cell expresses low or no Cre-recombinase. Grow cells for several passages in the selection medium (e.g., G418). (2) One loxP site in the helper virus is lost due to recombination with vector genomes. Test for the presence of both loxP sites using PCR and primers corresponding to sequences flanking both loxP sites. (3) The helper virus turned into RCA in case HEK293-based producer cells have been used. Test for presence of RCA using PCR
Helper and vector particle bands do not separate clearly	Especially if vector yield is low, the vector or helper virus band is hardly visible. Using a fiber-optic gooseneck lamp when drawing off the band from the CsCl gradient simplifies this procedure significantly

III. COMPLETE CHARACTERIZATION OF HC-Ad VECTOR PREPARATIONS

Materials

First Day: Seeding Cells

Cell lines: A549 (ATCC number: CCL-185) or, alternatively, HeLa cells (ATCC number: CCL-2)
Supplies: 24-well tissue culture plates

Second Day: Transduction of Cells with HC-Ad Vectors

Reagents and solutions: Tris-buffered saline (TBS, see Section II.)

Third Day: Slot Blotting of Cell and Vector Lysates; Hybridization

Equipment

Shaker (e.g., Scientific Industries: Vortex Genie Model G-560E)

Vacuum pump (e.g., Biometra: Typ PM12640-026.3)

Slot-blot apparatus (e.g., Amersham Biosciences: Hoefer PR648)

Hybridisation oven (e.g., Biometra: Duo-Thermo-Oven OV5)

Supplies

Positively charged nylon membrane (Pall: Biodyne B, 0.45 µm)

50 µCi [^{32}P] dCTP

RediPrime II DNA labeling kit (Amersham-Biosciences)

Reagents and Solutions

Phosphate-buffered saline (PBS): 6.46 mM Na$_2$HPO$_4$, 1.47 mM KH$_2$PO$_4$, 137 mM NaCl, 2.7 mM KCl, pH 7.4, autoclave

PBS/20 mM EDTA: Add EDTA from aqueous stock solution (0.5 M, pH 8, see Section II) to PBS

0.8 N NaOH: Dissolve 32 g NaOH pellets in 1 liter deionized H$_2$O, always prepare fresh

(Pre-)hybridisation solution: 2 × SSC (300 mM NaCl, 30 mM Na-citrate, pH 7.0) containing 10% dextran sulfate, 1% SDS, 0.5% milk powder (low fat), and 0.5 mg/ml salmon sperm DNA

DNA Templates for Generation of Probes

Template for generation of HC-Ad vector-specific probe: This should be a purified PCR fragment of 600–1200 bp length. Usually the PCR fragment is derived from the stuffer DNA or the transgene expression cassette of the HC-Ad vector plasmid. This probe is used for determination of both the number of infectious units and total particles. Alternatively, an Ad5-ITR probe can be used that is generated by PCR with the primers P1 (5'-CAT-CATCAATAATATACCTTATTTTG-3') and P2 (5'-AACGCCAACTTTGACCCGGAACGCGG-3') from a plasmid containing at least the left Ad5 ITR.

Template for generation of helper virus-specific probe: This is a PCR fragment comprising Ad5 nucleotides 31042–32390 (Ad5 fiber gene) obtained from a plasmid containing the fiber gene of Ad5 with the primers P3 (5'-ATGAAGCGCGCAAGACCGTCTG-3') and P4 (5'-CCAGATATTGGAGCCAAACTGCC-3').

Plasmids for Generating Standard Curves

Comment: It is recommended to prepare a large stock of standard plasmids at appropriate concentrations (1–3E+06 copies/µl) and keep it frozen (−20°C).

The plasmid used to generate the standard curve for determining infectious and total particle contents of HC-Ad vector preparations is usually the HC-Ad vector shuttle plasmid that was used to generate the vector and/or the PCR template for the probe. The concentration of this standard plasmid should be determined as accurately as possible.

The plasmid used to generate the standard curve for determining the helper virus content of HC-Ad vector preparations is usually a shuttle plasmid for the generation of E1-deleted vectors and should contain the Ad5 fiber gene. The concentration of this standard plasmid should be determined as accurately as possible.

Fourth Day: Exposition of PhosphorScreen and Signal Quantification

Equipment

PhosphorScreen (e.g., Amersham-Biosciences/Molecular Dynamics: Kodak Storage Phosphor Screen SO230)

PhosphorImager (e.g., Amersham-Biosciences/Molecular Dynamics: Storm 860)

Quantification Software (e.g., Amersham-Biosciences/Molecular Dynamics: ImageQuaNT)

Supplies

Saran wrap

Reagents and Solutions

Wash buffer I: 2 × SSC (300 mM NaCl, 30 mM Na-citrate, pH 7.0) containing 0.1% SDS

Wash buffer II: 0.1 × SSC (15 mM NaCl, 1.5 mM Na-citrate, pH 7.0) containing 0.1% SDS

Procedure

First Day: Seeding Cells

1. Calculate the amount of wells needed: [No. of vectors to be titered × 6] +10.
2. Seed 1–2E+05 A549 or HeLa cells per well.
3. Let cells attach overnight in 1 ml of medium (37°C).

Second Day: Transduction of Cells with HC-Ad Vectors

1. Dilute the vectors 1:20 with TBS.
2. Aspirate medium from the cells and replace with 300 µl fresh medium.
3. Transduce the cells with 2, 10, and 20 µl of the vector dilutions in duplicates (six wells per vector).
4. Leave 10 wells untransduced to generate a standard curve for infectious particles.
5. Incubate cells for 12–14 h at 37°C.

Third Day: Slot Blotting of Cell and Particle Lysates; Hybridization

Preparation of Cell Lysates from Transduced Cells for Determining the Number of Infectious Particles

1. Aspirate the medium from transduced cells.
2. Wash the cells once with 1 ml prewarmed PBS (37°C) per well.
3. Incubate the cells with 1 ml prewarmed PBS for 5 min at 37°C.
4. Aspirate PBS and add 200 µl of PBS/EDTA per well. Incubate for 10 min at 37°C.
5. Detach the cells by pipetting up and down several times and transfer them into 1.5-ml reaction tubes.
6. Add 200 µl 0.8N NaOH per reaction tube, mix thoroughly, and incubate at room temperature for 30 min. During incubation mix thoroughly every 10 min.

Preparation of Cell Lysates for Generation of a Standard Curve for Infectious Particles

Comment: The debris in the cell lysate that is slot blotted onto the membrane will influence the signal intensity due to partial blocking of the membrane. Therefore, to obtain reliable standard curves for determining the number of infectious HC-Ad vector particles, it is absolutely required to mix the standard plasmid with untransduced cells and prepare lysates in the same way as for transduced cells before blotting.

7. Aspirate the medium from untransduced cells.
8. Wash the cells once with 1 ml prewarmed PBS (37°C) per well.
9. Incubate the cells with 1 ml prewarmed PBS for 5 min at 37°C.
10. Aspirate PBS and add 200 µl of PBS/EDTA per well. Incubate for 10 min at 37°C.
11. Detach the cells by pipetting up and down several times and transfer them into 1.5-ml reaction tubes.
12. Add 1E+06, 5E+06, 1E+07, 5E+07, and 1E+08 copies of the standard plasmid in duplicates to the cells in the reaction tubes. Do not to add more than 40 µl.
13. Add 200 µl 0.8N NaOH per reaction tube, mix thoroughly, and incubate at room temperature for 30 min. During incubation mix thoroughly every 10 min.

Preparation of Particle Lysates for Determining the Number of Total Particles

Comment: Particle lysates for determining the number of total and helper virus particles do not contain any debris that blocks the membrane. Therefore, standard plasmids are prepared in PBS/EDTA and treated the same way as particle lysates.

14. Dilute the vectors 1:600 with TBS.
15. Transfer in duplicates 2, 10, and 20 µl of the vector dilutions into 1.5-ml reaction tubes and fill up to 200 µl with PBS/EDTA.
16. Add 200 µl 0.8N NaOH per reaction tube, mix thoroughly, and incubate at room temperature for 30 min. During incubation mix thoroughly every 10 min.

Preparation of a Standard for Determining the Number of Total Particles

17. Prepare 1E+06, 5E+06, 1E+07, 5E+07, and 1E+08 copies of the standard plasmid in 200 µl PBS/EDTA in duplicates in 1.5-ml reaction tubes.
18. Add 200 µl 0.8N NaOH per reaction tube, mix thoroughly, and incubate at room temperature for 30 min. During incubation mix thoroughly every 10 min.

Preparation of Particle Lysates for Determining the Number of Helper Virus Particles

19. Transfer 2, 10, and 20 µl of the undiluted vector stocks in duplicates into 1.5-ml reaction tubes and fill up to 200 µl with PBS/EDTA.
20. Add 200 µl 0.8N NaOH per reaction tube, mix thoroughly, and incubate at room temperature for 30 min. During incubation mix thoroughly every 10 min.

Preparation of a Standard for Determining the Number of Helper Virus Particles

21. Prepare 1E+06, 5E+06, 1E+07, 5E+07, and 1E+08 copies of the standard plasmid in 200 µl PBS/EDTA in duplicates in 1.5-ml reaction tubes.
22. Add 200 µl 0.8N NaOH per reaction tube, mix thoroughly, and incubate at room temperature for 30 min. During incubation mix thoroughly every 10 min.

Slot Blotting of Lysates and Standard onto Positively Charged Nylon Membranes

Comment: The cell lysates and particle lysates for determining the number of infectious and total HC-Ad-vector particles can be blotted onto the same membrane, as the same probe is used for hybridization. Particle lysates for determining the number of helper virus particles must be blotted onto a separate membrane and will be hybridized with a different probe.

23. Soak an appropriately sized membrane 2 min in 0.4N NaOH.
24. Assemble the slotbot apparatus with the membrane in correct position.

25. Connect the slot blot apparatus to the vacuum pump and apply vacuum (max –400 mbar).
26. Check tightness by applying 200 μl deionized water to one slot.
27. Release vacuum.
28. Pipet two-thirds of each lysate into separate slots.
29. Apply vaccum starting with –400 mbar and slowly inrease up to –800 mbar until lysates pass the membrane completely.
30. Release vacuum and disassemble slot blot apparatus.
31. Rinse the membrane twice in 2 × SSC (5 min).
32. Bake the membrane at 80°C for 20 min.
33. Boil prehybridization buffer for 10–15 min in a water bath. Cool down to room temperature on ice. Use 20 ml of prehybridization buffer per membrane.
34. Prehybridize membrane at 68°C in 20 ml prehybridization buffer (1–2 h).
35. Label 100 ng of the DNA probe with [^{32}P]dCTP and add it to the prehybridization buffer.
36. Hybridize for 12–15 h.

Fourth Day: Exposition of PhosphorScreen and Signal Quantification

1. Prewarm wash buffers I and II to 68°C in a water bath.
2. Wash membrane twice with prewarmed wash buffer I for 10 min at 68°C.
3. Wash membrane with prewarmed wash buffer II for 10 min at 68°C.
4. Wash membrane with wash buffer II for 5 min at room temperature.
5. Air dry membrane.
6. Start exposition of the PhosphorScreen (usually 2–3 h).
7. Read the phosphor screen with PhosphorImager and quantify signals.
8. Calculate standard curve from signal intensities and use linear regression to quantify the signal intensities of the vector samples. The coefficient of correlation for the standard should be above 0.99.
9. Calculate the inverse bioactivity [No. of total particles/No. of infectious particles].

IV. RESULTS

Figure 4a shows a typical result for determination of the number of infectious particles of different HC-Ad vector preparations. Figure 4b shows the standard curve obtained from the signal intensities of the standard plasmid. The parameters of the standard curve

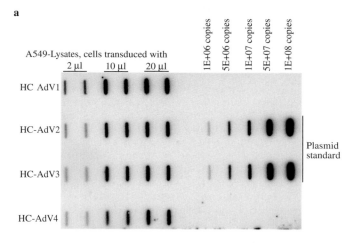

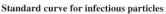

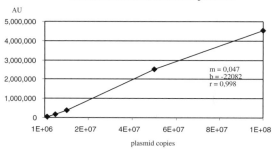

FIGURE 4 (a) Slot blot to determine the number of infectious units of different HC-Ad-vector preparations (HC-AdV1 to HC-AdV4). (b) The standard curve obtained from signal intensities of the standard plasmid. r, coefficient of correlation; b, intercept, and m, slope. These parameters were used to calculate the number of infectious particles for HC-AdV1 to HC-AdV4.

TABLE I   Number for Infectious Units Calculated from Signal Intensities for HC-AdV1 to HC-AdV4 as Shown in Fig. 4

Vector	Titer
HC-AdV1	$1.4 \pm 0.2 \times 10^7$ ip/μl
HC-AdV2	$7.8 \pm 1.2 \times 10^6$ ip/μl
HC-AdV3	$6.6 \pm 0.4 \times 10^6$ ip/μl
HC-AdV4	$5.5 \pm 0.9 \times 10^6$ ip/μl

(slope *m* and intercept *b*) were used to calculate the number of infectious particles of the vector preparations HC-AdV1 to HC-AdV4. These titers are given in Table I. Slot blots to determine the number of total particles and helper virus particles look similar and are omitted for brevity. The inverse bioactivity (number of total particles divided by number of infectious particles) usually is between 10 and 200, and the helper virus contamination is between 0.1 and 2%.

References

Fallaux, F. J., Bout, A., van der Velde, I., van den Wollenberg, D. J., Hehir, K. M., Keegan, J., Auger, C., Cramer, S. J., van Ormondt, H., van der Eb, A. J., Valerio, D., and Hoeben, R. C. (1998). New helper cells and matched early region 1-deleted adenovirus vectors prevent generation of replication-competent adenoviruses. *Hum. Gene Ther.* **9**, 1909–1917.

Gao, G. P., Engdahl, R. K., and Wilson, J. M. (2000). A cell line for high-yield production of E1-deleted adenovirus vectors without the emergence of replication-competent virus. *Hum. Gene Ther.* **11**, 213–219.

Graham, F. L., Smiley, J., Russel, W. C., and Nairn, R. (1977). Characteristics of a human cell line transformed by DNA from human adenovirus type 5. *J. Gen. Virol.* **36**, 59–74.

Hartigan-O'Connor, D., Barjot, C., Crawford, R., and Chamberlain, J. S. (2002). Efficient rescue of gutted adenovirus genomes allows rapid production of concentrated stocks without negative selection. *Hum. Gene Ther.* **13**, 519–531.

Hilgenberg, M., Schnieders, F., Löser, P., and Strauss, M. (2001). System for efficient helper-dependent minimal adenovirus construction and rescue. *Hum. Gene. Ther.* **12**, 643–657.

Kreppel, F., Biermann, V., Kochanek, S., and Schiedner, G. A DNA-based method to assay total and infectious particle contents and helper virus contamination in high-capacity adenoviral vector preparations. *Hum. Gene Ther.* **13**, 1151–1156.

Ng, P., Beauchamp, C., Evelegh, C., Parks, R., and Graham, F. L. (2001). Development of a FLP/frt system for generating helper-dependent adenoviral vectors. *Mol. Ther.* **3**, 809–815.

Palmer, D., and Ng, P. (2003). Improved system for helper-dependent adenoviral vector production. *Mol. Ther.* **11**, 504–511.

Parks, R. J., Bramson, J. L., Wan, Y., Addison, C. L., and Graham, F. L. (1999). Effects of stuffer DNA on transgene expression from helper-dependent adenovirus vectors. *J. Virol.* **73**, 8027–8034.

Parks, R. J., Chen, L., Anton, M., Sankar, U., Rudnicki, M. A., and Graham, F. L. (1996). A helper-dependent adenovirus vector system: Removal of helper virus by Cre-mediated excision of the viral packaging signal. *Proc. Natl. Acad. Sci. USA* **93**, 13,565–13,570.

Parks, R. J., and Graham, F. L. (1997). A helper-dependent system for adenovirus vector production helps define a lower limit for efficient DNA packaging. *J. Virol.* **7**, 3293–3298.

Sakhuja, K., Reddy, P. S., Ganesh, S., Cantaniag, F., Pattison, S., Limbach, P., Kayda, D. B., Kadan, M. J., Kaleko, M., and Connelly S. (2003). Optimization of the generation and propagation of gutless adenoviral vectors. *Hum. Gene Ther.* **14**, 243–254.

Sandig., V., Youil, R., Bett, A. J., Franlin, L. L., Oshima, M., Maione, D., Wang, F., Metzker, M. L., Savino, R., and Caskey, C. T. (2000). Optimization of the helper-dependent adenovirus system for production and potency *in vivo. Proc. Natl. Acad. Sci. USA* **97**, 1002–1007.

Schiedner, G., Hertel, S., and Kochanek, S. (2000). Efficient transformation of primary human amniocytes by E1 functions if Ad5: Generation of new cell lines for adenoviral vector production. *Hum. Gene Ther.* **11**, 2105–2116.

Schmid, S. I., and Hearing, P. (1998). Cellular components interact with adenovirus type 5 minimal DNA packaging domains. *J. Virol.* **72**, 6339–6347.

Thomas, C. E., Schiedner, G., Kochanek, S., Castro, M. G., and Löwenstein., P. R. (2000). Peripheral infection with adenovirus causes unexpected long-term brain inflammation in animals injected intracranially with first-generation, but not with high-capacity, adenovirus vectors: Toward realistic long-term neurological gene therapy for chronic diseases. *Proc. Natl. Acad. Sci. USA* **97**, 7482–7487.

Umana, P., Gerdes, C. A., Stone, D., Davis, J. R., Ward, D., Castro, M. G., and Löwenstein, P. R. (2001). Efficient FLPe recombinase enables scalable production of helper-dependent adenoviral vectors with negligible helper-virus contamination. *Nature Biotechnol.* **19**, 582–585.

55

Novel Approaches for Production of Recombinant Adeno-Associated Virus

Angelique S. Camp, Scott McPhee, and R. Jude Samulski

I. INTRODUCTION

This article describes how to produce a recombinant adeno-associated virus (rAAV) stock of high purity. The adeno-associated virus (AAV) is a human parvovirus that was first engineered as a recombinant viral vector for gene delivery by Hermonat and Muzyczka in 1984. The unique biology and life cycle of rAAV make it a popular choice as a gene delivery system as it satisfies the main criteria for successful gene vectors. These criteria include but are not limited to (1) efficient transduction of the target cell; (2) stable and long-term expression of the transgene of interest, especially with the use of promoters and enhancers that are not inactivated in the transduced cell; and (3) a lack of stimulation of a cytotoxic immune response to the vector or transduced cell, resulting in a very good safety profile for clinical application. Wild-type AAV has not previously been associated with disease in healthy adult humans and is classified as a risk group 1 agent under the NIH's recombinant DNA guidelines (rev. 04/02). Recombinant AAV vectors retain only the inverted terminal repeat (ITRs) sequences from the wild-type AAV genome, with 96% of the DNA genome removed. This includes all viral coding genes. Normally recombinant AAV is considered nonpathogenic, noninfectious, and nonhazardous. However, the incorporation of oncogenes or toxin encoding genes into the vector genome may alter this status. Therefore, laboratory facilities used to produce rAAV may be required by local institutions to operate in accordance with biosafety level 2 guidelines, despite rAAV not being a risk group 2 agent in typical circumstances. Wild-type AAV requires helper functions provided in *trans* by a helper virus such as adenovirus or herpes virus for AAV replication. Early generation rAAV preparations were produced using a helper adenovirus that was then almost completely eliminated in the purification process. More recently, trace levels of helper virus in rAAV stocks have been shown to elicit a cellular immune response to the AAV-transduced tissue (Monohan *et al.*, 1998). Soon after this observation, efforts were made to improve the procedure for generating rAAV vectors, with our laboratory developing a packaging procedure that uses nonoverlapping plasmid constructs to produce rAAV vectors free of contamination by wild-type AAV or helper adenovirus (Xiao *et al.*, 1998). All AAV vectors utilize a plasmid substrate carrying the viral ITR sequences flanking the therapeutic gene of interest. For efficient packaging, the rAAV insert size must be ~4.6 kb or smaller, consistent with wild-type genome size ~4.7 kb. The AAV plasmid is then cotransfected into human embryonic kidney (HEK) 293 cells, along with a plasmid(s) that provides AAV and adenovirus helper functions. HEK 293 cells contain an adenovirus 5 E1A gene integrated into the genome that activates the AAV Rep and Cap, as well as other essential Ad genes required for productive AAV infection. In this setting, only the gene insert along with the flanking ITRs is then packaged into rAAV virions. Major advances in AAV production have been directly related to better understanding the unique biology of this virus. For example, Summerford and Samulski (1998) identified the primary receptor for AAV type 2 as heparin sulfate proteoglycan. As a result, a novel purification procedure using affinity chromatography was developed to generate virus stocks with a very high level of purity. In addition to the affinity chromatography step, this protocol also uses an iodixanol gradient in place of the cesium chloride step used in

b. Balance the tubes and fill them completely by slowly adding 1X PBS dropwise to the tube to the uppermost layer. Insert a plug and centrifuge at 4°C for 1 h using a Beckman Ti-70 rotor at 70,000 rpm (350,000 g).

c. Carefully remove the Optiseal tubes from the rotor. In a viral hood, remove the plug from the top of the tube. Use a 5-ml syringe with an 18-gauge needle to puncture the tube at the 40–60% iodixanol interface. Remove 75% of the 40% iodixanol layer.

6. Purification Using a Heparin Column

Note: A viral preparation made from twenty 15-cm dishes of 293 cells can be purified on a 1-ml column with a single injection.

a. Sterilize the lines of the AKTA FPLC with $0.5 M$ NaOH and 1X PBS.
b. Run the "pump wash" program.
c. Set up the heparin column and check it for leaks.
d. Inject the sample and autozero the UV.
e. Run the 1-ml heparin column program.
f. Collect 0.5-ml fractions.
g. Sterilize the lines again using $0.5 M$ NaOH and 1X PBS.
h. Discard the column.
i. Test the fractions for the presence of virus using the dot blot hybridization assay and combine the fractions with the highest concentration of virus.

7. Delivery of Recombinant Virus in Vitro

a. Determination of rAAV titer by Dot Blot Assay

1. Place 5 μl of each fraction collected from the heparin sulfate column into a well of a 96-well microtiter plate. Assay duplicate samples of each fraction.

2. Add 50 μl of DNase 1 digestion mixture and incubate for 1 h at 37°C. This treatment digests any viral DNA that has not been packaged into capsids.

3. After 1 h, stop the digestion by adding 10 μl of $0.1 M$ EDTA to each reaction. Mix well.

4. Add 60 μl of proteinase K digestion mixture to each sample to release the viral DNA from the capsid. Incubate for 30 min at 50°C.

5. To create a set of DNA hybridization standards, use plasmid DNA that was used for the transfection. Linearize the plasmid and do serial dilutions in 10 mM Tris–HCl (pH 8.0) and 1 mM EDTA. A volume of 25 μl is convenient for each standard in wells of a 96-well microtiter plate. A suitable standard working range is 500 ng to 10 fg.

6. Denature the samples and control DNA by adding 100 μl of $0.5 M$ NaOH to each.

7. Equilibrate a nylon membrane and one piece of Whatman blotting paper in $0.4 M$ Tris–HCl (pH 7.5) and place them between the upper and lower blocks of a dot blot manifold apparatus; membrane should be on top of the Whatman paper.

8. Add the denatured DNA from the 96-well microtiter plate to wells of the dot blot manifold apparatus in the absence of a vacuum. After all the DNA has been transferred into the manifold, apply a vacuum for 3–5 min.

9. Radiolabel a transgene cassette-specific probe. (**Note:** The probe should not contain plasmid backbone or ITR sequences.)

10. In a hybridization bottle, prehybridize the nylon membrane with 5 ml of Church buffer for 5 min at 65°C. Discard the prehybridization Church buffer and replace with 5 ml of fresh Church buffer. Place at 65°C.

11. Boil the ^{32}P-dCTP-radiolabeled probe for 5 min, place on ice, and add to the hybridization bottle containing the dot blot. Hybridize overnight at 65°C.

12. Remove the hybridization solution and add 10 ml of low-stringency wash solution. Wash for 10 min at 65°C. Repeat the wash with 10 ml of fresh solution.

13. Wash the dot blot for 10 min at 65°C with the medium-stringency wash solution and discard the wash solution.

14. Monitor the dot blot with a Geiger counter. Continue the washes if needed using the high-stringency wash solution. Do not let the membrane dry out or the probe will permanently adhere to the membrane.

15. To quantitate each spot on the dot blot, expose the filter to a Phosphoimager cassette. Alternatively, employ X-ray film to identify labeled regions on the nylon membrane, excise each sample, and quantitate using a scintillation counter.

16. Plot a standard curve of DNA concentration vs integrated intensity per counts per minute for the DNA standards and employ the curve to determine the concentration of DNA in fractions obtained from the heparin sulfate column. (**Note:** The replication center assay is also a useful method to calculate the rAAV titer. The rAAV particle number of each fraction can be calculated. Remember to take into consideration that plasmid standards are double stranded whereas rAAV virions harbor only a single strand.)

IV. NOTES

1. *Hind*III is used in the digest to cut the *rep* and *cap* fragment in half for easy isolation of the plasmid backbone.

2. DNA preparation should be pure. Purify your viral fragment by agarose gel separation and running onto Whatman DEAE-8 1 paper or a preparation of equivalent high quality.

3. Alternatively, blunt-end ligation may be used to construct the rAAV vector plasmid.

4. For efficient packaging into AAV capsids, the size of the rAAV construct (including the 190-bp ITRs) must be 4.6 kb or less.

5. Plasmids are grown in the Sure strain of *Escherichia coli*. The literature shows that AAV ITRs are unstable in bacteria. To avoid deletion, restrict bacterial growth in the stationary phase. If you still obtain deletions, grow the plasmids at 30°C for only 12 h. The integrity of the plasmids can be assayed by restriction enzyme digests.

6. Polypropylene and glass attract ionic strength-dependent aggregates more than polystyrene. For this reason, mixing containers made of polystyrene are preferred for transfections.

7. The total DNA is equal to 37.5 µg/plate, and the ratio of rAAV construct to the pXX2 and pXX6 is equal to a molecular ratio of about 1:1:1.

8. If a coarse precipitate forms, decrease the incubation time. If a precipitate forms too quickly (i.e., less than 1 min) check the pH of the 2X HeBS. The pH range of 2X HeBs should be between 7.05 and 7.12.

9. Only use sterile ddH$_2$O. Do *not* autoclave!

10. After 24 h, the 293 cells (when viewed through a microscope) should have a rounded appearance, indicating viral replication. If detached cells are noted, the incubation was too long.

11. Cell suspensions or cell precipitates may be stored at −20°C for up to 6 months.

12. The NaCl in the 15% iodixanol layer will separate viral aggregates that may form due to the high concentration of virus.

13. Collecting fractions from the flow-through and washing steps ensures that all the virus was bound to the column and eluted as the salt gradient increased.

14. Using a new heparin column for each purification ensures that you will not cross-contaminate viral preparations.

References

Amiss, T. J., and Samulski, R. J., (2000). Methods for adeno-associated virus-mediated gene transfer into muscle. *In "Methods in Molecular Biology"* (M. P. Starkey, R. Elaswarapu, ed.), Vol. 175, pp. 455–469. Humana Press, Totowa, NJ.

Berns, K. I., and Giraud, C. (1995). Adeno-associated virus (AAV) vectors in gene therapy. *Curr. Topics Microbiol. Immun.* 218, 1–25.

Bartlett, J. S., and Samulski, R. J. (1996). Production of recombinant adeno-associated viral vectors. *In "Current Protocols in Human Genetics"*, pp. 12.1.1–12.1.24. Wiley, Philadelphia.

Daly, T. M., Okuyama, T., Vogler, C., Haskins, M. E., Muzyczka, N., and Sands, M. S. (1999). Neonatal intramuscular injection with recombinant adeno-associated virus results in prolonged beta-glucuronidase expression in situ and correction of liver pathology in mucopolysaccharidosis in type VII mice. *Hum.Gene Ther.* 10, 85–94.

Ferrari, F. K., Xiao, X., McCarty, D., and Samulski, R. J. (1997). New developments in the generaion of Ad-free, high-titer rAAV gene therapy vectors. *Nature med.* 3, 1295–1296.

Hermonat, P. L., Mendelson, E., and Carter, B. J. (1984). Use of adeno-associated virus as a mammalian DNA cloning vector: Transduction of neomycin resistance into mammalian tissue culture cells. *Proc. Natl. Acad. Sci. USA* 81, 6466–6470.

Kotin, R. M., Siniscalco, M., Samulski, R. J., Zhu, X. D., Hunter, L., Laughlin, C. A., Mclaughlin, S., Muzyczka, N., Rocchi, M., and Berns, K. I. (1990). Site-specific integration by adeno-associated virus. *Proc. Natl. Acad. Sci. USA* 87, 2211–2215.

Monahan, P. E., Samulski, R. J., Tazelaar, J., Xiao, X., Nichols, T. C., Bellinger, D. A., and Read, M. S. (1998). Direct intramuscular injection with recombinant AAV vectors results in sustained expression in a dog model of hemophilia. *Gene Ther.* 5, 40–49.

Samulski, R. J. (1993). Adeno-associated virus: Integration at a specific chromosomal location. *Curr. Opin. Gen. Dev.* 3, 74–80.

Samulski, R. J., Sally, M., and Muzycka, N. (1999). Adeno-associated viral vectors. *In "The Development of Human Gene Therapy"* (T. Friedmann, ed.), pp. 131–172. Cold Spring Harbor Laboratory, Cold Spring Harbor, NY.

Samulski, R. J., Zhu, X., Xiao, X., Brook, J. D., Housman, D. E., Epstein, N., and Hunter, L. A. (1991). Targeted integration of adeno-associated virus (AAV) into human chromosome 19 [published erratum appears in *EMBO J* 11(3), 1228 (1992)]. EMBO J. 10, 3941–3950.

Song, S., Morgan, M., Ellis, T., Poirier, A., Chesnut, K., Wang, J., Brantly, M., Muzyczka, N., Byrne, B. J., Atkinson, M., and Flotte, T. R. (1998). Sustained secretion of human alpha-1-antitrypsin from murine muscle transduced with adeno-associated virus vectors. *Proc. Natl. Acad. Sci. USA* 95, 14,384–14,388.

Summerford, C., and Samulski, R. J. (1998). Membrane-associated heparin sulfate proteoglycan is a receptor for adeno-associated virus type 2 virions. *J. Virol.* 72, 1438–1445.

Summerford, C., and Samulski, R. J. (1999). Viral receptors and vector purification: New approaches for generating clinical-grade reagents. *Nature Med.* 5, 587–588.

Xiao, X., Li, J., and Samulski, R. J. (1996). Efficient long-term gene transfer into muscle tissue of immunocompetent mice by adeno-associated virus vector. *J. Virol.* 11, 8098–8108.

Xiao, X., Li, J., and Samulski, R. J. (1998). Production of high-titer recombinant adeno-associated virus vectors in the absence of helper adenovirus. *J. Virol.* 72, 2224–2232.

Zolotukhin, S., Byrne, B. J., Mason, E., Zolotuknin, I., Potter, M., Chesnut, K., Summerford, C., Samulski, R. J., and Muzyczka, N. (1999). Recombinant adeno-associated virus purification using novel methods improves infectious titer and yield. *Gene Ther.* 6, 973–985.

Production and Purification of Antibodies

sequence for antibody production in the animals. The conjugated peptide (or MAP peptide) is then used as antigen in rabbits as described (Huet, 1998).

High-titered **antiserum** can be obtained within about 2–3 months and can be subsequently affinity purified using the synthetic peptide coupled to a suitable matrix. This process allows to get rid of all the unrelated antibodies present in the antiserum (endogenous antibodies from the rabbits and the antibodies produced against the carrier protein) that otherwise may give rise to background signals.

II. MATERIALS AND INSTRUMENTATION

15–25 mg of custom-made peptide (or MAP peptide) at >70% purity, stored lyophilized at 4°C under inert gas and protected from moisture

KLH (Sigma Cat. No. H7017), stored lyophilized at 4°C

Glutaraldehyde (Sigma Cat. No. G6257), 25% solution in water, stored at room temperature. Because this product is toxic, always work under a hood and wear gloves.

EDC (Pierce Cat. No. 22980)

MBS (Pierce Cat. No. 22311)

Dimethyl formamide (DMF) (Sigma Cat. No. D8654), stored at room temperature

Glycine (Sigma Cat. No. G7126), stored at room temperature

Boric acid (Vel Cat. No. 1021), phosphate-buffered saline (PBS) 10× (Sigma Cat. No. P7059), tromethamine (Tris) (Sigma Cat. No. T1503), sodium hydrogen carbonate trihydrate (NaHCO$_3$·3H$_2$O) (Vel Cat. No. 7958), sodium chloride (NaCl) (Vel Cat. No. 1723), disodium hydrogen phosphate dodecahydrate (Na$_2$HPO$_4$·12H$_2$O) (Vel Cat. No. 1773), glacial acetic acid (Vel Cat. No. 1005), 0.1 M hydrochloric acid (Sigma Cat. No. 210-4), sodium hydroxide (Sigma Cat. No. S5881), and guanidine hydrochloride (Sigma Cat. No. G4505), all stored at room temperature

PD10 column (Pharmacia Cat. No. 17-0851-01), activated CH-Sepharose 4B (Pharmacia Cat. No. 17-0490-01), EAH-Sepharose 4B (Pharmacia Cat. No. 17-0569-01), stored at room temperature

Dialysis membrane Spectra/Por (Spectrum Cat. No. 132678), molecular weight cutoff 12-14000

pH meter MP220 (Mettler-Toledo) or similar

Spectrophotometer Ultraspec 3300P20 (Pharmacia) or similar

FPLC Akta with UV detector (Pharmacia) or similar

III. PROCEDURES

A. Selection of Suitable Peptide Sequences and Definition of Coupling Strategy

1. Selection of Antigenic Peptides

Whenever possible, use a software for predicting the "antigenicity" of the **peptide**. We use the Lasergene software from DNASTAR, Inc., particularly the EditSeq (sequence edition), Protean (antigenicity prediction), MegAlign (sequence alignments), and GeneMan (homology searches) modules.

Other useful free software can be used on the Expasy Molecular Biology Server of the Swiss Institute of Bioinformatics (http://www.expasy.org/); e.g., the Protscale program can be used to assess hydrophilicity and other parameters. Links to software for predicting signal peptides, transmembrane domains, potential phosphorylation, and glycosylation sites can be found there too. Another good source is the protein analysis software contained in the GCG Wisconsin package from Accelrys (http://www.accelrys.com/products/gcg_wisconsin_package/program_list.html#Protein) and there may be many more available through the web. All these computer programs can be very helpful in selecting a well-suited peptide. A few hints are given here to facilitate the selections based on the most used algorithms.

Check in an international database if the protein has been described. There is often helpful information concerning different domains of known proteins in their description (e.g., transmembrane regions, signal peptides, potential glycosylation or phosphorylation sites, or even a link to the three-dimensional structure if this has been resolved). If the three-dimensional structure is known, use the structure information to define a surface-located region (e.g., using the Swiss Viewer software available at the Expasy server: http://www.expasy.org/spdbv/). After pasting the sequence in the EditSeq module, open it with the Protean program and check the following parameters over the whole length of the protein sequence: hydrophilicity (Hopp-Woods and Kyte-Doolittle), antigenic index (Jameson-Wolf), surface probability (Emini), flexibility (Karplus-Schulz), turn, α, β, and coil plot (Garnier-Robson), as well as positive and negative regions in terms of charge.

If the N or C terminus of the protein is hydrophilic, it is often a good choice. Some proteins contain signal sequences, which are cleaved in the mature proteins. Those signal peptides are very often hydrophobic parts of about 20–30 amino acids, which are not well suited. If doubt persists, use a software (e.g., SignalP program available at http://www.cbs.dtu.dk/services/SignalP/) to predict its cleavage site.

If internal peptides have to be used, they should be hydrophilic, present a high surface probability, and should not be predicted as strongly α helical. Short synthetic peptides will hardly adopt the same α-helical structure and therefore the antibodies will not recognize such structured regions in the native protein. Good hydrophilicity and surface probability are essential if the antibodies have to recognize the native protein (e.g., immunohistochemistry experiments).

Choose a peptide in a flexible region whenever possible. Avoid potential phosphorylation (serine, threonine, and tyrosine) and glycosylation sites (not predicted by these algorithms, but can be predicted by other software if no better information is available). Antibodies made against a simple peptide will off course not fix at the corresponding region in the protein if this contains a large sugar moiety.

Avoid peptides containing several cysteines, as they may participate to disulfide bridges impossible to mimic with a short synthetic peptide. If the protein contains transmembrane regions, avoid loops between two transmembrane domains that are shorter than 30–35 amino acids. Short loops are not well mimicked by flexible and linear peptides.

Choose several well-suited peptides of about 12–15 amino acids length and rank them from best to less good based on the aforementioned parameters. Shorter peptides (9–11 amino acids) can be used if the antibodies have to recognize only a very small region, e.g., a phosphorylation site, which should then be placed about in the middle of the peptide.

2. Homology Searches

In order to avoid unspecific or background signals, it is important to check the chosen peptide sequences for possible homologies with known proteins other than the target one. Depending on the application, interspecies cross-reactivities may not be disturbing or may even be desired.

For every chosen peptide sequence, run a BLAST search (available at the NCBI: http://www.ncbi.nlm.nih.gov/BLAST/) using the Protein Blast and the "Search for short nearly exact matches" option. A standard BLAST with a short peptide will hardly return any result, while the option here above gives matches for short parts of proteins.

Discard those peptides for which stretches of more than five identical amino acids are found in another protein(s). Because one epitope generally contains more than five amino acids, specific antibodies can be

expected if less or equal than five amino acids one following the other are present. The following example illustrates these issues:

peptide: EHRTPRGKEDSSVP
               | | | | | | | | | | |
other protein: EHRTPRGKYQSIVP

antibodies may cross-react (8 identical amino acids in a stretch)

peptide: EHDTDRGMIDSSVP
               | | | | | | | | |
other protein: EHRTPRGKYQSSVP

antibodies will be specific (no stretch >4 identical amino acids)

Care should also be taken with very similar amino acids, such as serine and cysteine, with the only difference being the functional group on the side chain (OH vs SH), which is only a very minor difference that may not be sufficient to avoid cross-reaction if the rest of the epitope is identical.

If specificity is the major goal (e.g., antibodies against one specific member out of a protein family), select those peptides with the lowest homologies as first choices. If the first goal is to obtain working antibodies, give a higher weight to antigenicity analysis than to the homology part.

3. Coupling Strategy

It is very important to consider the following rules for the coupling of peptides to the carrier protein. After coupling to the carrier, the peptide should mimic the original structure in the best possible way, which means that the peptide should be linked to the carrier protein at one end and not in the middle of the peptide. A **conjugation** in the middle of the sequence would only display two small parts useful for antibody production (considering that the antibodies containing the conjugation residue in their epitope will not recognize the protein of interest in which this conjugation reagent is not present of course).

If the chosen peptide corresponds to the real N terminus of the protein, coupling must be done on the C-terminal part of the peptide. If the peptide sequence contains lysine, aspartate, or glutamate, add a cysteine C-terminally for conjugation using MBS and synthesize the peptide as C-terminal amide (C terminus: —CONH$_2$), which better mimics the normal amide bond and does not introduce an unnatural negative charge. If it does not contain lysine, aspartate, or three times by 2 ml water, pH 4.5. Resuspend the gel in 1 ml of water and adjust pH to 4.5 by adding some 0.1 M hydrochloric acid solution.

Dissolve 10 mg of peptide in 1 ml of water and adjust pH to 4.5 by adding some 0.1 M hydrochloric acid solution.

Add the peptide solution to the gel suspension. Add 70 mg EDC and adjust the pH again to 4.5 by adding some 0.1 M hydrochloric acid solution.

Incubate for 4 h at room temperature. During the first hour of reaction, the pH decreases and it is very important to readjust the pH to 4.5 regularly by adding small volumes of 0.1 M sodium hydroxide solution.

Centrifuge and wash the gel with 3 ml of 0.1 M acetate buffer, 0.5 M salt, pH 4.0 buffer and then 3 ml of phosphate buffer, 0.5 M salt, pH 7.4. Repeat this washing at least three times.

After the last centrifugation, discard the supernatant, resuspend the gel in 1 ml PBS, and transfer the gel in the column adapted to the FPLC instrument.

3. Conjugation of Peptide to EAH-Sepharose 4B Using MBS

This procedure is to be used for peptides that have undergone a MBS coupling to a carrier protein.

Buffer preparation: 50 mM phosphate buffer, pH 6.0; dissolve 17.9 g of disodium hydrogen phosphate dodecahydrate ($Na_2HPO_4 \cdot 12H_2O$) in 1 liter of water

Take 3 ml of EAH-Sepharose 4B suspension, centrifuge, and discard the supernatant. Wash the gel three times by 2 ml of 50 mM phosphate buffer, pH 6.0. Discard the supernatant each time.

Resuspend the gel in 1 ml of 50 mM phosphate buffer, pH 6.0

Dissolve 15 mg MBS in 1 ml of DMF. Use this solution within 1 h!

Add 140 μl of the MBS solution and 500 μl DMF to the gel suspension and let react for 30 min at room temperature under gentle stirring.

Centrifuge and discard the supernatant. Wash with 2 ml of PBS buffer, pH 7.4, centrifuge, and discard the supernatant. Repeat this washing three times and then resuspend in 1 ml of PBS buffer.

Dissolve 10 mg of peptide in 1 ml of PBS buffer, pH 7.4, and add this solution to the gel suspension. Incubate under gentle stirring for 3 h at room temperature.

Centrifuge, discard the supernatant, resuspend in 1 ml of PBS buffer, pH 7.4, and transfer the gel in the column adapted to the FPLC instrument.

4. Affinity Purification of Antiserum

Buffer preparation: PBS high salt; dissolve 20 g of sodium chloride in 1 liter of PBS buffer

100 mM glycine buffer, pH 2.5; dissolve 7.5 g of glycine in 1 liter of water

1 M Tris buffer, pH 9.0; dissolve 121.14 g of Tris in 1 liter of water

1.5 M guanidine hydrochloride: dissolve 143.30 g of guanidine hydrochloride in 1 liter of water

Thaw the serum to be purified (we use generally 50 ml serum in one run), place it in a graduated cylinder, and add 1 g of sodium chloride per milliliter of serum.

Mount the column on the FPLC instrument and wash the column with 10 ml of 1.5 M guanidine hydrochloride followed by 10 ml of PBS high salt.

Prepare a Falcon tube for antibody recovery and place 400 μl of 1 M Tris buffer in the tube.

Charge the serum slowly on the column (maximum flow rate should be 0.5 ml/min) and take care of the counterpressures that may arise (in case the counterpressure gets too high, decrease the flow rate). Recover the flow through of the column in the original serum tube. Never discard any liquids before making sure the affinity purification has worked correctly!

Once all the serum has been charged, wash the column with PBS high salt until the absorbance (measured at 280 nm) decreases to zero. Stop the recovery in the original serum tube as soon as the absorbance starts to decrease.

Elute the antibodies by passing the glycine buffer, pH 2.5, on the column and recover the antibodies (increase in absorbance) in a Falcon tube containing Tris buffer. Stop the recovery as soon as the absorbance gets back close to zero.

Measure immediately the pH in the Falcon and adjust to pH 7–7.5 if necessary (by Tris buffer or glycine buffer). Dialyze the antibody solution against 5 liters PBS buffer for 4 h at 4°C; repeat the dialysis with 5 fresh liters of PBS.

Divide the antibody solution in working aliquots and store the aliquots at −20°C. If the antibodies are to be stored a long time, it is best to add 0.1% sodium azide as a preservative (except if it is planned to label the antibodies afterward, in this case 0.1% thimerosal as a preservative is preferred).

Wash the affinity column for 5 min with guanidine hydrochloride solution, check the absorbance during this process, and recover eventual peaks in a Falcon tube. Dialyze immediately (see Section IV) against 5 liters PBS buffer (two times).

Wash the column for 10 min by PBS, recover the gel material, and store it at 4°C after the addition of one equivalent volume of ethanol.

Quantify the antibody solution by measuring the absorbance of an aliquot at 280 nm. One optical density unit corresponds to about 0.75 mg of pure antibody.

IV. PITFALLS

Depending on the sequence, peptides are sometimes difficult to get into aqueous solution. In case a peptide does not dissolve in water, dissolve a dry quantity in a minimum amount of dimethyl sulfoxide (DMSO); once it is dissolved completely, dilute by water to the appropriate concentration. The DMSO does not influence the coupling process.

Coupling may sometimes give rise to precipitates, depending on the peptide sequence. These precipitates should not be discarded for the antibody production; vortex the solution well before aliquoting the injection amounts to include the peptide–carrier aggregates. Some antigens are known to be phagocytosed even better than soluble material. In case the aggregates are too large, crush them first with a spatula to get them as fine as possible. Never freeze the affinity gels; these are best kept at 4°C in PBS buffer containing sodium azide or PBS/50% ethanol.

It may happen that the glycine buffer does not elute some high-affinity antibodies. These antibodies will come off the column during the guanidine hydrochloride wash. Dialyze *immediately* the recovered solution two times against PBS as described earlier, as the antibodies will generally still be active.

References

Huet, C. (1998). Production of polyclonal antibodies in rabbits. *In* "*Cell Biology: A Laboratory Handbook*" (J. Celis, ed.), 2rd Ed., Vol. 2, pp.381–391. Academic Press, New York.

Tam, J. P. (1988). Synthetic peptide vaccine design: Synthesis and properties of a high-density multiple antigen peptide system. *Proc. Natl. Acad. Sci. USA* 85, 5409–5413.

tion. Complete Freund's adjuvant (CFA, Cat. No. 77140) and incomplete Freund's adjuvant (IFA, Cat. No. 77145) are from Pierce Biotechnology Inc. Hybridoma fusion cloning supplement (HFCS, Cat. No. 592247800) and polyethylene glycol 1500 in 75 mM HEPES, sterile and fusion tested (PEG, Cat. No. 783641), are from Roche Diagnostics. Red blood cell lysing buffer (8.3 g/liter NH_4Cl in 0.01 M Tris–HCl) (Cat. No. R-7757) and trypan blue solution (Cat. No. T-8154) are from Sigma. Dimethyl sulfoxide (DMSO, Cat. No. 102950) is from Merck. ELISA-based isotyping kits are available from a number of manufacturers, including Roche Diagnostics (Cat. No. 1183117) and Pierce Biotechnology Inc (Cat. No. 37502). Store all stock reagents as recommended by the manufacturer.

C. Cell Culture Materials

Tissue culture plates, 24-well (Cat. No. 143982) and 96-well (Cat. No. 167008) flat-bottom plates with lids, 25-cm^2 (Cat. No. 136196) and 80-cm^2 (Cat. No. 178891) tissue culture flasks, cryotubes (Cat. No. 375418), 10-ml disposable pipettes (Cat. No. 159633) and petri dishes (Cat. No. 150350) are from Nalge Nunc Int. V-bottom 30-ml tubes (Cat. No. 128A) and wide-bore graduated transfer pipettes (Cat. No. PP88SA) are from Bibby-Sterilin Ltd.

A container (Cat. No. 5100 Nalgene Nunc) or a foam box with walls of about 1 cm thick is required for the control rate freezing of cells.

In addition, we use Gilson Pipetman 5- to 20- and 20- to 200-µl pipettes and a Gilson Distriman repeating pipette for cloning, all with sterile tips. Hamilton 500-µl glass luer lock syringes are used for immunization (Cat. No. 1750 Hamilton, Reno. Nev). Sterile forceps and scissors are also required.

D. Myeloma Cell Line

Several myeloma cell lines are commonly used for fusion with murine spleen cells; most have been derived from the P3X63Ag8.653 murine myeloma line (Kearney *et al.*, 1979). The myeloma clones SP2/0 (Shulman *et al.*, 1978) and FO (de St Groth and Scheidegger, 1980), derived from P3X63Ag8.65, are commonly used. These cell lines are available from the American Type Culture Collection (ATCC) (http://www.atcc.org) as CRL1580, CRL1581, and CRL1646, respectively. We use P3X63Ag8.653 cells that have been maintained in our laboratory for many years, which are screened regularly for mycoplasma, and we select batches that have been previously successful in hybridoma production.

III. PROCEDURES

A. General

The preparation of monoclonal antibodies depends on a series of steps that require attention to detail and careful laboratory management. The major requirement is time; Table I gives an approximate time frame for monoclonal antibody preparation. Phase 1 is largely taken up by the immunization protocol, ensuring that the screening method is working, growing the myeloma cells, and finally performing the fusion. Phase 2 encompasses testing, cloning, and freezing and is more labor-intensive. However, the workload is variable and unless the screening method is cumbersome or many fusions are carried out, it is not a full-time occupation.

In our experience the secret to success in monoclonal antibody preparation is the screening method. Ideally the method should be in place before the animals are immunized. The method needs to be robust and reliable and, as there may be 30 to 50 samples to screen a day, reagents and equipment need to be readily available. The method should be appropriate to the intended use of the monoclonal antibod-

TABLE I Time Frame for Monoclonal Antibody Preparation

Phase One	
Light workload	
Prepare antigen and develop screening assay	
Immunize animals	4–8 weeks
Test bleed	Week 7
Thaw myeloma line	Fourth or 8th week
Fuse spleen and myeloma line HAT selection medium	Day 1
Feed with HT medium	Day 7
Feed with HT medium	Day 14
Colonies become visible, screening begins	Day 14 to about day 32
Phase Two	
Heavy workload	
Screen wells as required	
Clone, freeze, feed, test	About 21-day cycle
Reclone, freeze, feed, test	About 21-day cycle
Third clone cycle	About 21-day cycle
Keep extensive records	
Phase Three	
Variable workload	
Isotype and scale up	
Characterization.	
Antibody purification, labeling, etc.	
Maintenance of frozen stock	

ies. Although many antibodies do perform in alternative protocols, there is no guarantee unless they have been screened appropriately. Some suggestions for screening methods are included in Table II, but detailed descriptions are outside the scope of this article.

B. Antigen Preparation and Immunization

A number of protocols that we have used to immunize mice to a variety of antigens are shown in Table II. Animals can make antibodies to a wide range of molecular structures with two important general exceptions: animals will not usually make antibodies to self-antigen (see Section V) and will not usually recognize small molecules. Small molecules need to be conjugated to a carrier protein. Selection of the carrier protein appears to be largely personal, but the ease of and position of conjugation are important criteria. When linking peptides to carrier proteins, Landsteiner's principle (Landsteiner, 1945) should be taken into account, that antibody specificity tends to be directed to epitopes of the hapten furthest removed from the functional group linked to the carrier protein.

Commonly used carrier proteins include keyhole limpet hemocyanin (KLH) and bovine serum albumin (BSA). If you are going to screen for monoclonal antibodies by ELISA, make both peptide–KLH and peptide–BSA combinations; immunize with one and screen against the other. There are a number of methods for the conjugation of peptides to carrier proteins and the protocol selected will depend on the amino acid sequence of the peptide and the epitope of interest. However, if you are having a peptide synthesized, consider adding a biotin tag; a streptavidin-conjugated carrier protein can then make a convenient carrier.

If high-affinity IgG monoclonal antibodies are required, the immunization protocol needs to invoke T-cell help. Hence, adjuvant and several booster injections may be needed and test bleeds should be examined for the presence of antigen-specific IgG. If IgG isotype monoclonal antibodies are essential, they should be screened for specifically.

The protein antigen immunization schedule used in our laboratory is shown in Table III. Most protocols require antigen in adjuvant for immunization. Complete Freund's adjuvant (CFA) is highly effective but may cause toxic side effects. A number of new adjuvants are becoming available, including Hunter's TiterMax (http://www.titermax.com), Pierce's AdjuPrime Immune Modulator (http://www. piercenet.com), and RIBI adjuvant systems (http://www.corixa.com) among others. Advice from the Institutional Animal Ethics Committee should be taken on which adjuvants are recommended.

If CFA is used, it should only ever be used subcutaneously for the primary immunization and Freund's incomplete used for subsequent boosts.

Complete Freund's adjuvant causes inflammation and granuloma formation in mice and will cause severe injury in humans. Take suitable precautions to prevent self-injection and employ eye protection.

Six-week-old female BALB/c mice are preferred for immunization as the myeloma lines used for fusion were derived from that strain. Routinely, we immunize six animals, in batches of two, with a 2-week interval between immunizing the next two animals.

TABLE III   Immunization Schedule for Protein Antigens

Primary immunization	Day 0
First boost	Day 14
Second boost	Day 28
Test bleed	Day 35
Third boost	Day 42
Fusion	Day 45

TABLE II   Suggested Immunization and Screening Protocols

Immunogen	Concentration	Route of injection	Screening assay
Cells, bacteria, virus, nematodes	5×10^6–5×10^7	Intraperitoneal (200 μl saline)	Flow cytometry Dot blot, ELISA Immunohistochemistry
Plasmid DNA	50 μg plasmid	Intramuscular (50 μl/site, both rear legs)[a]	ELISA
Glycoprotein peptides/carrier Proteins	2–50 μg	Subcutaneous[b] (100 μl/site, CFA, IFA) or intraperitoneal (200 μl saline)	Western blot ELISA

[a] Anesthesia is essential or muscle contraction will eject the immunogen.
[b] Typically at the nape of the neck and at the base of the tail.

Steps

Wear eye protection.

1. Draw an equal volume of antigen in saline and CFA into two separate Hamilton 500-μl glass luer lock syringes.
2. Connect the syringes with a three-way stopcock and emulsify the mixture slowly by passing material from one syringe to the other.
3. Check that an emulsion has been formed by taking a drop of the mixture from the stopcock and placing on water, an emulsion will float and not dissipate.
4. Using the Hamilton syringe and a 21-gauge needle, immunize the animals with 100 μl of emulsion subcutaneously at the base of the tail and nape of the neck.

C. Test Bleeds

Analysis of a test bleed can allow the fusion to be postponed and the animals reimmunized if the antibody titer is low or nonexistent. Test bleeds should only be carried out by experienced personnel. Test bleeds should preferably be from the lateral tail vein using a 26-gauge needle and syringe. A maximum of 1% of the body weight of the animal should be collected, about 100–200 μl. Larger volumes of blood can be collected at the time of euthanasia and can provide a valuable positive control sample for the screening protocol.

D. Myeloma Cell Preparation

About 1 week before the fusion, thaw a vial of the myeloma cell line from the liquid nitrogen stock (see Section IV,B).

The myeloma cells should be of known pedigree; it is vital to maintain the myeloma line carefully. Mycoplasma-contaminated stock or stock that has been overgrown will result in poor production of hybridomas.

Steps

1. Transfer the cells into 24-well plates at about 2×10^5 per well, check cell numbers daily, and scale up as they double in number into 25-cm^2 flasks and then into 80-cm^2 flasks containing 50 ml of RF10.
2. Maintain vigorous growth rates during the scale-up period. The day prior to the fusion the myeloma cells should be given a one-to-one split so that the cells are in an exponential growth phase at the time of fusion.
3. The number of myeloma cells required depends on how many spleen cells are to be fused. Typically,

we use a spleen to myeloma cell fusion ratio of 10:1. Normal BALB/c mice produce about 10^8 leukocytes per spleen, hence 10^7 myeloma cells are required for each spleen.

4. On the day of the fusion and before the animals are euthanized, examine the myeloma cells carefully to check that the culture is not contaminated.

IV. CELL FUSION PROCEDURE

Solutions

1. *RF10 medium:* To 500 ml of RPMI 1640 add 50 ml of FBS and 5 ml of PSGx100. Store at 4°C for up to 14 days; after 14 days replace the PSG. Warm to 37°C in a water bath before use.

2. *HAT selection medium:* Reconstitute 100× lyophilized HAT supplement with 10 ml of sterile distilled water and store 1-ml aliquots at 4°C protected from light. Add 1 ml HAT and 2 ml of HFCS to 100 ml of RF10 medium. Make only as much as required for the fusion. Warm to 37°C in a water bath before use.

3. *HT medium:* To 500 ml of RF10 medium add 5 ml of HT supplement. Warm to 37°C in a water bath before use.

4. *HT medium +2% HFCS:* Add 2 ml of HFCS to 100 ml HT medium. Warm to 37°C in a water bath before use.

A. Background Notes

HAT medium is used to selectively grow hybrids following fusion. Aminopterin selects against unfused myeloma cells and myeloma:myeloma-fused cells by blocking the main synthetic pathway for DNA. Unfused spleen cells do not have the capacity to survive for more than a few days. In hybridomas, hypoxanthine and thymidine supply purines and pyrimidines for DNA synthesis via HGPRTase salvage pathways derived from the spleen cells. The HT medium acts as a rescue medium while the aminopterin is being diluted out. Although the HT medium can be withdrawn once the nucleoside biosynthesis pathways are re-established, in our experience it is rarely worth the effort.

Cells under pressure often require "supplements" to maintain growth. In many monoclonal antibody preparation methods, these are supplied by feeder cells, usually mouse thymocytes or peritoneal washout cells. In the method described, feeder cells are replaced with hybridoma fusion cloning supplement HFCS, which in addition to being preferred for reasons of animal ethics, is more convenient and, in our experience, more effective.

B. Fusion Protocol

Steps

1. For each spleen to be fused, place a 1-ml aliquot of PEG mixture and a 3- and a 7-ml aliquot of RF10 in a 37°C water bath.

2. After euthanasia, submerge the mouse in 70% alcohol for a few minutes; this prevents hair and bedding material from contaminating the laminar flow hood.

3. Place the animal on its right-hand side, grasp the skin posterior to the rib cage with sterile forceps, and make a small incision to cut the skin but not to penetrate the subcutaneous fascia. With two pairs of forceps, grasp the skin at either side of the incision and pull. The skin should peel away, revealing the subcutaneous fascia. Flush the area with 70% alcohol to remove any stray hair. Using forceps and scissors, cut through the subcutaneous fascia into the peritoneal cavity and locate the spleen and remove by cutting away the connective tissue.

4. Place the spleen into a petri dish containing about 15 ml of RF10. At this point, blood can be collected by opening the thoracic cavity, quickly puncturing the heart, and collecting blood; this can be useful as a positive control.

5. Half-fill two 10-ml syringes with RF10 from the petri dish and, using 21-gauge needles, gently disrupt the spleen by injecting the medium. Inject the media slowly, a little at a time; you should see clouds of cells going into the medium and the spleen turning lighter in color. Finally, when the spleen looks like a limp sack, gently tease apart the spleen with the needle and flat tweezers to remove any remaining cells. Discard the spleen connective tissue.

6. Gently pipette the cell suspension to break up clumps and transfer into a 30-ml V-bottom tube, ignoring large clumps and the remains of the spleen connective tissue; let the tube stand for about 5 min.

7. Remove cells from the tube and transfer into another 30-ml V-bottom tube, leaving behind the debris and connective tissue that has settled to the bottom.

8. Pellet the spleen cells at $400\,g$ for 5 min.

9. Resuspend the cell pellet in 5 ml of red cell lysing solution, leave for 5 min, and then fill the tube with media and pellet at $400\,g$ for 5 min.

10. Resuspend the spleen cells in 5 ml of media and perform a cell count. The number of leucocytes derived from a single spleen should be approximately 10^8 cells total.

11. Wash and count the myeloma cells; for 10^8 spleen cells, 10^7 myeloma cells will be required.

12. Add the myeloma cells to the spleen cell suspension to give a spleen/myeloma cell ratio of 10:1. *Note:* **This is the critical bit, for the next 20 min you will be committed to fusing the cells.**

13. Pellet the cells ($400\,g$ for 5 min) and remove *all* of the supernatant; this is best done with a Pasteur pipette attached to a vacuum line.

14. Tap the pellet to loosen the cells.

15. With a wide-bore transfer pipette, add 1 ml of PEG solution per spleen. Gently mix with the pipette for 10 s and then continue to mix the cells gently by tapping the tube for a further 50 s.

16. Retrieve the 3-ml RF10 sample and slowly add dropwise over a 10-min period continually mixing the cells gently by tapping the tube. This can be accomplished (for a right-handed person) by holding the tube in the left hand between thumb and forefinger, tapping the tube with the second or third finger, depending on comfort and reach, and adding the medium with a transfer pipette held in the right hand. With a little practice a steady rhythm can be produced.

17. After the first 3 ml, retrieve the 7-ml aliquot and add over the next 10-min period.

18. Pellet the cells at $400\,g$ for 5 min and resuspend in about 10 ml of RF10; loosen the tube cap to allow for CO_2 transfer and place the cells in the tissue culture incubator for about 1 h.

19. Pellet the cells at $400\,g$ for 5 min.

20. Gently resuspend the cell pellet in a small volume of the HAT selection medium.

21. Add this suspension back into HAT selection medium; 50 ml of HAT selection medium will be needed for every 10^7 myeloma cells used in the fusion.

22. Add about 1 ml of the cell suspension to each well of the 24-well plates. This is the equivalent of 2×10^5 myeloma cells per well. We prefer plating the fusion into 24-well plates, as evaporation is less of a problem. The larger volume also provides enough supernatant for any screening assay.

23. Label and place the plates in the tissue culture incubator.

With the exception of the 1-h incubation period, the entire fusion protocol should take no more then 2 to 3 h.

C. Maintaining Hybridomas

Examine the wells using the inverted microscope on days following the fusion. Do not keep the plates out of the incubator for more than about 10 min, as media in the wells will cool and the pH may also change. Large numbers of dead cells may be seen, this is normal. You may also see some small moving particles; this is due to Brownian movement of debris and does not indicate bacterial infection.

After 7 days, media needs to be replenished, this is also the beginning of diluting out the aminopterin from the HAT selection medium. To each well add 1 ml of HT medium +2% HFCS. Add the medium slowly so as not to break up any colonies; this is not critical, but it is easier to judge when to test a well if you can see the size of intact colonies.

From now on it is a matter of observing colony growth and changes in the pH of the medium. Screen for antibody from wells in which the medium is yellow (acidic) and colonies are about 25% confluent. Keep feeding wells on a 7-day cycle for slow growers and as needed for fast growers. For the second and subsequent feeds we use HT medium +1% HFCS.

D. Testing and Cloning

When testing for antibody, ensure that appropriate negative controls are used. A positive control can be a dilution of the test bleed or of serum collected when the spleen was removed. If screening by ELISA, false-positive "antiplastic responses" can be a nuisance. The use of antigen-negative wells as a control can help screen these out. It is a waste of time cloning cells from wells that display an antiplastic response.

Cells from wells that are positive in the screening assay need to be cloned and stored in liquid nitrogen as soon as possible. Cloning is performed not only to produce a monoclonal population, but also to stabilize cell growth and eliminate antibody nonproducers. Nonproducers often grow faster than cells producing antibody and can quickly outgrow the antibody-producing population.

We routinely clone at 3 cells per well and observe about two-thirds of the wells with cell growth. This approximates to Poisson's distribution of a probable cell cloning number of 1 cell per well. If you find more wells with cell growth, clone at a lower cell number per well, 1 or even 0.5 cells per well.

E. Cloning

Solutions

1. *HT medium +2% HFCS*: Add 2 ml of HFCS to 100 ml of HT medium. Store at 4°C. Warm to 37°C in a water bath before use.
2. *Trypan blue solution* (Cat. No. T8154 Sigma)

Steps

1. Gently resuspend the cells in a positive well of the 24-well plate using a sterile transfer pipette; avoid creating air bubbles. Transfer the cells to a new 24-well plate.
2. Perform a viable cell count at a 1:2 dilution in trypan blue solution.

In this example we will work from a viable cell count of 6×10^5 cells/ml

3. To obtain 3 cells/well, perform a series of dilutions: 1/1000 dilution = 600 cells/ml; add 10 μl cells into 10 ml of HT medium. From this make a 1/20 dilution = 30 cells/ml or 3 cells/100 μl.
4. Conveniently make this dilution by taking 500 μl of cells from the 1/1000 dilution and place into 9.5 ml of HT medium +2% HFCS. This gives 10 ml of cells, which is sufficient for one cloning plate at 100 μl per well.
5. Using a Gilson Distriman, add 100 μl of cell suspension to each well of a flat-bottomed 96-well plate. Label the new plate with the date and code of the original well and plate that it came from, e.g., P1D3 means plate 1 row D well 3.
6. Examine the cloning plates regularly and replenish with HT medium +2% HFCS on at least day 7 but beforehand if noticeable evaporation of media occurs. Some methods suggest wrapping 96-well plates with plastic film to slow down evaporation. This can be successful but makes viewing the wells difficult and can also give a false sense of security, care also needs to be taken when removing the film so as not to contaminate the plate.
7. Test wells that display cell growth as required, reclone, and freeze cells from positive wells using the same procedure. We clone a minimum of three times in order to achieve monoclonality.

When storing positive wells in liquid nitrogen, we routinely store two vials from the initial positive 24-well plate. From subsequent 96-well cloning plates we keep no more than three or four positive wells and store about two cryotubes of each and the same at each subsequent cloning step.

F. Isotyping and Scaling up of Hybridomas

Isotyping of the monoclonal antibody can be performed with a commercial ELISA kit, this can help confirm monoclonality and determine the strategy for antibody purification from the culture supernatant.

Care should be taken in scaling up the cultures from wells to flasks, ensure that cell growth is stable, and split cells into a number of wells of the 24-well plate;

we establish confluent growth in at least four wells before transferring the cells to a 25-cm² flask.

V. LIQUID NITROGEN STORAGE PROCEDURE

The only means by which long-term survival of hybridomas can be achieved is by storing cells in liquid nitrogen. It is essential that an up-to-date catalogue of stocks is maintained. Valuable hybridoma stocks should be kept in more than one freezer, preferably at separate sites.

A. Cell Freezing Protocol

Solutions

1. *Solution A*: Gently mix 10 ml of RF10 and 10 ml FBS. This solution may be stored at 4°C indefinitely if kept sterile. Warm to 37°C in a water bath before use.
2. *Solution B*: Add 6 ml of DMSO to 14 ml of RF10. This mixture causes an exothermic reaction and may be stored at 4°C for several days if kept sterile. Warm to 37°C in a water bath before use.

Steps

1. Resuspend cells in solution A.
2. To this add an equal volume of solution B. Mix the suspension gently and dispense into labeled sterile cryotubes.
3. After sealing the tubes, place them into a Nunc freezing container or a foam box.
4. Put the container into a −80°C freezer for 24 h, after which the tubes should be transferred into a liquid nitrogen store. The location and contents of the tubes have to be catalogued carefully.

B. Cell Thawing Protocol

We strongly recommend that full-face protection be worn when retrieving cells from liquid nitrogen.

1. Remove vials using cryogloves and forceps.
2. Transfer vials *immediately* to a shatter-proof container.
3. Using a beaker, transfer warm water from the 37°C water bath into the container.
4. The lid of the container need only be partially removed to allow access with the beaker and replaced as soon as water is added. *The use of the shatter-proof container prevents injury from tubes that explode on contact with warm water. This is a comparatively rare event but has caused severe injury.*
5. After a few minutes when the contents have thawed, remove and wash the vials in alcohol.
6. Remove the cells with a transfer pipette and place in a sterile 30-ml tube.
7. Slowly add an equal volume of RF10 medium dropwise.
8. Let stand for 5 min.
9. Add 10 ml of RF10 and let stand for a further 5 min.
10. Centrifuge at 400 g for 5 min and resuspend in HT medium for hybridomas and in RF10 for myeloma.
11. We routinely start cells growing at about 2×10^5 cells/well in a 24-well plate. The addition of 1% HFCS will speed up cell growth. Do not let cells overgrow and scale up to other wells and flasks as required.
12. The first priority is to maintain frozen stock, so ensure that more tubes are stored in liquid nitrogen.

V. COMMENTS

We have overcome tolerance to self-antigen by linking the conserved peptide to a fusion protein and using the construct as the immunogen. Using this strategy, we have been successful in making antibodies to a human peptide with 95% homology to the murine protein (Cavill *et al.*, 1999).

If purchasing a tissue culture incubator, ensure that it can be cleaned easily, has readily accessible filters that are changed easily, and does not have "hidden" tubing that cannot be cleaned and acts as a harbor for infections.

VI. PITFALLS

1. Make sure that the marker pen ink used to label the cryotubes tubes is resistant to alcohol.
2. We cannot overemphasize the importance of storing positive hybridomas in liquid nitrogen for subsequent retrieval.

References

Campbell, A. M. (1991). Monoclonal antibody and immunosensor technology. In "Laboratory Techniques in Biochemistry and Molecular Biology" (van der Vliet, P. C., ed.) Elsevier, Amsterdam.

Cavill, D., Macardle, P. J., Beroukas, D., Kinoshita, G., Stahl, J., McCluskey, J., and Gordon, T. P. (1999). Generation of a monoclonal antibody against human calreticulin by immunization with a recombinant calreticulin fusion protein: Application in paraffin-embedded sections. *App. Immunohistochem. Mol. Morphol.* 7, 150–155.

de St Groth, S. F., and Scheidegger, D. (1980). Production of monoclonal antibodies: Strategy and tactics. *J. Immunol. Methods* 35, 1–21.

Donohoe, P. J., Macardle, P. J., and Zola, H. (1994). Making and using conventional mouse monoclonal antibodies. *In "Monoclonal Antibodies: The Second Generation"* (H. Zola, ed.), *Biological Sciences*, Coronet Books, Philadelphia, PA.

Freshney, R. I. (2002). "Culture of Animal Cells: A Manual of Basic Technique," 4th Ed. Wiley-Liss, New York.

Goding, J. W. (1986). "Monoclonal Antibodies: Principles and Practice," 2nd Ed. Academic Press, London.

Grillo-Lopez, A. J. (2000). Rituximab: An insider's historical perspective. *Semin. Oncol.* 27(Suppl. 12), 9–16.

Harrison, M. A., and Rae, I. F. (2003). General Techniques of Cell Culture. *In Handbooks in Practical Animal Cell Biology*

(M. Harrison, I. E. Roe, A. Harris, Series Eds.) Cambridge Univ. Press, Cambridge, U.K.

Kearney, J., Radbruch, A., Liesegang, B., and Rajewsky, K. (1979). A new mouse myeloma cell line that has lost immunoglobulin expression but permits the construction of antibody-secreting cell lines. *J. Immunol.* 123, 1548–1550.

Kohler, G., and Milstein, C. (1975). Continuous cultures of fused cells secreting antibody of predefined specificity. *Nature* 256, 495–497.

Landsteiner, K. (1945). "The Specificity of Serological Reactions." Harvard Univ. Press, Boston, MA.

Shulman, M., Wilde, C. D., and Kohler, G. (1978). A better cell line for making hybridomas secreting specific antibodies. *Nature* 276, 269–270.

58

Rapid Development of Monoclonal Antibodies Using Repetitive Immunizations, Multiple Sites

Eric P. Dixon, Stephen Simkins, and Katherine E. Kilpatrick

I. INTRODUCTION

An in-depth understanding of both the formation of germinal centers and the immunoregulatory processes involved in T-cell-dependent B-cell responses (Levy *et al.*, 1989; Berek *et al.*, 1991; Jacob *et al.*, 1991; Kroese *et al.*, 1990; Nossal, 1992; MacLennan, 1994; Kelsoe, 1996) led us to initially explore the feasibility of modifying immunization and fusion time lines used for developing monoclonal antibodies. Our studies demonstrated that monoclonal antibodies could be generated quickly using an immunization and somatic fusion strategy, which we refer to as repetitive immunizations, multiple sites (RIMMS) (Kilpatrick *et al.*, 1997). Immunizations, somatic fusion, screening, and isolation of affinity-matured IgG-secreting hybridoma cell lines can be achieved within 1 month. RIMMS capitalizes on somatic fusion of immune B cells undergoing germinal center maturation in draining lymph nodes (Kilpatrick *et al.*, 1997, 2003). The immunization sites used for RIMMS are proximal to easily accessible regional lymph nodes. RIMMS involves the use of P3X63/Ag8.653 murine myeloma cells stably transfected with human BCL-2 (Kilpatrick *et al.*, 1997). Fusions can be performed as early as 7 days (Bynum *et al.*, 1999) out to 14 days after the onset of immunization using recombinant protein, conjugated synthetic peptides, or drug haptens (Kilpatrick *et al.*, 1997; Ignar *et al.*, 1998; Kinch *et al.*, 1998; Wring *et al.*, 1999; Alligood *et al.*, 2000; Ellis *et al.*, 2000; Lindley *et al.*, 2000). RIMMS has also been used successfully to generate high-affinity antibodies using DNA-based immunizations in conjunction with the PowderJect gene gun (Kilpatrick *et al.*, 1997, 1998, 2000, 2002, 2003; Kinch *et al.*, 2002).

This article describes the RIMMS procedure, including the preparation of adjuvant, immunization, isolation of lymph nodes, and our high-efficiency polyethylene glycol (PEG)-induced somatic fusion process. We also provide the reader with protocols for developing a BCL-2-modified myeloma cell line.

II. MATERIALS AND INSTRUMENTATION

Fine curved forceps (Cat. No. 1-23-20) and microdissecting scissors (Cat No. 11-250) are from Biomedical Research Instruments. The following items are from Corning Costar: 0.2-μm vacuum filter/storage units (Cat. No. 431205), T 25-cm^2 tissue culture flasks (Cat. No. 3056), and 96-well tissue culture plates (Cat. No. 3595), as well as 24-well tissue culture plates (Cat. No. 3524). Items purchased from Sigma include Freund's complete adjuvant (FCA) (Cat. No. F-5881), Hybri-MAX azaserine hypoxanthine, 50× (Cat. No. A9666), Hybri-MAX dimethyl sulphoxide (DMSO) (Cat. No. D2650), deoxycholic acid sodium salt (Cat. No. D-6750), and Igepal CA 630 (Cat. No. I-3021, used in place of NP-40). The following products are from InVitrogen: 0.45-μm nitrocellulose (Cat. No. LC2001), RPMI 640 (Cat. No. 11875-119), 10 m*M*, 100× nonessential amino acids (Cat. No. 11140-050), L-glutamine, 29.2 mg/ml with 10,000 units penicillin, 10,000 μg/ml streptomycin

(Cat. No. 10378-016), gentamycin reagent, 50 mg/ml (Cat. No. 10131-035), 8–16% Tris–glycine gels (Cat. No. EC6045), and SeeBlue Plus2 markers (Cat. No. LC5925), as well as pcDNA3.1+ (Cat. No. V790-20), Topo TA cloning (Cat. No. K4500-01), imMedia Amp (Cat. No. Q600-20), imMedia Amp Agar (Cat. No. Q601-20), and ethidium bromide (Cat. No. 15582-018). NUNC 1.8-ml freezer vials (Cat. No. 66021-986), as well as 10× phosphate-buffered saline (PBS) (Cat. No. EX-6506), are from VWR. Defined fetal bovine serum (FBS) (Cat. No. SH30070.03) is from Hyclone, and Origen cloning factor (Cat. No. 210001) is from Igen. JRH EX-Cell 610 HSF medium (Cat. No. 14610-1000M) is from JRH Bioscience. The P3-X63Ag8.653 cells (ATCC CRL-1580) and PEG 1450, MW 1300–1600, 2-g bottle are from ATCC. The RIBI adjuvant (Cat. No. R-700) is from Corixa. The anti-human BCL-2 monoclonal antibody (Cat. No. M0887) is from Dako. The following items are from Becton Dickinson: 1-ml Leur Lok syringes (Cat. No. 309626), 5-ml Leur Lok syringes (Cat. No. 309603), and 26G1/2 (Cat. No. 305111) and 16G1 (Cat. No. 305197) needles. BCIP/NBT color development substrate (Cat. No. S3771) is from Promega.

A PTC-200 Peltier thermal cycler is from MJ Research, Inc. (Cat. No. ALD-1244). DNA encoding human bcl-2 is from Genecopoeia (Cat. No. GC-B0284). Enzymes and lipid transfection reagents, Asp718 (Cat. No. 814 245), XbaI (Cat. No. 674 257), T4 DNA ligase (Cat. No. 481 220), and FuGENE 6 (Cat. No. 1 815 091) are from Roche. Software to analyze chromatographs is from Gene Codes Corporation (Sequencer Cat. No. SWC4.0). DNA isolation and cleanup kits are from Qiagen, Inc. (QiaPrep spin DNA kit, Cat. No. 27106 and QiaQuick gel extraction kit, Cat. No. 28704). Oligonucleotides are generated by Integrated DNA Technologies, and Microspin S400 columns are from Amersham Bioscience (Cat. No. 27-5140-01) for buffer exchange.

Eight- to 12-week-old female SJL mice are from Jackson Laboratories, and BALB/c mice are from Charles Rivers. Isoflurane (Iso Flo, Cat. No. 06-8550-2/R1) is from Abbott Labs and is administered to mice using a Vapomatic (Model 2) from AM Bickford, Inc.

III. PROCEDURES

A. Preparation of Culturing Media

Solutions

1. *Fusion selection medium, 1 liter*: Combine 500 ml ExCell-610 HSF media, 260 ml RPMI 1640, 100 ml Origen hybridoma cloning factor, 100 ml FBS, 10 ml L-glutamine/pen-strep, 10 ml nonessential amino acids, 2 vials of azaserine hypoxanthine 50×, each reconstituted with 10 ml RPMI 1640. Sterile filter media in a Corning 0.2-μm vacuum filter storage unit and store at 4°C.

2. *Fusion selection medium without Origen cloning factor, 1 liter*: Follow the directions in solution 1, but eliminate the Origen cloning factor and increase FBS from 10 to 20%. This medium is used to eliminate background from unfused B cells following somatic fusion.

3. *Fusion cloning medium, 1 liter:* This medium is used for limit dilution cloning of hybridomas. Combine 500 ml ExCell-610 HSF media, 280 ml RPMI 1640, 100 ml Origen hybridoma cloning factor, 100 ml FBS, 10 ml L-glutamine/pen-strep, and 10 ml nonessential amino acids. Sterile filter medium in a Corning 0.2-μm vacuum filter storage unit and store at 4°C.

4. *Culturing medium for BCL-2-transfected P3-X63Ag8.653 (ATCC CRL-1580) cells, 1 liter*: Combine 890 ml RPMI 1640, 100 ml FBS, 10 ml L-glutamine/pen-strep solution, and 200 μg/ml geneticin (G418). Sterile filter medium in a Corning 0.2-μm vacuum filter storage unit and store at 4°C. One week before using the cells for fusion, pass the cells into the just-described medium without G418. The same medium without G418 is used to culture the parental P3-X63Ag8.653 myeloma cell line.

5. *Serum-free wash medium, 500 ml*: Add 5 ml L-glutamine/pen-strep solution to 495 ml RPMI 1640, sterile filter medium in a Corning 0.2-μm vacuum filter storage unit, and store at 4°C.

B. Preparation of Antigen in Adjuvant

Steps

1. *Dose per mouse*: In a 1.5-ml Eppendorf tube, add 100 μl of antigen diluted in sterile PBS to a final concentration of 15 μg for the primary immunization. Add 100 μl of RIBI adjuvant to the tube containing the antigen and then vortex. Using a 1-ml Leur Lok syringe outfitted with a 16G1 needle, remove 100 μl of Freund's complete adjuvant (vortex the vial of FCA right before use).

2. To make an emulsion for the primary immunization, place the bevel of the needle against the inner wall at the bottom of the Eppendorf tube containing the antigen/RIBI mixture. Expel the Freund's complete adjuvant from the syringe into the tube and then draw the solution back up into the syringe holding the bevel of the needle against the inner wall of the tube. Repeat the process until a milky, slightly thickened

emulsion is formed. Draw the emulsion back up into the syringe. Carefully remove the 16 G needle and replace it with a 26 G 1/2 needle. The 300-μl volume is then delivered to the six subcutaneous sites indicated in Fig. 1, 50 μl per site, as detailed later.

3. The final concentration of antigen used for the secondary immunization is 5 μg, and 2–5 μg of the antigen is used for the tertiary immunization. For secondary and tertiary immunizations, dilute the antigen in a 100-μl volume of sterile PBS, increase the RIBI adjuvant volume to 200 μl, and eliminate FCA.

C. Immunization of Mice

Steps

1. For each antigen, immunize two 8- to 12-week-old female SJL or two BALB/c mice (one mouse will serve as a backup) at the sites indicated in Fig. 1 on days 0, 7, and 10. Using this immunization time line, fusion can be performed on day 11, 12, or 13. Alternatively, mice can be immunized on days 0, 4, and 8, and fusions can be performed on day 11. Fusions can also be performed as early as day 7 by immunizing mice on days 0, 2, and 4 (Bynum *et al.*, 1999). The backup mouse can be boosted every 2–3 weeks using conventional protocols (see previous article).

2. Anesthetize the mice with isoflourane for all immunization time points.

3. Inject 50 μl of the antigen/adjuvant emulsion into six sites proximal to axillary and brachial lymph nodes

(thoracic region), superficial inguinal lymph nodes (abdominal region), and popliteal lymph nodes (located behind the knee) as indicated in Fig. 1.

D. Harvesting Lymph Nodes

Steps

1. Perform euthanasia of mice using carbon dioxide.

2. Wet down the fur of the mouse with 70% ethanol.

3. Lay the mouse on its back and lift up the skin in the lower groin using 70% ethanol forceps. Using 70% ethanol-rinsed microdissecting scissors, make a small incision to open up the skin (do not cut open the abdominal wall). To make a midsection incision, insert the scissors under the skin and then make an incision starting from the lower groin region up to the neck using forceps to lift up the skin. Make incisions across the top of the shoulders, down to the front feet on the left and right sides. Then make incisions from the lower groin area across to the top of the legs continuing down to the hind feet. To expose the lymph nodes, peel the skin back from the midsection incision, pull back the skin from the hind legs, and secure with pins as shown in Fig. 1.

4. Using curved forceps that have been rinsed in 70% ethanol, remove the lymph nodes by placing curved microforceps under each node and then pull up gently to separate lymph nodes from surrounding tissue.

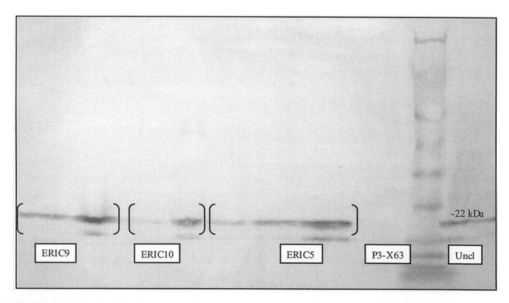

FIGURE 1 Immunization sites used for RIMMS are indicated by arrows. Following immunization, the axillary, brachial, superficial inguinal, and popliteal lymph nodes are exposed and then removed aseptically. Lymphocytes are isolated from the lymph nodes and are then used in a PEG-induced fusion.

5. Place the lymph nodes into a 60-mm sterile tissue culture dish containing 5 ml of serum-free wash medium.

E. Isolation of Lymphocytes from Lymph Nodes

Steps

1. In a laminar flow hood, remove the lymph nodes from the 60-mm tissue culture dish and transfer the nodes into a new 60-mm dish containing 5 ml of serum-free wash medium. Fill a 5-ml syringe with serum-free wash medium and add a 26 G needle. Gently hold individual lymph nodes with 70% ethanol-rinsed curved microforceps. Insert the needle into the node and then profuse with medium in order to flush out the lymphocytes. Repeat process on each node. Using two curved microforceps, gently tease remaining cells from the capsules.

2. Pipette the lymphocyte cell suspension into a 15-ml conical tube and allow the debris from the capsules of the nodes to settle to the bottom of the tube (less than 1 min). Pipette the cells away from the debris and then transfer the cell suspension into a new 15-ml conical tube. Count the cells (see article by Hoffman). For each fusion, use 3×10^7 lymphocytes (see step 3 in Section IIIF).

Note: The number of lymphocytes isolated from pooled lymph nodes will range from $1-2 \times 10^7$ (weak immunogen) to 1×10^8 when one female SJL mouse is immunized using RIMMS. You can expect one-half of this range when one female BALB/c mouse is used.

3. Remaining immune lymphocyte cells not used for fusion can be frozen back by pelleting the cells by centrifugation at $400\,g$ for 5 min. Resuspend the cells in 1 ml of 90% FCS, 10% DMSO freezing media. Place cells into 1×1.8-ml NUNC freezing vials labeled with the antigen designation and the date. Place the vial overnight at $-80°C$ in a styrofoam container and then transfer the cells to liquid nitrogen storage the following day. These cells can be subsequently thawed and used for somatic fusion.

F. PEG-Induced Somatic Fusion

Steps

1. To affect a 1:1 lymphocyte-to-myeloma ratio, harvest 3×10^7 P3-X63Ag8.653 (ATCC CRL-1580) or P3X-BCL-2-transfected myeloma cells to be used for the fusion in a 50-ml conical tube (see protocols on how to generate a bcl-2-modified fusion partner in Section IIIH). Check the viability of the myeloma cells using trypan blue; ideally you want 90–98% viability.

Wash the cells in serum-free wash medium by centrifuging at $400\,g$ for 5 min. Resuspend the cells in 2 ml serum-free wash medium.

2. To prepare the PEG for fusion (ATCC PEG 1450, MW 1300–1600, 2-g bottle), very slowly heat the bottle on a hot plate on a low setting, just enough to melt the PEG. Do not boil the PEG. Using a 5-ml syringe outfitted with a 16G1 needle, immediately add 3 ml of serum-free wash medium that has been prewarned to 37°C to make a 40% stock solution. Keep the PEG solution in a 37°C incubator until you are ready to perform the fusion. Use the PEG solution within 1.5 h, as the efficiency of the fusion will drop off significantly if you use the PEG beyond 2 h after preparation.

3. In a 15-ml conical tube, mix 3×10^7 immune lymphocyte cells isolated from lymph nodes with 3×10^7 myeloma cells (from step 1). Centrifuge the cell mixture at $400\,g$ for 5 min. Decant the media to leave a "dry pellet" by removing as much media as possible. Gently disrupt the pellet of mixed cells by tapping the bottom of the tube.

4. Using a 1-ml pipette, slowly add 300 µl of the PEG solution to the "dry pellet" of the myeloma and lymph node cell mixture in the bottom of the 15-cc tube. Mix gently and then allow the tube to incubate in the hood for 5 min. Gently mix the cell suspension again by tapping the tube and then allow the suspension to incubate in the hood for another 5 min.

5. Using a 5-ml pipette add 4 ml of the fusion selection media to the fusion, resuspend the cells gently, and then transfer the fusion suspension into a 250-ml sterile bottle containing 196 ml of fusion selection media. Swirl gently to mix cell suspension. Plate out the fusion in 10×96-well tissue culture plates by adding 200-µl per well.

G. Postfusion Care and Handling

Steps

1. Avoid removing the fusion plates from the incubator for microscopic observation during the first few days after the fusion.

2. Within 7 days, remove one-half of the hybridoma selection media and then replace with fresh hybridoma selection media (containing azaserine hypoxanthine). If you observe a high number of unfused lymphocytes still growing, change the media to hybridoma selection media containing 20% FBS, without Origen cloning factor. Antibody from unfused B cells will give misleading results in primary screening assays.

3. Within 5–10 days you will be able to observe the outgrowth of hybridomas in the fusion plates. Harvest supernatant for ELISA analysis. ELISA positives

are further tested in Western blot and immuno-precipitation.

4. For limit dilution cloning for the isolation and identification of monclonal antibody-producing cell lines, please refer to the previous article.

H. Generation of a P3X-BCL-2 Myeloma Cell Line

Cloning of Human BCL-2

Solutions

1. *10× TBE*: 108 g Tris base, 55 g boric acid, 40 ml 0.5 M EDTA (pH 8.0)
2. *Ethidium bromide*: 10 mg/ml in dH$_2$O

Steps

1. *PCR reaction mix*: Mix 20 pmol of primers (sense 5' cgg ggt acc gcc acc atg gcg cac gct ggg aga ac 3' and anti-sense 5' ccg tct aga tca ctt gtg gcc cag ata ggc a 3'), human BCL-2 cDNA, 10× Advantage PCR buffer, 200 μM dNTP mix, dH$_2$O, and Advantage HF polymerase.
2. Set parameters for human bcl-2 in a PTC-200 Peltier thermal cycler with the following conditions. Cycle steps: a denaturation step of 3 min at 95°C to generate a hot start, 30 cycles at 95°C for 15 s, 65°C for 15 s, and then 72°C for 1 min, followed by a soak step at 4°C.
3. After amplification of the cDNA, separate the PCR products and primers electrophoretically on a 1% agarose/TBE ethidium gel and purify the 642-bp band using the QiaQuick gel purification kit.
4. Perform buffer exchange over a Sephadex-400 column.
5. Insert the cloned cDNA into pCR2.1 using the TOPO TA cloning kit and transform into Top10 *Escherichia coli*.
6. Prepare 6–10 cultures with 2 ml LB-amp media in a 15-ml sterile culture tube. Pick a single isolated colony with a sterile loop and inoculate each mini-culture with the isolated bacteria. Incubate and shake overnight at 37°C.
7. Isolate recombinants using the QiaPrep spin DNA kit and digest 1 μg of the plasmid DNA with the restriction endonucleases Asp718 and *Xba*I for 1 h at 37°C.
8. Isolate the human BCL-2 cDNA fragment by excising the 642-bp band and then subclone into a linear pcDNA3.1(+) vector (digested with Asp718 and *Xba*I) using T4 DNA ligase.
9. Transform 2.5 μl of the ligation reaction into chemically competent Top 10 *E. coli* and spread onto a LB-amp agar plate. Incubate overnight at 37°C.

10. Culture, isolate, and digest another 6–10 plasmid DNA recombinants. Determine correct recombinants by restriction endonuclease digests and electrophoresis.
11. Transfect pcDNA3.1(+) human BCL-2 plasmid DNA into the P3-X63Ag8.653 cell line.

I. Transfection, Isolation, and Identification of BCL-2-Transfected Myeloma Cells

Solutions

1. *RIPA buffer*: 150 mM NaCl, 50 mM Tris, 1% Igepal, 0.25% deoxycholate, pH 7.5. To make 1 liter, add 8.76 g of sodium chloride, 6.35 g Tris–HCl, 1.18 g Tris base, 10 ml Igepal, and 2.5 g deoxycholate to a total volume of 1 liter. Sterile filter in a Corning 0.2-μm vacuum filter storage unit and store at 4°C.
2. *Culturing medium for P3X63/Ag8.653 murine myeloma cells*: Combine 890 ml RPMI 1640, 100 ml FBS, and 10 ml L-glutamine/pen-strep solution. Sterile filter in a Corning 0.2-μm vacuum filter storage unit and store media at 4°C.
3. *G418 selection medium, 1 liter, used for selection of BCL-2- transfected P3-X63Ag8.653 cells (parental cells P3-X63Ag8.653, ATCC CRL-1580)*: Combine 890 ml RPMI 1640, 100 ml FBS, and 10 ml L-glutamine/pen-strep solution. Add geneticin (G418) from 50-mg/ml stock (potency is 600 μg/mg) to final concentrations of 1 mg/ml, 500 μg/ml, and 250 μg/ml. Sterile filter in a Corning 0.2-μm vacuum filter storage unit and store media at 4°C.
4. *Culturing medium for BCL-2-transfected P3-X63Ag8.653 cells, 1 liter*: Combine 890 ml RPMI 1640, 100 ml FBS, 10 ml L-glutamine/pen-strep solution, and a final concentration of G418 at 200 μg/ml. Sterile filter in a Corning 0.2-μm vacuum filter storage unit. One week before using the cells for fusion, pass the cells into the just-described media without G418.
5. *Phosphate-buffered saline*: 10× PBS from VWR (137 mM NaCl, 2.7 mM potassium chloride, 10 mM phosphate buffer). To 100 ml of 10× PBS stock, add 900 ml distilled water. Sterile filter in a Corning 0.2-μm vacuum filter storage unit.
6. *Hybridoma freezing medium*: 90 ml fetal bovine serum, 10 ml Hybri-Max DMSO, sterile filter, and store at 4°C.
7. *Alkaline phosphatase developing buffer, pH 9.5 (0.1 M Tris HCL, 0.1 M NaCl, 5 mM MgCl$_2$)*: 1.52 g Tris–HCl, 10.94 g Tris–OH, 5.85 g NaCl, 1.15 g MgCl$_2$·6H$_2$O dissolved in 1 liter of distilled water, pH to 9.5, sterile filter, and store at 4°C.
8. *5% PBST Blotto*: To 100 ml of 1× PBST (0.05% Tween-20), add 5 g powdered milk, mix well, and then store at 4°C.

Steps

1. The day before transfection, plate P3X63/Ag8.653 cells into 12 wells of a 24-well tissue culture plate at a density of 4×10^5 cells per well in 2 ml of culturing media. Six of the wells will be used for transfection, and 6 wells will be used for controls for the G418 selection (see step 5). Incubate the cells overnight in a 37°C, 5% CO_2 humidified incubator. The cells should be approximately 50% confluent.

2. The DNA ratio used for transfection is 3:1. For each well, use 0.5 ml of serum-free RPMI 1640 medium containing 0.6 µl of FuGENE 6 with 0.2 µg of DNA encoding human BCL-2.

3. The following steps are taken to prepare the FuGENE 6 reagent:DNA complex for transfection of P3X63/Ag8.653 murine myeloma with the human BCL-2 gene. Add 50 µl of serum-free RPMI 1640 media to a small sterile tube. Add 3 µl of FuGENE 6 transfection reagent directly to the serum-free medium. Mix by tapping the tube gently. Add 0.2 µg of the DNA in a volume of 1 µl to the tube. Mix the contents by tapping the tube gently (do not vortex). Incubate the reaction at room temperature for 15–30 min.

4. Remove the medium from six wells of the plated P3X63/Ag8.653 cells. Add 0.6 ml culture medium to each well. Using a sterile tip, add dropwise 53 µl of the complex mixture from step 3. Gently mix the cell suspension to disperse the reagent:DNA complex evenly. Return the cells to the incubator.

5. Within 24 h, carefully remove medium from wells that underwent transfection. To duplicate wells add 2 ml of the 1-mg/ml, 500-µg/ml, and 250-µg/ml G418 selection media. To the remaining six wells that were not transfected, add to duplicate wells 2 ml of the 1-mg/ml, 500-µg/ml, and 250-µg/ml G418 selection media. These cells will serve as controls for the G418 selection process (cells should die).

6. Within 4 days, remove selection media and replace wells with fresh selection media. As the selected cells grow, continue to maintain the cells in selection media. Change the media every 3 to 5 days.

7. Expand the selected cell lines into T25-cm² flasks in the respective G418 selection media. Grow the cells to $7-9 \times 10^5$/ml. Spin down the cells at 400 g for 5 min. Decant the supernatant, resuspend the cells in hybridoma freezing medium, and aliquot 1 ml per 1.8-ml Nunc freezing vials. Place the cells into a styrofoam rack and then freeze overnight at −80°C. The next day, transfer the cells to LN_2. Freeze back stocks of the transfected cell lines at 1×10^6 cells per vial in freezing media. Continue to maintain the cells in culture in respective selection media in order to prepare RIPA

extracts to determine the presence of the human BCL-2 protein (see later).

J. Western Blot Detection of Human BCL-2

Steps

1. Following the transfections and selection of P3-X63Ag8.653 cells with plasmid encoding human BCL-2, maintain the myeloma cells in G418 selection media containing final concentrations of 1 mg/ml, 500 µg/ml, and 250 µg/ml in T25-cm² flasks. Also maintain the parental P3-X63Ag8.653 cells, which will serve as negative controls.

2. Count the parental P3-X63Ag8.653 and the BCL-2 transfected cell lines using trypan blue (see article by Hoffman). For each cell line, centrifuge 1×10^6 cells in a 15-ml conical tube at 400 g for 5 min.

3. Remove all of the culture supernatant. To make cell extracts in order to determine the presence of human BCL-2 by Western blot, add 10 µl of chilled RIPA buffer to the cell pellet, resuspend the cells, and then transfer the respective cell suspensions to 1.5-ml Eppendorf tubes. Incubate tubes on ice for 15 min.

4. Microfuge the sample tubes in a microfuge set on high for 15 min at 4°C. Remove the supernatant (discard the pellet) and then mix the supernatant 1:1 with 2× sample buffer. Heat samples at 96°C for 5 min.

5. Load a 20-µl volume of the P3-X63Ag8.653 RIPA extract (parental cell line) and each P3XBCL-2-transfected cell extract onto individual lanes of a 1×10 well 8–16% Tris–glycine gel (InVitrogen) using one lane for SeeBlue Plus2 markers. Run the gel for 90 min at 125 V and then transfer the gel to nitrocellulose. Block the nitrocelluose in 5% PBST Blotto overnight at 4°C.

6. For detection of human BCL-2, add 10 ml of anti-human BCL-2 (Dako). Dilute 160 µl of the antibody per 10 ml in 5% PBST Blotto blocking buffer. Incubate the blot for 1 h at room temperature on a shaker or rocker platform.

7. Wash the blot four times for 5 min with PBST. Add 10 ml of a 1:1000 dilution of goat anti-mouse IgG-alkaline phosphatase-labeled conjugate diluted in 5% PBST Blotto blocking buffer to the blot. Incubate for 1 h on a rocker platform.

8. Wash the blot four times with PBST. Develop the blot using 10 ml alkaline phosphatase buffer containing 66 µl of NBT and 33 µl of BCIP. Immerse the blot into the developing substrate, place on a rocking platform, and incubate until a purple color develops. Stop the reaction by removing the developing substrate and rinsing the blot with ddH_2O.

9. Figure 2 demonstrates a Western blot indicating the presence of BCL-2 in RIPA extracts made from

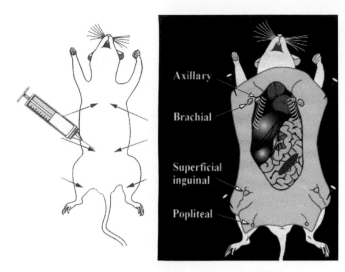

FIGURE 2 Titration of RIPA extracts of P3-X63Ag8.653 clones (ERIC5, ERIC9, and ERIC10) stably expressing human BCL-2, as well as uncloned parental BCL-2 transfected cells, indicates the presence of human BCL-2 by Western blot using an anti-human bcl-2-specific monoclonal antibody. Control untransfected P3-X63Ag8.653 cells serve as a negative control.

transfected, G418 selected P3-X63Ag8.653 myeloma cells.

10. Expand the uncloned cells from wells that demonstrate the presence of BCL-2 into a 75-cm² tissue culture flask containing 35 ml of media supplemented with 250 μg/ml G418. Grow the cells to $7–9 \times 10^5$/ml. Spin down the cells at $400\,g$ for 5 min. Decant the supernatant, resuspend the cells in hybridoma freezing media, and aliquot 1 ml per 1.8-ml Nunc freezing vials. Place the cells into a styrofoam rack and then freeze overnight at −80°C. The next day, transfer the cells to LN_2.

K. Limit Dilution Cloning of P3Xbcl-2 Myeloma Cells

Steps

1. Count the P3XBCL-2 cells (see article by Hoffman) and then dilute the cells to 330 cells in a total of 200 ml of culturing media containing 250 μg/ml G418. To increase the limit dilution cloning efficiency, the Origen cloning factor can be added to the G418 selection medium.

2. Add 200 μl of the cell suspension per well to 10 × 96-well sterile tissue culture plates. Incubate the cells at 37°C in a humidified incubator.

3. Microscopically scan the wells to identify one cell per well within 48 h of plating. Mark the wells that have clones derived from single cells.

4. Once the clones have grown to cover 50–75% of the well, expand the cells into 24-well plates containing 2 ml per well culturing media containing 200 μg/ml G418. Allow the cells to grow to 80–90% confluency in 6 wells. Remove cells from 3 wells and make RIPA lysates using the aforementioned procedure. Test for expression of BCL-2 using the Western blotting procedure. (Fig. 2).

5. Expand the cells that exhibit high BCL-2 expression levels to T25-cm² tissue culture flasks. Freeze back master stocks as detailed earlier.

6. Test the P3XBCL-2-transfected cells for their ability to negatively select (die) by placing the cells in fusion selection media containing HybriMax azaserine hypoxanthine. Due to the presence of BCL-2, the cells will take approximately 2–3 days longer to die as compared to the parental untransfected P3-X63Ag8.653 cell line.

IV. PITFALL

Unfused B cells producing antigen-specific immunoglobulin can contribute to background, which may indicate false positives in a primary ELISA screen. It is imperative to change the media on fusion plates one to two times before screening (see Section IIIG, step 2).

References

Alligood, K. J., Milla, M., Rhodes, N., Ellis, B., Kilpatrick, K. E., Lee, A., Gilmer, T. J., and Lansing, T. L. (2000). Monoclonal antibodies generated against recombinant ATM support kinase activity. *Hybridoma* 19, 317–321.

Berek, C., Berger, A., and Apel, M. (1991). Maturation of the immune response in germinal centers. *Cell* 67, 1121–1129.

Bynum, J., Andrews, J. L., Ellis, B., Kull, F. C., Austin, E. A., and Kilpatrick, K. E. (1999). Development of class-switched, affinity matured monoclonal antibodies following a 7 day immunization schedule. *Hybridoma* 18, 407–411.

Ellis, J. H., Ashman, C. A., Burden, M. N., Kilpatrick, K. E., Morse, M. A., and Hamblin, P. A. (2000). GRID, a novel Grg-2-related adapter protein which interacts with the activated T cell costimulatory receptor CD28. *J. Immunol.* 164, 5805–5814.

Ignar, D. M., Andrews, J. L., Witherspoon, S. M., Leray, J. D., Clay, W. C., Kilpatrick, K. E., Onori, J., Kost, T. A., and Emerson, D. L. (1998). Inhibition of establishment of primary and micrometastatic tumors by a urokinase plasminogen activator receptor antagonist. *Clin. Exp. Metast.* 16, 9–20.

Jacob, J., Kelsoe, G., Rajewsky, K., and Weiss, U. (1991). Interclonal generation of antibody mutants in germinal centres. *Nature* 354, 389–392.

Kelsoe, G. (1996) Life and death in germinal centers (Redux). *Immunity* 4, 107–111.

Kilpatrick, K. E., Cutler, T., Whitehorn, E., Drape, R. J., Macklin, M. D., Witherspoon, S. M., Singer, S., and Hutchins, J. T. (1998). Gene gun delivered DNA-based immunizations mediate rapid production of murine monoclonal antibodies to the Flt-3 receptor. *Hybridoma* 17, 569–576.

Kilpatrick, K. E., Danger, D. P., Hull-Ryde, E. A., and Dallas, W. (2000). High affinity monoclonal antibodies generated in less than 30 days using 5 μg of DNA. *Hybridoma* 19, 297–302.

Kilpatrick, K. E., Kerner, S., Dixon, E. P., Hutchins, J. T., Parham, J. H., Condreay, J. P., and Pahel, G. (2002). *In vivo* expression of a GST-fusion protein mediates the rapid generation of affinity matured monoclonal antibodies using DNA-based immunizations. *Hybridoma Hybridomi.* 21, 237–243.

Kilpatrick, K., Sarzotti, M., and Kelsoe, G. (2003). Induction of B cells by DNA vaccines. *In "DNA Vaccines"* (H. C. J. Ertl, ed.), pp 66–81. Eurekah Publ.

Kilpatrick, K. E., Wring, S. A., Walker, D. H., Macklin, M. D., Payne, J. A., Su, J.-L., Champion, B. R., Caterson, B., and McIntyre, G. D. (1997). Rapid development of affinity matured monoclonal antibodies using RIMMS. *Hybridoma* 16, 381–389.

Kinch, K. C., Kilpatrick, K. E., Stewart, J. C., and Kinch, M. S. (2002). Antibody targeting of the EphA2 receptor tyrosine kinase on malignant carcinomas. *Cancer Res.* 62, 2840–2847.

Kinch, M. S., Kilpatrick, K. E., and Zhong, C. (1998). Identification of tyrosine phosphorylated adhesion proteins in breast cancer. *Hybridoma* 17, 227–235.

Kroese, F. G. M., Timens, W., and Nieuwenhuis, P. (1990). Germinal center reactions and B lymphocytes: Morphology and function. *Curr. Topics. Pathol.* 84, 103–148.

Levy, N. S., Malipiero, U. V., Lebecque, S. G., and Gearhart, P. J. (1989). Early onset of somatic mutations in immunoglobulin V_H genes during the primary immune response. *J. Exp. Med.* 169, 2007–2019.

Lindley, K. M., Su, J.-L., Hodges, P. K., Wisely, C. B., Bledsoe, R. K., Condreay, J. P., Wineager, D. A., Hutchins, J. T., and Kost, T. A. (2000). Production of monoclonal antibodies using recombinant baculovirus displaying gp64-fusion proteins. *J. Immunol. Methods* 234, 123–135.

MacLennan, I. C. M. (1994). Germinal centers. *Annu. Rev. Immunol.* 12, 117–139.

Nossal, G. J. V. (1992). The molecular and cellular basis of affinity maturation in the antibody response. *Cell* 68, 1–2.

Wring, S. A., Kilpatrick, K. E., Hutchins, J. T., Witherspoon, S. M., Ellis, B., Jenner, W. N., and Serabjit-Singh, C. (1999). Shorter development of immunoassay for drugs: Application of the novel RIMMS technique enables rapid production of monoclonal antibodies to Ranitidine. *J. Pharma. Biomed. Anal.* 19, 695–707.

Phage-Displayed Antibody Libraries

Antonietta M. Lillo, Kathleen M. McKenzie, and Kim D. Janda

I. INTRODUCTION

Phage-displayed antibody libraries consist of large repertoires of Fab fragments (Barbas *et al.*, 1991), single chain variable regions (scFv) (McCafferty *et al.*, 1990), or diabodies (dimer scFv) (McGuinness *et al.*, 1996) cloned into genetically engineered phage or phagemid vectors (Smith, 1985; Smith and Petrenko, 1997), and expressed on the surface of a bacteriophage. This phenotype–genotype linkage enables phage-displayed antibodies to be selected by multiple rounds of antigen-based affinity purification and amplification. In addition, phage-displayed libraries can be constructed bypassing the immune system and therefore can be targeted to self-antigens (Zeidel *et al.*, 1995), as well as to nonimmunogenic and even toxic substances (Vaugham *et al.*, 1996). Furthermore, being an *in vitro* technique, phage display technology makes it easier to build fully human antibody libraries (Holt *et al.*, 2000). This article illustrates a general protocol for the generation of scFv libraries displayed on the surface of the pCGMT phage vector as part of either surface protein pIII (Gao *et al.*, 1997; Mao *et al.*, 1999; Gao *et al.*, 1999) or pIX (Gao *et al.*, 2002).

II. MATERIALS AND INSTRUMENTATION

First-strand cDNA synthesis kit is from Amersham Pharmacia (Cat. No. 27-9261-01). Carbenicillin, tetracycline, kanamycin, and IPTG are from Research Products International (Cat. No. C46000-1.0, T17000/1.0, K22000-1.0, and I56000-5.0). *Pfu* DNA polymerase, *Escherichia coli* XL-1 blue, VCSM13 interference-resistant helper phage, and total RNA purification kit are from Stratagene (Cat. No. 6000154, 200249, 200251, and 400790). Ultrapure agarose, electroporation cuvettes, 1-kb plus DNA ladder, T4 DNA ligase, bovine serum albumin (BSA), and 100 m*M* dNTP mix are from Invitrogen (Cat. No. 15510027, 15224017, P450-50, 10787018, 15561012, and 10216018). Dimethyl sulfoxide (DMSO) is from Aldrich (Cat. No. 27,043-1). *Sfi*I is from New England Biolabs (Cat. No. R0123S). QIAquick gel extraction kit is from Qiagen (Cat. No. 28704). Tryptone is from Becton Dickinson (Cat. No. 211043). Yeast extract is from obtained EM Science (Cat. No. 1.03753.5007). Nonfat dry milk is from Bio-Rad (Cat. No. 170-6404). All the salts, glycerol, polyethylene glycol, Tween 20, ethanol, and glucose are from Sigma (Cat. No. G5516, P2139, P1379, 27,074-1, G5767). Petri dishes and Nunc MaxiSorb Immuno-tubes are from Fisher (Cat. No. 08-757-1000, and 12-565-144). Polymerase chain reaction (PCR) tubes are from USA Scientific (Cat. No. 1402-4300). Nalgene 500-ml centrifuge bottles are from VWR (Cat. No. 21020-050). The thermocycler (Mastercycler Gradient) and benchtop centrifuge (5415C) are from Eppendorf. The electrophoresis chamber (Easy-Cast) is from Electrophoresis System. The electroporation apparatus (Gene Pulser II) is from Bio-Rad. The J2-H2 centrifuge is from Beckman.

III. PROCEDURES

A. Construction of a scFv Library

Solutions

1. *2xYT*: To make 1 liter, dissolve 16 g tryptone, 10 g yeast extract, and 5 g NaCl in 900 ml distilled

water. After adjusting the pH to 7.0 with NaOH, bring the volume to 1 liter. Sterilize by autoclaving.

2. *SB medium*: To make 1 liter, dissolve 30 g tryptone, 20 g yeast extract, and 10 g MOPS in 1 liter deionized water. Sterilize by autoclaving.

3. *LB-agar*: To make 1 liter, dissolve 10 g tryptone, 5 g yeast, and 10 g NaCl in 900 ml distilled water. Add 15 g agar and adjust the pH to 7.0 with NaOH. Fill to 1 liter and autoclave. Allow to cool to a reasonable temperature and supplement with 1 ml carbenicillin stock, 10 μg/ml tetracycline, and 2% glucose. Pour in petri dishes.

4. *SOC*: To make 1 liter, dissolve 20 g tryptone, 5 g yeast, and 0.5 g NaCl in 900 ml distilled water. Add 10 ml 250 mM KCl. Adjust the pH to 7.0, fill to 975 ml, and autoclave. Once cooled, add 5 ml 2M MgCl$_2$ and 20 ml 1M glucose

5. *Phosphate-buffered saline (PBS)*: To make 1 liter, dissolve 1.44 g sodium phosphate, 0.24 g potassium phosphate, 0.2 g potassium chloride, and 8 g NaCl in 900 ml distilled water. Adjust the pH to 7.4. Fill to 1 liter and autoclave.

6. *Blotto*: To make 100 ml, dissolve 4 g nonfat dry milk in enough PBS to make 100 ml final volume.

7. *3 M NaOAc*: Dissolve 24.61 g NaOAc in 90 ml distilled water. Adjust the pH to 5.2. Fill to 100 ml and autoclave.

8. *250 mM KCl*: Dissolve 1.86 g KCl in 100 ml distilled water. Sterilize by autoclaving.

9. *2 M MgCl$_2$*: Dissolve 40.7 g magnesium chloride hexahydrate in 100 ml distilled water. Sterilize by autoclaving.

10. *1 M glucose*: Dissolve 90 g glucose in 500 ml distilled water. Sterile filter.

11. *Carbenicillin stock*: Dissolve 1 g carbenicillin in 10 ml deionized water. Sterile filter. Store at −20°C.

12. *Tetracycline stock*: Dissolve 50 mg tetracycline in 10 ml 70% ethanol. Store at −20°C

13. *Kanamycin stock*: Dissolve 500 mg kanamycin in 10 ml deionized water. Sterile filter. Store at −20°C.

14. *0.5 M IPTG stock*: Dissolve 5 g isopropyl-β-D-thiogalactopyranoside (IPTG) in 42 ml deionized water sterile filter. Store at −20°C.

15. *VCSM13 helper phage solution*: Prepare this solution according to the vendor's instructions.

Steps

1. *Preparation of mRNA*. Extract mRNA from either human peripheral blood lymphocytes (human library) or mouse spleen cells (murine library) using the RNA Purification Kit (Stratagene) according to the vendor's instructions. Prepare first-strand cDNA from the total RNA by using the First-strand cDNA Synthesis Kit (Pharmacia) and dT18 primer according to the manufacturer's recommendations.

2. *PCR amplification of antibody variable region genes (see Fig. 1 and Appendix)*. In a 250-μl PCR tube, combine 2 μl of cDNA template, 2 μl of one forward primer (100 pmol/μl), 2 μl of an equimolar mixture of the respective reverse primers (100 pmol/μl), 200 μM dNTPs, 5% DMSO, 10 μl 10× *Pfu* DNA polymerase buffer, and 5 units *Pfu* DNA polymerase (final volume: 100 μl). For example, for the human heavy chain, set up 12 PCR reactions total, in which each HVH forward primer is combined with a mixture of HJH reverse primers. Set the temperature program as follows: denaturation at 94°C for 5 min; 30 cycles of amplification, including denaturation, 1 min, 94°C; annealing, 1 min, 50°C; extension, 1 min, 72°C; and final extension at 72°C for 10 min. Run a 1% agarose gel and cut out the appropriate bands. Combine the heavy chain bands into one pool and the light chain bands (including V$_\lambda$ and V$_\kappa$) into a separate pool. Purify using a Qiagen gel extraction kit.

3. *Construction of the scFv library (see Fig. 1 and Appendix)*. Assemble the scFv library by overlap PCR. In a 250-μl PCR tube, combine ~20 ng of each scFv fragment pool, 200 μM dNTPs, 2 μl 10× *Pfu* polymerase buffer, and 1 unit *Pfu* polymerase (final volume: 20 μl). Set the temperature program as follows: denaturation at 94°C for 5 min; 5 cycles of amplification, including denaturation, 1 min, 94°C; annealing, 1 min; 55°C; and extension, 1.5 min, 72°C. Add the outer primers HVH(Sfi) and HLJ(Sfi) [or MVH(Sfi) and MLJ(Sfi)] to a final concentration of 2 mM and bring the final volume to 50 μl. Set the thermocycler for 30 more cycles of denaturation, 30 s at 94°C; annealing, 30 s at 60°C; extension, 1.5 min at 72°C; and final extension at 72°C for 10 min. Gel purify the full-length scFv library on a 1% agarose gel using a Qiagen gel extraction kit.

4. *Digestion of the scFv library and vector*. Digest both the scFv library and either pCGMT (pIII library; Gao *et al.*, 1997) or pCGMT9 (pIX library; Gao *et al.*, 2002)

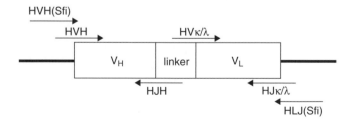

FIGURE 1 scFv diagram indicating design and primer overlap. HVH, human variable heavy chain; HJH, human constant heavy chain; HVκ/λ, human variable light chain; HJκ/λ, human constant light chain; HVH(Sfi), human forward *Sfi*I primer; HLJ(Sfi), human reverse *Sfi*I primer.

vector by combining 46 μl DNA, 6 μl 10× BSA, 6 μl 10× NE buffer 2, and 2 μl SfiI. Incubate the mixture for 2 h or overnight at 50°C. Gel purify each digestion on a 1% agarose gel using a Qiagen gel extraction kit.

5. *Ligation into vector.* Combine 5 μl SfiI digested vector, 10 μl scFv library, 4 μl 5X ligase buffer, and 1 μl T4 DNA ligase for a final volume of 20 μl. Incubate at 16°C for 2–10 h. Add 2 μl of 3 M NaOAc, pH 5.2, and mix well. Add 40 μl absolute ethanol and store at –20°C for >20 min. Centrifuge at ≥12,000 rpm for 10 min. Decant supernatant and wash the pellet with 1 ml 70% ethanol, mixing well. Spin at ≥12,000 rpm for 1–2 min and aspirate the solution. Air dry the pellet and resuspend in 10 μl distilled water.

6. *Transformation into electrocompetent E. coli XL-1 Blue.* Add 10 μl of the ligation mixture to 500 μl of a suspension of *E. coli* XL-1 Blue. Add 50-μl aliquots to 10 prechilled electroporation cuvettes. Electroporate at 2.5 kV per manufacture's instructions. Immediately add 200 μl SOC medium to the cuvette and incubate at 37°C for 30 min. Repeat as many times as necessary. Plate the cells on LB-agar plates supplemented with 2% glucose, 50 μg/ml carbenicillin, and 10 μg/ml tetracycline. Grow each plate overnight at 30°C.

7. *Library titer and storage.* Scrape the clones off each plate with 10 ml SB medium supplemented with 10% glycerol and store at -70°C. Determine the size of the library by counting the number of independent transformants. To titer the library, serially dilute the library (1:10) in SB out to 10⁻⁹. Starting with the lowest dilution, add 100 μl onto a LB-carb plate and spread evenly. Incubate overnight at 37°C and count the colony-forming units for each dilution.

8. *Rescue of the scFv phage library.* Inoculate ~5 × 10⁵ cells obtained from the previously described glycerol stock into 100 ml SB medium containing 2% glucose, 50 μg/ml carbenicillin, and 10 μg/ml tetracycline. Incubate the culture at 37°C on a reciprocal shaker until OD₆₀₀ ~ 0.6 is reached. Add ~4 × 10¹³ plaque-forming units of VCSM13 helper phage and let incubate for 30 min at room temperature and then for 90 min on the shaker at 37°C. Add 200 μl of 0.5 M IPTG and 140 μl kanamycin and allow the culture to grow over-night at 30°C.

9. *Phage precipitation.* Pellet the cells by centrifuging at 7000 rpm for 20 min in a 400-ml centrifuge tube. Decant the supernatant into clean tubes. Dissolve 3 g NaCl and 4 g PEG-8000 in each tube and place on ice for 30 min. Centrifuge at 7000 rpm for 20 min to pellet phage. Resuspend in an appropriate volume of PBS.

10. *Panning.* Add 1 ml of ~5–50 μg/ml antigen/hapten to an immunotube and incubate overnight at 4°C. Remove the antigen solution and block the tube

with 1 ml Blotto for 1 h at 37°C. Grow a culture of *E. coli* XL-1 Blue in 20 ml SB until the OD₆₀₀ ~ 0.6. Save for phage titration and rescue. Remove the blocking solution and add 10¹²–10¹³ colony-forming units of phage library in PBS containing 1% nonfat dry milk and 3% BSA. Incubate at 37°C for 2 h. Aspirate unbound phage solution and remove weakly bound phage by washing with 0.05% Tween 20 in PBS. Elute the tightly bound phage by either adding 1 ml of 0.1 M glycine, pH 2.5, or adding a solution of free antigen. Incubate at room temperature for 10 min. Remove the solution and neutralize with concentrated Tris base (usually 60 μl 2 M Tris, pH 8.0). Titrate the eluted phage by serially diluting (1:10) in SB out to 10⁻⁹. Starting with the lowest dilution, add 10 μl diluted phage to 90 μl *E. coli* XL-1 Blue. Plate 50 μl onto a LB-carb plate and spread evenly. Incubate overnight at 37°C and count the colony-forming units for each dilution. Rescue the remaining eluted phage by adding to 20 ml *E. coli* XL-1 Blue, mixing gently, and letting sit for 10 minutes at room temperature. Centrifuge at 3000 rpm for 10 minutes and resuspend the pellet in approximately 500 μl of the SB medium. Spread evenly on a LB-carb plate and incubate overnight at 37°C. Rescue the library as outlined in step 7. This will be the input library for the next round of panning. Phage libraries are usually subjected to four to six rounds of panning. The number of rounds is dependent on enrichment as determined by library titer.

IV. COMMENTS

This procedure outlines the construction of a human or murine scFv library, including a list of suitable primers (see **Appendix**). The technique is also applicable to the formation of a Fab library with suitable primers (Barbas *et al.*, 1991).

The stringency of the phage selection can be increased at the end of each round of panning in several ways. The first is to decrease the amount of antigen immobilized on the immunotubes. Similarly, the incubation time of the library with the antigen can be decreased. Alternatively, one can increase the number of washing steps and/or the amount of detergent in the wash buffer. Finally, one can use a progressively lower concentration of free antigen during the elution step. Note that when selecting for interactions influenced by the ionic strength of the medium, buffers other than PBS might be used for panning.

The number of transformations needed to obtain an appropriate library size has to be determined experimentally. Upon titration of the first transformation mixture, 30–50 colonies must be picked and the corresponding plasmids digested to determine the percentage of those containing inserts. Based on this number and the overall titer, one can determine the number of transformations needed to obtain ~10^9 independent transformants.

The quality of a library depends not only on the quantity of independent transformants, but also on diversity. In order to test such diversity, a set of 30–50 random clones should be sequenced so that the percentage of repeats can be determined. It is customary to define a library as satisfactory if it contains less than 10% repeated sequences.

V. PITFALLS

1. Take extreme care when working with RNA to prevent degradation by nucleases. This includes wearing gloves and using nuclease-free solutions.
2. Failure in the primary PCR reaction is typically due to a low concentration of cDNA. Repeat the first-strand synthesis if necessary. It may also be necessary to adjust the annealing temperature. Note that lowering the annealing temperature allows for less fidelity in primer overlap.
3. If overlap fails in step 3, check the concentrations of the various V_H, V_κ, and V_λ fragments by running a small amount on a gel and examining the intensity of the bands. Adjust the amounts accordingly to give approximately equimolar concentrations.
4. Scale the amount of phage preparation as necessary for library size.

References

Barbas, C. F., Kang, A. S., Lerner, R. A., and Benkovic, S. J. (1991). Assembly of combinatorial antibodies libraries on phage surfaces: the gene III site. *Proc. Natl. Acad. Sci. USA* **88**, 7978–7982.

De Haard, H. J., van Neer, N., Reurs, A., Hufton, S. E., Roovers, R. C., Henderikx, P., de Bruine, A. P., Arends, J. W., and Hoogemboom, H. R. (1999). A large non-immunized human Fab fragment phage library that permits rapid isolation and kinetic analysis of high affinity antibodies. *J. Biol. Chem.* **274**, 18218–18230.

Gao, C., Lin, C. H., Lo, C. H., Mao, S., Wirsching, P., Lerner, R. A., and Janda, K. D. (1997). Making chemistry selectable by linking it to infectivity. *Proc. Natl. Acad. Sci. USA* **94**, 11777–11782.

Gao, C. S., Brummer, O., Mao, S. L., and Janda, K. D. (1999). Selection of human metalloantibodies from a combinatorial phage single-chain antibody library. *J. Am. Chem. Soc.* **121**, 6517–6518.

Gao, C. S., Mao, S., Kaufmann, G., Wirsching, P., Lerner, R. A., and Janda, K. D. (2002). A method for the generation of combinatorial antibody libraries using pIX phage display. *Proc. Natl. Acad. Sci. USA* **99**, 12612–12616.

Haidaris, C. G., Malone, J., Sherrill, L. A., Bliss, J. M., Gaspari, A. A., Insel, R. A., and Sullivan, M. A. (1999). Recombinant human antibody single chain variable fragments reactive with *Candida albicans* surface antigens. *J. Immunol. Methods* **257**, 185–202.

Holt, L. J., Enever, C., de Wildt, R. M., and Tomlinson, I. M. (2000). The use of recombinant antibodies in proteomics. *Curr. Opin. Biotech.* **11**, 445–449.

Jahn, S., Grunow, R., Kiessig, S. T., Specht, U., Matthes, H., Hiepe, F., Heinak, A., and Von Baehr, R. (1988). Establishment of human Ig producing heterohybridomas by fusion of mouse myeloma cells with human lymphocytes derived form peripheral blood, bone marrow, spleen, lymph nodes and synovial fluids. *J. Immun. Methods* **107**, 59–66 and references therein.

Mao, S., Gao, C., Lo, C. H., Wirsching, P., Wong, C.-H., and Janda, K. D. (1999). Phage-display library selection of high-affinity human single-chain antibodies to tumor-associated carbohydrate antigens sialyl Lewis X and Lewis X. *Proc. Natl. Acad. Sci. USA* **96**, 6953–6958.

Marks, J. D., Tristem, M., Karpas, A., and Winter, G. (1991). Oligonucleotides primers for polymerase chain reaction amplification of human immunoglobulin variable genes and design of family-specific oligonucleotide probes. *Eur. J. Immunol.* **21**, 985–991.

McCafferty, J., Griffiths, A. D., Winter, G., and Chiswell, D. J. (1990). Phage antibodies: Filamentous phage displaying antibody variable domains. *Nature* **348**, 552–554.

McGuinness, B. T., Walter, G., FitzGerald, K., Schuler, P., Mahoney, W., Duncan, A. R., and Hoogenboom, H. R. (1996). Phage diabody repertoires for selection of large numbers of bispecific antibody fragments. *Nature Biotech.* **14**, 1149–1154.

Smith, G. P. (1985). Filamentous fusion phage: Novel expression vectors that display cloned antigens on the virion surface. *Science* **228**, 1315–1317.

Smith, G. P., and Petrenko, V. A. (1997). Phage display. *Chem. Rev.* **97**, 391–410.

Vaughan, T. J., Williams, A. J., Pritchard, K., Osbourn, J. K., Pope, A. R., Earnshaw, J. C., McCafferty, J., Hodits, R. A., Wilton, J., and Johnsons, K. S. (1996). Human antibodies with sub-nanomolar affinities isolated from a large non-immunized phage display library. *Nature Biotech.* **14**, 309–314.

Welschof, M., Terness, P., Kolbinger, F., Zewe, M., Dubel, S., Dorsam, H, Hain, C., Finger, M., Jung, M., Moldenhauer, G., Hayashi, N., Little, M., and Opelz, G. (1995). Amino acid sequence based PCR primers for amplification of rearranged human heavy and light chain immunoglobulin variable region genes. *J. Immunol. Methods* **179**, 203–214.

Zeidel, M., Rey, E., Tami, J., Fischbach, M., and Sanz, I. (1995). Genetic and functional characterization of human autoantibodies using combinatorial phage display libraries. *Ann. N. Y. Acad. Sci.* **764**, 559–564.

APPENDIX

The following primers were designed based on those published previously and the most recent genes segments entered in the V-Base sequence directory (de Haard *et al.*, 1999; Haidaris *et al.*, 1999; Welschof *et al.*, 1995; Marks *et al.*, 1991; Jahn *et al.*, 1988).

Primers for Human scFv Library

Primers for Amplification of Human V_H Genes

Forward primers
HVH(Sfi): TTGTTATTACTCGCGGCCCAGCCGGCC
ATGGCACAGGT;

HVH-1: CAGCCGGCCATGGCACAGGTNCAGCTG
GTRCAGTCTGG;

HVH-2: CAGCCGGCCATGGCACAGGTCCAGCTG
GTRCAGTCTGGGG;

HVH-3: CAGCCGGCCATGGCACAGGTKCAGCTG
GTGSAGTCTGGG;

HVH-4: CAGCCGGCCATGGCACAGGTCACCTTG
ARGGAGTCTGGTCC;

HVH-5: CAGCCGGCCATGGCACAGGTGCAGCTG
GTGGAGWCTGG;

HVH-6: CAGCCGGCCATGGCACAGGTGCAGCTG
GTGSAGTCYGG;

HVH-7: CAGCCGGCCATGGCACAGGTGCAGCTG
CAGGAGTCGG;

HVH-8: CAGCCGGCCATGGCACAGGTGCAGCTG
TTGSAGTCTG;

HVH-9: CAGCCGGCCATGGCACAGGTGCAGCTG
GTGCAATCTG;

HVH-10: CAGCCGGCCATGGCACAGGTGCAGCT
GCAGGAGTCCGG;

HVH-11: CAGCCGGCCATGGCACAGGTGCAGCTA
CAGCAGTGGG;

HVH-12: CAGCCGGCCATGGCACAGGTACAGCT
GCAGCAGTCAG.

Reverse primer

HJH: GGAGCCGCCGCCGCCAGAACCACCACCAC
CTGAGGAGACGGTGACCAKKGTBCC

Primers for Amplification of Human V$_\kappa$ Genes

Forward primers

HVκ-a: GGCGGCGGCGGCTCCGGTGGTGGTGGTT
CTGACATCSWGATGACCCAGTCTCC

HVκ-b: GGCGGCGGCGGCTCCGGTGGTGGTGGTT
CTGAAATTGTGYTGACKCAGTCTCC

HVκ-c: GGCGGCGGCGGCTCCGGTGGTGGTGGTT
CTGATGTTGTGATGACTCAGTCTCC

HVκ-d: GGCGGCGGCGGCTCCGGTGGTGGTGGTT
CTGAAACGACACTCACGCAGTCTCC

Reverse primers

HJκ-1: TGGAATTCGGCCCCCGAGGCCACGTTTG
ATTTCCACCTTGGTCCC;

HJκ-2: TGGAATTCGGCCCCCGAGGCCACGTTTG
ATCTCCAGCTTGGTCCC;

HJκ-3: TGGAATTCGGCCCCCGAGGCCACGTTTG
ATATCCACTTTGGTCCC;

HJκ-4: TGGAATTCGGCCCCCGAGGCCACGTTTG
ATCTCCACCTTGGTCCC;

HJκ-5: TGGAATTCGGCCCCCGAGGCCACGTTTAA
TCTCCAGTCGTGTCCC.

Primers for Amplification of Human V$_\lambda$ Genes

Forward primers

HVλ-a: GGCGGCGGCGGCTCCGGTGGTGGTGGTT
CTCAGTCTGTGTTGACGCAGCCGCC

HVλ-b: GGCGGCGGCGGCTCCGGTGGTGGTGGTT
CTCAGTCTGCCCTGACTCAGCCTGC

HVλ-c: GGCGGCGGCGGCTCCGGTGGTGGTGGTT
CTTCCTATGTGCTGACTCAGCCACC

HVλ-d: GGCGGCGGCGGCTCCGGTGGTGGTGGTT
CTTCTTCTGAGCTGACTCAGGACCC

HVλ-e: GGCGGCGGCGGCTCCGGTGGTGGTGGTT
CTCACGTTATACTGACTCAACCGCC

HVλ-f: GGCGGCGGCGGCTCCGGTGGTGGTGGTT
CTCAGGCTGTGCTCACTCAGCCGTC

HVλ-g: GGCGGCGGCGGCTCCGGTGGTGGTGGTT
CTAATTTTATGCTGACTCAGCCCCA

Reverse primers

HLJ(Sfi): GTCCTCGTCGACTGGAATTCGGCCCCC
GAGGCCAC;

HJλ-1: TGGAATTCGGCCCCCGAGGCCACCTAGGA
CGGTGACCTTGGTCCC;

HJλ-2: TGGAATTCGGCCCCCGAGGCCACCTAGGA
CGGTCAGCTTGGTCCC;

HJλ-3: TGGAATTCGGCCCCCGAGGCCACCTAAAA
CGGTGAGCTGGGTCCC.

Primers for Mouse scFv Library

Primers for Amplification of Mouse V$_H$ Gene

Forward primers

MVH(Sfi): TTGTTATTACTCGCGGCCCAGCCGGC
CATGGCA

MVH-1: GCCCAGCCGGCCATGGCAGAGGTRMAG
CTTCAGGAGTCAGGAC

MVH-2: GCCCAGCCGGCCATGGCAGAGGTSCAG
CTKCAGCAGTCAGGAC

MVH-3: GCCCAGCCGGCCATGGCACAGGTGCAG
CTGAAGSASTCAGG

MVH-4: GCCCAGCCGGCCATGGCAGAGGTGCAG
CTTCAGGAGTCSGGAC

MVH-5: GCCCAGCCGGCCATGGCAGARGTCCAG
CTGCAACAGTCYGGAC

MVH-6: GCCCAGCCGGCCATGGCACAGGTCCAG
CTKCAGCAATCTGG

MVH-7: GCCCAGCCGGCCATGGCACAGSTBCAG
CTGCAGCAGTCTGG

MVH-8: GCCCAGCCGGCCATGGCACAGGTYCAG
CTGCAGCAGTCTGGRC

MVH-9: GCCCAGCCGGCCATGGCAGAGGTYCAG
CTYCAGCAGTCTGG

MVH-10: GCCCAGCCGGCCATGGCAGAGGTCCA
RCTGCAACAATCTGGACC

MVH-11: GCCCAGCCGGCCATGGCACAGGTCCA
CGTGAAGCAGTCTGGG

MVH-12: GCCCAGCCGGCCATGGCAGAGGTGAA
SSTGGTGGAATCTG

MVH-13: GCCCAGCCGGCCATGGCAGAVGTGAA
GYTGGTGGAGTCTG

MVH-14:   GCCCAGCCGGCCATGGCAGAGGTGC
AGSKGGTGGAGTCTGGGG
MVH-15:   GCCCAGCCGGCCATGGCAGAKGTGCA
MCTGGTGGAGTCTGGG
MVH-16:   GCCCAGCCGGCCATGGCAGAGGTGA
AGCTGATGGARTCTGG
MVH-17:   GCCCAGCCGGCCATGGCAGAGGTGCA
RCTTGTTGAGTCTGGTG
MVH-18:   GCCCAGCCGGCCATGGCAGARGTRAA
GCTTCTCGAGTCTGGA
MVH-19:   GCCCAGCCGGCCATGGCAGAAGTGAA
RSTTGAGGAGTCTGG
MVH-20:   GCCCAGCCGGCCATGGCΛGAAGTGAT
GCTGGTGGAGTCTGGG
MVH-21:   GCCCAGCCGGCCATGGCACAGGTTA
CTCTRAAAGWGTSTGGCC
MVH-22:   GCCCAGCCGGCCATGGCACAGGTCCA
ACTVCAGCARCCTGG
MVH-23:   GCCCAGCCGGCCATGGCACAGGTYCA
RCTGCAGCAGTCTG
MVH-24:   GCCCAGCCGGCCATGGCAGATGTGAA
CTTGGAAGTGTCTGG
MVH-25:   GCCCAGCCGGCCATGGCAGAGGTGAA
GGTCATCGAGTCTGG
Reverse primers
MJH-1:   GGAGCCGCCGCCGCCAGAACCACCACC
ACCTGAGGAAACGGTGACCGTGGT
MJH-2:   GGAGCCGCCGCCGCCAGAACCACCACC
ACCTGAGGAGACTGTGAGAGTGGT
MJH-3:   GGAGCCGCCGCCGCCAGAACCACCACC
ACCTGCAGAGACAGTGACCAGAGT
MJH-4:   GGAGCCGCCGCCGCCAGAACCACCACC
ACCTGAGGAGACGGTGACTGAGGT

Primers for Amplification V_κ and V_λ Gene

Forward primers
MVκ-1: GGCGGCGGCGGCTCCGGTGGTGGTGGTT
CTGACATTGTTCTCACCCAGTCTCC
MVκ-2: GGCGGCGGCGGCTCCGGTGGTGGTGGTT
CTGACATTGTGCTSACCCAGTCTCC
MVκ-3: GGCGGCGGCGGCTCCGGTGGTGGTGGTT
CTGACATTGTGATGACTCAGTCTCC
MVκ-4: GGCGGCGGCGGCTCCGGTGGTGGTGGTT
CTGACATTGTGCTMACTCAGTCTCC
MVκ-5: GGCGGCGGCGGCTCCGGTGGTGGTGGTT
CTGACATTGTGYTRACACAGTCTCC
MVκ-6: GGCGGCGGCGGCTCCGGTGGTGGTGGTT
CTGACATTGTRATGACACAGTCTCC
MVκ-7: GGCGGCGGCGGCTCCGGTGGTGGTGGTT
CTGACATTMAGATRACCCAGTCTCC
MVκ-8: GGCGGCGGCGGCTCCGGTGGTGGTGGTT
CTGACATTCAGATGAMCCAGTCTCC

MVκ-9: GGCGGCGGCGGCTCCGGTGGTGGTGGTT
CTGACATTCAGATGACDCAGTCTCC
MVκ-10: GGCGGCGGCGGCTCCGGTGGTGGTGGT
TCTGACATTCAGATGACACAGACTAC
MVκ-11: GGCGGCGGCGGCTCCGGTGGTGGTGGT
TCTGACATTCAGATGATTCAGTCTCC
MVκ-12: GGCGGCGGCGGCTCCGGTGGTGGTGGT
TCTGACATTGTTCTCAWCCAGTCTCC
MVκ-13: GGCGGCGGCGGCTCCGGTGGTGGTGGT
TCTGACATTGTTCTCTCCCAGTCTCC
MVκ-14: GGCGGCGGCGGCTCCGGTGGTGGTGGT
TCTGACATTGWGCTSACCCAATCTCC
MVκ-15: GGCGGCGGCGGCTCCGGTGGTGGTGGT
TCTGACATTSTGATGACCCARTCTC
MVκ-16: GGCGGCGGCGGCTCCGGTGGTGGTGGT
TCTGACATTKTGATGACCCARACTCC
MVκ-17: GGCGGCGGCGGCTCCGGTGGTGGTGGT
TCTGACATTGTGATGACTCAGGCTAC
MVκ-18: GGCGGCGGCGGCTCCGGTGGTGGTGGT
TCTGACATTGTGATGACBCAGGCTGC
MVκ-19: GGCGGCGGCGGCTCCGGTGGTGGTGGT
TCTGACATTGTGATAACYCAGGATG
MVκ-20: GGCGGCGGCGGCTCCGGTGGTGGTGGT
TCTGACATTGTGATGACCCAGTTTGC
MVκ-21: GGCGGCGGCGGCTCCGGTGGTGGTGGT
TCTGACATTGTGATGACACAACCTGC
MVκ-22: GGCGGCGGCGGCTCCGGTGGTGGTGGT
TCTGACATTGTGATGACCCAGATTCC
MVκ-23: GGCGGCGGCGGCTCCGGTGGTGGTGGT
TCTGACATTTTGCTGACTCAGTCTCC
MVκ-24: GGCGGCGGCGGCTCCGGTGGTGGTGGT
TCTGACATTGTAATGACCCAATCTCC
MVκ-25: GGCGGCGGCGGCTCCGGTGGTGGTGGT
TCTGACATTGTGATGACCCACACTCC
MVλ: GGCGGCGGCGGCTCCGGTGGTGGTGGT
TCTCAGGCTGTTGTGACTCAGGAATC
Reverse primers
MJκ-1: TGGAATTCGGCCCCCGAGGCCACGTTTG
ATTTCCAGCTTGG
MJκ-2: TGGAATTCGGCCCCCGAGGCCACGTTTT
ATTTCCAGCTTGG
MJκ-3: TGGAATTCGGCCCCCGAGGCCACGTTTT
ATTTCCAACTTTG
MJκ-4: TGGAATTCGGCCCCCGAGGCCACGTTTC
AGCTCCAGCTTGG
MJλ: TGGAATTCGGCCCCCGAGGCCACCTAGG
ACAGTCAGTTTGG
MLJ(Sfi): GTCCTCGTCGACTGGAATTCGGCCCCC-
GAGGCCAC

60

Ribosome Display: *In Vitro* Selection of Protein–Protein Interactions

Patrick Amstutz, Hans Kaspar Binz, Christian Zahnd, and Andreas Plückthun

I. INTRODUCTION

Ribosome display is an *in vitro* technology to identify and evolve proteins or peptides binding to a given target (Fig. 1) (Hanes *et al.*, 2000a). While most selection technologies need living cells to achieve the essential coupling of genotype and phenotype, ribosome display uses the ribosomal complexes formed during *in vitro* translation to generate the physical coupling between polypeptide (phenotype) and mRNA (genotype) (Amstutz *et al.*, 2001). Hence, no transformation step limiting the size of the usable library is necessary, allowing the selection from very large combinatorial libraries. In addition, the rapid selection cycles require an integral polymerase chain reaction (PCR) step, which can be used for randomization, making this method ideal for directed evolution experiments. The fact that the ribosomal complex used for selection is not covalent allows an uncomplicated separation of the mRNA from the selected ribosomal complexes, even if the selected molecules bind the target with very high affinity or are even trapped covalently (Amstutz *et al.*, 2002; Jermutus *et al.*, 2001). All these benefits make ribosome display a good alternative to other selection techniques, such as phage display (Smith, 1985).

Ribosome display has been applied successfully for the selection of peptides (Matsuura and Plückthun, 2003; Mattheakis *et al.*, 1994), as well as folded proteins such as antibody fragments (Hanes and Plückthun, 1997; He and Taussig, 1997; Irving *et al.*, 2001). Ribosome display can also be considered for the screening of cDNA libraries for interaction partners. Ribosome display ultimately selects always for a specific binding event. However, by designing the selection pressure carefully, molecules can be selected for many other parameters, such as enzymatic turnover (by selection with a suicide inhibitor, or active site ligand) (Amstutz *et al.*, 2002; Takahashi *et al.*, 2002), protein stability (by selecting for binding under conditions where most library members will not fold) (Jermutus *et al.*, 2001), or protein biophysical properties (resistance to proteases and nonbinding to hydrophobic surfaces) (Matsuura and Plückthun, 2003). It is the combination of this array of selection pressures with the convenient PCR-based randomisation techniques that makes ribosome display a powerful and versatile technology.

II. MATERIALS AND INSTRUMENTATION

A. Reagents

The following chemicals and enzymes are necessary to prepare the extract and to perform ribosome-display selections: Luria broth base (GibcoBRL 12795-084); agarose (Invitrogen 30391-023); glucose (Fluka 49150); potassium dihydrogen phosphate (KH_2PO_4, Fluka 60230); dipotassium hydrogen phosphate ($K_2HPO_4 \cdot 3H_2O$, Merck 1.05099.1000); yeast extract (GibcoBRL 30393-037); thiamine (Sigma T-4625); Tris (Serva 37190); magnesium acetate (MgAc, Sigma M-0631); potassium acetate (KAc, Fluka 60034); L-glutamic acid monopotassium salt monohydrate (KGlu, Fluka 49601); 20 natural amino acids (Sigma LAA-21 kit); adenosinetriphosphate (ATP, Roche Diagnostics 519 987); phosphoenolpyruvate trisodium salt (PEP, Fluka 79435); pyruvate kinase (Fluka 83328);

Indirect Immunsorption at
CNBr Activated Sepharose Beads

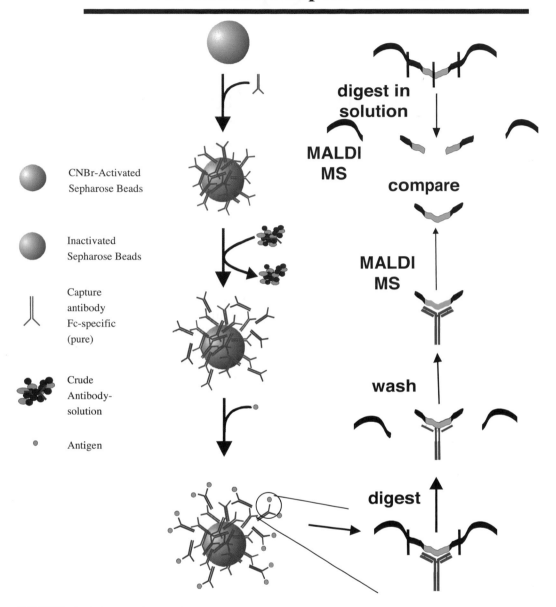

FIGURE 1 Scheme of proteolytic footprinting incorporating capture antibodies with cross-linking. [Adapted with permission from Peter and Tomer (2001). Copyright 2001 American Chemical Society.]

disulfide bonds; peptides originating from proteolysis at sites that do not disrupt antigen–antibody binding are washed off prior to analysis; and, when the primary and secondary antibodies are cross-linked, many potential proteolysis sites for the most commonly used enzymes are blocked. The capabilities of this approach are illustrated by mapping the epitope on adrenocorticotropin recognized by a mouse anti-ACTH IgG antibody (Peter and Tomer, 2001).

II. MATERIALS AND INSTRUMENTATION

Adrenocorticotropin is from Bachem, California (Cat. No. H1160). Monoclonal IgG1 anti-ACTH (clone 58) is from Biodesign International (Cat. No. E54008M). Polyclonal IgG goat antimouse Fc specific is from Sigma (Cat. No. M-4280), as are Trizma–HCl, (Cat. No. T-6666), Trizma–base (Cat. No. T-6791),

Tween 20 (Cat. No. P-8942, 10% in water), NaH_2PO_4-H_2O (Cat. No. S-9638), leucine aminopeptidase M (Cat. No. L-0632), and EDTA (Cat. No. E-5513). Endoproteinases are obtained from Roche Diagnostics (Lys-C Cat. No. 1 047 825; Glu-C Cat. No. 1 047 817; trypsin Cat. No. 1 418 025; and carboxypeptidase Y Cat. No. 1 111 914). Formic acid, 96% (Cat. No. 25,136-4), α-cyano-4-hydroxycinnamic acid (Cat. No. 47,687-0), and NH_4HCO_3 (Cat. No. 28,509-9) are from Aldrich Chemical Co. Bis(sulfosuccinimidyl) suberate, BS^3 (Cat. No. 21580), is from Pierce. Cyanogen bromide-activated Sepharose 4B beads are from Amersham Biosciences (Cat. No.17-0820-01). Compact reaction columns, CRC (Cat. No. 13928), and $35\,\mu M$ compact column filters (Cat. No. 13912) are from USB. HCl, $1M$ (Cat. No. 920-1), and NaOH, $1M$ (Cat. No. 930-65), are from Sigma Diagnostics. $CaCl_2 \cdot 2H_2O$ (Cat. No. 4160), $Na_2HPO_4 \cdot 3H_2O$ (Cat. No. 7914), $NaHCO_3$ (Cat. No. 7412), NH_4OAc (Cat. No. 3272), and $NaOAc-7H_2O$ (Cat. No. 7364) are from Mallinckrodt. NaCl is from Baker (Cat. No. 3624-05), and ethyl alcohol is from Pharmco Products. Becton Dickinson Falcon tubes, 15 ml, No. 352059, are from Fisher Scientific (Cat. No. 14-959-11B). The rotating incubator is a Lab Line hybridization incubator or similar. Deionized water is obtained from a Hydro Picopure 2 system (Hydro Systems, Research Triangle Park, NC).

MALDI mass spectra are obtained on a Voyager DE-STR mass spectrometer (Applied Biosystems). Similar results should be obtained from similar instrumentation.

III. PROCEDURES

A. Immobilization of Secondary Antibody to Cyanogen Bromide-Activated Sepharose Columns

Solutions

1. *Phosphate buffer, pH 7.2*: Prepare 100 ml $0.1M$ Na_2HPO_4 by dissolving 2.68 g $Na_2HPO_4 \cdot 7H_2O$ in water and 100 ml $0.1M$ NaH_2PO_4 by dissolving 1.38 g $NaH_2PO_4 \cdot H_2O$ in water. Use 100 ml of the $0.1M$ Na_2HPO_4 solution and titrate to pH 7.2 with the $0.1M$ NaH_2PO_4 solution (approximately 37 ml necessary).

2. *Phosphate-buffered Saline (PBS), pH 7.2*: To make 100 ml of PBS, pH 7.2, containing $0.1M$ phosphate buffer and $150\,mM$ NaCl, dissolve 0.88 g NaCl in previously prepared 100 ml $0.1M$ phosphate buffer.

3. *1 mM HCl*: To make 1 liter of $1\,mM$ HCl, dilute 1 ml $1M$ HCl to 1000 ml.

4. *Coupling buffer*: To prepare 50 ml of coupling buffer, add 0.42 g sodium bicarbonate to sufficient

deionized water to make 50 ml of solution. The pH will be between 8.2 and 8.35 and does not need further adjustment.

5. *$0.1M$ $NaHCO_3$/$0.15M$ NaCl*: To make 50 ml of solution, add 0.42 g sodium bicarbonate and 0.44 g NaCl to sufficient deionized water to make 50 ml of solution. Again, the pH does not need to be adjusted any further.

6. *Antibody solution*: Mix 20 µl of goat anti-mouse Fc-specific IgG (supplied as a 2.0-mg/ml solution in 10 mM phosphate buffered saline, pH 7.4, containing $15\,mM$ sodium azide) with 80 µl $0.1M$ $NaHCO_3$/$0.15M$ NaCl.

7. *$0.1M$ Tris, pH 8.0*: To make 50 ml of solution, add 0.79 g Trizma–HCl to approximately 45 ml deionized water and adjust pH with $1M$ NaOH. Add sufficient deionized water to yield 50 ml final volume.

8. *$0.1M$ NaOAc/$0.5M$ NaCl, pH 4.0*: To prepare 50 ml of solution, add 0.68 g $NaOAc \cdot 3H_2O$ and 1.46 g NaCl to approximately 45 ml deionized water and adjust pH with HCl. Add sufficient deionized water to make 50 ml final volume.

Steps

1. Place *ca.* 0.2 g of dry CNBr-Sepharose beads into a Falcon tube.
2. Add 10 ml of $1\,mM$ HCl. Mix gently and equilibrate for 15 min.
3. Place *ca.* 20 µl of the wet beads in a CRC and then drain the column.
4. Wash the column six times with 0.8 ml $1\,mM$ HCl and then six times with 0.4 ml coupling buffer ($0.1M$ $NaHCO_3$).
5. Add 100 µl of antibody solution with a final concentration of 400 µg/ml to the beads and incubate for 1 h at room temperature.
6. Remove the antibody solution and then, to block unreacted binding positions, add 200 µl of $0.1M$ Tris, pH 8.0, and react for 2 h at room temperature with slow rotation.
7. Wash alternately three times with $0.1M$ NaOAc/ $0.5M$ NaCl, pH 4.0, $0.1M$ Tris, pH 8.0, and PBS, pH 7.2.
8. Check immobilization of the secondary antibody by MALDI/MS using a 1-µl aliquot. Ions attributable to the antibody should be observed in low abundance only.

B. Binding of Primary Antibody to Immobilized Secondary Antibody

Solutions

1. *PBS/0.1% Tween 20*: To make 5 ml of solution, add 50 µl of 10% Tween 20 in water to 5 ml of previously prepared PBS, pH 7.2.

2. *10 mM BS³ in PBS, pH 7.2*: For a 1-ml solution, dissolve 5.72 mg BS³ in 1 ml previously prepared PBS, pH 7.2.

3. *0.1 M Tris, pH 8.0*: To make 50 ml of solution, add 0.79 g Trizma–HCl to approximately 45 ml deionized water and adjust pH with NaOH. Add sufficient deionized water to yield 50 ml final volume.

4. *Anti-ACTH IgG solution*: Prepare initial solution by dissolving 200 μg anti-ACTH IgG in 1 ml previously prepared PBS, pH 7.2.

Steps

1. Add 50 μl of mouse anti-ACTH IgG (200 μg/ml in PBS, pH 7.2) to the beads from step 7 and react for 1 h at room temperature.

2. Drain solution and wash the beads three times with 0.5 ml PBS/0.1% Tween 20 and three times with 0.5 ml PBS, pH 7.2.

3. Check that the primary antibody is bound to the immobilized secondary antibody by obtaining a MALDI/MS from a 1-μl aliquot. Ions arising from the antibody should be observed (Fig. 2A). If no ions are detected the procedure has not worked.

4. To cross-link the primary antibody to the secondary antibody, add a 10-μl aliquot of a 10 mM solution of BS³ in PBS, pH 7.2, to the beads. Incubate the mixture at RT in the dark with rotation for 45 min.

5. Drain and then quench the cross-linker with 400 μl 0.1 M Tris, pH 8.0, for 15 min at room temperature.

6. Wash the beads twice with 100 μl of 0.1 M Tris, pH 8.0.

7. Wash three times with 400 μl PBS, pH 7.2.

8. Set aside one-fourth of the beads for use as a control. *Note*: All further experiments should be performed on the control except that no ACTH should be bound to the immunocomplex. This will provide information about chemical background for subsequent MALDI analyses.

9. Check that the primary antibody has been cross-linked to the secondary antibody by MALDI/MS using a 1-μl aliquot. Ions attributable to the antibody should be of low abundance. (Fig. 2B, cf. Fig. 2A).

C. Binding of Antigen (ACTH) to Secondary Antibody

Solutions

1. *Phosphate buffer, pH 7.2*: Prepare 100 ml 0.1 M Na_2HPO_4 by dissolving 2.68 g $Na_2HPO_4 \cdot 7H_2O$ in water and 100 ml 0.1 M NaH_2PO_4 by dissolving 1.38 g $NaH_2PO_4 \cdot H_2O$ in water. Use 100 ml of the 0.1 M Na_2HPO_4 solution and titrate to pH 7.2 with

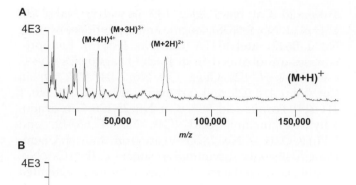

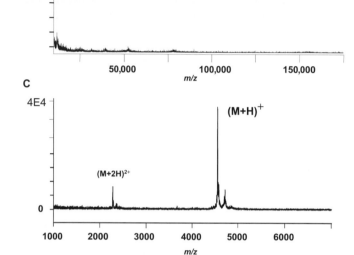

FIGURE 2. (A) Direct MALDI/MS of mouse anti-ACTH antibody from immunocomplex with Sepharose-bound antimouse Fc-specific IgG without cross-linking. (B) Direct MALDI/MS of mouse anti-ACTH antibody from immunocomplex with Sepharose-bound antimouse Fc-specific IgG with cross-linking. (C) Direct MALDI/MS of ACTH bound to cross-linked immunocomplex. [Adapted with permission from Peter and Tomer (2001). Copyright 2001 American Chemical Society.]

the 0.1 M NaH_2PO_4 solution (approximately 37 ml necessary).

2. *PBS*: To make 100 ml of PBS, pH 7.2, containing 0.1 M phosphate buffer and 150 mM NaCl, dissolve 0.88 g NaCl in 100 ml previously prepared 0.1 M phosphate buffer.

3. *10 μg/ml ACTH in PBS*: Dissolve 10 μg ACTH in 1 ml PBS from solution 2.

Steps

1. Drain the beads.

2. Add 100-μl aliquot of a 10-μg/ml solution of ACTH in PBS to the remainder of the beads from step 7. Allow reaction at room temperature for 1 h with slow rotation.

3. Incubate the control aliquot from step 8 in a CRC with 100 μl PBS, pH 7.2, for 1 h at room temperature with rotation.

4. Drain the two CRCs and wash the beads three times with 400 µl PBS and store in sufficient PBS to keep the beads moist.

5. Remove a 1-µl aliquot of the beads for MALDI/MS analysis (Fig. 2C).

D. Proteolytic Footprinting (Epitope Excision)

Successive enzymatic digestions can be performed. In this example, the immunocomplex is treated successively with Lys-C, Glu-C, trypsin, carboxypeptidase Y, and aminopeptidase M. Prepare all proteinase-containing solutions just prior to use to limit autolysis.

Solutions

1. *Lys-C solution*: To prepare 1 µg/µl Lys-C stock solution, dissolve 50 µg endoproteinase Lys-C in 50 µl deionized water. To make 50 ml stock buffer solution, dissolve 0.30 g Trizma–base and 18.6 mg EDTA disodium salt dihydrate in approximately 45 ml deionized water and adjust pH to 8.5 with HCl. Add sufficient deionized water to yield 50 ml final volume. To 90 µl of the buffer stock, add 10 µl of the Lys-C stock solution. Take 10 µl of this 0.1-µg/µl Lys-C solution and dilute it further with 990 µl of the buffer stock.

2. *Phosphate buffer, pH 7.2*: Prepare 100 ml 0.1 M Na_2HPO_4 by dissolving 2.68 g $Na_2HPO_4 \cdot 7H_2O$ in water and 100 ml 0.1 M NaH_2PO_4 by dissolving 1.38 g $NaH_2PO_4 \cdot H_2O$ in water. Use 100 ml of the 0.1 M Na_2HPO_4 solution and titrate to pH 7.2 with the 0.1 M NaH_2PO_4 solution (approximately 37 ml necessary).

3. *PBS, pH 7.2*: To make 100 ml of PBS, pH 7.2, containing 0.1 M phosphate buffer and 150 mM NaCl, dissolve 0.88 g NaCl in 100 ml 0.1 M phosphate buffer.

4. *Glu-C solution*: To prepare 1 µg/µl Glu-C stock solution, dissolve 50 µg Glu-C in 50 µl deionized water. To make 50 ml stock buffer solution, dissolve 98.8 mg ammonium bicarbonate in 45 ml deionized water and adjust pH to 7.8 with NaOH. Add sufficient deionized water to yield 50 ml final volume. To 90 µl of the buffer stock, add 10 µl of the Glu-C stock solution. Take 10 µl of this 0.1-µg/µl Glu-C solution and dilute it further with 990 µl of the buffer stock.

5. *Trypsin solution*: To prepare 1 µg/µl trypsin stock solution, dissolve 25 µg trypsin in 25 µl deionized water. To make 50 ml stock buffer solution, add 0.30 g Trizma–base and 5.55 mg $CaCl_2$ to approximately 45 ml deionized water and adjust pH to 8.5 with HCl. Add sufficient deionized water to yield 50 ml final volume. To 90 µl of the buffer stock, add 10 µl of the trypsin stock solution. Take 10 µl of this 0.1-µg/µl trypsin solution and dilute it further with 990 µl of the buffer stock.

6. *Carboxypeptidase Y solution*: To prepare 0.4 µg/µl carboxypeptidase Y stock solution, dissolve 20 µg carboxypeptidase Y in 50 µl deionized water. Buffer stock solution, 50 ml, is prepared by adding 193 mg ammonium acetate to 45 ml deionized water. Adjust the pH to 4.5 with acetic acid and add sufficient deionized water to yield 50 ml final volume. Add 10 µl carboxypeptidase Y to 90 µl of the stock solution. Take 10 µl of this 0.04-µg/µl carboxypeptidase Y solution and dilute it further with 90 µl of the buffer stock.

7. *Phosphate buffer, pH 7.0*: Prepare 100 ml 0.1 M Na_2HPO_4 by dissolving 2.68 g $Na_2HPO_4 \cdot 7H_2O$ in water and 100 ml 0.1 M NaH_2PO_4 by dissolving 1.38 g $NaH_2PO_4 \cdot H_2O$ in water. Use 100 ml of the 0.1 M Na_2HPO_4 solution and titrate to pH 7.0 with the 0.1 M NaH_2PO_4 solution (approximately 58 ml necessary).

8. *Leucine aminopeptidase solution*: To prepare 1 µg/µl aminopeptidase M stock solution, dissolve 50 µg aminopeptidase M in 50 µl deionized water. Dilute 10 µl of the aminopeptidase stock solution with 90 µl phosphate buffer, pH 7.0. Take 10 µl of this 0.1-µg/µl aminopeptidase M solution and dilute it further with 990 µl of the buffer stock.

Steps

1. Lys-C:
   a. Prepare a solution of Lys-C (1 ng/µl) in a solution of 50 mM Tris and 1 mM EDTA, pH 8.5. Add the enzyme solution (50 µl) (approximately 20:1 substrate-to-enzyme ratio) to the immunocomplex in the CRC.
   b. Incubate the beads overnight at 37°C with slow rotation.
   c. Remove the enzyme solution and wash the beads three times with 0.4 ml PBS, pH 7.2.
   d. Add sufficient buffer solution to keep beads moist.
   e. Remove a 1-µl aliquot for MALDI/MS analysis (Fig. 3A).

2. Glu-C:
   a. Prepare a solution of Glu-C (1 ng/µl) in a solution of 25 mM NH_4HCO_3, pH 7.8. Add the enzyme solution (50 µl) (approximately 20:1 substrate-to-enzyme ratio) to the immunocomplex in the CRC.
   b. Incubate the beads overnight at 37°C with slow rotation.
   c. Remove the enzyme solution and wash the beads three times with 0.4 ml PBS, pH 7.2.
   d. Add sufficient buffer solution to keep beads moist.
   e. Remove a 1-µl aliquot for MALDI/MS analysis (Fig. 3B).

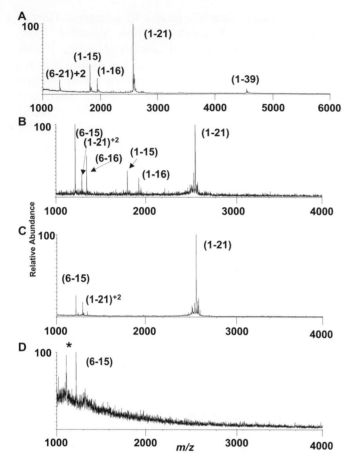

FIGURE 3. (A) Direct MALDI/MS of ACTH bound to the cross-linked immunocomplex after digestion with Lys-C. (B) Direct MALDI/MS of ACTH bound to the cross-linked immunocomplex after digestion with Lys-C followed by Glu-C. (C) Direct MALDI/MS of ACTH bound to the cross-linked immunocomplex after digestion with Lys-C followed by Glu-C and then trypsin. (D) Direct MALDI/MS of ACTH bound to the cross-linked immunocomplex after digestion with Lys-C followed by Glu-C, trypsin, and carboxypeptidase Y. Peak marked with an asterisk is background as observed in the control. [Adapted with permission from Peter and Tomer (2001). Copyright 2001 American Chemical Society.]

3. Trypsin:
   a. Prepare a solution of trypsin (1 ng/μl) in a solution of 50 mM Tris, 1 mM CaCl₂, pH 8.5. Add the enzyme solution (50 μl) (approximately 20:1 substrate-to-enzyme ratio) to the immuno-complex in the CRC.
   b. Incubate the beads overnight at 37°C with slow rotation.
   c. Remove the enzyme solution and wash the beads three times with 0.4 ml PBS, pH 7.2.
   d. Add sufficient buffer solution to keep beads moist.
   e. Remove a 1-μl aliquot for MALDI/MS analysis (Fig. 3C).

4. Carboxypeptidase Y:
   a. Prepare a solution of carboxypeptidase Y (4 ng/μl) in a 50 mM NH₄OAc, pH 4.5, solution. Add the enzyme solution to the immunocomplex in the CRC at a 3:1 substrate-to-enzyme ratio (83 μl).
   b. Incubate the beads overnight at 37°C with slow rotation.
   c. Remove the enzyme solution and wash the beads three times with 0.4 ml PBS, pH 7.2.
   d. Add sufficient buffer solution to keep beads moist.
   e. Remove a 1-μl aliquot for MALDI/MS analysis (Fig. 3D).

5. Leucine aminopeptidase:
   a. Prepare a solution of leucine aminopeptidase (1 ng/μl) in a 100 mM sodium phosphate, pH 7.0, solution. Add the enzyme solution to the immunocomplex in the CRC at a 20:1 substrate-to-enzyme ratio.
   b. Incubate the beads overnight at 37°C with slow rotation.
   c. Removed the enzyme solution and wash the beads three times with 0.4 ml PBS, pH 7.2.
   d. Add sufficient buffer solution to keep beads moist.
   e. Remove a 1-μl aliquot for MALDI/MS analysis. In this experiment, no additional cleavages are observed.

E. Preparation of Samples for MALDI/MS Analysis

1. Mix approximately 1 μl of the affinity beads on the MALDI target with 0.5–1.0 μl of saturated α-cyano-4-hydroxycinnamic acid in ethanol/water/formic acid (45/45/10 v/v/v).
2. Let the target air dry.
3. Obtain spectra as indicated by the instrument's user guide.

IV. COMMENTS

This procedure works well for linear epitopes and many continuous conformational epitopes. Problems can be encountered for conformational epitopes that are conformationally constrained by the presence of disulfide bonds when the disulfide bond is lost during proteolysis. For discontinuous epitopes, a combination of limited proteolysis and chemical modification of surface accessible amino acids may provide the required structural information (Hochleitner et al., 2000).

[1]SYSMEHFRWGKPVGKKRRPVKVYPNGAEDESAEAFPLEF[39]

Lys-C peptides observed

1-15 SYSMEHFRWGKPVGK

1-16 SYSMEHFRWGKPVGKK

1-21 SYSMEHFRWGKPVGKKRRPVK

Lys-C/Glu-C peptides observed

6-15 HFRWGKPVGK

6-16 HFRWGKPVGKK

1-15 SYSMEHFRWGKPVGK

1-16 SYSMEHFRWGKPVGKK

1-21 SYSMEHFRWGKPVGKKRRPVK

Lys-C/Glu-C/Tryp peptides observed

6-15 HFRWGKPVGK

1-21 SYSMEHFRWGKPVGKKRRPVK

Lys-C/Glu-C/Tryp/CPY peptides observed

6-15 HFRWGKPVGK

FIGURE 4. Proteolytic peptides observed for the various digestion steps involved in obtaining spectra shown in Fig. 3.

The aforementioned order of proteolytic enzymes used was used to provide initially large proteolytic peptides, followed by enzymes that would cleave within the sequence of the smallest peptide observed (Fig. 4).

V. PITFALLS

Do not use dithiothreitol because it can denature the antibody as well as the antigen.

Do not use too high a concentration of Ca^{2+} as this can interfere with the MALDI/MS analysis.

Although we have been able to perform four consecutive proteolysis experiments on one aliquot of antigen–antibody beads, the levels of background ions increase significantly while the abundance of analyte ions decreases.

References

Hochleitner, E. O., Borchers, C., Parker, C., Bienstock, R. J., and Tomer, K. B. (2000). Characterization of a discontinuous epitope of the human immunodeficiency virus (HIV) core protein p24 by epitope excision and differential chemical modification followed by mass spectrometric peptide mapping analysis. *Protein Sci.* 9, 487–496.

Jemmerson, R., and Paterson, Y. (1986). Mapping epitopes on a protein antigen by the proteolysis of antigen-antibody complexes. *Science* 232, 1001–1004.

Jeyarajah, S., Parker, C. E., Sumner, M. T., and Tomer, K. B. (1998). MALDI/MS mapping of HIV-gp120 epitopes recognized by a limited polyclonal antibody. *J. Am. Soc. Mass Spectrom.* 9, 157–165.

Legros, V., Jolivet-Reynaud, C., Battail-Poirot, N., Saint-Pierre, C., and Forest, E. (2000). Characterization of an anti-Borrelia burgdorferi OspA conformational epitope by limited proteolysis of monoclonal antibody-bound antigen and mass spectrometric peptide mapping. *Protein Sci.* 9, 1002–1010.

Macht, M., Fiedler, W., Kuerzinger, K., and Przybylski, M. (1996). Mass spectrometric mapping of protein epitope structures of myocardial infarct markers myoglobin and troponin T. *Biochemistry* 35, 15633–15639.

Papac, D. I., Hoyes, J., and Tomer, K. B. (1994). Epitope mapping of the gastrin releasing peptide/anti-bombesin monoclonal antibody complex by proteolysis followed by matrix-assisted laser desorption mass spectrometry. *Protein Sci* 3, 1488–1492.

Parker, C. E., Deterding, L. J., Hager-Braun, C., Binley, J. M., Schülke, N., Katinger, H., Moore, J. P., and Tomer, K. B. (2001). Fine definition of the epitope on the gp41 glycoprotein of human immunodeficiency virus type 1 for the neutralizing monoclonal antibody 2F5. *J. Virol* 75, 10906–10911.

Parker, C. E., Papac, D. I., Trojak, S. K., and Tomer, K. B. (1996). Epitope mapping by mass spectrometry: Determination of an epitope on HIV-1$_{IIIB}$ p26 recognized by a monoclonal antibody. *J. Immunol.* 15(1), 198–206.

Peter, J. F., and Tomer K. B. (2001). A general strategy for epitope mapping by direct MALDI-TOF mass spectrometry using secondary antibodies and crosslinking. *Anal. Chem.* 73, 4012–4019.

Suckau, D., Kohl, J., Karwath, G., Schneider, K., Casaretto, M., Bitter-Suermann, D., and Przybylski, M. (1990). Molecular epitope identification by limited proteolysis of an immobilized antigen-antibody complex and mass-spectrometric peptide-mapping. *Proc. Natl. Acad. Sci. USA* 87, 9848–9851.

Yu, L., Gaskell, S. J., and Brookman, J. L. (1998). Epitope mapping of monoclonal antibodies by mass spectrometry: Identification of protein antigens in complex biological systems. *J. Am. Soc. Mass Spectrom.* 9, 208–215.

Zhao, Y., Muir, T. M., Kent, S. B. H., Tischer, E., Scardina, J. M., and Chait, B. T. (1996). Mapping protein-protein interactions by affinity-directed mass spectrometry. *Proc. Natl. Acad. Sci. USA* 93, 4020–4024.

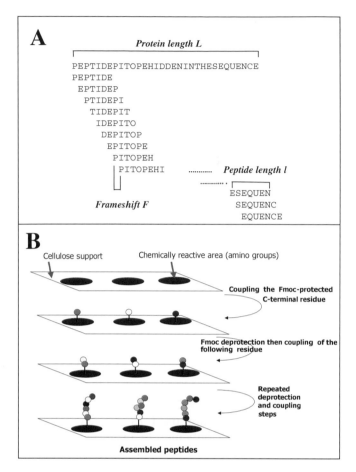

FIGURE 1 Principle of epitope identification using the SPOT method. (A) Design of overlapping peptides from a putative protein sequence. (B) Principle of the SPOT method.

antigen can be synthesized in the form of a series of overlapping peptides (Fig. 1) and further simultaneously probed for reactivity with antibody. The straightforward identification of peptides bound by the antibody has many advantages over methods using antigen modification or cleavage, mutagenesis, and so on to disclose the epitope. The identified peptide can be further prepared in large amounts and used as antigen, e.g., in diagnostic kits. Once the epitope of a monoclonal antibody has been mapped, the SPOT method can conveniently be used to determine key residues in the interaction by performing alanine scanning of the peptide. The reactivity of polyclonal antibodies with a set of immobilized peptides can also be assessed. In this case, however, it is only possible to identify antigenic regions and not precise epitopes. Nevertheless, the information derived from such an analysis is certainly more valuable than any epitope prediction method (Van Regenmortel and Pellequer, 1994) to design peptides suitable for raising antipeptide antibodies cross-reactive with the cognate protein.

However, as all methods relying on peptides, only continuous epitopes can be mapped by the SPOT approach. The identification of conformation-dependent epitopes using the SPOT technique has been reported but requires slightly more sophisticated methods for revealing low-affinity binding and discarding the background signal (Reineke *et al.*, 1999b).

The techniques used to synthesize peptides on a cellulose membrane and to probe the membrane with an antibody are described in the following sections.

II. MATERIALS AND INSTRUMENTATION

A. Spotter

The ASP 222 robot from Intavis (http://www.intavis.com) is used. The protein sequence file is submitted to the specific software together with the requirements for peptide length and frameshift, and the software automatically generates the files for synthesis. The computer then drives the spotting of the activated Fmoc amino acid by the robot according to the sequence of the individual peptides. The reader is encouraged to read the instructions carefully for use of the ASP 222 robot for efficient setup of the robot. Verification of the correct functioning of the robot is recommended before starting the synthesis: a mock synthesis is initiated using a sheet of paper instead of the membrane and is then stopped once accurate spotting has occured.

B. Chemicals

Recommended most common Fmoc amino acids (from Novabiochem, http://www.merckbiosciences. de) are as follows.

Fmoc-L-Ala; Fmoc-L-Arg(Pbf); Fmoc-L-Asn(Trt); Fmoc-L-Asp(OtBu); Fmoc-L-Cys(Trt); Fmoc-L-Gln(Trt); Fmoc-L-Glu(OtBu); Fmoc-Gly; Fmoc-L-His(Trt); Fmoc-L-Ile; Fmoc-L-Leu; Fmoc-L-Lys(Boc); Fmoc-L-Met; Fmoc-L-Phe; Fmoc-L-Pro; Fmoc-L-Ser(tBu); Fmoc-L-Thr(tBu); Fmoc-L-Trp(Boc); Fmoc-L-Tyr(tBu); Fmoc-L-Val; Fmoc-L-Cys(Acm).

C. Solvents and Reagents

N, N'-Dimethylformamide (DMF), ref. 0343549 from SDS (http://www.sds.tm.fr/)
N-Methylpyrrolidone-2 (NMP), ref. 0873516 from SDS
Piperidine, ref. 0663516 from SDS
N, N'-Diisopropylcarbodiimide (DIPC), ref. 38370F from Sigma-Aldrich (www.sigma-aldrich.com)

N-Hydroxybenzotriazole (HOBT), ref. 02–62-0008 from Novabiochem

Methanol, ref. 20847320 from VWR (http://www.vwr.com)

Bromphenol blue, ref. B-8026 from Sigma-Aldrich

Acetic anhydride, ref. 0140216 from SDS

Trifluoroacetic acid, ref. 80203 from SDS

Dichloromethane, ref. 029337E21 from SDS

Acetic acid, ref. 20103295 from VWR

Triethylsilane, ref. 90550 from Sigma-Aldrich

Dimethyl sulfoxide (DMSO), ref. 41640 from Sigma-Aldrich

D. Membranes for Spot Synthesis

Cellulose membranes (ref. 30100) are available from Intavis (http://www.intavis.com). They can be stored for months at −20°C. They consist of cellulose paper derivatized with amino polyethylene glycol. SPOT synthesis membranes should be identified. At the end of the first synthesis cycle, when spots are colored, the first and last spot number of each row of peptides should be noted in pencil, out of the arrayed area.

III. PROCEDURES

A. Chemical Synthesis

Reagents

1. *Fmoc amino acid solutions:* Weigh out each Fmoc amino acid derivative and prepare the corresponding stock solution in NMP using Table I. Use polypropylene tubes. Be sure that each derivative is completely dissolved, as at 0.6 M, most of the Fmoc-protected amino acids are at their solubility limit. Distribute, in 1.5 ml Eppendorf tubes, each stock solution into N 300 μl aliquots except for arginine (150 μl) (where N corresponds to twice the number of days the synthesis is planned to last. For example, four synthesis cycles can be performed in one working day. Supposing four membranes have to be prepared and if the peptides are 12 amino acids long, it will take 3 days to synthesize them. In this case therefore N = 8). The N series of amino acids are disposed on N tube racks and kept at −20°C.

2. *Bromphenol blue stock solution:* Prepare 10 ml of a 10 mg/ml solution of bromphenol blue in pure DMF

3. *Activators*: Prepare 1.2 M HOBT (648 mg/4 ml NMP) and 1.2 M DIPC (0.746 ml in 3.254 ml NMP) and store at −20°C

4. *Fmoc deprotection reagent:* Prepare daily 500 ml of a 20% piperidine solution in DMF

TABLE I Preparation of Fmoc Amino Acid Solutions[a]

One letter code	Amino acid	Molecular weight	Weight for a final volume of		
			1.6 ml	2.4 ml	3.2 ml
A	Fmoc-L-Ala	329	316	474	632
N	Fmoc-L-Asn(Trt)	597	574	861	1148
R	Fmoc-L-Arg(Pbf)	649	624	936	1248
D	Fmoc-L-Asp(OtBu)	412	396	594	792
C	Fmoc-L-Cys(Acm)	415	399	598.5	798
Q	Fmoc-L-Gln(Trt)	611	587	880.5	1174
E	Fmoc-L-Glu(OtBu)	444	427	640.5	854
G	Fmoc-Gly	297	286	429	572
H	Fmoc-L-His(Trt)	620	596	894	1192
I	Fmoc-L-Ile	354	340	510	680
L	Fmoc-L-Leu	354	340	510	680
K	Fmoc-L-Lys(Boc)	487	468	702	936
M	Fmoc-L-Met	372	358	537	716
F	Fmoc-L-Phe	388	373	559.5	746
P	Fmoc-L-Pro	338	325	487.5	650
S	Fmoc-L-Ser(tBu)	384	369	553.5	738
T	Fmoc-L-Thr(tBu)	398	383	574.5	766
W	Fmoc-L-Trp(Boc)	527	506	759	1012
Y	Fmoc-L-Tyr(tBu)	460	442	663	884
V	Fmoc-L-Val	340	327	490.5	654
O	Fmoc-L-Cys(Trt)	588	565	847.5	1130

[a] According to the desired final volume of solution, weigh out the indicated quantities, dissolve in the selected volume of NMP, and aliquot under 300 μl.

[b] Aliquot under 150 μl.

5. *Acetylation reagent:* Prepare daily 500 ml of 10% acetic anhydride in DMF

6. *Coloration reagent:* Add 5 ml bromphenol blue stock solution to 395 ml DMF preface translight.

7. *Side chain deprotection reagent:* Mix 15 ml trifluoroacetic acid, 15 ml dichloromethane, and 0.75 ml triethylsilane. *Warning!* Trifluoroacetic acid is extremely corrosive. Gloves and protective mask should be used when preparing and using this reagent.

Procedures

1. *Purity control for DMF and NMP:* Solvents used in Fmoc peptide synthesis should not contain contaminating free amines that could untimely deprotect Fmoc amino acid. To verify the absence of amines, pipette 10 μl of the bromphenol blue stock solution and add 990 μl of the solvent to be tested in a 1.5 ml Eppendorf tube. The solution should be light yellow. A greenish or blueish color indicates contamination with free amines, thus precluding the use of such a solvent.

2. *Synthesis:* Each synthesis cycle involves incorporation of a defined amino acid in the sequence of all peptides at the same time. There are two coupling

3. It is important to use an optimised the amount of blocking agent. We usually use the ECL Advance blocking agent, but other blocking agents (e.g., SuperBlock, Pierce Biotechnology Inc.) can be used and, in certain circumstances, perform better. As signal emission is stable for several hours, maximal reduction of background noise and nonspecific binding allows detection of even minute amounts of the target protein.

References

Celis, J. E., *et al.* (1991). The master two-dimensional gel database of human AMA cells proteins: Towards linking protein and genome sequence and mapping information. *Electrophoresis* **12**, 765–801.

Figeys, D. (2002). Functional proteomics: Mapping protein-protein interactions and pathways. *Curr. Opin. Mol. Ther.* **4**, 210–215.

Gerling, I. C., Solomon, S. S., and Bryer-Ash, M. (2003). Genomes, transcriptomes, and proteomes: Molecular medicine and its impact on medical practice. *Arch. Intern. Med.* **163**, 190–198.

Harlow, E., and Lane, D. (1988). "Antibodies: A Laboratory Manual." Cold Spring Harbor Laboratory, Cold Spring Harbor, NY.

Laemmli, U. K. (1970). Cleavage of structural proteins during the assembly of the head of bacteriophage T4. *Nature (London)* **227**, 680–685.

O'Farrell, P. H. (1975). High resolution two-dimensional electrophoresis of proteins. *J. Biol. Chem.* **250**, 4007–4021.

O'Farrell, P. Z., Goodman, H. M., and O'Farrell, P. H. (1977). High resolution two-dimensional electrophoresis of basic as well as acidic proteins. *Cell* **12**, 1133–1142.

Otto, J. J. (1993). Immunoblotting. *In* "Antibodies in Cell Biology" (D. J. Asai, ed.), pp 105–117. Academic Press, San Diego.

Panisko, E. A., Conrads, T. P., Goshe, M. B., and Veenstra, T. D. (2002). The postgenomic age: Characterization of proteomes. *Exp. Hematol.* **30**, 97–107.

Santoni, V., Molloy, M., and Rabilloud T. (2000). Membrane proteins and proteomics: un amour impossible? *Electrophoresis* **21**, 1054–1070.

Symington, J. (1984). *In* "Two-Dimensional Gel Electrophoresis of Proteins: Methods and Applications" (J. E. Celis, R. Bravo, eds.), pp. 126–168, Academic Press, New York.

Tastet, C., Charmount, S., Chevallet, M., Luche, S., and Rubilloud, T. (2003). Structure-efficiency relationships of zwitterionic detergents a protein solubilizers in two-dimensional electrophoresis. *Proteomics* **3**, 111–121.

Towbin, H., Staehelin, T., and Gordon, J. (1979). Electrophoretic transfer of proteins from polyacrylamide gels to nitrocellulose sheets: Procedure and some application. *Proc. Natl. Acad. Sci. USA* **76**, 4350–4354.

Valle, R. P., Jendoubi, M. (2003). Antibody-based technologies for target discovery. *Curr. Opin. Drug. Disco v. Dev.* **6**, 197–203.

64

Enzyme-Linked Immunosorbent Assay

Staffan Paulie, Peter Perlmann, and Hedvig Perlmann

I. INTRODUCTION

The enzyme-linked immunosorbent assay (ELISA) (Engvall and Perlmann, 1971) is a highly versatile and sensitive technique that can be used for qualitative or quantitative determinations of practically any antigen or antibody (Berzofsky *et al.*,1999). Reagents are stable, nonradioactive, and, in most cases, available commercially. Its use ranges from testing of individual samples to fully automated systems for high throughput screening. In one of its simplest forms, the assay involves immobilization of one reagent (e.g., antigen) on a plastic surface, followed by the addition of test antibodies specific for the antigen and, after washing, enzyme-conjugated secondary antibodies against the test antibodies. The addition of substrate giving coloured, fluorescent, or luminescent reaction products makes it possible to determine the concentrations of the reactants at very low levels (Butler, 1994). Depending on the quality of the reagents used and the choice of substrate, sensitivities in the picogram or subpicogram per milliliter range can be obtained.

Of several enzymes suitable for ELISA, alkaline phosphatase (ALP) and horseradish peroxidase (HRP) are the most commonly used. Various methods of enhancing sensitivity may be employed, most of which, like the commonly used biotin–streptavidin system, are designed to amplify the signal by increasing the amount of enzyme bound (Ternynck and Avrameas, 1990).

This article provides three examples of ELISA protocols: an indirect ELISA to determine antibodies (the prototype for many serological assays) and two sandwich or catcher ELISAs designed for the detection of antigen and antibodies, respectively. For the many other variants and applications of ELISA, the literature should be consulted (e.g., see Butler, 1994; Maloy *et al.*, 1991; Mark-Carter, 1994; Ravindranath *et al.*, 1994; Zielen *et al.*, 1996).

One important modification of the ELISA is the ELISpot, which instead of measuring an analyte in solution measures it at the site of a producing cell. This is made possible by using a precipitating rather than a soluble substrate, with the result being a visible imprint or spot, each representing an individual, producing cell. Enumeration of the spots gives the frequency of producing cells, which may be as low as 1 in 100,000 cells. Due to its very high sensitivity, the ELISpot is particularly well suited for measuring specific immune responses and it was originally developed for the detection of immunoglobulin production by specifically stimulated B cells (Czerkinsky *et al.*, 1983; Sedgwick and Holt, 1983). However, today it is mainly used for the analysis of specific T-cell responses where the induced production of cytokines by antigen-triggered T-cells is exploited (Lalvani *et al.*, 1997; Larsson *et al.*, 2002). Depending on the cytokine analysed the test may, apart from the number of responding cells, also give information about the type of responding cell (e.g., CTL, Th-1, or Th-2).

II. MATERIALS AND INSTRUMENTATION

A. Elisa

Flat-bottomed microtiter plates: Maxisorp from Nunc A/S or High Binding from Costar (Cat. No. 3590)
Round-bottomed microtiter plates for preparation of dilutions

III. PROCEDURES

A. Radioiodination with Iodine-125 Using Chloramine-T as Oxidant

This procedure is adapted from that originally published by Hunter and Greenwood (1962).

Solutions

1. *0.1 M phosphate buffer, pH 7.4*: Add 19 ml of 0.1 M sodium dihydrogen phosphate solution to 81 ml of 0.1 M disodium hydrogen phosphate solution. Check the pH and adjust with monobasic or dibasic solutions as required. Store at room temperature for 4 weeks.

2. *Chloramine-T*: 0.5 mg/ml in 0.1 M phosphate buffer, pH 7.4. Weigh 5 mg of chloramine-T in a universal container and add 10 ml of 0.1 M posphate buffer, pH 7.4. Swirl gently to dissolve. Keep cool until required. Prepare fresh.

3. *Sodium metabisulphite*: 0.5 g/ml weigh 5 mg of sodium metabisulphite in a universal container and add 10 ml of 0.1 M phosphate buffer, pH 7.4. Swirl gently to dissolve. Keep cool until required. Prepare fresh.

4. *0.5 M phosphate buffer, pH 7.4*: Add 19 ml of 0.5 M sodium dihydrogen phosphate solution to 81 ml of 0.15 disodium hydrogen phosphate solution. Check the pH and adjust with monobasic or dibasic solutions as required. Store at room temperature for 4 weeks.

Steps

1. Into a 1- to 2-ml polypropylene tube pipette 10–100 µg of the antibody and 50 µl of 0.5 M phosphate buffer, pH 7.4. Add the desired amount of iodine-125, typically 100 µCi–1 mCi, and mix by gently drawing the solution up and down in the pipette tip.
2. Add 20 µl of chloramine-T solution and mix again.
3. Cap the tube and leave for 5 min.
4. Add 40 µl of sodium metabisulphite solution and mix.
5. If desired, check labelling efficiency by ITLC (see Section III,E). This will typically range from 50 to 80%.
6. Separate labelled antibody from free iodine (see Section III,D).
7. If desired, determine the immunoreactive fraction of the radiolabelled antibody (see Section III,F).

B. Radioiodination with Iodine-125 Using Iodogen as Oxidant

This procedure is adapted from that originally published by Fraker and Speck (1978).

Solutions

1. *Iodogen tubes*: Dissolve 1 mg of Iodogen in 10 ml of dichloromethane in a glass or polypropylene container. Pipette 500 µl into as many 2-ml glass or polypropylene test tubes as required. Evaporate the solvent in a Speed-Vac, with a stream of nitrogen, or by leaving in a laminar flow hood for 2–4 h with the lights turned off. Cap the tubes and store in a closed container at −20°C for up to a year until required.

2. *0.1 M phosphate buffer, pH 7.4*: Add 19 ml of 0.1 M sodium dihydrogen phosphate solution to 81 ml of 0.1 M disodium hydrogen phosphate solution. Check the pH and adjust with monobasic or dibasic solutions as required. Store at room temperature for 4 weeks.

3. *0.5 M phosphate buffer, pH 7.4*: Add 19 ml of 0.5 M sodium dihydrogen phosphate solution to 81 ml of 0.15 M disodium hydrogen phosphate solution. Check the pH and adjust with monobasic or dibasic solutions as required. Store at room temperature for 4 weeks.

Steps

1. Into an Iodogen tube pipette 10–100 µg of the antibody and 50 µl of 0.5 M phosphate buffer, pH 7.4. Add the desired amount of iodine-125, typically 100 µCi–1 mCi, and mix by gently drawing the solution up and down in the pipette tip. Wait for 10 min mixing gently every 2–3 min.
2. Transfer the reaction mixture to a fresh test tube, wash the Iodogen tube with 0.5 ml of 0.1 M phosphate buffer, pH 7.4, and add to the mixture.
3. If desired, check labelling efficiency by ITLC (see Section III,E). This will typically range from 50 to 80%.
4. Separate labelled antibody from free iodine (see Section III,D).
5. If desired, determine the immunoreactive fraction of the radiolabelled antibody (see Section III,F).

C. Iodination of Antibody with "Bolton and Hunter" Reagent

This procedure is adapted from that originally published by Bolton and Hunter (1973).

Solutions

1. *Antibody solution*: 10–100 µg at a concentration of 2–5 mg/ml in 0.1 M HEPES, phosphate, or borate buffer, pH 8.0.
2. *Glycine solution*: 0.2 M in 0.1 M phosphate buffer, pH 8.0.
3. *N*-Succimidyl-3-(4-hydroxy-3-[125I]iodophenyl) propionate. Bolton and Hunter reagent [IM5861, Amersham Pharmacia Biotech, or equivalent].

4. *0.1 M phosphate buffer, pH 7.4*: Add 19 ml of 0.1 M sodium dihydrogen phosphate solution to 81 ml of 0.1 M disodium hydrogen phosphate solution. Check the pH and adjust with monobasic or dibasic solutions as required. Store at room temperature for 4 weeks.

5. *0.1 M phosphate buffer, pH 7.4 containing 0.05% polysorbate 20 (PBS/Tween)*: Add 50 μl of Tween 20 to 100 ml of 0.1 M phosphate buffer, pH 7.4. Mix gently but thoroughly. Store at room temperature for 4 weeks.

Steps

1. Into a 1.5-ml microcentrifuge tube (e.g., Eppendorf) pipette the required radioactivity of Bolton and Hunter reagent. Evaporate the solvent, ideally with a Speed-vac or, alternatively, under a gentle stream of nitrogen.
2. Add the required amount of antibody to the vial. Mix briefly and incubate for 30 min at room temperature.
3. Add 0.5 ml of 0.2 M glycine solution. Mix and incubate for a further 10 min.
4. If desired, check labelling efficiency by ITLC (see Section III,E and use ITLC paper and 20% trichloracetic acid as mobile phase). This will typically range from 30 to 50%.
5. Separate labelled antibody from free iodine (see Section III,D).
6. If desired, determine the immunoreactive fraction of the radiolabelled antibody (see Section III,F).

D. Separation of Radiolabelled Antibody from Free Iodide

Solutions

1. *Sephadex gel*: Weigh out 1 g of Sephadex G-50 powder and add 15 ml of deionised water. Mix well and either leave overnight or heat in a boiling water bath for 1 h to allow the gel to swell. Keep at 4°C for 4 weeks.

2. *0.1 M phosphate buffer, pH 7.4*: Add 19 ml of 0.1 M sodium dihydrogen phosphate solution to 81 ml of 0.1 M disodium hydrogen phosphate solution. Check the pH and adjust with monobasic or dibasic solutions as required. Store at room temperature for 4 weeks.

3. *Bovine serum albumin solution*: 1% BSA in phosphate-buffered saline, pH 7.4 (1% BSA/PBS). Weigh out 1 g of BSA and dissolve by gentle continuous mixing in 100 ml of 0.1 M phosphate buffer, pH 7.4. Keep at 4°C for 4 weeks.

Steps

1. Use either a prepacked PD-10 gel-filtration column or, if not available, prepare one as follows: Remove the barrel from a 10-ml disposable syringe and cap the luer tip. Plug the end of the syringe with a small circle of filter paper or lint dressing. Clamp the syringe vertically in a retort stand. Swirl the swollen Sephadex gel and pour as much as possible into the syringe. Allow the gel to settle for a few minutes. Then remove the luer cap and allow the liquid supernatant to run through into a waste container. Gently layer 10 ml of deionised water on top of the gel and allow to run through to waste. Replace the luer cap and use the prepared column as soon as possible.

2. Clamp either a prepacked or the home-made column vertically in a retort stand and remove the luer cap.

3. Wash the column with 30 ml of cold 1% BSA/PBS.

4. Apply the labelled antibody reaction mixture to the surface of the column and allow it to run into the gel. Gently pipette 1 ml of cold 1% BSA/PBS onto the gel and collect the eluate in a test tube.

5. Repeatedly elute the column with ten 1-ml aliquots of 1% BSA/PBS and collect each 1 ml of eluate in a fresh, numbered test tube.

6. Pipette 10-μl samples from each of the eluate fractions into counting tubes and count them in a gamma counter in order to identify tubes containing the labelled antibody fractions (typically tubes 3–5). Use or store the contents as required.

E. Determination of Radiochemical Purity by TLC

Solution

85% methanol solution: Pour 85 ml of methanol into a measuring cylinder and make up to 100 ml with deionised water. Store in a tightly closed container at room temperature for up to 4 weeks.

Steps

1. Cut a piece of chromatographic support material: Whatman 3-mm chromatography paper, silica gel-coated plastic TLC sheets, or silica gel-impregnated glass fibre (ITLC) approximately 1 × 10 cm in size. Make a faint pencil mark 1.5 cm from one end.
2. Pour enough 85% methanol into a 10- to 15-cm-tall glass beaker or similar container until it is 0.5 cm deep. Cover the beaker with a petri dish lid, aluminium foil, or similar.
3. Place a 1-μl spot of the sample to be analysed onto the centre of the pencil mark on the chromatographic strip and allow to dry.

use an entirely different chemistry for radioiodination. The most well-established alternative is the Bolton and Hunter method described in Section III,C, which results in antibody labelling at the site of lysine residues.

The other type of mechanism responsible for loss of immunoreactivity is oxidation. In addition to its desired role in oxidising the radioiodide to a reactive species, the oxidant may potentially oxidise critical residues, particularly methionine, in the antibody molecule. A way to find out which of the two possible mechanisms may in fact be the cause of a loss in immunoreactivity is to perform the labelling procedure without the addition of the radioiodine and to perform an ELISA assay. If an oxidative mechanism is responsible, then immunoreactivity will still be lost, as all the antibody molecules will be affected, not only those substituted with radioiodine. If this is found to be the case then either a Bolton and Hunter approach can be pursued or an alternative electrophilic substitution method that does not subject the antibody to such strong oxidising conditions can be employed. Two approaches may work. The first is to use a milder oxidising agent, such as the lactoperoxidase system (Morrison and Bayse, 1970). The alternative is to use a modification of the Iodogen system in which the radioiodine is first oxidised in the iodogen tube but is then transferred from the oxidising environment to another tube containing the antibody (van der Laken et al., 1997). It is likely that both of these procedures will result in a lower labelling efficiency but either may solve the problem of oxidative damage to the antibody.

The shelf life of radioiodinated antibodies is limited by radiolysis, which causes a gradual loss in both purity and immunoreactivity. The rate of deterioration can be reduced by the addition of carrier proteins or antioxidants that scavenge the radiolytic free radicals (Chakrabarti et al., 1996). A concentration of 0.1–1% albumin or 0.5% ascorbic acid is commonly used and antibodies may be stored in these solutions at either 4 or −20°C for at least a month without a significant loss of quality. If stored below 0°C, then the preparation should be divided into aliquots to save repeated freezing and thawing, which tends to favour aggregation of the antibody. If stored above 0°C, provided it does not interfere with the ultimate application, sodium azide can be added to a final concentration of 0.05% to limit microbial growth.

V. PITFALLS

The most likely cause of failure of any of the labelling methods described here is the presence of impurities in the antibody solution. The best solution is to repurify the antibody by either dialysis or gel filtration into freshly prepared buffers.

The most common problem experienced with the immunoreactive fraction assay described in Section III,F is that a curve, rather than a straight line, is obtained when data are plotted. This is nearly always caused by inaccuracies in diluting and losses in washing the cells. With practice and care, the problem usually goes away.

References

Bolton, A. E., and Hunter, W. M. (1973). The labelling of proteins to high specific activities by conjugation to a 125-I-containing acylating agent. *Biochem. J.* **133**, 529–538.

Chakrabarti, M. C., Le, N., Paik, C. H., De Graff, W. G., and Carrasquillo, J. A. (1996). Prevention of radiolysis of monoclonal antibody during labeling. *J. Nuclear Med.* **37**(8), 1384–1388.

Dewanjee, M. K. (1992). Radioiodination: Theory, Practice and Biomedical Applications. Kluwer Academic, Dordrecht.

Fraker, P. J., and Speck, J. C. (1978). Protein and cell membrane iodinations with a sparingly soluble chloramide 1,3,4,6-tetrachloro 3a.6a diphenylglycoluril. *Biochem. Biophys. Res. Commun.* **80**, 849.

Hunter, W. M., and Greenwood, F. C. (1962). Preparation of iodine-131 labelled human growth hormone of high specific activity. *Nature* **194**, 495–496.

Lindmo, T., Boven, E., and Cuttita, F. (1984). Determination of the immunoreactive fraction of radiolabelled monoclonal antibody by linear extrapolation to binding at infinite antigen excess. *J. Immunol Methods* **27**, 77–89.

Mather, S. (2000). Radiolabelling of monoclonal antibodies. *In* "Monoclonal Antibodies, a Practical Approach" (P. Shepherd and C. Dean, eds.), pp. 207–236. Oxford Univ. Press, Oxford.

Morrison, M., and Bayse, G. S. (1970). Catalysis of iodination by lactoperoxidase. *Biochemistry* **9**, 2995–3000.

van der Laken, C. J., Boerman, O. C., Oyen, W. J., van de Ven, M. T., Chizzonite, R., Corstens, F. H., and van der Meer, J. (1997). Preferential localization of systemically administered radiolabeled interleukin1 alpha in experimental inflammation in mice by binding to the type II receptor. *J. Clin. Invest.* **100**(12), 2970–2976.

IMMUNOCYTOCHEMISTRY

Immunofluorescence

FIGURE 1 Actin stress fibers in the RMCD rat mammary cell line revealed by staining with rhodamine-labeled phalloidin (×400).

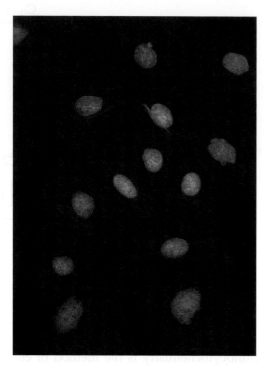

FIGURE 3 Keratin filaments in the rat kangaroo PtK2 cell line revealed by staining with antibodies to keratin and a rhodamine-labeled second antibody. DNA has been counterstained with Hoechst dye (×150).

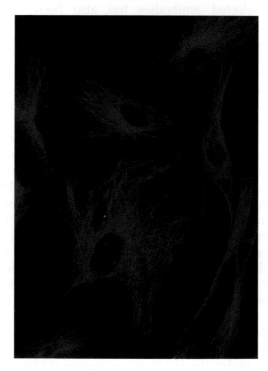

FIGURE 2 Microtubules in cells in culture revealed by staining with antibodies to tubulin followed by a rhodamine-labeled second antibody (×400).

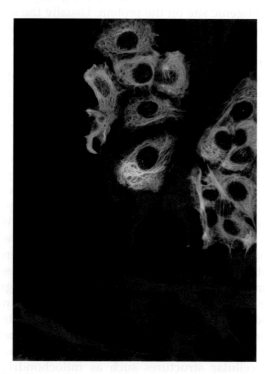

FIGURE 4 Artificial mixture of cells from the human breast carcinoma cell line MCF-7 and the human fibroblast cell line HS27 stained with an antibody to keratin (in green) and with the V9 antibody to vimentin (in red). Note that each cell type contains only a single type of intermediate filament. The yellow color results from MCF-7 and HS27 cells that lie over each other (×150).

Other reviews of immunofluorescence of cultured cells that concentrate on methods include those of Osborn (1981), Osborn and Weber (1981), and Wheatley and Wang (1981) and for live cells, Wang and Taylor (1989) and Prast *et al.* (2004). Alternatively, see Allan (2000). For an overview of the different cytoskeletal and motor proteins, see Kreis and Vale (1999). For interesting collections of color immunofluorescence micrographs from a wide variety of organisms, see Haugland *et al.* (2004) or the BioProbes newsletter (http://www.probes.com).

II. MATERIALS AND INSTRUMENTATION

A. Antibodies

Antibodies to many cellular proteins can be purchased commercially. Firms offering a variety of antibodies to cytoskeletal and other proteins include Amersham, Biomakor, Dako, Novocastra, Sigma-Aldrich, and Transduction Laboratories; other firms have specialized collections emphasizing narrower areas.

Primary antibodies today are usually monoclonal antibodies made in mice, although polyclonal antibodies made in species such as guinea pigs and rabbits are sometimes also available commercially. The appropriate dilution is established by a dilution series. Monoclonal antibodies supplied as hybridoma supernatants can often be diluted 1:1 to 1:20 for immunofluorescence or even more if other more sensitive immunocytochemical procedures are used (see also article by Osborn and Brandfass for additional information). Monoclonal antibodies supplied as ascites fluid can be diluted in the range of 1:100 to 1:1000. Use of ascites fluid at a dilution of less than 1:100 is not recommended as there are usually high titers of autoantibodies in such fluids. Polyclonal antibodies supplied as sera should be diluted in the range of 1:20 to 1:100. Note that many rabbits have relatively high levels of autoantibodies against keratins and/or other cellular proteins, so check presera. Affinity purification in which the antigen is coupled to a support and the polyclonal antibody is then put through the column usually results in a dramatic improvement in the quality of the staining patterns. Affinity-purified antibodies should work in the range of 5–20 µg/ml.

Secondary antibodies directed against IgGs of the species in which the first antibody is made are usually purchased already coupled to a fluorophore (e.g., from Jackson Laboratories http://jacksonimmuno.com, Molecular Probes http://www.probes.com and other companies listed earlier). Originally, only FITC and rhodamine-labeled antibodies were available commercially. Today there is a very wide choice of commercial antibodies coupled to different fluorophores, including AMCA, Cy2, FITC, Cy3, TRITC, phycoerythrin, RRX, Texas red, and Cy5 (see Haugland, 2004 and http://jacksonimmuno.com). Cy2 and Cy3 fluoresce in the green region of the visible spectrum, as does FITC. They are more photostable, less sensitive to pH, and are reported to give less background than many other fluorophores. Cy2 can be visualized with FITC filters and Cy3 with TRITC filters. Cy5 has an emission maximum at 670 nm and cannot be seen well by eye. Choice among the red fluorophores depends in part on the application. In addition, Molecular Probes has produced a series of Alexa Fluor dyes that cover the visible spectrum. Alexa Fluor 488 is claimed to be the best green fluorescent dye available. Alexa Fluor 555 spectra match those of Cy3, but are more fluorescent and more photostable. Alexa Fluor 647 spectra are very similar to those of Cy5. The working dilution for the secondary antibody is established by running a dilution series. Usually 1:50 to 1:150 dilutions of the commercial products are appropriate. An essential control is to check that the second antibody is negative when used alone. If nonspecific staining is present, it can sometimes be removed by absorbing the antibody on fixed monolayers of cells or on an acetone cell powder.

Antibodies other than IgMs should be stored in the freezer (−70°C for valuable primary antibodies and affinity-purified antibodies, −20°C for the rest). Antibodies should be stored in small aliquots and repeated freezing/thawing should be avoided. IgMs may be inactivated by freezing/thawing and are best kept in 50% glycerol in a freezer set at −20 to −25°C. If dilutions are made in a suitable buffer [e.g., phosphate-buffered saline (PBS), 0.5 mg/ml bovine serum albumin, $10^{-3} M$ sodium azide], diluted antibodies are stable for several months at 4°C.

B. Reagents and Other Useful Items

Methanol is of reagent grade. Formaldehyde can be diluted 1:10 from a concentrated 37% solution (e.g., Analar grade BDH Chemicals). As such solutions usually contain 11% methanol, it may be better to make the formaldehyde solution from paraformaldehyde. In this case, heat 18.5 g paraformaldehyde in 500 ml PBS on a magnetic stirrer to 60°C and filter through a 0.45-µm filter. Store at room temperature. PBS contains per liter 8 g NaCl, 0.2 g KCl, 0.2 g KH_2PO_4, and 1.15 g Na_2HPO_4, adjusted to pH 7.3 with NaOH.

Polyvinyl alcohol-based mounting media have the advantage in that although they are liquid when the sample is mounted, they solidify within hours of application. In addition, the fluorescence is stable if the sample is held in the dark and at 4°C. Samples can be reexamined and photographed after months or even years. Commonly used mounting media include Mowiol 4–88 (Calbiochem Cat. No. 475904). To make the mounting medium, place 6 g analytical grade glycerol in a 50-ml plastic conical centrifuge tube, add 2.4 g of Mowiol 4.88, and stir for 1 h to mix. Add 6 ml distilled water and stir for a further 2 h. Add 12 ml of 0.2 M Tris buffer (2.42 g Tris/100 ml water, pH adjusted to 8.5 with HCl as FITC has maximal fluorescence emission at this pH) and incubate in a water bath at 50°C for 10 min, stirring occasionally to dissolve the Mowiol. Clarify by centrifugation at 1200 g for 15 min and aliquot. Store at −20°C; unfreeze as required. Once unfrozen, the solution will be stable for several months at room temperature. While some laboratories add antifade reagents to mounting media, we have never found this to be necessary for our applications.

Other useful items include round (12 mm) or square (12 × 12 mm) glass coverslips (thickness 1.5). Ten round coverslips fit in a petri dish of 5.5 cm diameter. For screening purposes or when a large number of samples is needed (e.g., for hybridoma screening), microtest slides that contain 10 numbered circles 7 mm in diameter (Flow Labs, Cat. No. 6041505) are useful. Tweezers (e.g., Dumont No. 7) are used to handle the coverslips. Ceramic racks into which coverslips fit (Thomas Scientific, Cat. No. 8542E40) and glass containers in which these racks fit are also needed. Glass beakers (30 ml) are used to wash the specimens. Cells growing in suspension can be firmly attached to microscope slides using a cytocentrifuge such as the Cytospin 2 (Shandon Instruments).

C. Equipment

The essential requirement is access to a microscope equipped with appropriate filters to visualize the fluorochromes in routine use. Microscopes with CCD or digital cameras so that results can be viewed directly on screen are available from several manufacturers, e.g., Zeiss. Epifluorescence, an appropriate high-pressure mercury lamp (HBO 50 or HBO 100) and appropriate filters (so that specimens doubly labeled with, e.g., fluorescein and rhodamine can be visualized) are basic requirements. Lenses should also be selected carefully. The depth of field of the lens will decrease as the magnification increases. Round cultured cells will be in focus only with a ×25 or ×40 lens, whereas flatter cells can be studied with a ×63 or ×100

lens. To enable phase and fluorescence to be studied on the same specimen, some lenses should have phase optics. Only certain lenses transmit the Hoechst DNA stain (e.g., Neofluar lenses), and this stain also requires a separate filter set.

Increased resolution particularly in the z direction can be obtained by confocal microscopy (see Mason, 1999). Other forms of microscopy allow a further increase in resolution, but are not widely available (see later). Some institutes are pooling their light microscopy facilities (e.g., the Advanced Light Microscope Facility at EMBL, http:/www.embl-heidelberg.de/ExternalInfo/EurALMF, which provides state-of-the-art light microscopy image analysis and support for internal groups as well as visitors to EMBL).

III. INDIRECT IMMUNOFLUORESCENCE PROCEDURE

Steps

1. Trypsinize cells 1–2 days prior to the experiment onto glass coverslips or on multitest slides that have been washed in 100% ethanol and oven sterilized. For most applications, choose coverslips or multitest slides on which cells are two-thirds or less confluent. Drain coverslip or touch to filter paper to remove excess medium, but do not allow it to dry.

2. Place coverslips in a ceramic rack and multitest slides in metal racks, and immerse in methanol precooled to −10 to −20°C. Leave for 6 min at room temperature.

3. Make a wet chamber by lining a 13-cm-diameter (for coverslips) or a 24 × 24-cm² petri dish (for slides) with two or three sheets of filter paper and add sufficient water to moisten the filter paper. Numbers identifying the samples can be written on the top sheet of filter paper prior to wetting it.

4. Wash the fixed specimens briefly in PBS, remove excess PBS by touching to dry filter paper, and place cell side up over the appropriate number in the wet chamber.

5. Add 5–10 µl of an appropriate dilution of the primary antibody with an Eppendorf pipette. Use the tip to spread the antibody over the coverslip without touching the cells. Replace the top of the wet chamber, transfer to a humidified incubator at 37°C, and incubate for 45 min.

6. Wash by dipping each coverslip individually three times into each of three 30-ml beakers containing PBS. Wash slides by replacing slides in metal rack and

transferring through three PBS washes (180 ml each, leave for 2 min in each). Remove excess PBS with filter paper.

7. Replace specimens in wet chamber. Add 5–10 μl of an appropriately diluted second antibody carrying a fluorescent tag. Return to 37°C incubator for a further 30–45 min.

8. Repeat step 6.

9. Identify microscope slides with small adhesive labels on which date, specimen number, antibody, or other information is written. Place slides in cardboard microslide folders (e.g., Thomas Scientific, Cat. No. 6708-M10). Mount two coverslips per slide by inverting each coverslip and placing cell side down on a drop of mounting medium placed on the slide, with a disposable ring micropipette. Cover with filter paper and press gently to remove excess mounting medium. For samples on multitest slides, use 6 × 2.5-cm glass coverslips on which a drop of mounting medium has been placed. Secure the coverslips with nail polish. Store samples in the dark in slide boxes at 4°C.

10. Documentation. Use a fast film (e.g., Kodak 35 mm Tri-X) and push the development, e.g., with Diafine (Acufine), or record the image digitally. Phase micrographs of specimens embedded in Mowiol should be made as soon as possible after mounting the specimens. The fluorescence decreases a little in the first few days, but is then stable for years if the samples are stored in the dark at 4°C.

IV. COMMENTS

Specimens should not be allowed to dry out at any stage in the procedure. If coverslips are dropped accidentally, the side on which the cells are can be identified by focusing on the cells under an upright microscope and scratching gently with tweezers.

A. Fixation

The procedure gives good results with many cytoskeletal and other antigens; however, the optimal fixation protocol depends on the specimen, the antigen, and the location of the antigen within the cell. Three requirements have to be met. First, the fixation procedure must retain the antigen within the cell. Second, the ultrastructure must be preserved as far as possible without destroying the antigenic determinants recognized by the antibody. Third, the antibody must be able to reach the antigen; i.e., the fixation and permeabilization steps must extract sufficient cytoplasmic components so that the antibodies can penetrate into the fixed cells. In the procedure just given, fixation and permeabilization are achieved in a single step, i.e., with methanol. Alternative fixation methods include the following:

1. Formaldehyde–methanol: 3.7% formaldehyde in PBS for 10 min (to fix the cells) and then methanol at –10°C for 6 min (to permeabilize the cells).

2. Formaldehyde–Triton: 3.7% formaldehyde in PBS for 10 min (to fix) and then PBS with 0.2% Triton X-100 for 1 min at room temperature (to permeabilize).

3. Glutaraldehyde: Fix in 1% glutaraldehyde (electron microscopic grade) in PBS for 15 min and then methanol at –10°C for 15 min. Immerse in sodium borohydride solution (0.5 mg/ml in PBS made minutes before use) for 3 × 4 min. Wash with PBS 2 × 3 min each. Note that the sodium borohydride step is necessary to reduce the unreacted aldehyde groups; without this step the background will be very high.

Note that formaldehyde treatment destroys the antigenicity of many antigens. Alternatively, in a very few cases, positive staining may be observed only after formaldehyde fixation. Very few antigens react after glutaraldehyde fixation.

B. Special Situations

1. Fluorescently labeled phalloidin (extremely poisonous), a phallotoxin that binds to filamentous actin, is available commercially and is usually used to reveal the distribution of filamentous actin in cells (Fig. 1). To obtain good staining patterns, fix cells for 10 min in 3.7% formaldehyde in PBS. Wash with PBS. Incubate for 1 min in 0.2% Triton X-100 in PBS and wash with PBS. Incubate with an appropriate dilution of rhodamine-labeled phalloidin (e.g., Sigma Cat. No. P-1951) for 30 min at 37°C, wash with PBS, and mount in Mowiol. Note that phalloidin staining will not work after methanol fixation (for further discussion, see article by Prast et al., and Small et al., 1999).

2. To stain endoplasmic reticulum, use either an antibody, e.g., ID3 against a sequence region of protein disulfide isomerase (Vaux et al., 1990), or the lipophilic, cationic fluorescent dye $DiOC_6$ (3,3-dihexyloxacarbocyanine iodide, Kodak Cat. No. 14414) (Terasaki et al., 1984). To stain with dye, fix for 5 min in 0.25% glutaraldehyde in 0.1 M cacodylate and 0.1 M sucrose buffer, pH 7.4. Wash. Stain for 80 s with dye, mount in buffer, and observe using a ×63 or ×100 lens and the fluorescein filter. Reticular structures should be apparent. Note that mitochondria will also be stained. To stain only mitochondria, use either an antibody, e.g., to cytochrome oxidase, or the dye rhodamine 123.

3. Special fixation procedures may also be needed for other membrane structures in cells. In addition, lectins can be used to stain carbohydrate-containing organelles, e.g., staining of Golgi apparatus with fluorescently labeled wheat germ or other agglutinins.

4. To stain DNA for fluorescent applications dyes that bind to the minor groove in DNA such as the Hoechst dyes and DAPI are usually used. To stain with Hoechst use either Hoechst 33242 (Sigma Cat. No. 2261) or Hoechst 33258 (Sigma Cat. No. B2883). Note that Hoechst 33242 can also be used to stain DNA in live cells. When bound to DNA, Hoechst dyes fluoresce bright blue (cf. Fig. 1C). To stain with Hoechst prepare a 1 mM stock solution in sterile water and store at 4°C. To stain cells dilute the stock solution 1:1000 in PBS to a final working concentration of 1 μM. After step 8 in the immunofluorescence procedure, pipette 20–100 μl of the working solution onto each coverslip and leave for 4 min at room temperature. Wash twice with PBS, drain the coverslip, and mount in Mowiol. DNA can also be stained with DAPI, which shows a 20-fold fluorescence enhancement on binding to DNA and also gives blue fluorescence.

5. Some cellular structures, such as microtubules, are sensitive to calcium. In this case, add 2–5 mM EGTA to the 3.7% formaldehyde solution in Section IV,A and to the methanol in Section III, step 2.

6. Blocking steps are usually not necessary if antibodies are diluted in BSA containing buffers. If a blocking step is used it should be performed after fixation and prior to adding the first antibody. Blocking to reduce nonspecific staining is performed with 10% serum from the same host species as the labeled antibody.

7. Sometimes for cell surface components it may be advantageous to stain live cells. Expose such cells to antibody for 25 min and proceed with steps 6–8 in Section III. Then fix cells in 5% acetic acid/95% ethanol for 10 min at −10°C (Raff et al., 1978).

8. Another application of immunofluorescence microscopy is to monitor directly the distribution of fluorescently labeled proteins in live cells by video microscopy (e.g., Sammak and Borisy, 1988).

C. Double or Triple Immunofluorescence Mircroscopy

It is often advantageous to visualize two or three antigens in the same cell (Fig. 4). Here it is important to choose fluorophores that give good color separation, e.g., a Texas red/FITC combination will give a better separation than TRITC/FITC (see also article by Prast et al.). Most important is that the microscope is optimized for the fluorophores in use by the selection of the appropriate filters so that there is no overlap between the channels used to observe each of the fluorophores.

D. Stereomicroscopy

Fluorescence microscopy gives an overview of the whole cell. With practice, specimens can be seen in three dimensions when looking through a conventional microscope. Stereomicrographs can be made using a simple modification of commercially available parts (Osborn et al., 1978a). Today, however, confocal microscopy is the method of choice and is particularly useful for round cells, which are not in focus with the higher-power ×63 or ×100 lenses, or to document arrangements and obtain greater resolution at multiple levels in the same cell or organism (cf. Fox et al., 1991).

E. Limit of Resolution

Theoretically, this is ~200 nm when 515-nm wavelength light and a numerical aperture of 1.4 are used. Objects with dimensions above 200 nm will be seen at their real size. Objects with dimensions below 200 nm can be visualized provided they bind sufficient antibody, but will be seen with diameters equal to the resolution of the light microscope (cf. visualization of single microtubules in Osborn et al., 1978b). Objects closer together than 200–250 nm cannot be resolved by conventional fluorescence microscopy, e.g., microtubules in the mitotic spindle or ribosomes. However, some increase in the resolution of fluorescent images can be obtained using new forms of microscopy (see Hell, 2003).

F. New Developments

Immunofluorescence microscopy is an important technique not only for fixed cells (this article), but also because of the possibility of expressing GFP vectors coupled to particular constructs in living cells (see article by Prast et al.) and following changes in distribution by video microscopy. FRET imaging techniques (see this volume) and other novel techniques, such as the use of quantum dot ligands (e.g., Lidke et al., 2004), are also opening up new possibilities for more quantitative fluorescence measurements on live cells.

V. PITFALLS

Occasionally no specific structures are visualized, even though the cell is known to contain the antigen. This may be because:

1. Antibodies can be species specific. This can be a particular problem with monoclonal antibodies, which, for instance, may work with human but not with other species. If in doubt, check the species specificity with the supplier before purchase.

2. The fixation procedure may inactivate the antigen. For instance, many intermediate filament antibodies no longer react after fixation protocols such as those in Section IV,A.

3. The antigen may be present only at very low concentrations and therefore it may be necessary to use more sensitive methods to detect the antigen.

4. The antigen can be poorly fixed or extracted by the fixation procedure.

5. The antibody may not be able to gain access to the antigen, e.g., antibodies to tubulin often do not stain the midbody of the intracellular bridge.

6. The specimens may be generally fluorescent and it can be hard to decide whether this is due to specific or nonspecific staining.

References

Allan, V. J. (2000). Basic immunofluorescence. In "Protein Localisation by Fluorescence Microscopy: A Practical Approach," pp. 1–26. Oxford Univ. Press, New York.

Fox, M. H., Arndt-Jovin, D. J., Jovin, T. M., Baumann, P. H., and Robert-Nicoud, M. (1991). Spatial and temporal distribution of DNA replication sites localized by immunofluorescence and confocal microscopy in mouse fibroblasts. J. Cell Sci. 99, 247–253.

Haugland, R. P. (2004). Molecular probes. In "Handbook of Fluorescent Probes and Research Products," 9th Ed.

Hell, S. W. (2003). Toward fluorescence nanoscopy. Nature Biotechnol. 21, 1347–1355.

Kreis, T., and Vale, R. (1999). "Guidebook to the Cytoskeletal and Motor Proteins," 2nd Ed. Oxford Univ. Press, London.

Lazarides, E., and Weber, K. (1974). Actin antibody: The specific visualization of actin filaments in non-muscle cells. Proc. Natl. Acad. Sci. USA 71, 2268–2272.

Lidke, D. S., Nagy, P., Heintzmann, R., Arndt-Jovin, D. J., Post, J. N., Grecco, H. E., Jares-Erijman, E. A., and Jovin, T. M. (2004). Quantum dot ligands provide new insights into erbB/HER receptor-mediated signal transduction. Nature Biotechnol. 22, 198–203.

Mason, W. T. (1999). "Fluorescent and Luminescent Probes for Biological Activity." Academic Press, San Diego.

Osborn, M. (1981). Localization of proteins by immunofluorescence techniques. Techniq. Cell. Physiol. P107, 1–28.

Osborn, M., Born, T., Koitzsch, H. J., and Weber, K. (1978a). Stereo immunofluorescence microscopy. I. Three-dimensional arrangement of microfilaments, microtubules and tonofilaments. Cell 13, 477–488.

Osborn, M., and Weber, K. (1981). Immunofluorescence and immunochemical procedures with affinity purified antibodies. In "Methods in Cell Biology," Vol. 23. Academic Press, New York.

Osborn, M., Webster, R. E., and Weber, K. (1978b). Individual microtubules viewed by immunofluorescence and electron microscopy in the same PtK2 cell. J. Cell Biol. 77, R27–R34.

Pruss, R. M., Mirsky, R., Raff, M. C., Thorpe, R., Dowding, A. J., and Anderton, B. H. (1981). All classes of intermediate filaments share a common antigenic determinant defined by a monoclonal antibody. Cell 27, 419–428.

Quentmeier, H., Osborn, M., Reinhardt, J., Zaborski, M., and Drexler, H. G. (2001). Immunocytochemical analysis of cell lines derived from solid tumors. J. Histochem. Cytochem. 49, 1369–1378.

Raff, M. C., Mirsky, R., Fields, K. L., Lisak, R. P., Dorfman, S. H., Pilbenberg, D. H., Gregeon, N. A., Leibowitz, S., and Kennedy, M. C. (1978). Galactocereboside is a specific cell surface antigenic marker for oligodendrocytes in culture. Nature 274, 813–816.

Sammak, P. J., and Borisy, G. G. (1988). Direct observation of microtubule dynamics in living cells. Nature 332, 724–726.

Small, J. V., Rottner, K., Hahne, P., and Anderson, K. I. (1999). Visualising the actin cytoskeleton. Microsc. Res. Tech. 47, 3–17.

Terasaki, M., Song, J., Wong, J. R., Weiss, M. J., and Chen, L. B. (1984). Localization of endoplasmic reticulum in living and glutaraldehyde fixed cells with fluorescent dyes. Cell 38, 101–108.

Vaux, D., Tooze, J., and Fuller, S. (1990). Identification by anti-idiotype antibodies of an intracellular membrane protein that recognizes a mammalian endoplasmic reticulum retention signal. Nature 345, 495–502.

Wang, Y. L., and Taylor, D. L. (1989). "Methods in Cell Biology," Vols. 29 and 30. Academic Press, New York.

Wheatley, S. P., and Wang, Y. L. (1998). Indirect immunofluorescence microscopy in cultured cells. Methods Cell Biol. 57, 313–332.

67

Immunofluorescence Microscopy of the Cytoskeleton: Combination with Green Fluorescent Protein Tags

Johanna Prast, Mario Gimona, and J. Victor Small

Immunofluorescence microscopy is now a standard procedure for the localisation of molecules in cells (for an introduction to the method, see article by Osborn). Nevertheless, the method has its pitfalls, not least in requiring the immobilisation of cells by chemical fixation and multiple manipulations during labelling. This can lead to the loss of antigen as well as to the distortion of cell structure. Unfortunately, published pictures still appear in which distortions in cell structure are overlooked and where the conclusions drawn are consequently tenuous. Advances in live cell imaging, combined with green fluorescent protein (GFP) tags (and analogues) now provide independent methods for assessing the localisation of molecules. Nevertheless, immunofluorescence microscopy remains an important technique when used with discretion. Indeed, when applied in combination with cells expressing fluorescently tagged probes, it adds a further dimension to characterise the relative localisation of multiple components. This article provides some recipes suitable for labeling the cytoskeleton and gives examples where GFP tags and immunofluorescence can be usefully combined.

II. MATERIALS AND REAGENTS

1. *Coverslips*: Round glass coverslips 12 or 15 mm in diameter, cleaned in 60% ethanol/40% HCl (10 min), rinsed with H_2O (2 × 5 min), drained, cleaned with lint-free paper, and sterilized for tissue culture by exposure to ultraviolet light in the culture dish.

2. *Humid chamber*: Large petri dish 14 cm in diameter, or similar container with lid, containing a glass plate (around 9 cm^2) coated with a layer of Parafilm and supported on a moistened piece of filter paper on the bottom of the dish. A few drops of water on the glass plate facilitate spreading and flattening of the Parafilm.

3. *Washing reservoir*: Two multiwell dishes, 24 wells each (e.g., Nunc).

4. *Filter paper*: Whatman No. 1, 9 cm in diameter.

5. *Forceps*: Dumont No. 4 or No. 5 or equivalent watchmaker forceps.

6. *Pipettes*: Set of automatic pipettes (0–20, 20–200, 50–1000 μl) or capillary pipettes for diluting and aliquoting antibodies. Pasteur pipettes.

7. *Phalloidin*: Alexa 568- and Alexa 488-coupled phalloidins from Molecular Probes and CPITC (coumarin) phalloidin from Sigma. Store as 0.1-mg/ml stocks in methanol at −20°C.

8. *Secondary antibodies*: Commercial secondary antibodies carrying Molecular Probes Alexa 488, 568, and 350 conjugates are, in our experience, of generally good quality.

9. *Gelvatol, Vinol*: The basic ingredient of the mounting medium is polyvinyl alcohol (MW 10.000, around 87% hydrolysed), which comes under various trade names: Elvanol, Mowiol, and Gelvatol. We use Vinol 203 from Air Products and Chemical Inc.

II. PROCEDURES

Solutions

1. *0.5 M EGTA stock solution*: For 500 ml stock, weigh out 95.1 g EGTA (Sigma E-4378) into 400 ml H_2O, adjust pH to 7.0 with 1 N NaOH, and make up to 500 ml with H_2O. Store at room temperature in a plastic bottle.

2. *1 M $MgCl_2$ stock solution*: For 500 ml stock, weigh out 101.6 g $MgCl_2$, add H_2O to 500 ml, dissolve, and store at 4°C.

3. *Cytoskeleton buffer (CB)*: 10 nM MES (Sigma M-8250), 150 mM NaCl, 5 mM EGTA, 5 mM $MgCl_2$, and 5 mM glucose. For 1 liter, add the following amounts to 800 ml H_2O: MES, 1.95 g; NaCL, 8.76 g; 0.5 M EGTA, 10 ml; 1 M $MgCl_2$, 5 ml; and glucose, 0.9 g. Adjust pH to 6.1 with 1 N NaOH and fill up to 1 liter. Store at 4°C. For extended storage, add 100 mg streptomycin sulfate (Sigma S-6501).

4. *Phosphate-buffered saline (PBS) working solution*: 137 mM NaCl, 2.7 mM KCl, 4.3 mM $Na_2HPO_4 \cdot 7H_2O$, and 1.4 mM KH_2PO_4, pH 7.4.

5. *Triton X-100*: Make up 10% aqueous stock and store at 4°C.

6. *Glutaraldehyde (GA) stock*: Make up 2.5% solution of glutaraldehyde by diluting 25% glutaraldehyde EM grade (Agar scientific Ltd., Cat. No. R 1020 or equivalent) in CB. Readjust pH to 6.1 and store at 4°C.

7. *Paraformaldehyde (PFA) stock*: Make up a stock 4% solution of paraformaldehyde in PBS or CB (see fixative mixtures) (analytical grade Merck Cat. No. 4005). To make 100 ml, heat 80 ml of CB (or PBS) to 60°C, add 3 g paraformaldehyde, and mix 30 min. Add a few drops of 10 M NaOH until the solution is clear, cool, adjust pH (see appropriate mixture), and make up to 100 ml. Store in aliquots at −20°C.

8. *Fixative mixtures*: Aldehyde fixative mixtures are made up using the stock solutions given earlier to give the combinations listed under step 2.

9. *Blocking solution*: 1% bovine serum albumin and 5% horse serum in PBS.

10. *Antibody mixtures*: These are made up in PBS or in the blocking solution without serum. To remove any unwanted particles, centrifuge (10,000 g for 10 min) the diluted mixture before use. The antibody combinations used for this article are listed in Table I.

11. *Mounting medium*: Mix 2.4 g of polyvinyl alcohol with 6 g glycerol (87%) and then with 6 ml H_2O. After at least 2 h at room temperature, add 0.2 ml 0.2 M Tris–HCl, pH 8.5, to the mixture and further incubate the solution for 10 min at 60°C. Remove any precipitate by centrifugation at 17,000 g for 30 min. Store in aliquots at −20°C. [Antibleach agents are available that considerably reduce bleaching and thus enable multiple pictures to be taken of the same cells. We use n-propyl gallate (Giloh and Sedat, 1982) at 5 mg/ml or phenylenediamine (Johnson *et al.*, 1982) at 1–2 mg/ml in the mounting medium. After dissolving the additive, degas mounting medium before storage.]

Steps

1. Seed the cells onto coverslips in the petri dish and allow them to attach and spread for 4–48 h in an incubator at 37°C.

2. Aspirate growth medium and rinse dish gently with PBS (warmed 37°C PBS); avoid shifting of the coverslips over each other. Aspirate PBS and replace with one of the following fixative solutions.
   a. *Fix 1*: 4% PFA/0.1% Triton X-100 in PBS for 2 min. Rinse three times with PBS. 4% PFA in PBS for 20 min. Wash 2 × 10 min with PBS.
   b. *Fix 2*: 0.5% GA/0.25% Triton X-100 in CB for 1 min. Rinse 3× with CB. 4% PFA in CB for 20 min. Wash 2 × 10 min with CB.
   c. *Fix 3*: 0.25% GA/0.5% Triton X-100 in CB for 1 min. Rinse 3× with CB. 1% GA for 15 min. Wash 2 × 10 min with CB.
   d. *Fix 4*: 4% PFA/0.3% Triton X 100/0.1% GA in CB for 15 min. Wash 2 × 10 min with CB.

3. Block: Invert each PBS coverslip onto a 30-µl drop of blocking solution on Parafilm in the humid chamber. Before transfer to drop, dry the back side of the coverslip by holding it briefly on filter paper with a pair of forceps, taking care not to allow the cell side to dry. Drain any excess solution from the cell side by touching the edge of the coverslip to the filter paper. Incubate on blocking solution for 15 min or until first antibody mixtures are prepared. (Back side of coverslip should not be wet or else coverslip will sink during the washing step.)

4. Apply drops (20–50 µl) of first antibody mixture to unused part of Parafilm and transfer coverslips to appropriate drops after draining excess blocking solution on filter paper. Replace lid on petri dish and leave at room temperature for 45–60 min.

5. Wash: To ease removal of coverslips for washing, pipette 100 µl PBS under their edge to lift them up from the Parafilm. Using forceps, transfer coverslips to a multiwell dish in which the wells are filled to the brim with PBS so that the liquid surface is flat. The coverslips will float well, cell side down, as long as the back side remains dry. (For efficient washing, transfer dish gently to a tilting rotating table for 10 min.) Repeat washing steps after transfer of coverslips to a second dish containing fresh PBS two or three times.

6. Change Parafilm in humid chamber and apply drops of second antibody mixture. Transfer coverslips

to drops after briefly draining excess with filter paper and incubate for 45 min.

7. Wash as described in step 6.

8. Mount: Add a small drop of mounting medium to a cleaned glass slide using, for example, a plastic disposable pipette tip. Drain excess (PBS) from coverslip and gently invert onto drop. *Note*: The mounting medium dries quite fast so the drops should be applied singly and not in batches. If necessary, remove excess medium after mounting by applying small pieces of torn filter paper to the coverslip edge.

9. Observe directly in a fluorescence microscope with a dry lens. An oil immersion lens can be used the next day when the mounting medium has solidified. Alternatively, the drying time can be shortened by transfer of slides to a 27°C oven.

III. CHOICE OF FIXATION FOR MULTIPLE LABELLING

Different antibodies commonly require different fixation protocols to give optimal labelling. It is therefore important to test different fixation conditions for each antibody to determine the best compromise fixation for multiple labelling. Table I lists the characerics of the commercial antibodies used in this article in terms of the intensity of label obtained with each of the four fixation protocols described for cultured smooth muscle cells (A7r5, American Type Culture Collection). In general, you should establish the strongest fixation protocol that your antibodies can tolerate and draw your conclusions accordingly.

IV. COMBINING IMMUNOFLUORESCENCE AND GFP TAGS

The ability to express fluorescent proteins as tags to gene products in living cells represents an important advance in localisation methods. In addition to facilitating the visualisation of protein dynamics *in vivo* with sensitive imaging systems, these tagging methods can be usefully combined with the imunofluorescence procedure. Additional flexibility is offered by the fact that the tagged protein is already fluorescent so that restrictions with the use of secondary antibodies are reduced and triple labelling is relatively straightforward. With the cytoskeleton, phalloidin is a probe of choice for actin; when applying this probe, four labels in one cell would not be

TABLE I   Characteristics of Commercial Antibodies[a]

Staining	Fix 1	Fix 2	Fix 3	Fix 4
Mc. Anti-α-actinin mouse IgM (Sigma BM-75.2) 1:500	+++	++	++	++
Mc. Antiphosphotyrosine (Santa Cruz pY99) Mouse IgG 1:1000	+++	+	++	++
Mc. Anti-α-Tubulin mouse IgG (Sigma DM1A) 1:1000	−	+++	+++	+++
Mc. Antivimentin mouse IgG (Sigma V9) 1:200	+++	+++	+++	+++
Mc. Antivinculin mouse IgG (Sigma vVin-1) 1:400	+	++	+++	+++
Phalloidin CPITC (Sigma) 1:25	+	+	++	++
Phalloidin Alexa 488 (Molecular Probes) 1:300	++	+++	+++	+++
Phalloindin Alexa 568 (Molecular Probes) 1:300	++	+++	+++	+++

[a] **Secondary antibodies** used in this screen were as follows: (1) Rhodamine Red-conjugated donkey anti-mouse **IgM** [Affini Pure F(ab')₂ Fragment-Immuno Research Laboratories, Inc.] and (2) Alexa Fluor 568- and 488-conjugated goat anti-mouse **IgG** (H+L) (Molecular Probes)

problematic, given a suitable choice of fluophores (including Cy-5 in the infrared, for example).

An interesting aspect of proteins that are washed away easily during fixation is that they are retained more readily when tagged with a GFP moiety (M. Gimona, unpublished observations). This property offers an unexpected advantage of GFP-tagged probes for localisation studies with fixed cells.

Figures 1–3 show examples of cells expressing GFP-tagged proteins and then fixed and labelled with phalloidin (for actin) and a single antibody. The fixation procedure was "fixation 1" in step 2.

V. COMMENTS

We have generally aimed for fixation protocols that best preserve the actin cytoskeleton. Although stress fibers are easily preserved with most fixative protocols, the delicate peripheral lamellipodia are normally

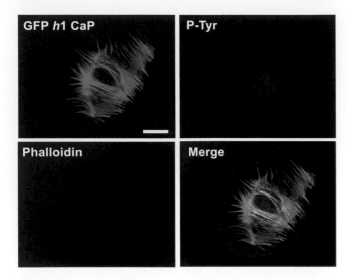

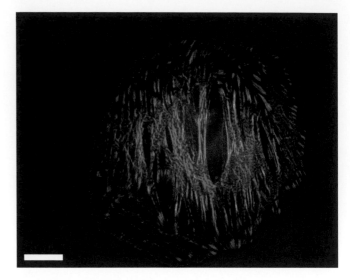

FIGURE 1 Flourescence images of an A7r5 smooth muscle cell transfected with GFP h1calponin (Gimona, 2003) and then counterstained with Alexa phalloidin 350 (dilution, 1:200) and mouse antiphosphotyrosine (PY99 Santa Cruz dilution 1:1000), followed by a GaM Alexa 568 as secondary antibody at a dilution of 1:750. Images were recorded on a Zeiss Axioskop fitted with an Zeiss Axiocam imaging system. All three images are combined in the merged image (bottom right). Bar: 20 μm.

FIGURE 3 Merged fluorescence images (as Fig. 1) of an A7r5 smooth muscle cell transfected with GFP-α-actinin (Gimona et al., 2003) and counterstained with mouse antiphosphotyrosine (as in Fig. 1). Bar: 20 μm.

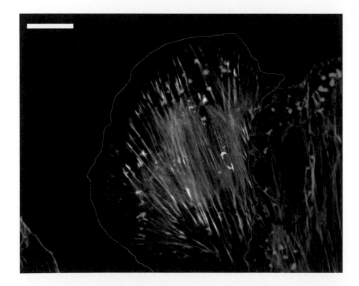

FIGURE 2 Merged fluorescence images (as Fig. 1) of an A7r5 smooth muscle cell transfected with Ds-Red zyxin (Bhatt et al., 2002) to mark focal adhesions and then labelled with Alexa 488 phalloidin (dilution, 1:300) and antiphosphotyrosine (see Fig. 1) with GaM Alexa 350 at a dilution of 500 as secondary antibody. The boundary of the transfected cell is indicated by a white line. Bar: 20 μm.

distorted or lost after methanol or formaldehyde fixation. This is why glutaraldehyde is included in two of the present mixtures. A stronger glutaraldehyde fixation than what we have used here is best for lamellipodia (see, e.g., Small, 1988) but cannot be used with

several of the antibodies described. So again we have had to compromise. Weber and colleagues (1978) introduced the use of sodium borohydride to reduce free aldehyde groups after glutaraldehyde fixation and thereby the autofluorescence introduced by this fixative. If autofluorescence is observed, for example, in the region of the nucleus, this can be eliminated by a brief treatment (3 × 10 min) of the coverslips in ice-cold cytoskeleton buffer containing freshly dissolved sodium borohydride (0.5 mg/ml). The coverslips are rinsed three times prior to immunolabelling.

VI. PITFALLS

If problems arise from sinking of coverslips during washing, use another washing protocol, e.g., immersion of coverslips cell side up in separate petri dishes containing PBS. Damaged cells normally arise from inadvertently allowing the coverslip to dry at any stage of the procedure or by touching the cell side with filter paper. Labelling with phalloidin can be improved by including this probe also in the first antibody. Successful double or triple immunofluorescence labelling requires that the individual antibody combinations each produce intense staining with a clean background, when used alone.

References

Bhatt, A., Kaverina, I., Otey, C., and Huttenlocher, A. (2002). Regulation of focal complex composition and disassembly by the calcium-dependent protease calpain. J. Cell Sci. **115**, 3415–3425.

Giloh, H., and Sedat, J. W. (1982). Fluorescence microscopy: Reduced photobleaching of rhodamine and fluorescein protein conjugates by *n*-propyl gallate. *Science* **217**, 1252–1255.

Gimona, M., Kaverina, I., Resch, G. P., Vignal, E., and Burgstaller, G. (2003). Calponin repeats regulate actin filament stabilty and formation of podosomes in A7r5 smooth muscle cells. *Mol. Biol. Cell.*

Johnson, G. D., Davidson, R. S., McNamee, K. C., Russell, G., Goodwin, D., and Holborow, E. J. (1982). Fading of immunoflu-orescence during microscopy: A study of the phenomenon and its remedy. *J. Immunol. Methods.* **55**, 231–242.

Small, J. V. (1988). The actin cytoskeleton. *Electron Microsc. Rev.* **1**, 155–174.

Weber, K., Rathke, P. C., and Osborn, M. (1978). Cytoplasmic micro-tubular images in glutaraldehyde-fixed tissue culture cells by electron microscopy and by immunofluorescence microscopy. *Proc. Natl. Acad. Sci. USA* **75**, 1820–1824.

more sensitive streptavidin–biotin method. The two latter methods have the advantage that nuclei can be counterstained with hemotoxylin and that only a simple light microscope is required to visualize the stain; however, fluorescence generally gives greater resolution.

As noted for cells, many antibodies that react well on cryostat sections may not react on the same tissue after it has been fixed in formaldehyde and embedded in paraffin. In such a case it may be advantageous to try fixing tissue, e.g., in B5, Bouin's, or Zenker's fixative or alcohol, prior to paraffin embedding. An interesting alternative is to use sections of formaldehyde-fixed, paraffin-embedded material that have been treated in a microwave oven.

For all methods mark the position of the section after fixation; either use a diamond pencil or circle the section with a water-repellent marker (Dako Cat. No. S2002). Remove excess buffer after rinsing steps with Q-tips. Use 10 µl of antibody per section. Apply with an Eppendorf pipette and use the pipette tip to spread the antibody over the section without touching the section. Several manufacturers (e.g., Dako) produce excellent protocol sheets for each immunocytochemical method.

1. Immunofluorescence

Steps

1. Fix cryostat sections or paraffin sections deparaffinized as described earlier for 10 min in acetone at –10°C. Air dry.

2. Use steps 2–9 of the protocol given in the article by Osborn in this volume for multitest slides. Nuclei can be counterstained with Hoechst dye (see Osborn's article). Positively stained cells will be green (see Fig. 3) if an FITC-labeled second antibody is used or red if a rhodamine-labeled second antibody is selected (see also the double label of a cytological sample in Fig. 4).

2. Peroxidase Staining

Steps

1. Fix cryostat sections for 10 min in acetone at –10°C, air dry, and wash in PBS.

2. Deparaffinize paraffin sections and incubate for 30 min at room temperature in 100 ml methanol containing 100 µl H_2O_2 to block endogenous peroxidase activity. Wash in PBS.

3. Incubate for 10 min at 37°C with normal rabbit serum. Drain, but do not wash after this step.

4. Incubate with primary antibody (e.g., mouse monoclonal) for 30 min at 37°C.

5. Wash three times in PBS.

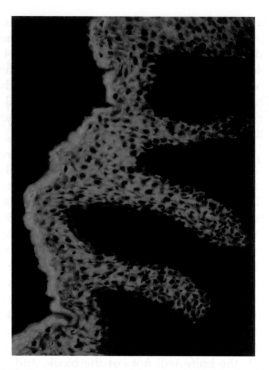

FIGURE 3 Frozen section of human skin stained with KL1 keratin antibody with FITC-labeled second antibody. Only the epidermis is stained (×150).

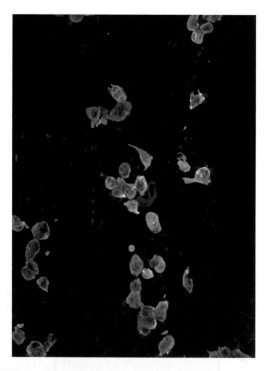

FIGURE 4 Cytological specimen from human breast carcinoma stained with a keratin antibody with FITC-labeled second antibody to show the tumor cells and with a vimentin antibody and a rhodamine-labeled second antibody to show the other cells in the specimen.

6. Incubate with second antibody coupled to peroxidase, e.g., rabbit antimouse for a monoclonal first antibody (Cat. No. P0260 from Dako diluted 1:10 to 1:20).

7. Wash three times in PBS and once in Tris buffer (6g NaCI, 6g Tris/liter, pH 7.4),

8. Develop for 10min at room temperature using freshly made solutions (e.g., 0.06g diaminobenzidine, Fluka Cat. No. 32750 in 100ml Tris buffer, 0.03ml H_2O_2). *Note:* Diaminobenzidine is a carcinogen; handle with care.

9. Wash in tap water.

10. Apply a light counterstain by immersing the slide in Hemalum for 1 to 10s. Remove when staining reaches the required intensity.

11. Wash in tap water.

12. Mount in Glycergel (Dako Cat. No. C0563, http://www.dako.com).

Notes

1. Structures that are positively stained will be dark brown, whereas nuclei will be light blue (Figs. 5–7).

2. The method can be made more sensitive by using an additional step with a peroxidase–antiperoxidase complex (see Sternberger, 1979).

3. Blocking should be performed with a 5% solution of normal serum from the same host species as the labeled antibody, i.e., in this example the tissue is

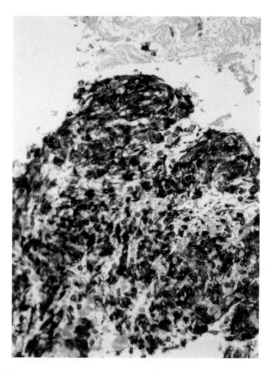

FIGURE 6 Frozen section of human rhabdomyosarcoma stained after microwave treatment with desmin antibody DEB5 and with peroxidase-labeled second antibody. Brown tumor cells are positive for desmin. Nuclei are counterstained blue (×160).

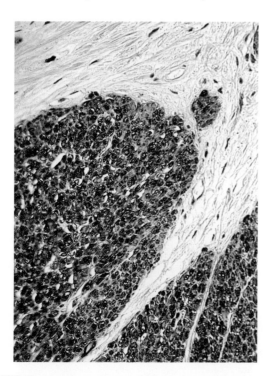

FIGURE 5 Paraffin section of human uterus stained with antibody after microwave fixation with the desmin DER 11 antibody in the peroxidase technique (×160).

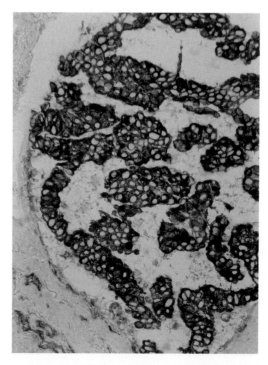

FIGURE 7 Frozen section of human breast carcinoma stained with the keratin KL1 antibody and with peroxidase-labeled second antibody. Brown tumor cells are positive for keratin. Nuclei are counterstained blue (×150).

blocked with normal rabbit serum because the labeled antibody is a rabbit antibody.

4. Bovine serum albumin (BSA) for diluting antibodies should be of high purity. Impure preparations can contain bovine IgGs that can cross-react with the labeled second antibodies.

3. Streptavidin–Biotin Stain

Buy the reagents separately or use the Histostain kit from Zymed Laboratories (http://www.zymed.com; Cat. No. 95-6543 for mouse primary antibody and Cat. No. 95-6143 for rabbit primary antibody). These kits are based on the strong binding between streptavidin, a 60,000-kDa protein isolated from *Streptomyces avidinii*, and biotin, a water-soluble vitamin (MW 244, $K_d = 10^{-15} M$). Instructions are given for the mouse kit.

Steps

1. Fix cryostat sections for 10 min in acetone at −10°C and air dry. Sections can be treated with 0.23% periodate for 45 s. Go to step 3.
2. Deparaffinize paraffin sections and air dry. Incubate 10 min in PBS at room temperature and then 10 min in H_2O_2 solution (nine parts methanol to one part 30% H_2O_2 in water). Wash 3 × 2 min in PBS.
3. Reduce nonspecific background staining by blocking for 10 min in 10% goat serum at room temperature (see Note 3 in Section III,D,2). Then drain but do not wash.
4. Incubate with primary antibody, e.g., mouse monoclonal antibody, in wet chamber for 30 min at 37°C.
5. Wash 3 × 2 min in PBS.
6. Add biotinylated second antibody, e.g., goat anti-mouse, for 10 min at room temperature and repeat step 5.
7. Add enzyme conjugate streptavidin–peroxidase diluted 1:20 for 5 min at room temperature. This binds to the biotin residues on the second antibody.
8. Develop by adding substrate–chromogen mixture for 5 min at 37°C or 15 min at room temperature. The enzyme peroxidase catalyzes the substrate hydrogen peroxide and converts the chromogen aminoethylcarbazole to a red, colored deposit.
9. Wash 3 × 2 min in distilled water.
10. Counterstain with Hemalum between 1 and 10 s.
11. Wash 7 min in tap water.
12. Mount in glycergel.

Notes

1. Structures that are positively stained will be red, whereas nuclei will be light blue (Figs. 8 and 9).

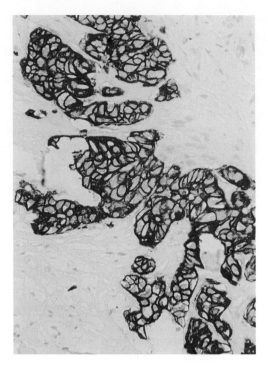

FIGURE 8 Paraffin section of human breast carcinoma stained after microwave treatment with keratin KL1 antibody in the streptavidin–biotin technique. Red tumor cells are keratin positive. Nuclei are counterstained blue (×150).

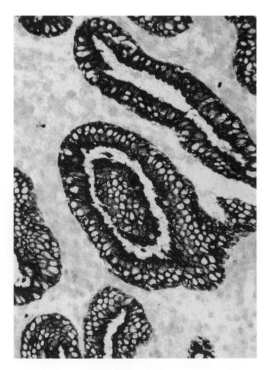

FIGURE 9 Frozen section of human uterus is stained with KL1 keratin antibody in the alkaline phosphatase–antialkaline phosphatase technique. Only epithelial cells are positive (×150).

2. The optimal dilution of primary antibody has to be determined for each antibody.

3. The streptavidin–peroxidase conjugate is the same for all species; the blocking serum and second antibody vary according to the species in which the first antibody is made.

Other Methods

1. The alkaline phosphatase–antialkaline phosphatase technique is also a very sensitive method (Fig. 8).

2. Both fluorescent and chromogenic signals can be enhanced up to 1000X with Tyramide Signal Amplification Technology (New England Nuclear). With this technique, primarily antibodies can be diluted in some instances more than 10,000-fold.

IV. COMMENTS

If possible, include (1) a positive control, i.e., a section of a tissue known to contain the antigen; (2) a negative control, i.e., a section known not to have the antigen; and (3) a reagent control, i.e., a section stained with nonimmune serum instead of the primary antibody. Ideally, the first control should be positive and the second and third controls negative. If high backgrounds are obtained, it may help to adjust the antibody concentrations by increasing the length of time for washing steps or by including $0.5\,M$ NaCl or 5% BSA in the antibody solutions.

References

Cattoretti, G., Backer, M. H. G., Key, G., Duchrow, M., Schlüter, C., Galle, J., and Gerdes, J. (1992). Monoclonal antibodies against recombinant parts of the Ki-67 antigen (MIB 1 and MIB 3) detect proliferating cells in microwave processed formalin-fixed paraffin sections. *J. Pathol.* **168**, 357–363.

Denk, H. (1987). Immunohistochemical methods for the demonstration of tumor markers. *In* "Morphological Tumor Markers" (G. Seifert, ed.), pp. 47–70. Springer-Verlag, Berlin.

Jenette, J. C. (1989). "Immunohistology in Diagnostic Pathology." CRC Press, Boca Raton, FL.

Ordonez, N. G., Manning, J. T., and Brooks, T. E. (1988). Effect of trypsinization on the immunostaining of formalin-fixed, paraffin-embedded tissues. *Am. J. Surg. Pathol.* **12**, 121–129.

Osborn, M., and Domagala, W. (1997). Immunocytochemistry. *In* "Comprehensive Cytopathology" (M. Bibbo, ed.), 2nd Ed., pp. 1033–1074. Saunders, Philadelphia.

Osborn, M., and Weber, K. (1983). Tumor diagnosis by intermediate filament typing: A novel tool for surgical pathology. *Lab. Invest.* **48**, 372–394.

Sternberger, L. A. (1979). "Immunohistochemistry," 2nd Ed. Wiley, New York.

Tubbs, R. R., Gephardt, G. N., and Petras, R. E. (1986). "Atlas of Immunohistology." American Society of Clinical Pathologists Press, Chicago.

Appendix

69

Representative Cultured Cell Lines and Their Characteristics

Robert J. Hay

Virtually thousands of different cell lines have been derived from human and other metazoan tissues. Many of these originate from normal tissues and exhibit a definable, limited doubling potential. Other cell lines may be propagated continuously, as they either have become immortalized from the normal by genetic changes or have been developed initially from tumor tissue. Finite lines of sufficient doubling potential and continuous lines can both be expanded to produce a large number of aliquots, frozen, and authenticated for widespread use in research.

Resources such as the American Type Culture Collection (ATCC) have been established to acquire, preserve, authenticate, and distribute reference cell lines and microorganisms for use by the academic and industrial scientific community (Hay *et al.*, 2000). The cell biology program at the ATCC performs these functions to include human and animal cell lines with over 4000 available in 2003.

The advantages of working with well-defined cell lines free from contaminating organisms may appear obvious. Unfortunately, however, the potential pitfalls associated with the use of cell lines casually obtained and processed still require emphasis (Stacey *et al.*, 2000). Numerous occasions where lines exchanged among cooperating laboratories have been contaminated with cells of other species have been recognized since the late 1960s (Nelson-Rees *et al.*, 1974, 1981; MacLeod *et al.*, 1999). The loss of time and research funds as a result of these problems is very extensive.

Although bacterial and fungal contaminations represent an added concern, in most instances they are overt and easily detected, and therefore have less serious consequences than the more insidious contaminations by mycoplasma. That the presence of these latter microorganisms in cultured cell lines often negates research findings entirely has been stated repeatedly over the years (Barile *et al.*, 1973; Hay *et al.*, 1989). However, the difficulties of detection and the prevalence of contaminated cultures in the research community suggest that the problem cannot be overemphasized. These and related difficulties associated with the use of cell lines obtained from different sources can be avoided if one acquires stocks from a centralized cell resource that applies appropriate quality control (Hay *et al.*, 2000).

Representative human cell lines from normal and tumor tissues available from the ATCC are listed in Table I with a selection of a few of the more important characteristics. Similar data on a variety of cell lines from other animals are included in Table II. More current information on these and other cell lines, their availability, and characteristics is available online via the ATCC website at www.atcc.org.

Cell Biology

TABLE I (*Continued*)

Tissue/tumor (number available)	Designation	ATCC No.	Age	Sex	Race	Diagnosis	Source	Medium[b]	Comments
	C-4I	CRL-1594	41	F	W	Cervical carcinoma	Primary tumor	Naymouth's MB752/ 1/10	Contains HPV-18 sequences and expresses HPV-18 RNA. Tumorigenic. One of a series
	ME-180	HTB-33	66	F	W	Invasive SCC	Metastasis to omentum	MC10	Tumorigenic; desmosomes hypotriploid, XXX
	Si Ha	HTB-35	55	F	Asian	Squamous cell CA	Primary tumor	MEM10	P53 and pRB + Integrated HPV-16
Colon (44)	SW480	CCL-228	50	M	W	Grade III–IV adenocarcinoma	Primary	L15/10	Tumorigenic; K-*ras* codon 12. One of an extensive series
	LoVo	CCL-229	56	M	W	Adenocarcinoma of colon	Metastasis to clevicula	F12/10	Tumorigenic; well characterized for growth kinetics and drug responses. Desmosomes shown. CEA produced
	T84	CCL-248	72	M	?	Colon carcinoma	Transplant tumor	F12DME/5	Desmosomes; receptor for peptide hormones and neurotransmitters; show vectorial transport
	NCI-H508	CCL-253	55	M	W	Cecum adenocarcinoma	Metastasis to abdominal wall	R10	High DOPA carboxylase; TAG-72, CA 19-9 and CEA antigens. One of a series
	FHC	RL 1831	Fetus 13 wk.	?	?	Normal	Colon	F12/DM10+	Epithelial-like; lacks keratin; limited doubling potential
	HT-29	HTB 38	44	F	W	Well-differentiated grade II adenocarcinoma	Primary	MC10	Tumorigenic; hypertriploid; 17 marker chromosomes
	Caco-2	HTB 37	72	M	W	Adenocarcinoma	Primary	EM20	Hypertetraploid; exhibits enterocyte differentiation; tumorigenic.
Duodenum (2)	FH 74	CCL-241	3- to 4- month fetus	F	?	Normal	Small intestine	DME10+	Epithelial-like; tonofibrils and microvilli; keratin –'ve
	HuTu80	HTB-40	53	M	W	Adenocarcinoma	Primary	EM10	Tumorigenic; forms well-differentiated papilloma; pseudodiploid
Embryonal carcinoma (5)	NTERA-2 cl-D1	CRL-1973	22	M	W	Testicular carcinoma	Metastasis to lung	DM10	Clone of Tera-2; pluripotent; differentiates on exposure to RA
	NCCIT	CRL-2073	Adult	M	J	Embryonal CA	Mediastinal mixed germ cell tumor	R10	Differentiates; -'ve for keratin; +'ve for vimentin and placental alkaline phosphatase
	Cates-1B	HTB-104	35	M	W	Testicular carcinoma	Metastasis to lymph node	MC15	Pseudodiploid; patient treated with cytoxan, vincristine, and actinomycin D

TABLE I (*Continued*)

Tissue/tumor (number available)	Designation	ATCC No.	Age	Sex	Race	Diagnosis	Source	Medium[b]	Comments
	Tera 1	HTB-105	47	M	W	Seminoma	Metastasis to lung	MC10	Not tumorigenic; bcl-1 mRNA
Endometrium (7)	AN3 CA	HTB-111	55	F	W	Adenocarcinoma	Metastasis to lymph node	EM10	Yields malignant, undifferentiated tumor
	HEC-1A	HTB-112	71	F	?	Adenocarcinoma stage IA	Primary tissue	MC10	Hyperdiploid; triploid variant available as HTC113
	SK-UT-1	HTB-114	75	F	W	Leiomyosarcoma grade III	Mixed Mesoderm-mal tumor	MEM10+	Hypodiploid to hyperdiploid. Diploid line selected and available as HTB-115
	KLE	CRL-1622	64	F	W	Poorly differentiated adenocarcinoma	Primary	DM/F12–10	Tumorigenic; forms microvilli and junctional complexes
	RL95–2	CRL-1671	65	F	W	Moderately differentiated adenosquamous			Estrogen receptors; *a*-keratin; microvilli
Kidney (30)	293 [HEK293]	CRL-1573	Fetus	?	?	Normal kidney	Early passage cells exposed to sheared adenovirus 5 DNA	MEM10HS	Used extensively in virus propagation. Many variants and engineered derivatives available
	ACHN	CRL-1611	22	M	W	Adenocarcinoma	Pleural effusion	EM10	Tumorigenic; invasive
	769-P	CRL-1933	63	F	W	Clear cell adenocarcinoma	Primary	R10	Tumorigenic; colonies in soft agar; microvilli and desmosomes; hypodiploid
	HK-2	CRL-2190	Adult	M	?	Normal proximal tubular line	E6/E7 trans-formed	Keratinocyte serum-free medium +	Positive for PT cell marker enzymes and other proteins
	CCD-1103 KIDTr	CRL-2304	Fetal	?	?	Normal kidney epithelia	E6/E7 trans-formed	F12K/10	Keratin and acid phosphatase positive; susceptible to coxsackie, measles, polio and syncytial virus infection
	SV7 tert	CRL-2461	63	F	?	Renal angiomyolipoma	Immor-talized with SV40 large T and h-tert	DME10	For studies on tuberous sclerosis
	A-498	HTB-44	52	F	?	Kidney carcinoma	Primary tumor	MEM10	Hypertriploid line; used in virus and tumorigenicity studies
	Caki-1	HTB-46	49	M	W	Renal carcinoma	Metastasis to skin	MC10	Tumorigenic; hypertriploid
Leukemia/ lymphoma (55)	CCRF-CEM	CCL-119	4	F	W	ALL	Peripheral blood	R20	T lymphoblast; malignant in newborn hamsters; model chromosome number 45–47

TABLE I (*Continued*)

Tissue/tumor (number available)	Designation	ATCC No.	Age	Sex	Race	Diagnosis	Source	Medium[b]	Comments
	HL-60	CCL-240	36	F	W	ALL	Peripheral blood	L15/10	Neutrophilic promyelocytes differentiate when exposed to RA and others; surface receptors for Fc
	MOLT 4	CRL-1582	19	M	?	ALL	Peripheral blood	R10	Stable T cell; high Tdt; model chromosome number 95
	TUR	CRL-2367	37	M	W	Histiocytic lymphoma	Effusion	R10	Stable transfectant of U-937 resistant to both G418 and TPA
	NK-92	CRL-2407	50	M	W	Hodgkin's lymphoma	Peripheral blood	Alpha MEM/125 H/FBS	Natural killer cell line, IL-2-dependent surface markers defined series of lines available
	Loucy	CRL-2629	38	F	W	T-ALL	Peripheral blood	R10	Useful for defining role of t(16; 20) and genetic studies
	Hut 78	TIB-161	50	M	W	Sezary syndrome	Peripheral blood	R10	Mature T cell with inducer/helper phenotype; yields and responds to interleukin-2; parent to H9 cell line
Liver (13)	Hep-3B	HB-8064	8	M	B	Hepatocellular carcinoma	Primary	EM10	Tumorigenic; produce haptoglobin; a-fetoprotein, albumin, a_2-macroglobulin; transferrin; fibrinogen, and other liver-specific proteins
	Hep-G2	HB-8065	15	M	W	Hepatocellular carcinoma	Primary	EM10	
	SNU-398	CRL-2233	42	M	A	Anaplastic hepatocellular carcinoma	Primary tumor	ACL4/5	Hyaline globules; HBV-DNA +'ve; HBV-RNA not expressed. One of a series of similar lines
	THLE-3	CRL-11233	Adult	?	?	Normal liver	Left lobe	BEGM	SV_{40} T antigen infected via PA317. Model for pharmacotoxicological studies
	SK-HEP-1	HTB-52	52	?	W	Adenocarcinoma	Ascites	EM10	Produces a-antitrypsin; has Weibel–Palade bodies and vimentin
Lung (203)	A-549	CCL-185	58	M	W	Carcinoma	Primary	F12K/15	Enzymes related to surfactant synthesis studied; lamellar inclusions but sparse.
	WI-38	CCL-75	Fetus	F	W	Normal	Primary	EM10	Euploid line, widely used in cell biology aging research, virology and vaccine manufacture. One of many normal fibroblast controls
	MRC-5	CCL-171	Fetus	M	W	Normal	Primary	EM10	Euploid line, widely used in cell biology aging research, virology and vaccine manufacture

TABLE I (*Continued*)

Tissue/tumor (number available)	Designation	ATCC No.	Age	Sex	Race	Diagnosis	Source	Medium[b]	Comments
	HLF-a	CCL-199	54	F	B	Normal	Primary	EM10	Diploid and stable; patient had epidermoid carcinoma of lung; tissue used was from site remote from cancer
	CCD-19Lu	CCL-210	20	F	W	Normal	Primary	EM10	Euploid; patient died of accidental head trauma
	NCI-H2126	CCL-256	65	M	W	Nonsmall cell lung CA	Plural effusion	HITES/5	One of an extremely large collection of human lung cancer lines and controls
	NCI-BL2126	CCL-256	65	M	W	Peripheral blood B lymphoblast line	EBV transformant	R10	Paired "control" for studies with NCI-H2126. Genomic DNA also available (ATCC# 45512 and 45513)
	NCI-H146	HTB-173	59	M	W	SCLC	Pleural fluid	R10	Tumorigenic; near triploid; high c-*myc* mRNA but no gene amplification; elevated biochemical markers for SCLC; keratin and vimentin positive
	NCI-H441	HTB-174	?	M	?	PapAC	Pericardial fluid	R10	Hyperdiploid; grows in soft agar; SP-A+' Clara and lamellar inclusions
	NCI-H82	HTB-175	40	M	W	SCLC	Pleural fluid	R10	Tumorigenic; near triploid, no Y, high c-*myc*; DNA and RNA, reduced amount and abnormal p53 mRNA (3.7 kb)
	NCI-H820	HTB-181	53	M	W	PapAC	Lymph node	A4	Near triploid; produces lamellar bodies and surfactants SP-A, -B and -C
	SK-LU-1	HTB-57	60	F	W	Adenocarcinoma	Primary	EM10	Tumorigenic in immunotolerant rats; hypotetraploid
Melanoma (32)	C32	CRL-1585	53	M	W	Amelanotic melanoma	Primary	EM10	Tumorigenic; hypodiploid with mode of 45
	Hs294T	HTB-140	56	M	W	Metastatic melanoma	Metastasis to lymph node	DM10	Nerve growth factor and interferon receptors; responsive to RA; tumorigenic and grows in soft agar
	MeWo	HTB-65	78	M	W	Malignant melanoma	Metastasis to lymph node	EMO	Well-studied line, XY karyottype, blood type A; HLA A2, A26, BW16, 18
	SK-MEL5	HTB-70	24	F	W	Metastatic melanoma	Metastasis to axillary node	EM10	Tumorigenic
	COLO 829	CRL-1974	45	M	W	Malignant melanoma	Subcutaneous metastasis	R10	Prior to therapy; some melanin produced; B-cell counterpart available

TABLE I (*Continued*)

Tissue/tumor (number available)	Designation	ATCC No.	Age	Sex	Race	Diagnosis	Source	Medium[b]	Comments
	COLO 829BL	CRL-1980	45	M	W	Malignant melanoma	Peripheral blood EBV transformed	R10	"Control" B-cell line to CRL 1974; DNA fingerprinting confirms identity
Myeloma/ plasmacytoma (18)	RPMI 8226	CCL-155	61	M	?	Multiple myeloma	Peripheral blood	R20	Produces λ light chains; no mature plasma cells
	U266B1	TIB-196	53	M	?	Myeloma	Perophiral	R15	IgE-A secreting
	HS-Sultan	CRL-1484	56	M	W	Plasmacytoma	Primary	R10	Produces IgG-*k*; hyperdiploid
	MC/CAR	CRL-8083	81	M	W	β-cell plasmacytoma	Peripheral blood	IDM20	Produce IgG1, have C3b complement receptors, and have normal diploid karyotype. Variant resistant to 8-aza-guanine available as CRL-8147
	NCI-H929	CRL-9068	62	F	W	Myeloma	Malignant effusion	R10+	Positive for PCA-1, transferrin receptor, and CD38. The 8q chromosomal anomaly is prevalent. Ras and C-myc RNA are expressed
Nasal septum (1)	RPMI 2650	CCL-30	52	M	?	Anaplastic SCC	Pleural effusion	EM10	Pseudodiploid with mode 46; keratin positive
Neuro-blastoma (13)	IMR-32	CCL-127	13 mo.	M	W	Neuroblastoma with organoid differentiation	Abdominal mass	EM10	Neuroblasts and large hyaline fibroblasts
	SK-N-MC	HTB-10	14	F	W	Neuroblastoma	Metastasis to supraorbital area	EM10 EM10	Pseudodiploid; dopamine hydroxlase positive
	SK-N-SH	HTB-11	4	F	?	Neuroblastoma	Metastasis to bone marrow		Hyperdiploid; dopamine hydroxylase
	BE(2)-M17	CRL-2267	2	M	?	Neuroblastoma	Metastasis to marrow	EM/F12 10	Grow in multilayer with long neuritic processes. Also as floating aggregates
	MC-IXC	CRL-2270	14	F	W	Neuroblastoma	Metastasis to supra-orbital area	EM/F12 10	High choline acetyltransferase clonal derivative of HTB-10
Ovary (11)	Caov-3	HTB-75	54	F	W	Adenocarcinoma	Primary	EM10	Extremely unusual chromosome morphology
	NIH: OVCAR-3	HTB-161	60	F	W	Progressive adenocarcinoma	Ascites	R20+	Tumorigenic; grows in soft agar; androgen and estrogen receptors
	PA-1	CRL1572	12	F	W	Teratocarcinoma	Ascites	EM10	Pseudodiploid, t(15q20q); highly malignant in nude mice
	TDV-112D	CRL-11731	42	F	W	Malignant adenocarcinoma	Primary ovary	MCDB104 and M199/15	Mutated p53, variants with deletions at 3p24 (CRL-11730 and 32)
Pancreas (19)	AsPC-1	CRL-1682	62	F	W	Metastatic carcinoma	Ascites	R20	Tumorigenic; CEA, PAA and PSA positive, hypotriploid

TABLE I (*Continued*)

Tissue/tumor (number available)	Designation	ATCC No.	Age	Sex	Race	Diagnosis	Source	Medium[b]	Comments
	SU86-86	CRL-1837	57	F	W	Pancreatic ductal adenocarcinoma	Liver metastasis	R10	Sensitivities in killer cell assays documented
	HPAF-11	CRL-1997	44	M	W	Pancreatic metastatic adenocarcinoma	Peritoneal fluid	EMIO	Highly metastatic
	MPanc-96	CRL-2380	67	M	W	Malignant adenocarcinoma	Lymph node	RI0	Transplantable to SCID mice. Recognized by autologous tumor-infiltrating leucocytes
	Capan-1	HTB-79	40	M	W	Metastatic carcinoma	Liver metastasis	R15	Tumorigenic; hypotriploid
	PANC-1	CRL-1469	56	M	W	Epitheloid ductal carcinoma	Primary	DM10	Hypertriploid
Pharynx (3)	Detroit 562	CCL-138	?	F	W	Metastatic carcinoma	Pleural fluid	EM10	Keratin positive
	FaDu	HTB-43	56	M	W	Squamous cell	Primary	EM10	Tumorigenic; desmosomes; 19 marker chromosomes
Placenta (5)	BeWo	CCL-98	Fetus	M	?	Malignant gestational choriocarcinoma	Hamster xenograft	F12-15	Secretes placental hormones, hCG, human placental lactogen, estrone, estradiol, progesterone
	JEG-3	HTB-36	Fetus	?	?	Choriocarcinoma, Edwin-Turner tumor	Hamster xenograft	EM10	Secretes hCG, human chorionic somatomammotropin, and progesterone; tumorigenic
	JAR	HTB-144	Fetus	?	?	Choriocarcinoma	Primary	R10	Secretes estrogen, progesterone, gonadotropin, and lactogen
	3A(tPA-30-1)	CRL-1583	Term	F	?	TsA30SV$_{40}$ + transformant	Normal term placenta	AMEM1O	Produce human chorionic gonadotrophin and are alkaline phosphatase positive at permissive (33) temperature. Postcrisis line available as CRL-1584
Prostate (10)	PC3	CRL-1435	60	M	W	Adenocarcinoma, grade IV	Primary	F12K-7	Tumorigenic and grows in soft agar; low acid phosphatase and steroid reductase
	LNCap.FGC	CRL-1740	50	M	W	Metastatic adenocarcinoma androgen-independent	Metastasis to supra-clavicular-lymph node	R10	Produces PSA, prostatic acid phosphatase; androgen receptors; tumorigenic
	PZ-HPV-7	CRL-2221	70	M	W	HPV-18 DNA transformed epithelium	Normal tissue peripheral prostate	KSFM	Keratins 5 and 8 positive. Not tumorigenic
	MDA Pca2b	CRL-2422	63	M	B	Prostatic adenocarcinoma	Bone	F12K/20	PSA and androgen receptor positive. Tumorigenic
	RWPE-1	CRL-11609	54	M	W	Transfected with plasmid carrying copy of HPV-18	Normal prostatic epithelia	KSFM	Nontumorigenic, positive for keratins 8 and 18. PSA develops on exposure to androgens. Transformed variants available as CRL-11610 and 11611

TABLE I (*Continued*)

Tissue/tumor (number available)	Designation	ATCC No.	Age	Sex	Race	Diagnosis	Source	Medium[b]	Comments
	DU145	HTB-81	69	M	W	Metastatic carcinoma	Metastasis to brain	EM10	Grows in soft agar; weak acid phosphatase; triploid; desmosomes
Rectum (5)	SW-837	CCL-235	53	M	W	Adenocarcinoma, grade IV	Primary	L15–10	Tumorigenic; hypodiploid
	SW-1463	CCL-234	66	F	W	Adenocarcinoma, grades II–III	Primary	L15–10	Tumorogenic; CEA positive; hypertriploid
Retinoblastoma (2)	WERI-Rb-1	HTB-169	1	F	W	Retinoblastoma	Primary	R10	Tumorigenic in rabbits; no colonies in soft agar; near diploid with 15 or 16 markers
	Y79	HTB-18	2.5	F	W	Retinoblastoma	Primary	R15	Reportedly reverse transcriptase positive
Rhabdomyo-sarcomas (5)	A204	HTB-82	1	F	?	Embryonal rhabdomyo-sarcoma	Primary	MC10	Tumorigenic; near diploid with abnormality on 22p
	HS729	HTB-153	74	M	W	Malignant rhabdomyosar-coma	Primary muscle	M5/10	Tumorigenic consistent with embryonal type
	RD	CCL-136	7	F	W	Malignant rhabdomyosar-coma	Pelvic tumor	DM10	No myofibrils but myoglobin and myosin ATPase activity; complex hyperdiploid karyology; also designated TE32 and 130T
Sarcoma (28 + genetic variants)	HT-1080	CCL-121	35	M	W	Fibrosarcoma	Acetabilum	EM10	Pseudodiploid tumorigenic. Susceptible to R0114 and FeLV. Contains activated N-ras
	HOS	CRL-1543	13	F	W	Osteogenic sarcoma	Primary	EM10	Flat morphology; sensitive to viral and chemical morphological transformation
	MES-SA	CRL-1976	56	F	W	Poorly differentiated sarcoma	Uterine tissue-hysterec-tomy	MCRO	Sensitive to chemotherapeutic agents. Resistant lines also available as CRL-1977 and CRL-2274
	MG-63	CRL-1427	14	M	W	Osteogenic sarcoma	Primary	EM10	Yields interferon on induction; hypotriploid with 18 or 19 markers
	SK-LMS-1	HTB-88	4	F	W	Leiomyosarcoma, grade II	Primary, vulva	EM10	Tumorigenic; hypertriploid with complex karyotype
	SW 1353	HTB-94	72	F	W	Chondrosarcoma	Primary, right humerus	L15/10	Hyperdiploid with trisomic N7 only
	U-205	HTB-96	15	F	W	Osteosarcoma	Tibia	MC15	Secretes osteosarcoma-derived growth factor similar to PDGFA chain
Skin (28)	Detroit-551	CCL-110	Fetus	F	W	Normal fibroblast	Skin	EM10	One of a series of lines from normal individuals and persons with genetic disorders

TABLE I (*Continued*)

Tissue/tumor (number available)	Designation	ATCC No.	Age	Sex	Race	Diagnosis	Source	Medium[b]	Comments
	A-431	CRL-1555	85	F	?	Epidermoid carcinoma	Primary	DM10	Tumorigenic and grows in soft agar; hypertriploid
	182-PFSK	CRL-1532	20	M	W	Normal	Skin	DM10	Apparently normal but carrying gene for hereditary adenomatosis of the colon
	MeKam	CRL-1279	10	M	W	Xeroderma pigmentosum	Skin	DM10	Xeroderma pigmentosum line from NIH. One of about 200 lines from humans with genetic disorders
	CCD 27Sk	CRL-1475	Fetus	M	B	Apparently normal fibroblasts	Skin	EM10	Established using skin from the chest area
	CD 977Sk	CRL-1900	20	F	W	Apparently normal fibroblasts	Skin	EM10	Established from skin of the breast
	CCD 966Sk	CRL-1881	78	F	B	Apparently normal fibroblasts	Skin	EM10	Established from skin of the breast
	CCD 1106 KERTr	CRL-2309	Fetus 19 wk.		W	E6/E7-transformed keratinocytes	Skin	KSFM	Epithelial specific and cytokeratin+; HLA not expressed, E6/E7 presence
	CCD 1102 KERTr	CRL-2310	Fetus 16 wk.		B	E6/E7-transformed keratinocytes	Skin	KSFM	Confirmed by PCR
Stomach (9)	AGS	CRL-1739	54	M	W	Adenocarcinoma	Primary	F12–10	No prior therapy; tumorigenic
	NCI-SNU-16	CRL-5974	33	F	A	Gastric carcinoma metastatic	Ascites	R10	DDC, CEA, and TAG-72 +'ve; muscarini and VIP receptors; c-myc amplified; one of a series of human cancer lines developed by J. Park and associates
	RF-48	CRL-1863	62	M	His-panic	Metastatic carcinoma	Primary metastasis	L15/10	Metastasis from CRL 1864 (RF-1); mucin and CEA negative
	RF-1	CRL-1864	62	M	His-panic	Metastatic carcinoma	Primary	L15/10	Stains for mucin; CEA positive Tuorigenic; hypertetraploid
	KATO-III	HTB-103	55	M	Mon-goloid		Pleural effusion	IDM10	
Submaxilla (1)	A-253	HTB-41	54	M	W	Epidermoid carcincoma	Primary	MC10	Hypotriploid; 14 marker chromosomes
Testes (3)	Cates-1B	HTB-104	34	M	W	Embryonal carcinoma	Metastasis to lymph node	MC10	Reportedly hypodiploid to diploid. See also HTB-105 and HTB-106
Thyroid (2)	SW579	HTB-107	59	M	W	SCC	Primary	L15/10	Tumorigenic
	TT	CRL-1803	77	F	W	Medullary thyroid carcinoma	Primary	F12K/10	Tumorigenic; neuropeptides produced
Tongue (4)	SCC—4	CRL-1624	55	M	?	SCC	Primary	F12/DM10	Tumorigenic; involucrin negative; hypopentaploid; positive for 40-kDa keratin. See CRL-1623 and 1629
	SCC-25	CRL-1628	70	M	?	SCC	Primary	F12/DM10	Tumorigenic, synthesizes low levels of involucrin; epidermal keratin; hypertriploid

TABLE I (*Continued*)

Tissue/tumor (number available)	Designation	ATCC No.	Age	Sex	Race	Diagnosis	Source	Medium[b]	Comments
	CAL27	CRL-2095	56	M	W	SCC	Primary	DME10	Tumorigenic; resistant to treatment with VDS, CDP, or ACTD. Cytokeratin+
Umbilicus (3)	HUV-EC-C	CRL-1730	Fetal-term			Apparently normal vascular endothelium	Umbilical vein	F12K/10+ECGS	Endothelial line; produces factor VIII; near diploid; limited life span.
	HUVE-12	CRL-2480	Fetal-term			Vascular endothelium apparently normal	Umbilical vein	F12K/10 + ECCS and heparin	Limited doubling potential. See "Wistar Special Collection" for more lines and detail
	HUVS-112D	CRL-2481	Fetal-term			Smooth muscle apparently normal	Umbilical vein	F12K/10 + ECCS and heparin	Limited doubling potential. See "Wistar Special Collection" for more lines and detail
Vulva (3)	SK-LMS-1	HTB-88	43	F	W	Leiomyosarcoma	Primary	EM10	Tumorigenic
	SW954	HTB-117	86	F	W	SCC	Primary	L15/10	Pseudodiploid
	SW962	HTB-118	64	F	W	SCC	Metastasis to lymph node	L15/10	Tumorigenic; hypertriploid with at least 15 marker chromosomes

[a] See Hay *et al.* (1992) for originators, references, and more details.

[b] Abbreviations for medium used (on the left-hand side) include (EM) Eagle's minimum essential medium; (R) RPMI 1640; (DM) Dulbecco's modification of Eagle's medium; (F12 and F12K) Ham's medium and Kaighn's modification, respectively; (IDM) Iscove's modification of Eagle's medium (L15) Leibovitz medium; (MC) McCoy's 5A medium; and (AMEM) α modification of Eagle's medium [see Hay *et al.* (1992) for formulas]. The number on the right-hand side indicates percentage of serum (usually fetal bovine) used. Additional recommended ingredients are indicated by a plus sign.

[c] SCC, squamous cell carcinoma; hGH, human growth hormone; Tdt, terminal deoxynucleotidyl transferase; hCG, human chorionic gonadotrpin; RA, retinoic acid; G6PD, glucose-6-phosphate dehydrogenase; ALL, acute lyphoblastic leukemia; SCLC, small cell lung cancer; CEA, carcinoembryonic antigen; PSA, pancreas-specific antigen; PAA, pancreas-associated antigen. See www.atcc.org for further details.

TABLE II Representative Cell Lines from Other Species (January 2003)

ATCC No.[a] (general category)	Designation	Species (No. of lines available)	Culture medium[b]	Comments
CCL-33	PK (15)	Pig (6)	MEM10	Adult kidney epithelial line. Used in virology. Keratin +. Produces plasminogen activator
CCL 34	MDCK	Canine (5)	EM10	Kidney epithelial line; model chromosomes number 78; forms domes in monolayer; used extensively in transport studies
CCL 70	CV1	African green monkey (14)	EM10	From male kidney; modal number 60; supports replication of SV40[c] and many other viruses
CCL-73	Ch1 Es (NBL-8)	Goat (3)	EM10	Continuous line from the esophagus of an adult animal. Used on studies of viruses affecting domestic animals
CCL 81	Vero	African green monkey	199-5	From adult kidney; hypodiploid (mode 58); used extensively in virus assay and production
CCL 92	3T3	Murine (129)	DM-CS10	From Swiss mouse embryo; hypertriploid (mode 68); contact sensitive; used for studies in oncogenic and viral transformation

TABLE II (*Continued*)

ATCC No.[a] (general category)	Designation	Species (No. of lines available)	Culture medium[b]	Comments
CCL-106	LLC-RK1	Rabbit (13)	199/10	Adult kidney, epithelial-like from New Zealand white animals. Viral susceptibility documented
CCL-208	4MBr-5	Rhesus monkey (8)	F12K/10	Normal, 2- to 3-year-old female animal. Bronchial epithelium. Required EGF, developed PAS-positive inclusions
CCL-209	CPAE	Bovine (101)	EM20	Enthothelial line from pulmonary artery; stable diploid karyotype; positive for angiotensin-converting activity
CRL-1476	A10	Rat (61)	DM20	Smooth muscle line from thoracic artery of a DB1X embryo; myokinase and creatine phosphokinase positive
CRL-1581	Sp2/0-Ag14	Murine	DM10	Myeloma used as fusion partner in hybridoma production; HAT sensitive; nonsecretor
CRL-1633	MDOK	Sheep (3)	EM10	From kidney of normal adult male. Epithelial-like cells support growth of several infectious viruses. Hypotetraploid
CRL-1651	COS-7	African green monkey	DM10	Transformed with origin-defective SV40; T antigen positive; suitable host in transfection studies
CRL-1711	Sf9	Fall armyworm (1)	G+10	Clonally derived from pupal ovarian tissue; susceptible to infection by baculovirus expression vectors
CRL-1721	PC12	Rat	R10HS-5	From pheochromocytoma; responsive to nerve growth factor; catecholamine; dopamine, and norepinephrine positive
CRL-1934	ES-B3	Murine	DME15 ± feeder }	Stem cell line. Propagate on STO feeder layer or permit differentiation in its absence. Populations express Oct-3, SSEA-1, EMA-1, or TROMA-1 under appropriate conditions. See www.stemcells@atcc.org for more information
Patented				
CRL-8002	OKT4	Murine	IDM20	One of a series of patented hybridomas; produces monoclonal to human helper T subset
CRL-8305	FRTL-5	Rat	F12K-5+	Thyroid epithelial cell line; produces thyroglobulin; responsive to thyroid-stimulating hormone
CRL-8509	528	Murine	R10	One of a series of patented hybridomas; produces monoclonal to epidermal growth factor receptor
CRL-8873	FHCR	Murine	R12+	Hybridoma FHCR-1-2624/FH6/FHOT-1-3019; secretes monoclonal to Le[x] ganglioside
CRL-10968	S4B6–1	Murine	R10	Produces monoclonal to murine interleukin-2
CRL-11422	MPRO clone 2–1	Murine	IDM + GMCSF/20	Infected with retroviral vector LRAR alpha 403SN. Neutrophil progenitor exhibits azurophilic granules, NSA7/4, chloroacetate esterase, and the RAR sequence
Hybridomas/tumor immunology Lines				
HB-55	L243	Murine	DM10	Secretes an IgG2a cytotoxic antibody to a nonpolymorphic determinant on human Ia
HB-95	WB/32	Murine	DM10	Secretes an IgG2a cytotoxic antibody that reacts with monomorphic determinants on HLA-A, -B, and -C
HB-170	R4–6A2	Murine/rat	DM10	Secretes IgG1 monoclonal to murine Interferon-γ

TABLE II (Continued)

ATCC No.[a] (general category)	Designation	Species (No. of lines available)	Culture medium[b]	Comments
HB-188	11B11	Murine/rat	R10	Secrets IgG1 monoclonal to murine B-cell stimulatory factor (BSF-1, interleukin-4)
HB-198 (215)	F4/80	Murine/rat	R5	Secretes rat IgG2b antibody to murine macrophages
TIB-63	P388D$_1$	Murine	R15	A phagocytic monocyte/macrophage line that produces interleukin-1 in response to PMA myelanonocyte LPS
TIB-68	WEHI-3	Murine	IDM10+	A myelomonocyte sensitive to LPS; produces colony-stimulating activity
TIB-71	RAW 264.7	Murine	DM10	A phagocytic monocyte/macrophage line; capable of antibody-dependent lysis of target cells
TIB-207	GK1.5	Murine	DM20	Secretes monoclonal to T-cell surface antigen L3T4 on a helper/inducer T-cell subset
TIB-214	CTLL-2	Murine	R10+	Clone of cytotoxic T cells dependent on interleukin-2 for proliferation

[a] Prefixes assigned to ATCC cell lines reflect the historical source of support for their addition to the collection and, to a certain extent, the degree of characterization applied. Representative certified cell lines (CCL); cell repository lines (CRL) in the general and patent collections; hybridomas in the collection supported by the National Institute for Allergy and Infectious Diseases (HB); and cells important for studies in tumor immunology (TIB) are listed. The CRL total does not include some 1900 additional cell lines in special collections. See www.atcc.org for names of originators, references, and more detail.

[b] Abbreviations for medium used (on the left-hand side) include (EM) Eagle's minimum essential medium; (R) RPMI 1640; (DM) Dulbecco's modification of Eagle's medium; (F12K) Kaighn's modification of Ham's F12; (IDM) Iscove's modification of Eagle's medium; (199) medium 199; and (G) Grace's insect medium [see Hay *et al.* (1992) for formulas]. The number on the right-hand side indicates percentage of serum used, usually fetal bovine, bovine calf (CS), or horse serum (HS). Additional recommended ingredients are indicated by a plus sign.

[c] SV40, simian virus 40; HAT, hypoxanthine aminopterin thymidine; PMA, phorbal myristic acetate; LPS, lipopolysaccharide. See www.atcc.org for further information.

References

Barile, M. F., Hopps, H. E., Grabowski, M. W., Riggs, D. B., and Del Giudice, R. A. (1973). The identification and sources of mycoplasmas isolated from contaminated cultures. *Ann. N.Y. Acad. Sci.* 225, 252–264.

Hay, R. J., Cleland, M. M., Durkin, S., and Reid, Y. A. (2000). Cell line preservation and authentication. *In* "Animal Cell Culture: A Practical Approach" (J. R. W. Masters, ed.), 3rd Ed., pp. 69–103. Oxford Univ. Press, Oxford.

Hay, R. J., Caputo, J., Chen, T. R., Macy, M. L., McClintock, P., and Reid, Y. A. (1992). "Catalogue of Cell Lines and Hybridomas," 7th Ed. American Type Culture Collection, Rockville, MD.

Hay, R. J., Macy, M. L., and Chen, T. R. (1989). Mycoplasma infection of cultured cells. *Nature (London)* 339, 487–488.

MacLeod, R. A. F., Dirks, W. G., Kaufmann M., Matsuo Y., Milch H., and Drexler H. G. (1999). Widespread intraspecies cross-contamination of human tumor cell line arising at source. *Int. J. Cancer* 83, 555–563.

Nelson-Rees, W., Daniels, W. W., and Flandermeyer, R. R. (1981). Cross-contamination of cells in culture. *Science* 212, 446–452.

Nelson-Rees, W. A., and Flandemeyer, R. R. (1977). Inter- and intraspecies contamination of human breast tumor cell lines HBC and BrCa5 and other cell cultures. *Science* 195, 1343–1344.

Nelson-Rees, W. A., Flandermeyer, R. R., and Hawthorne, P. K. (1974) Banded marker chromosomes as indicators of intraspecies cellular contamination. *Science* 184, 1093–1096.

Stacey, G. N., Masters, J. R. W., Hay, R. J., Drexler, H. G., MacLeod, R. A. F., and Freshney, R. I. (2000). Cell contamination leads to inaccurate data: We must take action now. *Nature* 403, 356.